Lecture Notes in Mathematics

2241

T0177965

More information about this series at http://www.springer.com/series/304

Ştefan Cobzaş • Radu Miculescu • Adriana Nicolae

Lipschitz Functions

 Springer

Ştefan Cobzaş
Faculty of Mathematics and Computer
Science
Babeş-Bolyai University
Cluj-Napoca, Romania

Radu Miculescu
Faculty of Mathematics and Computer
Science
Transilvania University of Braşov
Brasov, Romania

Adriana Nicolae
Faculty of Mathematics and Computer
Science
Babeş-Bolyai University
Cluj-Napoca, Romania

ISSN 0075-8434   ISSN 1617-9692   (electronic)
Lecture Notes in Mathematics
ISBN 978-3-030-16488-1   ISBN 978-3-030-16489-8   (eBook)
https://doi.org/10.1007/978-3-030-16489-8

Library of Congress Control Number: 2019936365

Mathematics Subject Classification (2010): Primary: 46-02, 26A16, 30L05, 46A22, 46B20, 46B22, 46B80, 46B85; Secondary: 46E15, 46E40, 47H09, 47B33, 47B07, 53C22, 54C20, 54C65

This Springer imprint is published by the registered company Springer Nature Switzerland AG
The registered company address is: Gewerbestrasse 11, 6330 Cham, Switzerland

Preface

The aim of this book is to give an essentially self-contained account of the main classical results in the theory of Lipschitz functions. In fact, this project originated as an outgrowth of a master course taught by the second-named author at the University of Bucharest. In time, we developed it and addressed additional relevant topics and recent research trends concerning this class of functions.

The prerequisites are basic results in real analysis, functional analysis, measure theory (including vector measures), and topology, which, for readers' convenience, are surveyed in the first chapter of the book, Chap. 1, "Prerequisites".

Lipschitz functions form a class of functions which appear not only in many branches of mathematics, as the theory of ordinary differential equations, partial differential equations, measure theory, nonlinear functional analysis, topology, metric geometry, and fractal theory, but also in computer science, as in image processing or in the study of Internet search engines.

Taking into account the classical theorem of H. Rademacher which states that a Lipschitz function $f : \Omega \to \mathbb{R}^m$, where Ω is an open subset of $\mathbb{R}^n$, is differentiable outside of a Lebesgue null subset of Ω, the condition of being Lipschitz could be viewed as a weakened version of differentiability, and therefore, these functions are a good substitute for smooth functions in the framework of metric spaces.

Chapter 2 contains some basic results concerning Lipschitz and locally Lipschitz functions—algebraic operations, sequences of Lipschitz functions, Lipschitz properties for differentiable functions (including a characterization in terms of Dini derivatives), or gluing Lipschitz functions together. The existence of Lipschitz and locally Lipschitz partitions of unity with applications to sandwich-type theorems, to Lipschitz selections of set-valued mappings, and to the Lipschitz separability of the Banach space $C(T)$ is also proved.

Chapter 3 starts with a detailed discussion on Lipschitz properties of convex functions, including vector functions. In the vector case, meaning convex functions defined on a locally convex space with values in a locally convex space ordered by a cone, we emphasize the key role played by the normality of the cone. Equi-Lipschitz properties of families of convex functions and Lipschitz properties of convex functions defined on metric linear spaces are discussed as well.

Other considered topics involve the existence of an equivalent metric making a given continuous function Lipschitz and metric spaces where every continuous function is Lipschitz. An old result of Fichtenholz (from 1922) on the relation between absolutely continuous and Lipschitz functions is included. The chapter ends with a discussion on the differentiability properties of Lipschitz functions— Rademacher-type theorems—in finite and in infinite dimension.

The possibility to extend a Lipschitz function from a subset of a metric space to the whole space with the preservation of the Lipschitz constant (a Hahn-Banach type result for Lipschitz functions) is studied in Chap. 4. This chapter contains several results on the existence of norm-preserving extensions of Lipschitz functions— Kirszbraun, McShane, Valentine, and Flett. A discussion on the corresponding property for semi-Lipschitz functions defined on quasi-metric spaces and for Lipschitz functions with values in a quasi-normed space is included as well.

Chapter 5 is concerned with Lipschitz functions on geodesic metric spaces, which are a natural generalization of Riemannian manifolds and provide a suitable setting for the study of problems from various areas of mathematics with important applications. We review in this chapter some selected properties of Lipschitz mappings in geodesic metric spaces, focusing mainly on certain extension theorems which generalize corresponding ones from linear contexts.

After introducing some definitions and results from the theory of geodesic metric spaces with an emphasis on the notion of curvature, we discuss Lipschitz extension results of Kirszbraun and McShane type in Alexandrov spaces with lower or upper curvature bounds considered globally. Even if the existence of a Lipschitz extension is guaranteed by an extension result, in general this extension is not unique. Here, we address the parameter dependence of extensions of Lipschitz mappings from the point of view of continuity (with respect to the supremum distance). This chapter additionally includes two counterparts of the Dugundji extension theorem for continuous or Lipschitz mappings with values in nonpositively curved spaces in the sense of Busemann.

Chapter 6 deals with the possibility to approximate various classes of functions (e.g., uniformly continuous) by Lipschitz functions, based on Lipschitz partitions of unity or on some extension results for Lipschitz functions. A result due to Baire on the approximation of semicontinuous functions by continuous ones, based on McShane's extension method, is also included. This chapter also contains a study of the homotopy of Lipschitz functions (two homotopic Lipschitz functions are Lipschitz homotopic) and an introduction to Lipschitz manifolds.

The main results of Chap. 7 are Aharoni's result (from 1974) on the bi-Lipschitz embeddability of separable metric spaces in the Banach space c_0 and a result of Väisälä (from 1992) on the characterization of the completeness of a normed space X by the non-existence of bi-Lipschitz surjections of X onto $X \setminus \{0\}$. Other related results are discussed in the final section of the chapter. The chapter offers only a glimpse of this very active area of research, the topic being treated at large in the books by Benyamini and Lindenstrauss [75] and in the two-volume treatise by Brudnyi and Brudnyi [126, 127].

The validity of an extension result of Hahn-Banach type for Lipschitz functions makes the space of Lipschitz functions on a Banach space X a good substitute for the linear dual X^*. This idea, combined with the method of Lipschitz free Banach spaces, made possible the extension of many results in functional analysis from the linear case to the Lipschitz one, a topic treated in Chap. 8.

We introduce several Banach spaces of Lipschitz functions (Lipschitz functions vanishing at a fixed point, bounded Lipschitz functions, little Lipschitz functions) on a metric space and present some of their properties. A detailed study of free Lipschitz spaces is carried out, including several ways to introduce them and corresponding duality results. The study of the Kantorovich-Rubinstein and Hanin norms is tightly connected with Lipschitz spaces, mainly via the weak convergence of probability measures, a topic treated here in detail. The case of functions with values in a Hilbert space is considered as well, the key tool for the treatment of this case being a sesquilinear integral for Hilbert space-valued functions.

Compactness and weak compactness properties of Lipschitz operators on Banach spaces and of composition operators on spaces of Lipschitz functions are also studied, emphasizing the key role played by Lipschitz free Banach spaces. Another theme presented here is the Bishop-Phelps property for Lipschitz functions, meaning density results for Lipschitz functions that attain their Lipschitz norm.

Applications to best approximation in metric spaces and in metric linear spaces X are given in the last section of this chapter, where it is shown how results from the linear case can be transposed to this situation by using as a dual space the space of Lipschitz functions defined on X.

There are two other books devoted to Lipschitz functions and spaces of Lipschitz functions—the book by Weaver [675] and that by Miculescu and Mortici [482]. We tried to keep the overlapping with these books at an inevitable minimum, making this book complementary to them. An important topic missing from this book is that of fixed points for Lipschitz mappings, but which is well treated in many books devoted to fixed point theory, as, for instance, in [570].

The bibliography (almost 700 items) contains references to the sources of the results included in the book as well as to further results mentioned in the final sections of each chapter, in order to help the potential reader to get acquainted with the current status of the subject and to find his own line of investigation. In spite of its wealth, the bibliography is far from being complete, but we strived to be as accurate as possible in attributing a result to the right person. We apologize in advance for any inadvertence. An exception is the chapter "Prerequisites", where references to some books rather than to original papers are given. At the beginning of each section in this chapter, the sources on which the presentation is based are indicated.

The book is accessible to graduate students (some parts even to undergraduates), but it also contains recent results of interest to researchers in various domains— metric geometry, mathematical analysis, and functional analysis. The book (or parts of it) is also suitable as a support for graduate or advanced undergraduate courses. We hope that it will be of interest to everyone whose domain of interest is mathematical analysis and its applications.

The authors express their thanks to the reviewers for the careful reading of the manuscript and for the remarks and suggestions that led to a substantial improvement of the presentation, in both style and contents. Our warmest thanks also go to the Springer Editors for the professional cooperation, especially to Ute McCrory whose support and encouragements helped us reach this final stage.

Cluj-Napoca, Romania   Ştefan Cobzaş
Brasov, Romania   Radu Miculescu
Cluj-Napoca, Romania   Adriana Nicolae
January 30, 2019

Contents

Chapter 1
Prerequisites

For the reader's convenience we collect in this chapter some notions and results used throughout the book. In this part we give references only to some appropriate books where the mentioned results can be found along with references to the original papers were they were first proved.

1.1 Ordered Sets

In this section we present some notions and results on ordered sets.

1.1.1 Preorder and Order

We shall use the following notation for numerical sets:

$$\mathbb{N} = \{1, 2, \dots\}\text{—the set of natural numbers;}$$

$$\mathbb{N}_0 = \{0, 1, 2, \dots\} = \mathbb{N} \cup \{0\};$$

$$\mathbb{Z} = \{-2, -1, 0, 1, 2, \dots\}\text{—the set of integers;}$$

$$\mathbb{Q}\text{—the set of rational numbers;}$$

$$\mathbb{R} = (-\infty, \infty)\text{—the set of real numbers;}$$

$$\mathbb{R}_+ = [0, \infty)\text{—the set of nonnegative real numbers;}$$

$$\mathbb{C}\text{—the set of complex numbers;}$$

$$\mathbb{K} = \mathbb{R} \text{ or } \mathbb{K} = \mathbb{C}.$$

© Springer Nature Switzerland AG 2019
Ş. Cobzaş et al., *Lipschitz Functions*, Lecture Notes in Mathematics 2241,
https://doi.org/10.1007/978-3-030-16489-8_1

A *preorder* on a nonempty set S is a binary relation on S, denoted by $\leq$, which satisfies the following properties:

(O1) $s \leq s$, for all $s \in S$;
(O2) if $s \leq s'$ and $s' \leq s''$, then $s \leq s''$.

If further

(O3) if $s \leq s'$ and $s' \leq s$, then $s = s'$,

then $\leq$ is called an *order*. Two elements $s, s' \in S$ are called *comparable* if $s \leq s'$ or $s' \leq s$. If any two elements $s, s' \in S$ are comparable, then the set S is called *totally preordered* (resp. *totally ordered*). Sometimes, to emphasize that an order is not total we say that it is a *partial order* and call the set $(S, \leq)$ *partially ordered* (some authors use the term *poset* for a partially ordered set). A totally ordered subset of an ordered set is also called a *chain*.

An element m in a partially ordered set S is called *maximal* (*minimal*) if $m \leq s$ ($s \leq m$) implies $s = m$, for every $s \in S$. One says that $m \in S$ is the *maximum* (*minimum*) of S if $s \leq m$ ($m \leq s$) for all $s \in S$. One also says that m is the *greatest* (*smallest* or *least*) element of S.

An *upper* (*lower*) *bound* of a subset T of a partially ordered set S is an element $s \in S$ such that $t \leq s$ (resp. $s \leq t$) for all $t \in T$. The lowest upper bound of T is called the *supremum* of T, while the greatest lower bound is called the *infimum* of T.

A partially ordered set S such that any pair of elements in S has a supremum and an infimum is called a *lattice*. If any pair of elements in S has an upper bound then one says that S is a *directed set*.

The set S is called *well-ordered* if every nonempty subset of S has a least element.

Theorem 1.1.1 (Zorn's Lemma) *Let S be a partially ordered set. If every totally ordered subset of S has an upper bound, then S contains a maximal element.*

A consequence of Zorn's lemma, actually a property equivalent to it (and to the Axiom of Choice) is the so-called *well-ordering principle*.

Corollary 1.1.2 *Every set can be well-ordered.*

1.1.2   Ordered Vector Spaces

In this subsection we briefly survey some basic results on ordered vector spaces. Details can be found in [34, 35, 113, 621], or [622]. Ordered topological vector spaces will be considered in Sect. 1.4.9.

A *preordered vector space* is a vector space X equipped with a preorder $\leq$ such that

(OVS1) $x \leq y \Rightarrow x+z \leq y+z$;
(OVS2) $x \leq y \Rightarrow tx \leq ty$,

for all $x, y, z \in X$ and all $t \geq 0$.

An order on a vector space satisfying (OVS1) and (OVS2) is called a *vector order* and the corresponding space, an *ordered vector space*.

In a preordered vector space one can add inequalities

$$x \leq y \text{ and } x' \leq y' \Rightarrow x+x' \leq y+y',$$

and multiply by positive numbers

$$x \leq y \iff tx \leq ty,$$

for all $x, x', y, y' \in X$ and $t > 0$. The multiplication by negative numbers reverses the inequalities

$$\forall t < 0, \quad (x \leq y \iff tx \geq ty).$$

A *wedge* in a vector space X is a nonempty subset C of X such that

$$(C1) \ C + C \subseteq C \quad \text{and} \quad (C2) \ \mathbb{R}_+ C \subseteq C.$$

It is clear that a wedge C is a convex set and

$$\alpha x + \beta y \in C,$$

for all $x, y \in C$, and all $\alpha, \beta \in \mathbb{R}_+$.

If $(X, \leq)$ is a preordered vector space, then

$$X_+ := \{x \in X : x \geq 0\}$$

is a wedge in X, called the wedge of positive elements. Conversely, if C is a wedge in a vector space X, then

$$x \leq_C y \iff y - x \in C$$

defines a vector preorder on X and $X_+ = C$.

The vector preorder $\leq_C$ is a vector order if and only if C is a *cone*, i.e., a wedge satisfying the condition

$$(C3) \quad C \cap (-C) = \{0\}.$$

We shall use the same term—preordered (ordered) vector space—for any pair (X, C), where X is a vector space and C is a wedge (cone) in X, inducing a vector preorder (order) $\leq_C$ as above.

The order is called *Archimedean* if, for any $x \in X$, the existence of an element $y \in X$ such that

$$nx \leq y \quad \text{for all } n \in \mathbb{N},$$

implies $x \leq 0$.

An *order-interval* in a preordered vector space (X, C) is a (possibly empty) set of the form

$$[x, y]_o := \{z \in X : x \leq z \leq y\} = (x + C) \cap (y - C), \tag{1.1.1}$$

for $x, y \in X$. If x, y are not comparable, then $[x, y]_o = \emptyset$. It is clear that an order-interval $[x, y]_o$ is a convex subset of X and that

$$[x, y]_o = x + [0, y - x]_o.$$

The notation $[x, y]$ will be reserved for an algebraic interval (or simply interval):

$$[x, y] := \{(1 - t)x + ty : t \in [0, 1]\}.$$

It is obvious that if $x \leq y$, then $[x, y] \subseteq [x, y]_o$, but the reverse inclusion may not hold, as the following example shows.

Example 1.1.3 Taking $X = \mathbb{R}^2$ with the coordinate order and $x = (0, 0)$, $y = (1, 1)$, then $[x, y]_o$ equals the (full) square with the vertices $(0, 0)$, $(0, 1)$, $(1, 1)$ and $(0, 1)$, so it is larger than the segment $[x, y]$.

A subset A of X is called *C-full* (or simply *full*) if $[x, y]_o \subseteq A$ for all $x, y \in A$. Since the intersection of an arbitrary family of C-full sets is C-full, we can define the *C-full envelope* $[A]_C$ of a nonempty subset A of X as the intersection of all C-full subsets of X containing A, i.e., the smallest C-full subset of X containing A. It follows that

$$[A]_C = \bigcup \{[x, y]_o : x, y \in A\} = (A + C) \cap (A - C).$$

Obviously, A is C-full if and only if $A = [A]_C$.

Sometimes we shall simply write $[A]$ instead of $[A]_C$.

We also mention the following result.

Proposition 1.1.4 *Let* $(X, \leq)$ *be an ordered vector space. Then the order* $\leq$ *is total if and only if every C-full subset of X is convex.*

Remark 1.1.5 Some authors use the terms *order-convex*, or *saturated*, to designate a *C*-full set.

An ordered vector space X is called a *vector lattice* (or a *Riesz space*) if every pair $x, y \in X$ admits a supremum, denoted by $x \vee y$. Since the infimum, denoted by $x \wedge y$, satisfies

$$x \wedge y = -[(-x) \vee (-y)],$$

it follows that every pair of elements in X admits an infimum. The property extends to finite subsets of X, i.e., every such subset has an infimum and a supremum.

For $x \in X$ one defines

$$x^+ = x \vee 0, \quad x^- = (-x) \vee 0, \quad |x| = x \vee (-x).$$

It follows that

(i) $x = x^+ - x^-$ and $x^+ \wedge x^- = 0$, $\quad |x| = x^+ + x^-$, $\quad |-x| = |x|$;

(ii) $||x| - |y|| \leq |x + y| \leq |x| + |y|$;

(iii) $|x| \leq a \iff (x \leq a$ and $-x \leq a)$ for any $a \geq 0$;

(iv) $|x| \vee |y| = \dfrac{1}{2}[|x + y| + |x - y|]$ and $|x| \wedge |y| = \dfrac{1}{2}||x + y| - |x - y||$;

(v) $x \leq y \leq z \Rightarrow |y| \leq |x| \vee |z|$.

$$(1.1.2)$$

We prove only the last assertion (v) from above which will be used in the proof of Theorem 3.1.26 (see also Remark 3.1.28). The others can be found in every book on ordered vector spaces (as, for instance, [35, Theorem 1.17]).

Observe that

$$x \leq y \leq z \Rightarrow 0 \leq y - x \leq z - x.$$

By (iv),

$$|x| \vee |z| = \frac{1}{2}[|z + x| + |z - x|] = \frac{1}{2}[|z + x| + z - x]$$

$$\geq \frac{1}{2}[z + x + y - x] = \frac{1}{2}[z + y] \geq y.$$

Since

$$x \leq y \leq z \Rightarrow -z \leq -y \leq -x,$$

it follows that

$$|x| \vee |z| = |-x| \vee |-z| \geq -y,$$

implying $|y| \leq |x| \vee |z|$.

In fact, the following general principles hold in vector lattices ([34, Theorem 8.6 and Corollary 8.7, p. 318]).

Theorem 1.1.6

1. *Every lattice identity that is true for real numbers is also true in every Archimedean Riesz space.*
2. *If a lattice inequality is true for real numbers, then it is true in any Riesz space.*

This is due to the fact that every Archimedean Riesz space is lattice isomorphic to an appropriate function space with the order defined pointwise.

1.1.3 Convex Sets and Convex Functions

Convexity is one of the most important notions in analysis.

A subset A of a vector space X is called *convex* if $[x, y] \subseteq A$ for all $x, y \in A$, where

$$[x, y] = \{(1 - t)x + ty : 0 \leq t \leq 1\}$$

denotes the segment determined by x and y.

Remark 1.1.7 It is easy to see that if the set A is convex then for every $n \in \mathbb{N}$,

$$\lambda_1 x_1 + \cdots + \lambda_n x_n \in A,$$

for all $x_1, \ldots, x_n \in A$ and all $\lambda_1, \ldots, \lambda_n \in \mathbb{R}_+$ with $\lambda_1 + \cdots + \lambda_n = 1$.

Also, the intersection of an arbitrary family of convex sets is convex.

By the above remark and the fact that the whole space X is convex, the following definition makes sense.

Let X be a vector space and $A \subseteq X$. The intersection of all convex subsets of X which contain A is called the *convex hull* of A and it is denoted by $\text{co}(A)$.

Remark 1.1.8 The following equality holds

$$\text{co}(A) = \{x \in X : \exists n \in \mathbb{N}, \ \exists x_1, \ldots, x_n \in A, \ \exists \lambda_1, \ldots, \lambda_n \in \mathbb{R}_+,$$

$$\text{with } \lambda_1 + \cdots + \lambda_n = 1 \text{ such that } x = \lambda_1 x_1 + \cdots + \lambda_n x_n\}.$$

Proposition 1.1.9 *Let X be a vector space.*

1. A subset A of X is convex if and only if

$$\alpha A + \beta A = (\alpha + \beta)A,$$

for all $\alpha, \beta > 0$.
2. If A, B are convex subsets of X, then

$$co(A \cup B) = \cup\{[a, b] : a \in A, \, b \in B\}.$$

Let K be a convex subset of a linear space X. A function $f : K \to \mathbb{R}$ is called
convex if

$$f((1 - \lambda)x + \lambda y) \le (1 - \lambda)f(x) + \lambda f(y) \tag{1.1.3}$$

for all $x, y \in K$ and $\lambda \in [0, 1]$.

Remark 1.1.10 By mathematical induction one can extend the inequality (1.1.3) to
the case of arbitrary convex combinations, namely the following inequality, known
as Jensen's inequality for convex functions, is valid:

$$f\left(\sum_{k=1}^{n} \lambda_k x_k\right) \le \sum_{k=1}^{n} \lambda_k f(x_k),$$

for all $n \in \mathbb{N}$, $x_1, x_2, \ldots, x_n \in K$ and $\lambda_1, \ldots, \lambda_n \in [0, 1]$ with $\sum_{k=1}^{n} \lambda_k = 1$.

We shall consider now some further properties of subsets of vector spaces.
A subset A of a vector space X is called:

- *balanced* if $\lambda A \subseteq A$ for every $|\lambda| \le 1$;
- *circled* if $\lambda A \subseteq A$ for every $|\lambda| = 1$;
- *symmetric* if $-A = A$;
- *absolutely convex* if it is convex and balanced;
- *absorbing* if $\{t > 0 : x \in tA\} \ne \emptyset$ for every $x \in X$;
- *radially bounded* if $\sup\{t > 0 : tx \in A\} < \infty$ for every $x \in X \setminus \{0\}$.

Remark 1.1.11 The following equivalences are immediate:

(i) A is absolutely convex;
(ii) $\forall a, b \in A$, $\forall \alpha, \beta \in \mathbb{R}$, with $|\alpha| + |\beta| = 1$, $\alpha a + \beta b \in A$;
(iii) $\forall a, b \in A$, $\forall \alpha, \beta \in \mathbb{R}$, with $|\alpha| + |\beta| \le 1$, $\alpha a + \beta b \in A$.

The above property can be extended to n points $x_1, \ldots, x_n \in A$ and n scalars
$\lambda_1, \ldots, \lambda_n \in \mathbb{K}$ with $|\lambda_1| + \cdots + |\lambda_n| \le 1$.

Notice that a balanced set is circled and a circled convex set is absolutely convex.
In the real case, circled is equivalent to symmetric.

Since the intersection of an arbitrary family of absolutely convex sets is absolutely convex, we can define the *absolutely convex hull* aco(A) of a subset A of a vector space X by

$$\mathrm{aco}(A) = \bigcap \{K : A \subseteq K \subseteq X,\ K \text{ absolutely convex}\}.$$

A formula, similar to that given in Remark 1.1.8 for the convex hull, holds for the absolutely convex hull too:

$$\mathrm{aco}(A) = \{x \in X : \exists n \in \mathbb{N},\ \exists x_1, \ldots, x_n \in A,\ \exists \lambda_1, \ldots, \lambda_n \in \mathbb{R},$$
$$|\lambda_1| + \cdots + |\lambda_n| \le 1 \text{ such that } x = \lambda_1 x_1 + \cdots + \lambda_n x_n\}. \tag{1.1.4}$$

The following properties are easily seen.

Proposition 1.1.12 *Let X be a vector space ordered by a cone C, $A \subseteq X$ nonempty and $[A] = (A - C) \cap (A + C)$.*

1. *If A is convex, then $[A]$ is also convex.*
2. *If A is balanced, then $[A]$ is also balanced.*
3. *If A is absolutely convex, then $[A]$ is also absolutely convex.*

Remark 1.1.13 Balanced, circled and absolutely convex sets can be also defined in the case of complex scalars simply by replacing $\mathbb{R}$ by $\mathbb{C}$ in the corresponding definitions. Formula (1.1.4) holds in the complex case too. A subset of $\mathbb{C}$ is circled if and only if it is invariant with respect to rotations around 0.

We end this subsection with a fundamental result in convexity with a large area of applications in various domains (see [175]).

Theorem 1.1.14 (Helly's Theorem) *Let $\mathscr{F}$ be a finite family of convex sets in $\mathbb{R}^n$ containing at least $n + 1$ members. Suppose that every $n + 1$ members of $\mathscr{F}$ have a non-empty intersection. Then $\mathscr{F}$ has a non-empty intersection.*

Remark 1.1.15 Helly's theorem also holds for infinite families of compact convex subsets of $\mathbb{R}^n$.

1.1.4   The Minkowski Functional, Norms and Seminorms

We define first some classes of functionals on vector spaces.
Let X be a vector space. A functional $p : X \to \mathbb{K}$ is called

- *homogeneous* if $p(\alpha x) = \alpha p(x)$ for all $x \in X$ and $\alpha \in \mathbb{K}$;
- *absolutely homogeneous* if $p(\alpha x) = |\alpha| p(x)$ for all $x \in X$ and $\alpha \in \mathbb{K}$;
- *additive* if $p(x + y) = p(x) + p(y)$ for all $x, y \in X$;
- *linear* if it is additive and homogeneous.

If $\mathbb{K} = \mathbb{R}$, then p is called

- *positive* if $p(x) \geq 0$ for all $x \in X$;
- *positively homogeneous* if $p(\alpha x) = \alpha p(x)$ for all $x \in X$ and $\alpha > 0$;
- *subadditive* if $p(x + y) \leq p(x) + p(y)$ for all $x, y \in X$;
- a *seminorm* if it subadditive and absolutely homogeneous;
- a *norm* if it is a seminorm and $p(x) > 0$ for every $x \neq 0$.

Remark 1.1.16 Notice that a positively homogeneous functional satisfies $p(0) = 0$ and a seminorm is positive.

Indeed $p(0) = p(2 \cdot 0) = 2p(0)$ implies $p(0) = 0$, and, if p is a seminorm, then

$$0 = p(0) = p(-x + x) \leq p(-x) + p(x) = 2p(x).$$

For an absorbing subset A of a vector space X the *Minkowski functional* p_A : $X \to [0, \infty)$ is given by

$$p_A(x) = \inf\{t > 0 : x \in tA\}, \quad x \in X. \tag{1.1.5}$$

We mention the following properties of this functional.

Proposition 1.1.17 *Let A be an absorbing subset of a vector space X.*

1. The functional p_A is positively homogeneous and

$$A \subseteq \{x \in X : p_A(x) \leq 1\}.$$

2. If A is convex, then p_A is also subadditive and

$$\{x \in X : p_A(x) < 1\} \subseteq A.$$

3. If A is absolutely convex, then p_A is a seminorm.
4. If A is absolutely convex and radially bounded, then p_A is a norm.

1.1.5 Limit Inferior and Limit Superior of Sequences of Real Numbers

For a sequence (x_n) in $\mathbb{R}$, put

$$\underline{x}_n = \inf\{x_k : k \geq n\} \quad \text{and} \quad \overline{x}_n = \sup\{x_k : k \geq n\}.$$

The inclusion $\{x_k : k \geq n+1\} \subseteq \{x_k : k \geq n\}$ implies $\underline{x}_{n+1} \geq \underline{x}_n$ and $\overline{x}_{n+1} \leq \overline{x}_n$ for all $n \in \mathbb{N}$. Consequently, there exist the limits (in $\overline{\mathbb{R}}$) $\underline{x} = \lim_{n \to \infty} \underline{x}_n$ and

$\bar{x} = \lim_{n\to\infty} \bar{x}_n$ called, respectively, the limit inferior and the limit superior of the sequence (x_n) and denoted by $\liminf_{n\to\infty} x_n$ and $\limsup_{n\to\infty} x_n$.

It can be shown that $\underline{x}$ is the infimum of all $x \in \mathbb{R}$ such that there exists a subsequence of (x_n) with limit x. Similarly, $\bar{x}$ is the supremum of all $x \in \overline{\mathbb{R}}$ such that there exists a subsequence of (x_n) with limit x.

The obvious inequalities $\underline{x}_n \leq \bar{x}_n$, $n \in \mathbb{N}$, imply $\underline{x} \leq \bar{x}$. We have $\underline{x} \geq \bar{x}$ if and only if the sequence (x_n) has a limit x in $\overline{\mathbb{R}}$, in which case $x = \underline{x} = \bar{x}$.

1.2 Topological Spaces

In this section we present some definitions and basic results in topology, referring for details and further results to [213, 223, 351].

1.2.1 The Notion of Topological Space

In this subsection we introduce topological spaces and mention some related notions.

Definition 1.2.1 A *topology* on a set X is a collection τ of subsets of X having the following properties:

  (i) $\emptyset$ and X are in τ;
 (ii) the union of the elements of any subcollection of τ is in τ;
(iii) the intersection of the elements of any finite subcollection of τ is in τ.

A *topological space* is a pair (X, τ) consisting of a set X and a topology τ on X.

Given a topological space X with topology τ, we say that a subset U of X is an *open set* of X if it belongs to the collection τ.

A subset A of a topological space X is said to be *closed* if the set $X \setminus A$ is open.

Remark 1.2.2 Taking into account this definition and de Morgan's rules from set theory, it follows that the family $\mathscr{F}$ of all closed subsets of the topological space (X, τ) satisfies the following conditions:

  (j) $\emptyset$ and X are in $\mathscr{F}$;
 (jj) the union of the elements of any finite subcollection of $\mathscr{F}$ is in $\mathscr{F}$;
(jjj) the intersection of the elements of an arbitrary subcollection of $\mathscr{F}$ is in $\mathscr{F}$.

Given a topological space X and a point x in X, we say that V is a *neighborhood* of x if V contains an open set containing x. The family of all neighborhoods of a point x of X (also called the neighborhood system of x), denoted by $\mathscr{V}(x)$ (or by

$\mathscr{V}_x$), satisfies the following properties:

(N1) $V \in \mathscr{V}(x) \Rightarrow x \in V$;
(N2) $U, V \in \mathscr{V}(x) \Rightarrow U \cap V \in \mathscr{V}(x)$;
(N3) $U \in \mathscr{V}(x)$ and $U \subseteq V \Rightarrow V \in \mathscr{V}(x)$;
(N4) $U \in \mathscr{V}(x) \Rightarrow \exists V \in \mathscr{V}(x)$ such that $U \in \mathscr{V}(y)$ for all $y \in V$.

Conversely, let X be a nonempty set. If to every $x \in X$ there corresponds a nonempty family $\mathscr{V}(x)$ of subsets of X satisfying the properties (N1)–(N4), then, defining τ by

$$\tau = \{Y \subseteq X : \forall y \in Y, \ Y \in \mathscr{V}(y)\},$$

it follows that τ is a topology on X. Furthermore, the neighborhood system of every point $x \in X$ with respect to τ agrees with $\mathscr{V}(x)$ (see Kelley [351] for details).

This approach to topology is convenient in topological vector spaces, because a vector topology on a vector space is determined by the neighborhoods of the origin—the neighborhoods of an arbitrary point are obtained by translations from the neighborhoods of the origin (see Sect. 1.4.1).

A *basis* for the topology τ is a family $\mathscr{B} \subseteq \tau$ such that any member of τ can be written as a union of sets in $\mathscr{B}$. A *neighborhood basis* (or a *basis of neighborhoods*) at a point $x \in X$ is a family $\mathscr{B}(x) \subseteq \mathscr{V}(x)$ such that for any $V \in \mathscr{V}(x)$ there exists $B \in \mathscr{B}(x)$ with $B \subseteq V$.

A topological space (X, τ) is called *first countable* if every point in x admits a countable basis of neighborhoods, and *second countable* if the topology τ admits a countable basis. One also says that X satisfies the first, respectively the second, axiom of countability.

The topological space X is called *separable* if it contains a countable dense subset. A subset Z of X is said to be *dense* in X if $\overline{Z} = X$.

If Y is a subset of a topological space (X, τ), then $\tau|_Y := \{Y \cap Z : Z \in \tau\}$ is a topology on Y called the *induced topology*.

Remark 1.2.3 Let Y be a subset of a topological space (X, τ) and $\tau|_Y$ the induced topology.

1. A subset Z of Y is $\tau|_Y$-closed (i.e., relatively closed in Y) if and only if $Z = W \cap Y$ for some τ-closed subset W of X.
2. A subset U of Y is a neighborhood of a point $y \in Y$ relative to $\tau|_Y$ if and only if $U = V \cap Y$ for some neighborhood V of y relative to τ.

Let $A \subseteq X$. A point $x \in X$ is said to be an *adherent point* of A if $V \cap A \neq \emptyset$ for all $V \in \mathscr{V}_x$, and an *accumulation point* of A if $V \cap (A \setminus \{x\}) \neq \emptyset$ for all $V \in \mathscr{V}_x$. A point $x \in X$ is called an *isolated point* of X if there exists a neighborhood V of x containing only the point x (in other words, if and only if $\{x\}$ is a neighborhood of x, or if and only if $\{x\}$ is open in X).

The set of all adherent points of A is denoted by $\overline{A}$ (or by $cl(A)$) and is called the *closure* of A. It is the smallest closed subset of X containing A. We have $A \subseteq \overline{A}$, $\overline{A} = \overline{\overline{A}}$ and $\overline{A} \subseteq \overline{B}$ if $A \subseteq B$. The set A is closed if and only if $A = \overline{A}$.

If Y is a subset of X and $A \subseteq Y$, then the closure of a subset A with respect to $\tau|_Y$ is equal to the intersection of the closure of A in (X, τ) with Y.

The set of all accumulation points of A, called the *derived set* of A, will be denoted by A'. We have $\overline{A} = A \cup A'$.

1.2.2 Separation Axioms

We shall present some separation properties of topological spaces.

Definition 1.2.4 (Separation Axioms) A topological space (X, τ) is called

- T_0 if for every pair x_1, x_2 of distinct points in X at least one of them has a neighborhood not containing the other one;
- T_1 if for every pair x_1, x_2 of distinct points in X each of them has a neighborhood not containing the other one;
- T_2 or *Hausdorff* if for every pair x_1, x_2 of distinct points in X, there exist the neighborhoods U_1 and U_2 of x_1 and x_2, respectively, such that $U_1 \cap U_2 = \emptyset$;
- *regular* if for every closed subset Y of X and every point $x \in X \setminus Y$ there exist the disjoint open sets $U, V \subseteq X$ such that $x \in U$ and $Y \subseteq V$; the space X is called T_3 if it is regular and T_1;
- *completely regular* if for every closed subset Y of X and every point $x \in X \setminus Y$ there exists a continuous function $f : X \to [0, 1]$ such that $f(x) = 1$ and $f(y) = 0$ for all $y \in Y$; the space X is called $T_{3\frac{1}{2}}$ if it is T_1 and completely regular;
- *normal* if every pair Y, Z of disjoint closed subsets of X can be separated by open sets, i.e., there exist the disjoint open sets $U, V \subseteq X$ such that $Y \subseteq U$ and $Z \subseteq V$; the space X is called T_4 if it is normal and T_1.

Remark 1.2.5 Let X be a topological space.

1. The space X is T_1 if and only if the one-point set $\{x\}$ is closed for every $x \in X$.
2. As suggested by their notation, these classes of topological spaces are related in the following way

$$T_4 \Rightarrow T_{3\frac{1}{2}} \Rightarrow T_3 \Rightarrow T_2 \Rightarrow T_1 \Rightarrow T_0.$$

3. Some authors (Engelking [223], for instance) define regular and normal spaces supposing that the axiom T_1 is satisfied. With this approach regular is equivalent to T_3 and normal with T_4. The separation axioms as given in Definition 1.2.4 are taken from [351].

1.2.3 Compactness

Let (X, τ) be a topological space and $A \subseteq X$. A family $\mathscr{B} = \{B_i : i \in I\}$ of subsets of X is called a *cover* of A if $A \subseteq \bigcup_{i \in I} B_i$. If I is finite (countable), then $\mathscr{B}$ is called a *finite (countable) cover*. If $J \subseteq I$ is such that $\{B_j : j \in J\}$ still covers A, then $\{B_j : j \in J\}$ is called a *subcover* of $\mathscr{B}$. A cover formed of open sets is called an *open cover*. A family $\{B_i : i \in I\} \subseteq 2^X$ is called *centered* (or a *centered system*) if $\bigcap_{j \in J} B_j \neq \emptyset$ for every nonempty finite subset J of I.

Definition 1.2.6 A subset A of a topological space (X, τ) is called

- *compact* if every open cover of A contains a finite subcover;
- *countably compact* (or σ-*compact*) if every countable open cover of A contains a finite subcover;
- *sequentially compact* if every sequence in A contains a convergent subsequence;
- *relatively compact* if its closure $\overline{A}$ is compact.

The following proposition is concerned with the relations between the compactness of a subset of a topological space and the compactness in the induced topology as well as with the relations between compactness and closedness.

Proposition 1.2.7 *Let (X, τ) be a topological space and $Y \subseteq X$.*

1. *The set Y is compact in (X, τ) if and only if the topological space $(Y, \tau|_Y)$ is compact.*
2. *If X is compact and $Y \subseteq X$ is closed, then Y is compact too.*
   If the topological space X is Hausdorff and Y is a compact subset of X, then Y is closed.

Again, de Morgan's rules yield the following characterization of compactness.

Theorem 1.2.8 *A topological space (X, τ) is compact if and only if every centered system of closed subsets of X has a nonempty intersection.*

1.2.4 Continuous Functions

Let X and Y be topological spaces and $f : X \to Y$. The function f is called:

- *continuous at* $x \in X$ if for each neighborhood V of $f(x)$, there exists a neighborhood U of x such that $f(U) \subseteq V$;
- *continuous on* X if it is continuous at each point x of X;
- a *homeomorphism* if f is continuous bijective and $f^{-1} : Y \to X$ is also continuous;
- a *homeomorphic embedding* of X in Y if f is a homeomorphism between X and $f(X)$ (for $f(X)$ endowed with the subspace topology).

The spaces X and Y are called *homeomorphic* if there exists a homeomorphism from X to Y. It is obvious that a homeomorphic embedding is a continuous function from X to Y (see Remark 1.2.3).

The following theorem contains characterizations of the global continuity.

Theorem 1.2.9 *Let* (X, τ) *and* (Y, v) *be topological spaces and* $f : X \to Y$. *Then the following are equivalent.*

1. *The mapping* f *is continuous on* X.
2. $f^{-1}(W) \in \tau$ *for every* $W \in v$.
3. $f^{-1}(Z)$ *is closed in* X *for every closed subset* Z *of* Y.

The separation axioms (see Sect. 1.2.2) are essential for the existence of continuous functions on topological spaces.

Theorem 1.2.10 (Urysohn's Lemma) *For every pair* A, B *of nonempty closed disjoint subsets of a normal topological space* X, *there exists a continuous function* $f : X \to [0, 1]$ *such that* $f(x) = 0$ *on* A *and* $f(x) = 1$ *on* B.

Theorem 1.2.11 (Tietze's Extension Theorem) *Let* X *be a normal topological space and* $A \subseteq X$ *nonempty and closed. Then every continuous function* $f : A \to [-1, 1]$ *admits a continuous extension* $F : X \to [-1, 1]$.

In fact, Tietze's extension theorem holds for every continuous function $f : A \to \mathbb{R}$. In the case of metric spaces Dugundji [212] (see also [80, Chapter II, §3]) extended Tietze's theorem to vector functions.

Theorem 1.2.12 (Dugundji's Extension Theorem) *Let* X *be a metric space,* $A \subseteq X$ *nonempty closed and* E *a Hausdorff LCS. Then every continuous function* $f : A \to E$ *admits a continuous extension* $F : X \to E$ *such that* $F(X) \subseteq \text{co}(f(A))$.

For locally convex spaces (LCS) see Sect. 1.4.2.

Denote by $C(Z, E)$ the space of all continuous functions from a topological space Z to a Hausdorff LCS E. On $C(Z, E)$ one can define several natural topologies, from which we mention the following three:

- the topology of pointwise convergence τ_p;
- the topology of uniform convergence on compacta τ_k;
- the topology of uniform convergence τ_u;

see [107, Chapter X].

Denote by $\mathscr{E}(f) = \{F \in C(X, E) : F|_A = f\}$ the set off all continuous extensions of a continuous function $f : A \to E$. A *selection* for the set-valued mapping $\mathscr{E}$ is a mapping $\Lambda : C(A, E) \to C(X, E)$ such that $\Lambda(f) \in \mathscr{E}(f)$ for every $f \in C(A, E)$.

The following result was proved by Dugundji [212] in the case $E = \mathbb{R}$, answering to a question of Borsuk [99] and, in this general form, by Michael [465] (see also [80, Chapter II, §3]).

Theorem 1.2.13 *Let X be a metric space, E a Hausdorff LCS and A a nonempty closed subset of X. Then the set-valued mapping $\mathscr{E} : C(A, E) \to 2^{C(X,E)} \setminus \{\emptyset\}$ admits a selection Λ which is a linear homeomorphic embedding of $C(A, E)$ into $C(X, E)$ when both the spaces $C(A, E)$ and $C(X, E)$ are equipped with one of the topologies τ_p, τ_k, τ_u mentioned above.*

1.2.5 Semicontinuous Functions

This is an important class of functions with applications in many fields of mathematics, as, for instance, optimization theory.

Definition 1.2.14 Let (X, τ) be a topological space. A function $f : X \to \mathbb{R}$ is called *upper semicontinuous* (*lower semicontinuous*) if for each $u \in \mathbb{R}$ the set $f^{-1}((-\infty, u)) = \{x \in X : f(x) < u\}$ $(f^{-1}((u, \infty)) = \{x \in X : f(x) > u\})$ is open.

The abbreviations usc (resp. lsc) are used to designate these properties.

Let us define $\liminf$ and $\limsup$ of a function $f : X \to \mathbb{R}$ by

$$\liminf_{x \to x_0} f(x) = \sup\{\inf f(V \setminus \{x_0\}) : V \in \mathscr{V}(x)\}, \quad \text{and}$$

$$\limsup_{x \to x_0} f(x) = \inf\{\sup f(V \setminus \{x_0\}) : V \in \mathscr{V}(x)\}.$$

The semicontinuity of the function f can be characterized in the following way.

Proposition 1.2.15 *Let (X, τ) be a topological space and $f : X \to \mathbb{R}$. The function f is lsc (usc) at $x_0 \in X$ if and only if*

$$f(x_0) \le \liminf_{x \to x_0} f(x) \quad (resp. \ \ f(x_0) \ge \limsup_{x \to x_0} f(x)).$$

Remark 1.2.16 The norm of a normed space X is weakly lsc and the dual norm is weakly*-lsc on X^* (see Proposition 1.4.15).

The semicontinuous functions share with continuous functions the following properties.

Proposition 1.2.17 *Let X be a Hausdorff compact space and $f : X \to \mathbb{R}$ a function.*

1. *If f is lsc on X, then f is bounded below on X and there exists $\underline{x} \in X$ such that $f(\underline{x}) = \inf f(X)$.*
2. *If f is usc on X, then f is bounded above on X and there exists $\overline{x} \in X$ such that $f(\overline{x}) = \sup f(X)$.*

1.2.6   Sequences and Nets in Topological Spaces

A sequence (x_n) in a topological space (X, τ) is called *convergent to* $x \in X$ is for every neighborhood V of x there exists $n_0 = n_0(V)$ such that $x_n \in V$ for all $n \geq n_0$. The element x is called a *limit* of the sequence (x_n).

Remark 1.2.18 It is obvious that if (x_n) is a sequence in a subset A of X converging to some $x \in X$, then $x \in \overline{A}$. The converse is not true in general, i.e., not every point in $\overline{A}$ is the limit of a sequence of points in A. This happens in first countable topological spaces (see Sect. 1.2.1).

Also, if X is Hausdorff, then any sequence in X has at most one limit.

In general, the topological properties can be characterized in terms of generalized sequences (or nets). The properties expressed in terms of usual sequences are called *sequential*, e.g. sequential compactness, sequential continuity, etc.

A preordered set $(I, \leq)$ is called *directed* if for every $i_1, i_2 \in I$ there exists $i \in I$ such that $i_1, i_2 \leq i$.

A *net* (or a *generalized sequence*) in a set X is a mapping φ from a directed set I to X. One also uses the notations $x_i = \varphi(i)$, $i \in I$, and $(x_i : i \in I)$, $(x_i)_{i \in I}$ or simply (x_i) to denote a net.

One says that a subset J of a directed set I is

- *cofinal* in I if for every $i \in I$ there exists $j_i \in J$ with $j_i \geq i$;
- *a section* of I if there exists $i_0 \in I$ such that $J = \{i \in I : i \geq i_0\}$.

It is well-known that the notion of subsequence of a sequence plays a key role in many questions of analysis. The extension of this notion to nets must be done in such a way that the properties of subsequences can be extended to this situation.

Let I, J be directed sets and $\varphi : I \to X$, $\psi : J \to X$ two nets. One says that ψ is a *subnet* of the net φ if the following conditions hold:

(i) there exists a mapping $h : J \to I$ such that $\psi = \varphi \circ h$;
(ii) $\forall i \in I$, $\exists j_i \in J$ such that $h(j) \geq i$ for all $j \in J$ with $j \geq j_i$.

An important example of a subnet is given in the following proposition.

Proposition 1.2.19 *Let* $h : J \to I$ *be an increasing mapping such that* $h(J)$ *is cofinal in* I. *Then for every net* $\varphi : I \to X$, $\psi = \varphi \circ h : J \to X$ *is a subnet of* φ.

In the treatise of Schechter [623, Section 7.14] this type of subnet is called a Willard subnet (see [682]), while the one considered before is referred to as a Kelley subnet (see [351]). Observe that the Willard subnet is closer to the notion of subsequence, where a subsequence (x_{n_k}) of a sequence (x_n) is determined by a strictly increasing sequence $n_1 < n_2 < \ldots$ of indices. In this case $h : \mathbb{N} \to \mathbb{N}$, $h(k) = n_k$, $k \in \mathbb{N}$, is the (strictly) increasing function determining the subsequence. In fact, this kind of subnets are sufficient to develop the whole theory of topological spaces.

One says that a net $(x_i : i \in I)$ in a topological space *converges to* $x \in X$ is for every neighborhood V of x there exists $i_0 \in I$ such that $x_i \in V$ for all $i \geq i_0$. The notations $\lim_i x_i = x$ or $x_i \to x$ are used to denote such a situation.

The following theorem illustrates the key role played by nets in topology.

Theorem 1.2.20 *Let* (X, τ), (Y, ν) *be topological spaces.*

1. *If a net* (x_i) *in* X *converges to* x, *then every subnet of* (x_i) *converges to* x.
2. *The topology of* X *is Hausdorff if and only if every net in* X *has at most one limit.*
3. *If* Z *is a subset of* X, *then* $x \in \overline{Z}$ *if and only if there exists a net* (z_i) *in* Z *such that* $\lim_i z_i = x$.
4. *A subset* Z *of* X *is compact if and only if every net in* Z *admits a subnet converging to some point in* Z.
5. *A subset* Z *of* X *is relatively compact if and only if every net in* Z *admits a subnet converging to some point in* X.
6. *A function* $f : X \to Y$ *is continuous at a point* $x \in X$ *if and only if* $\lim_i f(x_i) = f(x)$ *for every net* $(x_i)_{i \in i}$ *in* X *converging to* x.
7. *A function* $f : X \to \mathbb{R}$ *is lower (upper) semicontinuous at a point* $x \in X$ *if and only if*

$$ f(x) \leq \liminf_i f(x_i) \quad (resp.\ f(x) \geq \limsup_i f(x_i)), $$

for every net $(x_i)_{i \in i}$ *in* X *converging to* x.

Remark 1.2.21 The notions of $\liminf$ and $\limsup$ for nets in $\mathbb{R}$ are defined by analogy with the corresponding notions for sequences (see Sect. 1.1.5). If $(\alpha_i : i \in I)$ is a net in $\mathbb{R}$, then

$$ \liminf_i \alpha_i = \sup_{i \in I} \inf\{\alpha_j : i \leq j\} \quad \text{and} \quad \limsup_i \alpha_i = \inf_{i \in I} \sup\{\alpha_j : i \leq j\}. $$

1.2.7 Products of Topological Spaces. Tihonov's Theorem

Let $(X_i, \mathcal{V}_i)$, $i \in I$, be a family of topological spaces where $\mathcal{V}_i : X \to \mathscr{P}(\mathscr{P}(X_i))$ is the neighborhood function, i.e., for every $x_i \in X_i$, $\mathcal{V}_i(x_i)$ is the family of neighborhoods of the point x_i.

Consider the Cartesian product $X = \prod_{i \in I} X_i$ whose elements are denoted by $x = (x_i)_{i \in I}$ (or simply by (x_i)) and let $\pi_i : X \to X_i$ be the canonical projections: $\pi_i(x) = x_i$, for $x = (x_i) \in X$.

Let us introduce a topology on X by defining the family $\mathcal{V}(x)$ of neighborhoods of a point $x = (x_i)$ from X in the following way.

Definition 1.2.22 A subset V of X is a *neighborhood* of the point $x = (x_i) \in X$ provided that there exist a nonempty finite subset J of I and the neighborhoods

$V_j \in \mathcal{V}_j(x_j)$, $j \in J$, such that

$$\bigcap_{j \in J} \pi_j^{-1}(V_j) \subseteq V.$$

It follows that the family of subsets of X

$$\mathcal{B}(x) = \Big\{ \bigcap_{j \in J} \pi_j^{-1}(V_j) : V_j \in \mathcal{V}_j(x_j), \ j \in J, \ J \subseteq I, \ J \text{ nonempty finite} \Big\},$$

is a basis for $\mathcal{V}(x)$. The so defined topological space $(X, \mathcal{V})$ is called the *topological product* of topological spaces X_i, $i \in I$, or the *Tihonov product*.

The neighborhood basis $\mathcal{B}(x)$ can also be described in the following way. A subset B of X belongs to $\mathcal{B}(x)$ if and only if there exist the neighborhoods $V_i \in \mathcal{V}_i(x_i)$, $i \in I$, such that the set

$$\{i \in I : V_i \neq X_i\}$$

is finite and

$$B = \prod_{i \in I} V_i.$$

A subset G of X is open in the product topology if and only if for every $x \in G$ there exist a nonempty finite subset J of I and the open sets $G_j \subseteq X_j$, $j \in J$, such that

$$x \in \bigcap_{j \in J} \pi_j^{-1}(G_j) \subseteq G.$$

If the set I is finite, say $I = \{1, 2, \ldots, n\}$, then, for $x = (x_1, x_2, \ldots, x_n) \in X$, the neighborhood basis $\mathcal{B}(x)$ is formed by all the sets of the form

$$V_1 \times V_2 \times \cdots \times V_n, \quad V_i \in \mathcal{V}_i(x_i), \ i = 1, 2, \ldots, n.$$

In the following proposition we collect some basic properties of the topological product.

Proposition 1.2.23 *Let* $(X_i, \mathcal{V}_i)$, $i \in I$, *be topological spaces,* $(X, \mathcal{V})$ *their topological product and* $\pi_i : X \to X_i$, $i \in I$, *the canonical projections.*

1. *For every* $i \in I$ *the canonical projection* π_i *is a continuous and open mapping.*
2. *Let* Y *be another topological space. A function* $f : Y \to X$ *is continuous if and only if* $\pi_i \circ f : Y \to X_i$ *is continuous for each* $i \in I$.

3. *If all the topological spaces X_i are Hausdorff, then the product X is also Hausdorff.*
4. *A net $(x_\alpha : \alpha \in A)$ in X converges to $x = (x_i) \in X$ if and only if, the net $(\pi_i(x_\alpha) : \alpha \in A)$ converges in X_i to x_i for each $i \in I$.*

One of the deepest results in mathematics, equivalent to the Axiom of Choice, is Tihonov's compactness theorem.

Theorem 1.2.24 (Tihonov's Compactness Theorem) *If all the topological spaces $(X_i, \mathcal{V}_i)$, $i \in I$, are compact, then their topological product is also compact.*

1.3 Metric Spaces

As the natural framework for the study of Lipschitz functions is that of metric spaces, we present some notions and results used in the sequel. Metric spaces are treated in many books on topology, mathematical analysis, functional analysis, etc. As dedicated exclusively to metric spaces we mention the books [283] and [527].

1.3.1 The Notion of Metric Space

We start with the formal definition.

Definition 1.3.1 Let X be a nonempty set. A function $d : X \times X \to [0, \infty)$ is called a *metric* on X if it satisfies the following conditions:

$$\text{(i)} \quad d(x, y) = 0 \iff x = y;$$

$$\text{(ii)} \quad d(x, y) = d(y, x);$$

$$\text{(iii)} \quad d(x, z) \leq d(x, y) + d(y, z),$$

for all $x, y, z \in X$.

A set X endowed with a metric d is called a *metric space* and is denoted by (X, d).

If instead of (i) d satisfies

$$\text{(i}') \quad d(x, x) = 0 \quad \text{for all } x \in X,$$

(ii) and (iii), then d is called a *semimetric* (*pseudo-metric* by some authors).

We define now balls in metric spaces. Let (X, d) be a metric space, $x \in X$ and $r > 0$. The *open ball* $B(x, r)$ and the *closed ball* $B[x, r]$ with center x and radius r

are defined by

$$B(x, r) = \{y \in X : d(x, y) < r\},$$

and

$$B[x, r] = \{y \in X : d(x, y) \leq r\},$$

respectively. The *sphere* $S(x, r)$ is defined by

$$S(x, r) = \{y \in X : d(x, y) = r\}.$$

Every metric space can be endowed with a topology.

If d is a metric on the set X, then the collection τ_d of subsets of X given by

$$\tau_d := \{U \subseteq X : \forall x \in U, \exists \varepsilon > 0, \ B(x, \varepsilon) \subseteq U\}$$

is a topology on X (called the topology induced by d and denoted by τ_d). In this way a metric space (X, d) can be viewed as a topological space and all the topological notions will be considered in this sense.

A subset V of X is a neighborhood of a point $x \in X$ if and only if there exists $\varepsilon > 0$ such that $B(x, \varepsilon) \subseteq V$.

Using closed balls $B[x, r]$ instead of open balls in the above definition one obtains the same topology.

A metric space is first countable, Hausdorff normal (i.e. T_4, see Sect. 1.2.2) with respect to the topology τ_d.

We present now some characteristics of convergent sequences in metric spaces. We mark the convergence to x of a sequence (x_n) in a metric space (X, d) by $\lim_{n \to \infty} x_n = x$, by $x_n \to x$, or, if we want to be more accurate, by $x_n \xrightarrow{d} x$.

A sequence $(x_n)_{n \in \mathbb{N}}$ of elements from a metric space (X, d) is convergent to $x \in X$ if and only if for every $\varepsilon > 0$ there exists $n_\varepsilon \in \mathbb{N}$ such that

$$d(x_n, x) < \varepsilon$$

for all $n \in \mathbb{N}$ with $n \geq n_\varepsilon$.

The following simple properties will be used in the book without notice:

$$x_n \xrightarrow{d} x \iff d(x, x_n) \to 0,$$

and the continuity of the metric d as a function of two variables

$$x_n \xrightarrow{d} x \text{ and } y_n \xrightarrow{d} y \implies d(x_n, y_n) \to d(x, y).$$

1.3.2   Uniformly Continuous, Lipschitz and Hölder Functions

Now we consider some classes of functions specific to metric spaces.

Let (X, d) and (Y, d') be metric spaces. A function $f : X \to Y$ is called:

- *uniformly continuous* if for each $\varepsilon > 0$ there exists $\delta_\varepsilon > 0$ such that

$$d'(f(x), f(y)) < \varepsilon \text{ for all } x, y \in X \text{ with } d(x, y) < \delta_\varepsilon; \qquad (1.3.1)$$

- a *uniform homeomorphism* if f is bijective and both f and f^{-1} are uniformly continuous;
- *Lipschitz* if there exists $L \geq 0$ such that

$$d'(f(x), f(y)) \leq L\, d(x, y) \quad \text{for all} \quad x, y \in X; \qquad (1.3.2)$$

- a *uniform embedding* if f is a uniform homeomorphism between X and $f(X)$;
- an *isometric embedding* (or an *isometry*) if $d'(f(x), f(y)) = d(x, y)$ for all $x, y \in X$.

A function $f : X \to Y$ is continuous if and only if for every $x \in X$ and every $\varepsilon > 0$ there exists $\delta = \delta(\varepsilon, x) > 0$ such that:

$$d(x, x') < \delta \implies d'(f(x), f(x')) < \varepsilon,$$

for all $x' \in X$. Comparing this with the definition of the uniform continuity, it is plain that a uniformly continuous function is continuous.

Also, it is easy to check that a Lipschitz function is uniformly continuous. Indeed, suppose that $f : X \to Y$ is L-Lipschitz, for some $L \geq 0$. For an arbitrary $\varepsilon > 0$ take $\delta = \varepsilon/(L + 1)$. Then

$$d'(f(x), f(y)) \leq Ld(x, y) < \varepsilon,$$

for all $x, y \in X$ with $d(x, y) < \delta$.

The number L in (1.3.2) is called a *Lipschitz constant* for f and we say that f is L-*Lipschitz* if we want to emphasize this constant.

We define the *Lipschitz norm* $L(f)$ of a Lipschitz function $f : X \to Y$ by

$$L(f) := \inf\{L \geq 0 : L \text{ is a Lipschitz constant for } f\}. \qquad (1.3.3)$$

The following proposition contains some simple properties of the Lipschitz norm.

Proposition 1.3.2 *Let $(X, d), (Y, d')$ be metric spaces and $f : X \to Y$ a Lipschitz function. Then*

$$L(f) = \sup \left\{ \frac{d'(f(x), f(y))}{d(x, y)} : x, y \in X, x \neq y \right\}. \tag{1.3.4}$$

The Lipschitz norm $L(f)$ is the least Lipschitz constant for f, i.e.,

(i) $d'(f(x), f(y)) \leq L(f)d(x, y)$ for all $x, y \in X$, and
(ii) if L is a Lipschitz constant for f, then $L(f) \leq L$.

Proof Denoting by $\mathscr{L}(f)$ the set of all Lipschitz constants for f it follows that $L(f) = \inf \mathscr{L}(f)$. Denote also by γ the supremum in the right hand side of (1.3.4).
  If $L \in \mathscr{L}(f)$, then $d'(f(x), f(y))/d(x, y) \leq L$ for all $x, y \in X$, $x \neq y$. It follows that $\gamma \leq L$ for every $L \in \mathscr{L}(f)$, and so $\gamma \leq L(f)$.
  By the definition of γ, $d'(f(x), f(y)) \leq \gamma d(x, y)$ for all $x, y \in X$, i.e., $\gamma \in \mathscr{L}(f)$ and so $L(f) \leq \gamma$.
  Consequently, $\gamma = L(f)$.
  The assertions (i) and (ii) follow from the relations $\gamma \in \mathscr{L}(f)$ and $\gamma = \inf \mathscr{L}(f)$. $\square$

Along with Lipschitz functions one can consider Hölder functions.

Definition 1.3.3 Let $0 < \alpha < 1$ and let $(X, d), (Y, d')$ be metric spaces. A function $f : X \to Y$ is called *Hölder* if there exists $L \geq 0$ such that

$$d'(f(x), f(y)) \leq L \left(d(x, y) \right)^{\alpha}, \tag{1.3.5}$$

for all $x, y \in X$.
  One also says that a function f satisfying (1.3.5) is Hölder (or L-Hölder) of order α.

The class of Hölder functions of order α is denoted by $\mathrm{Lip}_{\alpha}(X, Y)$ for $0 < \alpha \leq 1$, with the convention that $\mathrm{Lip}_1(X, Y) = \mathrm{Lip}(X, Y)$.
  The *α-Hölder norm* of $f \in \mathrm{Lip}_{\alpha}(X, Y)$ is

$$L_{\alpha}(f) = \inf\{L \geq 0 : f \text{ is } L\text{-Hölder of order } \alpha\}$$

$$= \sup_{x \neq y} \frac{d'(f(x), f(y))}{(d(x, y))^{\alpha}}, \tag{1.3.6}$$

with the convention that $L_1(f) = L(f)$.
  Define $d^{\alpha} : X \times X \to [0, \infty)$ by

$$d^{\alpha}(x, y) = (d(x, y))^{\alpha}, \quad x, y \in X.$$

The inequality

$$(a+b)^\alpha \le a^\alpha + b^\alpha,$$

valid for all $a, b \in \mathbb{R}_+$, shows that d^α is a metric on X. We also have

$$f \text{ is } L\text{-Hölder of order } \alpha \iff f \text{ is } L\text{-}(d^\alpha, d')\text{-Lipschitz},$$

that is,

$$\mathrm{Lip}_\alpha((X, d), (Y, d')) = \mathrm{Lip}((X, d^\alpha), (Y, d')),$$

with the preservation of the α-Hölder and Lipschitz norms. For this reason their study reduces to that of Lipschitz spaces. In the case $Y = \mathbb{K}$ one uses the notation

$$\mathrm{Lip}_\alpha(X) \quad \text{or} \quad \mathrm{Lip}(X, d^\alpha).$$

It is natural to ask what happens in the case $\alpha > 1$. A metric space (X, d) is called *midpoint convex* if for every pair x, y of distinct points in X there exists $z \in X$ such that $d(x, z) = d(z, y) = \frac{1}{2}d(x, y)$.

Proposition 1.3.4 ([15], Lemma 2.7) *Let (X, d) be a midpoint convex metric space and (Y, d') an arbitrary metric space. If, for some $L > 0$ and $\alpha > 1$, the function $f : X \to Y$ satisfies the condition*

$$d'(f(x), f(y)) \le Ld^\alpha(x, y),$$

for all $x, y \in X$, then f is a constant function.

Proof We show first that, for all $n \in \mathbb{N}$,

$$d'(f(u), f(v)) \le \frac{L}{2^{n(\alpha-1)}} d^\alpha(u, v), \quad \text{for all } u, v \in X \text{ with } u \ne v. \quad (1.3.7)$$

We proceed by mathematical induction.
The case $n = 1$. For $u \ne v$ in X let w be such that $d(u, w) = d(w, v) = \frac{1}{2}d(u, v)$. Then

$$d'(f(u), f(v)) \le d'(f(u), f(w)) + d'(f(w), f(v))$$

$$\le L\left(d^\alpha(u, w) + d^\alpha(w, v)\right) = \frac{2L}{2^\alpha}d^\alpha(u, v) = \frac{L}{2^{\alpha-1}}d^\alpha(u, v).$$

The induction step. Suppose that (1.3.7) holds for $n = k$ and prove it for $n = k + 1$. For $u \ne v$ in X let w be such that $d(u, w) = d(w, v) = \frac{1}{2}d(u, v)$.

Applying (1.3.7) to the pairs u, w and w, v it follows

$$d'(f(u), f(w)) \leq \frac{L}{2^{k(\alpha-1)}} d^{\alpha}(u, w) = \frac{L}{2^{k(\alpha-1)}} \frac{1}{2^{\alpha}} d^{\alpha}(u, v) \quad \text{and}$$

$$d'(f(w), f(v)) \leq \frac{L}{2^{k(\alpha-1)}} d^{\alpha}(w, v) = \frac{L}{2^{k(\alpha-1)}} \frac{1}{2^{\alpha}} d^{\alpha}(u, v) .$$

But then

$$d'(f(u), f(v)) \leq d'(f(u), f(w)) + d'(f(w), f(v)) \leq \frac{L}{2^{(k+1)(\alpha-1)}} d^{\alpha}(u, v) .$$

Let now $x \neq y$ be two arbitrary points in X. Then, by (1.3.7),

$$d'(f(x), f(y)) \leq \frac{L}{2^{n(\alpha-1)}} d^{\alpha}(x, y) \to 0 \quad \text{as } n \to \infty ,$$

because $2^{\alpha-1} > 1$.

Consequently, $f(x) = f(y)$ for all $x, y \in X$, showing that f is a constant function.   □

Remark 1.3.5 If $f : I \to \mathbb{R}$, I an interval in $\mathbb{R}$, is Hölder with $\alpha > 1$, then

$$\left| \frac{f(x+h) - f(x)}{h} \right| \leq L|h|^{\alpha-1}$$

yields for $h \to 0$, $f'(x) = 0$, so that f is a constant function.

1.3.3 The Distance Function

For a nonempty subset A of a metric space (X, d), the *distance function* $d_A(x) = d(x, A)$ is given by

$$d(x, A) = \inf\{d(x, y) : y \in A\}, \quad x \in X .$$

The *distance* $d(A, B)$ between two nonempty subsets A, B of X is given by

$$d(A, B) = \inf\{d(x, y) : x \in A, y \in B\} .$$

Remark 1.3.6 Sometimes, for a metric space (X, δ), these distance functions will be denoted by $\text{dist}_\delta(x, A)$ and $\text{dist}_\delta(A, B)$, respectively.

The distance functions play a crucial role in many problems involving metric spaces, as e.g., in Lipschitz analysis which is our main concern.

In the following proposition we collect some of the basic properties of the distance function.

Proposition 1.3.7 *Let (X, d) be a metric space and A a nonempty subset of X with distance function d_A.*

1. The function d_A is 1-Lipschitz, that is,

$$|d(x, A) - d(x', A)| \le d(x, x'), \tag{1.3.8}$$

for all $x, x' \in X$.
2. The following equivalence holds

$$d(x, A) = 0 \iff x \in \overline{A}.$$

Consequently, if A is closed, $d(x, A) > 0$ for every $x \in X \setminus A$.

Proof The proof of 1 is simple. For $x, x' \in X$,

$$d(x, C) \le d(x, c) \le d(x, x') + d(x', c),$$

for all $c \in C$. Taking the infimum with respect to $c \in C$ one obtains

$$d(x, C) \le d(x, x') + d(x', C) \iff d(x, C) - d(x', C) \le d(x, x').$$

By symmetry $d(x', C) - d(x, C) \le d(x, x')$, so that $|d(x, C) - d(x', C)| \le d(x, x')$.

2. If $d(x, A) = 0$, then there exists a sequence (y_n) in A with $d(x, y_n) \to 0$. It follows that $y_n \to x$ and so $x \in \overline{A}$.

If $x \in \overline{A}$, then there exists a sequence (y_n) in A with $y_n \to x$. It follows that

$$0 \le d(x, A) \le d(x, y_n) \to 0 \text{ as } n \to \infty,$$

and so $d(x, A) = 0$. □

1.3.4 The Pompeiu-Hausdorff Metric

Let (X, d) be a metric space and $A, B \subseteq X$. The quantity

$$e(A, B) = \sup\{d(x, B) : x \in A\}$$

is called the *excess* of A over B and

$$d_H(A, B) = \max\{e(A, B), e(B, A)\}$$

is called the *Pompeiu-Hausdorff distance* (or *metric*) between the sets A, B.

For $A \subseteq X$ nonempty and $\varepsilon > 0$ let

$$A_\varepsilon = \{x \in X : d(x, A) < \varepsilon\} \quad \text{and} \quad A_{\bar\varepsilon} = \{x \in X : d(x, A) \le \varepsilon\}.$$

By the continuity of the distance function it follows that the set A_ε is open and $A_{\bar\varepsilon}$ is closed.

The Pompeiu-Hausdorff distance between $A, B \subseteq X$ satisfies the equalities

$$d_H(A, B) = \inf\{\varepsilon > 0 : A \subseteq B_\varepsilon \text{ and } B \subseteq A_\varepsilon\}$$

$$= \inf\{\varepsilon > 0 : A \subseteq B_{\bar\varepsilon} \text{ and } B \subseteq A_{\bar\varepsilon}\}$$

$$= \sup\{|d(x, A) - d(x, B)| : x \in X\}.$$

It follows that

$$d_H(A, B) = 0 \iff \overline{A} = \overline{B};$$

$$d_H(A, B) = d_H(B, A);$$

$$d_H(A, C) \le d_H(A, B) + d_H(B, C),$$

for all $A, B, C \subseteq X$.

Since, as defined, it is possible that $d_H(A, B) = \infty$ for some subsets A, B of X, it follows that d_H is an extended semimetric on 2^X. By an *extended metric* (or *extended semimetric*) on a set X one understands a mapping $d : X \times X \to [0, \infty]$ satisfying the axioms (i) (resp. (i′)), (ii) and (iii) from Definition 1.3.1.

We shall call the mapping d_H on $\mathscr{P}(X) \times \mathscr{P}(X)$ a Pompeiu-Hausdorff metric, although, it is not always a metric.

The *diameter* of a subset A of a metric space (X, d) is

$$\text{diam } A = \sup\{d(x, y) : x, y \in A\},$$

with the convention that diam $\emptyset = 0$. The subset A is called *bounded* if diam $A < \infty$. The metric space X is called *bounded* if diam $X < \infty$.

For a metric space (X, d) we shall consider the following classes of sets:

$\mathscr{P}_f(X) = \{Z \subseteq X : Z \text{ nonempty closed}\}$,

$\mathscr{P}_b(X) = \{Z \subseteq X : Z \text{ nonempty bounded}\}$,

$\mathscr{P}_k(X) = \{Z \subseteq X : Z \text{ nonempty compact}\}$.

If $(X, \|\cdot\|)$ is a normed space then we shall consider further the classes:

$\mathscr{P}_c(X) = \{Z \subseteq X : Z \text{ nonempty convex}\}$,

$\mathscr{P}_{wk}(X) = \{Z \subseteq X : Z \text{ nonempty, weakly-compact}\}$,

$\mathscr{P}_{w^*k}(X) = \{Z \subseteq X^* : Z \text{ nonempty, weakly*-compact}\}$.

The families obtained by taking intersections of these families are denoted by:

$\mathscr{P}_{bf}(X) = \mathscr{P}_b(X) \cap \mathscr{P}_f(X) = \{Z \subseteq X : Z \text{ nonempty, bounded, closed}\}$,

$\mathscr{P}_{kc}(X) = \mathscr{P}_k(X) \cap \mathscr{P}_c(X) = \{Z \subseteq X : Z \text{ nonempty, compact and convex}\}$,

etc.

The main classes in the case of a normed space will be

$$\mathscr{P}_{fc}(X), \quad \mathscr{P}_{bfc}(X), \quad \mathscr{P}_{kc}(X), \quad \mathscr{P}_{wk,c}(X)$$

whose definitions are clear from the mentioned notational convention.

Also, for a class $\mathscr{P}_{\alpha\beta\gamma}(X)$ we shall use a tilde to denote the corresponding class plus the empty set

$$\widetilde{\mathscr{P}}_{\alpha\beta\gamma}(X) = \mathscr{P}_{\alpha\beta\gamma}(X) \cup \{\emptyset\}.$$

It follows that d_H is a semimetric on $\mathscr{P}_b(X)$, an extended metric on $\mathscr{P}_f(X)$, and a metric on $\mathscr{P}_{bf}(X)$. We shall use the term Pompeiu-Hausdorff distance (or metric) and the notation d_H in all the cases.

Theorem 1.3.8 *Let (X, d) be a metric space.*

1. *If the space X is complete, then $\mathscr{P}_f(X)$ is complete with respect to the extended Pompeiu-Hausdorff metric d_H.*
2. *The classes $\mathscr{P}_{bf}(X)$ and $\mathscr{P}_k(X)$ are closed in $(\mathscr{P}_f(X), d_H)$.*
3. *If the metric space (X, d) is complete, then $(\mathscr{P}_{bf}(X), d_H)$ and $(\mathscr{P}_k(X), d_H)$ are complete metric spaces.*

In the case of normed spaces we also have good completeness results.

Theorem 1.3.9 *Let $(X, \|\cdot\|)$ be a normed space. The class $\mathscr{P}_{fc}(X)$ of all nonempty closed convex subsets of X, is closed in $(\mathscr{P}_f(X), d_H)$.*

Consequently, if X is a Banach space, then $\mathscr{P}_{fc}(X)$ is complete with respect to d_H.

1.3.5   Characterizations of Continuity in the Metric Case

In the following proposition we give some characterizations of continuity and uniform continuity in the metric case.

Proposition 1.3.10 *Let (X, d) and (Y, d') be metric spaces and $f : X \to Y$ a function.*

1. *The function f is continuous at a point $x_0 \in X$ if and only if*

$$x_n \xrightarrow{d} x_0 \implies f(x_n) \xrightarrow{d'} f(x_0), \tag{1.3.9}$$

for each sequence (x_n) in X.
2. *The function f is continuous on X if and only if*

$$d(x, A) = 0 \implies d'(f(x), f(A)) = 0,$$

for all $x \in X$ and nonempty subsets A of X.

3. *The function f is uniformly continuous on X if and only if*

$$d(x_n, y_n) \to 0 \;\Rightarrow\; d'(f(x_n), f(y_n)) \to 0\,,$$

for all sequences (x_n), (y_n) *in X.*
4. *The function f is uniformly continuous on X if and only if*

$$d(A, B) = 0 \;\Rightarrow\; d'(f(A), f(B)) = 0\,,$$

for all nonempty subsets A, B of X.

Remark 1.3.11

1. The condition (1.3.9) from property 1 in Proposition 1.3.10, called *sequential continuity*, holds in arbitrary topological spaces, i.e., any continuous function between two arbitrary topological spaces is sequentially continuous. The converse does not hold in general, i.e., there are sequentially continuous functions that are not continuous. The characterization of continuity in general topological spaces is given in terms of nets (see Sect. 1.2.6).
2. While the assertions 1–3 of Proposition 1.3.10 are easy to prove, the proof of 4 is more intricate. This property was discovered by V. A. Efremovich [220] in his study of proximity spaces (see also [45, Problem 276, p. 76]).

In the metric case semicontinuity (see Sect. 1.2.5) admits the following characterization.

Proposition 1.3.12 *Let* (X, d) *be a metric space. A function* $f : X \to \mathbb{R}$ *is lsc (usc) at* $x_0 \in X$ *if and only if*

$$f(x_0) \le \liminf_{n \to \infty} f(x_n) \qquad (resp. \; f(x_0) \ge \limsup_{n \to \infty} f(x_n))\,,$$

for every sequence (x_n) *in X converging to* x_0.

A metric space is normal (see Definition 1.2.4) and Urysohn's lemma (Theorem 1.2.10) admits in this case a very simple proof.

Theorem 1.3.13 (Urysohn's Lemma—the Metric Case) *Let A and B be disjoint closed subsets of a metric space X. Then there exists a continuous mapping* $f : X \to [0, 1]$ *such that* $f(x) = 0$ *for each* $x \in A$ *and* $f(x) = 1$ *for each* $x \in B$.

Proof In this case the function f is given by the formula

$$f(x) = \frac{d(x, A)}{d(x, A) + d(x, B)}\,, \qquad x \in X\,.$$

□

1.3.6 Completeness and Baire Category

Let (X, d) be a metric space (X, d). A sequence $(x_n)_{n \in \mathbb{N}}$ in X is called *Cauchy* (or *fundamental*) if for every $\varepsilon > 0$ there exists $n_\varepsilon \in \mathbb{N}$ such that

$$d(x_n, x_m) < \varepsilon \quad \text{for all} \quad m, n \in \mathbb{N} \text{ with } m, n \geq n_\varepsilon,$$

or, equivalently,

$$d(x_n, x_{n+k}) < \varepsilon \quad \text{for all} \quad n \in \mathbb{N} \text{ with } n \geq n_\varepsilon \text{ and all } k \in \mathbb{N}.$$

The metric space (X, d) is called *complete* if every Cauchy sequence in (X, d) is convergent.

Remark 1.3.14 The following properties hold true.

(i) Any convergent sequence is a Cauchy sequence. The converse is not generally true.
(ii) Every Cauchy sequence is bounded.
(iii) Every subsequence of a Cauchy sequence is Cauchy.
(iv) A Cauchy sequence which has a convergent subsequence is convergent.

Proposition 1.3.15 *Any complete metric space without isolated points is uncountable.*

Baire Category
An important tool in mathematics is that of Baire category.

Definition 1.3.16 Let T be a topological space. A subset S of T is called:

- *nowhere dense* if $\text{int}(\overline{S}) = \emptyset$;
- *of first Baire category* if there exists a family S_n, $n \in \mathbb{N}$, of nowhere dense subsets of T such that $S = \bigcup_{n=1}^{\infty} S_n$;
- *of second Baire category* if it is not of first Baire category;
- *residual* if $T \setminus S$ is of first Baire category.
  We consider also the following types of sets in topological spaces:
- the intersection of a countable family of open sets is called a G_δ-set and the union of a countable family of G_δ-sets a $G_{\delta\sigma}$-set;
- the union of a countable family of closed sets is called an F_σ-set and the intersection of a countable family of F_σ-sets an $F_{\sigma\delta}$-set.

The space T is called a *Baire space* if every nonempty open subset of T is of second Baire category.

The following theorem contains several characterizations of Baire spaces.

Theorem 1.3.17 *Let T be topological space. The following are equivalent.*

1. *T is a Baire space.*
2. *For every family G_n, $n \in \mathbb{N}$, of open dense subsets of T, the intersection $\bigcap_{n=1}^{\infty} G_n$ is dense in T.*
3. *For every family F_n, $n \in \mathbb{N}$, of closed subsets of T such that $\mathrm{int}(F_n) = \emptyset$ for all $n \in \mathbb{N}$, it follows that $\mathrm{int}(\bigcup_n F_n) = \emptyset$.*
4. *Any residual subset of T is dense in T.*
5. *Every subset of first Baire category of T has empty interior.*

The following consequences are often used in applications.

Corollary 1.3.18 *Let T be a Baire space. Then the following results hold.*

1. *If $T \neq \emptyset$, then for every family F_n, $n \in \mathbb{N}$, of closed subsets of T such that $T = \bigcup_n F_n$ there exists an $n \in \mathbb{N}$ such that $\mathrm{int}(F_n) \neq \emptyset$.*
2. *Every residual subset of T contains a subset of G_δ-type, dense in T.*

Two important classes of Baire spaces are presented in the following theorem.

Theorem 1.3.19

1. *Every complete metric space is a Baire space.*
2. *Every locally compact Hausdorff topological space is a Baire space.*

A topological space T is called *locally compact* if every point of T has a compact neighborhood. If T is Hausdorff, then this implies that every point of T has a basis of compact neighborhoods.

1.3.7 Compactness in Metric Spaces

A subset A of metric space (X, d) is called *totally bounded* if for each $\varepsilon > 0$ there exist $n_\varepsilon \in \mathbb{N}$ and $x_1, \dots, x_{n_\varepsilon} \in X$ such that

$$A \subseteq B(x_1, \varepsilon) \cup B(x_2, \varepsilon) \cup \cdots \cup B(x_{n_\varepsilon}, \varepsilon).$$

The family $x_1, \dots, x_{n_\varepsilon}$ is called a finite ε-net for A.

Notice that one obtains the same notion if instead of the open balls $B(x_i, \varepsilon)$ one takes closed balls $B[x_i, \varepsilon]$.

Theorem 1.3.20 *Let (X, d) be a metric space and $A \subseteq X$.*

1. *The set A is totally bounded if and only if every sequence in A contains a Cauchy subsequence.*
2. *The set A is compact if and only if it is totally bounded and complete.*

Corollary 1.3.21 *A subset of a complete metric space is relatively compact if and only if it is totally bounded.*

In a metric space the first three notions of compactness considered in Definition 1.2.6 agree.

Theorem 1.3.22 *For a subset A of a metric space (X, d) the following are equivalent.*

1. *The set A is compact.*
2. *The set A is countably compact.*
3. *The set A is sequentially compact.*

1.3.8   Equivalent Metrics

Definition 1.3.23 Let d_1 and d_2 be two metrics on a set X.
The metrics d_1, d_2 are called:

- *topologically equivalent* if they induce the same topology on X (i.e., $\tau_{d_1} = \tau_{d_2}$);
- *uniformly equivalent* if the identity mapping $\mathrm{Id} : X \to X$, $\mathrm{Id}(x) = x$, $x \in X$, is uniformly continuous both as a mapping from (X, d_1) to (X, d_2) and as a mapping from (X, d_2) to (X, d_1);
- *Lipschitz equivalent* if there exist two numbers $\alpha, \beta > 0$ such that

$$\alpha d_1(x, y) \leq d_2(x, y) \leq \beta d_1(x, y),$$

for all $x, y \in X$.

Proposition 1.3.24 *Let d_1 and d_2 be two metrics on a set X.*

1. *If the metrics d_1, d_2 are Lipschitz equivalent they are uniformly equivalent. If they are uniformly equivalent, then they are topologically equivalent.*
2. *The metrics d_1, d_2 are topologically equivalent if and only if*

$$x_n \xrightarrow{d_1} x \iff x_n \xrightarrow{d_2} x,$$

for every sequence (x_n) in X.
3. *The metrics d_1, d_2 are uniformly equivalent if and only if*

$$d_1(x_n, y_n) \to 0 \iff d_2(x_n, y_n) \to 0,$$

for every sequence $((x_n, y_n))_{n \in \mathbb{N}}$ in $X \times X$.
4. *If the metrics d_1, d_2 are uniformly equivalent then the metric space (X, d_1) is complete if and only if the metric space (X, d_2) is complete.*

Remark 1.3.25 If $\| \cdot \|_1, \| \cdot \|_2$ are norms on a vector space X, then these three notions of equivalence agree, that is, if $\| \cdot \|_1, \| \cdot \|_2$ are topologically equivalent, then they are Lipschitz equivalent.

Remark 1.3.26 The function $d(x, y) = |\arctan x - \arctan y|$, $x, y \in \mathbb{R}$, is a metric on $\mathbb{R}$ which is topologically equivalent to the usual metric $d_{|\cdot|}(x, y) = |x - y|$ on $\mathbb{R}$, but the sequence $x_n = n$, $n \in \mathbb{N}$, is d-Cauchy and has no limit in $\mathbb{R}$. This shows that the metric space $(\mathbb{R}, d)$ is not complete and so the metrics $d_{|\cdot|}$ and d are not uniformly equivalent. The metrics $d_1, d_2 : \mathbb{R} \times \mathbb{R} \to \mathbb{R}_+$ given by

$$d_1(x, y) = |x - y| \quad \text{and} \quad d_2(x, y) = |x - y|^{1/2} \quad \text{for } x, y \in \mathbb{R},$$

are uniformly equivalent (by Proposition 1.3.24(3)), but not Lipschitz equivalent, because the inequality $|x - y| \leq \beta |x - y|^{1/2}$ leads to the contradiction $|x - y|^{1/2} \leq \beta$ for all $x, y \in \mathbb{R}$.

Proposition 1.3.27 *Let (X, d) be a metric space. Then the functions $d', d^* : X \times X \to [0, \infty)$ given by*

$$d'(x, y) = \min\{1, d(x, y)\} \quad \text{and} \quad d^*(x, y) = \frac{d(x, y)}{1 + d(x, y)},$$

for $x, y \in X$, are metrics on X, topologically equivalent to d. They satisfy

$$d'(x, y) \leq d(x, y) \quad \text{and} \quad d^*(x, y) \leq d(x, y),$$

for all $x, y \in X$, that is, each of them is 1-Lipschitz with respect to d and so uniformly continuous with respect to d.

Both of the metrics d' and d^* are uniformly equivalent to d, but they are not Lipschitz equivalent to d, provided that the metric space (X, d) is unbounded, i.e.,

$$\text{diam } X := \sup\{d(x, y) : x, y \in X\} = \infty. \tag{1.3.10}$$

Proof The uniform equivalence of the metrics d' and d^* with d is an immediate consequence of the easily verified conditions

$$d(x_n, y_n) \to 0 \iff d'(x_n, y_n) \to 0, \quad \text{and}$$

$$d(x_n, y_n) \to 0 \iff d^*(x_n, y_n) \to 0,$$

for all sequences (x_n), (y_n) in X.

Let us show now that d is not Lipschitz, neither with respect to d' nor with respect to d^*, provided that the metric space (X, d) is unbounded.

In the first case, the inequalities $d(x, y) \leq cd'(x, y) \leq c$, valid for all $x, y \in X$, contradict (1.3.10).

In the second case, if for some $c > 0$,

$$d(x, y) \leq c \frac{d(x, y)}{1 + d(x, y)},$$

then $1 + d(x, y) \leq c$ for all $x, y \in X$ with $x \neq y$, again in contradiction to (1.3.10). $\qquad\square$

1.3.9 Ultrametric Spaces

A metric d on a set X is called an *ultrametric* if

$$d(x, z) \le \max\{d(x, y), d(y, z)\} \tag{1.3.11}$$

for all $x, y, z \in X$. The pair (X, d) is called an *ultrametric space*. The inequality (1.3.11) is called the *strong triangle inequality* (or the *ultrametric inequality*).

A valued field $(K, |\cdot|)$ such that $|a + b| \le \max\{|a|, |b|\}$, for all $a, b \in K$, is called a non-Archimedean valued field. Functional analysis over such fields is called non-Archimedean functional analysis, a well established domain of research (see, for instance, [551]). A norm $\|\cdot\|$ on a vector space X over a non-Archimedean valued field is called non-Archimedean provided that $\|x + y\| \le \max\{\|x\|, \|y\|\}$ for all $x, y \in X$.

Ultrametric spaces have some strange properties.

Proposition 1.3.28 *Let (X, d) be an ultrametric space.*

1. *If $d(x, y) \ne d(y, z)$, then*

$$d(x, z) = \max\{d(x, y), d(y, z)\}.$$

2. *The balls in an ultrametric space have the following properties:*

   (i) $y \in B(x, r) \implies B(y, r) = B(x, r)$;

   (ii) $y \in B[x, r] \implies B[y, r] = B[x, r]$;

   (iii) $y \in S(x, r) \implies B(y, r) \subseteq S(x, r)$;

   (iv) $B(x, r) \cap B(y, r') \ne \emptyset \implies [B(x, r) \subseteq B(y, r')$ *or* $B(y, r') \subseteq B(x, r)]$;

   (v) $B[x, r] \cap B[y, r'] \ne \emptyset \implies [B[x, r] \subseteq B[y, r']$ *or* $B[y, r'] \subseteq B[x, r]]$.

3. *The ball $B[x, r]$ and the sphere $S(x, r)$ are closed and, at the same time, open subsets of X.*[1]

Proof

1. Suppose that $d(x, y) < d(y, z)$. Then $\max\{d(x, y), d(y, z)\} = d(y, z)$, so $d(x, z) \le d(y, z)$. But $d(y, z) \le d(x, y) \vee d(x, z) \le d(y, z)$, so that $d(x, y) \vee d(x, z) = d(y, z)$. Since $d(x, y) < d(y, z)$ this implies $d(x, z) = d(y, z) = \max\{d(x, y), d(y, z)\}$.

---

[1] *Nature is an infinite sphere of which the center is everywhere and the circumference nowhere* (Blaise Pascal, *Pensées*). Probably the famous French philosopher and mathematician had in mind a non-Archimedean world. Initially he wrote "A frightful (*effroyable*) sphere" (see the essay on this topic by J. L. Borges at http://www.filosofiaesoterica.com/pascals-sphere/)

2. (i) Suppose that $d(x, y) < r$. If $d(y, z) < r$, then

$$d(x, z) \leq d(x, y) \vee d(y, z) < r,$$

showing that $B(y, r) \subseteq B(x, r)$. The inclusion $B(x, r) \subseteq B(y, r)$ follows by symmetry.

The equality from (ii) can be proved similarly.

(iii) Suppose $d(x, y) = r$. If $d(y, z) < r = d(x, y)$, then, using 1,

$$d(x, z) = d(x, y) \vee d(y, z) = d(x, y) = r,$$

that is, $z \in S(x, r)$.

(iv) Let $z \in B(x, r) \cap B(y, r')$ and suppose that $r \leq r'$. Then, by (i),

$$B(x, r) = B(z, r) \subseteq B(z, r') = B(y, r').$$

The implication from (v) can be proved similarly.

3. The ball $B[x, r]$ and the sphere $S(x, r)$ are closed sets in any metric space. Let us show that they are also open.

If $y \in S(x, r)$, i.e., $d(x, y) = r$, then, by (iii),

$$B(y, r) \subseteq S(x, r),$$

proving that $S(x, r)$ is open.

Since

$$B[x, r] = B(x, r) \cup S(x, r),$$

and the set $B(x, r)$ is open in any metric space, it follows that $B[x, r]$ is open too.   □

Remark 1.3.29 A subset of a topological space that is at the same time closed and open is called *clopen* (obtained by contraction from closed-open). The assertion 3 from above shows that in an ultrametric space every point has a basis of neighborhoods formed of clopen sets.

1.3.10 Paracompact Spaces

This is an important class of topological spaces that we briefly present in what follows.

Definition 1.3.30 Let $\mathscr{A}$, $\mathscr{B}$ be covers of a topological space X. We say that

- $\mathscr{B}$ *refines* $\mathscr{A}$ (or $\mathscr{B}$ is a *refinement* of $\mathscr{A}$) if for each $B \in \mathscr{B}$ there exists $A \in \mathscr{A}$ such that $B \subseteq A$, i.e., each set in $\mathscr{B}$ is contained in some set in $\mathscr{A}$;
- $\mathscr{A}$ is *point-finite* if each $x \in X$ belongs to finitely many members of $\mathscr{A}$;
- $\mathscr{A}$ is *locally finite* if for each $x \in X$ there exists a neighborhood of x that intersects only finitely many members of $\mathscr{A}$;
- $\mathscr{A}$ is *discrete* if each point $x \in X$ has a neighborhood intersecting at most one member of $\mathscr{A}$;
- $\mathscr{A}$ is *σ-discrete* (*σ-point-finite*, *σ-locally-finite*) if $\mathscr{A} = \bigcup_{k=1}^{\infty} \mathscr{A}_k$ where, for every $k \in \mathbb{N}$, $\mathscr{A}_k$ is a discrete (point-finite, locally-finite) family of subsets of X.

It is obvious that any locally finite family $\mathscr{A}$ of subsets of a topological space X is point-finite.

Definition 1.3.31 A topological space is called *paracompact* if every open cover of it admits a locally finite open refinement.

Remark 1.3.32 Since any subcover of $\mathscr{A}$ is a refinement of $\mathscr{A}$, it follows that a compact space is paracompact.

Theorem 1.3.33 (A. H. Stone) *Every metric space is paracompact.*

A proof of this result will be given in Sect. 2.6.3 with application to the Lipschitz partition of unity.

Theorem 1.3.34 ([213], Theorem 1.4, p. 162) *Let $(U_i)_{i \in I}$ be an open cover of a topological space X which has an open locally finite refinement (this is the case if X is a metric space). Then there exists an open locally finite refinement $(V_i)_{i \in I}$ of $(U_i)_{i \in I}$ such that $V_i \subseteq U_i$ for each $i \in I$.*

Theorem 1.3.35 *Let $(U_i)_{i \in I}$ be an open cover of a metric space X. Then there exists an open cover $(V_i)_{i \in I}$ of X such that $\overline{V_i} \subseteq U_i$ for each $i \in I$. If $(U_i)_{i \in I}$ is locally finite, then $(V_i)_{i \in I}$ is locally finite too.*

For the proof of the above theorem one can consult [213], Theorem 6.1, p. 152, taking into account that every paracompact space is normal (see [213] Theorem 2.2, p. 163).

Theorem 1.3.36 *Let $(U_i)_{i \in I}$ be an open cover of a metric space X. Then there exists an open locally finite refinement $(V_i)_{i \in I}$ of $(U_i)_{i \in I}$ such that $\overline{V_i} \subseteq U_i$ for each $i \in I$.*

Proof By Theorems 1.3.33 and 1.3.34, there exists an open locally finite refinement $(W_i)_{i \in I}$ of $(U_i)_{i \in I}$ such that $W_i \subseteq U_i$ for each $i \in I$. Taking into account Theorem 1.3.35, there exists an open cover $(V_i)_{i \in I}$ of X such that $\overline{V_i} \subseteq W_i$ for each $i \in I$. Then $(V_i)_{i \in I}$ is an open locally finite refinement of $(U_i)_{i \in I}$ such that $\overline{V_i} \subseteq U_i$ for each $i \in I$. $\square$

1.3.11 Partitions of Unity

Paracompact spaces are normal, so Urysohn's lemma guarantees the existence of continuous functions that separate disjoint closed convex sets, but, in fact, paracompact spaces can be characterized by a stronger property, namely the existence of partitions of unity.

Let (X, τ) be a topological space. The *support* $\mathrm{spt}(f)$ of a function $f : X \to \mathbb{R}$ is defined by

$$\mathrm{spt}(f) = \mathrm{cl}(\{x \in X : f(x) \neq 0\}).$$

Definition 1.3.37 A family $\mathscr{F} = \{f_i : i \in I\}$ of continuous functions $f_i : X \to [0, 1]$ is called a *partition of unity* if

(i) for every $x \in X$ there exists a neighborhood U of x with $f_i|_U = 0$ for all but finitely many $i \in I$;
(ii) $\sum_{i \in I} f_i(x) = 1$, for all $x \in X$, and $\sup f_i(X) > 0$ for every $i \in I$.

One says that a partition of unity $\{f_i : i \in I\}$ is *subordinated* to an open cover $\mathscr{A}$ of X if the cover $\{\mathrm{spt}(f_i) : i \in I\}$ of X refines $\mathscr{A}$. It is called *locally finite* if the family $\{\mathrm{spt}(f_i) : i \in I\}$ is locally finite.

Theorem 1.3.38 ([223], Theorem 5.1.9) *For a T_1 topological space (X, τ) the following are equivalent.*

1. *The space X is paracompact.*
2. *Every open cover of X admits a locally finite partition of unity subordinated to it.*

Remark 1.3.39 One can prove that if $\{A_i : i \in I\}$ is a locally finite open cover of X then there exists a partition of unity $\{f_i : i \in I\}$ such that $\mathrm{spt}(f_i) \subseteq A_i$ for all $i \in I$.

1.3.12 Sandwich and Approximation Results
for Semicontinuous Functions

A topological space (X, τ) is called *countably paracompact* if every countable open cover of X admits a (necessarily countable) locally finite refinement. A topological space X is called *perfectly normal* (the term *fully normal* is also used by some authors) if it is normal and every closed subset of X is G_δ (i.e., the intersection of a countable family of open sets).

The following theorem was proved independently by Dowker [202] and Katětov [346] (see also [223, Exercise 5.5.20]).

Theorem 1.3.40 *For a T_1 topological space X the following are equivalent.*

1. *The space X is normal and countably paracompact.*
2. *For any pair of functions $g, h : X \to \mathbb{R}$ such that g is usc, h is lsc and $g(x) < h(x)$ for all $x \in X$ there exists a continuous function $f : X \to \mathbb{R}$ such that $g(x) < f(x) < h(x)$ for all $x \in X$.*

The following result was obtained by Tong [652] (announced in [651]), see also [223, Problem 1.7.15.(b)].

Theorem 1.3.41 *For a T_1 topological space X the following are equivalent.*

1. *The space X is normal.*
2. *For any pair of functions $g, h : X \to \mathbb{R}$ such that g is usc, h is lsc and $g(x) \leq h(x)$ for all $x \in X$ there exists a continuous function $f : X \to \mathbb{R}$ such that $g(x) \leq f(x) \leq h(x)$ for all $x \in X$.*

Michael [466] (see also [223, Problem 1.7.15.(d)]) refined this result in the following way.

Theorem 1.3.42 *For a T_1 topological space X the following are equivalent.*

1. *The space X is perfectly normal.*
2. *For any pair of functions $g, h : X \to \mathbb{R}$ such that g is usc, h is lsc and $g(x) \leq h(x)$ for all $x \in X$ there exists a continuous function $f : X \to \mathbb{R}$ such that $g(x) \leq f(x) \leq h(x)$ for all $x \in X$ and $g(x) < f(x) < h(x)$ whenever $g(x) < h(x)$.*

H. Tong, *op. cit.* (see also [223, Problem 1.7.15.(c)]), showed as well that the possibility to approximate semicontinuous functions by continuous ones can be characterized in terms of perfect normality.

Theorem 1.3.43 *For a T_1 topological space X the following are equivalent.*

1. *The space X is perfectly normal.*
2. *For any lsc function $f : X \to \mathbb{R}$ there exists a sequence $f_n : X \to \mathbb{R}$, $n \in \mathbb{N}$, of continuous functions such that $f_n(x) \leq f_{n+1}(x)$, for all $x \in X$ and $n \in \mathbb{N}$, and $f(x) = \lim_{n \to \infty} f_n(x)$ for all $x \in X$.*
3. *For any usc function $f : X \to \mathbb{R}$ there exists a sequence $f_n : X \to \mathbb{R}$ of continuous functions such that $f_n(x) \geq f_{n+1}(x)$, for all $x \in X$ and $n \in \mathbb{N}$, and $f(x) = \lim_{n \to \infty} f_n(x)$ for all $x \in X$.*

1.4 Functional Analysis

For topological vector spaces we recommend [622]. Good expositions of functional analysis (both linear and nonlinear) in Banach spaces are given in the books [22, 232, 233, 456] (but, of course, the reader may have his own preferences).

1.4.1 Topological Vector Spaces

A *topological vector space* (TVS for short) is a vector space X endowed with a topology τ such that the operations $+ : X \times X \to \mathbb{K}$, $(x, y) \mapsto x + y$, and $\cdot :$ $\mathbb{K} \times X \to X$, $(\alpha, y) \mapsto \alpha x$, are continuous. A topology satisfying these conditions is called a *vector topology*.

In the following proposition we collect some useful properties of TVS.

Proposition 1.4.1 *Let* (X, τ) *be a TVS over* $\mathbb{K}$.

1. *For any* $x_0 \in X$ *and* $\alpha_0 \in \mathbb{K} \setminus \{0\}$, *the mappings* $x \mapsto x_0 + x$ *and* $x \mapsto \alpha_0 x$ *are homeomorphisms of* X *onto* X.
2. *For every* $x \in X$,

$$\mathscr{V}(x) = \{x + V : V \in \mathscr{V}(0)\}.$$

   If $\mathscr{B}$ *is a basis of 0-neighborhoods, then* $\mathscr{B}(x) := \{x + B : B \in \mathscr{B}\}$ *is a basis of neighborhoods of* x.
3. *Each point in* X *admits a basis formed of closed neighborhoods.*
4. *There exists a basis* $\mathscr{B}$ *of 0-neighborhoods such that:*

   (i) *each* $B \in \mathscr{B}$ *is absorbing and balanced;*
   (ii) *for every* $B \in \mathscr{B}$ *there exists* $C \in \mathscr{B}$ *such that* $C + C \subseteq B$;
   (iii) *for every* $B \in \mathscr{B}$ *there exists* $C \in \mathscr{B}$ *such that* $\lambda C \subseteq B$ *for all* $\lambda \in \mathbb{K}$, $|\lambda| \le 1$.

5. *The closure of a subset* A *of* X *is given by*

$$\overline{A} = \bigcap \{A + V : V \in \mathscr{V}(0)\} = \bigcap \{A + B : B \in \mathscr{B}\},$$

   where $\mathscr{B}$ *is a basis of 0-neighborhoods.*
6. *The closure of a balanced set* A *is balanced, as well as its interior provided that* $0 \in \text{int}(A)$.
7. *If* A *is a convex subset of* X, *then* $\overline{A}$ *and* $\text{int}(A)$ *are convex and* $\text{int}(A) = \text{int}(\overline{A})$. *Also,* $\overline{A} = \overline{\text{int}(A)}$ *if* $\text{int}(A) \ne \emptyset$.
8. *A linear operator between two TVS* X, Y *is continuous on* X *if and only if it is continuous at some point* $x_0 \in X$ *(usually* $x_0 = 0$*).*
9. *For a seminorm* $p : X \to [0, \infty)$ *the following are equivalent:*

   (i) p *is continuous at* $0 \in X$;
   (ii) *the set* $\{x \in X : p(x) < 1\}$ *is open in* X;
   (iii) p *is uniformly continuous on* X.

10. *Let* A *be an absorbing absolutely convex subset of* X. *If its Minkowski functional* p_A *is continuous, then*

$$\text{int}(A) = \{x \in X : p_A(x) < 1\} \subseteq A \subseteq \{x \in X : p_A(x) \le 1\} = \overline{A}.$$

Vector topologies can be introduced through some special bases of 0-neighborhoods.

Proposition 1.4.2 *Let X be a vector space and $\mathscr{B}$ a nonempty family of subsets of X satisfying the properties*

(B1) *each $B \in \mathscr{B}$ is absorbing and balanced;*
(B2) *for each $B \in \mathscr{B}$ there exists $C \in \mathscr{B}$ such that $C + C \subseteq B$.*

Defining the neighborhood system of an arbitrary point $x \in X$ by

$$\mathscr{V}(x) = \{V \subseteq X : \exists B \in \mathscr{B},\ x + B \subseteq V\},$$

one obtains a vector topology τ on X.

Definition 1.4.3 The *closed convex hull* $\overline{\mathrm{co}}(A)$ of a subset A of a TVS X is the intersection of all closed convex subsets of X containing A.

The *closed absolutely convex hull* of A, $\overline{\mathrm{aco}}(A)$, is defined similarly.

The following equalities hold

$$\overline{\mathrm{co}}(A) = \mathrm{cl}\,(\mathrm{co}(A)) \quad \text{and} \quad \overline{\mathrm{aco}}(A) = \mathrm{cl}\,(\mathrm{aco}(A))\,,$$

where $\mathrm{co}(A)$ and $\mathrm{aco}(A)$ denote the convex and the absolutely convex hull of A, respectively (see Sect. 1.1.3).

1.4.2 Locally Convex Spaces

A topological vector space X is called *locally convex* (LCS for short) if every point in X has a neighborhood basis formed of convex sets.

Proposition 1.4.4 *A TVS X is locally convex if and only if $0 \in X$ admits a neighborhood basis formed of absolutely convex sets.*

Locally Convex Topologies Generated by Families of Seminorms
Let P be a family of seminorms defined on a vector space X. For $p \in P$, $x \in X$ and $r > 0$, let

$$B_p(x, r) = \{y \in X : p(y - x) < r\} \quad \text{and} \quad B_p[x, r] = \{y \in X : p(y - x) \le r\}.$$

For $F \subseteq P$ nonempty finite, let

$$B_F(x, r) = \bigcap\{B_p(x, r) : p \in F\} \quad \text{and} \quad B_F[x, r] = \bigcap\{B_p[x, r] : p \in F\}.$$

Putting

$$B_p = B_p[0, 1], \quad B'_p = B_p(0, 1),$$
$$B_F = B_F[0, 1], \quad B'_F = B_F(0, 1),$$

the following equalities are satisfied

$$B_p[x, r] = x + r B_p, \quad B_p(x, r) = x + r B'_p;$$
$$B_F[x, r] = x + r B_F, \quad B_F(x, r) = x + r B'_F.$$

Proposition 1.4.5 *Let P be a family of seminorms defined on a vector space X.*

1. The family

$$\mathscr{B} = \{r B_F : F \subseteq P, \ F \ finite \ nonempty, \ r > 0\}$$

satisfies the conditions (B1), (B2) from Proposition 1.4.2, so it defines a vector topology τ_P on X. Since every B_F is absolutely convex this topology is locally convex.

   Each seminorm $p \in P$ is continuous with respect to τ_P.
   The same is true for the family

$$\mathscr{B}' = \{r B'_F : F \subseteq P, \ F \ finite \ nonempty, \ r > 0\}.$$

2. The topology τ_P is Hausdorff if and only if for every $x \in X$ there exists $p \in P$ with $p(x) > 0$.

3. A net $(x_i : i \in I)$ in X is τ_P-convergent to $x \in X$ if and only if

$$\lim_i p(x_i - x) = 0,$$

for every $p \in P$.

A family P of seminorms on a vector space X is called *directed* if for each $p_1, p_2 \in P$ there exists $p_3 \in P$ such that $p_1, p_2 \leq p_3$.

If the family P of seminorms is directed, then for every neighborhood V of $x \in X$ there exist $p \in P$ and $r > 0$ such that $B_p[x, r] \subseteq V$ (or, equivalently, there exist $p \in P$ and $r' > 0$ such that $B_p(x, r') \subseteq V$).

   Putting

$$p_F(x) = \max\{p(x) : p \in F\}, \ x \in X,$$

it follows that

$$\tilde{P} := \{p_F : F \subseteq P, \ F \ finite \ nonempty\}$$

is a directed family of seminorms generating the same topology as P.

Indeed, this follows from the equalities

$$p_{F_1}, p_{F_2} \le p_F, \quad \text{where } F = F_1 \cup F_2;$$
$$B_F(x, r) = B_{p_F}(x, r) \quad \text{and} \quad B_F[x, r] = B_{p_F}[x, r].$$

The following result shows that locally convex topologies are always generated by families of seminorms in the way described above.

Proposition 1.4.6 *Let X be a locally convex space and $\mathscr{B}$ a basis of 0-neighborhoods formed of absolutely convex sets. Then the topology of X is generated, in the way described above, by the family of seminorms*

$$P = \{p_B : B \in \mathscr{B}\},$$

formed of Minkowski functionals corresponding to the sets in $\mathscr{B}$.

Continuity of Linear Operators and the Dual of a LCS

The continuity of linear operators between locally convex spaces can be characterized in the following way.

Proposition 1.4.7 *Let (X, P) and (Y, Q) be LCS, where P, Q are families of seminorms generating the topologies. A linear operator $T : X \to Y$ is continuous if and only if for every $q \in Q$ there exist a nonempty finite subset F of P and $L > 0$ such that*

$$q(Tx) \le L \cdot \max\{p(x) : p \in F\},$$

for all $x \in X$.

If the family P is directed, then T is continuous if and only if for every $q \in Q$ there exist $p \in P$ and $L > 0$ such that

$$q(Tx) \le L \cdot p(x),$$

for all $x \in X$.

The space of all linear operators between two vector space X, Y is denoted by $L(X, Y)$. The space of all continuous linear operators between two LCS (X, P), (Y, Q) is denoted by $\mathscr{L}(X, Y)$. The space $X^* := \mathscr{L}(X, \mathbb{K})$ is called the *dual* (or *conjugate*) *space* of X. Sometimes we say that X^* is the *dual* (or *conjugate*) of X. The elements of X^* will be denoted by $x^*, y^*, \ldots$. The value of a functional $x^* \in X^*$ at $x \in X$ is also denoted by $\langle x, x^* \rangle$, i.e.,

$$\langle x, x^* \rangle = x^*(x).$$

Various topologies can be defined on $\mathscr{L}(X, Y)$ and on X^*. We shall present some of them in the case of normed spaces.

Remark 1.4.8 The following result holds for seminorms. If p, q are two seminorms defined on a vector space X and, for some $\alpha, \beta > 0$,

$$p(x) \le \alpha \implies q(x) \le \beta,$$

for all $x \in X$, then

$$q(x) \le \frac{\beta}{\alpha} p(x),$$

for all $x \in X$.

Separation of Convex Sets

The following theorem contains separation results for convex sets in TVS.

Theorem 1.4.9 (Separation of Convex Sets) *Let (X, τ) be a TVS and A, B disjoint nonempty convex subsets of X.*

1. *If A is open, then there exist a continuous linear functional $x^* \in X^*$ and $\alpha, \beta \in \mathbb{R}$ such that*

$$\operatorname{Re} x^*(x) < \alpha \le \operatorname{Re} x^*(y),$$

   for all $x \in A$ and $y \in B$.
2. *If A is compact, B closed and X is locally convex, then there exist a continuous linear functional $x^* \in X^*$ and $\alpha, \beta \in \mathbb{R}$ such that*

$$\operatorname{Re} x^*(x) < \alpha < \beta < \operatorname{Re} x^*(y),$$

   for all $x \in A$ and $y \in B$.

An important consequence of the separation results is the so-called *support property*.

Theorem 1.4.10 (Support Property of Convex Sets) *Let (X, τ) be a TVS and $A \subseteq X$ closed convex with nonempty interior. Then, for every boundary point x_0 of A there exists $x^* \in X^*$, $x^* \ne 0$, such that*

   (i) $\operatorname{Re} x^*(x_0) = 1;$

   (ii) $\operatorname{Re} x^*(x) \le 1$ *for all* $x \in A$.

The functional x^* is called a *support functional* of the set A and x_0 a *support point* of A.

Remark 1.4.11 It follows that $\operatorname{Re} x^*(x_0) < 1$ for all $x \in \operatorname{int}(A)$. Theorem 1.4.10 can be reformulated as:

For every boundary point x_0 of A there exists $x^* \in X^* \setminus \{0\}$ such that $\operatorname{Re} x^*(x_0) = \sup\{\operatorname{Re} x^*(x) : x \in A\}.$

In the above formulation, supremum can be replaced with infimum.

A remarkable result of Bishop and Phelps [84] says that, if A is a bounded closed convex subset of a real Banach space X, then the set of support functionals of A is dense in X^* and the support points are dense in the boundary of A.

Taking $A = B_X$, the closed unit ball of a normed space X, it follows that there exists a functional $x^* \in X^* \setminus \{0\}$ attaining its norm at x_0, i.e., $|x^*(x_0)| = \|x^*\|$. By the above mentioned result, it follows that the functionals attaining their norms on B_X are dense in X^*, a property called *subreflexivity* (compare with James' theorem, Theorem 1.4.20).

See Sect. 8.8 for further results.

1.4.3 Normed Spaces

A *normed space* is a pair $(X, \|\cdot\|)$, where X is a vector space and $\|\cdot\|$ a norm on X. The topology of X is generated by the metric $d_{\|\cdot\|}(x, y) = \|y - x\|$, $x, y \in X$. A complete normed space is called a *Banach space*.

For a normed space $(X, \|\cdot\|)$ one uses the following notations:

- $B_X = \{x \in X : \|x\| \le 1\}$—the closed unit ball;
- $B'_X = \{x \in X : \|x\| < 1\}$—the open unit ball;
- $S_X = \{x \in X : \|x\| = 1\}$—the unit sphere of X.

It follows that

$$B[x_0, r] = x_0 + rB_X \quad \text{and} \quad B(x_0, r) = x_0 + rB'_X,$$

for $x_0 \in X$ and $r > 0$.

When X is a concrete Banach space, one uses also the notation $B(X)$ for its closed unit ball, for instance $B(\operatorname{Lip}(X, \mathbb{K}))$, etc.

Let X, Y be normed spaces and $T : X \to Y$ a linear operator. Then T is continuous if and only if there exists a nonnegative real number L such that

$$\|T(x)\| \le L \|x\|, \tag{1.4.1}$$

for all $x \in X$. A number $L \ge 0$ satisfying (1.4.1) is called a Lipschitz constant for T. Notice that this definition agrees with the previous one given for the Lipschitz condition in metric spaces (see Sect. 1.3.2) because, by the linearity of T,

$$\|Tx - Ty\| = \|T(x - y)\| \le L\|x - y\|.$$

The *norm* of an operator $T \in \mathscr{L}(X, Y)$ is given by

$$\|T\| = \sup\{\|T(x)\| : x \in X \text{ and } \|x\| \le 1\}. \tag{1.4.2}$$

One shows that $\|T\|$ is the smallest Lipschitz constant for T, i.e.,

- $\|T\|$ is a Lipschitz constant for T, and
- $\|T\| \le L$ for every Lipschitz constant L for T.

For this reason continuous linear operators are also called *bounded linear operators* (they are bounded on the closed unit ball B_X).

A linear operator $T : X \to Y$ between two normed spaces X, Y is called a *linear isometry* if

$$\|T(x)\| = \|x\|$$

for each $x \in X$. If further T is bijective, then we say that X and Y are *linearly isometric* (or *isometrically isomorph*) and use the notation $X \simeq Y$.

We mention the following result, showing the universality of the Banach space $C[0, 1]$ in the class of separable Banach spaces.

Theorem 1.4.12 (Banach-Mazur) *Every separable Banach space is linearly isometric to a subspace of* $C[0, 1]$.

This theorem has the following important corollary.

Corollary 1.4.13 (Banach-Fréchet-Mazur) *Every separable metric space can be isometrically embedded in* $C[0, 1]$.

Proofs of Theorem 1.4.12 and Corollary 1.4.13 can be found in [233, pp. 240–241].

Proposition 1.4.14 *Let X, Y be normed spaces.*

1. *The functional $\| \cdot \| : \mathscr{L}(X, Y) \to [0, \infty)$ given by (1.4.2) is a norm on the space $\mathscr{L}(X, Y)$.*
2. *The normed space $\mathscr{L}(X, Y)$ is complete (i.e., is a Banach space) if and only if Y is a Banach space. In particular, the dual space $X^* = \mathscr{L}(X, \mathbb{K})$ of X is always a Banach space.*

1.4.4   The Best Approximation Problem

Let X be a normed space, M a nonempty subset of X and $x \in X$. Denote by

$$d(x, M) = d_M(x) = \inf\{\|x - y\| : y \in M\}$$

the distance from x to M and let

$$P_M(x) = \{y \in M : \|x - y\| = d(x, M)\}$$

the *metric projection*.

The elements of the set $P_M(x)$ are called *best approximation elements* of x by elements in M (or *nearest points* to x in M) and the set-valued mapping $P_M : X \to 2^M$ is called *the metric projection* of X onto M. When $P_M(x)$ is formed by a single element, say $y_0 \in M$, we write $y_0 = P_M(x)$.

The set M is called *proximinal* if $P_M(x) \neq \emptyset$ for every $x \in X$ and *Chebyshevian* (or *Chebyshev*) if $P_M(x)$ contains exactly one element for every $x \in X$.

If $H = \{x \in X : x^*(x) = c\}$ is a hyperplane in X determined by a nonzero continuous linear functional x^* and by a number c, then the distance from an arbitrary point $x \in X$ to H can be calculated by the following formula, called Ascoli's formula

$$d(x, H) = \frac{|x^*(x) - c|}{\|x^*\|}. \tag{1.4.3}$$

1.4.5 Weak Topologies

Let X be a normed space. The *weak topology* w (or the topology $\sigma(X, X^*)$) on X is the locally convex topology generated by the family $\{p_{x^*} : x^* \in X^*\}$ of seminorms, where, for $x^* \in X^*$, $p_{x^*} : X \to [0, \infty)$ is the seminorm defined by

$$p_{x^*}(x) = |x^*(x)|, \quad x \in X.$$

The *weak* * topology* w^* (or the topology $\sigma(X^*, X)$) on X^* is the locally convex topology generated by the family $\{p_x : x \in X\}$ of seminorms, where, for $x \in X$, $p_x : X^* \to [0, \infty)$ is the seminorm defined by

$$p_x(x^*) = |x^*(x)|, \quad x^* \in X^*.$$

The convergence of nets with respect to these topologies can be characterized in the following way:

- a net $(x_i : i \in I)$ in X is w-convergent to $x \in X$ if and only if $\lim_i x^*(x_i) = x^*(x)$ for every $x^* \in X^*$;
- a net $(x_i^* : i \in I)$ in X^* is w^*-convergent to $x^* \in X^*$ if and only if $\lim_i x_i^*(x) = x^*(x)$ for every $x \in X$.

On the account of the above characterization of w^*-convergence, the w^*-topology is called sometimes the *pointwise topology* of X^* (as it induces the pointwise convergence of nets of functionals).

In contrast to this "weak" topologies the topology generated by the norm is called sometimes the *strong topology* of the normed space $(X, \| \cdot \|)$.

A useful property of normed spaces is the semicontinuity of the norm.

Proposition 1.4.15 *Let* $(X, \| \cdot \|)$ *be a normed space with dual* $(X^*, \| \cdot \|^*)$. *Then the norm* $\| \cdot \|$ *is weakly lsc on* X *and the dual norm* $\| \cdot \|^*$ *is weakly*-lsc on* X^*.

There are several deep results concerning weak topologies of Banach spaces.

Theorem 1.4.16 (Eberlein-Shmulian, see [456], Theorem 2.8.6) *Let* A *be a subset of a normed space* X. *Then the following are equivalent.*

1. *The set* A *is weakly compact.*
2. *The set* A *is countably weakly compact.*
3. *The set* A *is sequentially weakly compact.*

   The same equivalences hold for the relative versions of the above notions of weak compactness.

Another strong result is James' theorem on the characterization of weakly compact sets.

Theorem 1.4.17 (R. C. James, see [456], Theorem 2.9.3) *A weakly closed subset* A *of a Banach space* X *is weakly compact if and only if for every* $x^* \in X^*$ *there exists* $x_0 \in A$ *such that*

$$|x^*(x_0)| = \sup\{|x^*(x)| : x \in A\}.$$

In other words, a weakly closed subset A of a Banach space is weakly compact if and only if the modulus of every continuous linear functional attains its supremum on A.

Concerning the weak* topology we mention the following fundamental result.

Theorem 1.4.18 (Alaoglu-Bourbaki, see [456], Theorem 2.6.18) *The closed unit ball* B_{X^*} *of the dual* X^* *of a normed space* X *is* w^*-*compact.*

Remark 1.4.19 Notice that the equivalences from Theorem 1.4.16 do not hold for the weak*-topology.

1.4.6 The Bidual and Reflexivity

Let X be a normed space and let $X^{**} = (X^*)^*$ its bidual space. Define the mapping $j_X : X \to X^{**}$ by

$$j_X(x)(x^*) = x^*(x), \quad x^* \in X^*.$$

One shows that the definition is correct, i.e., $j_X(x) : X^* \to \mathbb{K}$ is linear and continuous. Furthermore, the operator j_X is linear and continuous and $\|j_X(x)\| = \|x\|$ for all $x \in X$, i.e., j_X is an isometric linear embedding of X into X^{**}. Taking into account this isometry the normed space X can be identified with its image in the bidual space X^{**}. Some authors identify the elements of X with their images in the bidual, i.e., they consider x as a functional on X^* acting by the rule $x(x^*) = x^*(x)$, $x^* \in X^*$.

The normed space X is called *reflexive* if $j_X(X) = X^{**}$. Since X^{**} is Banach, a reflexive normed space is necessarily Banach (i.e., complete).

Examples of reflexive Banach spaces are the spaces ℓ^p, L^p, for $1 < p < \infty$, and Hilbert spaces.

One says that a functional $x^* \in X^*$ *attains its norm* on the closed unit ball B_X of a normed space X if there exists $x \in B_X$ such that $|x^*(x)| = \|x^*\|$. If $x^* \neq 0$, then x is necessarily of norm one. One also says that x^* *supports* the unit ball B_X at x and that x is a *support point* for x^* in B_X.

A famous result of R. C. James characterizes reflexive Banach spaces in terms of the support functionals of the closed unit ball.

Theorem 1.4.20 (James' Characterization of Reflexivity) *A Banach space X is reflexive if and only if every continuous linear functional attains its norm on the closed unit ball of X.*

By the Alaoglu-Bourbaki theorem (Theorem 1.4.18) the unit ball of the dual X^* of a normed space X is weakly*-compact. The weak compactness of the unit ball of X characterizes reflexivity.

Theorem 1.4.21 *A Banach space X is reflexive if and only if its closed unit ball is weakly compact.*

For proofs of these results, see [456, Sections 2.8 and 2.9].

For further results on support functionals and support points, see Sect. 8.8.

1.4.7 Series and Summable Families in Normed Spaces

Let X be a normed space and $\sum x_n$ a (formal) series in X.

Definition 1.4.22 The series $\sum x_n$ is called:

- *convergent* (or *norm convergent*) if there exists $x \in X$ such that

$$\lim_{n \to \infty} \Big\| \sum_{k=1}^{n} x_k - x \Big\| = 0 \,,$$

i.e., the sequence $\left(\sum_{k=1}^{n} x_k \right)_{n \in \mathbb{N}}$ of partial sums is norm-convergent to x;

- *absolutely convergent* if the series $\sum_{n=1}^{\infty} \|x_n\|$ is convergent;
- *weakly convergent* if there exists $x \in X$ such that $x^*(x) = \sum_{n=1}^{\infty} x^*(x_n)$ for every $x^* \in X^*$;
- *unconditionally convergent* if the series $\sum_{k=1}^{\infty} x_{\sigma(k)}$ is convergent for every permutation σ (meaning a bijection $\sigma : \mathbb{N} \to \mathbb{N}$) of $\mathbb{N}$;
- *weakly unconditionally convergent* if the series $\sum_{k=1}^{\infty} x^*(x_{\sigma(k)})$ is convergent for every permutation σ of $\mathbb{N}$ and every $x^* \in X^*$;
- *subseries convergent* if the series $\sum_{k=1}^{\infty} x_{\sigma(k)}$ is convergent for every strictly increasing mapping $\sigma : \mathbb{N} \to \mathbb{N}$.

Remark 1.4.23 If the series $\sum_n x_n$ is convergent, then $\lim_{n \to \infty} \sum_{k=n}^{\infty} x_k = 0$. Also, the uniqueness of the sum follows for any of the considered kinds of convergence (except the absolute one).

Concerning the absolute convergence we mention the following result.

Proposition 1.4.24 *A normed space $(X, \|\cdot\|)$ is a Banach space if and only if every absolutely convergent series in X is convergent.*

The types of convergence considered in Definition 1.4.22 are related in the following way.

Theorem 1.4.25 *Let X be a Banach space.*

1. *A series $\sum x_n$ in X is (weakly) unconditionally convergent if and only if every subseries of $\sum x_n$ is (weakly) convergent.*
2. *Any absolutely convergent series is unconditionally convergent.*
3. *(Mazur-Orlicz) A series in X is weakly unconditionally convergent if and only if it is unconditionally convergent.*
4. *(Dvoretzky-Rogers) The unconditional and absolute convergence of series in X agree if and only if X is finite dimensional.*

We refer to [193] for proofs and further results.

Summable Families

These are the analogs of series defined for arbitrary families. Let $(X, \| \cdot \|)$ be a normed space and $\mathscr{A} = \{x_i : i \in I\} \subseteq X$ a family of elements in X. One says that $\mathscr{A}$ is *summable* to $x \in X$ if for every $\varepsilon > 0$ there exists a finite subset F_ε of I such that

$$\left\| x - \sum_{j \in F} x_j \right\| < \varepsilon ,$$

for every finite subset F of I containing F_ε. It follows that the set $J = \{i \in I : x_i \neq 0\}$ is at most countable and

$$x = \sum_{j \in J} x_j,$$

where the series is unconditionally convergent. Indexing J, $J = \{j_n : n \in \mathbb{N}\}$ this is equivalent to the fact that the series

$$\sum_{k=1}^{\infty} x_{j_{\sigma(k)}}$$

is convergent for every permutation (= bijection) $\sigma : \mathbb{N} \to \mathbb{N}$.

1.4.8 Inner Product Spaces

An *inner product* on a vector space X is a mapping $\langle \cdot, \cdot \rangle : X \times X \to \mathbb{K}$ satisfying the conditions

(IP1) $x \neq 0 \implies \langle x, x \rangle > 0$;

(IP2) $\langle y, x \rangle = \overline{\langle x, y \rangle}$;

(IP3) $\langle x + y, z \rangle = \langle x, z \rangle + \langle y, z \rangle$;

(IP4) $\langle \alpha x, y \rangle = \alpha \langle x, y \rangle$,

for all $x, y, z \in X$ and all $\alpha \in \mathbb{K}$. (The bar represents the conjugation operation in $\mathbb{C}$).

Remark 1.4.26 The above conditions imply

(IP3′) $\langle x, y + z \rangle = \langle x, y \rangle + \langle x, z \rangle$;

(IP4′) $\langle x, \alpha y \rangle = \overline{\alpha} \langle x, y \rangle$,

for all $x, y, z \in X$ and $\alpha \in \mathbb{K}$. One says that the inner product is linear in the first variable and conjugate linear with respect to the second one and the mapping $\langle \cdot, \cdot \rangle$ is called sesquilinear.

If $\langle \cdot, \cdot \rangle$ is an inner product on a vector space X, then

$$\|x\| = \sqrt{\langle x, x \rangle}, \quad x \in X, \tag{1.4.4}$$

is a norm on X.

An *inner product space* is a vector space X equipped with an inner product $\langle \cdot, \cdot \rangle$. The metric properties of X are understood in the sense of the norm (1.4.4). A complete inner product space is called a *Hilbert space*.

The inner product is continuous with respect to the norm (1.4.4), i.e.,

$$x_n \to x \quad \text{and} \quad y_n \to y \; \Rightarrow \; \langle x_n, y_n \rangle \to \langle x, y \rangle.$$

The triangle inequality for the norm is obtained as a consequence of the *Cauchy-Schwarz inequality*

$$|\langle x, y \rangle| \leq \sqrt{\langle x, x \rangle} \cdot \sqrt{\langle y, y \rangle}, \tag{1.4.5}$$

for all $x, y \in X$.

Some examples of Hilbert spaces are:

- $\mathbb{K}^n$ with the inner product

$$\langle x, y \rangle = \sum_{k=1}^{n} x_k \overline{y_k}, \tag{1.4.6}$$

for $x = (x_1, \ldots, x_n)$ and $y = (y_1, \ldots, y_n)$ in $\mathbb{K}^n$;
- the space ℓ^2 with the inner product

$$\langle x, y \rangle = \sum_{k=1}^{\infty} x_k \overline{y_k},$$

for $x = (x_k)_{k \in \mathbb{N}}$ and $y = (y_k)_{k \in \mathbb{N}}$ in ℓ^2;
- the space L^2 with the inner product

$$\langle f, g \rangle = \int f \overline{g} d\mu,$$

for $f, g \in L^2(\mu)$.

In $\mathbb{R}^n$ with the inner product (1.4.6) the inequality (1.4.5) becomes

$$\left| \sum_{k=1}^{n} x_k y_k \right| \leq \left(\sum_{k=1}^{n} x_k^2 \right)^{1/2} \left(\sum_{k=1}^{n} y_k^2 \right)^{1/2},$$

the usual Cauchy-Schwarz inequality.

Two elements x, y of an inner product space X are called *orthogonal*, denoted by $x \perp y$, if

$$\langle x, y \rangle = 0.$$

Pythagoras' rule is valid for any pair x, y of orthogonal elements,

$$\|x + y\|^2 = \|x\|^2 + \|y\|^2 .$$

The *orthogonal complement* of a nonempty subset Z of X is the set

$$Z^\perp = \{y \in X : \langle y, z \rangle = 0 \text{ for all } z \in Z\} .$$

It follows that $Z^\perp$ is always a closed subspace of X and $Z^\perp = \{0\}$ if and only if Z is dense in X.

Theorem 1.4.27 *Let X be a Hilbert space. If Y is a closed subspace of X, then*

$$X = Y \oplus Y^\perp,$$

i.e., every $x \in X$ can be uniquely written as $x = y + z$, where $(y, z) \in Y \times Y^\perp$ satisfy the equality $\|x\|^2 = \|y\|^2 + \|z\|^2$.

The above result characterizes Hilbert spaces in the sense that if every closed subspace of a Banach space $(X, \| \cdot \|)$ is complemented, then X is isomorphic to a Hilbert space, that is, there exists an inner product on X generating a norm equivalent to $\| \cdot \|$.

A family $\{x_i : i \in I\}$ of elements in X is called *orthogonal* if $\langle x_i, x_j \rangle = 0$ for all $i \neq j$ in I. An *orthonormal system* is an orthogonal system such that $\langle x_i, x_i \rangle = 1$ for all $i \in I$.

The following result is essential in the theory of Hilbert spaces.

Theorem 1.4.28 *For any Hilbert space X there exists an orthonormal system $\{e_i : i \in I\}$ such that every $x \in X$ can be written as*

$$x = \sum_{i \in I} \langle x, e_i \rangle e_i . \tag{1.4.7}$$

If X is separable then there exists a countable orthonormal system such that every $x \in X$ admits a representation of the form (1.4.7).

The expansion (1.4.7) is called the *Fourier expansion* of the element x and $\langle x, e_i \rangle$ its *Fourier coefficients*. A classical example is that of the Hilbert space $L^2[-\pi, \pi]$ and the orthonormal system $e_n(t) = \frac{1}{2\pi} \exp(int)$, $n \in \mathbb{Z}$. Every $f \in L^2[-\pi, \pi]$ can be written as

$$f = \sum_{n=-\infty}^{\infty} c_n e_n , \tag{1.4.8}$$

where

$$c_n = \frac{1}{2\pi} \int_{-\pi}^{\pi} f(t) \exp(-int) dt , \quad n \in \mathbb{Z} ,$$

are the Fourier coefficients of the expansion.

The convergence in (1.4.8) is understood in the L^2-norm. A more delicate subject of Fourier analysis is that of pointwise convergence in (1.4.8).

Another important result is the Riesz representation theorem.

Theorem 1.4.29 (Riesz) *Let X be a Hilbert space. For every linear continuous functional $x^* \in X^*$ there exists a unique $u \in X$ such that*

$$x^*(x) = \langle x, u \rangle \quad \text{for all } x \in X .$$

The equality $\|x^*\| = \|u\|$ is satisfied and the mapping $\Phi : X \to X^*$ given for $u \in X$ by $\Phi(u)(x) = \langle x, u \rangle$, $x \in X$, is a conjugate linear isometry of X onto X^*, that is, Φ is bijective and

$$\Phi(u + v) = \Phi(u) + \Phi(v), \quad \Phi(\alpha u) = \overline{\alpha} \Phi(u) \quad \text{and} \quad \|\Phi(u)\| = \|u\| ,$$

for all $u, v \in X$ and $\alpha \in \mathbb{K}$.

It follows that, in some sense, the dual of a Hilbert space X can be identified with X. In fact, the dual X^* is also a Hilbert space with respect to the inner product $\langle x^*, y^* \rangle^* = \langle \Phi^{-1}(y^*), \Phi^{-1}(x^*) \rangle$, and the dual of $(X^*, \langle \cdot, \cdot \rangle^*)$, given by Riesz' representation theorem, is X, implying the reflexivity of Hilbert spaces.

Corollary 1.4.30 *A Hilbert space is reflexive (viewed as a Banach space).*

1.4.9 Ordered Topological Vector Spaces

In the case of an ordered TVS (X, τ) some connections between order and topology hold. Let (X, τ) be a TVS with a preorder or an order, $\leq$ generated by a cone C.

We start with a simple result.

Proposition 1.4.31 *The cone C is closed if and only if the inequalities are preserved by limits, meaning that for all nets $(x_i : i \in I)$, $(y_i : i \in I)$ in X,*

$$(\forall i \in I, \ x_i \leq y_i \ \text{and} \ \lim_i x_i = x, \ \lim_i y_i = y) \Rightarrow x \leq y .$$

Other results are contained in the following proposition.

Proposition 1.4.32 ([35], Lemmas 2.3 and 2.4) *Let (X, τ) be a TVS ordered by a τ-closed cone C. Then the following hold.*

1. *The topology τ is Hausdorff.*
2. *The cone C is Archimedean.*
3. *The order-intervals are τ-closed.*
4. *If $(x_i : i \in I)$ is an increasing net which is τ-convergent to $x \in X$, then $x = \sup_i x_i$.*

Conversely, if the topology τ is Hausdorff, $\mathrm{int}(C) \neq \emptyset$ and C is Archimedean, then C is τ-closed.

Remark 1.4.33 In what follows by an ordered TVS we shall understand a TVS ordered by a closed cone. Also, in an ordered TVS (X, τ, C) we have some parallel notions—with respect to topology and with respect to order. To make a distinction between them, those referring to order will have the prefix "order-", as, for instance, "order-bounded", "order-complete", etc., while for those referring to topology we shall use the prefix "τ-", or "topologically-", e.g., "τ-bounded", "τ-complete" (resp. "topologically-bounded", "topologically-complete"), etc.

Normed Lattices

A seminorm p on a vector lattice X is called a *lattice seminorm* if $|x| \leq |y|$ implies $p(x) \leq p(y)$ for all $x, y \in X$. A norm with this property is called a *lattice norm*. A vector lattice equipped with a lattice norm is called a *normed lattice*. If it is also complete, then it is called a *Banach lattice*. A good presentation of various problems concerning Banach lattices and applications is given in [464].

1.4.10 Spaces of Continuous Functions

Let T be a Hausdorff topological space. Recall that the support $\mathrm{spt}(f)$ of a continuous function $f : T \to \mathbb{K}$ is the closure of the set $\{t \in T : f(t) \neq 0\}$. One says that the continuous function f *vanishes at infinity* if for every $\varepsilon > 0$ there exists a compact set $K_\varepsilon \subseteq T$ such that

$$|f(t)| \leq \varepsilon \text{ for all } t \in T \setminus K_\varepsilon.$$

A function f vanishing at infinity is bounded. Indeed, if K_1 is a compact subset of T corresponding to $\varepsilon = 1$, then

$$|f(t)| \leq \max\left\{1, \sup_{t \in K_1} |f(t)|\right\},$$

for all $t \in T$.

Consider the following spaces of continuous functions:

- $B(T)$—the space of all bounded functions $f : T \to \mathbb{K}$;
- $C(T)$—the space of all continuous functions $f : T \to \mathbb{K}$;
- $C_b(T)$—the space of all bounded continuous functions $f : T \to \mathbb{K}$;
- $C_0(T)$—the space of all continuous functions $f : T \to \mathbb{K}$ vanishing at infinity;
- $C_{00}(T)$—the space of all continuous functions $f : T \to \mathbb{K}$ with compact support.

It is obvious that all these four sets are vector spaces with respect to the pointwise operations of addition and multiplication by scalars and that

$$C_{00}(T) \subseteq C_0(T) \subseteq C_b(T) \subseteq C(T) \subseteq B(T).$$

If T is compact then $C_{00}(T) = C(T)$ so that all spaces of continuous functions from above coincide.

For $f \in C_b(T)$ put

$$\|f\|_\infty = \sup\{|f(t)| : t \in T\}. \tag{1.4.9}$$

Theorem 1.4.34 *Let T be a Hausdorff topological space. The functional $\| \cdot \|_\infty$ given by (1.4.9) is a norm on $B(T)$ and so on the subspaces $C_{00}(T)$, $C_0(T)$, $C_b(T)$ too. The space $B(T)$ is complete with respect to this norm as well as its subspaces $C_0(T)$ and $C_b(T)$.*

Remark 1.4.35 The norm (1.4.9) is called the *supremum norm* (or the *sup-norm*). It can be defined on $B(T)$ for an arbitrary set T and the convergence of a sequence (f_n) to a function f in $B(T)$ with respect to this norm is the uniform convergence, denoted by $f_n \overset{u}{\to} f$ (or by $f_n \overset{u}{\underset{S}{\to}} f$ if we want to emphasize the set $S \subseteq T$ on which the uniform convergence takes place). This means that, for every $\varepsilon > 0$ there exists $n_0 \in \mathbb{N}$ such that, for all $n \geq n_0$,

$$|f_n(t) - f(t)| \leq \varepsilon \text{ for all } t \in T \quad (\text{resp. } t \in S).$$

Since the normed space $(B(T), \| \cdot \|_\infty)$ is complete, to prove the completeness of the subspaces $C_0(T)$ and $C_b(T)$ it suffices to show that they are closed in $B(T)$.

The normed space $(C_{00}(T), \| \cdot \|_\infty)$ might not be complete if T is not compact. If T is locally compact, then the following density result holds.

Theorem 1.4.36 ([166], Proposition 7.3.1) *If T is a locally compact Hausdorff space, then $C_{00}(T)$ is dense in $C_0(T)$ with respect to the sup-norm.*

1.4.11 The Stone-Weierstrass Theorem

An *algebra* is a vector space X equipped with a product $\cdot : X \times X \to X$ satisfying the conditions

(A1) $(x \cdot y) \cdot z = x \cdot (y \cdot z)$ (associativity);

(A2) $\begin{aligned} &x \cdot (y + z) = x \cdot y + x \cdot z \quad \text{and} \\ &(y + z) \cdot x = y \cdot x + z \cdot x \end{aligned}$ (distributivity);

(A3) $1 \cdot x = x$ and $\alpha(\beta x) = (\alpha\beta)x$,

for all $x, y, z \in X$ and all $\alpha, \beta \in \mathbb{K}$.

If there exists $e \in X$ such that

$$\text{(A4)} \qquad e \cdot x = x \cdot e = x ,$$

for all $x \in X$, then e is called a *unit* and the algebra X is called an *algebra with unit* (or a *unital algebra*). If the multiplication is commutative then the algebra X is called *commutative*.

A normed algebra is an algebra equipped with a norm $\| \cdot \|$ such that

$$\text{(A5)} \qquad \|x \cdot y\| \le \|x\| \, \|y\| ,$$

for all $x, y \in X$. If X has a unit e, then one asks that

$$\text{(A6)} \qquad \|e\| = 1 .$$

A complete normed algebra is called a *Banach algebra*. A typical example of a Banach algebra is the algebra $C(T)$ of continuous functions on a compact Hausdorff space T with values in $\mathbb{K}$. The multiplication is the pointwise multiplication defined by

$$(f \cdot g)(t) = f(t) \, g(t), \quad t \in T ,$$

for $f, g \in C(T)$.

This algebra is commutative and has unit $\mathbf{1}(t) = 1$, $t \in T$. If one considers real scalars, then the notation $C(T, \mathbb{R})$ is used.

Another example is the algebra $\mathscr{L}(X)$ of continuous linear operators on a Banach space X with the composition of operators as multiplication. This is a non-commutative Banach algebra having as unit the identity operator $\mathrm{Id}_X(x) = x$, $x \in X$.

The space $C(T, \mathbb{R})$ is also a Banach lattice with respect to the pointwise order defined by

$$f \leq g \iff f(t) \leq g(t) \quad \text{for all } t \in T,$$

for $f, g \in C(T, \mathbb{R})$. In this case, the lattice operation $\vee$ is given by

$$(f \vee g)(t) = \max\{f(t), g(t)\}, \quad t \in T,$$

and $\wedge$ by

$$(f \wedge g)(t) = \min\{f(t), g(t)\}, \quad t \in T.$$

We say that a family $\mathscr{A} \subseteq C(T)$ is *separating* (or that it *separates the points* in T) if for every $t_1 \neq t_2$ in T there exists $f \in \mathscr{A}$ such that $f(t_1) \neq f(t_2)$.

Theorem 1.4.37 (Stone-Weierstrass Theorem—Lattice Version) *Let T be a compact Hausdorff space. Any sublattice Y of $C(T, \mathbb{R})$ that separates the points of T and contains the constant functions is dense in $C(T, \mathbb{R})$.*

Remark 1.4.38 The conditions imposed on Y mean that

(i) Y is a subspace of $C(T, \mathbb{R})$;
(ii) $\mathbf{1} \in Y$;
(iii) $f, g \in Y \implies f \vee g \in Y$,
(iv) Y separates the points of T.

Theorem 1.4.39 (Stone-Weierstrass Theorem—Real Algebraic Version) *Let T be a compact Hausdorff space and Y a subalgebra of $C(T, \mathbb{R})$. If*

(i) *Y separates the points of T;*
(ii) *$\mathbf{1} \in Y$,*

then Y is dense in $C(T, \mathbb{R})$.

In the complex case a supplementary condition should be added.

Theorem 1.4.40 (Stone-Weierstrass Theorem—Complex Algebraic Version) *Let T be a compact Hausdorff space and let Y be a subalgebra of $C(T)$ such that*

(i) *Y separates the points of T;*
(ii) *$\mathbf{1} \in Y$;*
(iii) *$f \in Y \implies \overline{f} \in Y$,*

where $\overline{f}(t) = \overline{f(t)}$, $t \in T$ (the bar is the complex conjugation).
   Then Y is dense in $C(T)$.

Remark 1.4.41 A subalgebra of an algebra X is a subset Y of X that is also an algebra with respect to the operations from X. This means that

(i) $y_1 + y_2 \in Y$;   (ii) $\alpha y \in Y$;   (iii) $y_1 \cdot y_2 \in Y$,

for all $y, y_1, y_2 \in Y$ and all $\alpha \in \mathbb{K}$.

1.4.12 Compactness in Spaces of Continuous Functions

For a compact Hausdorff space T and a Banach space E denote by $C(T, E)$ the vector space of all continuous functions from T to E. Equipped with the norm

$$\|f\|_\infty = \sup\{\|f(t)\| : t \in T\}, \quad f \in C(T, E),$$

$C(T, E)$ becomes a Banach space.

A subset Y of $C(T, E)$ is called:

- *equicontinuous* if for every $t \in T$ and every $\varepsilon > 0$ there exists a neighborhood V of t such that

$$\|f(t) - f(t')\| \le \varepsilon \text{ for all } t' \in V \text{ and all } f \in Y;$$

- *pointwise bounded* if, for every $t \in T$,

$$\sup\{\|f(t)\| : f \in Y\} < \infty,$$

i.e., the set $\{f(t) : f \in Y\}$ is bounded in E for every $t \in T$;
- if T is a compact metric space with metric δ, then Y is called *uniformly equicontinuous* if for every $\varepsilon > 0$ there exists $\delta > 0$ such that

$$\|f(t) - f(t')\| \le \varepsilon \text{ for all } t, t' \in T \text{ with } \delta(t, t') < \delta \text{ and all } f \in Y.$$

It is obvious that every norm-bounded subset of $C(T, E)$ is pointwise bounded.

The following compactness criteria will be used several times throughout this book.

Theorem 1.4.42 (Arzela-Ascoli Theorem—The Vector Case) *A subset Y of $C(T, E)$ is relatively compact if and only if*

(i) *Y is equicontinuous;*
(ii) *for every $t \in T$ the set $\{f(t) : f \in Y\}$ is totally bounded in E.*

The subset Y is compact if and only if it is closed and satisfies (i) and (ii).

Remark 1.4.43 If T is a compact metric space, then every relatively compact subset of $C(T, E)$ is uniformly equicontinuous and bounded.

Since a bounded subset of $\mathbb{K}^n$ is totally bounded, the above criterion has the following corollary, also known as the Arzela-Ascoli theorem.

Corollary 1.4.44 (Arzela-Ascoli Theorem—The Scalar Case) *A subset Y of $C(T, \mathbb{K}^n)$ is relatively compact if and only if*

(i) *Y is equicontinuous;*
(ii) *Y is pointwise bounded.*

The subset Y is compact if and only if it is closed and satisfies (i) and (ii).

1.4.13   Extreme Points of Convex Sets

Let A be a subset of a linear space X. A subset B of A is said to be an *extremal subset* of A if x, y belong to B whenever $\lambda x + (1 - \lambda)y \in B$ for $x, y \in A$ and some $\lambda \in (0, 1)$. If an extremal subset of A consists of a single point, then that point is called an *extreme point* of A. The set of all extreme points of A is denoted by $\mathrm{ext}(A)$.

Remark 1.4.45 A point $a \in A$ belongs to $\mathrm{ext}(A)$ if and only if $x = a = y$ for any $x, y \in A$ such that $a = \lambda x + (1 - \lambda)y$ for some $\lambda \in (0, 1)$. Geometrically this means that $a \in \mathrm{ext}(A)$ if and only if a does not lie on any open segment whose endpoints are in A.

The most important and useful result concerning extreme points is the Krein-Milman theorem.

Theorem 1.4.46 (Krein-Milman) *If K is a nonempty compact convex subset of a locally convex space then $K = \overline{\mathrm{co}}(\mathrm{ext}(K))$.*

In the finite dimensional case the result can be strengthened.

Theorem 1.4.47 (Minkowski) *Let $X = \mathbb{R}^n$.*

1. *Any point in the convex hull of a subset A of X can be written as a convex combination of at most $n + 1$ points in A.*
2. *Let K be a compact convex subset of X. Then every boundary point of K is a convex combination of at most n points in $\mathrm{ext}(K)$, and every interior point of K is a convex combination of at most $n + 1$ points in $\mathrm{ext}(K)$.*

1.4.14   Differentiability of Vector Functions

For vector-valued functions there are two basic notions of differentiability, namely Gâteaux and Fréchet.

Let X and Y be normed spaces, U an open subset of X and $x_0 \in U$. A function $f : U \to Y$ is said to be *Gâteaux differentiable* at x_0 if there exists a bounded linear operator $A : X \to Y$ such that

$$\lim_{t \to 0} \frac{f(x_0 + tu) - f(x_0)}{t} = A(u) \qquad (1.4.10)$$

for each $u \in X$.

The operator A is called the *Gâteaux differential* of f at x_0 and it is denoted by $Df(x_0; \cdot)$.

If Y is the dual of a Banach space Z, then one can define the w^*-*differential* of a function $f : U \to Z^*$ at $x_0 \in U$ as a continuous linear operator $A : X \to Z^*$ for which the limit (1.4.10) exists in weak*-sense, that is

$$\lim_{t \to 0} \left(\frac{f(x_0 + tu) - f(x_0)}{t} - A(u) \right)(z) = 0,$$

for every $z \in Z$. We shall use the notation $A = D_{w^*} f(x_0; \cdot)$.

If the limit in the definition of Gâteaux differential exists uniformly with respect to u in the unit sphere of X, we say that f is *Fréchet differentiable* at x_0 and A is called the *Fréchet differential* of f at x_0. Equivalently, f is Fréchet differentiable at x_0 if there exists a bounded linear operator $A : X \to Y$ such that

$$\lim_{\|u\| \to 0} \frac{f(x_0 + u) - f(x_0) - A(u)}{\|u\|} = 0.$$

The bounded linear operator A is denoted in this case by $Df(x_0)$, or by $f'(x_0)$.

It is obvious that if f is Fréchet differentiable at x_0 with Fréchet differential $A \in \mathscr{L}(X, Y)$, then it is Gâteaux differentiable at x_0 and the Gâteaux differential is also A.

In the case of a function $f : U \to Y$, where U is an open subset of $\mathbb{K} (= \mathbb{R}$ or $\mathbb{C})$, the limit

$$f'(t_0) = \lim_{s \to 0} \frac{f(t_0 + s) - f(t_0)}{s}$$

is called the *derivative* of f at $t_0 \in \mathbb{K}$ (provided it exists, as an element of Y).

Notice that every $\varphi \in \mathscr{L}(\mathbb{K}, Y)$ can be written as $\varphi(t) = t\varphi(1)$, $t \in \mathbb{K}$, and the mapping $\varphi \mapsto \varphi(1)$ is an isometric isomorphism between $\mathscr{L}(\mathbb{K}, Y)$ and Y.

Based on this remark, it follows that the function f is Fréchet differentiable at $t_0 \in \mathbb{K}$ if and only if it has a derivative at t_0, the differential $Df(t_0) \in \mathscr{L}(\mathbb{K}, Y)$ and the derivative $f'(t_0) \in Y$ being related by the equality

$$Df(t_0)(s) = sf'(t_0), \quad s \in \mathbb{K}.$$

1.4.15 Some Geometric Properties of Normed Spaces

We shall present some geometric properties of normed spaces. Good presentations of various aspects of the geometry of Banach spaces are given in the books [22, 71, 191, 192, 232, 233, 395, 456]. The volumes [311, 312] contain expert surveys on various aspects of the geometry of Banach spaces.

Strictly Convex Normed Spaces

A normed space X is called *strictly convex* (or *rotund*) if $\|2^{-1}(x+y)\| < 1$ for every pair x, y of distinct elements in S_X, that is, the unit sphere of X does not contain nontrivial line segments. In fact, for $x \neq y$ in S_X, the function $t \mapsto \|x + t(y - x)\|$, $t \in [0, 1]$, is convex, so that if $\|(1 - t_0)x + t_0 y\| = 1$ for some $t_0 \in (0, 1)$, then $\|(1 - t)x + ty\| = 1$ for all $t \in [0, 1]$.

An equivalent formulation: the normed space X is strictly convex if and only if for all $x, y \neq 0$ the equality $\|x + y\| = \|x\| + \|y\|$ implies $y = \alpha x$ for some $\alpha > 0$. For this reason, strictly convex spaces are called by some authors "strictly normed".

The strict convexity of a normed space can be characterized in terms of the uniqueness of best approximation.

Proposition 1.4.48 *A normed space X is strictly convex if and only if every nonempty convex subset C of X is a uniqueness set for best approximation, i.e., every point in X has at most one nearest point in C.*

Uniformly and Locally Uniformly Convex Normed Spaces

A normed space X is called *uniformly convex* (or *uniformly rotund*) if for every $\varepsilon > 0$ there exists $\delta > 0$ such that for all $x, y \in S_X$

$$\|2^{-1}(x + y)\| > 1 - \delta \implies \|x - y\| < \varepsilon, \tag{1.4.11}$$

or, equivalently,

$$\|x - y\| \geq \varepsilon \implies \|2^{-1}(x + y)\| \leq 1 - \delta.$$

The *modulus of convexity* δ_X of the space X is given by

$$\delta_X(\varepsilon) = \inf\{1 - \|2^{-1}(x + y)\| : x, y \in S_X, \|x - y\| \geq \varepsilon\},$$

for $\varepsilon > 0$.

The normed space X is uniformly convex if and only if $\delta_X(\varepsilon) > 0$ for all $\varepsilon \in (0, 2]$.

If for every $x \in S_X$ and $\varepsilon > 0$ there exists $\delta > 0$ such that (1.4.11) holds for all $y \in S_X$, then the normed space X is called *locally uniformly convex* (or *locally uniformly rotund*).

The *local modulus of convexity* δ_X of the space X is given by

$$\delta_X(x, \varepsilon) = \inf\{1 - \|2^{-1}(x + y)\| : y \in S_X, \|x - y\| \geq \varepsilon\},$$

for $x \in S_X$ and $\varepsilon > 0$.

The normed space X is locally uniformly convex if and only if $\delta_X(x, \varepsilon) > 0$ for all $\varepsilon \in (0, 2]$ and $x \in S_X$.

Theorem 1.4.49 (B. J. Pettis) *A uniformly convex Banach space is reflexive.*

Remark 1.4.50 The spaces L^p and ℓ^p are uniformly convex, so reflexive. But there are locally uniformly convex Banach spaces which are not reflexive. In fact, on every separable Banach space there exists an equivalent locally uniformly convex norm, see, for instance, [178, Theorem 1, p. 160].

A Banach space is called *superreflexive* if it admits a uniformly convex equivalent norm (see [233, Ch. 9]).

The following characterizations are useful in applications.

Proposition 1.4.51 *Let X be a normed space.*

1. *The space X is uniformly convex if and only if for all sequences (x_n), (y_n) in B_X,*

$$\|x_n + y_n\| \to 2 \implies \|x_n - y_n\| \to 0.$$

2. *The space X is locally uniformly convex if and only if for all $x \in S_X$ and all sequences (y_n) in B_X,*

$$\|x + y_n\| \to 2 \implies \|x - y_n\| \to 0.$$

Smooth Normed Spaces
A normed space X is called *smooth* if the norm $\|\cdot\|$ is Gâteaux differentiable for every $x \in X \setminus \{0\}$. Taking into account the convexity of the mapping $x \mapsto \|x\|$, this is equivalent to the fact that the limit

$$\lim_{t \to 0} \frac{\|x + th\| - \|x\|}{t} \tag{1.4.12}$$

exists for every $h \in X$. In this case there exists $x^* \in X^*$ such that

$$\lim_{t \to 0} \frac{\|x + th\| - \|x\|}{t} = x^*(h)$$

for every $h \in X$. In its turn, this is equivalent to the fact that

$$\lim_{t \to 0+} \frac{\|x + th\| + \|x - th\| - 2\|x\|}{t} = 0 \tag{1.4.13}$$

for every $h \in X$.

One shows that the normed space X is smooth if and only if for every $x \in S_X$ there exists exactly one $x^* \in S_{X^*}$ attaining its norm at x, that is, such that $x^*(x) = 1$.

The following duality results between strict convexity and smoothness hold.

Proposition 1.4.52 *Let X be a normed space and X^* its conjugate.*

1. *If X^* is strictly convex, then X is smooth.*
2. *If X^* is smooth, then X is strictly convex.*
3. *If X is reflexive, then*

   (i) *X is strictly convex $\iff$ X^* is smooth, and*
   (ii) *X is smooth $\iff$ X^* is strictly convex.*

Uniformly Smooth Normed Spaces

One defines the *uniformly smooth* normed spaces as those spaces for which the limit (1.4.12) exists uniformly with respect to $x, h \in S_X$. One can also say that the norm is *uniformly Fréchet differentiable* on S_X.

Motivated by the relation (1.4.13), the modulus of smoothness $\delta_X : (0, \infty) \to [0, \infty)$ of a normed space X is defined by the formula

$$\delta_X(t) = \sup \left\{ \frac{1}{2}(\|x + th\| + \|x - th\|) - 1 : x, h \in S_X \right\}, \quad 0 < t < \infty.$$

The following theorem illustrates the role the modulus of smoothness plays in the characterization of smoothness.

Theorem 1.4.53 *A normed space X is uniformly smooth if and only if*

$$\lim_{t \to 0+} \frac{\delta_X(t)}{t} = 0.$$

There are two remarkable formulae, proved by Lindenstrauss in 1963 (see [395, Propositions 1.e.2 and 1.e.3]), relating the moduli of convexity and of uniform smoothness, having as consequence the duality between uniform convexity and uniform smoothness.

Theorem 1.4.54 *Let X be a normed space and X^* its conjugate.*

1. *The modulus of convexity δ_X, δ_{X^*} and the modulus of smoothness δ_X, δ_{X^*} of the spaces X and X^* are related by the following formulae*

$$\delta_{X^*}(t) = \sup\{ \frac{t\epsilon}{2} - \delta_X(\epsilon) : 0 \le \epsilon \le 2\}$$

and

$$\delta_X(t) = \sup\{\frac{t\epsilon}{2} - \delta_{X^*}(\epsilon) : 0 \leq \epsilon \leq 2\}$$

for every $t > 0$.
2. *A Banach space X is uniformly convex if and only if X^* is uniformly smooth.*
3. *A uniformly smooth Banach space is reflexive.*

1.4.16 Quasi-Normed Spaces

We present now some classes of non-locally convex topological vector spaces. Good references for spaces of this kind are [328, 335], [366, pp. 156–166], [603].

Throughout this subsection all topological vector spaces (TVS) will be supposed to be Hausdorff.

A *quasi-norm* on a vector space X is a functional $\|\cdot\| : X \to \mathbb{R}_+$ for which there exists a real number $k \geq 1$ so that

(QN1) $\|x\| = 0 \iff x = 0$;
(QN2) $\|\alpha x\| = |\alpha|\,\|x\|$;
(QN3) $\|x + y\| \leq k(\|x\| + \|y\|)$,

for all $x, y \in X$ and $\alpha \in \mathbb{K}$.

If $k = 1$, then $\|\cdot\|$ is a norm. The smallest constant k for which the inequality (QN3) is satisfied for all $x, y \in X$ is called the *modulus of concavity* of the quasi-normed space X.

A *quasi-normed space* is a pair $(X, \|\cdot\|)$ where X is a vector space and $\|\cdot\|$ is quasi-norm on X. One defines a topology $\tau_{\|\cdot\|}$ on a quasi-normed space $(X, \|\cdot\|)$ in the usual way: the balls $B[x, r]$, $r > 0$, form a neighborhood basis at the point $x \in X$. It turns out that this is a metrizable vector topology, so that a quasi-normed space is a TVS (usually non-locally convex). The convergence of a sequence (x_n) in X to $x \in X$, with respect to this topology, is equivalent to $\lim_{n\to\infty} \|x_n - x\| = 0$. The sequence (x_n) is called *Cauchy* if $\lim_{m,n\to\infty} \|x_n - x_m\| = 0$ and one says that X is a *quasi-Banach* space if every Cauchy sequence in X converges to some $x \in X$.

For a linear operator T from a quasi-normed space $(X, \|\cdot\|_X)$ to a normed space $(Y, \|\cdot\|_Y)$ put

$$\|T\| = \sup\{\|Tx\|_Y : x \in X, \|x\|_X \leq 1\}.$$

In particular,

$$\|x^*\| = \sup\{|x^*(x)| : x \in X, \|x\|_X \leq 1\}, \quad x^* \in X^*, \tag{1.4.14}$$

is a norm on the dual space $X^* = (X, \|\cdot\|_X)^*$.

It follows that T is continuous if and only if $\|T\| < \infty$ and, in this case,

$$\|Tx\|_Y \leq \|T\|\|x\|_X, \quad x \in X,$$

$\|T\|$ being the smallest number $L \geq 0$ for which the inequality $\|Tx\|_Y \leq L\|x\|_X$ holds for all $x \in X$. If Y is also a quasi-normed space, then $\|\cdot\|$ is only a quasi-norm on the space $\mathscr{L}(X, Y)$ of all continuous linear operators from X to Y.

Two quasi-norms $\|\cdot\|_1, \|\cdot\|_2$ on a vector space X are called *equivalent* if they generate the same topology, or equivalently, if

$$\|x_n - x\|_1 \to 0 \iff \|x_n - x\|_2 \to 0,$$

for all sequences (x_n) in X and $x \in X$. As in the case of norms, the equivalence of two quasi-norms $\|\cdot\|_1, \|\cdot\|_2$ on a vector space X is equivalent to the existence of two numbers $\alpha, \beta > 0$ such that

$$\alpha\|x\|_1 \leq \|x\|_2 \leq \beta\|x\|_1,$$

for all $x \in X$.

A subset A of a TVS (X, τ) is called *bounded* if it is absorbed by any 0-neighborhood, i.e. for every $V \in \mathscr{V}_\tau(0)$ there exists $t > 0$ such that $A \subseteq tV$. A TVS is called *locally bounded* if it has a bounded 0-neighborhood. A quasi-normed space $(X, \|\cdot\|)$ is locally bounded, as the closed unit ball $B_X = \{x \in X : \|x\| \leq 1\}$ is a bounded neighborhood of 0. One shows that, conversely, the topology of every locally bounded TVS is generated by a quasi-norm.

A quasi-normed space $(X, \|\cdot\|)$ is normable (i.e., there exists a norm $\|\cdot\|_1$ on X equivalent to the quasi-norm $\|\cdot\|$) if and only if 0 has a bounded convex neighborhood (implying that X is locally convex).

Definition 1.4.55 An *F-norm* on a vector space X is a mapping $\|\cdot\| : X \to \mathbb{R}_+$ satisfying the conditions

(F1) $\|x\| = 0 \iff x = 0$;

(F2) $\|\lambda x\| \leq \|x\|$ for all $\lambda \in \mathbb{K}$ with $|\lambda| \leq 1$;

(F3) $\|x + y\| \leq \|x\| + \|y\|$;

(F4) $\|x_n\| \to 0 \implies \|\lambda x_n\| \to 0$;

(F5) $\lambda_n \to 0 \implies \|\lambda_n x\| \to 0$,

for all $x, y, x_n \in X$ and all $\lambda, \lambda_n \in \mathbb{K}$.

It follows that $d(x, y) = \|y - x\|$, $x, y \in X$, is a translation-invariant metric on X defining a vector topology. It is known that the metrizability of a TVS (X, τ) is equivalent to the existence of a countable basis of 0-neighborhoods, and, in this case, there exists a translation-invariant metric d on X generating the topology τ.

One shows, see [366, p. 163], that the topology of a metrizable TVS can be always given by an F-norm. If (X, τ) is a TVS, then the topology τ generates a uniformity $\mathscr{W}_\tau$ on X, a basis of it being given by the sets

$$W_U = \{(x, y) \in X^2 : y - x \in U\},$$

where U runs over a 0-neighborhood basis in X. Any translation-invariant metric generating the topology τ generates the same uniformity $\mathscr{W}_\tau$, so that if X is complete with respect to $\mathscr{W}_\tau$, then it is complete with respect to any translation-invariant metric generating the topology τ.

An F-*space* is a vector space equipped with a complete F-norm.

Typical examples of quasi-Banach spaces are the spaces $L^p[0, 1]$ and ℓ^p with $0 < p < 1$ equipped with the quasi-norms

$$\|f\|_p = \left(\int_0^1 |f(t)| dt \right)^{1/p} \quad \text{and} \quad \|x\|_p = \left(\sum_{k=1}^\infty |x_k|^p \right)^{1/p}, \tag{1.4.15}$$

for $f \in L^p[0, 1]$ and $x = (x_k)_{k\in\mathbb{N}} \in \ell^p$, respectively.

The quasi-norms $\| \cdot \|_p$ satisfy the inequalities

$$\|f + g\|_p \le 2^{(1-p)/p}(\|f\|_p + \|g\|_p) \text{ and}$$

$$\|x + y\|_p \le 2^{(1-p)/p}(\|x\|_p + \|y\|_p), \tag{1.4.16}$$

for all $f, g \in L^p[0, 1]$ and all $x, y \in \ell^p$.

The constant $2^{(1-p)/p} > 1$ is sharp, i.e., the moduli of concavity of the spaces $L^p[0, 1]$ and ℓ^p are both equal to $2^{(1-p)/p}$.

To show this, we start with the elementary inequalities

$$(a + b)^p \le a^p + b^p \le 2^{1-p}(a + b)^p, \tag{1.4.17}$$

valid for all $a, b > 0$.

Let $f, g \in L^p[0, 1]$. The first inequality from above implies

$$|f(t) + g(t)|^p \le (|f(t)| + |g(t)|)^p \le |f(t)|^p + |g(t)|^p,$$

for almost all $t \in [0, 1]$, so that

$$\|f + g\|_p^p = \int_0^1 |f(t) + g(t)|^p dt \le \int_0^1 |f(t)|^p dt + \int_0^1 |g(t)|^p dt$$

$$= \|f\|_p^p + \|g\|_p^p. \tag{1.4.18}$$

This inequality and the second inequality from (1.4.17) yield

$$\|f + g\|_p = \left(\|f + g\|_p^p\right)^{1/p} \le \left(\|f\|_p^p + \|g\|_p^p\right)^{1/p}$$
$$\le 2^{(1-p)/p}(\|f\|_p + \|g\|_p).$$

Similar calculations can be done to show that

$$\|x + y\|_p \le 2^{(1-p)/p}(\|x\|_p + \|y\|_p),$$

for all $x, y \in \ell^p$.

To show that the constant $2^{(1-p)/p}$ is sharp take $x = (1, 0, 0 \ldots)$ and $y = (0, 1, 0 \ldots)$ in the case of the space ℓ^p. Then

$$\|x + y\|_p = 2^{1/p} \quad \text{and} \quad 2^{(1-p)/p}(\|x\|_p + \|y\|_p) = 2^{(1-p)/p} \cdot 2 = 2^{1/p},$$

that is, we have equality in the second inequality from (1.4.16). In the case of the space $L^p[0, 1]$ take $f = \chi_{[0, \frac{1}{2})}$ and $g = \chi_{[\frac{1}{2}, 1]}$ to obtain equality in the first inequality from (1.4.16).

Remark 1.4.56 Apparently similar, the quasi-normed spaces ℓ^p and $L^p[0, 1]$ drastically differ. For instance, the space $L^p[0, 1]$ has trivial dual, $(L^p[0, 1])^* = \{0\}$, while $(\ell^p)^* = \ell^\infty$ (see [366, pp. 156–158]).

The duality $\alpha \mapsto \varphi_\alpha \in (\ell^p)^*$ for $\alpha = (\alpha_k) \in \ell^\infty$, is given by the formula

$$\varphi_\alpha(x) = \sum_{k=1}^{\infty} \alpha_k x_k, \quad \text{for} \ x = (x_k) \in \ell^p,$$

(see [417, p. 110]).

Pallaschke [535] and Turpin [659] proved that every compact endomorphism of L^p, $0 < p < 1$, is null. Kalton and Shapiro [334] showed that there exists a quasi-Banach space with trivial dual admitting non-trivial compact endomorphisms. The example is a quotient space of the Hardy space H^p, $0 < p < 1$.

A *p-norm*, where $0 < p \le 1$, is a mapping $\| \cdot \| : X \to \mathbb{R}_+$ satisfying (QN1) and (QN2) and

$$(QN3') \qquad \|x + y\|^p \le \|x\|^p + \|y\|^p,$$

for all $x, y \in X$.

The inequality from (1.4.18) shows that the quasi-norm of the space $L^p[0, 1]$ is a p-norm. Similarly, the quasi-norm of ℓ^p is a p-norm too.

A famous result of Aoki [39] and Rolewicz [601] says that on any quasi-normed space $(X, \| \cdot \|)$ there exists a p-norm equivalent to $\| \cdot \|$, where p is determined from the equality $2^{1/p} = k$, k being the constant from (QN3).

Let $0 < p \leq 1$. A subset A of a vector space X is called p-*convex* if $\alpha x + \beta y \in A$ for all $x, y \in A$ and all $\alpha, \beta \geq 0$ with $\alpha^p + \beta^p = 1$, and p-*absolutely convex* if $\alpha x + \beta y \in A$ for all $x, y \in A$ and all $\alpha, \beta \in \mathbb{K}$ with $|\alpha|^p + |\beta|^p \leq 1$. For $p = 1$ one obtains the usual convex and absolutely convex sets, respectively.

A TVS X is p-normable if and only if it has a bounded p-convex 0-neighborhood, see [366, p. 161]. One shows first that under this hypothesis there exists a bounded p-absolutely convex neighborhood V of 0 and one defines the p-norm as the Minkowski functional corresponding to V (see (1.1.5)).

Remark 1.4.57 In [366] by a p-norm on a vector space X one understands a mapping $\| \cdot \|' : X \to \mathbb{R}_+$ such that

$$\|x\|' = 0 \iff x = 0, \quad \|\alpha x\|' = |\alpha|^p \|x\|' \quad \text{and} \quad \|x + y\|' \leq \|x\|' + \|y\|',$$

for all $x, y \in X$ and $\alpha \in \mathbb{K}$. In this case the "p-norm" corresponding to a bounded absolutely p-convex 0-neighborhood is given by $\|x\|' = \inf\{t^p : t > 0, x \in tV\}$.

It follows that $\| \cdot \|$ is a p-norm in the sense given here if and only if $\| \cdot \|^p$ is a p-norm in the sense given in [366].

The Banach Envelope

Let $(X, \| \cdot \|)$ be a quasi-Banach space and $B_X = \{x \in X : \|x\| \leq 1\}$ its closed unit ball. Denote by $\| \cdot \|_C$ the Minkowski functional of the set $C = \mathrm{co}(B_X)$. It is obvious that $\| \cdot \|_C$ is a seminorm on X and a norm on the quotient space X/N, where $N = \{x \in X : \|x\|_C = 0\}$. Since, for $x \neq 0$, $x' := x/\|x\| \in B_X \subseteq C$, it follows that $\|x'\|_C \leq 1$, that is, $\|x\|_C \leq \|x\|$. Denote by $\widehat{X}$ the completion of X/N with respect to the quotient-norm $\| \cdot \|_{\widehat{X}}$ corresponding to $\| \cdot \|_C$, whose (unique) extension to $\widehat{X}$ is also denoted by $\| \cdot \|_{\widehat{X}}$. It follows that $\|\widehat{x}\|_{\widehat{X}} \leq \|x\|$ for all $x \in X$, hence the embedding $j : X \to \widehat{X}$ is continuous and one shows that $j(X)$ is dense in $\widehat{X}$. The space $\widehat{X}$ is called the *Banach envelope* of the quasi-Banach space X.

We distinguish two situations.

I. *X has trivial dual*: $X^* = \{0\}$.

In this case $C = \mathrm{co}(B_X) = X$ (see [335, Proposition 2.1, p. 16]) and so $\| \cdot \|_C \equiv 0$, $N = X$ and $X/X = \{\widehat{0}\}$. It follows that $\widehat{X} = \{0\}$ and $\widehat{X}^* = \{0\} = X^*$. In particular $\widehat{L^p} = \{0\}$, where $L^p = L^p[0, 1]$.

II. *X has a separating dual.*

This means that for every $x \neq 0$ there exists $x^* \in X^*$ with $x^*(x) \neq 0$ (e.g. $X = \ell^p$ with $0 < p < 1$). In this case $\| \cdot \|_C$ is a norm on X which can be calculated by the formula

$$\|x\|_C = \sup\{|x^*(x)| : x^* \in X^*, \|x^*\| \leq 1\}, \tag{1.4.19}$$

where the norm of $x^* \in X^*$ is given by (1.4.14).

Consequently, $N = \{0\}$, $X/N = X$ and we can consider X as a dense subspace of $\widehat{X}$ (in fact, continuously and densely embedded in $\widehat{X}$).

It follows that:

(i) every continuous linear functional on $(X, \| \cdot \|)$ has a unique norm-preserving extension to $(\widehat{X}, \| \cdot \|_{\widehat{X}})$;

(ii) every continuous linear operator T from $(X, \| \cdot \|)$ to a Banach space Y has a unique norm-preserving extension $\widehat{T} : (\widehat{X}, \| \cdot \|_{\widehat{X}}) \to Y$.

Consequently, $(X, \| \cdot \|)^*$ can be identified with $(\widehat{X}, \| \cdot \|_{\widehat{X}})^*$ and the norm $\| \cdot \|_{\widehat{X}}$ can also be calculated by the formula (1.4.19) for all $x \in X$.

One shows that the Banach envelope of ℓ^p is ℓ^1, for every $0 < p < 1$.

Another way to define the Banach envelope in the case of a quasi-Banach space with separating dual is via the embedding j_X of X into its bidual X^{**} (see [335]). Since X^* separates the points of X, it follows that j_X is injective of norm $\|j_X\| \leq 1$ (in this case one cannot prove that $\|j_X\| = 1$ because the Hahn-Banach extension theorem may fail for non-locally convex spaces).

By (1.4.19),

$$\|x\|_C = \sup\{|x^*(x)| : \|x^*\| \leq 1\} = \|j_X(x)\|_{X^{**}},$$

so we can identify $\widehat{X}$ with the closure of $j_X(X)$ in $(X^{**}, \| \cdot \|_{X^{**}})$.

1.5 Elements of Measure Theory and Integration

In this section we present some elements of measure theory and integration that we shall use throughout the book. Our presentation essentially follows [73, 166, 222, 613]. In spite of the time passed from its publication, the book by Saks [615] is still a valuable source of information, containing fine classical results on the differentiability and integrability of functions.

1.5.1 Algebras and σ-Algebras

Let Ω be a set and $\mathscr{P}(\Omega)$ the family of all subsets of Ω. Sometimes this set will be denoted by 2^Ω. For $A \subseteq \Omega$ denote by $\complement(A)$ the complement of the set A in Ω, $\complement(A) = \Omega \setminus A$. For $A, B \subseteq \Omega$ one denotes by $A \triangle B$ their symmetric difference:

$$A \triangle B = (A \setminus B) \cup (B \setminus A) = (A \cup B) \setminus (A \cap B).$$

For a family $\mathscr{A}$ of subsets of Ω consider the following properties:

(A1) $\emptyset \in \mathscr{A}$;
(A2) $A \in \mathscr{A} \Rightarrow \complement(A) \in \mathscr{A}$;
(A3) $A, B \in \mathscr{A} \Rightarrow A \cup B \in \mathscr{A}$;
(A4) $A_n \in \mathscr{A}, n \in \mathbb{N} \Rightarrow \bigcup_{n=1}^{\infty} A_n \in \mathscr{A}$.

A family $\mathscr{A}$ of subsets of Ω is called an *algebra* if it satisfies the conditions (A1)–(A3), and a σ-*algebra* if it satisfies the conditions (A1), (A2) and (A4).

It is obvious that a σ-algebra is an algebra. Trivial examples of σ-algebras are $\mathscr{A} = \{\emptyset\}$ and $\mathscr{A} = \mathscr{P}(\Omega)$.

The following properties are immediate consequences of the definitions and of de Morgan's rules in set theory.

Proposition 1.5.1 *Let $\mathscr{A}$ be a family of subsets of Ω.*

1. If $\mathscr{A}$ is an algebra, then for every $n \in \mathbb{N}$,

$$A_1, \ldots, A_n \in \mathscr{A} \;\Rightarrow\; A_1 \cup \cdots \cup A_n \in \mathscr{A} \; \text{ and } \; A_1 \cap \cdots \cap A_n \in \mathscr{A}.$$

2. If $\mathscr{A}$ is a σ-algebra, then

$$A_n \in \mathscr{A}, \, n \in \mathbb{N} \;\Rightarrow\; \bigcap_{n=1}^{\infty} A_n \in \mathscr{A}.$$

3. If $\mathscr{A}_i, i \in I$, is a family of algebras (σ-algebras), then $\bigcap_{i \in I} \mathscr{A}_i$ is an algebra (a σ-algebra).

Based on Proposition 1.5.1(3) one can define the algebra (respectively σ-algebra) generated by a family of subsets of Ω.

Let $\mathscr{B}$ be family of subsets of Ω. The intersection of all algebras (σ-algebras) of subsets of Ω containing $\mathscr{B}$ is an algebra (σ-algebra) called the *algebra (σ-algebra) generated* by $\mathscr{B}$. It is the smallest algebra (σ-algebra) of subsets of Ω containing $\mathscr{B}$. In the case of a σ-algebra we use the notation $\sigma(\mathscr{B})$.

Also, for $A \in \mathscr{A}$ the family $\mathscr{A}|_A := \{A \cap B : B \in \mathscr{A}\}$ is a σ-algebra of subsets of A, called the σ-algebra induced by $\mathscr{A}$ on A.

1.5.2 Measures

Let $\mathscr{A}$ be an algebra of subsets of Ω and $\mu : \mathscr{A} \to \overline{\mathbb{R}}$ a mapping which, besides finite values, can take only one of the values $-\infty, \infty$.

Definition 1.5.2 The mapping μ is called a *finitely additive measure* if

$$\text{(i) } \mu(\emptyset) = 0,$$

$$\text{(ii) } \mu\Big(\bigcup_{k=1}^{n} A_k\Big) = \sum_{k=1}^{n} \mu(A_k),$$

for any family $A_1, \ldots, A_n$ of pairwise disjoint sets in $\mathscr{A}$ and all $n \in \mathbb{N}$.

If $\mathscr{A}$ is a σ-algebra then μ is called a *measure* on $\mathscr{A}$ if it satisfies (i) and

$$\text{(iii)} \quad \mu\Big(\bigcup_{k=1}^{\infty} A_k\Big) = \sum_{k=1}^{\infty} \mu(A_k),$$

for any family A_n, $n \in \mathbb{N}$, of pairwise disjoint sets in $\mathscr{A}$.

A measure μ is called *finite* if $\mu(\mathscr{A}) \cap \{-\infty, \infty\} = \emptyset$, i.e., it takes only finite values, $\mu(\mathscr{A}) \subseteq \mathbb{R}$. It is called σ-*finite* if $\Omega = \bigcup_{n=1}^{\infty} A_n$ with $A_n \in \mathscr{A}$ and $|\mu(A_n)| < \infty$ for all $n \in \mathbb{N}$. Also, a set $A \in \mathscr{A}$ is called σ-*finite* if it can be written as a countable union of sets of finite measure μ.

A mapping $\mu : \mathscr{A} \to \overline{\mathbb{R}}$ satisfying (ii) or (iii) is called *finitely additive* or *countably additive* (σ-*additive*, sometimes), respectively.

A *positive finitely additive measure* (*positive measure*) is a finitely additive (countably additive) measure $\mu : \mathscr{A} \to [0, \infty]$.

Remark 1.5.3 A positive finitely additive measure μ on an algebra $\mathscr{A} \subseteq \Omega$ is monotone and finitely subadditive, i.e.,

$$A \subseteq B \implies \mu(A) \le \mu(B), \text{ and}$$

$$\mu(A \cup B) \le \mu(A) + \mu(B),$$

for all $A, B \in \mathscr{A}$.

If μ is positive and countably additive then it countably subadditive, i.e.,

$$\mu\left(\bigcup_{n=1}^{\infty} A_n\right) \le \sum_{n=1}^{\infty} \mu(A_n),$$

for any countable family $\{A_n\}$ of sets in $\mathscr{A}$.

Also, if μ is a measure on $\mathscr{A}$ and the series $\sum_{k=1}^{\infty} \mu(A_k)$ in the right hand side of (iii) from Definition 1.5.2 converges (i.e., it has finite sum), then it converges unconditionally, and so absolutely (by Theorem 1.4.25(4)).

Indeed, $\mu(B) = \mu(A) + \mu(B \setminus A) \ge \mu(A)$ for any $A, B \in \mathscr{A}$ with $A \subseteq B$. Hence, writing $A \cup B = (A \setminus B) \cup B$, it follows that $\mu(A \cup B) = \mu(A \setminus B) + \mu(B) \le \mu(A) + \mu(B)$ for arbitrary $A, B \in \mathscr{A}$.

The countable subadditivity is proved as in the finite case. Putting $A = \bigcup_{n=1}^{\infty} A_n$, $B_1 = A_1$ and $B_{n+1} = A_{n+1} \setminus (A_1 \cup \cdots \cup A_n)$ it follows that $A = \bigcup_{n=1}^{\infty} B_n$ where the sets B_n are mutually disjoint. Consequently,

$$\mu(A) = \mu\Big(\bigcup_{n=1}^{\infty} B_n\Big) = \sum_{n=1}^{\infty} \mu(B_n) \le \sum_{n=1}^{\infty} \mu(A_n).$$

Also, since the union $\bigcup_{k=1}^{\infty} A_k$ does not depend on the order of the sets A_k, the series $\sum_{k=1}^{\infty} \mu(A_k)$ converges unconditionally.

Remark 1.5.4 In many books on measure theory by a measure one understand a positive measure, while measures with values in $\overline{\mathbb{R}}$ are called signed measures. Since we are mainly working with measures with values in $\mathbb{R}$ we adopt the convention to call *measures* those taking values in $\overline{\mathbb{R}}$ and *positive measures* those taking values in $[0, \infty]$. One can also consider measures taking complex values as mappings $\mu :$ $\mathscr{A} \to \mathbb{C}$ satisfying (i) and (iii), called *complex measures*.

A pair $(\Omega, \mathscr{A})$, where $\mathscr{A}$ is a σ-algebra of subsets of a set Ω is called a *measurable space*. If $\mu : \mathscr{A} \to \overline{\mathbb{R}}$ is a measure on $\mathscr{A}$, then we call the triple $(\Omega, \mathscr{A}, \mu)$ a *measure space*. If μ is positive (or complex-valued), the one says that $(\Omega, \mathscr{A}, \mu)$ is a *positive measure space* (resp. a *complex measure space*). If the measure μ is finite, then we say that $(\Omega, \mathscr{A}, \mu)$ is a *finite measure space*. By definition, a complex measure space is always finite.

Let $(\Omega, \mathscr{A}, \mu)$ be a measure space. One says that μ is *positive* (*negative*) on $A \in \mathscr{A}$ if $\mu(B) \geq 0$ ($\mu(B) \leq 0$) for every $B \in \mathscr{A}$, $B \subseteq A$. The *positive* (*negative*) *part* μ^+, μ^- of the measure μ are defined by

$$\mu^+(A) = \sup\{\mu(B) : B \in \mathscr{A}, B \subseteq A\} \text{ and}$$
$$\mu^-(A) = -\inf\{\mu(B) : B \in \mathscr{A}, B \subseteq A\}, \tag{1.5.1}$$

for all $A \in \mathscr{A}$.

Since $\emptyset \subseteq A$ it follows that $\mu^+(A) \geq \mu(\emptyset) = 0$ and $-\mu^-(A) \leq \mu(\emptyset) = 0$, so that $\mu^+(A), \mu^-(A) \geq 0$, for every $A \in \mathscr{A}$. Actually, stronger properties are satisfied by μ^+ and μ^-.

A set $A \in \mathscr{A}$ is called μ-*null* if $\mu(B) = 0$ for every $B \in \mathscr{A}$, $B \subseteq A$, or, equivalently, if

$$\mu^+(A) = 0 = \mu^-(A).$$

An *atom* in a positive measure space $(\Omega, \mathscr{A}, \mu)$ is a set $A \in \mathscr{A}$ such that $\mu(A) > 0$ and $\mu(B) = \mu(A)$ for every nonempty $B \in \mathscr{A}$ contained in A. If a one-point set $\{t\}$ belongs to $\mathscr{A}$ and $\mu(\{t\}) > 0$, then $\{t\}$ is obviously an atom.

A positive measure space $(\Omega, \mathscr{A}, \mu)$ is called *complete* if for every $A \in \mathscr{A}$ with $\mu(A) = 0$, $B \subseteq A$ implies $B \in \mathscr{A}$ (and so $\mu(B) = 0$). One shows that any positive measure space admits a completion, i.e., a minimal complete positive measure space $(\Omega, \tilde{\mathscr{A}}, \tilde{\mu})$ such that $\mathscr{A} \subseteq \tilde{\mathscr{A}}$ and $\tilde{\mu}|_{\mathscr{A}} = \mu$. For instance the σ-algebra $\Lambda(\mathbb{K}^n)$ of Lebesgue measurable subsets of $\mathbb{K}^n$ with the Lebesgue measure λ_n is the completion of the measure space $(\mathbb{K}^n, \mathscr{B}(\mathbb{K}^n), \gamma_n)$ where γ_n is the Lebesgue measure defined on the σ-algebra $\mathscr{B}(\mathbb{K}^n)$ of Borel subsets of $\mathbb{K}^n$.

Theorem 1.5.5 *Let $(\Omega, \mathscr{A}, \mu)$ be a measure space.*

1. *The positive part μ^+ and the negative one μ^- of the measure μ are positive measures on $\mathscr{A}$.*
2. *If μ is positive (negative) on $A \in \mathscr{A}$, then $\mu(A) = \mu^+(A)$ (resp. $\mu(A) = -\mu^-(A)$). Conversely, if $\mu(A) = \mu^+(A) < \infty$ ($\mu(A) = -\mu^-(A) > -\infty$), then μ is positive (resp. negative) on A.*
3. *(Hahn decomposition) There exists a pair P, N of sets in $\mathscr{A}$ such that*

$$\text{(i)} \quad P \cup N = \Omega, \ P \cap N = \emptyset,$$
$$\text{(ii)} \quad \mu \text{ is positive on } P \text{ and negative on } N. \tag{1.5.2}$$

4. *If the sets $P, N \in \mathscr{A}$ satisfy (1.5.2), then*

$$\mu^+(A) = \mu(A \cap P) \ and \ \mu^-(A) = \mu(A \cap N),$$

for every $A \in \mathscr{A}$.

Remark 1.5.6 A pair P, N of sets in $\mathscr{A}$ satisfying (1.5.2) is called a *Hahn decomposition* of the measure space $(\Omega, \mathscr{A}, \mu)$ (or, shorter, for μ). It is unique excepting a μ-null set. This means that if P_1, N_1 is another Hahn decomposition for μ, then the sets $P_1 \triangle P$ and $N_1 \triangle N$ are both μ-null.

The *variation* of a measure $\mu : \mathscr{A} \to \overline{\mathbb{R}}$ over a subset $A \in \mathscr{A}$ is defined by

$$|\mu|(A) = \sup \sum_{k=1}^{n} |\mu(B_k)|, \tag{1.5.3}$$

where the supremum is taken over all partitions $A = B_1 \cup \cdots \cup B_n$ of A into pairwise disjoint sets $B_k \in \mathscr{A}$, $k = 1, \ldots, n$, $n \in \mathbb{N}$. It follows

$$|\mu(A)| \le |\mu|(A),$$

for all $A \in \mathscr{A}$.

The *total variation* of μ is the number

$$\|\mu\| = |\mu|(\Omega). \tag{1.5.4}$$

We define now an important class of measures.

Definition 1.5.7 A measure $\mu : \mathscr{A} \to \overline{\mathbb{R}}$ is called of *bounded variation* if $\|\mu\| < \infty$. We denote by $M(\mathscr{A}, \overline{\mathbb{R}})$ the set of all measures on $\mathscr{A}$ having bounded variation. The set of all measures $\mu : \mathscr{A} \to \mathbb{C}$ with bounded variation is denoted by $M(\mathscr{A}, \mathbb{C})$. We write $M(\mathscr{A}, \mathbb{K})$ for $\mathbb{K} = \mathbb{R}$ or $\mathbb{C}$.

The following result holds.

Theorem 1.5.8 *Let $(\Omega, \mathscr{A}, \mu)$ be a measure space. Then the following equalities*

$$\text{(i) } \mu = \mu^+ - \mu^-, \quad \text{(the Jordan decomposition);}$$

$$\text{(ii) } |\mu| = \mu^+ + \mu^-, \tag{1.5.5}$$

hold, where μ^+ and μ^- are given by (1.5.1).

Remark 1.5.9 The decomposition (i) from (1.5.5) is called the *Jordan decomposition* of the measure μ. It is minimal in the sense that if μ_1, μ_2 are positive measures such that $\mu = \mu_1 - \mu_2$, then $\mu^+ \le \mu_1$ and $\mu^- \le \mu_2$.

Remark 1.5.10 The equality $\mu = \mu^+ - \mu^-$ implies that at least one of the measures μ^+, μ^- is finite, while from $|\mu| = \mu^+ + \mu^-$ it follows that the measure μ is with bounded variation if and only if both measures μ^+ and μ^- are finite. One shows that this happens if and only if the measure μ is finite.

Consequently, a measure $\mu : \mathscr{A} \to \overline{\mathbb{R}}$ is with bounded variation if and only if it is finite.

The Lattice Structure of $M(\mathscr{A}, \mathbb{R})$
Let $\mathscr{A}$ be a σ-algebra of subsets of a set Ω and $M(\mathscr{A}, \mathbb{R})$ the set of all finite measures $\mu : \mathscr{A} \to \mathbb{R}$. It follows that each $\mu \in M(\mathscr{A}, \mathbb{R})$ is of bounded variation, i.e., $\|\mu\| = |\mu|(\Omega) < \infty$.

Define an order on $M(\mathscr{A}, \mathbb{R})$ by

$$\mu \le \nu \iff \mu(A) \le \nu(A) \text{ for all } A \in \mathscr{A}, \tag{1.5.6}$$

and let $M^+(\mathscr{A}, \mathbb{R})$ be the set of all positive measures on $\mathscr{A}$. Then $M^+(\mathscr{A}, \mathbb{R})$ is a cone in $M(\mathscr{A}, \mathbb{R})$ inducing the order (1.5.6).

Defining pointwise the operation of addition and multiplication by scalars, it follows that $M(\mathscr{A}, \mathbb{R})$ is a vector space.

Theorem 1.5.11 *Let $\mathscr{A}$ be a σ-algebra of subsets of a set Ω.*

1. *The total variation $\|\cdot\|$ is a complete norm on $M(\mathscr{A}, \mathbb{R})$, i.e., $(M(\mathscr{A}, \mathbb{R}), \|\cdot\|)$ is a Banach space.*
2. *$(M(\mathscr{A}, \mathbb{R}), \|\cdot\|)$ is a Banach lattice with respect to the order defined by (1.5.6). The ordered vector space $M(\mathscr{A}, \mathbb{R})$ is Dedekind complete with respect to this order, i.e., every order-bounded above nonempty subset of $M(\mathscr{A}, \mathbb{R})$ has a supremum (and, consequently, every order-bounded below nonempty subset has an infimum).*
3. *For $\mu \in M(\mathscr{A}, \mathbb{R})$, $\mu^+ = \mu \vee 0$, $\mu^- = -(\mu \wedge 0)$ and $\mu^+ \wedge \mu^- = 0$. This means that the positive and the negative part of a measure, as defined by (1.5.1), agree with those given by the lattice operations, that is, there is no ambiguity in notation.*

Complex Measures

A *complex measure space* is a triple $(\Omega, \mathscr{A}, \mu)$, where Ω is a set, $\mathscr{A}$ is a σ-algebra of subsets of Ω and $\mu : \mathscr{A} \to \mathbb{C}$ is a countably additive measure, called a *complex measure*. The variation $|\mu|$ of a complex measure μ is defined by the same formula (1.5.3), as in the real case, with the absolute values taken in the complex sense. Any complex measure μ can be written in the following forms

$$\mu = \mu_1 + i\mu_2 = \mu_1^+ - \mu_1^- + i(\mu_2^+ - \mu_1^-), \qquad (1.5.7)$$

where $\mu_1(A) = \operatorname{Re}\mu(A)$, $\mu_2(A) = \operatorname{Im}\mu(A)$, $A \in \mathscr{A}$, are real measures having the Jordan decompositions $\mu_1 = \mu_1^+ - \mu_1^-$, $\mu_2 = \mu_2^+ - \mu_1^-$.

Proposition 1.5.12 *Let $(\Omega, \mathscr{A}, \mu)$ be a complex measure space.*

1. $(\Omega, \mathscr{A}, |\mu|)$ is a positive measure space and

$$|\mu|(A) \le \mu_1^+(A) + \mu_1^-(A) + \mu_2^+(A) + \mu_2^-(A),$$

for all $A \in \mathscr{A}$.
2. Also,

$$\sup\{|\mu(B)| : B \in \mathscr{A}, B \subseteq A\} \le |\mu|(A) \le 4\sup\{|\mu(B)| : B \in \mathscr{A}, B \subseteq A\},$$

for all $A \in \mathscr{A}$.

1.5.3 Measurable Functions and Integration

Let $(\Omega, \mathscr{A})$ be a measurable space. For $f : \Omega \to \overline{\mathbb{R}}$ and $\alpha \in \mathbb{R}$ put

$$[f \ge \alpha] = \{t \in \Omega : f(t) \ge \alpha\}.$$

We define now the important class of measurable functions. A function $f : \Omega \to \overline{\mathbb{R}}$ is called $\mathscr{A}$-*measurable* if $[f \ge \alpha]$ belongs to $\mathscr{A}$ for every $\alpha \in \mathbb{R}$. If there is no danger of ambiguity we say that f is a *measurable function*.

Defining the sets $[f > \alpha]$, $[f \le \alpha]$, $[f < \alpha]$ in a similar way, it follows that each of the following conditions is equivalent to the measurability of f:

- $[f \ge \alpha]$ belongs to $\mathscr{A}$ for every $\alpha \in \mathbb{R}$;
- $[f > \alpha]$ belongs to $\mathscr{A}$ for every $\alpha \in \mathbb{R}$;
- $[f \le \alpha]$ belongs to $\mathscr{A}$ for every $\alpha \in \mathbb{R}$;
- $[f < \alpha]$ belongs to $\mathscr{A}$ for every $\alpha \in \mathbb{R}$.

In the following proposition we collect some properties of measurable functions.

Proposition 1.5.13 *Let $(\Omega, \mathscr{A})$ be a measurable space.*

1. *If $f, g : \Omega \to \overline{\mathbb{R}}$ are measurable functions, then the functions*

$$(f \vee g)(t) := \max\{f(t), g(t)\} \quad and \quad (f \wedge g)(t) := \max\{f(t), g(t)\}, \quad t \in \Omega,$$

are also measurable.

2. *If the functions $f_n : \Omega \to \overline{\mathbb{R}}$, $n \in \mathbb{N}$, are measurable, then the functions, defined for $t \in \Omega$ by*

$$(\sup_n f_n)(t) := \sup\{f_n(t) : n \in \mathbb{N}\}, \qquad (\inf_n f_n)(t) := \inf\{f_n(t) : n \in \mathbb{N}\},$$

$$(\limsup_n f_n)(t) := \limsup_n f_n(t), \qquad (\liminf_n f_n)(t) := \liminf_n f_n(t),$$

$$(\lim_n f_n)(t) := \lim_n f_n(t), \quad (provided\ that\ the\ limit\ exists),$$

are all measurable.

3. *If $f, g : \Omega \to [0, \infty]$ are measurable functions and $\alpha \geq 0$, then the functions*

$$(f + g)(t) := f(t) + g(t) \quad and \quad (\alpha f)(t) := \alpha f(t), \quad t \in \Omega,$$

are also measurable (with the convention $0 \cdot \infty = 0$).

A function $f : \Omega \to \overline{\mathbb{R}}$ is called *simple* if it takes only finitely many values. If $f(\Omega) = \{\alpha_1, \ldots, \alpha_n\}$, then $f = \sum_{i=1}^{n} \alpha_i \chi_{A_i}$, where $A_i := \{t \in \Omega : f(t) = \alpha_i\}$. The simple function f is $\mathscr{A}$-measurable if and only if $A_i \in \mathscr{A}$ for $i = 1, \ldots, n$.

The following result is essential for the definition of the integrals of measurable functions.

Proposition 1.5.14 *Let $(\Omega, \mathscr{A})$ be a measurable space.*

1. *For every $\mathscr{A}$-measurable function $f : \Omega \to [0, \infty]$ there exists a nondecreasing sequence (f_n) of $[0, \infty)$-valued measurable simple functions such that*

$$\lim_{n \to \infty} f_n(t) = f(t) \quad for\ all\ t \in \Omega. \tag{1.5.8}$$

2. *If the measurable function f takes values in $\overline{\mathbb{R}}$ (or $\mathbb{C}$), then there exists a sequence (f_n) of $\mathbb{R}$-valued ($\mathbb{C}$-valued) measurable simple functions satisfying (1.5.8) and such that*

$$|f_n(t)| \leq |f_{n+1}(t)|, \tag{1.5.9}$$

for all $t \in \Omega$ and $n \in \mathbb{N}$.

If the function f is bounded on Ω, then there exists a sequence of measurable simple functions satisfying (1.5.9) and converging to f uniformly on Ω.

The Integration of Measurable Functions

Let $(\Omega, \mathscr{A}, \mu)$ be a positive measure space. The *integral* of an $\mathscr{A}$-measurable simple function $f : \Omega \to [0, \infty)$, $f = \sum_{i=1}^{n} \alpha_i \chi_{A_i}$, $\alpha_i \geq 0$, $A_i \in \mathscr{A}$, $i = 1, \ldots, n$, is defined by

$$\int f d\mu = \sum_{i=1}^{n} \alpha_i \mu(A_i).$$

Sometimes, in order to emphasize the set Ω, one uses the notation $\int_{\Omega} f d\mu$, or even $\int_{\Omega} f(t) d\mu(t)$ if we also want to specify the variable of integration.

One shows that $\int f d\mu$ does not depend on the representation of the simple function f as a linear combination of characteristic functions of sets in $\mathscr{A}$. It is also obvious that $\int f d\mu \in [0, \infty]$.

Now based on Proposition 1.5.14, one can define the integrals of nonnegative measurable functions.

Definition 1.5.15 Let $(\Omega, \mathscr{A}, \mu)$ be a positive measure space. The integral of a $\mathscr{A}$-measurable function $f : \Omega \to [0, \infty]$ is defined by

$$\int f d\mu = \lim_{n \to \infty} \int f_n d\mu, \qquad (1.5.10)$$

where (f_n) is a nondecreasing sequence of positive $\mathscr{A}$-measurable simple functions which is pointwise convergent to f (whose existence is guaranteed by Proposition 1.5.14).

One shows that the definition is correct, i.e., that the limit (1.5.10) does not depend on the particular choice of the sequence (f_n). Since the integral of positive simple functions is monotone, it follows that $\int f_n d\mu \leq \int f_{n+1} d\mu$, so that in (1.5.10) we have

$$\lim_{n \to \infty} \int f_n d\mu = \sup_{n \in \mathbb{N}} \int f_n d\mu.$$

One shows further that,

$$\int f d\mu = \sup \left\{ \int g d\mu : g : \Omega \to [0, \infty) \text{ is } \mathscr{A}\text{-measurable, simple and } g \leq f \right\}.$$

Writing an $\mathscr{A}$-measurable function $f : \Omega \to \overline{\mathbb{R}}$ as $f = f^+ - f^-$, where $f^{\pm}(t) = \max\{\pm f(t), 0\}$, $t \in \Omega$, set

$$\int f d\mu = \int f^+ d\mu - \int f^- d\mu, \qquad (1.5.11)$$

provided that at least one of the integrals in the left hand side of (1.5.11) is finite. In this case we say that f has (or admits) an integral. If both these integrals are finite, then we say that f is *integrable* (or μ-*integrable*).

If f is μ-integrable, then $|f| = f^+ + f^-$ is also integrable and

$$\left| \int f d\mu \right| \le \int |f| d\mu . \qquad (1.5.12)$$

The integral of an $\mathscr{A}$-measurable complex-valued function $f : \Omega \to \mathbb{C}$ is defined by

$$\int f d\mu = \int \text{Re } f d\mu + i \int \text{Im } f d\mu .$$

If both the integrals of Re f and Im f are finite, then we say that the function f is μ-*integrable*. The inequality (1.5.12) is satisfied in this case, too.

The Integration with Respect to Arbitrary Measures

Let $(\Omega, \mathscr{A})$ be a measurable space and $\mu : \mathscr{A} \to \overline{\mathbb{R}}$ a measure. An $\mathscr{A}$-measurable function f from Ω to $\overline{\mathbb{R}}$ (or to $\mathbb{C}$) is called μ-*integrable* if both of the integrals $\int f d\mu^+$ and $\int f d\mu^-$ are finite, where $\mu = \mu^+ - \mu^-$ is the Jordan decomposition of the measure μ. The integral of f with respect to μ is defined by

$$\int f d\mu = \int f d\mu^+ - \int f d\mu^- . \qquad (1.5.13)$$

If only one of the integrals in the right side of (1.5.13) is finite, then one says that the integral of f with respect to μ exists.

If μ is a complex measure, then the integral of f is defined by

$$\int f d\mu = \int f d\mu_1 + i \int f d\mu_2 ,$$

where $\mu(A) = \mu_1(A) + i \mu_2(A)$, $A \in \mathscr{A}$.

Integration on a Set

Let $(\Omega, \mathscr{A})$ be a measurable space and $A \in \mathscr{A}$. The measurability of a function f on A is defined as the measurability with respect to the σ-algebra $\mathscr{A}|_A$ induced by $\mathscr{A}$ on A. The integral of f with respect to a measure μ (extended real-valued or complex-valued) on $\mathscr{A}$ is given by

$$\int_A f d\mu = \int f \chi_A d\mu \quad \left(= \int_\Omega f \chi_A d\mu \right).$$

It is obvious that

$$\left|\int_A f d\mu\right| \le \int_A |f| \cdot d|\mu|, \quad A \in \mathscr{A},$$

and that

$$\int_A f d\mu = 0, \tag{1.5.14}$$

whenever $|\mu|(A) = 0$.

The integral is countably additive both with respect to the set on which the integral is taken and the integrated functions.

If $\{A_n\}$ is a family of pairwise disjoint sets in $\mathscr{A}$ and $A := \bigcup_{n=1}^{\infty} A_n$, then

$$\int_A f d\mu = \sum_{n=1}^{\infty} \int_{A_n} f d\mu.$$

If (f_n) are μ-integrable functions such that $\sum_{n=1}^{\infty} |f_n|$ is μ-integrable, then $f := \sum_{n=1}^{\infty} f_n$ is μ-integrable as well and

$$\int f d\mu = \sum_{n=1}^{\infty} \int f_n d\mu.$$

For a measure space $(\Omega, \mathscr{A}, \mu)$, the set of $\mathbb{K}$-valued integrable functions will be denoted by $\mathscr{L}^1(\mu, \mathbb{K})$ ($\mathscr{L}^1(A, \mu, \mathbb{K})$ for $A \in \mathscr{A}$). It is a vector space with respect to the pointwise operations of addition and multiplication by scalars.

We mention the following result.

Proposition 1.5.16 ([73], Proposition 5.1.10) *Let $(\Omega, \mathscr{A}, \mu)$ be a positive measure space, $f : \Omega \to \overline{\mathbb{R}}$ an $\mathscr{A}$-measurable function such that $\int_\Omega f d\mu$ exists and $v : \mathscr{A} \to \overline{\mathbb{R}}$ defined by*

$$v(A) = \int_A f d\mu, \quad A \in \mathscr{A}.$$

Then v is a measure on $\mathscr{A}$ and its Jordan decomposition and total variation are given by

$$v^{\pm}(A) = \int_A f^{\pm} d\mu \text{ and } |v|(A) = \int_A |f| d\mu,$$

for all $A \in \mathscr{A}$, respectively.

The Hahn decomposition corresponding to the measure v is given by $P = \{t \in \Omega : f(t) > 0\}$ and $N = \Omega \setminus P$.

1.5.4   The Radon-Nikodým Theorem

Let $(\Omega, \mathscr{A})$ be a measurable space and μ, ν arbitrary measures on $\mathscr{A}$ (i.e., extended real-valued or complex-valued measures). One says that ν is *absolutely continuous* with respect to μ if

$$|\mu|(A) = 0 \implies \nu(A) = 0, \tag{1.5.15}$$

for all $A \in \mathscr{A}$.

One uses the notation $\nu \ll \mu$ to indicate that ν is absolutely continuous with respect to μ.

Remark 1.5.17 There is a notion of absolute continuity for functions $f : [a, b] \to \mathbb{R}$. This notion and its relation with the Lipschitz property will be discussed in Sect. 3.3.1.

The following characterization of absolute continuity holds.

Proposition 1.5.18 ([73], Theorem 5.2.8, and [166], Lemma 4.2.1) *Let* $(\Omega, \mathscr{A}, \mu)$ *be a positive measure space.*

1. *If ν a finite countably additive measure on $\mathscr{A}$ (with values in $\mathbb{R}$ or in $\mathbb{C}$), absolutely continuous with respect to μ, then for every $\varepsilon > 0$ there exists $\delta > 0$ such that*

$$\mu(A) < \delta \implies |\nu(A)| < \varepsilon, \tag{1.5.16}$$

for all $A \in \mathscr{A}$.
2. *Conversely, if the measure μ is finite and ν is a finitely additive measure on $\mathscr{A}$ (with values in $\overline{\mathbb{R}}$ or $\mathbb{C}$) such that μ and ν satisfy the condition (1.5.16), then ν is countably additive and absolutely continuous with respect to μ.*

From this theorem one obtains the following characterization of absolute continuity.

Corollary 1.5.19 *Let $(\Omega, \mathscr{A})$ be a measurable space and μ, ν measures on $\mathscr{A}$, with ν finite in the real case. Then ν is absolutely continuous with respect to μ if and only if for every $\varepsilon > 0$ there exists $\delta > 0$ such that*

$$|\mu|(A) < \delta \implies |\nu(A)| < \varepsilon,$$

for all $A \in \mathscr{A}$.

Remark 1.5.20 It can be shown that for arbitrary measures μ, ν,

$$\nu \ll \mu \iff |\nu| \ll \mu \iff |\nu| \ll |\mu|.$$

Remark 1.5.21 The finiteness of the measure v in Proposition 1.5.18(1) is essential for the validity of (1.5.16). The measure $v(A) = \int_A |t| d\lambda(t)$, $A \in \mathscr{B}(\mathbb{R})$, is absolutely continuous with respect to the Lebesgue measure λ, but for no positive ε does there exist $\delta > 0$ such that $A \in \mathscr{B}(\mathbb{R})$ and $\lambda(A) < \delta$ imply $v(A) < \varepsilon$.

Also, if $v \ll \mu$ and μ is finite, then v is not necessarily finite. This can be seen from the following example: $\Omega = (0, 1)$, $\mu = \lambda$ (the Lebesgue measure), and $v(A) = \int_A t^{-1} d\lambda(t)$ for $A \subseteq (0, 1)$ Lebesgue measurable.

Indeed, for any $A \subseteq [n, \infty)$ with $0 < \lambda(A) < \delta$,

$$\int_A |t| d\lambda(t) \geq n\lambda(A) \to \infty \text{ as } n \to \infty.$$

Example 1.5.22 If the function f from Ω to $\overline{\mathbb{R}}$, or to $\mathbb{C}$, is integrable with respect to a positive measure μ, then, by (1.5.14), the measure

$$v(A) = \int_A f d\mu, \quad A \in \mathscr{A},$$

is absolutely continuous with respect to μ.

The remarkable Radon-Nikodým theorem tells us that the converse of this result is also true.

Theorem 1.5.23 (Radon-Nikodým) *Let $(\Omega, \mathscr{A})$ be a measurable space, $\mu : \mathscr{A} \to [0, \infty]$ a σ-finite measure on $\mathscr{A}$ and v a $\mathbb{R}$-valued ($\mathbb{C}$-valued) measure defined on $\mathscr{A}$. If v is absolutely continuous with respect to μ, then there exists an integrable function $g : \Omega \to \mathbb{R}$ ($g : \Omega \to \mathbb{C}$) such that*

$$v(A) = \int_A g d\mu \text{ for all } A \in \mathscr{A}.$$

1.5.5 Borel Measures

A *Borel subset* of a locally compact Hausdorff topological space T is an element of the σ-algebra $\mathscr{B}(T)$ generated by the family of open subsets of T. A *Baire subset* of a locally compact Hausdorff topological space T is an element of the σ-algebra $\mathscr{B}_0(T)$ generated by the family of all compact G_δ (intersection of a countable family of open sets) subsets of T. Note that this is the smallest σ-algebra $\mathscr{B}_0$ of subsets of T making all real-valued continuous functions with compact support $\mathscr{B}_0$-measurable. If T is further σ-compact, then this is the smallest σ-algebra of subsets of T making $\mathscr{B}_0$-measurable all real-valued continuous functions on T. Since a compact G_δ-set belongs to $\mathscr{B}(T)$, it follows that $\mathscr{B}_0(T) \subseteq \mathscr{B}(T)$. If T is further metrizable, then $\mathscr{B}_0(T) = \mathscr{B}(T)$.

A measure $\mu : \mathscr{B}(T) \rightarrow \overline{\mathbb{R}}$ is called a *Borel measure*. If μ takes values in $[0, \infty]$ or in $\mathbb{C}$, then we call it a *positive Borel measure* or a *complex Borel measure*, respectively.

One says that a measure $\mu : \mathscr{A} \rightarrow [0, \infty]$ is *supported* by a set $S \in \mathscr{A}$ if $\mu(T \setminus S) = 0$. The *support* spt(μ) of a positive Borel measure μ is the complement to the largest open subset G of T for which $\mu(G) = 0$. It follows that the support is a closed set and a positive Borel measure μ is supported by S if and only if spt$(\mu) \subseteq S$.

The *Dirac measure* corresponding to $t \in T$ is the measure $\delta_t : \mathscr{B}(T) \rightarrow \mathbb{R}_+$ defined for $A \in \mathscr{B}(T)$ by $\delta_t(A) = 1$ if $t \in A$ and $\delta_t(A) = 0$ otherwise. A linear combination $\mu = \sum_{i=1}^{n} \alpha_i \delta_{t_i}$ of Dirac measures is called a *discrete measure*. Obviously, its support is the set $\{t_1, \ldots, t_n\}$.

A positive Borel measure μ is called *regular* if $\mu(K) < \infty$ for every compact subset K of T and

(i) $\mu(B) = \inf\{\mu(U) : U$ open and $B \subseteq U\}$,   for every $B \in \mathscr{B}(T)$, and

(ii) $\mu(U) = \sup\{\mu(K) : K$ compact and $K \subseteq U\}$,   for every $U \subseteq T$ open.

It follows that

$$\mu(B) = \sup\{\mu(K) : K \text{ compact and } K \subseteq B\},$$

for every σ-finite Borel set B. Also, for every such B and every $\varepsilon > 0$ there exist an open set U and a compact set K such that

$$K \subseteq B \subseteq U \text{ and } \mu(U \setminus K) < \varepsilon.$$

A Borel measure μ with values in $\overline{\mathbb{R}}$ or $\mathbb{C}$ is called *regular* if $|\mu|$ is a regular positive Borel measure. In the real case this is equivalent to the fact that both μ^+ and μ^- are regular positive Borel measures. In the complex case all the measures from the representations (1.5.7) must be regular. Denote by $\mathscr{M}(T)$ the space of all regular Borel signed measures. If T is a compact Hausdorff space and μ is a regular Borel measure on T, then $\|\mu\| := |\mu|(T) < \infty$ is a complete norm on $\mathscr{M}(T)$, that is, $\mathscr{M}(T)$ is a Banach space with respect to this norm. This follows from Riesz' representation theorem.

There is a more general approach to measurability. Consider two measurable spaces $(\Omega_1, \mathscr{A}_1)$ and $(\Omega_2, \mathscr{A}_2)$. A function $f : \Omega_1 \rightarrow \Omega_2$ is called $(\mathscr{A}_1, \mathscr{A}_2)$-*measurable* if $f^{-1}(B) \in \mathscr{A}_1$ for all $B \in \mathscr{A}_2$.

Since, for an arbitrary function $f : \Omega_1 \rightarrow \Omega_2$, the set $\{f^{-1}(B) : B \in \mathscr{A}_2\}$ is a σ-algebra of subsets of Ω_1, the following result holds.

Proposition 1.5.24 *Let $(\Omega, \mathscr{A})$ be a measurable space, T a topological space, $\mathscr{B}(T)$ the σ-algebra of Borel subsets of T and $f : \Omega \to T$ a mapping. Then the following are equivalent.*

1. *The function f is $(\mathscr{A}, \mathscr{B}(T))$-measurable.*
2. *$f^{-1}(G)$ belongs to $\mathscr{A}$ for every open subset G of T.*
3. *$f^{-1}(S)$ belongs to $\mathscr{A}$ for every closed subset S of T.*

In particular, for scalar-valued functions one obtains.

Corollary 1.5.25 *Let $(\Omega, \mathscr{A})$ be a measurable space. A function $f : \Omega \to \mathbb{K}$ is measurable (in the sense defined in Sect. 1.5.3) if and only if it is $(\mathscr{A}, \mathscr{B}(\mathbb{K}))$-measurable.*

1.5.6 Riesz' Representation Theorem

Let T be a topological space. A functional $I : C(T, \mathbb{R}) \to \mathbb{K}$ is called *positive* if $I(f) \geq 0$ for every $f \geq 0$ in $C(T, \mathbb{R})$, where the order in $C(T, \mathbb{R})$ is the pointwise order (see Sect. 1.4.11). One denotes by $C_0^+(T, \mathbb{R})$ the positive cone of $C_0(T, \mathbb{R})$, i.e., $C_0^+(T, \mathbb{R}) = \{ f \in C_0(T, \mathbb{R}) : f \geq 0 \}$.

All the integral representation results for linear functionals on $C(T)$ are known under the generic name of "*Riesz' representation theorem*". In fact, Riesz [597] proved in 1909 the result in the case $C[a, b]$, when the integral representation is done via the Riemann-Stieltjes integral. In 1938, Markov [436] extended it to some noncompact intervals and, finally, Kakutani [325] gave in 1941 a proof for general compact spaces. For these reasons the theorem is called sometimes "*the Riesz-Markov-Kakutani representation theorem*".

The first result is the following one.

Theorem 1.5.26 *Let T be a locally compact Hausdorff topological space. For any positive linear functional $I : C_{00}(T, \mathbb{R}) \to \mathbb{R}$ there exists a unique regular positive Borel measure μ such that*

$$I(f) = \int_T f \, d\mu \quad \text{for all } f \in C(T, \mathbb{R}). \qquad (1.5.17)$$

Concerning the case of continuous positive linear functionals we mention the following result.

Theorem 1.5.27 ([222], Kapitel VIII, Satz 2.8) *Consider the space $C_{00}(T, \mathbb{R})$ equipped with the sup-norm, where T is a locally compact Hausdorff topological space. Let $I : C_{00}(T, \mathbb{R}) \to \mathbb{R}$ be a positive linear functional and μ the* *unique regular positive Borel measure satisfying* (1.5.17). *Then the following are equivalent.*

(i) *the functional* I *is continuous;*
(ii) *the Borel measure* μ *representing* I *is finite;*
(iii) $\|I\| = \mu(T)$.

In the case of the space $C_0(T)$ any positive linear functional is continuous.

Theorem 1.5.28 *Let* T *be a locally compact Hausdorff topological space.*

1. *Any positive linear functional* $I : C_0(T, \mathbb{R}) \to \mathbb{R}$ *is continuous with respect to the sup-norm on* $C_0(T, \mathbb{R})$.
2. *If* $I : C_0(T, \mathbb{R}) \to \mathbb{R}$ *is a continuous linear functional, then there exists two (continuous) positive linear functionals* $I^+, I^- : C_0(T, \mathbb{R}) \to \mathbb{R}$ *such that* $I = I^+ - I^-$, *where, for any* $f \in C_0^+(T, \mathbb{R})$, I^+ *is given by*

$$I^+(f) = \sup\{I(h) : 0 \le h \le f\}.$$

The decomposition $I = I^+ - I^-$ *is minimal, in the sense that if* $I = J - L$ *with* J, L *positive linear functionals on* $C_0(T, \mathbb{R})$, *then* $I^+ \le J$ *and* $I^- \le L$.

The general representation result is the following one. Denote by $\mathscr{M}(T, \mathbb{K})$ the space of all $\mathbb{K}$-valued regular Borel measures. With respect to the total variation norm $\|\mu\| = |\mu|(T)$ $\mathscr{M}(T, \mathbb{K})$ is a Banach space.

Theorem 1.5.29 *Let* T *be a locally compact Hausdorff topological space and* $C_0(T)$ *the Banach space (with respect to the sup-norm) of all vanishing at infinity* $\mathbb{K}$-*valued continuous functions on* T *and* $C_0(T)^*$ *its dual space. The mapping* $\Phi : \mathscr{M}(T, \mathbb{K}) \to C_0(T)^*$ *given for* $\mu \in \mathscr{M}(T, \mathbb{K})$ *by*

$$\Phi(\mu)(f) = \int_T f \, d\mu, \quad \text{for all } f \in C_0(T),$$

is an isometric isomorphisms between the Banach spaces $\mathscr{M}(T, \mathbb{K})$ *and* $C_0(T)^*$.

In the case $\mathbb{K} = \mathbb{R}$, Φ *is also an order isomorphism and* $\Phi(\mu^+) = I^+$, $\Phi(\mu^-) = I^-$. *Here* $\mu = \mu^+ - \mu^-$ *is the Jordan decomposition of the measure* $\mu \in \mathscr{M}(T, \mathbb{R})$, I *is given by* $I = \Phi(\mu)$ *and* $I = I^+ - I^-$ *is the decomposition of* I *given by Theorem 1.5.28.*

A proof of Riesz' representation theorem for the dual of $C_0(T)$, T locally compact Hausdorff, is also given in [613, p. 131].

1.5.7 Radon Measures

In the three volume book [149] the theory of integration is developed à la Bourbaki. For a locally compact Hausdorff space T one equips $C_{00}(T)$ with the inductive limit topology τ_{ind} with respect to the family $C_K(T) = \{f \in C_{00}(T) : \text{spt}(f) \subseteq K\}$, $K \subseteq T$, K compact, of subspaces of $C_{00}(T)$. It follows that a sequence (f_n) in $C_{00}(T)$ converges to $f \in C_{00}(T)$ with respect to τ_{ind} if and only if there exists a compact subset K of T such that $\text{spt}(f_n) \subseteq K$, $n \in \mathbb{N}$, $\text{spt}(f) \subseteq K$, and

$$f_n|_K \xrightarrow{u} f|_K \qquad \text{(uniform convergence on } K\text{)}.$$

A *Radon measure* is a continuous linear functional $I : (C_{00}(T), \tau_{ind}) \to \mathbb{R}$. The family of all Radon measures is denoted by $\mathscr{R}ad(T)$ and the family of all positive Radon measures is denoted by $\mathscr{R}ad^+(T)$. The continuity of a linear functional $I : (C_{00}(T), \tau_{ind}) \to \mathbb{R}$ is equivalent to the following property: for every compact $K \subseteq T$ there exists a number $\beta_K > 0$ such that

$$|I(f)| \leq \beta_K \|f\|_\infty,$$

for all $f \in C_{00}(T)$ with $\text{spt}(f) \subseteq K$.

Note that a positive linear functional $I : (C_{00}(T), \tau_{ind}) \to \mathbb{R}$ is automatically continuous. Riesz' representation theorem tells us that in the case when T is compact, $\mathscr{R}ad(T)$ can be identified with the set $\mathscr{M}(T, \mathbb{R})$ of all regular Borel measures, and $\mathscr{R}ad^+(T)$ with the set $\mathscr{M}^+(T, \mathbb{R})$ of all regular positive Borel measures.

Ordering $X = C_{00}(T)$, as usual, by the pointwise order

$$f \leq g \iff \forall t \in T, \ f(t) \leq g(t),$$

it follows that $\mathscr{R}ad^+(T)$ is the dual cone $X_+^* \subseteq X^*$ of X_+. We mention the following order properties.

Theorem 1.5.30 ([149], Vol. I, Theorem 11.2) *Let T be a locally compact Hausdorff space. Then the following assertions are true.*

1. *$C_{00}(T)$ is a vector lattice.*
2. *Every positive linear functional is continuous and $\mathscr{R}ad(T)$ agrees with the space of all order-bounded linear functionals on $C_{00}(T)$.*
3. *$\mathscr{R}ad(T)$ is a complete lattice.*

A real-valued linear functional f on an ordered vector space X is called *order-bounded* if it maps order-bounded subsets of X to (order) bounded subsets of $\mathbb{R}$.

Remark 1.5.31 Let T be a locally compact Hausdorff space. In [222, Kapitel VIII] the dual of the space $C_{00}(T)$ equipped with the topology τ_k of uniform convergence on compacta is also discussed. The topology τ_k is the locally convex topology generated by the family p_K of seminorms, where for $K \subseteq T$, K compact, p_K is defined by

$$p_K(f) = \sup\{|f(t)| : t \in K\}, \quad f \in C_{00}(T).$$

In the same book the duals of $C_b(T)$ (the Banach space of bounded continuous functions on T with the sup-norm) for more general topological spaces (completely regular, normal) are presented. In this case the representing measures are, in general, only finitely additive.

1.6   Vector Measures

In this section we shall present some results on vector measures and the integration of vector functions. Details and further results can be found in [75, Chapter 5], [112, 194, 196], [295, Chapter 1].

A good part of the scalar measure and integration theory can be transposed to Banach space-valued functions with similar proofs. Everything goes smoothly (roughly speaking, by replacing the absolute value with the norm) until smoothness enters the scene, meaning by this Lebesgue's theorem on the a.e. differentiability of absolutely continuous functions, the fundamental theorem of calculus and the Radon-Nikodým theorem, when some geometric properties of the Banach space are required. The study of these properties in the vector case led to a rich geometrical and topological theory of Banach spaces and established connections between apparently unrelated notions. The pioneers of this direction of research were Clarkson [153], who introduced uniformly convex Banach spaces to obtain a vector analog of the above mentioned differentiability theorem, and Gelfand [249], who studied similar problems in reflexive and in separable Banach spaces (which are spaces satisfying the Radon-Nikodým property, see Sect. 1.6.3)

Concerning this matter we include the following quotation from [194, p. 44].

Indeed, some have said that the Bochner integral is only the Lebesgue integral with absolute value signs replaced by norm signs. We shall see that often this is the case, and sometimes it is a totally ignorant appraisal of the Bochner integral. In fact, as we shall see later, the failure of the Radon-Nikodým theorem for the Bochner integral lies at the base of some of the most intriguing results in the theory of vector measures and the structure theory of Banach spaces.

1.6.1   The Integration of Vector Functions

In this subsection we shall present some results on the integration of vector functions with respect to scalar measures, with emphasis on Bochner's integral. Let $(\Omega, \mathscr{A})$ be a measurable space, E a Banach space with dual E^* and $\mathscr{B}(E)$ the σ-algebra of Borel subsets of E.

For a function $f : \Omega \to E$ and $x^* \in E^*$, we denote by $\langle f, x^* \rangle$ the function $x^* \circ f$, i.e., $t \mapsto \langle f(t), x^* \rangle$, $t \in \Omega$, and, for $g : \Omega \to E^*$ and $x \in E$, one denotes by $\langle x, g \rangle$ the function $t \mapsto \langle x, g(t) \rangle$, $t \in \Omega$.

A *measurable simple function* is a function of the form $\sum_{i=1}^{n} \chi_{A_i} x_i$, where $A_i \in \mathscr{A}$, $x_i \in E$, $i = 1, \ldots, n$, $n \in \mathbb{N}$.

Definition 1.6.1 A function $f : \Omega \to E$ is called

- *Borel measurable* if it is $(\mathscr{A}, \mathscr{B}(E))$-measurable;
- *strongly measurable* if there exists a sequence (f_n) of measurable simple functions that is pointwise convergent to f;
- *weakly measurable* if the function $\langle f, x^* \rangle$ given by $t \mapsto \langle f(t), x^* \rangle$, $t \in \Omega$, is measurable for every $x^* \in E^*$. More general, one can consider a subset Γ of E^* and call f Γ-measurable if $\langle f, x^* \rangle$ is measurable for all $x^* \in \Gamma$. For $\Gamma = E$, considered as a subset of E^{**}, one says that $f : \Omega \to E^*$ is weak*-measurable. This means that the mapping $t \mapsto \langle x, f(t) \rangle$, $t \in \Omega$, is measurable for every $x \in E$.

Apparently, a natural definition of the measurability of a vector function f would be that of Borel measurability, i.e., $f^{-1}(B) \in \mathscr{A}$ for every $B \in \mathscr{B}(E)$. The main disadvantage of such a definition is that the σ-algebra $\mathscr{B}(E)$ could be too large. For instance, the σ-algebra $\sigma(E^*)$ generated by all continuous linear functionals on E can be strictly smaller than $\mathscr{B}(E)$ if the space E is nonseparable, and this restricts the possibility to use tools specific to functional analysis, such as the Hahn-Banach theorem. For this reason, one introduces the notion of strong measurability, inspired by a property of measurable scalar functions (see Proposition 1.5.14).

By the σ-algebra generated by a subset G of E^* one understands the σ-algebra generated by the sets

$$\{x \in E : (x_1^*(x), \ldots, x_n^*(x)) \in B\},$$

for $n \in \mathbb{N}$, $x_1^*, \ldots, x_n^* \in G$, $B \in \mathscr{B}(\mathbb{K}^n)$.

If E is separable, then the situation is better.

Proposition 1.6.2 *Let E be a separable Banach space.*

1. *The following equalities hold*

$$\sigma(G) = \sigma(E^*) = \mathscr{B}(E),$$

for every w^*-dense subspace G of E^*.
2. *A function $f : \Omega \to E$ is Borel measurable if and only if the function $\langle f, x^* \rangle$ is measurable for every $x^* \in E^*$.*

Notice that a subspace G of E^* is w^*-dense in E^* if and only if it separates the points in E, that is, for every $x \neq 0$ there exists $x^* \in G$ with $x^*(x) \neq 0$.

The relation between weak and strong measurability is given by the following result of B. J. Pettis. A function $f : \Omega \to E$ is called *separably valued* if $f(\Omega)$

is a norm-separable subset of E (i.e., it is separable with respect to the topology generated by the norm). This is equivalent to the fact that the closed linear subspace E_0 of E generated by $f(\Omega)$ is separable.

Theorem 1.6.3 (Pettis' Measurability Theorem I) *Let $(\Omega, \mathscr{A})$ be a measurable space, E a Banach space, G a w^*-dense subspace of E and $f : \Omega \to E$ a vector function. Then the following are equivalent.*

1. *The function f is strongly measurable.*
2. *The function f is separably valued and Borel measurable.*
3. *The function f is separably valued and weakly measurable.*
4. *The function f is separably valued and $\langle f, x^* \rangle$ is measurable for every $x^* \in G$.*

Remark 1.6.4

1. One can show that if f is strongly measurable and takes values in a closed subspace E_0 of E, then the sequence (f_n) of measurable simple functions converging to f can be chosen in such a way that each function f_n takes values in E_0. Also, the strong measurability of f as a function from Ω to E is the same as the strong measurability of f as a function from Ω to E_0.
2. If f is strongly measurable, then there exists a sequence (f_n) of measurable simple functions such that

$$\|f_n(x)\| \le \|f(x)\| \text{ and } \|f_n(x) - f(x)\| \to 0 \text{ as } n \to \infty \,,$$

   for all $x \in E$.
3. The pointwise limit of a sequence (f_n) of strongly measurable functions is strongly measurable.
4. If E_1, E_2 are Banach spaces, $f : \Omega \to E_1$ is strongly measurable and $\varphi : E_1 \to E_2$ is Borel-to-Borel measurable, then $\varphi \circ f : \Omega \to E_2$ is strongly measurable. In particular, if $f : \Omega \to E$ is strongly measurable, then the function

$$\|f\|(t) = \|f(t)\|, \ t \in \Omega \,,$$

   is measurable.

μ-Measurable Functions

Suppose now that $(\Omega, \mathscr{A}, \mu)$ is a positive measure space and E is a Banach space. A function $f : \Omega \to E$ is called *strongly μ-measurable* if there exists a sequence (f_n) of measurable simple functions such that $(f_n(t))$ converges to $f(t)$ μ-a.e. $t \in \Omega$, i.e., there exists a set $A \in \mathscr{A}$ with $\mu(A) = 0$ such that $f_n(t) \to f(t)$ for every $t \in \Omega \setminus A$.

All the notions considered for strongly measurable functions have their μ-a.e. analogs. For instance, a function $f : \Omega \to E$ is called *μ-a.e. separably valued* if there exists $A \in \mathscr{A}$ with $\mu(A) = 0$ such that $f(\Omega \setminus A)$ is a norm separable subset of E.

Pettis measurability criteria given in Theorem 1.6.3 have their analogs for μ-measurable functions.

Theorem 1.6.5 (Pettis' Measurability Theorem II) *Let* $(\Omega, \mathscr{A}, \mu)$ *be a positive measure space,* E *a Banach space and* $f : \Omega \to E$ *a vector function. Then the following are equivalent.*

1. *The function* f *is strongly* μ-*measurable.*
2. *The function* f *is* μ-*a.e. separably valued and weakly measurable.*
3. *The function* f *is* μ-*a.e. separably valued and there exists a* w^*-*dense subspace* G *of* E *such that* $\langle f, x^* \rangle$ *is measurable for every* $x^* \in G$.

The above theorem has the following consequence.

Corollary 1.6.6 *A vector function* $f : \Omega \to E$ *is strongly measurable if and only if there exist a sequence* (f_n) *of measurable countable-valued functions and a* μ-*null set* $A \in \mathscr{A}$ *such that the sequence* (f_n) *converges to* f *uniformly on* $\Omega \setminus A$.

Remark 1.6.7

1. Let $(\Omega, \mathscr{A}, \mu)$ be a positive measure space, E a Banach space and $f : \Omega \to E$.

   (i) If f is strongly μ-measurable, then f is μ-a.e. equal to a strongly measurable function.
   (ii) If the measure μ is σ-finite, then the converse is also true. Namely, if f is μ-a.e. equal to a strongly measurable function, then f is strongly μ-measurable.

2. One can also show that if f is strongly μ-measurable, then there exists a sequence (f_n) of measurable simple functions such that

$$\|f_n(x)\| \leq \|f(x)\| \quad \text{and} \quad \|f_n(x) - f(x)\| \to 0 \quad \text{as } n \to \infty,$$

   for μ-almost all $x \in E$.
3. The μ-a.e. limit of a sequence (f_n) of strongly μ-measurable functions is strongly μ-measurable.
4. If E_1, E_2 are Banach spaces, $f : \Omega \to E_1$ is strongly μ-measurable and $\varphi : E_1 \to E_2$ is Borel-to-Borel measurable, then $\varphi \circ f : \Omega \to E_2$ is strongly μ-measurable, provided that μ is σ-finite or $\varphi(0) = 0$.

   In particular, if $f : \Omega \to E$ is strongly μ-measurable, then the function

$$\|f\|(t) = \|f(t)\|, \quad t \in \Omega,$$

   is μ-measurable.

The Bochner Integral

A measurable simple function $f = \sum_{i=1}^{n} \chi_{A_i} x_i$ is called μ-*integrable* if all the sets A_i are of finite measure μ. The *integral* of f is defined by

$$\int_{\Omega} f d\mu = \sum_{i=1}^{n} \mu(A_i) x_i .$$

The integral over a set $A \in \mathscr{A}$ is the integral of the simple function $\chi_A \cdot f = \sum_{i=1}^{n} \chi_{A \cap A_i} x_i$:

$$\int_{\Omega} f d\mu = \sum_{i=1}^{n} \mu(A \cap A_i) x_i .$$

One shows that the integral is well-defined, i.e., it does not depend on the representation of the simple function and that

$$\left\| \int_A f d\mu \right\| \le \int_A \|f\| d\mu .$$

Definition 1.6.8 A strongly μ-measurable function $f : \Omega \to E$ is called *Bochner integrable* if there exists a sequence (f_n) of μ-integrable simple functions such that

$$\lim_{m,n \to \infty} \int_{\Omega} \|f_n - f_m\| d\mu = 0 . \tag{1.6.1}$$

The integral is defined by

$$\int_{\Omega} f d\mu = \lim_{n \to \infty} \int_{\Omega} f_n d\mu .$$

The integral over a set $A \in \mathscr{A}$ is defined by

$$\int_A f d\mu = \int_{\Omega} (\chi_A \cdot f) d\mu .$$

One shows that $\left(\int_{\Omega} f_n d\mu \right)_{n \in \mathbb{N}}$ is a Cauchy sequence in the Banach space E, so it has a limit and that this limit is independent of the sequence of simple μ-integrable functions satisfying (1.6.1).

We mention the following properties of the Bochner integral.

Proposition 1.6.9 *Let* $(\Omega, \mathscr{A}, \mu)$ *be a positive measure space,* E *a Banach space and* $f : \Omega \to E$ *a strongly μ-measurable function.*

1. *The function is Bochner integrable if and only if* $\int_{\Omega} \|f\| d\mu < \infty$.

2. *If f is Bochner integrable, then the following hold:*

(i) $\left\| \int_{\Omega} f d\mu \right\| \le \int_{\Omega} \|f\| d\mu$;

(ii) $\lim_{\mu(A)\to 0} \int_A f d\mu = 0$;

(iii) *the integral is countably additive with respect to the set of integration, i.e., if $A = \bigcup_{n=1}^{\infty} A_n$, where $A_n \in \mathscr{A}$ are pairwise disjoint, then*

$$\int_A f d\mu = \sum_{n=1}^{\infty} \int_{A_n} f d\mu ;$$

(iv) $\frac{1}{\mu(A)} \int_A f d\mu \in \overline{co}(f(A))$ *for every $A \in \mathscr{A}$ with $\mu(A) > 0$;*

(v) *if F is another Banach space and $T : E \to F$ is a continuous linear operator, then $T \circ f : \Omega \to F$ is Bochner integrable and*

$$\int_A (T \circ f) d\mu = T\left(\int_A f d\mu \right) ,$$

for every $A \in \mathscr{A}$.

The Dunford, Pettis and Gelfand Integrals

These can be viewed as weak forms of integration.

Proposition 1.6.10 *Let $(\Omega, \mathscr{A}, \mu)$ be a positive measure space, E a Banach space and $f : \Omega \to E$. If f is weakly μ-measurable and for every $x^* \in E^*$, $\langle f, x^* \rangle \in L^1(\mu)$, then for every $A \in \mathscr{A}$ there exists a unique element $x_A^{**} \in E^{**}$ such that*

$$x_A^{**}(x^*) = \int_A (\langle f, x^* \rangle) d\mu , \tag{1.6.2}$$

for all $x^* \in E^*$.

Starting from this result we define the corresponding integrals.

Definition 1.6.11 A weakly μ-integrable function $f : \Omega \to E$ such that $\langle f, x^* \rangle \in L^1(\mu)$ for every $x^* \in E^*$ is called *Dunford integrable*. The *Dunford integral* of f over a set $A \in \mathscr{A}$ is the unique element $x_A^{**} \in E^{**}$ satisfying (1.6.2).

If for every $A \in \mathscr{A}$ there exists $x_A \in E$ such that

$$x^*(x_A) = \int_A (\langle f, x^* \rangle) d\mu , \tag{1.6.3}$$

for all $x^* \in E^*$, then one says that f is *Pettis integrable* and the unique element $x_A \in E$ satisfying (1.6.3) is called the *Pettis integral* of f over A.

If $f : \Omega \to E^*$ is such that the function $t \mapsto \langle x, f(t) \rangle$, $t \in \Omega$, belongs to $L^1(\mu)$ for every $x \in E$, then one says that f is *Gelfand integrable* if for every $A \in \mathscr{A}$ there exists $x_A^* \in E^*$ such that

$$x_A^*(x) = \int_A \langle x, f \rangle d\mu, \qquad (1.6.4)$$

for all $x \in E$. The functional x_A^* given by (1.6.4) is called the *Gelfand integral* of f over A.

Good presentations of Pettis integration theory are given in [488, 489].

1.6.2 Vector Measures

In this subsection we shall give a quick introduction to vector measures. Let $(\Omega, \mathscr{A})$ be a measurable space and E a Banach space. A mapping $v : \mathscr{A} \to E$ is called a *vector measure* if

$$\begin{aligned} &\text{(i)} \quad v(\emptyset) = 0, \\[4pt] &\text{(ii)} \quad v(A) = \sum_{n=1}^{\infty} v(A_n), \end{aligned} \qquad (1.6.5)$$

whenever $A = \bigcup_{n=1}^{\infty} A_n$, where $\{A_n\}$ is a countable family of mutually disjoint sets in $\mathscr{A}$. One can also consider finitely additive vector measures by supposing that (ii) holds only for finite families of mutually disjoint sets in $\mathscr{A}$ (in this case it suffices to suppose that $\mathscr{A}$ is only an algebra of subsets of Ω).

Remark 1.6.12 As in the case of scalar measures (see Remark 1.5.3), it follows that the series in the right hand side of the equality (ii) converges unconditionally, but in this case it does not follow that the convergence is also absolute (see Sect. 1.4.7).

A mapping $v : \mathscr{A} \to E$ is called *weakly countably additive* if it satisfies (i) (from (1.6.5)) and the series in (ii) is only weakly convergent. By the Mazur-Orlicz theorem (see Theorem 1.4.25), the series is norm unconditionally convergent, that is, the following result holds.

Proposition 1.6.13 *A weakly countably additive vector measure is (norm) countably additive.*

The total *variation* of a vector measure v is defined by replacing the absolute value in the definition of the variation of a scalar measure (see Definition 1.5.3) by the norm:

$$|v|(A) = \sup \sum_{k=1}^{n} \| v(B_k) \|,$$

where the supremum is taken over all partitions $A = B_1 \cup \cdots \cup B_n$ of $A \in \mathscr{A}$ into pairwise disjoint sets $B_k \in \mathscr{A}$, $k = 1, \ldots, n$, $n \in \mathbb{N}$.

Define also the *total variation* of v by

$$\|v\| = |v|(\Omega),$$

and call the vector measure v of *bounded variation* provided that $\|v\| < \infty$.

One shows that if the vector measure v is of bounded variation, then its total variation is a positive measure on $\mathscr{A}$ (countably additive, by the definition of a measure).

The composition $(x^* \circ v)(A) = x^*(v(A))$ of the vector measure v with a continuous linear functional $x^* \in E^*$ is a scalar measure on $\mathscr{A}$. Using this fact one can consider a weaker form of variation, called *semivariation*, defined for $A \in \mathscr{A}$ by

$$|v|_w(A) = \sup \left\{ |x^* \circ v|(A) : x^* \in B_{E^*} \right\}, \tag{1.6.6}$$

where $|x^* \circ v|$ is the variation of the scalar measure $x^* \circ v$. If $|v|_w(\Omega) < \infty$, then one says that v is of *bounded semivariation* (compare with (1.5.4) and Definition 1.5.7).

The following proposition shows that the semivariation (1.6.6) can be calculated without appealing to the dual space.

Proposition 1.6.14 *Let $(\Omega, \mathscr{A})$ be a measurable space and $v : \mathscr{A} \to E$ a vector measure. The semivariation of v on $A \in \mathscr{A}$ can be calculated by the formula*

$$|v|_w(A) = \sup \left\{ \Big\| \sum_{k=1}^{n} \varepsilon_k v(A_k) \Big\| \right\},$$

where the supremum is taken over all finite partitions $A_1, \ldots, A_n$ of A and all scalars satisfying $|\varepsilon_k| \leq 1$, $k = 1, \ldots, n$, $n \in \mathbb{N}$.

Also, the following inequalities hold

$$\sup \{\|v(B)\| : B \in \mathscr{A}, \ B \subseteq A\} \leq |v|_w(A)$$

$$\leq 4 \sup \{\|v(B)\| : B \in \mathscr{A}, \ B \subseteq A\}.$$

Consequently, v is of bounded semivariation if and only if its range is bounded in E, i.e., $v(\mathscr{A})$ is a bounded subset of E.

Remark 1.6.15 One can show further that the measure v has a relatively weakly compact range, i.e., the set $v(\mathscr{A})$ is relatively weakly compact in E.

The Integral with Respect to a Vector Measure

Let $(\Omega, \mathscr{A})$ be a measurable space and $v : \mathscr{A} \to E$ a vector measure. Denote by $S(\Omega, \mathscr{A})$ the vector space of $\mathscr{A}$-measurable simple functions on Ω. For $f =$

$\sum_{i=1}^n \alpha_i \chi_{A_i} \in S(\Omega, \mathscr{A})$ define the integral of f with respect to v by

$$\int f\, dv = \sum_{i=1}^n \alpha_i v(A_i).$$

It follows that the mapping $T_v : S(\Omega, \mathscr{A}) \to E$ given by $T_v(f) = \int f\, dv$ is a linear operator. By Proposition 1.6.14, the following inequality

$$\|T_v(f)\| \le \|f\|_\infty |v|_w(\Omega),$$

holds for every $f \in S(\Omega, \mathscr{A})$, implying that T_v is also continuous, with $\|T_v\| \le |v|_w(\Omega)$. It is easy to check that, in fact, $\|T_v\| = |v|_w(\Omega)$. Here $\|f\|_\infty = \sup\{|f(t)| : t \in \Omega\}$. Denote by $B(\Omega, \mathscr{A})$ the closure with respect to the sup-norm of the space $S(\Omega, \mathscr{A})$ in the Banach space $B(\Omega)$ of all bounded functions on Ω. By Proposition 1.5.14, $B(\Omega, \mathscr{A})$ agrees with the space of all bounded $\mathscr{A}$-measurable functions on Ω. It follows that T_v has a unique continuous linear extension $\tilde{T}_v$ to $B(\Omega, \mathscr{A})$, of the same norm as T_v, which will be called the integral with respect to v of functions in $B(\Omega, \mathscr{A})$, denoted by $\int f\, dv$. It follows that

$$\int f\, dv = \lim_{n\to\infty} \int f_n\, dv,$$

where (f_n) is a sequence of μ-integrable simple functions uniformly convergent to f (the limit does not depend on the chosen sequence).

1.6.3 The Radon-Nikodým Property

Since the Radon-Nikodým theorem, as presented in Sect. 1.5.4, does not hold in general for vector measures and vector functions, one imposes a study of those Banach spaces for which this property holds. Further properties of Banach spaces with the Radon-Nikodým property are considered in Sect. 8.8.4.

Let $(\Omega, \mathscr{A}, \mu)$ be a finite positive measure space, E a Banach space and $v : \mathscr{A} \to E$ a vector measure. One says that v is absolutely continuous with respect to μ if

$$\mu(A) = 0 \Rightarrow v(A) = 0.$$

It turns out that the characterization of absolute continuity given in Proposition 1.5.18 holds in this case too. The vector measure v is absolutely continuous with respect to μ if and only if the following condition holds: for every $\varepsilon > 0$ there exists $\delta > 0$ such that

$$\mu(A) < \delta \implies \|\nu(A)\| < \varepsilon,$$

for all $A \in \mathcal{A}$.

If $f : \Omega \to E$ is a Bochner integrable function, then $\nu(A) = \int_A f d\mu$, $A \in \mathcal{A}$, is a vector measure on $\mathcal{A}$ which is absolutely continuous with respect to μ (see Proposition 1.6.9). We have seen that for a scalar measure ν, by the Radon-Nikodým theorem (Theorem 1.5.23), the converse is also true, a property that is no longer true for vector measure as shown by the following simple example.

Example 1.6.16 ([194]) The failure of the Radon-Nikodým theorem for a c_0-valued vector measure.

Let $([0, 1], \mathcal{L}, \lambda)$ be the Lebesgue measure space on $[0, 1]$ and let $\nu : \mathcal{L} \to c_0$ be defined by

$$\nu(A) = \left(\int_A \sin(n\pi t)dt \right)_{n \in \mathbb{N}}, \quad A \in \mathcal{L}. \tag{1.6.7}$$

By the Riemann-Lebesgue lemma, ν takes values in c_0. Since $|\sin s| \le 1$, $s \in \mathbb{R}$,

$$\|\nu(A)\| = \sup_n \left| \int_A \sin(n\pi t)dt \right| \le \lambda(A).$$

It follows that ν is countably additive, λ-absolutely continuous and of bounded variation. Suppose that there exists a Bochner integrable function $g = (g_n)_{n \in \mathbb{N}} : [0, 1] \to c_0$ such that $\nu(A) = \int_A g(t)dt$, $A \in \mathcal{L}$. Denoting by P_n the projection operator on the n-th coordinate in c_0, it follows that

$$P_n \nu(A) = \int_A P_n g(t)dt = \int_A g_n(t)dt. \tag{1.6.8}$$

By (1.6.7) and (1.6.8), $g_n(t) = \sin(n\pi t)$ for λ-a.e. $t \in [0, 1]$, $n \in \mathbb{N}$. But the sequence $(\sin(n\pi t))_{n \in \mathbb{N}}$ belongs to c_0 only for $t \in \mathbb{Z}$.

For this reason one introduces the class of Banach spaces for which this property holds.

Definition 1.6.17 One says that the Banach space E has the *Radon-Nikodým Property* (RNP for short) for the finite positive measure space $(\Omega, \mathcal{A}, \mu)$ if for every vector measure $\nu : \mathcal{A} \to E$ of bounded variation and absolutely continuous with respect to μ there exists a Bochner integrable function $g : \Omega \to E$ such that

$$\nu(A) = \int_A g d\mu,$$

for all $A \in \mathscr{A}$. One says that the Banach space E *has the RNP* if it has the RNP for every finite positive measure space.

The Radon-Nikodým Property admits characterizations in terms of functions defined on $[0, 1]$.

Following [194, p. 217] we list some properties equivalent to the RNP.

Theorem 1.6.18 *Each of the following conditions is necessary and sufficient for a Banach space E to have the RNP.*

1. *Every closed linear subspace of E has the RNP.*
2. *Every separable closed linear subspace of E has the RNP.*
3. *Every function $f : [0, 1] \to E$ of bounded variation is differentiable off a fixed set of Lebesgue measure zero.*
4. *Every function $f : [0, 1] \to E$ of bounded variation is weakly differentiable off a fixed set of Lebesgue measure zero.*
5. *Every Lipschitz function $f : [0, 1] \to E$ is differentiable off a fixed set of Lebesgue measure zero.*
6. *Every absolutely continuous function $f : [0, 1] \to E$ is differentiable off a fixed set of Lebesgue measure zero. In this case*

$$f(b) - f(a) = \int_a^b f'(t)dt$$

for every $a, b \in [0, 1]$.
7. *Every absolutely continuous function $f : [0, 1] \to E$ is weakly differentiable off a fixed set of Lebesgue measure zero. In this case*

$$x^*(f(b) - f(a)) = \int_a^b \langle f'(t), x^* \rangle dt$$

for all $a, b \in [0, 1]$ and $x^* \in E^*$.

Remark 1.6.19 A Banach space E such that every absolutely continuous function $f : [0, 1] \to E$ is a.e. differentiable is called a *Gelfand space* in [194, p. 106] (see [249]). A Banach space E such that every Lipschitz function $f : [0, 1] \to E$ is a.e. differentiable is called a Banach space with the *Gelfand-Fréchet property* in [21]. By Theorem 1.6.18, both these properties are equivalent to E having the RNP.

Example 1.6.20

1. The function $f : [0, 1] \to L^1[0, 1]$ given by $f(t) = \chi_{[0,t]}$, $t \in [0, 1]$, is Lipschitz but nowhere differentiable on $[0, 1]$.
2. The function $g : [0, 1] \to c_0$ given by $g(t) = (g_n(t))_{n=1}^\infty$, where

$$g_n(t) = \int_0^t \sin(n\pi s)ds, \quad 0 \le t \le 1, \ n \in \mathbb{N},$$

is Lipschitz but not a.e. differentiable on $[0, 1]$.

3. The function $h : [0, 1] \to c_0$ given by $h(t) = (h_n(t))_{n=1}^{\infty}$, where

$$h_n(t) = \frac{1}{n}\sin(nt), \quad 0 \le t \le 1, \ n \in \mathbb{N},$$

is Lipschitz and nowhere differentiable on $[0, 1]$.

1. Indeed for $0 \le s < t \le 1$ we have

$$\|f(s) - f(t)\|_1 = \|\chi_{(s,t]}\|_1 = |s - t|,$$

that is, f is Lipschitz.
For $t_0 \in [0, 1)$ and $t_0 < s < t \le 1$, one has

$$\left\| \frac{f(s) - f(t_0)}{s - t_0} - \frac{f(t) - f(t_0)}{t - t_0} \right\|_1 = 2\frac{t - s}{t - t_0} \to 2 \text{ if } s \searrow t_0,$$

showing that the limit

$$\lim_{t \to t_0} \frac{f(t) - f(t_0)}{t - t_0}$$

does not exist. (If $t_0 = 1$ one works with $0 \le t < s < 1$ to obtain a similar result.)

2. By the Lebesgue-Riemann lemma (see [287, p. 249]) $\lim_{n \to \infty} g_n(t) = 0$, so that $g(t) \in c_0$ for all $t \in [0, 1]$. Also,

$$|g_n(t) - g_n(s)| = \left| \int_t^s \sin(n\pi u) du \right| \le \left| \int_t^s du \right| = |t - s|,$$

for all $n \in \mathbb{N}$, implying $\|g(t) - g(s)\|_{c_0} \le |t - s|$ for all $s, t \in [0, 1]$.
If g were differentiable, then $g'(t) = (g_n'(t))_{n=1}^{\infty} = (\sin(n\pi t))_{n=1}^{\infty}$.
But $\lim_{n \to \infty} \sin(n\pi t) = 0$ only for t belonging to a subset of measure 0 of $[0, 1]$, so that the function g is not a.e. differentiable on $[0, 1]$.

3. Concerning the function h, it is obvious that $h(t)$ belongs to c_0 for every $t \in [0, 1]$ and

$$|h_n(t) - h_n(s)| = \left| \int_t^s \cos(nu) du \right| \le \left| \int_t^s du \right| = |t - s|,$$

for all $n \in \mathbb{N}$, implying $\|h(s) - h(t)\|_{c_0} \le |s - t|$ for all $s, t \in [0, 1]$.
If h were differentiable, then $h'(t) = (g_n'(t))_{n=1}^{\infty} = (\cos(nt))_{n=1}^{\infty}$. But the sequence $(\cos(nt))_{n=1}^{\infty}$ belongs to c_0 for no $t \in [0, 1]$, so that the function h is nowhere differentiable on $[0, 1]$.

Remark 1.6.21 In the light of the characterizations of RNP given in Theorem 1.6.18(5), it follows that the spaces $L^1[0, 1]$ and c_0 do not have the RNP. A fairly complete treatment of the differentiation, Bochner integration and Radon-Nikodým property for functions defined on intervals in $\mathbb{R}$ and with values in a Banach space is given in Chapter 1 of the book [43]. In Example 1.6.20 the functions f and g are taken from [554], while h is from [43, Proposition 1.2.9] (see also [194]).

In [194, p. 218] there is a list of spaces having or not having the RNP.
Among the Banach spaces **having the RNP** we mention:

- reflexive Banach spaces and separable dual spaces;
- weakly compactly generated (WCG) dual spaces and dual subspaces of WCG spaces;
- locally uniformly convex and weakly locally uniformly convex dual spaces;
- dual spaces with Fréchet differentiable norm;
- the space $\ell^1(\Gamma)$ for arbitrary Γ;
- the Hardy space $H^p(D)$, $1 \leq p < \infty$, in the unit disk D;
- the space $\mathscr{L}(\ell^p, \ell^q)$, $1 \leq p < q < \infty$;
- the space $L^p(\nu, X)$ if X has the RNP.

Some Banach spaces that **do not have the RNP**:

- the spaces $L^1[0, 1]$ and $BV_0[0, 1]$;
- the space $L^1(\nu)$ if ν is not purely atomic;
- the spaces c_0, c, ℓ^∞, $L^\infty[0, 1]$;
- the space $C(T)$, T infinite compact Hausdorff;
- the space of compact operators $\mathscr{K}(X)$ for $X = \ell^p$, L^p or $C(T)$;
- the Hardy space $H^\infty(D)$ and the disk algebra $A(D)$, D the open unit disk in $\mathbb{C}$.

Chapter 2
Basic Facts Concerning Lipschitz Functions

In this chapter we prove the existence of some Lipschitz functions (the analog of Urysohn's lemma for Lipschitz functions) and of Lipschitz partitions of unity. We also study algebraic operations with Lipschitz functions, sequences of Lipschitz functions, Lipschitz properties for differentiable functions (including a characterization in terms of Dini derivatives), and the possibility of gluing together Lipschitz functions. Applications are given to a sandwich type theorem, to Lipschitz selections for set-valued mappings and to the separability of the space $C(T)$.

2.1 Lipschitz and Locally Lipschitz Functions

As the following proposition shows, there is a good supply of Lipschitz functions on a (reasonably rich) metric space.

Proposition 2.1.1 *Let (X, d) be a metric space.*

1. *If A is a nonempty subset of X such that $\overline{A} \neq X$, then the distance function $f(x) = d(x, A)$, $x \in X$, is Lipschitz with $L(f) = 1$ and $f(x) = 0$ for $x \in \overline{A}$.*
2. *If C, D are nonempty closed subsets of X such that $d(C, D) > 0$, then the function $f : X \to [0, 1]$ given by*

$$f(x) = \frac{d(x, C)}{d(x, C) + d(x, D)}, \quad x \in X, \tag{2.1.1}$$

is Lipschitz with $L(f) \leq 1/d(C, D)$, and

$$f(x) = 0 \iff x \in C \quad and \quad f(x) = 1 \iff x \in D. \tag{2.1.2}$$

In particular, this holds if the sets C, D are nonempty, closed, disjoint and one of them is compact.

© Springer Nature Switzerland AG 2019
Ş. Cobzaş et al., *Lipschitz Functions*, Lecture Notes in Mathematics 2241,
https://doi.org/10.1007/978-3-030-16489-8_2

Proof

1. We have seen in Proposition 1.3.7 that $|f(x) - f(x')| \leq d(x, x')$, for all $x, x' \in X$, so that $L(f) \leq 1$.
   Let $x \in X \setminus \bar{A}$. Then $d(x, A) > 0$ and there exists a sequence (y_n) in A such that $\lim_{n \to \infty} d(x, y_n) = d(x, A)$. It follows that

$$L(f) \geq \frac{|f(x) - f(y_n)|}{d(x, y_n)} = \frac{d(x, A)}{d(x, y_n)} \to 1 \text{ as } n \to \infty,$$

   showing that $L(f) \geq 1$, and so $L(f) = 1$.

2. It is clear that the function f given by (2.1.1) satisfies the conditions from (2.1.2). It remains to show that it is Lipschitz. Observe that for every $x \in X$

$$d(C, D) \leq d(x, C) + d(x, D). \tag{2.1.3}$$

   This follows from the inequalities

$$d(C, D) \leq d(c, d) \leq d(c, x) + d(x, d),$$

   valid for all $(c, d) \in C \times D$.
   Observe also that

$$\left| \frac{u}{u + v} - \frac{u'}{u' + v'} \right| = \frac{|uv' - u'v|}{(u + v)(u' + v')} \leq \frac{u|v' - v| + v|u - u'|}{(u + v)(u' + v')},$$

   for all $u, v, u', v' \in \mathbb{R}_+$ with $u + v > 0$ and $u' + v' > 0$.
   Putting $u = d(x, C)$, $u' = d(x', C)$, $v = d(x, D)$, $v' = d(x', D)$, and taking into account (1.3.8) and (2.1.3), one obtains

$$|f(x) - f(x')| \leq \frac{d(x, C)|d(x', D) - d(x, D)| + d(x, D)|d(x', C) - d(x, C)|}{(d(x, C) + d(x, D))(d(x', C) + d(x', D))}$$

$$\leq \frac{d(x, x')}{d(x', C) + d(x', D)} \leq \frac{1}{d(C, D)} \cdot d(x, x'),$$

   for all $x, x' \in X$.

$$\square$$

Remark 2.1.2 The second assertion from Proposition 2.1.1 is an Urysohn-type result for Lipschitz functions (see Theorem 1.3.13).

Remark 2.1.3 If C, D are arbitrary nonempty closed and disjoint subsets of a metric space X, then, as the following example shows, it is not sure that there exists a Lipschitz function f on X such that $f \equiv 1$ on C and $f \equiv 0$ on D. Consequently, the corresponding assertion from [675, p. 4] holds only under this supplementary hypothesis (namely $d(C, D) > 0$).

Indeed, take $X = \mathbb{R}^2$ with the Euclidean metric, $C = \{(x, y) \in \mathbb{R}^2_+ : xy \geq 1\}$ and $D = \{(x, y) \in \mathbb{R}^2 : y \leq 0\}$. If $f = 1$ on C and $f = 0$ on D, then, taking $\xi_n = (n, n^{-1}) \in C$ and $\eta_n = (n, 0) \in D$ for $n \in \mathbb{N}$, it follows that

$$\frac{|f(\xi_n) - f(\eta_n)|}{d(\xi_n, \eta_n)} = n \to \infty \quad \text{as} \quad n \to \infty.$$

Definition 2.1.4 Let (X, d) and (Y, d') be two metric spaces. A function $f :$ $(X, d) \to (Y, d')$ is called *locally Lipschitz* if for every $x \in X$ there exists a neighborhood V_x of x such that $f|_{V_x}$ is Lipschitz.

From the intuitive point of view one can imagine a locally Lipschitz function as one that obeys (temporarily) speed limits.

Remark 2.1.5 A locally Lipschitz function is continuous.

The following proposition shows that, in the case of a compact metric space (X, d), any locally Lipschitz function is Lipschitz. As we shall see in Remark 2.2.8 there exist locally Lipschitz functions which are not Lipschitz.

Theorem 2.1.6 *Let (X, d) and (Y, d') be metric spaces. If (X, d) is compact, then each locally Lipschitz function $f : (X, d) \to (Y, d')$ is Lipschitz.*

Proof Let $f : X \to Y$ be locally Lipschitz. By the continuity of f, the set $f(X)$ is compact, so bounded, implying that its diameter

$$\text{diam } f(X) := \sup\{d'(f(x), f(y)) : x, y \in X\}$$

is finite. Put $M := \text{diam } f(X)$.

If f is not Lipschitz on X, then for every $n \in \mathbb{N}$ there exist $x_n, y_n \in K$ such that

$$d'(f(x_n), f(y_n)) > nd(x_n, y_n). \tag{2.1.4}$$

By the compactness of X there exists a subsequence (x_{n_k}) of (x_n) converging to some $x \in X$. By (2.1.4)

$$M \geq d'(f(x_{n_k}), f(y_{n_k})) > n_k d(x_{n_k}, y_{n_k}),$$

for all $k \in \mathbb{N}$, implying $d(x_{n_k}, y_{n_k}) \to 0$, an so $y_{n_k} \to x$ for $k \to \infty$. Let V be a neighborhood of x on which f is L-Lipschitz. By (2.1.4), $x_{n_k} \neq y_{n_k}$ for every k, and $x_{n_k}, y_{n_k} \in V$ for k sufficiently large (say $k \geq k_0$). Consequently,

$$L \geq \frac{d'(f(x_{n_k}), f(y_{n_k}))}{d(x_{n_k}, y_{n_k})} > n_k,$$

for all $k \geq k_0$, in contradiction to the fact that $n_k \to \infty$ for $k \to \infty$. $\qquad\square$

Remark 2.1.7 Theorem 2.1.6 shows in fact that any locally Lipschitz function f : $X \to Y$, where X, Y are arbitrary metric spaces, is Lipschitz on every compact subset of X.

2.2 Lipschitz Properties of Differentiable Functions

In this section we show that if a function (scalar or vectorial) has some differentiability property, then the Lipschitz property can be characterized in terms of its differentials. The proofs are based on mean value theorems (MVT).

2.2.1 Differentiable Functions

In the case of differentiable functions we have the following simple characterizations of the Lipschitz property.

Proposition 2.2.1 *Let I be an interval in $\mathbb{R}$ and let $f : I \to \mathbb{R}$ be continuous on I and differentiable on the interior* $\mathrm{int}(I)$ *of I. Then f is Lipschitz if and only if the derivative f' is bounded on* $\mathrm{int}(I)$.
  In this case

$$L(f) = \sup\{|f'(x)| : x \in \mathrm{int}(I)\}. \tag{2.2.1}$$

Proof Let $\gamma := \sup\{|f'(x)| : x \in \mathrm{int}(I)\}$.
  If f is L-Lipschitz, then, for every $x \in \mathrm{int}(I)$

$$\left|\frac{f(x+h) - f(x)}{h}\right| \le L,$$

for sufficiently small $h \neq 0$. Letting $h \to 0$ one obtains $|f'(x)| \le L$ for every $x \in \mathrm{int}(I)$, implying $\gamma \le L$. Since this holds for every Lipschitz constant L, it follows that $\gamma \le L(f)$.
  If f' is bounded on $\mathrm{int}(I)$ (i.e., $\gamma < \infty$), then, by the Mean Value Theorem, for every $x < y$ in I there exists c, $x < c < y$, such that

$$\left|\frac{f(x) - f(y)}{x - y}\right| = |f'(c)| \le \gamma,$$

showing that f is γ-Lipschitz, hence $L(f) \le \gamma$.
  The proposition is proved as well as the equality (2.2.1) ∎

One can formulate an analogous criterion in terms of one-sided derivatives

$$f'_+(x) = \lim_{h \searrow 0} \frac{f(x+h) - f(x)}{h} \quad \text{and} \quad f'_-(x) = \lim_{h \nearrow 0} \frac{f(x+h) - f(x)}{h},$$

the right and left derivative of f at x, respectively. In this case the mean value theorem takes the following form.

Proposition 2.2.2 (Lagrange) *Let $f : [a, b] \to \mathbb{R}$.*

1. If

    (i) *f is continuous on $[a, b]$,*
    (ii) *$\forall x \in (a, b]$ there exists $f'_-(x)$ finite,*

then there exist $\xi_1, \xi_2 \in (a, b)$ such that

$$f'_-(\xi_1) \leq \frac{f(b) - f(a)}{b - a} \leq f'_-(\xi_2). \tag{2.2.2}$$

2. If

    (i) *f is continuous on $[a, b]$,*
    (ii) *$\forall x \in [a, b)$ there exists $f'_+(x)$ finite,*

then there exist $\xi'_1, \xi'_2 \in [a, b)$ such that

$$f'_+(\xi'_1) \leq \frac{f(b) - f(a)}{b - a} \leq f'_+(\xi'_2).$$

Proof We shall prove 1, the proof of 2 being similar. Consider the function $\varphi : [a, b] \to \mathbb{R}$ given by

$$\varphi(x) = f(x) - \frac{f(b) - f(a)}{b - a} x, \quad x \in [a, b].$$

Then φ is continuous, $\varphi(a) = \varphi(b)$ and

$$\varphi'_-(x) = f'_-(x) - \frac{f(b) - f(a)}{b - a}, \quad x \in (a, b]. \tag{2.2.3}$$

Let $\xi_1, \xi_2 \in [a, b]$ be such that $\varphi(\xi_1) = \min \varphi([a, b])$ and $\varphi(\xi_2) = \max \varphi([a, b])$. We can suppose that φ is not constant, so that $\varphi(\xi_1) < \varphi(\xi_2)$, hence $\xi_1 \neq \xi_2$. Taking into account the equality $\varphi(a) = \varphi(b)$ we can further suppose $\xi_1, \xi_2 \in (a, b]$. But then

$$\varphi'_-(\xi_1) \leq 0 \quad \text{and} \quad \varphi'_-(\xi_2) \geq 0,$$

which, by (2.2.3), yield (2.2.2). $\qquad\qquad\qquad\qquad\qquad\qquad\qquad\square$

Proposition 2.2.3 *Let* $f : [a, b] \to \mathbb{R}$ *be continuous. If* $f'_-(x)$ *exists and is bounded on* $(a, b]$, *then* f *is Lipschitz on* $[a, b]$ *and*

$$L(f) = \sup\{|f'_-(x)| : x \in (a, b]\}.$$

A similar result holds for the right derivative $f'_+(x)$ *on* $[a, b)$.

Proof For $x, y \in [a, b]$, $x \neq y$, there exists $\xi_1, \xi_2 \in (x, y]$ such that

$$f'_-(\xi_1) \leq \frac{f(y) - f(x)}{y - x} \leq f'_-(\xi_2),$$

implying

$$|f(y) - f(x)|/|y - x| \leq \max\{|f'_-(\xi_1)|, |f'_-(\xi_2)|\} \leq \sup_{z \in (a,b]} |f'_-(z)|.$$

Hence $L(f) \leq \sup_{z \in (a,b]} |f'_-(z)|$. On the other hand, for every $x \in (a, b]$ and $a - x < h < 0$,

$$\left| \frac{f(x + h) - f(x)}{h} \right| \leq L(f),$$

implies $|f'_-(x)| \leq L(f)$ for all $x \in (a, b]$, and so $\sup_{x \in (a,b]} |f'_-(x)| \leq L(f)$. $\square$

In order to present similar results for the n-dimensional and vector case, we need mean value theorems in this framework, such as the following well-known result. Denote by $B_\infty(a, r)$ the open ball with respect to the ℓ^∞-norm on $\mathbb{R}^n$, $\|x\|_\infty = \max_{1 \leq k \leq n} |x_k|$, for $x = (x_1, \ldots, x_n) \in \mathbb{R}^n$. In fact $B_\infty(a, r)$ is an open hypercube of center a, with each face perpendicular on a coordinate axis and edges of size $2r$.

The following result can be obtained by successive applications of the mean value theorem for functions of one variable.

Lemma 2.2.4 *Let* $f : B_\infty(a, r) \subseteq \mathbb{R}^n \to \mathbb{R}$ *be a function that has partial derivatives at each point of* $B_\infty(a, r)$. *Then, for* $x = (x_1, \ldots, x_n)$, $y = (y_1, \ldots, y_n) \in B_\infty(a, r)$ *there exist* ξ_k *between* x_k *and* y_k, $k = 1, 2, \ldots, n$, *such that*

$$f(x) - f(y) = \sum_{k=1}^{n} \frac{\partial f}{\partial x_k}(x_1, \ldots, x_{k-1}, \xi_k, y_{k+1}, \ldots, y_n)(x_k - y_k).$$

Proposition 2.2.5 *Each function* $f : B_\infty(a, r) \subseteq \mathbb{R}^n \to \mathbb{R}$ *having bounded partial derivatives on* $B_\infty(a, r)$ *is a Lipschitz function.*

Proof Let us choose $M > 0$ such that

$$\left| \frac{\partial f}{\partial x_i}(x) \right| \leq M,$$

for all $i \in \{1, 2, \ldots, n\}$ and all $x \in B_\infty(a, r)$.

Using Lemma 2.2.4, we have:

$$|f(x) - f(y)| \leq \sum_{k=1}^{n} \left| \frac{\partial f}{\partial x_k}(x_1, \ldots, x_{k-1}, \xi_k, y_{k+1}, \ldots, y_n) \right| |x_k - y_k|$$

$$\leq M \sum_{k=1}^{n} |x_k - y_k| \leq M\sqrt{n} \, \|x - y\|,$$

for all $x, y \in B_\infty(a, r)$, where the norm is the Euclidean one.   □

In the vector case the Mean Value Theorem takes the following form.

Proposition 2.2.6 *Let* $(X, \| \cdot \|), (Y, \| \cdot \|)$ *be normed spaces,* $\emptyset \neq U \subseteq X$ *open and* $f : U \to Y$ *Gâteaux differentiable on* U. *If* $x, y \in U$ *are such that the segment* $[x, y] := \{x + t(y - x) : t \in [0, 1]\}$ *is contained in* U, *then there exists* $\theta \in (0, 1)$ *such that*

$$f(x) - f(y) = f'(x + \theta(x - y))(x - y). \tag{2.2.4}$$

The following version of Proposition 2.2.5 can be proved using this Mean Value Theorem.

Proposition 2.2.7 *If* U *is an open subset of* $\mathbb{R}^n$ *(or of a normed space* $(X, \| \cdot \|)$), K *is a convex subset of* U *and* $f : U \to \mathbb{R}$ *is Gâteaux differentiable on* K *with bounded partial derivatives (resp. with the Gâteaux differential bounded on* K), *then* f *is Lipschitz on* K.

Proof Let $\|f'(x)\| \leq M$ for all $x \in K$. Then, by (2.2.4),

$$\|f(x) - f(y)\| = \|f'(x + \theta(x - y))(x - y)\|$$

$$\leq \|f'(x + \theta(x - y))\| \cdot \|(x - y)\| \leq M\|x - y\|,$$

for all $x, y \in K$.   □

Remark 2.2.8 There exist uniformly continuous functions which are not Lipschitz. Also, there exist locally Lipschitz functions which are not Lipschitz.

Consider the function $f : [0, 1] \to \mathbb{R}$ given by

$$f(x) = \begin{cases} x \cos \dfrac{\pi}{2x}, & \text{if } x \in (0, 1] \\ 0, & \text{if } x = 0 \end{cases}$$

and let g be the restriction of f to $(0, 1)$.

Then the function f is uniformly continuous but not Lipschitz and g is locally Lipschitz but not Lipschitz.

Indeed, the function f is uniformly continuous since it is continuous on the compact set $[0, 1]$.

Its derivative $f'(x) = \cos \dfrac{\pi}{2x} + \dfrac{\pi}{2x} \sin \dfrac{\pi}{2x}$, $x \in (0, 1)$, is unbounded on $(0, 1)$. Indeed, for $x_k = 1/(4k + 1)$,

$$f'(x_k) = \frac{(4k + 1)\pi}{2} \to \infty \quad \text{as } k \to \infty,$$

implying $\sup\{|f'(x)| : x \in (0, 1)\} = \infty$. By Proposition 2.2.1, the function f is not Lipschitz on $[0, 1]$ and neither is the function $g = f|_{(0,1)}$.

We claim that g is locally Lipschitz.

Indeed, for each $x \in (0, 1)$ there exist $\alpha_x, \beta_x \in (0, 1)$ such that $\alpha_x < x < \beta_x$ and $[\alpha_x, \beta_x] \subseteq (0, 1)$. The derivative f' of f is continuous on $(0, 1)$, so bounded on $V_x := [\alpha_x, \beta_x]$, implying that f is Lipschitz on the neighborhood V_x of x.

Using the more general MVT proved in Dieudonné [195, Theorem 8.5.1], one obtains more general criteria for the Lipschitz property. Recall first the MVT.

Theorem 2.2.9 Let $I = [a, b]$ be an interval in $\mathbb{R}$, X a normed space and $f : I \to X$, $g : I \to \mathbb{R}$ be continuous mappings. Suppose that there is a denumerable set $N \subseteq I$ such that, for each $x \in I \setminus N$, f and g have both a derivative at x and that $\|f'(x)\| \le g'(x)$. Then

$$\|f(b) - f(a)\| \le g(b) - g(a).$$

In the case of one function one obtains the following corollary, [195, Theorems 8.5.2 and 8.5.3].

Corollary 2.2.10 Let X be a normed space, $I = [a, b] \subseteq \mathbb{R}$ and $N \subseteq I$ a denumerable set.

1. If $f : I \to X$ is continuous and differentiable at every $x \in I \setminus N$ with $\|f'(x)\| \le L$, then $\|f(b) - f(a)\| \le L(b - a)$.
2. If $g : I \to \mathbb{R}$ is continuous and differentiable at every point $x \in I \setminus N$, with $m \le g'(x) \le M$, then

$$m(b - a) \le g(b) - g(a) \le M(b - a).$$

In fact

$$m(b - a) < g(b) - g(a) < M(b - a),$$

except when $g(x) = f(a) + m(x - a)$ *or* $g(x) = f(a) + M(x - a)$ *for all* $x \in I$.

2.2.2 Characterizations in Terms of Dini Derivatives

Dini derivatives (also called Dini numbers) play a crucial role in the study of differentiability of functions of one real variable. For the convenience of the reader, we shall present, following [337, Chapter 3], some results on Dini derivatives. A good presentation of Dini derivatives and of their use in the study of real-valued functions is also given in [615].

Definition 2.2.11 Let $f : [a, b] \to \mathbb{R}$ be a function. For $a \leq x_0 < b$ the quantities

$$D^+ f(x_0) = \limsup_{x \searrow x_0} \frac{f(x) - f(x_0)}{x - x_0} \quad \text{and} \quad D_+ f(x_0) = \liminf_{x \searrow x_0} \frac{f(x) - f(x_0)}{x - x_0}$$

are called the *upper* (resp. *lower*) *right Dini derivatives* of f at x_0 (infinite values are admitted).

Similarly, for $a < x_0 \leq b$ the quantities

$$D^- f(x_0) = \limsup_{x \nearrow x_0} \frac{f(x) - f(x_0)}{x - x_0} \quad \text{and} \quad D_- f(x_0) = \liminf_{x \nearrow x_0} \frac{f(x) - f(x_0)}{x - x_0}$$

are called the *upper* (resp. *lower*) *left Dini derivatives* of f at x_0.

Let also

$$\overline{D} f(x_0) = \limsup_{x \to x_0} \frac{f(x) - f(x_0)}{x - x_0} \quad \text{and} \quad \underline{D} f(x_0) = \liminf_{x \to x_0} \frac{f(x) - f(x_0)}{x - x_0}$$

be the *upper* (resp. *lower*) *derivatives* of f at x_0.

Remark 2.2.12 We mention some simple properties of these derivatives.

$$D^+ f(x_0) = D_+ f(x_0) = \alpha \iff \exists f'_+(x_0) = \alpha;$$
$$D^- f(x_0) = D_- f(x_0) = \beta \iff \exists f'_-(x_0) = \beta;$$
$$D^+ f(x_0) = D^- f(x_0) = \gamma \iff \exists \overline{D} f(x_0) = \gamma;$$
$$D_+ f(x_0) = D_- f(x_0) = \delta \iff \exists \underline{D} f(x_0) = \delta.$$

Also,

$$D^+(-f) = -D_+f \quad \text{and} \quad D^-(-f) = -D_-f,$$
$$D^+f \geq D_+f \qquad \text{and} \quad D^-f \geq D_-f.$$

Proposition 2.2.13 *Let $f : [a, b] \to \mathbb{R}$.*

1. *If, for $x_0 \in [a, b]$, both $\overline{D}f(x_0)$ and $\underline{D}f(x_0)$ are finite, then f is continuous at x_0.*
2. *If f is continuous on $[a, b]$ and one of the four Dini derivatives is zero on $[a, b]$, excepting an at most countable set, then the function f is constant on $[a, b]$.*

An important result is the following characterization of monotonicity.

Theorem 2.2.14 ([337]) *Let $f : [a, b] \to \mathbb{R}$ be such that*

(a) $\limsup\limits_{x' \nearrow x} f(x') \leq f(x)$ *for all* $a < x \leq b$, *and*

(b) $\limsup\limits_{x' \searrow x} f(x') \leq f(x)$ *for all* $a \leq x < b$.

If one of the following conditions

(i) $D^+ f(x) \geq 0$ *for all* $x \in [a, b]$ *excepting an at most countable set, or*

(ii) $D^- f(x) \geq 0$ *for all* $x \in [a, b]$ *excepting an at most countable set,*

holds, then f is nondecreasing on $[a, b]$.

Remark 2.2.15 A function f satisfying the condition (a) (or (b)) from Theorem 2.2.14 could be called left (resp. right) upper semicontinuous.

Based on these results, one can obtain another characterization of the Lipschitz property.

Theorem 2.2.16 *Let $f : [a, b] \to \mathbb{R}$ be a continuous function. Then*

(i) $\sup\left\{ \dfrac{f(x) - f(x')}{x - x'} : x, x' \in [a, b], x \neq x' \right\} = \sup\{\Delta f(x) : x \in (a, b)\},$

(ii) $\inf\left\{ \dfrac{f(x) - f(x')}{x - x'} : x, x' \in [a, b], x \neq x' \right\} = \inf\{\Delta f(x) : x \in (a, b)\},$

$$(2.2.5)$$

for all $\Delta \in \{D^+, D_+, D^-, D_-\}$.

Proof Let

$$M = \sup \left\{ \frac{f(x) - f(x')}{x - x'} : x, x' \in [a, b], \; x \neq x' \right\} \quad \text{and} \quad m = \sup\{D_+ f(x) : x \in (a, b)\}.$$

It is obvious that $m \leq M$. Suppose $m < M$ and let $m < k < M$. The function $g(x) = kx - f(x)$ is continuous and

$$D^+ g(x) \geq D_+ g(x) = k - D_+ f(x) > 0,$$

for all $x \in (a, b)$, so that, by Theorem 2.2.14, it is nondecreasing. If $x < x'$ are two points in $[a, b]$, then

$$g(x') - g(x) \geq 0 \iff \frac{f(x) - f(x')}{x - x'} \leq k.$$

The same inequality holds in the case $x' < x$, i.e., $(f(x) - f(x'))/(x - x') \leq k$, for all $x, x' \in [a, b]$, $x \neq x'$, yielding the contradiction $M \leq k$. Consequently,

$$\sup \left\{ \frac{f(x) - f(x')}{x - x'} : x, x' \in [a, b], \; x \neq x' \right\} = \sup\{D_+ f(x) : x \in (a, b)\}.$$

The inequality $D^+ f(x) \geq D_+ f(x)$ implies

$$M \geq \sup\{D^+ f(x) : x \in (a, b)\} \geq \sup\{D_+ f(x) : x \in (a, b)\} = M,$$

so that $M = \sup\{D^+ f(x) : x \in (a, b)\}$.

Similar reasonings show that

$$M = \sup_{x \in (a,b)} D^- f(x) = \sup_{x \in (a,b)} D_- f(x).$$

Replacing f with $-f$ one obtains the equalities with infimum, i.e., those from (2.2.5).(ii). □

Corollary 2.2.17 *Let* $f : [a, b] \to \mathbb{R}$ *be continuous. If one of the four Dini derivatives given in Definition 2.2.11, call it* Δ, *is bounded on* (a, b), *then the function* f *is Lipschitz and*

$$L(f) = \sup_{x \in (a,b)} |\Delta f(x)|.$$

Proof The proof is based on the equalities (2.2.5) and on the formula

$$\sup_{x \in A} |g(x)| = \max \left\{ \left| \sup_{x \in A} g(x) \right|, \left| \inf_{x \in A} g(x) \right| \right\}, \tag{2.2.6}$$

valid for any function $g : A \to \mathbb{R}$, where A is an arbitrary nonempty set.

Indeed,

$$\sup_{x,x'}\left|\frac{f(x)-f(x')}{x-x'}\right| = \max\left\{\left|\sup_{x,x'}\frac{f(x)-f(x')}{x-x'}\right|,\left|\inf_{x,x'}\frac{f(x)-f(x')}{x-x'}\right|\right\}$$

$$= \max\left\{\left|\sup_{x\in(a,b)}\Delta f(x)\right|,\left|\inf_{x\in(a,b)}\Delta f(x)\right|\right\} = \sup_{x\in(a,b)}|\Delta f(x)|.$$

Here, one understands that the corresponding suprema and infima are taken over all $x, x' \in [a, b]$ with $x \neq x'$.   □

Another consequence of Theorem 2.2.16 is the following one.

Corollary 2.2.18 *If any one of the four Dini derivatives of a continuous function f is continuous at the point x_0, so are the other three. In this case, all four Dini derivatives are equal and f is differentiable at x_0.*

For completeness, let us prove the formula (2.2.6).

Proof of (2.2.6) Suppose that $g : A \to \mathbb{R}$, $g \neq 0$, is bounded and put

$$\|g\| = \sup_{x\in A}|g(x)|.$$

Then, the inequalities

$$g(x) \leq |g(x)| \leq \|g\| \quad \text{and} \quad -g(x) \leq |g(x)| \leq \|g\|,$$

valid for all $x \in A$, imply

$$\sup g(A) \leq \|g\|, \quad -\sup g(A) = \inf(-g)(A) \leq \|g\|, \quad \text{and}$$
$$\inf g(A) \leq \|g\|, \quad -\inf g(A) = \sup(-g)(A) \leq \|g\|,$$

so that

$$\max\{|\sup g(A)|, |\inf g(A)|\} \leq \|g\|. \tag{2.2.7}$$

Let now $0 < \varepsilon < \|g\|$ and let $x \in A$ be such that $\|g\| - \varepsilon < |g(x)| \leq \|g\|$.
If $g(x) > 0$, then $\|g\| - \varepsilon < g(x) \leq \|g\|$ implies $|\sup g(A)| = \sup g(A) \geq g(x) > \|g\| - \varepsilon$. If $g(x) < 0$, then $\|g\| - \varepsilon < -g(x) \leq \|g\|$ or, equivalently, $-\|g\| \leq g(x) < -\|g\| + \varepsilon < 0$. It follows that

$$|\inf g(A)| = -\inf g(A) = \sup(-g)(A) \geq -g(x) > \|g\| - \varepsilon,$$

so that

$$\max\{|\sup g(A)|, |\inf g(A)|\} > \|g\| - \varepsilon .$$

Letting $\varepsilon \searrow 0$, one obtains $\max\{|\sup g(A)|, |\inf g(A)|\} \geq \|g\|$, which, combined with (2.2.7), yields (2.2.6).   □

V. Bogdan [91] extended the mean value theorem, Theorem 2.2.9, to Dini derivatives that are finite a.e. (in the Lebesgue sense), and to a special derivative of a vector function, called Lipschitz derivative.

Definition 2.2.19 Let Y be a Banach space, $f : (a, b) \rightarrow Y$ a vector function, where $a, b \in \mathbb{R}$, $a < b$. The *left* (*right*) *Lipschitz derivative* of f at $x \in (a, b)$ is given by

$$D_\ell^-(f)(x) = \limsup_{h \nearrow 0} \|\frac{1}{h}(f(x+h) - f(x))\|, \quad \text{respectively}$$

$$D_\ell^+(f)(x) = \limsup_{h \searrow 0} \|\frac{1}{h}(f(x+h) - f(x))\| ,$$

while the quantity

$$D_\ell(f)(x) = \limsup_{h \to 0} \|\frac{1}{h}(f(x+h) - f(x))\|$$

is called the *Lipschitz derivative* of f at x.

Let Ω be a subset of a Banach space X and $f : \Omega \rightarrow Y$ a mapping to another Banach space Y. One says that f is *Lipschitz at the point* $x \in \Omega$ (or *pointwise Lipschitz at x*) if there exist $\beta > 0$ and $\delta > 0$ such that

$$\|f(x) - f(x')\| \leq \beta \|x - x'\| ,$$

for all $x' \in \Omega \cap B_X(x, \delta)$. If $\Omega = (a, b) \subseteq \mathbb{R}$, then one says that f is *Lipschitz from the right* (*left*) *at x* provided that there exist $\beta > 0$ and $\delta > 0$ such that

$$\|f(x) - f(x')\| \leq \beta |x - x'| ,$$

for all $x' \in (a, b)$ with $x < x' < x + \delta < b$ (resp. $x - \delta < x' < x$).

It is obvious that f has a finite (left, right) Lipschitz derivative at x if and only if it is Lipschitz (from the left, from the right) at x.

The mean value theorem in this case reads as follows.

Theorem 2.2.20 (MVT for Dini Derivatives, [91]) *Let $[a, b]$ be a closed bounded interval in $\mathbb{R}$, Y a Banach space, $f : [a, b] \rightarrow Y$ and $g : [a, b] \rightarrow \mathbb{R}$ continuous functions, and $E \subseteq (a, b)$ a set of Lebesgue measure zero. If the right-sided*

Lipschitz derivative $D_\ell^+ f(x)$ and the right-sided lower Dini derivative $D_+g(x)$ exist, are finite and

$$D_\ell^+ f(x) \le D_+g(x),\qquad\qquad(2.2.8)$$

for all $x \in (a, b) \setminus E$, then

$$\|f(b) - f(a)\| \le g(b) - g(a).\qquad\qquad(2.2.9)$$

For the proof we shall need the following result.

Theorem 2.2.21 ([120], Theorem 7.6, p. 271) *Let E be a subset of Lebesgue measure zero of the interval $[a, b] \subseteq \mathbb{R}$. Then there exists a strictly increasing continuous function $\varphi : [a, b] \to \mathbb{R}$ such that $\varphi'(x) = \infty$ for all $x \in E$.*

Corollary 2.2.22 *Under the hypotheses of Theorem 2.2.21 there exists a strictly increasing continuous function $\psi : [a, b] \to (0, \infty)$ such that $\psi'(x) = \infty$ for all $x \in E$.*

Indeed, take $\psi(x) = \varphi(x) + M + 1$, where $M = \sup\{|\varphi(x)| : x \in [a, b]\}$.

Proof of Theorem 2.2.20 Let $\psi : [a, b] \to (0, \infty)$ be a function as in Corollary 2.2.22. We shall show that for every $\varepsilon > 0$ the inequality

$$\|f(x) - f(a)\| \le g(x) - g(a) + \varepsilon[(x - a) + \psi(x)]\qquad\qquad(2.2.10)$$

holds for all $x \in [a, b]$. Taking $x = b$ in this inequality and letting $\varepsilon \searrow 0$, one obtains (2.2.9).

Suppose, by contradiction, that (2.2.10) does not hold. Then there exists $\varepsilon > 0$ such that the set

$$U := \{x \in [a, b] : \varphi(x) > 0\}$$

is nonempty, where $\varphi : [a, b] \to \mathbb{R}$ is given by

$$\varphi(x) := \|f(x) - f(a)\| - [g(x) - g(a)] - \varepsilon[(x - a) + \psi(x)], \quad x \in [a, b].$$

Since the function φ is continuous the set U is relatively open in $[a, b]$. Let $c := \inf U \ge a$. Because $\varphi(a) = -\varepsilon\psi(a) < 0$, it follows that, for some $\gamma > 0$, $\varphi(x) < 0$ for all x in the neighborhood $[a, a + \gamma)$ of a. Hence $[a, a + \gamma) \cap U = \emptyset$ and so $c \ge a + \gamma > a$. If $c = b$, then $U = \{b\}$, in contradiction to the openness of U.

Consequently, $a < c < b$ and $c \notin U$ (because U is open).
Observe that $c \notin U$ is equivalent to

$$\|f(c) - f(a)\| \le g(c) - g(a) + \varepsilon[(c - a) + \psi(c)].\qquad\qquad(2.2.11)$$

Put $\alpha = D_\ell^+ f(c)$ and $\beta = D_+ g(c)$. Since

$$\alpha = \inf_{0 < \delta < b-c} \sup \left\{ \frac{1}{h} \| f(c+h) - f(c) \| : 0 < h < \delta \right\}, \quad \text{and}$$

$$\beta = \sup_{0 < \delta < b-c} \inf \left\{ \frac{1}{h} [g(c+h) - g(c)] : 0 < h < \delta \right\},$$

it follows that there exists $0 < \delta_1 < b - c$ such that

(i) $\dfrac{1}{h} \| f(c+h) - f(c) \| < \alpha + \dfrac{\varepsilon}{2}$ and

(ii) $\dfrac{1}{h} [g(c+h) - g(c)] > \beta - \dfrac{\varepsilon}{2}$,

(2.2.12)

for all $0 < h < \delta_1$.

We distinguish two situations.

Case I. $c \notin E$.

By hypothesis, $\alpha \le \beta$ so that

$$\frac{1}{h} \| f(c+h) - f(c) \| < \alpha + \frac{\varepsilon}{2} \le \beta + \frac{\varepsilon}{2} < \frac{1}{h} [g(c+h) - g(c)] + \varepsilon,$$

implying

$$\| f(x) - f(c) \| < g(x) - g(c) + \varepsilon(x - c),$$

for all $x \in (c, c + \delta_1)$.

By this inequality, the inequality (2.2.11) and the fact that the function ψ is increasing, one obtains

$$\begin{aligned}
\| f(x) - f(a) \| &\le \| f(x) - f(c) \| + \| f(c) - f(a) \| \\
&\le [g(x) - g(c)] + \varepsilon(x - c) + [g(c) - g(a)] \\
&\quad + \varepsilon[(c - a) + \psi(c)] \\
&\le [g(x) - g(a)] + \varepsilon[(x - a) + \psi(x)]
\end{aligned}$$

for all $x \in (c, c + \delta_1)$.

This means that $(c, c + \delta_1) \cap U = \emptyset$, in contradiction to the fact that $c = \inf U$ and $c \notin U$.

Case II. $c \in E$.

In this case $\psi'(c) = \infty$ so there exists $0 < \delta_2 < \min\{c - a, b - c\}$ such that

$$\varepsilon \frac{1}{h} [\psi(c+h) - \psi(c)] > \alpha - \beta + \varepsilon,$$

or, equivalently,

$$\alpha + \frac{\varepsilon}{2} < \beta - \frac{\varepsilon}{2} + \varepsilon \frac{1}{h}[\psi(c+h) - \psi(c)]\,,$$

for all $0 < |h| < \delta_2$.
But then, by this inequality and the inequalities (2.2.12), one obtains

$$\frac{1}{h}\|f(c+h) - f(c)\| < \alpha + \frac{\varepsilon}{2} < \beta - \frac{\varepsilon}{2} + \frac{1}{h}[\psi(c+h) - \psi(c)]$$

$$< \frac{1}{h}[g(c+h) - g(c)] + \varepsilon \frac{1}{h}[\psi(c+h) - \psi(c)]\,,$$

for all $0 < h < \delta$, where $\delta := \min\{\delta_1, \delta_2\}$.
This can be rewritten as

$$\|f(x) - f(c)\| < [g(x) - g(c)] + \varepsilon[\psi(x) - \psi(c)]\,,$$

for all $c < x < c + \delta$.
Taking into account this inequality and (2.2.11), one obtains

$$\|f(x) - f(a)\| \leq \|f(x) - f(c)\| + \|f(c) - f(a)\|$$

$$\leq [g(x) - g(c)] + \varepsilon[\psi(x) - \psi(c)] + [g(c) - g(a)]$$

$$+ \varepsilon[(c - a) + \psi(c)]$$

$$= g(x) - g(a) + \varepsilon[(c - a) + \psi(x)]$$

$$< g(x) - g(a) + \varepsilon[(x - a) + \psi(x)]\,,$$

for all $c < x < c + \delta$. It follows that $(c, c + \delta) \cap U = \emptyset$ and, as in Case I, this
leads to a contradiction. $\square$

This theorem has as a consequence the following mean value theorem.

Corollary 2.2.23 (Cartan [136], Theorem 1.3.3, see also [137]) *Let f, g be as in
Theorem 2.2.20 and let N be an at most countable subset of (a, b). Suppose that the
right derivatives $f'_+(x), g'_+(x)$ exist, are finite and*

$$\|f'_+(x)\| \leq g'_+(x)\,,$$

for all $x \in (a, b) \setminus N$. Then

$$\|f(b) - f(a)\| \leq g(b) - g(a)\,.$$

Remark 2.2.24 Based on Theorem 2.2.14 and on the Hahn-Banach theorem, one can give a simpler proof of Theorem 2.2.20, under the hypothesis that the inequality (2.2.8) holds excepting an at most countable subset N of (a, b).

Indeed, by a consequence of the Hahn-Banach theorem, there exists $y^* \in Y^*$ such that $\|y^*\| = 1$ and $y^*(f(b) - f(a)) = \|f(b) - f(a)\|$. Consider the function $\varphi : [a, b] \to \mathbb{R}$, $\varphi(x) = g(x) - y^*(f(x))$. Then for every $x \in (a, b)$ and $0 < h < b - x$,

$$
\begin{aligned}
\frac{\varphi(x + h) - \varphi(x)}{h} &= \frac{g(x + h) - g(x)}{h} - y^* \left(\frac{f(x + h) - f(x)}{h} \right) \\
&\geq \frac{g(x + h) - g(x)}{h} - \|y^*\| \frac{\|f(x + h) - f(x)\|}{h} \\
&= \frac{g(x + h) - g(x)}{h} - \frac{\|f(x + h) - f(x)\|}{h}.
\end{aligned}
$$

Thus,

$$
D_+\varphi(x) \geq D_+g(x) - D_l^+ f(x) \geq 0,
$$

for all $x \in (a, b) \setminus N$. Applying Theorem 2.2.14 we get that φ is nondecreasing, so

$$
0 \leq \varphi(b) - \varphi(a) = g(b) - g(a) - y^*(f(b) - f(a)) = g(b) - g(a) - \|f(b) - f(a)\|,
$$

from where

$$
\|f(b) - f(a)\| \leq g(b) - g(a).
$$

From Theorem 2.2.20 one also obtains criteria of monotonicity, constancy and of the Lipschitz property.

Corollary 2.2.25 *Let $[a, b]$ be an interval in $\mathbb{R}$, $E \subseteq (a, b)$ of Lebesgue measure zero and Y a Banach space.*

1. *If the continuous function $g : [a, b] \to \mathbb{R}$ has a finite Dini derivative $D_+g(x) \geq 0$ for all $x \in (a, b) \setminus E$, then g is nondecreasing on (a, b).*
2. *If the continuous function $f : [a, b] \to Y$ has the right derivative $f_+'(x) = 0$ for all $x \in (a, b) \setminus E$, then f is constant on $[a, b]$.*
3. *If the continuous function $f : [a, b] \to Y$ has a finite right Lipschitz derivative satisfying $D_l^+ f(x) \leq \beta$ for all $x \in (a, b) \setminus E$, then*

$$
\|f(x) - f(y)\| \leq \beta|x - y|,
$$

for all $x, y \in [a, b]$.

Proof

1. Take $f \equiv 0$ and let $x < x'$ be two points in $[a, b]$. Then the hypotheses of Theorem 2.2.20 are satisfied on the interval $[x, x']$ implying

$$0 = \|f(x) - f(x')\| \le g(x) - g(x').$$

The proof of 2 is similar, by taking this time $g \equiv 0$.
3. Take $g(x) = \beta x$ for $x \in [a, b]$. Applying Theorem 2.2.20 on an interval $[x, y] \subseteq [a, b]$ with $x < y$, one obtains $\|f(y) - f(x)\| \le \beta(y - x)$. It follows that

$$\|f(x) - f(y)\| \le \beta|x - y|,$$

for all $x, y \in [a, b]$.

$\square$

If Ω is an open subset of a Banach space X and $f : \Omega \to Y$, where Y is another Banach space, one defines the Lipschitz derivative $D_\ell f(x)$ of f at a point $x \in \Omega$ by

$$D_\ell f(x) = \limsup_{h \to 0} \frac{\|f(x + h) - f(x)\|}{\|h\|}.$$

Obviously, this definition agrees with that given in Definition 2.2.19 in the case $\Omega = (a, b)$.

Proposition 2.2.26 *Let X, Y be Banach space, $\Omega \subseteq X$ open, $K \subseteq \Omega$ convex and N an at most countable subset of K. If the Lipschitz derivative of f exists and satisfies $D_\ell f(x) \le \beta$ for all $x \in K \setminus N$, for some $\beta > 0$, then*

$$\|f(x) - f(y)\| \le \beta\|x - y\|,$$

for all $x, y \in K$.

2.3   Algebraic Operations with Lipschitz Functions

The aim of this section is to establish which algebraic operations preserve the Lipschitz property of functions.

Proposition 2.3.1 *Let (X, d), (Y, d') and (Z, d'') be metric spaces. If $f : X \to Y$ and $g : Y \to Z$ are Lipschitz, then $g \circ f : X \to Z$ is Lipschitz. Moreover, the following inequality is valid:*

$$L(g \circ f) \le L(g)L(f).$$

If f and g are locally Lipschitz, then $g \circ f$ is locally Lipschitz.

Proof Since, for f and g as in the statement of the proposition, we have

$$d''((g \circ f)(x), (g \circ f)(y)) \leq L(g)d'(f(x), f(y)) \leq L(g)L(f)d(x, y),$$

for all $x, y \in X$, we obtain that $g \circ f$ is Lipschitz and $L(g \circ f) \leq L(g)L(f)$. $\square$

The next result shows that the study of Lipschitz functions taking values in $\mathbb{R}^n$ reduces to the case $n = 1$.

Proposition 2.3.2 *Let (X, d) be a metric space, $\mathbb{R}^n$ endowed with an arbitrary norm and $f : X \to \mathbb{R}^n$ a function having the components $f_1, f_2, \ldots, f_n : X \to \mathbb{R}$. Then f is Lipschitz (locally Lipschitz) if and only if $f_1, f_2, \ldots, f_n$ are all Lipschitz (locally Lipschitz).*

Proof Since all the norms on $\mathbb{R}^n$ are Lipschitz equivalent, it is sufficient to prove the result for the ℓ^∞-norm (the sup-norm) $\|x\|_\infty = \max\{|x_1|, \ldots, |x_n|\}$ for $x = (x_1 \ldots, x_n) \in \mathbb{R}^n$.

The assertion follows from the equivalence

$$\|f(x) - f(y)\|_\infty \leq L\,d(x, y) \iff |f_k(x) - f_k(y)| \leq L\,d(x, y),$$

$$k = 1, \ldots, n,$$

valid for all $x, y \in \mathbb{R}^n$. $\square$

Proposition 2.3.3 *Let (X, d) be a metric space, Y a normed space and $f, g : X \to \mathbb{R}$ two Lipschitz functions and $\alpha \in \mathbb{R}$. Then $f + g$ and αf are Lipschitz functions, and the following relations are valid:*

$$L(f + g) \leq L(f) + L(g) \quad and \quad L(\alpha f) = |\alpha|L(f).$$

If f and g are locally Lipschitz, then $f + g$ and αf are locally Lipschitz.

Proof Since, for f and g as in the statement of the proposition, we have

$$\|(f+g)(x)-(f+g)(y)\| \leq \|f(x)-f(y)\|+\|g(x)-g(y)\| \leq (L(f)+L(g))d(x, y)$$

and

$$\|(\alpha f)(x) - (\alpha f)(y)\| = |\alpha|\|f(x) - f(y)\| \leq |\alpha|L(f)d(x, y),$$

for all $x, y \in X$, we obtain that $f+g$ and αf are Lipschitz, $L(f+g) \leq L(f)+L(g)$ and $L(\alpha f) = |\alpha|L(f)$. $\square$

Proposition 2.3.4 *Let (X, d) be a metric space. If the functions $f, g : X \to \mathbb{R}$ are bounded and Lipschitz, then fg is Lipschitz. Moreover, the following inequality is valid:*

$$L(fg) \leq \sup_{x \in X} |f(x)|\,L(g) + \sup_{x \in X} |g(x)|\,L(f).$$

If $f, g : X \to \mathbb{R}$ are locally Lipschitz, then fg is locally Lipschitz.

Proof Since

$$|(fg)(x) - (fg)(y)| \leq |f(x)|\,|g(x) - g(y)| + |g(y)|\,|f(x) - f(y)|$$

$$\leq \left[\sup_{x \in X} |f(x)|\, L(g) + \sup_{x \in X} |g(x)|\, L(f) \right] d(x, y),$$

for all $x, y \in X$, we infer that fg is Lipschitz and

$$L(fg) \leq \sup_{x \in X} |f(x)|\, L(g) + \sup_{x \in X} |g(x)|\, L(f).$$

□

Remark 2.3.5 If (X, d) is a metric space of finite diameter, then every Lipschitz function $f : X \to \mathbb{R}$ is bounded.

Indeed, taking a point $x_0 \in X$, it follows that

$$|f(x)| \leq |f(x_0)| + |f(x) - f(x_0)| \leq |f(x_0)| + L(f)d(x, x_0) \leq |f(x_0)| + L(f) \operatorname{diam} X,$$

for all $x \in X$.

Remark 2.3.6 Let us note that if the functions f and g are not bounded, then the conclusion of Proposition 2.3.4 may not be true.

Indeed, the functions $f = g = \operatorname{Id}_{\mathbb{R}} : \mathbb{R} \to \mathbb{R}$, given by $\operatorname{Id}_{\mathbb{R}}(x) = x$ for all $x \in \mathbb{R}$, are Lipschitz:

$$|\operatorname{Id}_{\mathbb{R}}(x) - \operatorname{Id}_{\mathbb{R}}(y)| = |x - y|$$

for all $x, y \in \mathbb{R}$. But their product $(fg)(x) = x^2$, $x \in \mathbb{R}$, is not Lipschitz because its derivative $(fg)'(x) = 2x$ is unbounded on $\mathbb{R}$ (see Proposition 2.2.1).

Proposition 2.3.7 *Let (X, d) be a metric space, $\mathbb{R}$ endowed with the usual Euclidean metric and $f : X \to \mathbb{R}$ a Lipschitz function having the property that there exists $m > 0$ such that $|f(x)| \geq m$, for all $x \in X$. Then $1/f$ is Lipschitz and*

$$L(\frac{1}{f}) \leq \frac{L(f)}{m^2}.$$

If the locally Lipschitz function $f : X \to \mathbb{R}$ has the property that $f(x) \neq 0$ for every $x \in X$, then $1/f$ is locally Lipschitz.

Proof Since

$$\left|(\frac{1}{f})(x) - (\frac{1}{f})(y)\right| = \frac{|f(x) - f(y)|}{|f(x)||f(y)|} \le \frac{L(f)}{m^2}d(x, y),$$

for all $x, y \in X$, we infer that $1/f$ is Lipschitz and $L(\frac{1}{f}) \le \frac{L(f)}{m^2}$.

Suppose that f is locally Lipschitz and $f(x) \ne 0$ for all $x \in X$. For $x_0 \in X$ there exists a neighborhood V of x_0 on which f is Lipschitz and $|f(x)| \ge |f(x_0)|/2 > 0$ for all $x \in V$. It follows that $1/f$ is Lipschitz on V.   □

Corollary 2.3.8 *Let* (X, d) *be a compact metric space and* $f : X \to \mathbb{R}$ *a locally Lipschitz function having the property that* $f(x) \ne 0$ *for every* $x \in X$. *Then* $1/f$ *is Lipschitz.*

Proof Since the metric space X is compact the locally Lipschitz function f is Lipschitz on X (by Theorem 2.1.6).

Now, by the continuity of f and the compactness of X there exists $x_0 \in X$ such that

$$\inf\{|f(x)| : x \in X\} = |f(x_0)| > 0,$$

so that, by Proposition 2.3.7, $1/f$ is Lipschitz.   □

Proposition 2.3.9 *Let* (X, d) *be a metric space and* $\mathscr{F}$ *a family of real-valued* K-*Lipschitz functions defined on* X *such that* $\varphi(x) := \sup\{f(x) : f \in \mathscr{F}\}$ *is finite for every* $x \in X$. *Then the function* φ *is* K-*Lipschitz. Similarly, if* $\psi(x) := \inf\{f(x) : f \in \mathscr{F}\}$ *is finite for every* $x \in X$, *then* ψ *is* K-*Lipschitz.*

In particular, for two Lipschitz functions $f, g : X \to \mathbb{R}$, *the functions*

$$(f \vee g)(x) := \max\{f(x), g(x)\} \quad and \quad (f \wedge g)(x) := \min\{f(x), g(x)\}, \quad x \in X,$$

are K-*Lipschitz with* $K = \max\{L(f), L(g)\}$.

The function $|f|$ *is also Lipschitz and* $L(|f|) \le L(f)$.

Proof Indeed, for $x, y \in X$

$$f(x) \le f(y) + Kd(x, y) \le \varphi(y) + Kd(x, y),$$

for all $f \in \mathscr{F}$. Taking the supremum with respect to $f \in \mathscr{F}$, one obtains

$$\varphi(x) \le \varphi(y) + Kd(x, y) \iff \varphi(x) - \varphi(y) \le Kd(x, y).$$

Interchanging the roles of x, y one obtains $\varphi(y) - \varphi(x) \le Kd(x, y)$ proving that φ is K-Lipschitz.

The fact that ψ is Lipschitz can be proved similarly.

The inequalities

$$\big||f(x)| - |f(y)|\big| \le |f(x) - f(y)| \le L(f)d(x, y),$$

show that $|f|$ is Lipschitz and $L(|f|) \le L(f)$.   □

2.4   Sequences of Lipschitz Functions

In this section we investigate when the limit of a sequence of Lipschitz functions is Lipschitz.

Proposition 2.4.1 *Let (X, d), (Y, d') be metric spaces and $f_n : X \to Y$, $n \in \mathbb{N}$, Lipschitz functions, such that there exists M with the following property:*

$$L(f_n) \le M,$$

for all $n \in \mathbb{N}$.

If for every $x \in X$ there exists the limit

$$f(x) := \lim_{n\to\infty} f_n(x), \tag{2.4.1}$$

then f is Lipschitz with $L(f) \le M$ and the convergence of (f_n) is uniform on every totally bounded subset of X.

In particular, if the metric space X is compact, then (f_n) converges to f uniformly on X.

Proof Since

$$d'(f_n(x), f_n(y)) \le L(f_n)d(x, y) \le Md(x, y),$$

for all $x, y \in X$, letting $n \to \infty$ we infer that

$$d'(f(x), f(y)) = \lim_{n\to\infty} d'(f_n(x), f_n(y)) \le Md(x, y),$$

for all $x, y \in X$, hence f is Lipschitz with $L(f) \le M$.

Suppose now that K is a totally bounded subset of X and let $\varepsilon > 0$.
Let $x_1, \dots, x_m$ be an ε-net for K, that is,

$$\forall x \in K, \ \exists k \in \{1, 2 \dots, m\} \quad \text{such that} \quad d(x, x_k) \le \varepsilon. \tag{2.4.2}$$

By (2.4.1) there exists $n_0 \in \mathbb{N}$ such that

$$|f_n(x_k) - f(x_k)| \le \varepsilon, \quad k = 1, 2 \dots, m, \tag{2.4.3}$$

for all $n \ge n_0$.

Let x be an arbitrary element of K. Choosing $k \in \{1, 2, \ldots, m\}$ according to (2.4.2) and taking into account (2.4.3), it follows that

$$|f_n(x) - f(x)| \leq |f_n(x) - f_n(x_k)| + |f_n(x_k) - f(x_k)| + |f(x_k) - f(x)|$$
$$\leq L(f_n)d(x, x_k) + \varepsilon + L(f)d(x, x_k) \leq (2M + 1)\varepsilon,$$

for all $n \geq n_0$. Since $x \in K$ was arbitrarily chosen, it follows that (f_n) converges to f uniformly on K. □

Corollary 2.4.2 *Let (X, d) be a metric space, $(Y, \|\cdot\|)$ a normed space and $f_n :$ $X \to Y$, $n \in \mathbb{N}$, Lipschitz functions such that the series $\sum_n L(f_n)$ is convergent.*
If for every $x \in X$ there exists the sum

$$f(x) := \sum_n f_n(x),$$

then the function $f : X \to Y$ is Lipschitz and

$$L(f) \leq \sum_n L(f_n).$$

Proof Since

$$\lim_{n \to \infty} S_n(x) = f(x),$$

for all $x \in X$, where $S_n = f_1 + f_2 + \cdots + f_n$ and

$$L(S_n) \leq \sum_{k=1}^{n} L(f_k) \leq \sum_{k=1}^{\infty} L(f_k),$$

for all $n \in \mathbb{N}$, we conclude, taking into account Proposition 2.4.1, that f is Lipschitz and

$$L(\sum_n f_n) \leq \sum_n L(f_n).$$

□

Remark 2.4.3 Proposition 2.4.1 may fail if we weaken our hypotheses by requiring only that each f_n is Lipschitz, even if the sequence $(f_n)_{n \in \mathbb{N}}$ converges uniformly.
Indeed, consider, for each $n \in \mathbb{N}$, the function $f_n : [0, \infty) \to \mathbb{R}$ given by

$$f_n(x) = \sqrt{x + \frac{1}{n}}, \quad x \in [0, \infty).$$

Then:

(i) the sequence $(f_n)_{n\in\mathbb{N}}$ converges uniformly to the function $f : [0, \infty) \to \mathbb{R}$ given by $f(x) = \sqrt{x}$, $x \in [0, \infty)$;

(ii) f_n is Lipschitz and

$$L(f_n) = \frac{\sqrt{n}}{2}$$

for each $n \in \mathbb{N}$;

(iii) f is not Lipschitz.

Let us proceed with the proofs.

(i) The inequalities

$$|f_n(x) - f(x)| = \left|\sqrt{x + \frac{1}{n}} - \sqrt{x}\right| = \frac{1}{n} \cdot \frac{1}{\sqrt{x + \frac{1}{n}} + \sqrt{x}} \leq \frac{1}{\sqrt{n}},$$

valid for all $x \in [0, \infty)$ and all $n \in \mathbb{N}$, show that $(f_n)_{n\in\mathbb{N}}$ converges uniformly to f.

(ii) The derivative of f_n is

$$f_n'(x) = \frac{1}{2\sqrt{x + \frac{1}{n}}}.$$

It follows that

$$\sup\{|f_n'(x)| : x \in (0, \infty)\} = \frac{\sqrt{n}}{2},$$

so that, by Proposition 2.2.1, f_n is Lipschitz and $L(f_n) = \sqrt{n}/2$.

(iii) The derivative of f is

$$f'(x) = \frac{1}{2\sqrt{x}}, \quad x \in (0, \infty),$$

implying $\sup\{|f'(x)| : x \in (0, \infty)\} = \infty$, so that, by the same proposition, f is not Lipschitz on $[0, \infty)$.

Example 2.4.4 There exists a sequence $(f_n)_{n\in\mathbb{N}}$ of Lipschitz functions converging uniformly to a Lipschitz function such that the sequence $(L(f_n))_{n\in\mathbb{N}}$ is unbounded.

Indeed, for each $n \in \mathbb{N}$, consider the function $f_n : [0, 1] \to \mathbb{R}$ given by

$$f_n(x) = nxe^{-n^n x}, \quad x \in [0, 1].$$

Then:

(i) the sequence $(f_n)_{n \in \mathbb{N}}$ converges uniformly to the function $f \equiv 0$ which is Lipschitz;

(ii) f_n is Lipschitz and

$$L(f_n) = n$$

for each $n \in \mathbb{N}$.

Simple calculations show that

$$f_n'(x) = n \left(1 - n^n x\right) e^{-n^n x} \quad \text{and} \quad f_n''(x) = n^{n+1} \left(n^n x - 2\right) e^{-n^n x} .$$

(i) The only root of f_n' is $x_n' = 1/n^n$, so that f_n attains its maximum value on $[0, 1]$ at the point x_n' and $f_n(x_n') = \frac{1}{e n^{n-1}}$. It follows that $0 \leq f_n(x) \leq \frac{1}{e n^{n-1}}$ for all $x \in [0, 1]$ and all $n \in \mathbb{N}$. Since $\lim_{n \to \infty} \frac{1}{e n^{n-1}} = 0$, this shows that the sequence (f_n) is uniformly convergent to the function $f \equiv 0$.

Now we prove (ii). The only root of f_n'' is $x_n'' = 2n^{-n}$, implying

$$\max\{|f_n'(x)| : x \in [0, 1]\} = \max\{|f_n'(0)|, |f_n'(\frac{2}{n^n})|, |f_n'(1)|\}$$

$$= \max \left\{n, \frac{n}{e^2}, \frac{(n^n - 1)n}{e^{n^n}}\right\} = n,$$

for all $n \geq 2$, because

$$\frac{(n^n - 1)n}{e^{n^n}} < n \iff e^{n^n} > n^n - 1 ,$$

and the last inequality is true in virtue of the inequality $e^x \geq x + 1$ valid for all $x \in \mathbb{R}$.

By Proposition 2.2.1,

$$L(f_n) = \max\{|f_n'(x)| : x \in [0, 1]\} = n ,$$

for all $n \in \mathbb{N}$, $n \geq 2$.

Example 2.4.5 There exists a sequence of functions which are not Lipschitz, converging uniformly to a Lipschitz function.

Indeed, for each $n \in \mathbb{N}$, let us consider the function $f_n : [0, 1] \to \mathbb{R}$ given by

$$f_n(x) = \frac{1}{n}\sqrt{x}, \quad x \in [0, 1] .$$

Then:

(i) the sequence $(f_n)_{n\in\mathbb{N}}$ converges uniformly to the function $f \equiv 0$ which is Lipschitz;

(ii) f_n is not Lipschitz for each $n \in \mathbb{N}$.

The assertion (i) is obvious. In Remark 2.4.3, (ii) it was shown that the function $g(x) = \sqrt{x}$ is not Lipschitz on $[0, \infty)$. The proof given there shows that g is not Lipschitz on $[0, 1]$, so neither is f_n and (ii) holds.

Remark 2.4.6 Chapter 6 contains some more results concerning limits of Lipschitz or locally Lipschitz functions. For instance, in Sect. 6.4 it will be shown that bounded from below lsc (resp. bounded from above usc) functions are pointwise limits of monotonic sequences of Lipschitz functions.

2.5   Gluing Lipschitz Functions Together

Let us consider the metric spaces $X = A \cup B$ and Y, and a function $f : X \to Y$ such that $f|_A$ and $f|_B$ are locally Lipschitz.

In general f is not locally Lipschitz, in fact it may not even be continuous. For example, if $A = [0, 1]$, $B = (1, 2]$, and

$$f(x) = \begin{cases} 1, & x \in [0, 1] \\ 2, & x \in (1, 2], \end{cases}$$

then $f|_A$ and $f|_B$ are constant, hence locally Lipschitz, but f is not continuous.
If, in addition,

$$\overline{A \setminus B} \cap (B \setminus A) = \overline{B \setminus A} \cap (A \setminus B) = \emptyset$$

(one says that $A \setminus B$ and $B \setminus A$ are separated), then f is continuous. In this case we call $A \cup B$ a *proper union* of A and B.

Even if $A \cup B$ is a proper union of A and B and $f|_A$ and $f|_B$ are locally Lipschitz, f may not be locally Lipschitz. Consider, for example, the sets

$$A = \{(t, 0) : 0 \le t \le 1\} \quad \text{and} \quad B = \{(t, t^2) : 0 \le t \le 1\}$$

endowed with the usual metric induced from $\mathbb{R}^2$, and the function $f : X = A \cup B \to \mathbb{R}$ given by

$$f(t, 0) = 0 \quad \text{and} \quad f(t, t^2) = t \quad \text{for all } t \in [0, 1].$$

Then $f|_A$ and $f|_B$ are Lipschitz, but f is not locally Lipschitz (so it is not Lipschitz).

Indeed, $(0, 0)$ has no neighborhood where f is Lipschitz. Otherwise there exists L such that

$$\left| f(t, t^2) - f(t, 0) \right| = t \leq L d((t, t^2), (t, 0)) = L t^2,$$

i.e.,

$$1 \leq Lt,$$

for all t in $(0, \delta)$, for some $0 < \delta < 1$. The last assertion is obviously false.

Next we present two results providing sufficient conditions for gluing together Lipschitz and locally Lipschitz functions, respectively.

Definition 2.5.1 Let (X, d) be a metric space. A *path* in X is a continuous mapping $\gamma : [a, b] \subseteq \mathbb{R} \to X$. If $\gamma(a) = x$ and $\gamma(b) = y$, then we say that γ *joins* x and y. The *length* of a path $\gamma : [a, b] \to X$ is

$$l(\gamma) = \sup \sum_{i=1}^{n} d(\gamma(t_i), \gamma(t_{i-1})),$$

where the supremum is taken over $n \in \mathbb{N}$ and all partitions $(t_i)_{0 \leq i \leq n}$ of $[a, b]$ with $a = t_0 \leq t_1 \leq \cdots \leq t_n = b$. A path is called *rectifiable* if its length is finite.

Remark 2.5.2 If $\gamma : [a, b] \to X$ is a path and $\psi : [c, d] \to [a, b]$ is surjective and monotonic (and hence continuous), then $\gamma \circ \psi : [c, d] \to X$ is a path satisfying $l(\gamma \circ \psi) = l(\gamma)$. Thus, we can work in the sequel with paths defined on $[0, 1]$.

Definition 2.5.3 Let (X, d) be a metric space. For $x, y \in X$ we denote by $C(x, y, X)$ the infimum of all $C \geq 1$ for which there exists a rectifiable path γ in X joining x and y such that $l(\gamma) \leq C d(x, y)$. If no such path exists we put $C(x, y, X) = \infty$.

For a subset A of X, we consider

$$C(A, X) = \sup\{C(x, y, X) : x, y \in A\}$$

and

$$C(X) = C(X, X).$$

Definition 2.5.4 A metric space X is called *quasiconvex* if $C(X) < \infty$ and C-*quasiconvex* (where $C \geq 1$) if $C(X) \leq C$. In case that each point of X has a neighborhood U such that $C(U, X) < \infty$, then we say that X is *locally quasiconvex*.

Remark 2.5.5 Each convex subset of a normed space is 1-quasiconvex and the unit sphere in $\mathbb{R}^{n+1}$ is $\frac{\pi}{2}$-quasiconvex.

Theorem 2.5.6 *Let* (X, d) *be a C-quasiconvex metric space,* $\mathscr{A}$ *a cover of X such that for each* $x \in X$ *the set* $\mathrm{st}(x, \mathscr{A}) := \bigcup\{A \in \mathscr{A} : x \in A\}$ *is a neighborhood of* x *and let* (Y, d') *be another metric space. If a function* $f : X \to Y$ *has the property that* $f|_A$ *is L-Lipschitz for every* $A \in \mathscr{A}$, *then* f *is CL-Lipschitz.*

Proof Let us choose $x, y \in X$ and $\varepsilon > 0$. Since $C(x, y, X) \leq C$, there exists a rectifiable path $\gamma : [0, 1] \to X$ such that $\gamma(0) = x$, $\gamma(1) = y$ and $l(\gamma) \leq (C + \varepsilon)d(x, y)$.

As $\mathrm{st}(\gamma(s), \mathscr{A})$ is a neighborhood of $\gamma(s)$ and γ is continuous, there exists an open ball $B(s)$ in $[0, 1]$ of center s such that

$$\gamma(B(s)) \subseteq \mathrm{st}(\gamma(s), \mathscr{A}),$$

for each $s \in [0, 1]$.

Since $[0, 1]$ is compact, there exist $k \in \mathbb{N}$ and $s_1, s_2, \dots, s_k \in [0, 1]$ such that $B(s_1), B(s_2), \dots, B(s_k)$ is a cover of $[0, 1]$ having no proper subcover. Without loss of generality we may assume that $s_i < s_{i+1}$ and $\emptyset \neq B(s_i) \cap B(s_{i+1}) \subseteq (s_i, s_{i+1})$ for each $i \in \{1, 2, \dots, k - 1\}$. Note that $0 \in B(s_1)$ and $1 \in B(s_k)$. Let us consider $s_i' \in B(s_i) \cap B(s_{i+1})$, where $i \in \{1, 2, \dots, k - 1\}$. Then

$$\gamma(s_i') \in \gamma(B(s_i) \cap B(s_{i+1})) \subseteq \mathrm{st}(\gamma(s_i), \mathscr{A}) \cap \mathrm{st}(\gamma(s_{i+1}), \mathscr{A})$$

for each $i \in \{1, 2, \dots, k - 1\}$ and therefore there exist $A_{2i}, A_{2i+1} \in \mathscr{A}$ such that

$$\{\gamma(s_i), \gamma(s_i')\} \subseteq A_{2i} \text{ and } \{\gamma(s_i'), \gamma(s_{i+1})\} \subseteq A_{2i+1}.$$

Likewise, there exist $A_1, A_{2k} \in \mathscr{A}$ such that $\{x, \gamma(s_1)\} \subseteq A_1$ and $\{\gamma(s_k), y\} \subseteq A_{2k}$. By relabeling the sequence $(0, s_1, s_1', \dots, s_{k-1}, s_{k-1}', s_k, 1)$, we get a new sequence $(t_0, t_1, \dots, t_{2k-1}, t_{2k})$ such that $0 = t_0 \leq t_1 \leq \cdots \leq t_{2k-1} \leq t_{2k} = 1$ and, with the notation $x_i = \gamma(t_i)$, where $i \in \{0, 1, \dots, 2k\}$, we have $\{x_{i-1}, x_i\} \subseteq A_i$ for all $i \in \{1, \dots, 2k\}$. Consequently,

$$d'(f(x_i), f(x_{i-1})) \leq Ld(x_i, x_{i-1}).$$

Hence,

$$d'(f(x), f(y)) \leq L \sum_{i=1}^{2k} d(x_i, x_{i-1}) \leq Ll(\gamma) \leq L(C + \varepsilon)d(x, y),$$

for each $\varepsilon > 0$ and therefore $d'(f(x), f(y)) \leq LCd(x, y)$.

As x and y were arbitrarily chosen, we infer that f is CL-Lipschitz. $\square$

Lemma 2.5.7 *Let* (X, d) *be a locally quasiconvex metric space and V an open subset of X. Then V is a locally quasiconvex metric space.*

Proof Since X is a locally quasiconvex metric space, for each $x \in V$ there exists a neighborhood U of x such that $C := C(U, X) < \infty$. Then $U \cap V$ is a neighborhood of x and therefore there exists $r > 0$ such that

$$B(x, (2C + 3)r) \subseteq U \cap V.$$

Let us consider

$$y, z \in B(x, r) \subseteq B(x, (2C + 3)r) \subseteq U.$$

Since $C + 1 > C = C(U, X) \geq C(y, z, X)$ there exist $1 \leq M < C + 1$ and a rectifiable path γ in X joining y and z such that

$$l(\gamma) \leq Md(y, z) < (C + 1)d(y, z).$$

We claim that $\operatorname{Im} \gamma \subseteq V$. Indeed, if $w \in \operatorname{Im} \gamma$, then, taking into account that $y, z \in B(x, r)$, we get

$$d(y, w) \leq l(\gamma) < (C + 1)d(y, z) \leq 2(C + 1)r$$

and therefore

$$d(x, w) \leq d(x, y) + d(y, w) \leq r + 2(C + 1)r = (2C + 3)r,$$

so

$$w \in B(x, (2C + 3)r) \subseteq U \cap V \subseteq V.$$

This proves our claim and hence $C(y, z, V) \leq C + 1$. As y and z were arbitrarily chosen in $B(x, r)$ we infer that

$$C(B(x, r), V) \leq C + 1 < \infty.$$

$\square$

Theorem 2.5.8 *Let (X, d) be a locally quasiconvex metric space, $\mathscr{A}$ a point-finite cover of X having the property that $\operatorname{st}(x, \mathscr{A}) = \cup \{A \in \mathscr{A} : x \in A\}$ is a neighborhood of x, for each $x \in X$ and let (Y, d') be a metric space. If the function $f : X \to Y$ has the property that $f|_A$ is locally Lipschitz for all $A \in \mathscr{A}$, then f is locally Lipschitz.*

Proof For $x \in X$ let us consider the set $\{A_1, A_2, \ldots, A_k\}$ of all the elements of $\mathscr{A}$ containing x. Let us choose an open set U containing x and $L \geq 0$ such that

$$U \subseteq \operatorname{st}(x, \mathscr{A}) = A_1 \cup A_2 \cup \cdots \cup A_k$$

and $f|_{U \cap A_i}$ is L-Lipschitz for all $i \in \{1, 2, \ldots, k\}$. Then, taking into account that f is continuous, it follows that $f|_{U \cap \overline{A_i}}$ is L-Lipschitz for all $i \in \{1, 2, \ldots, k\}$.

According to Lemma 2.5.7, U is locally quasiconvex, so there exists a neighborhood V of x such that $V \subseteq U$ and $C := C(V, U) < \infty$. We claim that $f|_V$ is CL-Lipschitz. Then, as x was arbitrarily chosen, we infer that f is locally Lipschitz.

Now let us prove the claim. For $y, z \in V$ and $\varepsilon > 0$, since

$$C(y, z, U) \le C(V, U) = C < C + \varepsilon,$$

it follows that there exist $1 \le M < C + \varepsilon$ and a rectifiable path $\gamma : [0, 1] \to U$ joining y and z such that

$$l(\gamma) \le Md(y, z) < (C + \varepsilon)d(y, z).$$

Now we choose $t_0, t_1, \ldots, t_n$ having the property that $0 = t_0 \le t_1 \le \cdots \le t_n = 1$ and for each $i \in \{1, 2, \ldots, n\}$ there exists $m_i \in \{1, 2, \ldots, k\}$ such that

$$\{\gamma(t_i), \gamma(t_{i-1})\} \subseteq U \cap \overline{A_{m_i}}.$$

Then

$$d'(f(y), f(z)) \le \sum_{i=1}^{n} d'(f(\gamma(t_{i-1})), f(\gamma(t_i))) \le L \sum_{i=1}^{n} d(\gamma(t_{i-1}), \gamma(t_i)) \le Ll(\gamma)$$

$$\le L(C + \varepsilon)d(y, z).$$

As ε was arbitrarily chosen, we infer that

$$d'(f(y), f(z)) \le LCd(y, z),$$

for all $y, z \in V$, i.e., $f|_V$ is CL-Lipschitz. $\square$

2.6 Lipschitz Partitions of Unity

It is well-known that (continuous, differentiable) partitions of unity are an extremely powerful tool for obtaining global results from local ones. In this section we shall prove the existence of locally Lipschitz and of Lipschitz partitions of unity.

Definition 2.6.1 A *locally Lipschitz partition of unity* subordinated to the open cover $(U_\alpha)_{\alpha \in A}$ of a metric space X is a family $(\varphi_\alpha)_{\alpha \in A}$ of functions $\varphi_\alpha : X \to [0, 1]$ such that:

(a) φ_α is locally Lipschitz for every $\alpha \in A$;
(b) the supports $\mathrm{spt}(\varphi_\alpha) := \varphi_\alpha^{-1}((0, 1])$, $\alpha \in A$, form a locally finite family;

(c) $\{\mathrm{spt}(\varphi_\alpha) : \alpha \in A\}$ refines $\{U_\alpha : \alpha \in A\}$;
(d) $\sum_{\alpha \in A} \varphi_\alpha(x) = 1$ for all $x \in X$.
   A particular case of (c) is
(c′) $\mathrm{spt}(\varphi_\alpha) \subseteq U_\alpha$ for all $\alpha \in A$.
   A *Lipschitz partition of unity* subordinated to the open cover $(U_\alpha)_{\alpha \in A}$ of the
   metric space X is a family $(\varphi_\alpha)_{\alpha \in A}$ of functions $\varphi_\alpha : X \to [0, 1]$ satisfying
(a′) φ_α is Lipschitz for every $\alpha \in A$,

and the conditions (b), (c), (d) from above.

2.6.1 The Locally Lipschitz Partition of Unity

Now we prove the existence of locally Lipschitz partitions of unity.

Theorem 2.6.2 *For each open cover $(U_i)_{i \in I}$ of a metric space X, there exists a locally Lipschitz partition of unity $\{\varphi_i\}_{i \in I}$ such that $\mathrm{spt}(\varphi_i) \subseteq U_i$, $i \in I$.*

Proof We may assume that $U_i \neq X$ for all $i \in I$.
   Let us choose a locally finite open refinement $(V_i)_{i \in I}$ of $(U_i)_{i \in I}$ and a locally finite open refinement $(W_i)_{i \in I}$ of $(V_i)_{i \in I}$ such that

$$\overline{W_i} \subseteq V_i \subseteq \overline{V_i} \subseteq U_i$$

for each $i \in I$ (see Theorem 1.3.36).
   Let us consider the 1-Lipschitz functions $\psi_i : X \to [0, \infty)$ given by

$$\psi_i(x) = d(x, X \setminus W_i)$$

for each $x \in X$.
   Since

$$d(x, X \setminus W_i) > 0 \iff x \notin \overline{X \setminus W_i} = X \setminus W_i \iff x \in W_i,$$

it follows that $\psi_i^{-1}((0, \infty)) \subseteq W_i$, hence

$$\mathrm{spt}(\psi_i) = \overline{\psi_i^{-1}((0, \infty))} \subseteq \overline{W_i} \subseteq V_i \subseteq \overline{V_i} \subseteq U_i$$

for each $i \in I$.
   Consequently, as $(V_i)_{i \in I}$ is a locally finite family, $(\mathrm{spt}(\psi_i))_{i \in I}$ has the same property.
   In addition, for each $i \in I$,

$$\psi_i(x) > 0 \iff x \in W_i,$$

for all $x \in X$.

Moreover, since $(\text{spt}(\psi_i))_{i \in I}$ is a locally finite family, we infer that the cardinal of the set $\{i \in I : \psi_i(x) \neq 0\}$ is finite and greater than or equal to 1 for each $x \in X$. Consequently, we can define the functions

$$\varphi_i = \frac{\psi_i}{\sum_{j \in I} \psi_j}$$

having the following properties:

(a) $\varphi_i : X \to [0, 1]$ is locally Lipschitz for each $i \in I$;
(b) $\text{spt}(\varphi_i) \subseteq U_i$ for all $i \in I$, (since $\text{spt}(\varphi_i) = \text{spt}(\psi_i)$);
(c) $(\text{spt}(\varphi_i))_{i \in I}$ is a locally finite family (since the family $(\text{spt}(\psi_i))_{i \in I}$ has the same property);
(d) $\sum_{i \in I} \varphi_i(x) = 1$, for all $x \in X$.

We show that the family $(\varphi_i)_{i \in I}$ satisfies the conditions from Definition 2.6.1.

Only the assertion (a) needs some motivation. For $x \in X$ consider a neighborhood N_x of x such that the set $I_x := \{i \in I : W_i \cap N_x \neq \emptyset\}$ is finite. Since $\{W_i\}$ is a cover of X, there exists i_x such that $x \in W_{i_x}$, implying $i_x \in I_x$ and $\psi_{i_x}(x) > 0$. Take now a neighborhood $G_x \subseteq N_x$ of x such that $\psi_{i_x}(x') \geq m := 2^{-1}\psi_{i_x}(x) > 0$ for all $x' \in G_x$. It follows that

$$\sum_{i \in I} \psi_i(x') = \sum_{i \in I_x} \psi_i(x') \geq m ,$$

for all $x' \in G_x$, so we can apply Propositions 2.3.3 and 2.3.7 to conclude that φ_i is Lipschitz on G_x. Consequently, φ_i is locally Lipschitz on X for every $i \in I$, and so $(\varphi_i)_{i \in I}$ is a locally Lipschitz partition of unity subordinated to the open cover $(U_i)_{i \in I}$.   □

By Theorem 2.1.6 every locally Lipschitz function on a compact metric space is Lipschitz, so that Theorem 2.6.2 has the following corollary.

Corollary 2.6.3 *Let (X, d) be a compact metric space. Then for each open cover $(U_i)_{i \in I}$ of X, there exists a Lipschitz partition of unity subordinated to this cover.*

Remark 2.6.4 In the next subsection we shall prove the existence of Lipschitz partitions of unity on arbitrary metric spaces.

2.6.2 The Lipschitz Partition of Unity

In fact, with an extra effort, the existence of Lipschitz partitions of unity can be proved.

Theorem 2.6.5 *For each open cover of a metric space there exists a Lipschitz partition of unity subordinated to this cover.*

The proof is based on the following result.

Lemma 2.6.6 (M. E. Rudin [612]) *For any open cover $\{U_i : i \in I\}$ of a metric space (X, d) there exist the open covers $\{V_{n,i} : n \in \mathbb{N}, i \in I\}$ and $\{W_{n,i} : n \in \mathbb{N}, i \in I\}$ of X satisfying the conditions:*

(1) $V_{n,i} \subseteq W_{n,i} \subseteq \overline{W}_{n,i} \subseteq U_i$ *for all $n \in \mathbb{N}$ and $i \in I$;*
(2) $d(V_{n,i}, X \setminus W_{n,i}) \geq 2^{-n-1}$ *for all $n \in \mathbb{N}$ and $i \in I$;*
(3) $d(W_{n,i}, W_{n,j}) \geq 2^{-n}$ *for all $n \in \mathbb{N}$ and $i \neq j$ in I;*
(4) *for each $x \in X$ there exist an open neighborhood U_x of x and $m_x \in \mathbb{N}$ such that*

    (a) $k > m_x \implies U_x \cap W_{k,i} = \emptyset$ *for all $i \in I$, and*
    (b) $1 \leq k \leq m_x \implies U_x \cap W_{k,i} \neq \emptyset$ *for at most one $i \in I$.*

It follows that the open covers $\{V_{n,i}\}$ and $\{W_{n,i}\}$ are locally finite σ-discrete refinements of the open cover $\{U_i\}$.

Proof of Theorem 2.6.5 Let (X, d) be a metric space and $\{U_i : i \in I\}$ an open cover of X and $\{V_{n,i} : (n, i) \in \mathbb{N} \times I\}$, $\{W_{n,i} : (n, i) \in \mathbb{N} \times I\}$ the refinements of $\{U_i\}$ given by Lemma 2.6.6.

For $(n, i) \in \mathbb{N} \times I$ let $\varphi_{n,i} : X \to [0, 1]$ be given by

$$\varphi_{n,i}(x) = \frac{d(x, X \setminus W_{n,i})}{d(x, V_{n,i}) + d(x, X \setminus W_{n,i})}, \quad x \in X.$$

By Proposition 2.1.1 and condition (2) from Lemma 2.6.6, $\varphi_{n,i}$ is Lipschitz with Lipschitz norm

$$L(\varphi_{n,i}) \leq \frac{1}{d(V_{n,i}, X \setminus W_{n,i})} \leq 2^{n+1},$$

and

$$\varphi_{n,i}(x) = 1 \quad \text{for all } x \in V_{n,i}, \text{ and}$$

$$\varphi_{n,i}(x) = 0 \quad \text{for all } x \in X \setminus W_{n,i}.$$

Define $\varphi_n : X \to [0, 1]$ by

$$\varphi_n(x) = \begin{cases} \varphi_{n,i}(x) & \text{if there exists } i \in I \text{ such that } x \in W_{n,i}, \\ 0 & \text{for } x \in X \setminus \bigcup_{j \in I} W_{n,j}. \end{cases}$$

Since the family $\{W_{n,i} : i \in I\}$ is discrete, there exists at most one $i \in I$ such that $x \in W_{n,i}$, so the function $\varphi_n(x)$ is well defined.

132 2 Basic facts

The function φ_n is also Lipschitz. To prove this we have to consider several situations:

(1) if $x, y \in W_{n,i}$, then

$$|\varphi_n(x) - \varphi_n(y)| = |\varphi_{n,i}(x) - \varphi_{n,i}(y)| \le 2^{n+1} \|x - y\|;$$

(2) if $x \in W_{n,i}$ and $y \in W_{n,j}$ for $j \ne i$, then, by condition (3) in Lemma 2.6.6, $\|x - y\| \ge 2^{-n}$ and so

$$|\varphi_n(x) - \varphi_n(y)| \le 1 \le 2^n \|x - y\|;$$

(3) if $x \in W_{n,i}$, $y \in X \setminus \bigcup_{j \in I} W_{n,j} \subseteq X \setminus W_{n,i}$, then $\varphi_n(x) = \varphi_{n,i}(x)$ and $\varphi_n(y) = 0 = \varphi_{n,i}(y)$, so that

$$|\varphi_n(x) - \varphi_n(y)| = |\varphi_{n,i}(x) - \varphi_{n,i}(y)| \le 2^{n+1} \|x - y\|;$$

(4) if $x, y \in X \setminus \bigcup_{j \in I} W_{n,j}$, then $\varphi_n(x) = 0 = \varphi_n(y)$.

It follows that φ_n is Lipschitz with $L(\varphi_n) \le 2^{n+1}$.
Consider now the functions

$$\psi_1 = \varphi_1 \text{ and } \psi_n = \varphi_n(1 - \varphi_{n-1})\ldots(1 - \varphi_1), \text{ for } n \ge 2.$$

Then, for every $n \in \mathbb{N}$, the function $\psi_n : X \to [0, 1]$ is Lipschitz (as a product of bounded Lipschitz functions) and

$$\psi_1 + \cdots + \psi_n = 1 - (1 - \varphi_1)\ldots(1 - \varphi_n). \tag{2.6.1}$$

We show that, for every $x \in X$, the series $\sum_n \psi_n(x)$ has only finitely many non-null terms and

$$\sum_n \psi_n(x) = 1.$$

Indeed, for $x \in X$ there exists $(m, i) \in \mathbb{N} \times I$ such that $x \in V_{m,i} \subseteq W_{m,i}$ implying $\varphi_m(x) = \varphi_{m,i}(x) = 1$ and so $\psi_k(x) = 0$ for all $k > m$. The identity (2.6.1) (for $n = m$) shows that $\psi_1(x) + \cdots + \psi_m(x) = 1$.
Finally, for $(n, i) \in \mathbb{N} \times I$ define $\psi_{n,i} : X \to [0, 1]$ by

$$\psi_{n,i}(x) = \begin{cases} \psi_n(x) & \text{if } x \in W_{n,i}, \\ 0 & \text{for } x \in X \setminus W_{n,i}. \end{cases}$$

To show that $\psi_{n,i}$ is Lipschitz we have to consider again the four cases from above. Let $x, y \in X$. In case (1)

$$|\psi_{n,i}(x) - \psi_{n,i}(y)| = |\psi_n(x) - \psi_n(y)| \le L(\psi_n)\|x - y\|.$$

Case (2) can be treated as the corresponding one for the function φ_n.

If x, y are as in (3), then $\psi_{n,i}(x) = \psi_n(x)$, $\psi_{n,i}(y) = 0$ and $\psi_n(y) = 0$ because $\varphi_n(y) = 0$, so that

$$|\psi_{n,i}(x) - \psi_{n,i}(y)| = |\psi_n(x) - \psi_n(y)| \le L(\psi_n)\|x - y\|.$$

Finally, in case (4) we have $\psi_{n,i}(x) = 0 = \psi_{n,i}(y)$ because $X \setminus \bigcup_{j\in I} W_{n,j} \subseteq X \setminus W_{n,i}$.

It follows that $\psi_{n,i}$ is Lipschitz with $L(\psi_{n,i}) \le \max\{2^n, L(\psi_n)\}$.

For every $n \in \mathbb{N}$, the functions $\psi_{n,i}$ satisfy the identity

$$\sum_{j\in I} \psi_{n,j} = \psi_n.$$

Let $x \in X$. If, for some $i \in I$, $x \in W_{n,i}$, then $\psi_{n,i}(x) = \psi_n(x)$ and $\psi_{n,j}(x) = 0$ for all $j \in I \setminus \{i\}$. If $x \in X \setminus \bigcup_{j\in I} W_{n,j}$, then $\psi_{n,j}(x) = 0$ for all $j \in I$ and $\psi_n(x) = 0$ as well (because $\varphi_n(x) = 0$).

It follows that

$$\sum_{(n,i)\in\mathbb{N}\times I} \psi_{n,i}(x) = \sum_{n\in\mathbb{N}} \sum_{i\in I} \psi_{n,i}(x) = \sum_{n\in\mathbb{N}} \psi_n(x) = 1,$$

for every $x \in X$.

Since in an arbitrary topological space

$$U \cap A = \emptyset \iff U \cap \overline{A} = \emptyset,$$

for any open set U, it follows that the family $\{\overline{W}_{n,i}\}$ also satisfies the condition (4) of Lemma 2.6.6, so it is locally finite and σ-discrete, since $\{\overline{W}_{n,i} : i \in I\}$ is discrete for every $n \in \mathbb{N}$.

Because $\mathrm{spt}(\psi_{n,i}) \subseteq \overline{W}_{n,i}$, the family $\{\psi_{n,i} : (n, i) \in \mathbb{N}\times I\}$ is a locally finite and σ-discrete Lipschitz partition of unity subordinated to the open cover $\{U_i : i \in I\}$.

$\square$

Remark 2.6.7 The proof given above is inspired from [275] and [303], where the existence of Lipschitz C^k-smooth (for $k \in \mathbb{N} \cup \{\infty\}$) partitions of the unity on some Banach spaces is proved, with applications to approximation for some classes of functions (continuous, uniformly continuous) by polynomials or by Lipschitz C^k-smooth functions (see also the book [276]).

Other proofs are given in [244, 245] and [600].

2.6.3   A Proof of Rudin's Lemma

As in M. E. Rudin [612] the result is stated in a slightly different form, and in [275, 276] and [303] Lemma 2.6.6 is given without proof, we shall present a proof following the ideas from [612].

Let $\mathcal{U} = \{U_i : i \in I\}$ be an open cover of the metric space (X, d). Since, by a consequence of Zorn's lemma (Theorem 1.1.1 and Corollary 1.1.2), any set can be well ordered, we suppose that the set I is well ordered by an order relation $\leq$.

Define, by induction with respect to $n \in \mathbb{N}$, the families of sets $E_{n,i}$, $V_{n,i}$, $W_{n,i}$, $i \in I$, in the following way.

Step I. $z \in E_{1,i}$ if

    ($*$) i is the first element in I such that $z \in U_i$;
    ($**$) $B(z, 3 \cdot 2^{-1}) \subseteq U_i$.

For $i \in I$ define the sets $V_{1,i}$ and $W_{1,i}$ by

$$V_{1,i} = \bigcup \left\{ B(z, 2^{-2}) : z \in E_{1,i} \right\},$$

$$W_{1,i} = \bigcup \left\{ B(z, 2^{-1}) : z \in E_{1,i} \right\}.$$

Let $n \geq 2$. Supposing that the sets $E_{k,i}, V_{k,i}, W_{k,i}$ are defined for $i \in I$ and $1 \leq k < n$, we define $E_{n,i}$ by the conditions.
Step II. $z \in E_{n,i}$ if

    (i) i is the first element in I for which $z \in U_i$;
    (ii) $z \in X \setminus \bigcup \{V_{k,i} : i \in I, 1 \leq k < n\}$;
    (iii) $B(z, 3 \cdot 2^{-n}) \subseteq U_i$.

For $i \in I$ put

$$V_{n,i} = \bigcup \{B(z, 2^{-n-1}) : z \in E_{n,i}\} \text{ and}$$

$$W_{n,i} = \bigcup \{B(z, 2^{-n}) : z \in E_{n,i}\} .$$

Let us show first that the families of sets $\{V_{n,i}\}$ and $\{W_{n,i}\}$ are open covers of X. To do this it is sufficient to show that $\{V_{n,i} : i \in I, n \in \mathbb{N}\}$ covers X.

For $x \in X$ let i be the first element in I for which $x \in U_i$ and let $n \in \mathbb{N}$ be such that $B(x, 3 \cdot 2^{-n}) \subseteq U_i$. If $x \in X \setminus \bigcup \{V_{k,i} : i \in I, 1 \leq k < n\}$, then $x \in E_{n,i} \subseteq V_{n,i}$. Otherwise, then there exists $j \in I$ and $1 \leq k < n$ such that $x \in V_{k,j}$, so that $x \in \bigcup \{V_{k,j} : j \in I, k \in \mathbb{N}\}$ in both cases.

By definitions, the sets $V_{n,i}$ and $W_{n,i}$ are open and the inclusions $V_{n,i} \subseteq W_{n,i} \subseteq U_i$, $i \in I$, show that they are refinements of $\mathcal{U}$.

Let us show now that these sets fulfill the conditions (1)–(4) from Lemma 2.6.6.

Proof of (2) Let $x \in V_{n,i}$ and $y \in X \setminus W_{n,i}$. Let $z \in E_{n,i}$ be such that $x \in B(z, 2^{-n-1}) \subseteq V_{n,i}$. Since $y \notin W_{n,i}$, it follows that $d(z, y) \geq 2^{-n}$, so that

$$\frac{1}{2^n} \leq d(z, y) \leq d(z, x) + d(x, y)$$

$$< \frac{1}{2^{n+1}} + d(x, y),$$

hence $d(x, y) > 2^{-n-1}$ which proves (2). □

Proof of (3) Let $x \in W_{n,j}$ and $x' \in W_{n,j'}$, where $j < j'$. Then there exist $z \in E_{n,j}$ and $z' \in E_{n,j'}$ such that

$$B(z, 3 \cdot 2^{-n}) \subseteq U_j, \quad x \in B(z, 2^{-n}) \text{ and}$$

$$B(z', 3 \cdot 2^{-n}) \subseteq U_{j'}, \quad x' \in B(z', 2^{-n}).$$

Since, by (i), $z' \notin U_j$, it follows that $d(z, z') \geq 3 \cdot 2^{-n}$ and so

$$\frac{3}{2^n} \leq d(z, z') \leq d(z, x) + d(x, x') + d(x', z')$$

$$< d(x, x') + \frac{2}{2^n}.$$

Hence $d(x, x') > \frac{1}{2^n}$, which proves (3). □

Proof of (4) For $x \in X$ let i be the first element in I for which there exists $n \in \mathbb{N}$ such that $x \in V_{n,i}$. Since $V_{n,i}$ is open, there exists $m \in \mathbb{N}$ such that

$$B(x, 2^{-m}) \subseteq V_{n,i}. \tag{2.6.2}$$

We show that

(α) $k > m + n - 1 \Rightarrow B(x, 2^{-m-n-1}) \cap W_{k,j} = \emptyset$ for all $j \in I$;

(β) $1 \leq k \leq m + n - 1 \Rightarrow B(x, 2^{-m-n-1}) \cap W_{k,j} \neq \emptyset$ for at most one $j \in I$.

Obviously, (α), (β) imply (a) and (b) from (4), with $U_x = B(x, 2^{-m-n-1})$ and $m_x = m + n - 1$.

Let $k \geq m + n$ and $y \in B(x, 2^{-m-n-1}) \cap W_{k,j}$.

It follows that

$$d(x, y) < \frac{1}{2^{m+n+1}},$$

and the existence of an element $z \in E_{k,j}$ such that $y \in B(z, 2^{-k}) \subseteq W_{k,j}$. Since $k > n$, the condition (ii) implies $z \notin V_{n,i}$, and so $d(x, z) \geq 2^{-m}$ (by (2.6.2)).

But then

$$\frac{1}{2^m} \leq d(x, z) \leq d(x, y) + d(y, z)$$

$$< \frac{1}{2^{m+n+1}} + \frac{1}{2^{m+n}} = \frac{3}{2^{m+n+1}},$$

which yields the contradiction $2^{n+1} < 3$ (recall that $\mathbb{N} = \{1, 2, \dots\}$).

Let now $1 \leq k < m + n$ and suppose that there exist $j < j'$ in I and $y, y' \in X$ such that

$$y \in B(x, 2^{-m-n-1}) \cap W_{k,j} \quad \text{and} \quad y' \in B(x, 2^{-m-n-1}) \cap W_{k,j'}.$$

It follows that

$$d(y, y') \leq d(y, x) + d(x, y') < \frac{1}{2^{m+n}}. \qquad (2.6.3)$$

Taking into account (3) and the hypothesis on k, one obtains

$$d(y, y') \geq d(W_{k,j}, W_{k,j'}) \geq \frac{1}{2^k} \geq \frac{1}{2^{m+n}},$$

in contradiction to (2.6.3). $\square$

Proof of (1) We have only to show that $\overline{W}_{n,i} \subseteq U_i$, which will be a consequence of the inclusions

$$W_{n,i} \subseteq \{x \in X : d(x, V_{n,i}) < \frac{1}{2^n}\} \subseteq \{x \in X : d(x, V_{n,i}) \leq \frac{1}{2^n}\}$$

$$\subseteq \{x \in X : d(x, V_{n,i}) < \frac{2}{2^n}\} \subseteq U_i.$$

Indeed, since $\{x \in X : d(x, V_{n,i}) \leq \frac{1}{2^n}\}$ is closed,

$$\overline{W}_{n,i} \subseteq \{x \in X : d(x, V_{n,i}) \leq \frac{1}{2^n}\} \subseteq U_i.$$

Let us show first that

$$W_{n,i} \subseteq \{x \in X : d(x, V_{n,i}) < \frac{1}{2^n}\}.$$

Let $x \in W_{n,i}$. Then there exists $z \in E_{n,i} \subseteq V_{n,i}$ such that $x \in B(z, 2^{-n}) \subseteq W_{n,i}$, so that

$$d(x, V_{n,i}) \leq d(x, z) < \frac{1}{2^n}.$$

We show now that

$$\{x \in X : d(x, V_{n,i}) < \frac{2}{2^n}\} \subseteq U_i.$$

If $d(x, V_{n,i}) < \frac{2}{2^n}$, then there exists an element $y \in V_{n,i}$ such that

$$d(x, y) < \frac{2}{2^n}.$$

From $y \in V_{n,i}$ one obtains the existence of $z \in E_{n,i}$ such that $d(z, y) < \frac{1}{2^{n+1}}$. Consequently,

$$d(x, z) \leq d(x, y) + d(y, z)$$
$$< \frac{2}{2^n} + \frac{1}{2^{n+1}} < \frac{3}{2^n}.$$

Taking into account (iii), it follows that $x \in B(z, 3 \cdot 2^{-n}) \subseteq U_i$.

For every $x \in X$ the neighborhood U_x of x given by (4) intersects only finitely many of $W_{k,i}$, and, for fixed k, at most one $W_{k,i}$. Consequently, the family $\{W_{n,i} : i \in I, n \in \mathbb{N}\}$ is locally finite and $\{W_{n,i} : i \in I\}$ is discrete for every $n \in \mathbb{N}$, so that

$$\{W_{n,i} : i \in I, n \in \mathbb{N}\} = \bigcup_{n \in \mathbb{N}} \{W_{n,i} : i \in I\}$$

is also σ-discrete.

Lemma 2.6.6 is completely proved. $\square$

2.7 Applications of Lipschitz Partitions of Unity

In this section we shall present some applications of the locally Lipschitz partition of unity. Another application will be presented in Sect. 6.2.

2.7.1 A Sandwich-Type Theorem

The first application is a sandwich type theorem for locally Lipschitz functions, extending to this setting some known results for continuous functions (see Sect. 1.3.12).

In the following theorem we use the notation

$$f < g \iff \forall x \in X, \quad f(x) < g(x).$$

Theorem 2.7.1 *Let X be a metric space. If $g : X \to \mathbb{R}$ is an upper semicontinuous function and $h : X \to \mathbb{R}$ is a lower semicontinuous function such that $g(x) < h(x)$ for all $x \in X$, then there exists a locally Lipschitz function $f : X \to \mathbb{R}$ satisfying the inequality*

$$g < f < h.$$

Proof For each $r \in \mathbb{Q}$, let us consider the set

$$U_r = \{x \in X : g(x) < r\} \cap \{x \in X : r < h(x)\}.$$

Since g is upper semicontinuous and h is lower semicontinuous, it follows that the set U_r is open, for each $r \in \mathbb{Q}$.

As $(U_r)_{r \in \mathbb{Q}}$ is an open cover of X, there exists a locally Lipschitz partition of unity $(\varphi_r)_{r \in \mathbb{Q}}$ subordinated to this cover.

Then the locally Lipschitz function $f : X \to \mathbb{R}$, given by

$$f(x) = \sum_{r \in \mathbb{Q}} r \varphi_r(x)$$

for all $x \in X$, is the desired function.

Indeed, for each $x \in X$ there exist $r_1, \ldots, r_n \in \mathbb{Q}$ such that $x \in \mathrm{spt}(\varphi_{r_i}) \subseteq U_{r_i}$ for each $i \in \{1, \ldots, n\}$ and $x \notin \mathrm{spt}(\varphi_r)$ for all $r \in \mathbb{Q} \setminus \{r_1, \ldots, r_n\}$. Then

$$g(x) = \sum_{i=1}^{n} g(x)\varphi_{r_i}(x) < \sum_{i=1}^{n} r_i \varphi_{r_i}(x) < \sum_{i=1}^{n} h(x)\varphi_{r_i}(x) = h(x),$$

which shows that

$$g(x) < f(x) < h(x),$$

for each $x \in X$.   □

2.7.2 Selections of Set-Valued Mappings

We shall use now the locally Lipschitz partitions of unity to show that there exist locally Lipschitz functions which take given values.

Theorem 2.7.2 *Let us consider a function* $\varphi : X \to \mathscr{P}(Y)$ *and* V *a convex open neighborhood of* 0_Y, *where* X *is a metric space and* Y *a real normed space. Suppose that*

(i) $\varphi(x)$ *is nonempty and convex, for each* $x \in X$;
(ii) *the set* $U_y := \{x \in X : y \in \varphi(x) + V\}$ *is open, for each* $y \in Y$.

Then there exists a locally Lipschitz function $f : X \to Y$ *having the property that*

$$f(x) \in \varphi(x) + V,$$

for each $x \in X$.

Proof Let

$$Z := \{y \in Y : U_y \neq \emptyset\}.$$

Observe that

$$U_y \neq \emptyset \iff \exists x \in X, \; y \in \varphi(x) + V \iff y \in \bigcup \{\varphi(x) + V : x \in X\} = \varphi(X) + V,$$

that is, $Z = \varphi(X) + V$.

Since, for every $x \in X$, $x \in U_y$, where y is an arbitrary point in $\varphi(x) + V$, it follows that

$$\mathscr{U} = \{U_y : y \in Z\}$$

is an open cover of X. Consider now the locally Lipschitz function $f : X \to Y$ given by

$$f(x) = \sum_{y \in Z} y f_y(x), \quad x \in X,$$

where $(f_y)_{y \in Z}$ is the locally Lipschitz partition subordinated to the cover $\mathscr{U}$.

Let show that f is locally Lipschitz. Since the family $\{\mathrm{spt}(f_y) : y \in Z\}$ is locally finite, any point $x \in X$ has a neighborhood U that intersects only finitely many of these supports, say $\mathrm{spt}(f_{y_1}), \ldots, \mathrm{spt}(f_{y_n})$. Since each f_{y_i} is locally Lipschitz, there exists a neighborhood $U' \subseteq U$ of x on which all the functions f_{y_i} are Lipschitz. But then

$$f(x') = \sum_{i=1}^{n} y_i f_{y_i}(x'),$$

for all $x' \in U'$, so that the function f is Lipschitz on U'.

If, for some $i \in \{1, \ldots, n\}$, $f_{y_i}(x) \neq 0$, then $x \in \mathrm{spt}(f_{y_i}) \subseteq U_{y_i}$, implying $y_i \in \varphi(x) + V$.

Consequently,

$$f(x) = \sum_{i=1}^{n} y_i f_{y_i}(x) \in \varphi(x) + V,$$

since $\varphi(x) + V$ is convex, as $\varphi(x)$ and V are convex sets and

$$\sum_{i=1}^{n} f_{y_i}(x) = \sum_{y \in Z} f_y(x) = 1.$$

$\square$

2.7.3   The Lipschitz Separability of the Space $C(T)$

It is known that the space $C(T)$ is separable if T is a compact metric space. We show that the separability can be done with Lipschitz functions.

Theorem 2.7.3 *Let (T, ϱ) be a compact metric space. Then the space $C(T)$ contains a countable dense subset formed of Lipschitz functions.*

Proof Consider first the real Banach space $C(T, \mathbb{R})$.

Since T is separable it contains a countable dense subset $S = \{t_n : n \in \mathbb{N}\}$.

The density of S implies that the open balls $U_i^n := B(t_i, 1/n)$, $i \in \mathbb{N}$, form an open cover of the compact space T. It follows that, for every $n \in \mathbb{N}$, there exists a finite subset J_n of $\mathbb{N}$ such that $T = \bigcup_{j \in J_n} U_j^n$.

Consider now a locally Lipschitz partition f_j^n, $j \in J_n$, subordinated to the open cover $\{U_j^n : j \in J_n\}$, that is,

(i) $f_j^n \geq 0,$ (ii) $\forall j \in J_n, \ \forall t \in T, \ t \in T \setminus U_j^n \ \Rightarrow \ f_j^n(t) = 0,$ and

(iii) $\displaystyle\sum_{j \in J_n} f_j^n(s) = 1$ for all $s \in T$.

$$(2.7.1)$$

Since T is a compact metric space the locally Lipschitz functions f_i^n are in fact Lipschitz (Theorem 2.1.6).

Put

$$F_n = \Big\{ \sum_{j \in J_n} a_j f_j^n : a_j \in \mathbb{Q} \ \text{for all} \ j \in J_n \Big\} \quad \text{and} \quad F = \bigcup_{n=1}^{\infty} F_n.$$

Every function in F is Lipschitz and the set F is obviously countable (as a countable union of countable sets), so that the proof will be complete if we show that F dense in $C(T, \mathbb{R})$.

Let $f \in C(T, \mathbb{R})$ and $\varepsilon > 0$.

By the uniform continuity of f there exists $\delta > 0$ such that

$$\varrho(s, s') < \delta \;\Rightarrow\; |f(s) - f(s')| < \varepsilon,$$

for all $s, s' \in T$.

Choose $n \in \mathbb{N}$ such that $1/n < \delta$. It follows that, for any $j \in J_n$,

$$s \in U_j^n \;\Rightarrow\; |f(s) - f(t_j)| < \varepsilon. \tag{2.7.2}$$

Finally, let $a_j \in \mathbb{Q}$ be such that

$$|f(t_j) - a_j| < \varepsilon \text{ for all } j \in J_n. \tag{2.7.3}$$

By (2.7.1).(iii),

$$\left| f(s) - \sum_{j \in J_n} a_j f_j^n(s) \right| = \left| \sum_{j \in J_n} (f(s) - a_j) f_j^n(s) \right|$$
$$\leq \sum_{j \in J_n} |f(s) - a_j| f_j^n(s), \tag{2.7.4}$$

for every $s \in T$.

Let $s \in T$. Then by (2.7.1).(ii), (2.7.2) and (2.7.3), for every $j \in J_n$,

$$s \notin U_j^n \;\Rightarrow\; f_j^n(s) = 0, \quad \text{and}$$
$$s \in U_j^n \;\Rightarrow\; |f(s) - a_j| \leq |f(s) - f(t_j)| + |f(t_j) - a_j| < 2\varepsilon.$$

Consequently, by (2.7.4) and (2.7.1).(iii),

$$\left| f(s) - \sum_{j \in J_n} a_j f_j^n(s) \right| < 2\varepsilon \sum_{j \in J_n} f_j^n(s) = 2\varepsilon.$$

Since this inequality holds for every $s \in T$, it follows that

$$\left\| f - \sum_{j \in J_n} a_j f_j^n \right\| < 2\varepsilon.$$

The complex case reduces to the real one. If g_n, $n \in \mathbb{N}$, is a countable family of $\mathbb{R}$-valued Lipschitz functions, dense in $C(T, \mathbb{R})$, then

$$F = \{g_n + ig_m : m, n \in \mathbb{N}\}$$

is a countable family of $\mathbb{C}$-valued Lipschitz functions dense in $C(T)$.   □

Remark 2.7.4 In fact, the result holds for the space $C(T, \mathbb{K}^n)$.

Indeed, if $A = \{f_k : k \in \mathbb{N}\}$ is a countable family of Lipschitz functions, dense in $C(T, \mathbb{K})$, then A^n is a countable family of $\mathbb{K}^n$-valued Lipschitz functions, dense in $C(T, \mathbb{K}^n)$.

2.8 Bibliographic Comments

The notion of Lipschitz function is due to the German mathematician Rudolph Lipschitz (1832–1903), who introduced it in order to prove the existence of solutions to some differential equations and to provide a sufficient condition for the convergence of Fourier series.

Part of the results in Sects. 2.6 and 2.7 are from [413] (namely Theorem 2.6.2 and Theorem 2.7.1) and the other ones from [473].

Chapter 3
Relations with Other Classes of Functions

We have seen (see Sect. 1.3.2) that any Lipschitz function is uniformly continuous. In this chapter we shall present Lipschitz properties of convex functions and convex operators, and equi-Lipschitz properties for families of convex vector-functions. In the vector case, meaning convex functions defined on a locally convex space with values in a locally convex space ordered by a cone, one emphasizes the key role played by the normality of the cone. Other considered topics involve the existence of an equivalent metric making a given continuous function Lipschitz, and metric spaces where every continuous function is Lipschitz. An old result of G. Fichtenholz (from 1922) on the relation between absolutely continuous and Lipschitz functions is included. The chapter ends with a discussion on the differentiability properties of Lipschitz functions—Rademacher-type theorems—in finite and in infinite dimension.

3.1 Lipschitz Properties of Convex Functions

This section is concerned with Lipschitz properties of convex functions.

3.1.1 Overview

As it is well-known every convex function defined on an open interval of the real axis is Lipschitz on each compact subinterval of its domain of definition (see, e.g. [281], Chapter 3, §18). This result can be extended to convex functions defined on convex open subsets Ω of $\mathbb{R}^n$—every such function is locally Lipschitz on Ω and Lipschitz on every compact subset of Ω (see, for instance [552]). Assuming the continuity of the convex function the result can be further extended to the case when

© Springer Nature Switzerland AG 2019
Ş. Cobzaş et al., *Lipschitz Functions*, Lecture Notes in Mathematics 2241,
https://doi.org/10.1007/978-3-030-16489-8_3

Ω is an open convex subset of a normed space (see, e.g., [221, 598, 599, 674]), or of a locally convex space (see [154, 155, 161, 163, 431, 694, 695]). Lipschitz functions in the setting of locally convex spaces were introduced by Mankiewicz [428] in 1972 (see also [429]).

Convex mappings (or convex operators), that is, mappings defined on a convex subset of a vector space and with values in an ordered vector space, have been intensively studied in the last years, mainly in connection with optimization problems and mathematical programming in ordered vector spaces, see [101, 102, 522], or the monographs [268, 370]. The normality of the cone is essential in the proofs of the continuity properties of convex vector-functions and, as remarked by Carioli and Veselý [135], the normality is, in some sense, also necessary for the validity of these properties (see Sect. 3.1.5).

Lipschitz properties of continuous convex vector functions defined on an open convex subset of a normed space and with values in a normed space ordered by a normal cone were proved in [655] and [539, 540].

Equicontinuity results (Banach–Steinhaus type principles) for pointwise bounded families of continuous convex mappings were proved in [365, 523]. Kosmol [364] proved that a pointwise bounded family of continuous convex mappings, defined on an open convex subset Ω of a Banach space X and with values in a normed space Y ordered by a normal cone, is locally equi-Lipschitz on Ω. The case of real-valued functions was considered in [363]. Jouak and Thibault [319] proved equicontinuity and equi-Lipschitz results for families of continuous convex mappings defined on open convex subsets of Baire topological vector spaces or of barrelled locally convex spaces and taking values in a topological vector space, respectively in a locally convex space ordered by a normal cone. New proofs of these results were given in [157]. Breckner and Trif [114] extended these results to families of rationally s-convex functions. Marinescu [433] considered a more general class of functions (called quasi-Lipschitz) and used them in the study of Lipschitz properties of convex functions with values in ordered locally convex spaces. Condensation of singularities principles for non-equicontinuous families of continuous convex mappings have been proved in [115].

Our presentation, essentially following [161], will show that some geometric properties (monotonicity of the slope, the normality of the seminorms) allow to extend the proofs from the scalar case to the vector one. In this way the proofs become more transparent and natural.

3.1.2 Normal Cones in Locally Convex Spaces

Since in the study of convex functions taking values in topological vector spaces ordered by a closed cone a key role is played by the normality of the cone, we present some results on normal cones, referring to [34, 113, 181, 622] for details. The basic properties of ordered vector spaces and of ordered topological vector spaces are presented in Sects. 1.1.2 and 1.4.9.

Let (X, τ) be a topological vector space ordered by a closed cone C.

The cone C is called *normal* if the space X admits a neighborhood basis at the origin formed of C-full sets.

Let $\gamma > 0$. A seminorm p on a vector space X is called:

- *γ-monotone* if $0 \leq x \leq y \ \Rightarrow \ p(x) \leq \gamma p(y)$;
- *γ-absolutely monotone* if $-y \leq x \leq y \ \Rightarrow \ p(x) \leq \gamma p(y)$;
- *γ-normal* if $x \leq z \leq y \ \Rightarrow \ p(z) \leq \gamma \max\{p(x), p(y)\}$.

The following characterizations of normal cones hold.

Theorem 3.1.1 ([113, 622]) *Let (X, τ) be a LCS ordered by a cone C. Then the following are equivalent.*

1. *The cone C is normal.*
2. *The LCS X admits a basis of 0-neighborhoods formed of C-full absolutely convex sets.*
3. *There exist $\gamma > 0$ and a family of γ-normal seminorms generating the topology τ of X.*
4. *There exist $\gamma > 0$ and a family of γ-monotone seminorms generating the topology τ of X.*
5. *There exist $\gamma > 0$ and a family of γ-absolutely monotone seminorms generating the topology τ of X.*

All the above equivalences also hold with $\gamma = 1$ in all places.

A subset Z of a topological vector space (X, τ) is called *bounded* (or *topologically bounded*) if it is absorbed by every neighborhood of 0, i.e., for every neighborhood V of 0, there exists $\lambda > 0$ such that $\lambda Z \subseteq V$.

If X is a locally convex space with the topology generated by a family P of seminorms, then $Z \subseteq X$ is topologically bounded if and only if

$$\sup\{p(z) : z \in Z\} < \infty,$$

for every $p \in P$. If, further, X is a normed space, then Z is topologically bounded if and only if

$$\sup\{\|z\| : z \in Z\} < \infty.$$

A subset Z of a vector space $(X, \leq)$ ordered by a cone C is called *upper (lower) o-bounded* (o comes from "order") if there exists $y \in X$ such that $z \leq y$ (resp. $y \leq z$) for all $z \in Z$, where $\leq := \leq_C$ is the order generated by the cone C. It is called *o-bounded* if it is both upper and lower bounded, i.e., there exist $x, y \in X$ such that $Z \subseteq [x, y]_o$, where $[x, y]_o$ denotes the order-interval determined by x and y (see (1.1.1)).

We mention the following result.

Proposition 3.1.2 *Let (X, τ) be a topological vector space ordered by a closed cone C.*

1. *If the cone C is normal, then every o-bounded subset of X is topologically bounded.*
2. *Let X be a Banach space. If the cone C is not normal, then there exists $w \geq 0$ in X such that the order-interval $[0, w]_o$ is norm-unbounded.*
   It follows that, in this case, the cone C is normal if and only if every o-bounded subset of X is topologically bounded.

Proof

1. Suppose that the cone C is normal and let Z be an o-bounded subset of X. Then there exist $x, y \in X$ such that $Z \subseteq [x, y]_o$. Let V be a C-full neighborhood of $0 \in X$. Since V is absorbing, there exists $\lambda > 0$ such that $\lambda x, \lambda y \in V$. It follows that $[\lambda x, \lambda y]_o \subseteq [V] = V$, so that $\lambda Z \subseteq [\lambda x, \lambda y]_o \subseteq V$.
2. Since C is not normal there exist two sequences (x_n) and (y_n) in X such that $0 \leq x_n \leq y_n$, $\|y_n\| = 1$ and $\|x_n\| = 3^n + 1$ for all $n \in \mathbb{N}$. One takes $w = \sum_{k=1}^{\infty} 2^{-k} y_k$ and

$$z_n = w - \sum_{k=1}^{n-1} 2^{-k} y_k - 2^{-n} x_n = w - \sum_{k=1}^{n} 2^{-k} y_k + 2^{-n}(y_n - x_n).$$

Then $0 \leq z_n \leq w$ and

$$\|z_n\| \geq \left(\frac{3}{2}\right)^n - \left\| w - \sum_{k=1}^{n-1} 2^{-k} y_k \right\| \to \infty \quad \text{as} \quad n \to \infty,$$

since $\|x_n - y_n\| \geq 3^n$.

$\square$

3.1.3 Some Properties of Convex Vector-Functions

We consider now convex mappings from a more general point of view, meaning mappings with values in an ordered vector space which are convex with respect to the vector order, and give some simple results that are essential for the proofs in the following sections.

Let X, Y be real vector spaces and suppose that Y is ordered by a cone C. If Ω is a convex subset of X, then a mapping $f : \Omega \to Y$ is called *convex* (or a *convex operator*, or *C-convex*) provided that

$$f((1 - \alpha)x_1 + \alpha x_2) \leq (1 - \alpha)f(x_1) + \alpha f(x_2),$$

for all $x_1, x_2 \in \Omega$ and $\alpha \in [0, 1]$, where $\leq := \leq_C$ stands for the order induced by the cone C, namely $x \leq_C y \iff y - x \in C$.

The following results are well-known in the case of real-valued convex functions.

Proposition 3.1.3 *Let I be an interval in $\mathbb{R}$, Y a vector space ordered by a cone C and $\varphi : I \to Y$ a C-convex function.*

1. The following equivalent inequalities hold:

(a) $\quad \varphi(t_2) \leq \dfrac{t_3 - t_2}{t_3 - t_1} \varphi(t_1) + \dfrac{t_2 - t_1}{t_3 - t_1} \varphi(t_3),$

(b) $\quad \dfrac{\varphi(t_2) - \varphi(t_1)}{t_2 - t_1} \leq \dfrac{\varphi(t_3) - \varphi(t_1)}{t_3 - t_1},$

$$(3.1.1)$$

(c) $\quad \dfrac{\varphi(t_3) - \varphi(t_1)}{t_3 - t_1} \leq \dfrac{\varphi(t_3) - \varphi(t_2)}{t_3 - t_2},$

(d) $\quad \dfrac{\varphi(t_2) - \varphi(t_1)}{t_2 - t_1} \leq \dfrac{\varphi(t_3) - \varphi(t_2)}{t_3 - t_2},$

where $\leq := \leq_C$ is the order induced by the cone C and $t_1 < t_2 < t_3$ are points in I.

2. For $t_0 \in I$ fixed, the slope of φ at t_0, defined by

$$\Delta_{t_0}(\varphi)(t) = \frac{\varphi(t) - \varphi(t_0)}{t - t_0}, \quad t \in I \setminus \{t_0\},$$

is an increasing function of t, i.e.,

$$\frac{\varphi(t) - \varphi(t_0)}{t - t_0} \leq \frac{\varphi(t') - \varphi(t_0)}{t' - t_0},$$

for all $t, t' \in I \setminus \{t_0\}$ with $t < t'$.

Proof

1. The proof is based on the identity

$$t_2 = \frac{t_3 - t_2}{t_3 - t_1} t_1 + \frac{t_2 - t_1}{t_3 - t_1} t_3, \tag{3.1.2}$$

valid for all points $t_1 < t_2 < t_3$ in I. The identity can be verified by a direct calculation.

The inequality (3.1.1).(a) follows from (3.1.2) and the convexity of φ.

Isolating in the left-hand side of the inequalities (b), (c), (d) the value $\varphi(t_2)$, one obtains in all cases the inequality from (a), proving their equivalence.

The statement 2 follows from 1. $\qquad \square$

For $x, y \in X$, $x \neq y$, the right line $D(x, y)$ and the algebraic segment determined by x, y are

$$D(x, y) = \{x + t(y - x) : t \in \mathbb{R}\} \quad \text{and} \quad [x, y] = \{x + t(y - x) : t \in [0, 1]\},$$

respectively.

Consider now a more general framework.

Proposition 3.1.4 *Let X be a vector space and p a seminorm on X. For fixed $x, y \in X$ such that $p(x - y) > 0$ put $z_t = x + t(y - x)$, $t \in \mathbb{R}$.*

1. For every $t, t' \in \mathbb{R}$

$$p(z_t - z_{t'}) = |t - t'| \, p(y - x).$$

2. Let $z_i = z_{t_i}$, $i = 1, 2, 3$, where $t_1 < t_2 < t_3$. Then

$$z_2 = \frac{p(z_3 - z_2)}{p(z_3 - z_1)} z_1 + \frac{p(z_2 - z_1)}{p(z_3 - z_1)} z_3 \quad \text{and}$$

$$p(z_3 - z_1) = p(z_3 - z_2) + p(z_2 - z_1).$$

3. Let Ω be a convex subset of X, Y a vector space ordered by a cone C and $f : \Omega \to Y$ a C-convex function. For $x_0 := x + t_0(y - x) \in D(x, y) \cap \Omega$, the p-slope of f is given by

$$\Delta_{p,x_0}(f)(z_t) = \frac{f(z_t) - f(x_0)}{p(z_t - x_0)},$$

for $t \in \mathbb{R}$ such that $z_t \in D(x, y) \cap \Omega \setminus \{x_0\}$.
Then $t_0 < t < t'$ or $t < t' < t_0$ implies

$$\Delta_{p,x_0}(f)(z_t) \leq \Delta_{p,x_0}(f)(z_{t'}),$$

and $t < t_0 < t'$ implies

$$\frac{f(x_0) - f(z_t)}{p(x_0 - z_t)} \leq \frac{f(z_{t'}) - f(x_0)}{p(z_{t'} - x_0)} \quad (\Longleftrightarrow -\Delta_{p,x_0}(f)(z_t) \leq \Delta_{p,x_0}(f)(z_{t'})).$$

$$(3.1.3)$$

Proof The equality from 1 follows by the definition of z_t.

For 2, observe that the equality

$$t_2 = \frac{t_3 - t_2}{t_3 - t_1} t_1 + \frac{t_2 - t_1}{t_3 - t_1} t_3$$

implies

$$z_2 = \frac{t_3 - t_2}{t_3 - t_1} z_1 + \frac{t_2 - t_1}{t_3 - t_1} z_3 .$$

By 1,

$$\frac{t_3 - t_2}{t_3 - t_1} = \frac{p(z_3 - z_2)}{p(z_3 - z_1)} \quad \text{and} \quad \frac{t_2 - t_1}{t_3 - t_1} = \frac{p(z_2 - z_1)}{p(z_3 - z_1)} ,$$

proving the representation formula for z_2.

The equality $p(z_3 - z_1) = p(z_2 - z_1) + p(z_3 - z_2)$ is equivalent to $t_3 - t_1 = (t_3 - t_2) + (t_2 - t_1)$.

3. Let $x_0 = x + t_0(y - x)$, $z = x + t(y - x)$ and $z' = x + t'(y - x)$. The function $\varphi(t) = f(x + t(y - x))$ is convex, so that, by Proposition 3.1.3, its slope is increasing. If $t_0 < t < t'$, then

$$\frac{f(z) - f(x_0)}{p(z - x_0)} = \frac{\varphi(t) - \varphi(t_0)}{(t - t_0)p(y - x)} \leq \frac{\varphi(t') - \varphi(t_0)}{(t' - t_0)p(y - x)} = \frac{f(z') - f(x_0)}{p(z' - x_0)} .$$

The case $t < t' < t_0$ can be treated similarly. If $t < t_0 < t'$, then

$$\frac{f(x_0) - f(z)}{p(x_0 - z)} = \frac{\varphi(t_0) - \varphi(t)}{(t_0 - t)p(y - x)} \leq \frac{\varphi(t') - \varphi(t_0)}{(t' - t_0)p(y - x)} = \frac{f(z') - f(x_0)}{p(z' - x_0)} .$$

$\square$

3.1.4 Continuity Properties of Convex Functions

In this section we prove some results on the continuity of convex functions.

We start with real-valued functions of one real variable, a typical case. Based on the monotonicity of the slope one can give a simple proof of the Lipschitz continuity of convex functions.

Proposition 3.1.5 Let $\varphi : I \to \mathbb{R}$ be a convex function defined on an interval $I \subseteq \mathbb{R}$. Then φ is continuous on $\mathrm{int}(I)$ and Lipschitz on every compact interval $[\alpha, \beta] \subseteq \mathrm{int}(I)$.

Proof It is obvious that it suffices to check the fulfillment of the Lipschitz condition. For $[\alpha, \beta] \subseteq \mathrm{int}(I)$ with $\alpha < \beta$, let $a, b \in \mathrm{int}(I)$ be such that $a < \alpha < \beta < b$.

Let $\alpha \leq t < t' \leq \beta$. By Proposition 3.1.3.2,

$$\frac{\varphi(t') - \varphi(t)}{t' - t} \leq \frac{\varphi(b) - \varphi(t)}{b - t} \leq \frac{\varphi(b) - \varphi(\beta)}{b - \beta} =: B ,$$

and

$$A := \frac{\varphi(\alpha) - \varphi(a)}{\alpha - a} \lesseqgtr \frac{\varphi(t') - \varphi(a)}{t' - a} \leq \frac{\varphi(t) - \varphi(t')}{t - t'} \, .$$

Taking $L := \max\{|A|, |B|\}$, it follows that

$$|\varphi(t) - \varphi(t')| \leq L \, |t - t'| \quad \text{for all} \quad t, t' \in [\alpha, \beta] \, .$$

$\square$

We also mention the following results.

Proposition 3.1.6 *The convex functions have the following properties.*

1. *Let I be an interval in $\mathbb{R}$, $\varphi : I \to \mathbb{R}$ a convex function and $a < b$ two points in I. If for some $0 < t_0 < 1$, $\varphi((1-t_0)a + t_0 b) = (1-t_0)\varphi(a) + t_0\varphi(b)$, then φ is an affine function on the interval $[a, b]$, that is, $\varphi((1-t)a + tb) = (1-t)\varphi(a) + t\varphi(b)$ for every $t \in [0, 1]$.*
2. *Under the assumptions from 1, if $a, b \in \text{int}(I)$ and $\varphi(a) < \varphi(b)$, then φ is strictly increasing on the interval $I_{b+} = \{\alpha \in I : \alpha \geq b\}$. If $\varphi(a) > \varphi(b)$, then φ is strictly decreasing on the interval $I_{a-} = \{\alpha \in I : \alpha \leq a\}$.*
3. *Any nonconstant convex function $\varphi : \mathbb{R} \to \mathbb{R}$ is unbounded, more exactly,*

$$\sup \varphi(\mathbb{R}) = \infty \, .$$

4. *Let $\varphi : [0, \infty) \to [0, \infty)$ be convex such that $\varphi(\alpha) = 0 \iff \alpha = 0$. Then φ is strictly increasing and superadditive, that is,*

$$\varphi(\alpha + \beta) \geq \varphi(\alpha) + \varphi(\beta) \, ,$$

for all $\alpha, \beta \in [0, \infty)$.
   If $\varphi : [0, \infty) \to [0, \infty)$ is concave and $\varphi(\alpha) = 0 \iff \alpha = 0$, then φ is increasing and subadditive, that is,

$$\varphi(\alpha + \beta) \leq \varphi(\alpha) + \varphi(\beta) \, ,$$

for all $\alpha, \beta \in [0, \infty)$.

Proof

1. Suppose that for some t, $t_0 < t < 1$, $\varphi(a + t(b - a)) < \varphi(a) + t(\varphi(b) - \varphi(a))$. Let $c = a + t_0(b - a)$ and $c_t = a + t(b - a)$. It follows that $0 < t_0/t < 1$, $c =$

$a + \frac{t_0}{t}(c_t - a)$, and

$$\varphi(c) = \varphi(a) + t_0(\varphi(b) - \varphi(a))$$

$$= \left(1 - \frac{t_0}{t}\right)\varphi(a) + \frac{t_0}{t}[\varphi(a) + t(\varphi(b) - \varphi(a))]$$

$$> \left(1 - \frac{t_0}{t}\right)\varphi(a) + \frac{t_0}{t}\varphi(c_t),$$

in contradiction to the convexity of f.

The case $0 < t < t_0$ can be treated similarly.

2. Suppose that $\varphi(a) < \varphi(b)$ and let $\alpha > b$ be a point in I. Then, by the monotonicity of the slope,

$$\frac{\varphi(\alpha) - \varphi(b)}{\alpha - b} \geq \frac{\varphi(b) - \varphi(a)}{b - a} > 0 \Rightarrow \varphi(\alpha) > \varphi(b).$$

If $b < \alpha < \alpha'$ belong to I, then $\varphi(\alpha) > \varphi(b)$, and applying the above reasoning to the points $b < \alpha < \alpha'$, it follows that $\varphi(\alpha) < \varphi(\alpha')$.

In the case $\varphi(a) > \varphi(b)$, a similar argument applied to points $\alpha \in I$ with $\alpha < a$ shows that φ is strictly decreasing on I_{a-}.

3. Suppose that there exists two points $a < b$ in $\mathbb{R}$ such that $\varphi(a) \neq \varphi(b)$.

Case I. $\varphi(b) - \varphi(a) > 0$

Let $\alpha_t = a + t(b - a)$, $t > 1$. The monotonicity of the slope implies

$$\frac{\varphi(\alpha_t) - \varphi(a)}{\alpha_t - a} \geq \frac{\varphi(b) - \varphi(a)}{b - a}.$$

Since $\alpha_t - a = t(b - a) > 0$, it follows that

$$\varphi(\alpha_t) - \varphi(a) \geq t(\varphi(b) - \varphi(a)) \to +\infty \text{ as } t \to \infty.$$

Case II. $\varphi(b) - \varphi(a) < 0$

Taking $\alpha_t = a + t(b - a)$ for $t < 0$, it follows that $\alpha_t < a < b$, so that, by the monotonicity of the slope,

$$\frac{\varphi(\alpha_t) - \varphi(a)}{\alpha_t - a} \leq \frac{\varphi(b) - \varphi(a)}{b - a}.$$

Since, in this case, $\alpha_t - a = t(b - a) < 0$, it follows that

$$\varphi(\alpha_t) - \varphi(a) \geq t(\varphi(b) - \varphi(a)) \to +\infty \text{ as } t \to -\infty.$$

4. By 2, φ is strictly increasing on $[0, \infty)$ because $\varphi(\alpha) > 0 = \varphi(0)$ for every $\alpha > 0$.

Let now $0 < \alpha < \beta$. Then, by the convexity of φ,

$$\varphi(\alpha) = \varphi\left(\left(1 - \frac{\alpha}{\beta}\right)0 + \frac{\alpha}{\beta}\beta\right) \leq \frac{\alpha}{\beta}\varphi(\beta),$$

so that

$$\alpha\varphi(\beta) - \beta\varphi(\alpha) \geq 0. \tag{3.1.4}$$

Again, by the convexity of φ,

$$\varphi(\beta) \leq \frac{\alpha}{\beta}\varphi(\alpha) + \frac{\beta - \alpha}{\beta}\varphi(\alpha + \beta)$$

implying

$$\varphi(\alpha + \beta) \geq \frac{\alpha\varphi(\beta) - \beta\varphi(\alpha)}{\beta - \alpha} + \varphi(\alpha) + \varphi(\beta)$$

$$\geq \varphi(\alpha) + \varphi(\beta) \qquad \text{(by (3.1.4))}.$$

Suppose now that φ is concave and not increasing on $[0, \infty)$. Then there exist two numbers $0 < \alpha < \beta$ such that $\varphi(\alpha) > \varphi(\beta)$. Let $\alpha_t = \alpha + t(\beta - \alpha)$ with $t > 1$. Since the slope of φ is decreasing, we have

$$\frac{\varphi(\alpha_t) - \varphi(\alpha)}{\alpha_t - \alpha} \leq \frac{\varphi(\beta) - \varphi(\alpha)}{\beta - \alpha},$$

implying

$$\varphi(\alpha_t) \leq \varphi(\alpha) + t(\varphi(\beta) - \varphi(\alpha)) \to -\infty \quad \text{as } t \to \infty.$$

Consequently, $\varphi(\alpha_t) < 0$ for t large enough, in contradiction to the hypothesis that $\varphi \geq 0$.

The proof of the subadditivity follows the same line (reversing the inequalities) as the proof of superadditivity in the case of a convex function.

$\square$

Remark 3.1.7 Geometrically, the property 1 from Proposition 3.1.6 says that if a point $(t_0, \varphi(t_0))$, with $a < t_0 < b$, belongs to the segment $[A, B]$ where $A(a, \varphi(a))$ and $B(b, \varphi(b))$ are points on the graph of φ, then the graph of φ for $t \in [a, b]$ agrees with the segment $[A, B]$.

The example of the function $\varphi(t) = t$ for $t \in [0, 1]$ and $\varphi(t) = 1$ for $t \geq 1$ shows that a concave function satisfying the hypotheses from Proposition 3.1.6.4, can be only increasing, without being strictly increasing.

We consider now a more general situation.

Proposition 3.1.8 *Let X be a TVS, $\Omega \subseteq X$ open and convex and $f : \Omega \to \mathbb{R}$ a convex function.*

1. *If the function f is bounded from above on a neighborhood of some point $x_0 \in \Omega$, then f is continuous at x_0.*
2. *If there exists a point $x_0 \in \Omega$ and a neighborhood $U \subseteq \Omega$ of x_0 such that f is bounded from above on U, then f is locally bounded from above on Ω, that is, every point $x \in \Omega$ has a neighborhood $V \subseteq \Omega$ such that f is bounded from above on V.*
3. *If the function f is bounded from above on a neighborhood of some point $x_0 \in \Omega$, then f is continuous on Ω.*

Proof

1. Let U be a balanced neighborhood of 0 such that $x_0 + U \subseteq \Omega$ and, for some $\beta > 0$, $f(x) \leq \beta$ for all $x \in x_0 + U$, or, equivalently, to $f(x_0 + u) \leq \beta$ for all $u \in U$.

   For $0 < \varepsilon < 1$, $\pm \varepsilon u \in U$ and, by the convexity of f,

   $$f(x_0 + \varepsilon u) - f(x_0) = f((1 - \varepsilon)x_0 + \varepsilon(x_0 + u)) - f(x_0)$$
   $$\leq (1 - \varepsilon)f(x_0) + \varepsilon f(x_0 + u) - f(x_0),$$

   so that

   $$f(x_0 + \varepsilon u) - f(x_0) \leq \varepsilon(f(x_0 + u) - f(x_0)) \leq \varepsilon(\beta - f(x_0)). \qquad (3.1.5)$$

   On the other hand,

   $$f(x_0) = f\left(\frac{x_0 + \varepsilon u + x_0 - \varepsilon u}{2}\right) \leq \frac{1}{2}f(x_0 + \varepsilon u) + \frac{1}{2}f(x_0 - \varepsilon u),$$

   implying

   $$f(x_0) - f(x_0 + \varepsilon u) \leq f(x_0 - \varepsilon u) - f(x_0) \leq \varepsilon(\beta - f(x_0)). \qquad (3.1.6)$$

   The last inequality from above follows by replacing u with $-u$ in (3.1.5). Now, by (3.1.5) and (3.1.6) it follows that

   $$|f(x_0 + \varepsilon u) - f(x_0)| \leq \varepsilon(\beta - f(x_0)) \quad \text{for all} \quad u \in U,$$

   which is equivalent to

   $$|f(x_0 + v) - f(x_0)| \leq \varepsilon(\beta - f(x_0)) \quad \text{for every} \quad v \in \varepsilon U,$$

   which shows that f is continuous at x_0.

2. The proof has a geometric flavor and can be nicely illustrated by a drawing. Let U be a balanced neighborhood of 0 such that $x_0 + U \subseteq \Omega$ and, for some $\beta > 0$, $f(x) \leq \beta$ for all $x \in x_0 + U$.

   Let $x \in \Omega$. Since the set Ω is open, there exists $\alpha > 1$ such that $x_1 := x_0 + \alpha(x - x_0) \in \Omega$, implying $x = \frac{\alpha - 1}{\alpha} x_0 + \frac{1}{\alpha} x_1$. Putting $t = 1/\alpha$ it follows that $x = (1-t)x_0 + t x_1$ with $0 < t < 1$. Consider the neighborhood $V := x + (1-t) U$ of x. We have $V \subseteq \Omega$, because, by the convexity of Ω,

$$x + (1 - t)u = t x_1 + (1 - t)(x_0 + u) \in t\Omega + (1 - t)\Omega \subseteq \Omega,$$

   for all $u \in U$.

   Also

$$f\left(x + (1 - t)u\right) = f\left(t x_1 + (1 - t)(x_0 + u)\right) \leq t f(x_1) + (1 - t) f(x_0 + u)$$
$$\leq t f(x_1) + (1 - t)\beta,$$

   for every $u \in U$.
3. The assertion from 3 follows from 1 and 2.

$$\square$$

Based on these results one can give a characterization of the continuity of a convex function in terms of its epigraph. Let X be a vector space, Ω a nonempty subset of X and $f : \Omega \to \mathbb{R}$ a function. Let

$$\mathrm{epi}(f) = \{(x, \alpha) \in X \times \mathbb{R} : f(x) \leq \alpha\} \quad \text{and}$$
$$\mathrm{epi}'(f) = \{(x, \alpha) \in X \times \mathbb{R} : f(x) < \alpha\},$$

be the epigraph and the strict epigraph of f, respectively.

The following result is a direct consequence of the definitions.

Proposition 3.1.9 *Let X be a vector space, $\Omega \subseteq X$ a convex set and $f : \Omega \to \mathbb{R}$ a function. The following equivalences hold:*

$$\text{the function } f \text{ is convex} \iff \mathrm{epi}(f) \text{ is a convex subset of } X \times \mathbb{R}$$
$$\iff \mathrm{epi}'(f) \text{ is a convex subset of } X \times \mathbb{R}.$$

We can characterize now the continuity of f.

Proposition 3.1.10 *Let X be a TVS, $\Omega \subseteq X$ nonempty open convex and $f : \Omega \to \mathbb{R}$ a convex function.*

1. (a) $\mathrm{int}(\mathrm{epi}(f)) \subseteq \mathrm{epi}'(f)$;
   (b) *if $(x, \alpha) \in \mathrm{int}(\mathrm{epi}(f))$, then f is continuous at x;*
   (c) *if f is continuous at $x \in \Omega$, then $(x, \alpha) \in \mathrm{int}(\mathrm{epi}(f))$ for all $\alpha > f(x)$.*

2. *The following are equivalent:*

   (i) *f is continuous on Ω;*
   (ii) *int(epi(f)) $\neq \emptyset$;*
   (iii) *epi$'(f)$ is an open subset of $X \times \mathbb{R}$.*

3. *If int(epi(f)) $\neq \emptyset$, then int(epi(f)) = epi$'(f)$.*

Proof

1. (a) If $(x, \alpha) \in$ int(epi(f)), then there exist a neighborhood U of $0 \in X$ and
    $\delta > 0$ such that $W := (x + U) \times (\alpha - \delta, \alpha + \delta) \subseteq$ epi(f). But then
    $(x, \alpha - \delta/2) \in W \subseteq$ epi(f) so that $f(x) \leq \alpha - \delta/2 < \alpha$, that is, $(x, \alpha) \in$
    epi$'(f)$.
  (b) Let $W \subseteq$ epi(f) be as above. Then, for every $u \in U$, $(x + u, \alpha) \in W \subseteq$
    epi(f), so that $f(x + u) \leq \alpha$ for all $u \in U$, which, by Proposition 3.1.8,
    implies the continuity of f at x.
  (c) Suppose that f is continuous at $x \in \Omega$ and let $\alpha > f(x)$. Then $\delta := (\alpha -$
    $f(x))/2 > 0$ and there exists a neighborhood U of $0 \in X$ such that

$$f(x + u) < f(x) + \delta = \alpha - \delta < \alpha,$$

  for all $u \in U$. It follows that the neighborhood $(x + U) \times (\alpha - \delta, \infty)$ of (x, α)
  is contained in epi(f), which implies that $(x, \alpha) \in$ int(epi(f)).

2. Notice that, by Proposition 3.1.8, the continuity of f at a point $x \in \Omega$ is
  equivalent to the continuity of f on Ω.
    (i) $\Longleftrightarrow$ (ii) follows from the assertions (b) and (c) of point 1 of the proposition.
    (i) $\Rightarrow$ (iii). Suppose that f is continuous on Ω. If $(x, \alpha) \in$ epi$'(f)$, then
  $f(x) < \alpha$ so that, by claims (b) and (a) from 1, $(x, \alpha) \in$ int(epi(f)) $\subseteq$ epi$'(f)$.
  It follows that int(epi(f)) is a neighborhood of (x, α) contained in epi$'(f)$,
  that is, $(x, \alpha) \in$ int(epi$'(f)$). Consequently, epi$'(f) \subseteq$ int(epi$'(f)$) and so
  epi$'(f) =$ int(epi$'(f)$) is open.
    (iii) $\Rightarrow$ (i). If epi$'(f)$ is open, then, $\emptyset \neq$ epi$'(f) \subseteq$ int(epi(f)) so that (ii)
  holds, which implies the continuity of f.

3. If int(epi(f)) $\neq \emptyset$, then f is continuous on Ω, so that epi$'(f)$ is open. The
  inclusion epi$'(f) \subseteq$ epi(f) implies epi$'(f) \subseteq$ int(epi(f)) and so, taking into
  account 1.(a), epi$'(f) =$ int(epi(f)).

□

The following proposition shows that in the finite dimensional case the convex
functions are continuous.

Proposition 3.1.11 *Let* $f : \Omega \subseteq \mathbb{R}^n \to \mathbb{R}$ *be a convex function, where the set* Ω *is open and convex. Then* f *is locally bounded from above on* Ω *and, consequently, continuous on* Ω.

Proof Let us choose $x_0 \in \Omega$ and $K \subseteq \Omega$ be a hypercube having the center in x_0.

We are going to prove that f is bounded from above on K.

If $v_1, \ldots, v_m$, where $m = 2^n$, are the vertices of K, then for each $x \in K$ there exist $\lambda_1, \ldots, \lambda_m \in [0, 1]$, with $\sum_{k=1}^{m} \lambda_k = 1$, such that $x = \sum_{k=1}^{m} \lambda_k v_k$.

Taking into account Jensen's inequality for convex functions, we obtain that

$$f(x) = f(\sum_{k=1}^{m} \lambda_k v_k) \leq \sum_{k=1}^{m} \lambda_k f(v_k) \leq \max_{k \in \{1,2,\ldots,m\}} f(v_k),$$

showing that f is bounded from above on K. Since K is a neighborhood of x, f is continuous at x, and so on Ω (see Proposition 3.1.8). $\qquad\qquad\square$

A convex function defined on an infinite dimensional normed linear space is not necessarily locally bounded as the following example shows.

Example 3.1.12 Let X be the space of polynomials endowed with the norm given by

$$\|P\| = \max_{x \in [-1,1]} |P(x)|.$$

Then the function $f : X \to \mathbb{R}$ given by

$$f(P) = P'(1)$$

for each $P \in X$ is convex (even linear) but it is not locally bounded.

To show this, consider for each $n \in \mathbb{N}$ the polynomial

$$P_n(x) = \frac{1}{\sqrt{n}} x^n.$$

Then

$$\|P_n\| = \frac{1}{\sqrt{n}} \to 0, \quad n \to \infty,$$

but

$$f(P_n) = \sqrt{n} \to \infty, \quad n \to \infty,$$

proving the discontinuity of the functional f.

Remark 3.1.13 In fact a normed space X is finite dimensional if and only if every linear functional on X is continuous. On the other hand there exist infinite dimensional locally convex spaces X such that every convex function on X is continuous.

Indeed, it is known that every linear functional on a finite dimensional topological vector space is continuous. If X is an infinite dimensional normed space then it contains a linearly independent set $D = \{e_n : n \in \mathbb{N}\} \subseteq S_X$. Consider a Hamel basis E of X containing this set and define $\varphi : E \to \mathbb{R}$ by $\varphi(e_n) = n$, $n \in \mathbb{N}$, and $\varphi(e) = 0$ for $e \in E \setminus D$, extended by linearity to whole X. Then $\sup\{\varphi(x) : x \in X, \|x\| \le 1\} \ge \sup\{\varphi(e_n) : n \in \mathbb{N}\} = \infty$, proving the discontinuity of φ.

Concerning the second affirmation, let X be an infinite dimensional vector space equipped with the finest locally convex topology τ. A neighborhood basis at 0 for this topology is formed by all absolutely convex absorbing subsets of X. A family of seminorms generating this topology is formed of the Minkowski functionals of these neighborhoods. Since every seminorm p on X is the Minkowski functional of the absolutely convex absorbing subset $B_p = \{x \in X : p(x) \le 1\}$, it follows that τ is generated by the family P of all seminorms on X. It is in fact characterized by this property: the finest locally convex topology on a vector space X is the locally convex topology τ on X such that every seminorm on X is τ-continuous. For the finest locally convex topology on a vector space, see [622, p. 56 and Exercise 7, p. 69] and [550, pp. 3–4]. It follows that every convex absorbing subset of X is a neighborhood of 0 and every linear functional is continuous on X. Also, every convex function defined on a nonempty open convex subset Ω of X is continuous on Ω.

For the convenience of the reader we sketch the proof following [163] (see also [161]), where further details can be found.

Fact 1. *If C is a convex subset of vector space such that $0 \in C$, then $\alpha C \subseteq \beta C$ for all $0 < \alpha < \beta$.*

Indeed, by the convexity of C and the fact that $0 \in C$,

$$\alpha c = \beta \left(\frac{\alpha}{\beta} c + \left(1 - \frac{\alpha}{\beta} \right) \cdot 0 \right) \in \beta C ,$$

for all $c \in C$.

Fact 2. *Let Y be a vector space equipped with the finest locally convex topology τ. Then every convex absorbing subset C of Y is a neighborhood of 0.*

The set $D := C \cap (-C)$ is absolutely convex and contains 0. For $x \in Y$ there exist $\alpha, \beta > 0$ such that $x \in \alpha C$ and $-x \in \beta C \iff x \in \beta(-C)$. Then, by Fact 1, $x \in \gamma C \cap \gamma(-C)$, where $\gamma = \max\{\alpha, \beta\}$. This implies that there exist $c, c' \in C$ such that $x = \gamma c$ and $x = \gamma(-c')$. But then $c = -c' \in -C$, that is, $x \in \gamma D$. Since D is absolutely convex and absorbing it is a neighborhood of 0 and $C \supset D$ as well.

Fact 3. *Let X be a vector space. Consider the space $X \times \mathbb{R}$ equipped with the finest locally convex topology and X with the induced topology. If Ω is an open convex subset of X, then every convex function $f : \Omega \to \mathbb{R}$ is continuous.*

For more clarity we denote by θ the null element in X.

We can suppose, passing, if necessary, to the set $\widetilde{\Omega} := \Omega - x_0$ and to the function $\tilde{f}(x) := f(x + x_0) - f(x_0) - 1$, $x \in \widetilde{\Omega}$, that $\theta \in \Omega$ and $f(\theta) < 0$.

The convex function f is continuous on Ω if and only if it is continuous at $\theta \in \Omega$. In its turn, by Proposition 3.1.10, this holds if the strict epigraph $\mathrm{epi}'(f) := \{(x, \alpha) \in X \times \mathbb{R} : f(x) < \alpha\}$ is a neighborhood of $(\theta, 0)$ in $X \times \mathbb{R}$. By Fact 2, $\mathrm{epi}'(f)$ is a neighborhood of $(\theta, 0)$ in $X \times \mathbb{R}$ if it is convex and absorbing in $X \times \mathbb{R}$.

The convexity of $\mathrm{epi}'(f)$ follows from the convexity of f.

Let us show that $\mathrm{epi}'(f)$ is absorbing. Consider first the case $(\theta, \alpha) \in X \times \mathbb{R}$. If $\alpha > f(\theta)$, then $(\theta, \alpha) \in \mathrm{epi}'(f)$. If $\alpha \leq f(\theta) < 0$, then, as $\lim_{\gamma \searrow 0} \gamma\alpha = 0$, it follows that $\gamma\alpha > f(\theta)$ for sufficiently small positive γ, that is, $\gamma(\theta, \alpha) = (\theta, \gamma\alpha) \in \mathrm{epi}'(f)$. Let now $(x, \alpha) \in X \times \mathbb{R}$ with $x \neq \theta$. Then $I := \{t \in \mathbb{R} : tx \in \Omega\}$ is an open interval in $\mathbb{R}$ and $g : I \to \mathbb{R}$, $g(t) := f(tx)$, $t \in I$, is convex, and so continuous. But then $\mathrm{epi}'(g)$ is an open convex subset of $\mathbb{R}^2$. Since $g(0) = f(\theta) < 0$, it follows that $(0, 0) \in \mathrm{epi}'(g)$, hence, by Proposition 3.1.10, $\mathrm{epi}'(g)$ is a neighborhood of $(0, 0)$, and so an absorbing set in $\mathbb{R}^2$. Let $\lambda > 0$ be such that $(\lambda, \lambda\alpha) = \lambda(1, \alpha) \in \mathrm{epi}'(g)$. The equivalences

$$(\lambda, \lambda\alpha) \in \mathrm{epi}'(g) \iff g(\lambda) < \lambda\alpha \iff f(\lambda x) < \lambda\alpha$$

$$\iff \lambda(x, \alpha) = (\lambda x, \lambda\alpha) \in \mathrm{epi}'(f),$$

show that $\lambda(x, \alpha) \in \mathrm{epi}'(f)$ and so $\mathrm{epi}'(f)$ is an absorbing subset of $X \times \mathbb{R}$.

3.1.5 Further Properties of Convex Vector-Functions

Now we shall present, following [539], some further results on C-convex mappings.

Let X be a TVS, Y a vector space ordered by a cone C and Ω an open subset of X. We say that a mapping $f : \Omega \to Y$ is *locally o-bounded* on Ω if every point in Ω has a neighborhood on which f is o-bounded.

The following proposition is the analog of Proposition 3.1.8 with boundedness replaced by o-boundedness.

Proposition 3.1.14 *Let X, Y be as above and suppose that $\Omega \subseteq X$ is open and convex, and $f : \Omega \to Y$ a C-convex mapping.*

1. *If f is upper o-bounded on a neighborhood of some point $x_0 \in \Omega$, then f is locally o-bounded on Ω.*
2. *If Y is a TVS ordered by a normal cone C and f is topologically bounded on a neighborhood of a point $x_0 \in \Omega$, then f is continuous at x_0.*

3. *If Y is a TVS ordered by a normal cone C and f is o-bounded on a neighborhood of a point $x_0 \in \Omega$, then f is continuous at x_0.*
4. *If Y is a TVS ordered by a normal cone C and f is upper o-bounded on a neighborhood of some point $x_0 \in \Omega$, then f is continuous on Ω.*

Proof

1. Let U be a balanced 0-neighborhood and let $y \in Y$ be such that $x_0 + U \subseteq \Omega$ and $f(x_0 + u) \le y$ for all $u \in U$. Then $-u \in U$ and

$$f(x_0) \le \frac{1}{2}[f(x_0 + u) + f(x_0 - u)]$$

implies

$$f(x_0) - f(x_0 + u) \le f(x_0 - u) - f(x_0) \le y - f(x_0).$$

It follows that

$$f(x_0 + u) \ge 2f(x_0) - y,$$

for all $u \in U$, showing that f is also lower o-bounded on $x_0 + U$.

The fact that f is locally o-bounded on Ω can be proved similarly to the proof of assertion 2 in Proposition 3.1.8.

2. Suppose first that $0 \in \Omega$ and $f(0) = 0$.

Let $U \subseteq \Omega$ be a balanced neighborhood of 0 such that f is topologically bounded on U, that is, the set $f(U)$ is topologically bounded in Y. Let V be a balanced C-full neighborhood of $0 \in Y$. The boundedness of $f(U)$ implies the existence of $\lambda > 0$ such that $\lambda f(U) \subseteq V$ and, as the set V is balanced, we can further suppose that $\lambda < 1$.

By the convexity of f

$$f(\lambda u) = f((1 - \lambda)0 + \lambda u) \le (1 - \lambda)f(0) + \lambda f(u) = \lambda f(u) \in V,$$

for all $u \in U$.

Also, $f(\lambda(-u)) \le \lambda f(-u)$ and

$$0 = f(0) \le \frac{1}{2}[f(-\lambda u) + f(\lambda u)]$$

implies

$$f(\lambda u) \ge -f(-\lambda u) = -f(\lambda(-u)) \ge -\lambda f(-u) \in V.$$

Consequently, $-\lambda f(-u) \le f(\lambda u) \le \lambda f(u)$, with $-\lambda f(-u), \lambda f(u) \in V$. Since V is C-full, this implies $f(\lambda u) \in V$ for all $u \in U$. Since λU is a neighborhood of $0 \in X$ and $f(\lambda U) \subseteq V$, this proves the continuity of f at 0.

In general, let $x_0 \in \Omega$ and $U \subseteq \Omega$ a neighborhood of x_0 such that $f(U)$ is topologically bounded in Y. Consider the set $\widetilde{\Omega} = -x_0 + \Omega$ and the function $\widetilde{f} : \widetilde{\Omega} \to Y$ given by $\widetilde{f}(z) = f(x_0 + z) - f(x_0)$, $z \in \widetilde{\Omega}$. It follows that $\widetilde{f}$ is topologically bounded on the neighborhood $\widetilde{U} := -x_0 + U \subseteq \widetilde{\Omega}$ of $0 \in X$, so that it is continuous at 0, implying the continuity of the mapping f at $x_0 \in \Omega$.
3. Let $U \subseteq \Omega$ be a neighborhood of $x_0 \in \Omega$ such that $f(U)$ is o-bounded in Y. Since the cone C is normal it follows that $f(U)$ is topologically bounded, so that by 2, the function f is continuous at x_0.
4. Follows from 1 and 3.

$\square$

Remark 3.1.15 The proof of the assertion 3 of Proposition 3.1.14, shows that it actually holds under a weaker hypothesis than the topological boundedness of $f(U)$ for some neighborhood of x_0. Namely it is sufficient to suppose that:

$(*)$ *For every neighborhood V of $0 \in Y$ there exist a neighborhood $U \subseteq \Omega$ of x_0 and $\lambda > 0$ such that $\lambda f(U) \subseteq V$.*

In the finite dimensional case one obtains the following extension of Proposition 3.1.11.

Corollary 3.1.16 *Let Ω be a nonempty open convex subset of $\mathbb{R}^n$ and Y a TVS ordered by a normal cone C. Then every C-convex function $f : \Omega \to Y$ is locally o-bounded, and so continuous, on Ω.*

Proof The proof of Proposition 3.1.11 can be transposed *mutatis mutandis* to this situation, replacing the order relation in $\mathbb{R}$ by the order relation $\leq_C$ generated by the normal cone C. $\square$

Carioli and Veselý [135] showed that the normality of the cone C is, in some sense, necessary for the continuity of upper o-bounded convex vector-functions.

Theorem 3.1.17 *Let $I \subseteq \mathbb{R}$ be an open interval, X a (nontrivial) Hausdorff locally convex space, $\Omega \subseteq X$ an open, convex set and Y a Banach space ordered by a closed cone C. The following assertions are equivalent.*

1. *The cone C is normal.*
2. *Every convex function $\varphi : I \to Y$ is continuous.*
3. *Every convex function $\varphi : I \to Y$ is locally norm bounded.*
4. *Every convex function $f : \Omega \to Y$, which is upper o-bounded on some open subset of Ω, is continuous on Ω.*
5. *Every convex function $f : \Omega \to Y$, which is upper o-bounded on some nonempty open subset of Ω, is locally norm bounded on Ω.*

Proof We prove the following implications:

$1 \Rightarrow 2$. Follows from Corollary 3.1.16.

The implication $2 \Rightarrow 3$ is obvious.

$3 \Rightarrow 1$. We proceed by contradiction. Suppose that the cone C is not normal and show that there exists a convex function $\varphi : \mathbb{R} \to Y$, locally upper o-bounded on $\mathbb{R}$ but norm-unbounded on every neighborhood of 0.

By Proposition 3.1.2 there exists $w \geq 0$ in Y such that the order-interval $[0, w]_o$ is norm-unbounded. Then the interval $[\alpha w, \beta w]_o$ is also norm-unbounded for all $0 \leq \alpha < \beta$. Take the numbers λ, α with $\lambda \in (0, 1)$ and $1 < \alpha < \lambda^{-1}(1 - \lambda + \lambda^2)$. Since $1 - \lambda + \lambda^2 > \lambda$, α is well-defined. Consider the intervals $\Delta_n := \left[\lambda^{2n} w, \alpha \lambda^{2n} w\right]_o$ for $n \in \mathbb{N}_0 := \mathbb{N} \cup \{0\}$. Since $\alpha \lambda < 1 - \lambda + \lambda^2 < 1$, it follows that $\alpha \lambda^{2n+2} w \leq \lambda^{2n} w$ and $\alpha \lambda^{2n+2} w \neq \lambda^{2n} w$, so that the intervals Δ_n are pairwise disjoint and $z' \leq z$ for $z \in \Delta_n$, $z' \in \Delta_{n'}$ with $n < n'$.

Choose $w_n \in \Delta_n$ such that $\|w_n\| > n$ and define the function $\varphi : \mathbb{R} \to Y$ by $\varphi(t) = 0$ for $t \in (-\infty, 0]$, $\varphi(\lambda^k) = w_k$, $k \in \mathbb{N}_0$, and affine on each interval $[\lambda^{n+1}, \lambda^n]$. Then $\varphi(t) = \varphi_n(t)$ for $t \in [\lambda^{n+1}, \lambda^n]$, where

$$\varphi_n(t) = \frac{\lambda^n w_{n+1} - \lambda^{n+1} w_n}{\lambda^n - \lambda^{n+1}} + \mu_n t, \quad \text{with } \mu_n = \frac{w_n - w_{n+1}}{\lambda^n - \lambda^{n+1}}.$$

Put also $\varphi(t) = \varphi_0(t)$ for $t > 1$. One shows that $\mu_{n+1} \leq \mu_n$ and that the so defined function φ is C-convex. Since $\|\varphi(\lambda^n)\| = \|w_n\| \to \infty$, it is norm-unbounded on every neighborhood of $0 \in \mathbb{R}$. Since it takes values in $[0, w]_o$, it is o-bounded, and so locally o-bounded on $\mathbb{R}$.

The implication $1 \Rightarrow 4$ follows from Proposition 3.1.14, while the implication $4 \Rightarrow 5$ is evident.

$5 \Rightarrow 1$. We proceed again by contradiction. Suppose that the cone C is not normal and show that there exists a continuous convex function $f : X \to Y$ which is locally upper o-bounded on some neighborhood of 0 and norm-unbounded on every neighborhood of 0.

Let $[0, w]_o$ be a norm-unbounded interval in Y and $\varphi : \mathbb{R} \to Y$ the convex function given in the proof of the implication $3 \Rightarrow 1$. For a fixed element $v \in X \setminus \{0\}$ there exists a continuous linear functional $x^* \in X^*$ such that $x^*(v) = 1$. Define the function $f : X \to Y$ by $f(x) = \varphi(x^*(x))$, $x \in X$. Then f is convex, continuous and

$$\|f(\lambda^n v)\| = \|\varphi(\lambda^n\| = \|w_n\| \to \infty \text{ as } n \to \infty.$$

The function f is order-bounded on every neighborhood V_ε of $0 \in X$ of the form $V_\varepsilon = \{x \in X : |x^*(x)| < \varepsilon\}$, $\varepsilon > 0$. $\qquad \square$

3.1.6   Lipschitz Properties of Convex Vector-Functions

In this section we shall prove some results on Lipschitz properties for convex vector-functions, meaning functions which are convex with respect to a cone. The order-Lipschitz property as defined by Papageorgiou [539, 540] will be considered as well.

3.1.7   Convex Functions on Locally Convex Spaces

We define first Lipschitz functions between locally convex spaces.

Definition 3.1.18 Let (X, P) and (Y, Q) be locally convex spaces, where P, Q are directed families of seminorms generating their topologies, and $A \subseteq X$. A function $f : A \to Y$ is said to satisfy the *Lipschitz condition* (or that f is a *Lipschitz function*) if for each $q \in Q$ there exist $p \in P$ and $L = L_q \geq 0$ such that

$$q(f(x) - f(y)) \leq Lp(x - y),$$

for all $x, y \in A$.

The function f is called *locally Lipschitz* on A if every point $x \in A$ has a neighborhood V such that f is Lipschitz on $V \cap A$.

Remark 3.1.19 It is easy to check that the definition does not depend on the (directed) families of seminorms P, Q generating the locally convex topologies on X and Y, respectively.

Remark 3.1.20 If X and Y are Banach spaces then the above definition coincides with the standard definition (with respect to the metrics generated by the norms).

If $Y = \mathbb{K}$, then $f : A \to \mathbb{K}$ is *Lipschitz* if there exist $p \in P$ and $L > 0$ such that

$$|f(x) - f(y)| \leq Lp(x - y),$$

for all $x, y \in A$.

The next theorem shows that continuous convex vector-functions defined on open convex subsets of locally convex spaces are locally Lipschitz. For a seminorm p on a vector space X we use the notations

$$B_p = \{x \in X : p(x) \leq 1\} \quad \text{and} \quad B'_p = \{x \in X : p(x) < 1\}.$$

Arbitrary balls satisfy the equalities

$$B_p[x_0, r] := \{x \in X : p(x - x_0) \leq r\} = x_0 + rB_p, \quad \text{and}$$
$$B_p(x_0, r) := \{x \in X : p(x - x_0) < r\} = x_0 + rB'_p,$$

for $x_0 \in X$ and $r > 0$.

Theorem 3.1.21 *Let (X, P), (Y, Q) be locally convex spaces, C a normal cone in Y and Ω an open convex subset of X.*

If $f : \Omega \to Y$ is a continuous convex mapping, then f is locally Lipschitz on Ω.

Furthermore, f is Lipschitz on every compact subset of Ω.

We start with the following proposition, the key tool in the proof of the theorem.

Proposition 3.1.22 *Let X be a vector space, $x_0 \in X$, p a seminorm on X, Y a vector space ordered by a cone C and let q be the Minkowski functional of an absolutely convex C-full absorbing subset W of Y.*

For $R > 0$ let $V = B_p[x_0, R]$ and let $f : V \to Y$ be a C-convex function.

If, for some $\beta > 0$, $q(f(x)) \le \beta$ for all $x \in V$, then for every $0 < r < R$,

$$q(f(x) - f(y)) \le \frac{2\beta}{R - r}\, p(x - y),$$

for all $x, y \in B_p[x_0, r]$.

We need the following simple result.

Lemma 3.1.23 ([113], Proposition 2.5.6) *Let Y be a vector space ordered by a cone C. If W is a C-full absolutely convex absorbing subset of Y then the Minkowski functional q of W is a seminorm, satisfying the condition*

$$q(y) \le \max\{q(x), q(z)\},$$

for all $x, y, z \in Y$ with $x \le y \le z$.

Proof Let $a := \max\{q(x), q(z)\}$. Then, for every $\varepsilon > 0$, $q(x), q(z) < a + \varepsilon$, so, by the definition of the Minkowski functional, there exist $b, c \in (0, a + \varepsilon)$ such that $x \in bW$ and $z \in cW$. Since W is balanced,

$$bW = (a + \varepsilon)\, \frac{b}{a + \varepsilon}\, W \subseteq (a + \varepsilon)W,$$

and

$$cW = (a + \varepsilon)\, \frac{c}{a + \varepsilon}\, W \subseteq (a + \varepsilon)W,$$

implying $(a + \varepsilon)^{-1}x$, $(a + \varepsilon)^{-1}z \in W$. Since W is C-full and

$$(a + \varepsilon)^{-1}x \le (a + \varepsilon)^{-1}y \le (a + \varepsilon)^{-1}z,$$

it follows that $(a+\varepsilon)^{-1}y \in W$ or, equivalently, $y \in (a+\varepsilon)W$. But then $q(y) \le a+\varepsilon$. Since $\varepsilon > 0$ was arbitrarily chosen, this implies

$$q(y) \le a = \max\{q(x), q(z)\}.$$

$\square$

Proof of Proposition 3.1.22 Let $x, y \in B_p[x_0, r]$, $x \ne y$.

Case I. $p(x - y) = 0$.
  We show that in this case $q(f(x) - f(y)) = 0$.
  Observe first that the line $D(x, y) := x + \mathbb{R}(y - x)$ is contained in $B_p[x_0, r]$.
  Indeed, for $z_t = x + t(y - x)$,

$$p(z_t - x_0) \le p(x - x_0) + |t| p(y - x) \le r,$$

for all $t \in \mathbb{R}$, proving that $D(x, y) \subseteq B_p[x_0, r]$.
  For $t > 1$ let $z_t = y + t(x - y)$ and $z'_t = x + t(y - x)$. Then $x = (1 - t^{-1})y + t^{-1}z_t$ and $y = (1 - t^{-1})x + t^{-1}z'_t$, so that, by the convexity of f,

$$f(x) \le (1 - t^{-1})f(y) + t^{-1}f(z_t)$$

implying

$$f(x) - f(y) \le t^{-1}(f(z_t) - f(y)).$$

Interchanging the roles of x and y one obtains

$$f(y) - f(x) \le t^{-1}(f(z'_t) - f(x)) \iff f(x) - f(y) \ge t^{-1}(f(x) - f(z'_t)).$$

But then, by Lemma 3.1.23,

$$q(f(x) - f(y)) \le \max\{t^{-1}q(f(z_t) - f(y)), t^{-1}q(f(x) - f(z'_t))\} \le \frac{2\beta}{t}.$$

Letting $t \to \infty$, one obtains $q(f(x) - f(y)) = 0$.
Case II. $p(x - y) > 0$.
  Putting $u := x + \frac{R-r}{p(x-y)}(x - y)$ and $v := y + \frac{R-r}{p(y-x)}(y - x)$, one obtains

$$p(u - x_0) \le p(x - x_0) + R - r \le R \quad \text{and} \quad p(v - x_0) \le p(x - x_0) + R - r \le R,$$

that is, $u, v \in B_p[x_0, R]$ and so $q(f(u)) \le \beta$, $q(f(v)) \le \beta$.
  Also,

$$p(u - x) = R - r = p(v - y).$$

Appealing to (3.1.3), it follows that

$$\frac{f(x) - f(u)}{R - r} = \frac{f(x) - f(u)}{p(x - u)} \leq \frac{f(y) - f(x)}{p(y - x)} \quad \text{and}$$

$$\frac{f(y) - f(x)}{p(y - x)} \leq \frac{f(v) - f(y)}{p(v - y)} = \frac{f(v) - f(y)}{R - r} .$$

By hypothesis, $q((f(x) - f(u))/(R - r) \leq 2\beta/(R - r)$ and $q((f(v) - f(y))/(R - r) \leq 2\beta/(R - r)$, so that, by Lemma 3.1.23,

$$q\left(\frac{f(y) - f(x)}{p(y - x)}\right) \leq \frac{2\beta}{R - r} \iff q(f(y) - f(x)) \leq \frac{2\beta}{R - r} p(y - x) .$$

$\square$

Remark 3.1.24 If $Y = \mathbb{R}$ the case $p(x - y) = 0$ can be treated appealing to Proposition 3.1.6. Indeed, as we have seen, in this case $D(x, y) \subseteq B_p[x_0, r]$, so we can consider the convex function $\varphi : \mathbb{R} \to \mathbb{R}$, $\varphi(t) = f(x + t(y - x))$, $t \in \mathbb{R}$. By hypothesis the function φ is bounded, so that by Proposition 3.1.6.2 it is constant. But then $f(x) = \varphi(0) = \varphi(1) = f(y)$.

Proof of Theorem 3.1.21 Suppose that P is directed and that the seminorms in Q are the Minkowski functionals of the members of a neighborhood basis of $0 \in Y$ formed of absolutely convex C-full sets (see Theorem 3.1.1).

Let $x_0 \in \Omega$ and $q \in Q$. The continuity of f at x_0 implies the existence of a seminorm $p \in P$ and of $R > 0$ such that $V := x_0 + R B_p \subseteq \Omega$ and

$$q(f(x)) \leq 1,$$

for all $x \in V$.

If $0 < r < R$ then, by Proposition 3.1.22,

$$q(f(x) - f(y)) \leq \frac{2}{R - r} p(x - y),$$

for all $x, y \in x_0 + r B_p$.

Let us show now that f is Lipschitz on every compact subset K of Ω. Let $q \in Q$ be the Minkowski functional of a C-full absolutely convex neighborhood of $0 \in Y$. By the first part of the proof, for every $x \in K$ there are $p_x \in P$, $L_x > 0$ and $r_x > 0$ such that $U_x := x + r_x B'_{p_x} \subseteq \Omega$ and

$$q(f(u) - f(v)) \leq L_x p_x(u - v),$$

for all $u, v \in U_x$.

The compactness of K implies the existence of a finite set $\{x_1, \ldots, x_n\} \subseteq K$ such that

$$K \subseteq \bigcup_{i=1}^{n} U_i,$$

where $U_i = U_{x_i}$. Put $p_i = p_{x_i}$, $r_i = r_{x_i}$, $L_i = L_{x_i}$, and let $p \in P$, $p \geq p_i$, $i = 1, \ldots, n$ and $L = \max\{L_1, \ldots, L_n\}$. We show that

$$q(f(x) - f(y)) \leq Lp(x - y)$$

for all $x, y \in K$.

Let x, y be distinct points in K. Suppose first that $p(x - y) > 0$. If $i, j \in \{1, \ldots, n\}$ are such that $x \in U_i$ and $y \in U_j$ then, since these sets are open, there exist $a < 0$ and $b > 1$ such that

$$u := x + a(y - x) \in U_i \quad \text{and} \quad v := x + b(y - x) \in U_j.$$

Then

$$q(f(x) - f(u)) \leq L_i p_i(x - u) \leq Lp(x - u) \quad \text{and}$$
$$q(f(v) - f(y)) \leq L_j p_j(v - y) \leq Lp(v - y).$$

Now, by (3.1.3),

$$\frac{f(x) - f(u)}{p(x - u)} \leq \frac{f(y) - f(x)}{p(y - x)} \leq \frac{f(v) - f(y)}{p(v - y)},$$

so that, by Lemma 3.1.23,

$$\frac{q(f(y) - f(x))}{p(y - x)} \leq \max\left\{ \frac{q(f(x) - f(u))}{p(x - u)}, \frac{q(f(v) - f(y))}{p(v - y)} \right\} \leq L.$$

If $p(x - y) = 0$, then

$$p(y - x_i) \leq p(y - x) + p(x - x_i) < r_i$$

implying $x, y \in U_i$ and

$$q(f(x) - f(y)) \leq L_i p_i(x - y) \leq Lp(x - y).$$

$\square$

Taking into account Proposition 3.1.11 and Theorem 3.1.21, one obtains the following consequence.

Corollary 3.1.25 *Let $f : \Omega \subseteq \mathbb{R}^n \to \mathbb{R}$ be a convex function, where the set Ω is open and convex. Then f is locally Lipschitz on Ω and Lipschitz on every compact subset of Ω.*

3.1.8 The Order-Lipschitz Property

Papageorgiou [539] considered a notion of Lipschitzness for convex vector functions related to the order. Let X be a normed space and Y a normed lattice, $\Omega \subseteq X$ and $f : \Omega \to Y$. One says that f is *o*-Lipschitz on a subset Z of Ω if there exists $y \geq 0$ in Y such that

$$|f(z) - f(z')| \leq y\|z - z'\|, \qquad (3.1.7)$$

for all $z, z' \in Z$.

Notice that an *o*-Lipschitz function is Lipschitz. Indeed, from (3.1.7),

$$\|f(z) - f(z')\| \leq \|y\|\|z - z'\|,$$

for all $z, z' \in Z$, because in a normed lattice $|x| \leq |x'|$ implies $\|x\| \leq \|x'\|$.

Theorem 3.1.26 *Let X be a normed space, Y a normed lattice, $\Omega \subseteq X$ open and convex and $f : \Omega \to Y$ a function, convex with respect to the order of Y. If f is upper o-bounded on a neighborhood of a point $x_0 \in \Omega$, then f is locally o-Lipschitz on Ω.*

The proof will follow from an analog of Proposition 3.1.22.

Lemma 3.1.27 *Under the hypotheses of Theorem 3.1.26, if $R > 0$ is such that $V = B[x_0, R] \subseteq \Omega$ and, for some $z \geq 0$ in Y,*

$$|f(x)| \leq z, \qquad (3.1.8)$$

for all $x \in V$, then for every $0 < r < R$

$$|f(x) - f(y)| \leq \frac{2z}{R - r}\|x - y\|, \qquad (3.1.9)$$

for all $x, y \in U := B[x_0, r]$.

Proof The proof is similar to that of Proposition 3.1.22, so we only sketch it.

Let $x \neq y$ in U. Since $\|x - y\| > 0$ we have to consider only Case 2 of the corresponding proof. As there, let

$$u := x + \frac{R - r}{\|x - y\|}(x - y) \quad \text{and} \quad v := y + \frac{R - r}{\|y - x\|}(y - x).$$

Then

$$\|u - x\| = R - r = \|v - y\|,$$

and

$$\|u - x_0\| \le \|x - x_0\| + R - r \le R,$$
$$\|v - x_0\| \le \|y - x_0\| + R - r \le R.$$

Appealing to (3.1.3), it follows that

$$\frac{f(x) - f(u)}{R - r} = \frac{f(x) - f(u)}{\|x - u\|} \le \frac{f(y) - f(x)}{\|y - x\|}$$

$$\frac{f(y) - f(x)}{\|y - x\|} \le \frac{f(v) - f(y)}{\|v - y\|} = \frac{f(v) - f(y)}{R - r}.$$

(3.1.10)

so that

$$\frac{|f(y) - f(x)|}{\|y - x\|} \le \frac{2z}{R - r} \iff |f(y) - f(x)| \le \frac{2z}{R - r}\|y - x\|.$$

$\square$

Proof of Theorem 3.1.26 By Proposition 3.1.14 the function f is locally o-bounded on Ω. Therefore, for any $x \in \Omega$ there exist $R > 0$ and $y \ge 0$ such that (3.1.8) holds. By Lemma 3.1.27 the function f satisfies (3.1.9), that is, it is o-Lipschitz on $B[x, r]$, for every $r \in (0, R)$. $\square$

Remark 3.1.28 We have used some properties of the order relation in a vector lattice (see Sect. 1.1.2). For instance, at the end of the proof of Lemma 3.1.27, we applied to the inequalities (3.1.10) the property

$$u \le v \le w \implies |v| \le |u| \vee |w|,$$

(see the proof following the relations (1.1.2)).

3.1.9 C-Bounded Functions

In this subsection (X, P) and (Y, Q) will be locally convex spaces, where P, Q are directed families of seminorms generating the corresponding locally convex topologies.

Suppose that (Y, Q) is ordered by a cone C. A subset Z of Y is called:

- *C-bounded from above* if for every $q \in Q$ there exists $\beta > 0$ such that $Z \subseteq \beta B_q - C$;
- *C-bounded from below* if for every $q \in Q$ there exists $\beta > 0$ such that $Z \subseteq \beta B_q + C$;
- *C-bounded* if for every $q \in Q$ there exists $\beta > 0$ such that

$$Z \subseteq [\beta B_q] = (\beta B_q + C) \cap (\beta B_q - C),$$

i.e., it is C-bounded both from below and from above.

These notions and the results from below are taken from [655], where they have been proved with different methods.

Remark 3.1.29

1. An order-bounded subset of Y is C-bounded.
2. If the cone C is normal, then a subset Z of Y is C-bounded if and only if it is topologically bounded.

Indeed if $Z \subseteq [y, y']_o$, then, for every $q \in Q$, $y, y' \in \beta B_q$ for some $\beta > 0$, so that

$$Z \subseteq [y, y']_o \subseteq [\beta B_q].$$

If the cone C is normal, then we can take Q to be the Minkowski functionals of a neighborhood basis at 0 formed of absolutely convex C-full sets, so that, for every $q \in Q$, $[B_q] = B_q$ and

$$Z \subseteq \beta[B_q] \iff Z \subseteq \beta B_q.$$

Let Ω be a subset of X. A function $f : \Omega \to Y$ is called:

- *C-bounded from above* on a set $Z \subseteq \Omega$ if $f(Z)$ is C-bounded from above in Y;
- *C-bounded from above around* $x \in \Omega$ if there exists a neighborhood U of x such that f is C-bounded from above on $U \cap \Omega$;
- *simply C-bounded from above around* $x \in \Omega$ if for every $q \in Q$ there exists a neighborhood U of x and $\beta > 0$ such that $f(U) \subseteq \beta B_q - C$;
- *locally C-bounded from above* on Ω if it is C-bounded from above around every point $x \in \Omega$;
- *locally simply C-bounded from above* on Ω if it is simply C-bounded from above around every point $x \in \Omega$.

Similar definitions can be given with respect to the notions "C-bounded from below" and "C-bounded".

Lemma 3.1.30 *Let X be a vector space, $\Omega \subseteq X$ a convex set, $x_0 \in X$ and p a seminorm on X. Let also Y be a vector space ordered by a cone C and let q be the Minkowski functional of an absolutely convex C-full absorbing subset of Y. Finally let $f : \Omega \to Y$ be a C-convex function such that, for some $R, \beta > 0$, $U := x_0 + RB_p \subseteq \Omega$ and*

$$f(U) \subseteq \beta B_q - C .$$

Then

(i) $f(U) \subseteq (2\alpha + \beta)B_q + C,$ *and so*

(ii) $f(U) \subseteq (2\alpha + \beta)[B_q] = (2\alpha + \beta)B_q ,$

(3.1.11)

where $\alpha = q(f(x_0))$.
  Also, for every $0 < r < R$,

$$q(f(x) - f(y)) \le \frac{2(2\alpha + \beta)}{R - r} p(x - y) , \qquad (3.1.12)$$

for all $x, y \in x_0 + rB_p$.

Proof By hypothesis, for every $u \in RB_p$ there exist $z, z' \in B_q$ and $c, c' \in C$ such that

$$f(x_0 + u) = \beta z - c \quad \text{and} \quad f(x_0 - u) = \beta z' - c' .$$

If $\alpha = q(f(x_0))$, then $f(x_0) \in \alpha B_q$ and, by the convexity of f,

$$f(x_0 + u) \ge 2f(x_0) - f(x_0 - u) = 2f(x_0) - \beta z' + c' ,$$

implying $f(x_0+u)-2f(x_0)+\beta z' \ge c' \ge 0$, so that $f(x_0+u)-2f(x_0)+\beta z' = d$, for some $d \in C$, that is,

$$f(x_0 + u) = 2f(x_0) + \beta(-z') + d \in 2\alpha B_q + \beta B_q + C$$
$$= (2\alpha + \beta)B_q + C .$$

Consequently, $f(x_0 + RB_p) \subseteq (2\alpha + \beta)B_q + C$, and so

$$f(x_0 + RB_p) \subseteq (\beta B_q - C) \cap ((2\alpha + \beta)B_q + C)$$
$$\subseteq ((2\alpha + \beta)B_q - C) \cap ((2\alpha + \beta)B_q + C)$$
$$= (2\alpha + \beta)[B_q] = (2\alpha + \beta)B_q .$$

The validity of (3.1.12) follows by the inclusion (ii) in (3.1.11) and Proposition 3.1.22.   □

Proposition 3.1.31 *Let (X, P), (Y, Q) be LCS, with Y ordered by a cone C, Ω an open convex subset of X and $f : \Omega \to Y$ a C-convex function.*

1. *If f is C-bounded from above (from below, C-bounded) on a neighborhood $U \subseteq \Omega$ of a point $x_0 \in \Omega$, then it is locally C-bounded from above (from below, C-bounded) on Ω.*
   The same is true for the "simply" version of C-boundedness.
2. *If f is C-bounded from above on a neighborhood $U \subseteq \Omega$ of a point $x_0 \in \Omega$, then it is also C-bounded from below on U, that is, it is C-bounded on U.*
3. *If the cone C is normal and f is simply C-bounded from above around $x_0 \in \Omega$, then it is continuous at x_0, and so on Ω.*
4. *If the cone C is normal and f is C-bounded from above on a neighborhood $U \subseteq \Omega$ of a point $x_0 \in \Omega$, then it is locally Lipschitz on Ω.*
5. *If the cone C is normal and f is locally simply C-bounded from above on Ω, then it is Lipschitz on every compact subset of Ω.*

Proof

1. Let $U := x_0 + rB_p \subseteq \Omega$, $q \in Q$ and $\beta > 0$ be such that $f(U) \subseteq \beta B_q - C$. For $x \neq x_0 \in \Omega$ let $s > 0$ be such that $x_1 = x + s(x - x_0) \in \Omega$. Then $x = tx_1 + (1 - t)x_0$, with $t = 1/(s + 1)$, and $V := tx_1 + (1 - t)U$ is a neighborhood of x contained in Ω. Let $\alpha = tq(f(x_1))$. Then, for any $u \in U$,

$$f(tx_1 + (1 - t)u) \leq tf(x_1) + (1 - t)f(u),$$

hence, there exists $c \in C$ such that

$$f(tx_1 + (1 - t)u) = tf(x_1) + (1 - t)f(u) - c$$
$$\in \alpha B_q + (1 - t)(\beta B_q - C) - c$$
$$= (\alpha + (1 - t)\beta)B_q - C,$$

showing that $f(V) \subseteq (\alpha + (1 - t)\beta)B_q - C$.

The assertions concerning other kinds of boundedness can be proved by a similar argument.

2. This follows from Lemma 3.1.30.
3. Suppose that the seminorms in Q are the Minkowski functionals of a neighborhood basis at $0 \in Y$ formed of absolutely convex C-full sets.

   Let $q \in Q$ and $0 < \varepsilon < 1$. By hypothesis and Lemma 3.1.30 there exist $p \in P, r > 0$ and $\beta > 0$ such that $x_0 + rB_p \subseteq \Omega$ and

$$f(x_0 + rB_p) \subseteq \beta B_q . \tag{3.1.13}$$

   For $u \in rB_p$, $\varepsilon u \in rB_p$ and, by the convexity of f,

$$f(x_0 + \varepsilon u) \leq (1 - \varepsilon)f(x_0) + \varepsilon f(x_0 + u),$$

that is,

$$f(x_0 + \varepsilon u) - f(x_0) \leq \varepsilon (f(x_0 + u) - f(x_0)).$$

Similarly

$$f(x_0 - \varepsilon u) - f(x_0) \leq \varepsilon (f(x_0 - u) - f(x_0)).$$

Appealing again to the convexity of f,

$$f(x_0 + \varepsilon u) - f(x_0) \geq f(x_0) - f(x_0 - \varepsilon u) \geq \varepsilon (f(x_0) - f(x_0 - u)).$$

Consequently,

$$\varepsilon (f(x_0) - f(x_0 - u)) \leq f(x_0 + \varepsilon u) - f(x_0) \leq \varepsilon (f(x_0 + u) - f(x_0)),$$

so that, by Lemma 3.1.23 and (3.1.13),

$$q(f(x_0 + \varepsilon u) - f(x_0)) \leq \varepsilon \max\{q(f(x_0 + u) - f(x_0)), q(f(x_0) - f(x_0 - u))\}$$
$$\leq \varepsilon (\beta + q(f(x_0))),$$

for all $u \in r B_p$, proving the continuity of f at x_0.
4. The fact that f is locally Lipschitz on Ω follows from 3 and Theorem 3.1.21.
5. The proof of the Lipschitz property on compact sets from the same theorem can be adapted to yield 5.

$$\square$$

3.1.10   Equi-Lipschitz Properties of Families of Continuous Convex Mappings

Let $(X, P), (Y, Q)$ be real locally convex spaces, where P, Q are directed families of seminorms generating the topologies, Ω an open convex subset of X and F a family of functions from Ω to Y. The family F is called *equi-Lipschitz* on a subset A of Ω if for every $q \in Q$ there are $p = p_q \in P$ and a number $L_q \geq 0$ such that

$$q(f(x) - f(y)) \leq L_q p(x - y)$$

for all $x, y \in A$ and all $f \in F$. The family F is called *locally equi-Lipschitz* on Ω if each point $x \in \Omega$ has a neighborhood $U_x \subseteq \Omega$ such that F is equi-Lipschitz on U_x.

The family F is called *pointwise bounded* on Ω if, for every $q \in Q$,

$$\sup\{q(f(x)) : f \in F\} < \infty$$

holds for each $x \in \Omega$.

A *barrel* in a locally convex space (X, P) is an absorbing absolutely convex and closed subset. The locally convex space X is called *barrelled* if each barrel is a neighborhood of 0 in X.

Notice the following properties:

- any Baire LCS, hence any complete semimetrizable LCS, is a barrelled space;
- there exist barrelled locally convex spaces and barrelled normed spaces that are not Baire, see [550, p. 100] and [618], respectively;
- there exist incomplete normed spaces that are Baire (Libor Veselý), see
  http://users.mat.unimi.it/users/libor/AnConvessa/Baire-incompleto.pdf

The following result was proved in [319]. The proof given here is adapted from [157] and [161].

Theorem 3.1.32 *Let (X, P) be a barrelled locally convex space, (Y, Q) a locally convex space ordered by a normal cone C and Ω an open convex subset of X.*

If $\mathscr{F}$ is a pointwise bounded family of continuous convex functions from Ω to Y then $\mathscr{F}$ is locally equi-Lipschitz on Ω.

Furthermore, the family $\mathscr{F}$ is equi-Lipschitz on every compact subset of Ω.

Proof Suppose that the seminorms in Q are the Minkowski functionals of members of a neighborhood basis $\mathscr{B}$ of $0 \in Y$ formed of absolutely convex C-full sets.

Let $x_0 \in \Omega$, $W \in \mathscr{B}$ and let $q \in Q$ be the Minkowski functional of the set W. We show that there are $p \in P$, $R > 0$ and $\beta > 0$ such that $V := x_0 + R B_p \subseteq \Omega$ and

$$q(f(x)) \le \beta \tag{3.1.14}$$

for all $x \in V$ and all $f \in \mathscr{F}$. Taking into account Proposition 3.1.22, the relation (3.1.14) yields that, for any $0 < r < R$, we have

$$q(f(x) - f(y)) \le \frac{2\beta}{R - r} p(x - y)$$

for all $x, y \in x_0 + r B_p$ and all $f \in \mathscr{F}$.

Let

$$B = \{u \in X : x_0 \pm u \in \Omega \text{ and } f(x_0 \pm u) - f(x_0) \in \frac{1}{2} W - C \text{ for all } f \in \mathscr{F}\}.$$

The notation with $\pm z$ means that the corresponding relation holds for both w and $-w$.

A simple verification shows that B is convex and symmetric, and so absolutely convex. We show that B is also absorbing. To this end for $x \in X$ let $\alpha > 0$ be such that $x_0 \pm \alpha x \in \Omega$ (possible since the set Ω is open). Then $x_0 \pm t\alpha x \in \Omega$, for any $0 < t < 1$ (since Ω is convex) and

$$f(x_0 \pm t\alpha x) = f((1-t)x_0 + t(x_0 \pm \alpha x)) \le (1-t)f(x_0) + tf(x_0 \pm \alpha x)$$

implying

$$f(x_0 \pm t\alpha x) - f(x_0) \le t(f(x_0 \pm \alpha x) - f(x_0)), \qquad (3.1.15)$$

for all $f \in \mathscr{F}$. Since the family $\mathscr{F}$ is pointwise bounded there exists t_0, $0 < t_0 < 1$, such that

$$t_0(f(x_0 \pm \alpha x) - f(x_0)) \in \frac{1}{2}W$$

for all $f \in \mathscr{F}$, so that by (3.1.15),

$$f(x_0 \pm t_0\alpha x) - f(x_0)$$
$$= [f(x_0 \pm t_0\alpha x) - f(x_0) - t_0(f(x_0 \pm \alpha x) - f(x_0))]$$
$$+ t_0(f(x_0 \pm \alpha x) - f(x_0)) \in -C + \frac{1}{2}W,$$

for all $f \in \mathscr{F}$, showing that $t_0\alpha x \in B$. Consequently, the set $\overline{B}$ is a barrel in X and, since X is barrelled, $\overline{B}$ is a neighborhood of $0 \in X$.

Take $R > 0$ and $p \in P$ such that $V := x_0 + RB_p \subseteq x_0 + \overline{B}$. For $f \in \mathscr{F}$ and $u \in RB_p \subseteq \overline{B}$, there exists a net $(u_i)_{i \in I}$ in B converging to u. The relations

$$f(x_0 \pm u_i) - f(x_0) \in 2^{-1}W - C, \quad i \in I,$$

and the continuity of f imply

$$f(x_0 \pm u) - f(x_0) = \lim_i(f(x_0 \pm u_i) - f(x_0)) \in \mathrm{cl}(\frac{1}{2}W - C)$$
$$\subseteq \frac{1}{2}W - C + \frac{1}{2}W = W - C \qquad \text{(by Proposition 1.4.1.5)}.$$

By the convexity of f

$$2f(x_0) \le f(x_0 + u) + f(x_0 - u) \iff f(x_0 + u) - f(x_0) \ge f(x_0) - f(x_0 - u)$$
$$\Rightarrow f(x_0 + u) - f(x_0) \in f(x_0) - f(x_0 - u) + C.$$

But then

$$f(x_0 + u) - f(x_0) \in -W + C + C = W + C.$$

Therefore

$$f(x_0 + u) - f(x_0) \in (W - C) \cap (W + C) = W \subseteq B_q,$$

i.e.,

$$q(f(x) - f(x_0)) \le 1,$$

for all $x \in V$ and $f \in \mathscr{F}$. Hence

$$q(f(x)) \le 1 + q(f(x_0)) \le 1 + \sup\{q(f(x_0)) : f \in \mathscr{F}\} =: \beta,$$

for all $x \in V$ and all $f \in \mathscr{F}$.

The proof of the fact that $\mathscr{F}$ is equi-Lipschitz on every compact subset of Ω proceeds as in the case of one function, taking into account that, by (3.1.14), we can add "for all $f \in \mathscr{F}$" to each of the relations used in the proof of the corresponding assertion of Theorem 3.1.21.   □

3.1.11 Convex Functions on Metrizable TVS

In this section we shall discuss the Lipschitz properties of convex functions on metrizable TVS. A metric d on a vector space X is called *translation invariant* if

$$d(x + z, y + z) = d(x, y),$$

for all $x, y, z \in X$.

As it was shown in [163] continuous convex functions are also locally Lipschitz with respect to some translation invariant metrics.

For $0 < p < 1$ consider the linear space ℓ^p of all sequences $x = (x_k)$ of real numbers such that $\sum_{k=1}^{\infty} |x_k|^p < \infty$. The function

$$d(x, y) = \sum_{k=1}^{\infty} |y_k - x_k|^p$$

is a translation invariant metric on ℓ^p generating a linear topology on ℓ^p. It satisfies the inequality

$$d(x + y, 0) \le d(x, 0) + d(y, 0),$$

for all $x, y \in \ell^p$.

Proposition 3.1.33 *Let Ω be an open convex subset of the space ℓ^p, $0 < p < 1$. If $f : \Omega \to \mathbb{R}$ is continuous and convex, then f is locally Lipschitz on Ω.*

Proof For $x_0 \in \Omega$ there exist $r > 0$ and $a > 0$ such that $|f(x)| \leq a$ for all $x \in U$, where $U := \{x \in \ell^p : d(x_0, x) \leq r\} \subseteq \Omega$ is a neighborhood of x_0. Let $V := \{x \in \ell^p : d(x_0, x) \leq r/4\} \subseteq U$. For $x, y \in V$, $x \neq y$, we have $d(x, y) \leq r/2$ and

$$d\left(\frac{r}{2d(x, y)}(y - x), 0\right) = \left(\frac{r}{2d(x, y)}\right)^p d(y - x, 0)$$

$$= \left(\frac{r}{2d(x, y)}\right)^p d(x, y) = \left(\frac{r}{2}\right)^p (d(x, y))^{1-p} \leq \frac{r}{2}.$$

The element $z := y + r\,(2d(x, y))^{-1}\,(y - x)$ belongs to U because

$$d(z - x_0, 0) \leq d(y - x_0, 0) + d\left(\frac{r}{2d(x, y)}(y - x), 0\right) \leq \frac{r}{4} + \frac{r}{2} < r.$$

It follows that

$$y = \frac{2d(x, y)}{2d(x, y) + r}z + \frac{r}{2d(x, y) + r}x,$$

so that, by the convexity of f,

$$f(y) \leq \frac{2d(x, y)}{2d(x, y) + r}f(z) + \frac{r}{2d(x, y) + r}f(x),$$

implying

$$f(y) - f(x) \leq \frac{2d(x, y)}{2d(x, y) + r}(f(z) - f(x)) \leq \frac{4a}{2d(x, y) + r}d(x, y) \leq \frac{4a}{r}d(x, y).$$

By symmetry

$$f(x) - f(y) \leq \frac{4a}{r}d(x, y),$$

so that

$$|f(y) - f(x)| \leq \frac{4a}{r}d(x, y).$$

Consequently, f is Lipschitz on V with $L = (4a)/r$.   □

Remark 3.1.34 As we have noticed in Remark 1.4.56, the dual of the space ℓ^p, $0 < p < 1$, is the space ℓ^∞ of all bounded sequences. Consequently, for

$0 < p < 1$ every space ℓ^p contains a good supply of nonempty open convex sets and nonconstant continuous convex functions.

In contrast, $(L^p[0,1])^* = \{0\}$ for every $0 < p < 1$, so that $L^p[0,1]$ does not contain nonempty open convex subsets (see [614, §1.47]). Also, the only continuous convex functions on $L^p[0,1]$ are the constant ones.

We shall justify only the affirmation concerning convex functions on $L^p[0,1]$. Let $L^p = L^p[0,1]$, where $0 < p < 1$, and suppose that $\varphi : L^p \to \mathbb{R}$ is a nonconstant continuous convex function. Let $f, g \in L^p$ be such that $\varphi(g) < \varphi(f)$. Then the subdifferential of φ at f is nonempty (see [221, Ch. 1, §5]), so there exists $x^* \in (L^p)^*$ such that

$$x^*(h) - x^*(f) \le \varphi(h) - \varphi(f),$$

for all $h \in L^p$. Since $x^* = 0$, this leads to the contradiction

$$0 \le \varphi(g) - \varphi(f) < 0.$$

A result similar to Proposition 3.1.33 holds in metrizable LCS. Let (X, τ) be a Hausdorff LCS with the topology generated by the countable directed family $(p_n)_{n \in \mathbb{N}}$ of seminorms. It is known that the topology of X is metrizable and

$$d(x, y) = \sum_{n=1}^{\infty} \frac{1}{2^n} \cdot \frac{p_n(x - y)}{1 + p_n(x - y)}, \quad x, y \in X, \tag{3.1.16}$$

is a translation invariant metric on X generating the topology τ.

Proposition 3.1.35 *Let X be a metrizable LCS and Ω an open convex subset of X. If $f : \Omega \to \mathbb{R}$ is a continuous convex function, then f is locally Lipschitz on Ω with respect to the metric (3.1.16).*

Proof Let $x_0 \in \Omega$. By Theorem 3.1.21 there exists a convex neighborhood $U \subseteq \Omega$ of x_0, $m \in \mathbb{N}$ and $L_m > 0$ such that

$$|f(x) - f(y)| \le L_m p_m(x - y),$$

for all $x, y \in U$. Let $r > 0$ be such that $V := \{x \in X : d(x_0, x) \le r\} \subseteq U \cap \{x \in X : p_m(x - x_0) \le 1\}$. Then, for any $x, y \in V$, $p_m(x - y) \le 2$ and

$$|f(x) - f(y)| \le L_m p_m(x - y) = 2^m L_m(1 + p_m(x - y)) \cdot \frac{1}{2^m} \cdot \frac{p_m(x - y)}{1 + p_m(x - y)}$$

$$\le 3 \cdot L_m \cdot 2^m \cdot \sum_{k=1}^{\infty} \frac{1}{2^k} \cdot \frac{p_k(x - y)}{1 + p_k(x - y)} = L \cdot d(x, y),$$

where $L := 3 \cdot L_m \cdot 2^m$. $\qquad\square$

Remark 3.1.36 The fact that the metric d is translation invariant is essential for the validity of Propositions 3.1.33 and 3.1.35.

Indeed, on $X = \mathbb{R}$ the metric $d(x, y) = |x^3 - y^3|$, $x, y \in \mathbb{R}$, generates the usual topology on $\mathbb{R}$. The function $f(x) = x$, $x \in \mathbb{R}$, is continuous and convex on $\mathbb{R}$, but it is not Lipschitz around 0, because

$$|f(x) - f(y)| = \frac{1}{x^2 + xy + y^2} \cdot |x^3 - y^3| \quad \text{for } (x, y) \neq (0, 0),$$

and

$$\lim_{(x,y) \to (0,0)} \frac{1}{x^2 + xy + y^2} = \infty.$$

3.2   Transforming Continuous Functions into Lipschitz Functions

The next result shows that each continuous function from a metric space into another metric space becomes Lipschitz by replacing the metric on the first space with an equivalent one.

Theorem 3.2.1 *Let $f : X_1 \to X_2$ be a continuous function, where (X_1, d_1) and (X_2, d_2) are metric spaces. Then there exists a metric d_1^* on X_1, topologically equivalent to d_1 such that f is Lipschitz with respect to d_1^* and d_2.*

Proof Let us consider the function $d_1^* : X_1 \times X_1 \to [0, \infty)$ given by

$$d_1^*(x, y) = d_1(x, y) + d_2(f(x), f(y))$$

for all $x, y \in X_1$.

It is easy to check that d_1^* is a metric on X_1. The metric d_1^* will be topologically equivalent to d_1 if and only if

$$x_n \xrightarrow{d_1} x \iff x_n \xrightarrow{d_1^*} x,$$

for every sequence (x_n) in X and every $x \in X$.

The obvious inequality $d_1 \leq d_1^*$ yields

$$x_n \xrightarrow{d_1^*} x \implies x_n \xrightarrow{d_1} x.$$

If $x_n \xrightarrow{d_1} x$, then $d_1(x_n, x) \to 0$ and, by the continuity of f, $f(x_n) \xrightarrow{d_2} f(x)$, which is equivalent to $d_2(f(x_n), f(x)) \to 0$. Consequently,

$$d_1^*(x_n, x) = d_1(x_n, x) + d_2(f(x_n), f(x)) \to 0,$$

as $n \to \infty$.

The inequality

$$d_2(f(x), f(y)) \leq d_1(x, y) + d_2(f(x), f(y)) = d_1^*(x, y),$$

valid for all $x, y \in X$, shows that f is 1-Lipschitz with respect to the metrics d_1^* on X_1 and d_2 on X_2. □

Remark 3.2.2 The following example shows that, under the hypotheses of Theorem 3.2.1, it is possible that no metric d_2^* exists on X_2, topologically equivalent to d_2 and such that f is Lipschitz with respect to d_1 and d_2^*.

Let

$$X_1 = [0, 2] \setminus \{1\} \quad \text{and} \quad X_2 = \{u, v\},$$

where $u \neq v$, X_1 endowed with the usual distance in $\mathbb{R}$ denoted by d_1, and X_2 endowed with the distance d_2 given by $d_2(u, v) = 1$, $d_2(u, u) = 0$, $d_2(v, v) = 0$.

The function $f : X_1 \to X_2$ given by

$$f(x) = \begin{cases} u, & \text{if } x \in [0, 1) \\ v, & \text{if } x \in (1, 2] \end{cases}$$

is obviously continuous.

Let d be an arbitrary metric on X_2. Since X_2 is finite, d is topologically equivalent to d_2. Suppose that for some $L > 0$, $d(f(x), f(y)) \leq Ld_1(x, y)$ for all $x, y \in X_1$.

Taking $0 < x < 1 < y < 2$ it follows that $d(u, v) = d(f(x), f(y)) \leq Ld(x, y)$. Letting $x \nearrow 1$ and $y \searrow 1$ one obtains the contradiction $0 < d(u, v) \leq 0$.

Remark 3.2.3 Theorem 3.2.1 is true if we consider not only one function, but a finite number of functions $f_1, f_2, \ldots, f_n$, by defining

$$d_1^*(x, y) = d_1(x, y) + d_2(f_1(x), f_1(y)) + \cdots + d_2(f_n(x), f_n(y)),$$

for all $x, y \in X$.

In fact, we can prove a stronger result, namely that a sequence of continuous functions from a metric space (X_1, d_1) to another metric space (X_2, d_2) can be turned into a sequence of Lipschitz functions if both metrics d_1 and d_2 are replaced by some equivalent ones.

Theorem 3.2.4 *Let* $f_n : X_1 \rightarrow X_2, n \in \mathbb{N}$, *be continuous functions, where* (X_1, d_1) *and* (X_2, d_2) *are metric spaces. Then there exist a metric* d_1^* *on* X_1, *topologically equivalent to* d_1, *and a metric* d_2^* *on* X_2, *topologically equivalent to* d_2, *such that each function* f_n *is Lipschitz with respect to* d_1^* *and* d_2^*.

Proof Let us consider a metric d_2^* on X_2, equivalent to d_2 and such that $d_2^* \leq 1$ (see Proposition 1.3.27) and let $d_1^* : X_1 \times X_1 \rightarrow [0, \infty)$ be given by

$$d_1^*(x, y) = d_1(x, y) + \sum_{n=1}^{\infty} \frac{d_2^*(f_n(x), f_n(y))}{2^n}$$

for all $x, y \in X_1$.

It is easy to check that d_1^* is well-defined and that it is a metric on X_1.
The obvious inequality $d_1 \leq d_1^*$ implies

$$x_n \xrightarrow{d_1^*} x \implies x_n \xrightarrow{d_1} x.$$

Let now (x_k) be a sequence in X_1 such that $d_1(x_k, x) \rightarrow 0$. By the continuity of f_n, $\lim_{k\to\infty} d_2^*(f_n(x_k), f_n(x)) = 0$ for every $n \in \mathbb{N}$.
Then, given $\varepsilon > 0$, let $n_0 \in \mathbb{N}$ be such that

$$\sum_{n > n_0} \frac{1}{2^n} < \frac{\varepsilon}{2}.$$

Since

$$\lim_{k\to\infty} \left(d_1(x_k, x) + \sum_{1 \leq n \leq n_0} \frac{d_2^*(f_n(x_k), f_n(x))}{2^n} \right) = 0,$$

there exists k_0 such that

$$d_1(x_k, x) + \sum_{1 \leq n \leq n_0} \frac{d_2^*(f_n(x_k), f_n(x))}{2^n} < \frac{\varepsilon}{2},$$

for all $k > k_0$.

It follows that $d_1^*(x_k, x) < \varepsilon$ for all $k > k_0$, proving that $x_k \xrightarrow{d_1^*} x$ as $k \to \infty$. $\square$

Actually, a more general result is valid.

Theorem 3.2.5 *Let* $(X_n, d_n)_{n\in\mathbb{N}}$ *be a sequence of metric spaces and*

$$f_{m,n,k} : (X_m, d_m) \rightarrow (X_n, d_n), \quad m, n, k \in \mathbb{N},$$

*a family of continuous functions. Then, for each $i \in \mathbb{N}$, there exists a metric d_i^*
on X_i, topologically equivalent to d_i, such that $f_{m,n,k} : (X_m, d_m^*) \to (X_n, d_n^*)$ is
Lipschitz for all $m, n, k \in \mathbb{N}$.*

A natural question is whether these results can be extended to the uncountable
case. More precisely, given two metric spaces (X_1, d_1) and (X_2, d_2), are there two
metrics d_1^* and d_2^*, topologically equivalent to d_1 and d_2, respectively, such that each
continuous function $f : (X_1, d_1) \to (X_2, d_2)$ is (d_1^*, d_2^*)-Lipschitz?

The example below shows that, in general, the answer is negative.

Example 3.2.6 Let $X_1 = \mathbb{Q}$ and $X_2 = [0, 1]$ endowed with the usual distances in
$\mathbb{R}$, denoted by d_1 and d_2, respectively. Then for every metric d_1^* on X_1, topologically
equivalent to d_1, and every metric d_2^* on X_2, topologically equivalent to d_2, there
exists a continuous function $f : (X_1, d_1^*) \to (X_2, d_2^*)$ which is not Lipschitz.

The space (X_1, d_1) has no isolated points and so, since d_1^* is topologically
equivalent to d_1, the space (X_1, d_1^*) will have the same property. Since X_1 is
countable, the metric space (X_1, d_1^*) is not complete (see Proposition 1.3.15),
implying the existence of a d_1^*-Cauchy sequence $(x_n)_{n \in \mathbb{N}}$ of distinct elements in
X_1 which is not convergent.

Consider the sets

$$A = \{x_1, x_3, \ldots, x_{2n+1}, \ldots\} \quad \text{and} \quad B = \{x_2, x_4, \ldots, x_{2n}, \ldots\}.$$

Then $A' = B' = \emptyset$. Indeed, if $x \in A'$, then there exists a subsequence
$(x_{2k_i-1})_{i \in \mathbb{N}}$ of $(x_{2k-1})_{k \in \mathbb{N}}$ such that $\lim_{i \to \infty} x_{2k_i-1} = x$. Since $(x_{2k_i-1})_{i \in \mathbb{N}}$ is a
convergent subsequence of the d_1^*-Cauchy sequence (x_n), it follows that $x_n \to x$
as $n \to \infty$, in contradiction to the choice of (x_n). The equality $B' = \emptyset$ follows
similarly.

But then $\overline{A} = A \cup A' = A$ and $\overline{B} = B \cup B' = B$. Consequently, A, B are
closed and disjoint subsets of X_1, so that by Urysohn's lemma (Theorem 1.3.13),
there exists a continuous function $f : (X_1, d_1^*) \to (X_2, d_2^*)$ such that $f(A) = \{0\}$
and $f(B) = \{1\}$.

The function $f : (X_1, d_1^*) \to (X_2, d_2^*)$ is not Lipschitz.

Indeed, suppose that there exists $L > 0$ such that

$$d_2^*(f(x), f(y)) \leq L d_1^*(x, y),$$

for all $x, y \in X_1$. In particular, for $x = x_{2n-1}$ and $y = x_{2n}$, we get

$$d_2^*(0, 1) = d_2^*(f(x_{2n-1}), f(x_{2n})) \leq L d_1^*(x_{2n-1}, x_{2n}) \tag{3.2.1}$$

for each $n \in \mathbb{N}$. As the sequence $(x_n)_{n \in \mathbb{N}}$ is d_1^*-Cauchy it follows that

$$\lim_{n \to \infty} d_1^*(x_{2n-1}, x_{2n}) = 0.$$

Letting $n \to \infty$ in (3.2.1) one obtains the contradiction

$$0 < d_2^*(0, 1) \leq 0.$$

Our last result of this section says that for each infinite metric space X there exists a continuous function $f : X \to \mathbb{R}$ which is not Lipschitz.

Theorem 3.2.7 *For a metric space (X, d) the following are equivalent.*

1. *The set X is finite.*
2. *Each function $f : X \to \mathbb{R}$ is Lipschitz.*
3. *Each continuous function $f : X \to \mathbb{R}$ is Lipschitz.*
4. *Each uniformly continuous function $f : X \to \mathbb{R}$ is Lipschitz.*

Proof The implications $1 \Rightarrow 2 \Rightarrow 3 \Rightarrow 4$ are obvious.

We shall prove the implications $4 \Rightarrow 2$ and $2 \Rightarrow 1$.

$4 \Rightarrow 2$. For the very beginning we prove that

$$X' = \emptyset.$$

Suppose that $X' \neq \emptyset$, that is, X has an accumulation point x_0. Consider the function $\varphi : X \to \mathbb{R}$ given by

$$\varphi(x) = \sqrt{d(x, x_0)}, \quad x \in X.$$

Since the function $g(t) = \sqrt{t}$ is uniformly continuous on $[0, \infty)$ and $h(x) = d(x, x_0)$ is 1-Lipschitz, and so uniformly continuous on X, it follows that $\varphi = g \circ h$ is uniformly continuous on X. By hypothesis it is Lipschitz on X, that is,

$$\left|\varphi(x) - \varphi(x')\right| \leq L(\varphi)d(x, x'),$$

for all $x, x' \in X$.

If (x_n) is a sequence in $X \setminus \{x_0\}$ converging to x_0, then

$$\sqrt{d(x_n, x_0)} = |\varphi(x_n) - \varphi(x_0)| \leq L(\varphi)d(x_n, x_0),$$

implying

$$1 \leq L(\varphi)\sqrt{d(x_n, x_0)},$$

for all $n \in \mathbb{N}$. Letting $n \to \infty$, one obtains the contradiction $1 \leq 0$.

Now, we prove that each function $f : X \to \mathbb{R}$ is uniformly continuous, and so, according to the hypothesis, each function $f : X \to \mathbb{R}$ is Lipschitz, completing the proof of the implication $4 \Rightarrow 2$.

Let us suppose that there exists a function $f : X \to \mathbb{R}$ which is not uniformly continuous. We shall arrive at a contradiction by constructing a function $g : X \to \mathbb{R}$ which is uniformly continuous, but not Lipschitz.

As f is not uniformly continuous there exist $\varepsilon_0 > 0$ and two sequences $(x_n)_{n \in \mathbb{N}}$ and $(y_n)_{n \in \mathbb{N}}$ in X such that

$$d(x_n, y_n) < \frac{1}{n}$$

and

$$|f(x_n) - f(y_n)| \geq \varepsilon_0,$$

for each $n \in \mathbb{N}$. Let us note that

$$x_n \neq y_n$$

for each $n \in \mathbb{N}$. Passing if necessary to a subsequence, we also can suppose that all the elements x_n are distinct.

Put $Z := \{x_n : n \in \mathbb{N}\}$ and define the function $g : X \to \mathbb{R}$ by

$$g(x) = \begin{cases} 0, & \text{if } x \notin Z \\ \sqrt{d(x_n, y_n)}, & \text{if } x = x_n \text{ for some } n \in \mathbb{N}. \end{cases}$$

Claim I. *The function g is not Lipschitz.*

Indeed, supposing the contrary, the relations

$$\sqrt{d(x_n, y_n)} = |g(x_n) - g(y_n)| \leq L(g)d(x_n, y_n)$$

yield

$$1 \leq L(g)\sqrt{d(x_n, y_n)} \leq L(g) \cdot \frac{1}{\sqrt{n}},$$

for all $n \in \mathbb{N}$, a contradiction.

Claim II. *The function g is uniformly continuous.*

Indeed, given $\varepsilon > 0$, let $n_0 \in \mathbb{N}$ be such that

$$d(x_n, y_n) < \varepsilon^2,$$

for all $n > n_0$.

Since X has no accumulation points there exists $\gamma > 0$ such that for every $k \in \{1, 2, \ldots, n_0\}$,

$$d(x_k, z) < \gamma \implies z = x_k,$$

for every $z \in Z$.

Put $\delta = \min\{\gamma, \varepsilon^2\}$ and let $x, y \in X$, $x \neq y$, be such that $d(x, y) < \delta$. We have to consider several cases.

(i) If $x, y \in X \setminus Z$, then $|g(x) - g(y)| = 0$.
(ii) Suppose that $x = x_n$ for some $n \in \mathbb{N}$ and $y \notin Z$.
    If $1 \leq n \leq n_0$, then $d(x_n, y) < \delta \leq \gamma$ would imply $y = x_n$, in contradiction to the hypothesis $y \neq x$. Consequently, $n > n_0$ and

$$|g(x_n) - g(y)| = \sqrt{d(x_n, y_n)} < \varepsilon.$$

The case $y \in Z$ and $x \notin Z$ follows by symmetry.
(iii) Suppose now that $x = x_n$ and $y = y_m$ for some $m, n \in \mathbb{N}$.
    If $n \leq n_0$ or $m \leq n_0$, then, as above, the inequalities $d(x_n, y_m) < \delta \leq \gamma$ would imply $y_m = x_n$, in contradiction to the hypothesis $x \neq y$. Consequently, $n > n_0$ and $m > n_0$, so that

$$|g(x_n) - g(y_m)| = |\sqrt{d(x_n, y_n)} - \sqrt{d(x_m, y_m)}| < \varepsilon.$$

It follows that g is uniformly continuous.

$2 \implies 1$. Let us suppose, by contradiction, that there exists a sequence $(x_n)_{n \in \mathbb{N}}$ consisting of distinct elements of X. Then the function $h : X \to \mathbb{R}$ given by

$$h(x) = \begin{cases} 2k \cdot d(x_{2k}, x_{2k+1}), & \text{if } x = x_{2k} \text{ for some } k \in \mathbb{N}, \\ 0, & \text{otherwise}, \end{cases}$$

is not Lipschitz.

Indeed, if h were Lipschitz, then

$$2k d(x_{2k}, x_{2k+1}) = |h(x_{2k}) - h(x_{2k+1})| \leq L(h) d(x_{2k}, x_{2k+1}),$$

would imply

$$2k \leq L(h),$$

for all $k \in \mathbb{N}$, a contradiction. $\square$

3.3 Lipschitz Versus Absolutely Continuous Functions

In this section we discuss some connections between Lipschitz and uniformly continuous functions.

3.3.1 Absolutely Continuous Functions

We first present some properties of Lipschitz functions related to bounded variation and absolute continuity.

Definition 3.3.1 Let $a, b \in \mathbb{R}$, $a < b$, and $f : [a, b] \to \mathbb{R}$. The function f is called:

(i) of *bounded variation* if

$$V_a^b(f) := \sup \Big\{ \sum_{i=1}^{n} |f(t_i) - f(t_{i-1})| : n \in \mathbb{N}, a = t_0 < t_1 < \cdots < t_{n-1} < t_n = b \Big\}$$

is finite; the number $V_a^b(f)$ is called the *total variation* of f on $[a, b]$;

(ii) *absolutely continuous* if for every $\varepsilon > 0$ there exists $\delta_\varepsilon > 0$ such that

$$\sum_{i=0}^{n} |f(t_i') - f(t_i)| < \varepsilon .$$

for any set

$$a \le t_0 < t_0' \le t_1 < t_1' \le \cdots \le t_{n-1} < t_{n-1}' \le t_n < t_n' \le b , \tag{3.3.1}$$

of points in $[a, b]$ satisfying $\sum_{i=0}^{n} |t_i' - t_i| < \delta_\varepsilon$.

By the Lebesgue theorem, a function $f : [a, b] \to \mathbb{R}$ is absolutely continuous if and only if it is differentiable a.e. on $[a, b]$, $f' \in L^1[a, b]$ and $f(x) = f(a) + \int_a^x f'(t)dt$, $x \in [a, b]$.

An absolutely continuous function is of bounded variation and

$$V_a^b(f) = \int_a^b |f'(t)|dt .$$

The relations between these classes of functions are clarified by a classical result of Banach and Zarecki.

Theorem 3.3.2 (Banach–Zarecki) *A necessary and sufficient condition for a function* $f : [a, b] \to \mathbb{R}$ *to be absolutely continuous is to satisfy the following three conditions:*

(i) f *is continuous on* $[a, b]$;
(ii) f *is of bounded variation on* $[a, b]$;
(iii) f *satisfies Lusin's condition (N); that is, the image by f of any set of Lebesgue measure zero is of Lebesgue measure zero.*

For this result and others concerning these classes of functions see, for instance, [120, Theorem 7.11] (or [287]).

We start with a simple remark.

Proposition 3.3.3 *Each Lipschitz function* $f : [a, b] \to \mathbb{R}$ *is absolutely continuous, and so of bounded variation.*

Proof This follows from the inequality

$$\sum_{i=0}^{n} |f(t_i') - f(t_i)| \le L(f) \sum_{i=0}^{n} |t_i' - t_i|,$$

valid for all families (t_i), (t_i') of points in $[a, b]$ as in (3.3.1).   □

Our next goal is to provide a necessary and sufficient condition for a function to be Lipschitz. Let us start by providing an alternate characterization of absolutely continuous functions.

Proposition 3.3.4 *Let* $[a, b]$, $a < b$, *be an interval in* $\mathbb{R}$ *and* $f : [a, b] \to \mathbb{R}$ *a function. Then f is absolutely continuous if and only if for every $\varepsilon > 0$ there exists $\delta_\varepsilon > 0$ such that* $\left| \sum_{k=1}^{n} (f(\beta_k) - f(\alpha_k)) \right| < \varepsilon$ *for any finite family* (α_k, β_k), $k = 1, 2, \ldots, n$, *of pairwise disjoint intervals contained in $[a, b]$ such that* $\sum_{k=1}^{n} (\beta_k - \alpha_k) < \delta_\varepsilon$.

Proof The implication $\Rightarrow$ is obvious, taking into account the inequality

$$\left| \sum_{k=1}^{n} (f(\beta_k) - f(\alpha_k)) \right| \le \sum_{k=1}^{n} |f(\beta_k) - f(\alpha_k)|.$$

The implication $\Leftarrow$. Given $\varepsilon > 0$ let $\delta_\varepsilon > 0$ be such that $\left| \sum_{k=1}^{n} (f(\beta_k) - f(\alpha_k)) \right| < \varepsilon/2$, for each family (α_k, β_k), $k = 1, 2, \ldots, n$, of pairwise disjoint subintervals of $[a, b]$ with $\sum_{k=1}^{n} (\beta_k - \alpha_k) < \delta_\varepsilon$.

If (α_k, β_k), $k = 1, 2, \ldots, n$, is such a family, let

$$M_+ := \{k : 1 \le k \le n, \ f(\beta_k) - f(\alpha_k) \ge 0\} \quad \text{and}$$
$$M_- := \{k : 1 \le k \le n, \ f(\beta_k) - f(\alpha_k) < 0\}.$$

Since

$$\sum_{k \in M_+} (\beta_k - \alpha_k) < \delta_\varepsilon \quad \text{and} \quad \sum_{k \in M_-} (\beta_k - \alpha_k) < \delta_\varepsilon ,$$

it follows that

$$A := \Big| \sum_{k \in M_+} (f(\beta_k) - f(\alpha_k)) \Big| < \frac{\varepsilon}{2} \quad \text{and} \quad B := \Big| \sum_{k \in M_-} (f(\beta_k) - f(\alpha_k)) \Big| < \frac{\varepsilon}{2} ,$$

so that

$$\sum_{k=1}^{n} |f(\beta_k) - f(\alpha_k)| = A + B < \varepsilon .$$

□

The next result provides a necessary and sufficient condition for a function to be Lipschitz. The condition is the same as the condition for absolute continuity, except that the subintervals are allowed to overlap. The result was proved by Fichtenholz [237] (see also [238]) and appears as an exercise in Natanson's book on real analysis [519], as well as in [287, Exercise (18.34)]. The superposition of absolutely continuous functions is also discussed in [615, §8, p. 286].

Theorem 3.3.5 *Let* $[a, b]$, $[c, d]$ *be nondegenerate intervals in* $\mathbb{R}$ *and* $f : [a, b] \rightarrow \mathbb{R}$ *a function. Then the following conditions are equivalent.*

1. *f is Lipschitz on $[a, b]$.*
2. *For every $\varepsilon > 0$ there exists $\delta > 0$ such that $\sum_{k=1}^{n} |f(b_k) - f(a_k)| < \varepsilon$ for any $a \le a_k < b_k \le b$, $k = 1, 2, \ldots, n$, with $\sum_{k=1}^{n} (b_k - a_k) < \delta$.*
3. *For every absolutely continuous function $g : [c, d] \rightarrow [a, b]$ the composition $f \circ g$ is absolutely continuous on $[c, d]$.*
4. *For every Lipschitz function $g : [c, d] \rightarrow [a, b]$ the composition $f \circ g$ is absolutely continuous on $[c, d]$.*

Proof We shall prove here only the equivalence 1 ⟺ 2, as the rest of equivalences will be proved in a more general context (see Theorem 3.3.8).

The proof that a Lipschitz function satisfies condition 2 is the same as the proof of Proposition 3.3.3.

To prove the reverse implication, observe that, by Theorem 2.1.6, the function f is Lipschitz if it is locally Lipschitz on $[a, b]$, so it suffices to prove that f is locally Lipschitz provided that f satisfies the condition 2.

In fact, we can suppose that a weaker condition is satisfied: there exist $\varepsilon > 0$ and $\delta > 0$ such that $\sum_{k=1}^{n} |f(b_k) - f(a_k)| < \varepsilon$ for every family (a_k, b_k), $1 \le k \le n$, of subintervals of $[a, b]$ such that $\sum_{k=1}^{n} (b_k - a_k) < \delta$.

Suppose, by contradiction, that f is not locally Lipschitz on $[a, b]$. Then there exists $x_0 \in [a, b]$ such that f is not Lipschitz on $(x_0 - \frac{\delta}{2}, x_0 + \frac{\delta}{2}) \subseteq [a, b]$. This implies that there exist two points $x, y \in (x_0 - \frac{\delta}{2}, x_0 + \frac{\delta}{2}) \cap [a, b]$ such that

$$|f(y) - f(x)| \geq \frac{2\varepsilon}{\delta} |y - x|. \tag{3.3.2}$$

Without loosing the generality, we can assume that $x < y$.

If $y - x \geq \frac{\delta}{2}$, then, by (3.3.2), we get the following contradiction

$$\varepsilon = \frac{\delta}{2} \cdot \frac{2\varepsilon}{\delta} \leq |y - x| \cdot \frac{2\varepsilon}{\delta} \leq |f(y) - f(x)| < \varepsilon$$

(the last inequality holds because $0 < y - x < \delta$).

Consequently, $y - x < \frac{\delta}{2}$. Let $n \in \mathbb{N}$ be such that

$$(n - 1)(y - x) < \delta \text{ and } n(y - x) \geq \delta. \tag{3.3.3}$$

It follows that $n \geq 3$. Putting $a_k = x$, $b_k = y$, $k = 1, 2, \ldots, n - 1$, the family (a_k, b_k), $1 \leq k \leq n - 1$, of intervals satisfies the condition $\sum_{k=1}^{n-1} (b_k - a_k) = (n - 1)(y - x) < \delta$, so that $(n - 1)|f(y) - f(x)| = \sum_{k=1}^{n-1} |f(b_k) - f(a_k)| < \varepsilon$.

Then, by (3.3.2) and the second inequality in (3.3.3), one obtains the contradiction

$$\varepsilon > (n - 1)|f(y) - f(x)| \geq (n - 1)|y - x|\frac{2\varepsilon}{\delta} = n(y - x)\frac{2(n - 1)\varepsilon}{n\delta}$$

$$\geq \frac{2(n - 1)}{n}\varepsilon > \varepsilon \quad \text{(because } 2(n - 1) > n \text{ for } n \geq 3\text{)}.$$

□

Remark 3.3.6 From Theorem 3.3.5 one can see again that a Lipschitz function $f : [a, b] \subseteq \mathbb{R} \to \mathbb{R}$ is absolutely continuous.

Remark 3.3.7 Since any two intervals $[c, d], [c', d'] \subseteq \mathbb{R}$ are bi-Lipschitz equivalent by the affine function $\varphi : [c, d] \to [c', d']$, $\varphi(t) = ((d' - c')t + dc' - cd')/(d - c)$, $t \in [c, d]$, it follows that a function $F : [c', d'] \to \mathbb{R}$ is absolutely continuous on $[c', d']$ if and only if $F \circ \varphi$ is absolutely continuous on $[c, d]$. A similar result holds for the notion of bounded variation. Consequently, if property 3 in Theorem 3.3.5 holds for an interval $[c, d]$, then it holds for any other interval $[c', d']$. The same is true for property 4 in Theorem 3.3.5, and properties 2 and 3 from Theorem 3.3.8.

Maligranda et al. [427] extended the equivalences $1 \Longleftrightarrow 3 \Longleftrightarrow 4$ from Theorem 3.3.5 by proving the following result.

Theorem 3.3.8 *Let* $[c, d]$ *be a nondegenerate interval in* $\mathbb{R}$, *C a convex subset of a normed space X and $f : C \to \mathbb{R}$ a function. The following conditions are equivalent.*

1. *For every compact set $K \subseteq C$ the restriction $f|_K$ is Lipschitz on K.*
2. *For every absolutely continuous function $g : [c, d] \to C$ the composition $f \circ g$ is absolutely continuous on $[c, d]$.*
3. *For every Lipschitz function $g : [c, d] \to C$ the composition $f \circ g$ is absolutely continuous on $[c, d]$.*

Proof $1 \Rightarrow 2$. Let $g : [c, d] \to C$ be absolutely continuous. Since the set $K = g([c, d])$ is compact as a continuous image of the compact set $[c, d]$, the function $f \circ g$ is absolutely continuous on $[c, d]$ as a composition of Lipschitz and absolutely continuous mappings.

The implication $2 \Rightarrow 3$ is obvious.

$3 \Rightarrow 1$. Suppose that there exists a compact $K \subseteq C$ such that $f|_K$ is not Lipschitz. We show that there exist $b > 0$ and a Lipschitz function $g : [0, b] \to C$ such that $f \circ g$ is not of bounded variation, and so not absolutely continuous. Taking into account Remark 3.3.7, this implies that 3 does not hold.

By the hypotheses made on K and f, there exist $x_n, y_n \in K$, $x_n \neq y_n$, $n \in \mathbb{N}$, such that

$$|f(x_n) - f(y_n)| \geq 2n^3 \|f\|_K d_n \,, \qquad (3.3.4)$$

for all $n \in \mathbb{N}$, where $d_n := \|x_n - y_n\| > 0$, $n \in \mathbb{N}$, and $\|f\|_K = \sup_{x \in K} |f(x)|$.

It follows that $d_n \leq 1/n^3$ for all n. The compactness of K implies the existence of a subsequence (x_{n_k}) of (x_n) convergent to some $x_0 \in K$. Since (x_{n_k}) is Cauchy there exists further a subsequence $(x_{n_{k_i}})$ of (x_{n_k}) with $\|x_{n_{k_i}} - x_{n_{k_{i+1}}}\| \leq 1/2^i$. Consequently, without restricting the generality, we can suppose that the sequence (x_n) satisfies the conditions

$$x_n \to x_0 \in K \quad \text{and} \quad \sum_{n=1}^{\infty} \|x_n - x_{n+1}\| < \infty \,.$$

It follows that $y_n \to x_0$ too.

Put $k_n := \left[\frac{1}{n^2 d_n}\right]$ (the integer part). Then $\frac{1}{n^2} - d_n \leq k_n d_n \leq \frac{1}{n^2}$, that is, $k_n d_n \sim \frac{1}{n^2}$, and so $\sum_{n=1}^{\infty} k_n d_n < \infty$.

Define inductively the sequences (a_n), (b_n) by

$$a_1 = 0 \qquad\qquad b_1 = a_1 + 2k_1 d_1$$

$$a_{n+1} = b_n + \|x_n - x_{n+1}\| \qquad b_{n+1} = a_{n+1} + 2k_{n+1} d_{n+1}, \ n \in \mathbb{N}.$$

An induction argument shows that

$$0 = a_1 < b_1 \leq a_2 < b_2 \leq \ldots,$$

$$a_n = \sum_{i=1}^{n-1} \|x_i - x_{i+1}\| + 2\sum_{i=1}^{n-1} k_i d_i \quad \text{and}$$

$$b_n = \sum_{i=1}^{n-1} \|x_i - x_{i+1}\| + 2\sum_{i=1}^{n} k_i d_i \,,$$

for all $n \geq 2$.

It follows that

$$b_n \to b := \sum_{i=1}^{\infty} \|x_i - x_{i+1}\| + 2\sum_{i=1}^{\infty} k_i d_i < \infty \,,$$

and $a_n \to b$ as well. Also,

$$[0, b] = \bigcup_{n=1}^{\infty} ([a_n, b_n] \cup [b_n, a_{n+1}]) \cup \{b\} \,.$$

Define a function $g : [0, b] \to C$ by

(i) $g(a_n + 2i d_n) = x_n$ for $0 \leq i \leq k_n$ and $g(a_n + (2i - 1)d_n) = y_n$ for $1 \leq i \leq k_n$;
(ii) g is linear on every interval $[a_n + (j - 1)d_n, a_n + j d_n]$, $1 \leq j \leq 2k_n$.

The length of each interval $[a_n + (j - 1)d_n, a_n + j d_n]$ is $d_n = \|x_n - y_n\|$ so that g is 1-Lipschitz on each of these intervals, and so on $[a_n, b_n]$.

Define g to be linear on the segment $[b_n, a_{n+1}]$. Since $a_{n+1} - b_n = \|x_n - x_{n+1}\|$ it follows that g is 1-Lipschitz on this interval too. Putting $g(b) = x_0$ it follows that g is continuous at b, and so one obtains a 1-Lipschitz function $g : I \to C$, where $I = [0, b]$.

Let $h = f \circ g$. Then, taking into account (3.3.4), the variation of $h = f \circ g$ is

$$V_{a_n}^{b_n}(h) \geq \sum_{i=1}^{2k_n} |h(a_n + i d_n) - h(a_n + (i - 1)d_n)|$$

$$= \sum_{i=1}^{2k_n} |f(x_n) - f(y_n)| = 2k_n |f(x_n) - f(y_n)| \geq 4n^3 k_n d_n \|f\|_K \sim n \,,$$

because $k_n d_n \sim \frac{1}{n^2}$. It follows that $V_0^b(h) \geq \sum_{n=1}^{\infty} V_{a_n}^{b_n}(h) = \infty$.

As a function with infinite variation, h is not absolutely continuous.   □

From Theorem 3.3.8 one obtains the following result.

Corollary 3.3.9 *Let $[a, b], [c, d]$ be nondegenerate intervals in $\mathbb{R}$, $Q = [a, b]^2$ and $f : Q \to \mathbb{R}$ a function. The following conditions are equivalent.*

1. *f is Lipschitz on Q.*
2. *For every absolutely continuous functions $g_1, g_2 : [c, d] \to [a, b]$, the function $f(g_1(x), g_2(x))$ is absolutely continuous.*
3. *The composition $f(g_1(x), g_2(x))$ is absolutely continuous for every Lipschitz functions $g_1, g_2 : [c, d] \to [a, b]$.*

Similar characterizations hold for compositions of functions with bounded variation [317], or with generalized bounded variation [571].

Maligranda et al. [427] showed that the result proved by Fichtenholz actually solves a problem posed by Eidelheit in 1940 in the famous *Scottish Book* from Lvow, see [451].

Problem (Max Eidelheit, Problem 188.1, p. 261 in [451]) Let a function $f : Q \to \mathbb{R}$, $Q = I^2$, $I = [0, 1]$, be absolutely continuous on every straight line parallel to the axes of the coordinate system and let $g_1, g_2 : I \to I$ be absolutely continuous functions. Is the function $f(g_1(t), g_2(t))$ also absolutely continuous? If not, then perhaps this holds under the additional assumptions that $\iint_Q |f_x'|^p dxdy < \infty$ and $\iint_Q |f_y'|^p dxdy < \infty$, where $p > 1$?

In this general form, Eidelheit's problem has a negative answer.

Example 3.3.10 Consider Schwarz's function $f(x, y) = xy(x^2 + y^2)^{-1}$ for $(x, y) \neq 0$ and $f(0, 0) = 1$. For fixed $x > 0$, $|f(x, u) - f(x, v)| \leq \frac{2}{x}|u - v|$ and $f(0, u) - f(0, v) = 0$, showing that the function $f(x, \cdot)$ is absolutely continuous on $[0, 1]$ for every $x \in [0, 1]$. Similarly, $f(\cdot, y)$ is absolutely continuous on $[0, 1]$ for every $y \in [0, 1]$. Taking $g_1(t) = g_2(t) = t$ one obtains $h(t) = f(g_1(t), g_2(t)) = 1$ for $t \in (0, 1]$ and $h(0) = 0$. The function h is discontinuous in $t = 0$, so it is not absolutely continuous on $[0, 1]$.

The integrals from Eidelheit's problem are both infinite.

The function $f(x) = x(\ln x - 1)$, $x \in (0, 1]$, $f(0) = 0$, is absolutely continuous because its derivative $f'(x) = \ln x$, $x \in (0, 1]$, is Lebesgue integrable on $[0, 1]$ and $f(x) = \int_0^x \ln t dt$, $x \in [0, 1]$. It is not Lipschitz because f' is unbounded on $(0, 1]$.

This function furnishes another counterexample to Eidelheit's problem with both integrals finite.

Example 3.3.11 Let $I = [0, 1]$ and $f(x) = \int_0^x \ln t dt$, $x \in I$. If $\varphi(x, y) = f(x)$, $(x, y) \in Q = I^2$, then the functions $\varphi(x, \cdot)$ and $\varphi(\cdot, y)$ are absolutely continuous on I for any $(x, y) \in Q$.

Consider the Lipschitz function $g : I \to I$ constructed in the proof of Theorem 3.3.8 and take $g_1(t) = g_2(t) = g(t)$. Then the function $h(t) = \varphi(g_1(t), g_2(t)) = f(g(t))$ is not absolutely continuous.

We have $\iint_Q |\varphi_x'|^p dxdy = \iint_Q |\ln x|^p < \infty$ and $\iint_Q |\varphi_y'|^p dxdy = 0$, for all $p > 1$.

3.3.2   Another Characterization of Lipschitz Functions

The characterization of the Lipschitz property given in Theorem 3.3.5 (the equivalence 1 $\iff$ 2) admits the following extension to metric spaces.

Theorem 3.3.12 ([320]) *Let (X, d) and (Y, ϱ) be metric spaces and $f : X \to Y$ a bounded function. Then the following conditions are equivalent.*

1. *The function f is Lipschitz.*
2. *For every $\varepsilon > 0$ there exists $\delta > 0$ such that $\sum_{k=1}^{n} \varrho(f(x_k), f(y_k)) < \varepsilon$ for every finite set $\{(x_k, y_k) : k = 1, 2, \dots, n\} \subseteq X \times X$ such that $\sum_{k=1}^{n} d(x_k, y_k) < \delta$.*
3. *For every sequence $(x_k, y_k)_{k \in \mathbb{N}}$ in $X \times X$ the condition $\sum_{k=1}^{\infty} d(x_k, y_k) < \infty$ implies $\sum_{k=1}^{\infty} \varrho(f(x_k), f(y_k)) < \infty$.*

The proof given in [320] appeals to methods of constructive analysis as presented, for instance, in the book by Bishop and Bridges [82]. We adapt the proof to the simpler methods of classical analysis. Also in [320] the result is proved under the hypotheses that the metric space X is σ-compact (countable union of compact subsets) and the function f is further continuous.

For the proof we shall need an auxiliary, rather technical, result concerning series of positive real numbers. We shall consider sequences as functions $x : \mathbb{N} \to \mathbb{R}$ and we shall use the notation $x(k)$ for their terms. For two subsets I, J of $\mathbb{N}$ we use the notation $I \gg J$ if $i > j$ for every $(i, j) \in I \times J$.

Proposition 3.3.13 *Let $\mu : \mathbb{N} \to [0, \infty)$ be an increasing sequence and let $t : \mathbb{N} \to [0, \infty)$ be such that $\sum_{k \in \mathbb{N}} t(k) = \infty$ and $t(k)\mu(k) < 1$ for all $k \in \mathbb{N}$. Then the following conditions are equivalent.*

1. *The sequence μ is bounded, i.e., $\sup\{\mu(k) : k \in \mathbb{N}\} < \infty$.*
2. *For every $\varepsilon > 0$ there exists $\delta > 0$ such that $\sum_{k \in A} t(k)\mu(k) < \varepsilon$ for every finite subset A of $\mathbb{N}$ such that $\sum_{k \in A} t(k) < \delta$.*
3. *For every subset I of $\mathbb{N}$ the condition $\sum_{k \in I} t(k) < \infty$ implies $\sum_{k \in I} t(k)\mu(k) < \infty$.*

In the proof of this proposition we shall use the following lemma.

Lemma 3.3.14 *Let $a, b \in \mathbb{R}_+$ be such that $2b < a$. If the sequence $t : \mathbb{N} \to [0, \infty)$ satisfies the conditions $\sum_{k \in \mathbb{N}} t(k) = \infty$ and $t(k) < a$ for all $k > k_0$, for some $k_0 \in \mathbb{N}$, then there exists a finite subset A of $\mathbb{N}$ such that $A \gg \{k_0\}$ and $b < \sum_{k \in A} t(k) < a$.*

Proof Since $\sum_{k \in \mathbb{N}} t(k) = \infty$ there exists a finite subset B of $\mathbb{N}$ such that $B \gg \{k_0\}$ and $\sum_{k \in B} t(k) > b$. We prove that there exists a subset A of B such that $b < \sum_{k \in A} t(k) < a$.

If there exists $k \in B$ such that $b < t(k) < a$, then we can take $A = \{k\}$.

Suppose now that $t(k) < b$ for all $k \in B$ and write B as $k_1 < \cdots < k_p$. If $t(k_1) + t(k_2) > b$, then $b < t(k_1) + t(k_2) < 2b < a$. Continuing in this manner we find i, $2 \le i \le p$ such that $t(k_1) + t(k_2) + \cdots + t(k_{i-1}) \le b$ and $t(k_1) + t(k_2) + \cdots +$

$t(k_{i-1})+t(k_i) > b$. It follows that $b < t(k_1)+t(k_2)+\cdots+t(k_{i-1})+t(k_i) < 2b < a$, so we can take $A = \{k_1, \ldots, k_i\}$. $\square$

Proof of Proposition 3.3.13 The implications $1 \Rightarrow 2$ and $1 \Rightarrow 3$ are obvious.

$2 \Rightarrow 1$. For $\varepsilon = 1$ take $\delta > 0$ according to 2. We show that $\mu(k) \leq \beta$ for all $k \in \mathbb{N}$, where $\beta = 3/\delta$.

Suppose that $\mu(k) > 3/\delta > 1/\delta$ for some $k \in \mathbb{N}$. Since $\mu(k) \leq \mu(k')$ for $k' > k$, it follows that

$$t(k') \cdot \frac{1}{\delta} < t(k')\mu(k) \leq t(k')\mu(k') < 1.$$

Hence $t(k') < \delta$ for all $k' > k$, so that, by Lemma 3.3.14, there exists a finite subset A of $\mathbb{N}$ such that $A \gg \{k\}$ and $\delta/3 < \sum_{k' \in A} t(k') < \delta$. But then

$$\frac{\delta}{3} \cdot \mu(k) \leq \mu(k) \sum_{k' \in A} t(k') \leq \sum_{k' \in A} t(k')\mu(k') < 1,$$

implying $\mu(k) < 3/\delta$, in contradiction to the choice of k.

Consequently, $\mu(k) \leq 3/\delta$ for all $k \in \mathbb{N}$.

$3 \Rightarrow 1$. First we construct inductively an increasing sequence $(\alpha(k))$ of nonnegative integers in the following way.

Put $\alpha(0) = 0$.

If $\mu(k) \leq 2^{\alpha(k-1)+2}$, then put $\alpha(k) = \alpha(k - 1)$.

If $\mu(k) > 2^{\alpha(k-1)+2}$, then take $\alpha(k)$ such that

$$2^{\alpha(k-1)+2} \leq 2^{\alpha(k)} < \mu(k) \leq 2^{\alpha(k)+1} < 2^{\alpha(k)+2}.$$

Let

$$s(k) = \sum_{i=\alpha(k-1)+1}^{\alpha(k)} 2^{-i} \quad \text{if} \quad \alpha(k) > \alpha(k - 1),$$

and

$$s(k) = 0 \quad \text{if} \quad \alpha(k) = \alpha(k - 1).$$

Now we consider some finite subsets $A(k)$ of $\mathbb{N}$ defined in the following way. If $\alpha(k) = 0$, then $A(k) = \emptyset$.

Suppose $s(k) > 0$, i.e., $\alpha(k) > \alpha(k - 1)$. Then, the inequalities $s(k) \geq 2^{-\alpha(k)}$ and $\mu(k) > 2^{\alpha(k)}$ imply

$$s(k)\mu(k) \geq 1. \tag{3.3.5}$$

Hence

$$s(k) \geq \frac{1}{\mu(k)} \geq \frac{1}{\mu(k')} > t(k'),$$

for every $k' > k$. By Lemma 3.3.14 there exists a finite subset $A(k)$ of $\mathbb{N}$ such that $A(k) \gg \{k\} \cup A(k-1)$ and

$$\frac{1}{3} \cdot s(k) \leq \sum_{i \in A(k)} t(i) \leq s(k). \tag{3.3.6}$$

Put $I = \bigcup\{A(k) : k \in \mathbb{N}\}$. Let A' be a finite subset of I and $p = \max A'$. Then, by the second inequality in (3.3.6), one obtains the inequalities

$$\sum_{i \in A'} t(i) \leq \sum_{k=1}^{p} \sum_{i \in A(k)} t(i) \leq \sum_{i=1}^{k} s(k) < 1,$$

which imply $\sum_{i \in I} t(i) < \infty$. By hypothesis, $\beta := \sum_{i \in I} t(i)\mu(i) < \infty$.

Let $p \in \mathbb{N}$ and $A' = \bigcup\{A(k) : 1 \leq k \leq p\}$. The condition $A(k) \gg \{k\}$ and the monotonicity of μ imply $\mu(i) \leq \mu(k)$ for all $i \in A(k)$, so that, taking into account the first inequality in (3.3.6), one obtains

$$\frac{1}{3} \sum_{k=1}^{p} s(k)\mu(k) \leq \sum_{k=1}^{p} \mu(k) \sum_{i \in A(k)} t(i) \leq \sum_{i \in A'} t(i)\mu(i) \leq \beta,$$

showing the convergence of the series $\sum_{k=1}^{\infty} s(k)\mu(k)$. Taking into account (3.3.5) it follows that only a finite number of terms can be nonzero, implying the existence of $k_0 \in \mathbb{N}$ such that $\alpha(k) = \alpha(k_0 - 1)$ for all $k \geq k_0$, which is equivalent to

$$\mu(k) \leq 2^{\alpha(k_0-1)+2},$$

for all $k \geq k_0$, proving the boundedness of the sequence μ. □

Now we are able to prove Theorem 3.3.12.

Proof of Theorem 3.3.12 The implications $1 \Rightarrow 2$ and $1 \Rightarrow 3$ are obvious. In fact, they hold for any Lipschitz function f, without the boundedness hypothesis.

We prove the implication $2 \Rightarrow 1$ in the equivalent form $\neg 1 \Rightarrow \neg 2$.

As f is bounded, we can assume that

$$\rho(f(x), f(y)) < 1 \quad \text{for all } x, y \in X, \tag{3.3.7}$$

(otherwise we rescale ρ).

Suppose that f is not Lipschitz. Then there exists a sequence $(x_n, y_n) \in X \times X$ such that

$$\rho(f(x_n), f(y_n)) > nd(x_n, y_n) \quad \text{for all } n \in \mathbb{N}. \tag{3.3.8}$$

It follows $d(x_n, y_n) > 0$ for all $n \in \mathbb{N}$.
For $k \in \mathbb{N}$, define the finite sets

$$E_k = \left\{ (x_n, y_n) : 1 \leq n \leq k \text{ and } d(x_n, y_n) \geq \frac{1}{k} d(x_1, y_1) \right\}.$$

These sets are nonempty (because $(x_1, y_1) \in E_k$) and $E_k \subseteq E_{k+1}$ for all $k \in \mathbb{N}$.
Define now the function

$$m(x, y) = \rho(f(x), f(y))/d(x, y),$$

for all $(x, y) \in X \times X$ with $x \neq y$. Put for $k \in \mathbb{N}$

$$\mu(k) = \max\{m(x, y) : (x, y) \in E_k\}, \tag{3.3.9}$$

and

$$t(k) = d(x_{n_k}, y_{n_k}),$$

where $1 \leq n_k \leq k$ is such that $(x_{n_k}, y_{n_k}) \in E_k$ satisfies $m(x_{n_k}, y_{n_k}) = \mu(k)$.

Let us show that the functions μ and t satisfy the hypotheses of Proposition 3.3.13.

Since $E_k \subseteq E_{k+1}$, we have $\mu(k) \leq \mu(k+1)$ for all $k \in \mathbb{N}$. Also, $(x_{n_k}, y_{n_k}) \in E_k$ implies $t(k) \geq d(x_1, y_1)/k$, so that $\sum_{k=1}^{\infty} t(k) = \infty$.

Taking into account (3.3.7),

$$t(k)\mu(k) = d(x_{n_k}, y_{n_k}) \frac{\rho(f(x_{n_k}), f(y_{n_k}))}{d(x_{n_k}, y_{n_k})} = \rho(f(x_{n_k}), f(y_{n_k})) < 1,$$

for all $k \in \mathbb{N}$.

For every $n \in \mathbb{N}$ there exists a sufficiently large $k \geq n$ such that $d(x_n, y_n) \geq d(x_1, y_1)/k$, i.e., $(x_n, y_n) \in E_k$. Hence, by (3.3.9) and (3.3.8),

$$\mu(k) \geq m(x_n, y_n) > n,$$

showing that μ is unbounded.

But then, by the equivalence 1 $\iff$ 2 from Proposition 3.3.13, there exists $\varepsilon_0 > 0$ such that for every $\delta > 0$ there exists a finite subset $A_\delta \subseteq \mathbb{N}$ satisfying

$$\sum_{k \in A_\delta} d(x_k, y_k) < \delta \quad \text{and} \quad \sum_{k \in A_\delta} \rho(f(x_k), f(y_k)) \geq \varepsilon_0.$$

Consequently, condition 2 of Theorem 3.3.12 fails.
The implication $\neg 1 \Rightarrow \neg 3$ follows in a similar way.   □

3.4 Differentiability of Lipschitz Functions: Rademacher's Theorem

In this section we shall discuss the differentiability properties of Lipschitz functions.

3.4.1 Rademacher's Theorem and Some Extensions

Let us recall that, on the one hand, according to Jordan's theorem (see [611, Theorem 6.27, p. 120]) a function of bounded variation can be represented as the difference of two monotonically increasing functions. On the other hand, according to Lebesgue's theorem (see [166, Theorem 6.3.3, p. 186]) each monotonically increasing function is differentiable almost everywhere with respect to the Lebesgue measure. Hence each Lipschitz function $f : [a, b] \to \mathbb{R}$ is differentiable almost everywhere with respect to Lebesgue measure. This is just a particular case of the famous Rademacher's theorem.

Theorem 3.4.1 (Rademacher [587]) *Let U be a nonempty open subset of $\mathbb{R}^n$. Then every Lipschitz function $f : U \to \mathbb{R}^m$ is differentiable almost everywhere on U.*

The proof of this theorem can be split into two independent parts:

(a) A Lipschitz function which is Gâteaux differentiable at a point is Fréchet differentiable at that point.
(b) Any Lipschitz function is Gâteaux differentiable almost everywhere with respect to Lebesgue measure.

The interested reader can find proofs of this theorem in [231, Section 3.1], [283, Theorem 6.15], [284, Section 3]. See also [521] and [693].

Remark 3.4.2 It is easy to check that Theorem 3.4.1 holds for locally Lipschitz functions $f : U \to \mathbb{R}^m$.

Indeed, let $x_0 \in U$, $x_0 \neq 0$, $a := d(x_0, \partial U) > 0$ and $b := \|x_0\| > 0$, provided that $\partial U \neq \emptyset$. Consider the sets

$$U_n = \left\{ x \in U : d(x, \partial U) > a(2n)^{-1} \right\} \cap B(0, 2nb) \quad \text{and}$$

$$K_n = \left\{ x \in U : d(x, \partial U) \geq a(2n)^{-1} \right\} \cap B[0, 2nb],$$

for every $n \in \mathbb{N}$. If $\partial U = \emptyset$ (which is equivalent to $U = \mathbb{R}^n$), then take $x_0 = 0$, $U_n = B(0, n)$ and $K_n = B[0, n]$ for $n \in \mathbb{N}$.

The sets U_n, K_n are nonempty ($x_0 \in U_n \subseteq K_n$), U_n is open, K_n compact and

$$U = \bigcup_{n=1}^{\infty} U_n = \bigcup_{n=1}^{\infty} K_n.$$

By Theorem 2.1.6 (see also Remark 2.1.7) the function f is Lipschitz on K_n, and so on U_n as well. Rademacher's theorem implies that there exists a Lebesgue null subset Λ_n of U_n such that f is differentiable on $U_n \setminus \Lambda_n$. It follows that f is differentiable on $U \setminus \Lambda$, where $\Lambda = \bigcup_{n=1}^{\infty} \Lambda_n$.

Rademacher's theorem (Theorem 3.4.1) was published in 1920, but as remarked V. Stepanov [648] (see also [649]), actually, Rademacher proved a more general result.

Theorem 3.4.3 (V. V. Stepanov) *Let U be a nonempty open subset of $\mathbb{R}^n$. Then every function $f : U \to \mathbb{R}^m$ is differentiable almost everywhere on the set*

$$\left\{ x \in U : \limsup_{y \to x} \frac{\|f(y) - f(x)\|}{\|y - x\|} < \infty \right\}.$$

A consequence of Stepanov's theorem is the following result.

Corollary 3.4.4 *Let U be a nonempty open subset of $\mathbb{R}^n$. A function $f : U \to \mathbb{R}^m$ is differentiable almost everywhere on U if and only if*

$$\limsup_{y \to x} \frac{\|f(y) - f(x)\|}{\|y - x\|} < \infty \qquad (3.4.1)$$

holds, excepting a Lebesgue null subset of U.

Notice that a function $f : U \to \mathbb{R}$ satisfies (3.4.1) if and only if there exist the numbers $r, L > 0$ such that

$$\|f(y) - f(x)\| \leq L\|y - x\|,$$

for all $y \in B[x, r]$. Such a function is called *pointwise Lipschitz at x*. It is called *pointwise Lipschitz* on a set $A \subseteq U$ if it is pointwise Lipschitz at every $x \in A$.

Remark 3.4.5 W. Stepanoff is the same as the Russian mathematician V. V. Stepanov, and so Theorem 3.4.3 is generally known as *"Stepanov's theorem"*.

A finer result, concerning the points where f has partial derivatives but is not differentiable, was obtained by Bessis and Clarke [79]:

> The set of those points at which a function $f : \mathbb{R}^n \to \mathbb{R}^m$ is not differentiable but it is differentiable in n linearly independent directions is σ-porous (and so of Lebesgue measure 0 and of first Baire category).

Definition 3.4.6 A subset E of $\mathbb{R}^n$ is called *porous at* $x \in E$ if there exist $c > 0$ and a sequence $y_n \to 0$ such that $E \cap B(x + y_n, c\|y_n\|) = \emptyset$ for all $n \in \mathbb{N}$. The set E is called *porous* if it is porous at each $x \in E$ and σ-*porous* it can be written as a countable union of porous sets.

Porous and σ-porous sets in $\mathbb{R}^n$ are of Lebesgue measure 0 and of first Baire category.

Remark 3.4.7 The set E is porous at $x \in E$ if and only if the distance function $d(\cdot, E)$ is not Fréchet differentiable at x. The Gâteaux differentiability of this function is related to a weaker version of porosity, called directional porosity (see [397]).

Indeed, since the function $d(\cdot, E)$ attains its minimum at every $x \in E$, if it is differentiable at $x \in E$ (in any sense), then the differential must be 0. If E is porous at x, then $E \cap B(x + y_n, c\|y_n\|) = \emptyset$, for some $c > 0$ and a sequence $y_n \to 0$. It follows $|d(x + y_n, E) - d(x, E)|/\|y_n\| = d(x + y_n, E)/\|y_n\| \geq c$, so that $d(\cdot, E)$ is not Fréchet differentiable at x. Conversely, if $d(\cdot, E)$ is not Fréchet differentiable at $x \in E$, then there exists $\varepsilon > 0$ and a sequence (y_n) with $\|y_n\| \to 0$ such that $d(x + y_n, E)/\|y_n\| > \varepsilon$, implying $E \cap B(x + y_n, \varepsilon\|y_n\|) = \emptyset$ for all $n \in \mathbb{N}$.

Remark 3.4.8 Rademacher's theorem says that the condition of being Lipschitz could be viewed as a weakened version of differentiability. Let us mention that in view of the compactness of balls in finite dimensional Banach spaces, the concepts of Gâteaux differentiability and Fréchet differentiability coincide for Lipschitz functions whose domain is a Banach space having finite dimension (see [75]). If the domain is infinite dimensional, there are examples which show that there is a big difference between these two concepts.

Alberti and Marchese [12] proved an extension of Rademacher's theorem to an arbitrary positive measure μ on $\mathbb{R}^n$. The differential of a Lipschitz function $f : \mathbb{R}^n \to \mathbb{R}^m$ with respect to a linear subspace W of $\mathbb{R}^n$ at a point $x \in \mathbb{R}^n$ is a linear functional $d_W(x)$ on W such that

$$f(x + h) - f(x) = d_W(x)(h) + o(h) \quad \text{for all} \quad h \in W .$$

(Here, $o(h)$ means that $\lim_{h \to 0} o(h)/\|h\| = 0$). They attached to every point $x \in \mathbb{R}^n$ a subspace $V(\mu, x)$ of $\mathbb{R}^n$ and proved that every Lipschitz function $f : \mathbb{R}^n \to \mathbb{R}^m$ is differentiable with respect to $V(x, \mu)$ μ-a.e. $x \in \mathbb{R}^n$. The result is optimal, in the sense that there exists a Lipschitz function $f : \mathbb{R}^n \to \mathbb{R}^m$ such that μ-a.e. $x \in \mathbb{R}^n$ and every $v \in \mathbb{R}^n \setminus V(\mu, x)$ the derivative of f in the direction v does not exist. The correspondence $x \mapsto V(\mu, x)$ (depending only on μ) is called the decomposability bundle corresponding to μ.

Another extension, in the case $n = m = 1$, was proved by Aldaz [25]. Let μ be a Borel measure on $\mathbb{R}$. A function $f : \mathbb{R} \to \mathbb{R}$ is called μ-differentiable at $x \in \mathbb{R}$ if the limits

$$\lim_{h \searrow 0} \frac{f(x+h) - f(x)}{\mu([x, x+h])} \quad \text{and} \quad \lim_{h \searrow 0} \frac{f(x+h) - f(x)}{\mu([x-h, x])}$$

exist and are equal. The common value, denoted by $\frac{df}{d\mu}(x)$, is called the μ-differential of f at x.

Aldaz [25] proved the following extension of Rademacher's theorem:

Given a Borel measure μ on $\mathbb{R}$, every Lipschitz function $f : \mathbb{R} \to \mathbb{R}$ is μ-a.e. μ-differentiable on $\mathbb{R}$.

Taking into account Remark 3.4.2 and Corollary 3.1.25, one obtains the following result:

Let $f : U \to \mathbb{R}$ be a convex function, where $U \subseteq \mathbb{R}^n$ is open and convex. Then f is differentiable almost everywhere on U.

In fact, a much stronger result, proved by Aleksandrov [26] (see [231] for a proof), holds:

A convex function, $f : U \to \mathbb{R}$, where U is an open convex subset of $\mathbb{R}^n$, is twice differentiable almost everywhere on U.

Extensions of Aleksandrov's theorem to infinite dimensions were given in [104, 105] and [86] and, for H-convex functions on stratified groups, by Magnani [418].

3.4.2 The Converse of Rademacher's Theorem

There also some results concerning the converse of Rademacher's theorem, meaning the existence of Lipschitz functions $f : \mathbb{R}^n \to \mathbb{R}^m$ nowhere differentiable on a given Lebesgue null subset of $\mathbb{R}^n$. We start with the following two results.

Theorem 3.4.9 ([14]) *For a given subset E of $\mathbb{R}$ there is a Lipschitz function $f : \mathbb{R} \to \mathbb{R}$ which is not differentiable at any point $x \in E$ if and only if E has Lebesgue measure 0.*

A set $E \subseteq \mathbb{R}^n$ is called the *nondifferentiability* set of a function $f : \mathbb{R}^n \to \mathbb{R}^m$ if f is differentiable on $\mathbb{R}^n \setminus E$ and nondifferentiable at every point in E. The exact characterization of the nodifferentiability sets of Lipschitz functions on $\mathbb{R}$ was given by Zahorski [692].

Theorem 3.4.10 (Z. Zahorski) *For every $G_{\delta\sigma}$ subset E of $\mathbb{R}$ of Lebesgue measure 0, there exists a Lipschitz function $f : \mathbb{R} \to \mathbb{R}$ which is differentiable on $\mathbb{R} \setminus E$ and nondifferentiable at every point of E.*

Since the nondifferentiability set of a Lipschitz function $f : \mathbb{R} \to \mathbb{R}$ is $G_{\delta\sigma}$ and of Lebesgue measure 0, it follows that the above theorem actually gives a characterization of the nondifferentiability sets of Lipschitz functions.

Zahorski [692] also gave a description of the nondifferentiability sets of continuous functions:

Let $E \subseteq \mathbb{R}$. For the existence of a continuous function $f : \mathbb{R} \to \mathbb{R}$ differentiable on $\mathbb{R} \setminus E$ and nowhere differentiable on E it is necessary and sufficient that $E = E_1 \cup E_2$, where E_1 is an arbitrary G_δ set and E_2 is $G_{\delta\sigma}$ and of Lebesgue measure 0.

Zahorski used the Lipschitz function whose existence is stated in Theorem 3.4.10 to construct a continuous function whose nondifferentiability set is E. Fowler and Preiss [241] gave a simpler proof of Theorem 3.4.10 based on the following result.

Theorem 3.4.11 *For every G_δ subset E of $\mathbb{R}$ of Lebesgue measure 0, there exists a Lipschitz function $f : \mathbb{R} \to \mathbb{R}$, with Lipschitz constant $L(f) \leq 1$, differentiable outside E and satisfying*

$$\liminf_{t \to 0} \frac{f(x+t) - f(x)}{t} = -1 \quad and \quad \limsup_{t \to 0} \frac{f(x+t) - f(x)}{t} = 1,$$

for all $x \in E$.

Preiss [579] (see also [14]) proved the following result:

For any subset E of $\mathbb{R}^2$ of Lebesgue measure 0 there is a Lipschitz function $f : \mathbb{R}^2 \to \mathbb{R}^2$ which is not differentiable at any point of E.

On the other hand, he constructed the following surprising example.

Example 3.4.12 (Preiss [579]) There exists a subset E of $\mathbb{R}^2$ of Lebesgue measure 0, such that every Lipschitz function $f : \mathbb{R}^2 \to \mathbb{R}$ is differentiable at a least one point in E.

Good presentations of this circle of problems is given in [13] and [14].

A set $E \subseteq \mathbb{R}^n$ having the property from Example 3.4.12 is called a *universal differentiability set*, a term coined by Doré and Maleva [201]. The set constructed by Preiss is rather large—it contains all the lines connecting the points in E with rational coordinates and is dense in $\mathbb{R}^2$, so a natural question was to find smaller universal differentiability sets. This was done by M. Doré, M. Dymond and O. Maleva who proved the existence of universal differentiability sets in $\mathbb{R}^n$ with the following properties:

- compact and of Lebesgue measure 0 in [199];
- compact, of Lebesgue measure 0, and of Hausdorff dimension 1 in [201];
- compact, of Lebesgue measure 0, and of Minkowski dimension 1 in [216].

The universal differentiability set constructed by Preiss (adapted to $\mathbb{R}^n$) has Hausdorff dimension n. Also, the Hausdorff dimension is smaller than Minkowski dimension, see [216] (or the book [448]). The result from [201] was extended in [200] by proving that every non-zero Banach space with separable dual contains a closed and bounded universal differentiability set of Hausdorff dimension one.

In general, the converse of Rademacher' theorem for Lipschitz functions from $\mathbb{R}^n$ to $\mathbb{R}^m$ holds if and only if $m \geq n$. The solution in the case $m \geq n$ is announced by M. Csörnyiei and P. Jones in http://www.math.sunysb.edu/Videos/dfest/PDFs/38-Jones.pdf

The remaining case, $m < n$ for $n > 1$, is treated in [580]:

Let $n > 1$. Then there exists a Lebesgue null set $E \subseteq \mathbb{R}^n$ containing a point of differentiability of every Lipschitz function $f : \mathbb{R}^n \to \mathbb{R}^{n-1}$.

For further discussions on this subject, see the paper [426].

Universal differentiability sets in the Heisenberg group are discussed in [572] and [575].

3.4.3 Infinite Dimensional Extensions

Since Rademacher's theorem is an important tool in mathematical analysis, many attempts have been made to extend it to infinite dimensional spaces.

The first (and easy) step is to consider functions from an open subset of $\mathbb{R}^n$ to a Banach space Y. The example of the function $f : [0, 1] \to L^1[0, 1]$ given by $f(t) = \chi_{[0,t]}$, $t \in [0, 1]$, which is a nowhere differentiable isometry on $[0, 1]$ (see Example 1.6.20) shows that some restrictions must be imposed on the space Y. It turns out that Y must satisfy the Radon–Nikodým property—given an open subset U of $\mathbb{R}^n$ and a Banach space Y, a Rademacher type theorem holds for every Lipschitz function $f : U \to Y$ if and only if Y has the Radon–Nikodým property (see, e.g., [51]). Concerning the Radon–Nikodým property (RNP), see Sect. 1.6.3.

An extension of Rademacher's theorem to Lipschitz functions with values in an arbitrary Banach space was given by Kirchheim [356] (see also [37]) by replacing the linear differential by a seminorm. More exactly, let $f : \mathbb{R}^n \to (Y, \|\cdot\|)$ be a Lipschitz function, where $(Y, \|\cdot\|)$ is a Banach space. Then, for a given $u \in \mathbb{R}^n \setminus \{0\}$, the limit

$$MDf(x; u) := \lim_{t \searrow 0} \frac{\|f(x + tu) - f(x)\|}{t} \tag{3.4.2}$$

exists for almost all (with respect to the Lebesgue measure on $\mathbb{R}^n$) $x \in \mathbb{R}^n$. One shows that, for almost every $x \in \mathbb{R}^n$, the function $MDf(x; \cdot)$ is a seminorm on $\mathbb{R}^n$ and

$$\|f(z) - f(y)\| - MDf(x; z-y) = o(\|x-z\| + \|y-z\|) \quad \text{as} \quad y, z \to x. \tag{3.4.3}$$

Based on these remarks, the *metric differential* of a function $f : \mathbb{R}^n \to Y$ at a point $x \in \mathbb{R}^n$ is defined as a seminorm $MDf(x; \cdot)$ on $\mathbb{R}^n$ satisfying the condition

$$\|f(y) - f(x)\| - MDf(x; y - x) = o(\|y - x\|) \quad \text{as} \quad y \to x .$$

It follows that the stronger condition (3.4.3) holds in this case. Also, if Y is the dual of a separable Banach space Z, and $f : \mathbb{R}^n \to Z^*$ is Lipschitz, then the w^*-differential $D_{w^*} f(x; \cdot)$ (see Sect. 1.4.14) and the metric differential $MDf(x; \cdot)$ exist almost everywhere and the equality

$$MDf(x; \cdot) = \|D_{w^*} f(x; \cdot)\|$$

holds for almost all $x \in \mathbb{R}^n$ (see [37]). For further developments, including a Stepanov type result for the metric differential, see [344] and [345].

Remark 3.4.13 Actually, Kirchheim's result holds for functions defined on $\mathbb{R}^n$ and with values in an arbitrary metric space (X, d). The space $\mathbb{R}^n$ can be also replaced by a measurable subset of $\mathbb{R}^n$.

The next step is to extend Rademacher's theorem to Lipschitz functions defined on open subsets of infinite dimensional Banach spaces. The main difficulty in obtaining such extensions is to find an infinite dimensional generalization of the notion of a set of Lebesgue measure zero.

Since we are interested in problems concerning continuous functions, we need a class $\mathcal{N}$ of Borel sets (which will be considered as the class of null sets) having the following properties:

1. $\mathcal{N}$ is closed with respect to countable unions and translations;
2. $\mathcal{N}$ is hereditary, i.e., if $A \in \mathcal{N}$, then $B \in \mathcal{N}$ for any Borel subset B of A;
3. $\mathcal{N}$ does not contain any nonempty open set, i.e., if A is nonempty and open, then $A \notin \mathcal{N}$;
4. $\mathcal{N}$ coincides with the Borel sets of Lebesgue measure zero in the finite dimensional case.

In infinite dimensional spaces there is no analog of the Lebesgue measure (see [639, p. 108]). Hence we cannot define the notion of "almost everywhere" in the usual way, by means of a measure.

There are several ways to define negligible sets in an infinite dimensional Banach space. Let us mention some of them:

- Haar null sets introduced by Christensen [150, 151] in 1970;
- cube null sets introduced by Mankiewicz [429] in 1973;
- Aronszajn null sets introduced by Aronszajn [51] in 1976;
- Gauss null sets introduced by Phelps [564] in 1978;
- Borel σ-directionally porous sets introduced by Preiss and Zajíček [582] in 2001;
- Γ-null sets introduced by Lindenstrauss and Preiss [393] in 2003 (see also the book [399]), whose definition involves both the notions of category and measure.

Each of these classes forms a proper σ-ideal of Borel subsets of the considered Banach space (a nonempty family $\mathscr{A}$ of subsets of a set X is called a (*proper*) σ-*ideal* provided that: (i) $X \notin \mathscr{A}$; (ii) $A \in \mathscr{A}$ and $B \subseteq A \ \Rightarrow \ B \in \mathscr{A}$; (iii) $A_n \in \mathscr{A}$, $n \in \mathbb{N}$, $\Rightarrow \ \bigcup_{n=1}^{\infty} A_n \in \mathscr{A}$). In 1999 Csörnyei [171] showed that in every separable Banach space, the classes of Gauss null, cube null and Aronszajn null sets agree and that they are properly contained in the class of Haar null sets. For more details one can consult the sixth chapter of [75], or the second chapter of [399].

It is known that, in finite dimensions, Lebesgue null sets are preserved by Lipschitz mappings. Two infinite-dimensional counterexamples of Haar null sets with Lipschitz image having Haar null complement are provided in [396].

An extension to Fréchet spaces of the class of Aronszajn null sets was given by Bongiorno in [96]. A class of null sets playing in Fréchet spaces a role similar to that of the class considered by Preiss and Zajíček was introduced by La Russa in [371]. In [89] a Rademacher type theorem is proved with respect to a quasi-invariant Radon measure on a linear space.

Preiss and Zajíček [582] defined several σ-ideals as, for example, the class, denoted by $\widetilde{\mathscr{A}}$, which is strictly smaller than the class of Gauss null sets and proved almost everywhere (with respect to this class) differentiability results for Lipschitz functions.

For other related results see [94–96].

The Gâteaux Differentiability
Concerning the Gâteaux differentiability, the following positive result is known:

If X is a separable Banach space and Y is a Banach space having the Radon–Nikodým property, then every Lipschitz function $f : U \to Y$, where U is a nonempty open subset of X, is Gâteaux differentiable outside a Gauss (Aronszajn) null set (see [51, 151, 429], or Theorem 6.24 in [75]).

The Gâteaux differentiability has some inconveniences as, for instance, that signaled by S. A. Shkarin [633]: there exists a nowhere continuous and everywhere Gâteaux differentiable function $f : \ell^2 \to \mathbb{R}$.

The Fréchet Differentiability
The question of Fréchet differentiability needs a different approach since there are examples of nowhere Fréchet differentiable Lipschitz functions for which the Gâteaux differentiability results mentioned above are valid (see [51], Example I, Section 3, Chapter II, pp. 167–169, or [559], pp. 981–982). Some relations between the Gâteaux and Fréchet differentiability of convex functions are also discussed in [103].

The function $f : L^1[0, \pi] \to \mathbb{R}$ given by

$$f(x) = \int_0^{\pi} \sin(x(t)) dt \,,$$

for each $x \in L^1[0, \pi]$ is an example of Lipschitz function that is everywhere Gâteaux differentiable but nowhere Fréchet differentiable (see [252]). Its restriction

g to $L^2[0, 1]$ is everywhere Fréchet differentiable (even uniformly) but some authors (see the comments in [252]) erroneously considered the function g as an example of a Gâteaux differentiable function that is nowhere Fréchet differentiable. Results on the existence of Fréchet differentials are rare and hard to prove, but they are necessary for providing local approximation of functions. For instance, Phelps [559] (see also [567]) gave an example of an equivalent norm on ℓ^1 that is Gâteaux-differentiable everywhere (except at the origin) and nowhere Fréchet-differentiable. An example of nowhere Fréchet-differentiable Lipschitz function on ℓ^1 was given by Aronszajn [51].

If X is a Banach space, the function $f : X \to \mathbb{R}$ given by

$$f(x) = \|x\|, \quad x \in X,$$

is Lipschitz. If $X = \ell^1$, then f is Fréchet differentiable at no point in ℓ^1 and, in general, a similar situation may occur whenever X is separable with nonseparable dual space X^*.

Matoušek and Matoušková [447] constructed an equivalent norm on ℓ^2 whose points of Fréchet differentiability form an Aronszajn null set, disproving so a conjecture by Borwein and Noll [105] on the non-existence of continuous convex functions with this property.

The non-differentiability sets of Lipschitz functions are also related with porosity (see Remark 3.4.7). D. Preiss and J. Tišer [581] proved that if E is a countable union of closed porous subsets of a separable Banach space X, then there exists a Lipschitz function $f : X \to \mathbb{R}$ nowhere Fréchet differentiable on E. In the same paper it is shown that any σ-porous set in a separable Banach space X belongs to the σ-ideal generated by the sets of points of Fréchet non-differentiability of real-valued Lipschitz functions on X.

Therefore, in order to obtain extensions of Rademacher's theorem involving Fréchet differentiability we have to restrict the class of Banach spaces that can serve as domain spaces to the class of well-behaved Banach spaces which is given by the following definition.

Definition 3.4.14 A Banach space X is said to be an *Asplund space* if the dual of every separable subspace of X is separable.

Asplund [54] proved that if X is a Banach space with separable dual, then every continuous convex function $f : \Omega \to \mathbb{R}$ is Fréchet differentiable on a dense G_δ-subset of Ω, for every open convex subset Ω of X. To honor this remarkable result, the Banach spaces with this property were called later *Asplund spaces*. One can show that this differentiability property actually characterizes Asplund spaces.

Among the classes of Asplund spaces we mention the Banach spaces whose duals are separable and reflexive Banach spaces.

One also shows that a Banach space X is an Asplund space if and only if X^* has the RNP. For an excellent introduction to the subject we recommend Phelps' book [567] (a good source is also the book by Giles [253]).

D. Preiss established in 1990 the following result:

Any real-valued Lipschitz function defined on a nonempty subset of an Asplund space is Fréchet differentiable on a dense subset of its domain (see [391, 579] and [398]).

Remark 3.4.15 Since Asplund spaces can be defined as those Banach spaces on which every equivalent norm has a point of Fréchet differentiability, it follows that Preiss' result is the best possible.

It can be shown that if X is separable and Y has the RNP, then every Lipschitz function $f : X \to Y$ satisfies $L(f) = \sup\{\|D(f;\cdot)\| : x \in G(f)\}$, where $G(f)$ denotes the points of Gâteaux differentiability of f. If there exists $x_0 \in G(f)$ such that $\|D(f;x_0)\| = L(f)$, then f is Fréchet differentiable at x_0 (at least in the case when X is uniformly smooth). This suggests an approach to obtain Fréchet differentiability from Gâteaux differentiability based on a variational principle, i.e., the maximum of an appropriate perturbation of a given function is attained. This approach is followed in the paper [398].

For other results in this direction one can consult [207, 208], [251], [254, 255, 269]. Mankiewicz [429], Bongiorno [96] and La Russa [371] proved Rademacher type theorems in the setting of separable Fréchet spaces.

Let us also mention a result which says that a Lipschitz function from $\mathbb{R}^n$ to $\mathbb{R}$ equals a C^1 function except on a "small" set (see [231], p. 251):

Let $f : \mathbb{R}^n \to \mathbb{R}$ be a Lipschitz function and $\varepsilon > 0$. Then there exists a C^1 function $\widetilde{f} : \mathbb{R}^n \to \mathbb{R}$ such that the Lebesgue measure of the set $\{x \in \mathbb{R}^n : f(x) \neq \widetilde{f}(x)$ or $Df(x) \neq D\widetilde{f}(x)\}$ is less than ε.

A famous open question is whether every finite or countable collection of real-valued Lipschitz functions on an Asplund space has a common point of Fréchet differentiability. In this direction, let us mention the following two results:

A. Every Lipschitz mapping from an open subset of a Hilbert space to a two-dimensional space has points of Fréchet differentiability. The result does not hold in the case of Lipschitz functions with values in $\mathbb{R}^3$.
In the case of the space ℓ^p, $1 < p < \infty$, the result is true for Lipschitz functions $f : \ell^p \to \mathbb{R}^n$ with $n \leq p$. (see [399], Corollaries 13.1.2 and 13.1.3).
B. If $X = C(K)$ with K countable compact, or X is a subspace of c_0, then every Lipschitz function from an open subset U of X to a Banach space with RNP is Fréchet differentiable on U excepting a Γ-null set (see J. Lindenstrauss and D. Preiss [393]).

The book [399] contains an excellent presentation of the above mentioned question and of other problems concerning the differentiability of Lipschitz functions (see also [392] and [397]).

The papers [390] and [316] provide sufficient conditions on a pair of Banach spaces X and Y under which each Lipschitz mapping from a domain in X to Y has, for every $\varepsilon > 0$, a point of ε-Fréchet differentiability. Given two Banach spaces X and Y, D an open subset of X, $x_0 \in D$ and $\varepsilon > 0$, a function $f : D \to Y$ is called

ε-Fréchet differentiable at x_0 if there exists a bounded linear operator $T : X \to Y$ and $\delta > 0$ such that $\| f(x_0 + u) - f(x_0) - T(u) \| \leq \varepsilon \| u \|$ for every $u \in X$ with $\| u \| \leq \delta$.

3.4.4   Metric Measure Spaces

Let (X, d) be a metric space. A positive Borel measure μ on X is called *doubling* if there exists a constant $D > 0$ such that

$$0 < \mu(B[x, 2r]) \leq D\mu(B[x, r]) < \infty, \tag{3.4.4}$$

for all $x \in X$ and $r > 0$. Let $C, s > 0$. The measure μ is called (C, s)-*homogeneous* if

$$0 < \mu(B[x, \lambda r]) \leq C\lambda^s \mu(B[x, r]) < \infty, \tag{3.4.5}$$

for all $x \in X$, $r > 0$ and $\lambda \geq 1$. It is obvious that if the measure μ is (C, s)-homogeneous, then it is doubling with $D = 2^s C$. Conversely, if μ satisfies (3.4.4), then it satisfies (3.4.5) with $(C, s) = (D, \log_2 D)$ (see [409, §6] or [38, Th. 5.2.2]).

We define now the metric notions of homogeneity and doubling.

Definition 3.4.16 Let $c, s \in [0, \infty)$. A metric space (X, d) is called (c, s)-*homogeneous* if for every $0 < r < R$,

$$\operatorname{card}(X_0) \leq c\, (R/r)^s ,$$

for any subset X_0 of X such that $r \leq d(x, y) \leq R$ for all distinct points $x, y \in X_0$.

The space (X, d) is called s-*homogeneous* if it is (c, s)-homogeneous for some $c \in [0, \infty)$.

The *Assouad dimension* of (X, d) is

$$\dim_A(X, d) = \inf \{ s \geq 0 : (X, d) \text{ is } s\text{-homogeneous} \}$$

if this infimum exists; otherwise put $\dim_A(X, d) = \infty$.

Alternatively, the Assouad dimension of a metric space can be defined in terms of covering by balls. The Assouad dimension of a metric space (X, d) is the infimum of all $\beta > 0$ for which there exists $c > 1$ such that every ball $B[x, R]$ in X can be covered by at most $c(R/r)^\beta$ closed balls of radius r, for any $0 < r < R$ and any $x \in X$ (see [283, pp. 81–82]). In fact, in [283] this property is defined in terms of the covering of sets of diameter δ by sets of diameter $\delta/2$, with the mention that this is equivalent to the definition with balls given above.

A good presentation of the properties of the Assouad dimension is given in the Appendix of the paper [409] and in the books [283] and [285].

A subset Y of a metric space (X, d) is called ε-*separated* if $d(y, y') \geq \varepsilon$ for all $y, y' \in Y$ with $y \neq y'$. The metric space X is called *doubling with constant m* if every ε-separated subset of X contains at most m elements. This property can also be characterized in terms of covering properties. The following result holds:

If (X, d) is a doubling metric space with constant m, then every open ball of radius $r > 0$ in X can be covered by m open balls of radius $r/2$. Conversely, if X is a metric space such that every open ball of radius $r > 0$ in X can be covered by m open balls of radius $r/2$ then X is doubling with constant m^2 (see [285, Lemma 4.1.11]).

It is easy to see that $\mathbb{R}^n$ is doubling with a constant depending only on n, and in fact the Assouad dimension of $\mathbb{R}^n$ is n. Thus, every subset of Euclidean space is doubling. Also, doubling spaces are precisely the spaces of finite Assouad dimension (see [283, p. 81–82]).

Remark 3.4.17 Homogeneous spaces were introduced by Coifman and Weiss [167] in connection with some problems in harmonic analysis. They considered the more general situation of a "metric" d defined on a set X satisfying, for some $K \geq 1$, the inequality

$$d(x, z) \leq K(d(x, y) + d(y, z)) \quad \text{for all } x, y, z \in X,$$

called the relaxed triangle inequality. They called d a quasi-metric (by analogy with quasi-norm, see Sect. 1.4.16). Since this term is used in the present book in another sense (see Sect. 4.3.1), we shall call d a *b-metric* (following the terminology of the fixed point community). The pair (X, d) is called a *b-metric space*. More general approaches, with "balls" defined axiomatically as nonempty bounded open subsets $B(x, r)$ of $\mathbb{R}^n$, for $x \in \mathbb{R}^n$ and $r > 0$, satisfying some appropriate conditions (which hold for balls in b-metric spaces), are considered as well (see, for instance, [647, p. 8]).

It turned out that the existence of a homogeneous measure on a metric space X is tightly connected with the metric homogeneity of X. Coifman and Weiss [167] remarked that the existence of a doubling measure on a metric space (X, d) implies that the metric d is doubling, a result that is true for (C, s)-homogeneous measures—if the measure μ is (C, s)-homogeneous, then the metric d is (c, s)-homogeneous, for some $c \geq 0$ depending on (C, s).

A. L. Vol'berg and S. V. Konyagin [673] proved the converse: if X is a compact subset of $\mathbb{R}^n$ which is (c, s)-homogeneous with respect to the induced Euclidean metric, then there exists a (c, s)-homogeneous measure on X. They introduced a new dimension, denoted by $\dim_{VK}(X)$ in [409], defined as the infimum of all $s \geq 0$ such that X carries a (C, s)-homogeneous measure for some $C \geq 0$. The result was extended in [412] to arbitrary complete metric spaces:

Let $c \geq 1$, $s > 0$ and $t > s$. Then there exists $C \geq 1$ such that every (c, s)-homogeneous complete metric space carries a (C, t)-homogeneous measure.

It follows that $\dim_A(X) = \dim_{VK}(X)$ for every complete metric space X. Also, for every $n \in \mathbb{N}$, there exists a constant $C_n \geq 1$ such that every closed subset of

$\mathbb{R}^n$ carries a (C_n, n)-homogeneous measure (see [412]). In [409, Theorem 5.2] one also shows that a subset X of $\mathbb{R}^n$ is porous (see Definition 3.4.6) if and only if $\dim_A(X) < n$.

The existence of a doubling measure μ on a metric space (X, d) imposes some restrictions, both on the measure μ and on the space X. For instance, $\mu(\{x\}) = 0$ for every non-isolated point x of X (one says that the measure μ is diffuse). Also the space X must be separable and every bounded subset of X is totally bounded. If X is further complete, then every closed bounded subset of X is compact.

However, as it was shown by E. Saksman [616], not any doubling metric space carries a doubling measure:

For each $n \geq 2$ there is a bounded Jordan domain Ω of $\mathbb{R}^n$ (even the image of $B_{\mathbb{R}^n}(0, 1)$ under a homeomorphism of $\mathbb{R}^n$) which does not carry a doubling measure with respect to the Euclidean metric.

It turned out that a lot of results in analysis can be transposed to metric spaces with a doubling measure. These can be done for those involving continuity notions. In order to extend results concerning differentiable functions (the so-called first-order calculus), some supplementary hypotheses are needed.

Let (X, d) be a metric space and μ a Borel measure on X. A key notion in constructing a first order differential calculus is that of upper gradient. An *upper gradient* for a function $f : X \to \mathbb{R}$ is a Borel measurable function $g : X \to [0, \infty]$ such that

$$|f(x) - f(y)| \leq \int_\gamma g ds, \qquad (3.4.6)$$

for every rectifiable curve γ joining x and y and for all $x, y \in X$. If $f : \mathbb{R}^n \to \mathbb{R}$ is smooth, then an upper gradient for f is $g(x) = \|\nabla f(x)\|$, $x \in \mathbb{R}^n$, showing that the notion of upper gradient is an extension of the norm of the gradient to arbitrary metric spaces. If f is locally Lipschitz, then

$$\mathrm{lip}(f)(x) := \liminf_{r \searrow 0} \left(\frac{1}{r} \sup_{x \in B(x,r)} d(f(x), f(y)) \right), \quad x \in X,$$

is also an upper gradient for f (see [285, Lemma 6.2.6]).

Another ingredient is the validity of a Poincaré-type inequality. One says that the metric measure space (X, d, μ) satisfies a *weak p-Poincaré inequality*, for some $1 < p < \infty$, if there exist $C > 0$ and $\lambda > 1$ such that

$$\frac{1}{\mu(B)} \int_B |f - f_B| d\mu \leq Cr \left(\frac{1}{\mu(\lambda B)} \int_{\lambda B} g^p d\mu \right)^{1/p} \qquad (3.4.7)$$

for any Borel measurable function f and any upper gradient g of f. Here B is a closed ball of radius $r > 0$ and λB is the ball with the same center and radius

λr, $\lambda > 1$. Also, by f_B one denotes the mean value of the function f on the ball B

$$f_B = \frac{1}{\mu(B)} \int_B f \, d\mu \, .$$

If (3.4.7) holds with $\lambda = 1$, then one says that (X, d, μ) satisfies a *strong p-Poincaré inequality*.

As it is written in [285, p. 3]:

> When coupled with the doubling condition for the measure μ, the Poincaré inequality becomes a powerful tool with both analytic and geometric consequences.

There are numerous examples of metric measure spaces supporting a Poincaré inequality such as Carnot groups, Heisenberg groups with the Carnot-Carathédori metric, Alexandrov spaces, etc., see [285, §14.2].

Cheeger [141] proved Rademacher-type differentiability theorems for real-valued Lipschitz functions defined on a metric space (X, d) carrying a doubling measure μ and satisfying some appropriate conditions (as, e.g., supporting a Poincaré inequality). Together with Kleiner [142, 143] he extended these results to Lipschitz functions with values in a Banach space with the Radon–Nikodým property. More exactly, J. Cheeger proved the existence of a chart $\{(U_i, \varphi_i) : i \in \mathbb{N}\}$, where U_i is a Borel subset of X, $X = \bigcup_i U_i$, and $\varphi_i : U_i \to \mathbb{R}^{m(i)}$, $i \in \mathbb{N}$, are Lipschitz. One proves that, under these circumstances, for every Lipschitz map $f : X \to \mathbb{R}$, for every $i \in \mathbb{N}$ and for μ-almost all $x \in U_i$ there exists a unique (co)vector $df(x)_i \in \mathbb{R}^{m(i)}$ such that

$$\limsup_{y \to x} \frac{|f(y) - f(x) - \langle df(x)_i, \varphi_i(y) - \varphi(x) \rangle|}{d(y, x)} = 0 \, .$$

Cheeger's results were extended to a more general setting, namely metric measure spaces admitting differentiable structures, by Keith [347, 348] (see also [349]). A conjecture from [141], on the absolute continuity of the push-forward $(\varphi_i)_\#(\mu|_{U_i})$ of the restriction of the measure μ to U_i with respect to Lebesgue measure on $\mathbb{R}^{m(i)}$, was positively solved in [184].

Rademacher type results for Lipschitz functions on Carnot groups were proved by Pansu [537]. The case of Lipschitz mappings from a subset of a Carnot group to a Banach homogeneous group, satisfying a suitably weakened Radon–Nikodým property, was treated by Magnani and Rajala [420]. Extensions to the more general framework of stratified groups were given by Magnani [419].

Kirchheim and Magnani [357] gave an example of a Lipschitz function on the Heisenberg group which is nowhere differentiable in the metric sense. This is in contrast to Pansu's classical theorem [537] of a.e. horizontal differentiability of Lipschitz functions on Carnot groups.

Porous sets and their relevance to the differentiability of Lipschitz functions on Carnot groups are discussed by Pinamonti and Speight [573, 574].

For an introduction to analysis on homogeneous spaces, including metric spaces with doubling measures, we recommend the introductory texts [38] and [283]. For a full treatment of more advanced topics and references to recent work, see [285] and [647].

3.5 Bibliographic Comments and Miscellaneous Results

Theorems 3.2.1, 3.2.4 and 3.2.5 are due to N. Levine (see [384, 385]), and Theorem 3.2.7 to Marino [434].

We also mention the following result from [435], related to the characterizations of finite metric spaces given in Theorem 3.2.7. The authors prove that a subset Y of a metric space X is X-finitely chainable if and only if $f(Y)$ is bounded for every uniformly locally Lipschitz mapping $f : X \to \mathbb{R}$. A function $f : X \to \mathbb{R}$ is called *uniformly locally Lipschitz* if there exists $r > 0$ such that for every $x \in X$, f is Lipschitz on $B(x, r)$ (with a Lipschitz constant depending on x).

An ε-chain of length n joining two points $x, y \in X$ is a finite set $x_0, x_1, \ldots, x_n$ of elements in X such that $x_0 = x, x_n = y$ and $d(x_i, x_{i+1}) \leq \varepsilon$ for all $i = 0, 1, \ldots, n-1$. One says that a subset Y of X is finitely ε-chainable in X if there exist a finite subset Z of X and a number m such that every point in Y can be joined with some point in Z by an ε-chain of length m. The set Y is called finitely chainable in X if it is finitely ε-chainable in X for every $\varepsilon > 0$.

Chapter 4
Extension Results for Lipschitz Mappings

In this chapter we present various extension results for Lipschitz functions obtained by Kirszbraun, McShane, Valentine and Flett—the analogs of Hahn-Banach and Tietze extension theorems. A discussion on the corresponding property for semi-Lipschitz functions defined on quasi-metric spaces and for Lipschitz functions with values in a quasi-normed space is included as well.

The problem of finding Lipschitz extensions of a Lipschitz function has diverse applications in geometry (see [470]), computer science, image processing (see [139, 140, 608] and [644]), elasticity and optimal design (see [173, 174] and the references therein), and medicine (see [457, 458]).

4.1 McShane Type Theorems

We prove the existence of extensions of Lipschitz functions preserving Lipschitz constants as well as an extension result for locally Lipschitz functions.

4.1.1 McShane's Theorem

Theorem 4.1.1 (McShane) *Let (X, d) be a metric space, S a subset of X and $f : S \to \mathbb{R}$ an L-Lipschitz function.*

Then the functions $F, G : X \to \mathbb{R}$, defined for $x \in X$ by

$$F(x) := \sup_{y \in S}[f(y) - Ld(x, y)] \quad and \quad G(x) := \inf_{y' \in S}[f(y') + Ld(x, y')],$$

$$(4.1.1)$$

© Springer Nature Switzerland AG 2019
Ş. Cobzaş et al., *Lipschitz Functions*, Lecture Notes in Mathematics 2241,
https://doi.org/10.1007/978-3-030-16489-8_4

are L-Lipschitz extensions of f and any other L-Lipschitz extension H of f satisfies the inequalities

$$F \leq H \leq G. \tag{4.1.2}$$

If f is L-Lipschitz and bounded, then it admits an L-Lipschitz extension $\tilde{H}$ satisfying the condition $\|\tilde{H}\|_\infty = \|f\|_\infty$.

Proof For arbitrary $x \in X$ and $y, y' \in S$, the inequalities

$$f(y) - f(y') \leq Ld(y, y') \leq Ld(x, y) + Ld(x, y'),$$

imply

$$f(y) - Ld(x, y) \leq f(y') + Ld(x, y'). \tag{4.1.3}$$

It follows that the functions $F, G : X \to \mathbb{R}$ given by (4.1.1) are well-defined and

$$F(x) \leq G(x), \tag{4.1.4}$$

for all $x \in X$.

Let us show that F, G are L-Lipschitz extensions of f and any other L-Lipschitz extension H satisfies $F \leq H \leq G$.

I. $F|_S = G|_S = f$.

If $x \in S$, then, putting $y = y' = x$ in (4.1.3) and taking into account the definitions (4.1.1) of the functions F, G and the inequality (4.1.4), one obtains

$$f(x) \leq F(x) \leq G(x) \leq f(x),$$

showing that $F(x) = G(x) = f(x)$.

II. *The functions F, G are L-Lipschitz.*

For $x, x' \in X$ we have

$$G(x) \leq f(y') + Ld(x, y') \leq f(y') + Ld(x', y') + Ld(x, x'),$$

for all $y' \in S$, implying

$$G(x) \leq G(x') + Ld(x, x') \iff G(x) - G(x') \leq Ld(x, x').$$

Interchanging the roles of x and x' one obtains $G(x') - G(x) \leq Ld(x, x')$, so that $|G(x') - G(x)| \leq Ld(x, x')$.

Starting from the inequalities

$$F(x) \geq f(y) - Ld(x, y) \geq f(y) - Ld(x', y) - Ld(x, x'),$$

a similar argument shows that the function F is L-Lipschitz too.

III. $F \leq H \leq G$ *for any L-Lipschitz extension H of f.*

For $x \in X$ and $y \in S$, $H(x) - H(y) = H(x) - f(y)$. Since H is L-Lipschitz, it follows that

$$-Ld(x, y) \leq H(x) - f(y) \leq Ld(x, y),$$

or, equivalently,

$$f(y) - Ld(x, y) \leq H(x) \leq f(y) + Ld(x, y).$$

Taking the supremum with respect to $y \in S$ in the left hand side of the above inequalities and the infimum in the right hand side, one obtains

$$F(x) \leq H(x) \leq G(x).$$

Suppose that, in addition, f is bounded, i.e., $a := \|f\|_\infty < \infty$ and let H be an L-Lipschitz extension of f. The function $\pi_a : \mathbb{R} \to [-a, a]$ given by

$$\pi_a(t) = \begin{cases} t & \text{for} & |t| \leq a \\ at/|t| & \text{for} & |t| > a \end{cases} \tag{4.1.5}$$

is 1-Lipschitz (it is the metric projection of $\mathbb{R}$ onto $[-a, a]$, see Proposition 1.3.7), so that the function $\tilde{H} = \pi_a \circ H : X \to [-a, a]$ is L-Lipschitz and satisfies the equality $\|\tilde{H}\|_\infty = \|f\|_\infty$. □

Remark 4.1.2 In the same way, one can show that if $f : S \to [a, b]$ is L-Lipschitz, then it admits an L-Lipschitz extension $\tilde{F}$ with values in $[a, b]$, i.e., $\tilde{F}(X) \subseteq [a, b]$.

Indeed, taking π_Δ to be the metric projection of the Hilbert space $\mathbb{R}$ onto the convex subset $\Delta = [a, b]$ it follows that π_Δ is 1-Lipschitz. If F is an L-Lipschitz extension of f to X, then $\tilde{F} = \pi_\Delta \circ F$ satisfies all the requirements.

In the complex case the extension with the same Lipschitz constant is not always possible, but one can prove the following result.

Proposition 4.1.3 *Let (X, d) be a metric space, S a subset of X and $f : S \to \mathbb{C}$ an L-Lipschitz function. Then f admits a Lipschitz extension F to X satisfying $L(F) \leq \sqrt{2}\,L$.*

If f is L-Lipschitz and bounded, then it admits a Lipschitz extension $\tilde{F}$ satisfying the conditions $L(\tilde{F}) \leq \sqrt{2}\,L$ and $\|\tilde{F}\|_\infty = \|f\|_\infty$.

Proof Writing f as $f = f_1 + if_2$, it follows that $f_1, f_2 : S \to \mathbb{R}$ are both L-Lipschitz, so that they have L-Lipschitz extensions $F_1, F_2 : X \to \mathbb{R}$. Putting $F = F_1 + iF_2$ it follows that

$$|F(x) - F(y)| = \sqrt{(F_1(x) - F_1(y))^2 + (F_2(x) - F_2(y))^2} \le \sqrt{2}\, Ld(x, y),$$

for all $x, y \in X$.

Supposing $a := \|f\|_\infty < \infty$, consider the function π_a given by (4.1.5) but for $t \in \mathbb{C}$. Then π_a is 1-Lipschitz, so that $\tilde{F} = \pi_a \circ F$ satisfies $L(\tilde{F}) \le L(F) \le L\sqrt{2}$ and $\|\tilde{F}\|_\infty = \|f\|_\infty$. □

Remark 4.1.4 Using the same ideas, one can show that an L-Lipschitz function f defined on a subset S of a metric space (X, d) and with values in the Euclidean space $\mathbb{R}^n$ admits a Lipschitz extension to X satisfying $L(F) \le \sqrt{n}L$.

The following example, taken from [675, p. 18], shows that norm-preserving extensions of complex valued Lipschitz functions do not always exist.

Example 4.1.5 Let $X = \{x_0, x_1, x_2, x_3\}$. Define a metric d on X by $d(x_0, x_i) = 1/2$, $i = 1, 2, 3$, and $d(x_i, x_j) = 1$ for $i, j = 1, 2, 3$, $i \ne j$. Let $S = \{x_1, x_2, x_3\}$ and let $f(x_i)$, $i = 1, 2, 3$, be the vertices of an equilateral triangle in $\mathbb{C}$ with sides of length 1. It follows that $L(f) = 1$ (in fact, f is an isometry) and the extension F of f to X with smallest Lipschitz number takes x_0 to the center of this triangle. Consequently,

$$|F(x_i) - F(x_0)| = \frac{1}{\sqrt{3}} = \frac{2}{\sqrt{3}} d(x_0, x_i), \ i = 1, 2, 3.$$

It follows that $L(F) = 2/\sqrt{3} > 1 = L(f)$.

For a further discussion on this matter, see the paper [609].

In order to state a McShane type result for Lipschitz functions from a locally convex space into $\mathbb{R}^n$, let us recall that given a locally convex Hausdorff space X, a directed family $(p_\alpha)_{\alpha \in A}$ of seminorms generating the topology of X and $S \subseteq X$, a function $f : S \to \mathbb{R}^n$ is called *Lipschitz* if there exist $\alpha \in A$ and $L \ge 0$ such that

$$\|f(x) - f(y)\| \le Lp_\alpha(x - y),$$

for all $x, y \in S$.

Theorem 4.1.6 *Let X be a locally convex Hausdorff space, S a subset of X and $f : S \to \mathbb{R}^n$ a Lipschitz function. Then there exists a Lipschitz function $F : X \to \mathbb{R}^n$ such that $F|_S = f$.*

4.1.2 The Extension of Locally Lipschitz Functions

This subsection is concerned with the extension of real-valued functions which are locally Lipschitz (see Definition 2.1.4) at the points of some subset of a metric space.

Theorem 4.1.7 *Let (X, d) be a metric space, S a closed subset of X and $f : S \to \mathbb{R}$ a function which is locally Lipschitz at each $x \in S$. Then there exists a function $F : X \to \mathbb{R}$ which is locally Lipschitz at each $x \in X$ such that $F|_S = f$.*

Proof For the very beginning, we assume that f is bounded, say

$$|f(x)| \le M$$

for each $x \in S$, for some real number M.

We claim that for each $x \in S$ there exists a real number K_x such that

$$|f(x) - f(y)| \le K_x d(x, y),$$

for all $y \in S$.

Indeed, for each $x \in S$, according to the hypothesis, there exist $k_1 \ge 0$ and $r > 0$ such that

$$|f(x) - f(y)| \le k_1 d(x, y)$$

for each $y \in S \cap B(x, r)$. Let

$$K_x \ge \max \left\{ 2Mr^{-1}, k_1 \right\}.$$

Then, for $y \in S \setminus B(x, r)$ we have

$$|f(x) - f(y)| \le |f(x)| + |f(y)| \le 2M = \frac{2M}{r} \cdot r \le K_x \, d(x, y),$$

so that

$$|f(x) - f(y)| \le K_x d(x, y),$$

for all $y \in S$.

We shall prove that the function $F : X \to \mathbb{R}$ given by

$$F(x) = \inf_{y \in S}[f(y) + K_y d(x, y)], \quad x \in X,$$

is the extension that we are looking for.

Since

$$f(y) + K_y d(x, y) \geq -M,$$

for each $y \in S$, F is well-defined and

$$F(x) \geq -M, \qquad (4.1.6)$$

for all $x \in X$.

Let $x \in S$. Then

$$F(x) = \inf_{y \in S}[f(y) + K_y d(x, y)] \leq f(x).$$

Also, for every $y \in S$,

$$f(x) - f(y) \leq K_y d(x, y) \iff f(x) \leq f(y) + K_y d(x, y),$$

implies $f(x) \leq F(x)$, and so $F(x) = f(x)$, that is, $F|_S = f$.

We have to prove that F is locally Lipschitz at every $x_0 \in X$. To this end we shall consider two cases.

I. $x_0 \in S$.

For $x \in X$ we have

$$F(x) \leq f(x_0) + K_{x_0} d(x, x_0),$$

so that

$$F(x) - F(x_0) = F(x) - f(x_0) \leq K_{x_0} d(x, x_0).$$

If $y \in S \setminus \{x_0\}$, then

$$f(y) - f(x_0) \geq -K_y d(y, x_0)$$

and

$$f(y) - f(x_0) \geq -K_{x_0} d(y, x_0).$$

Multiplying the first inequality by $d(x, y)/\left(d(x, y) + d(x, x_0)\right)$, the second one by $d(x, x_0)/\left(d(x, y) + d(x, x_0)\right)$ and adding them, one obtains

$$f(y) - f(x_0) \geq -[K_y d(x, y) + K_{x_0} d(x, x_0)]\frac{d(y, x_0)}{d(x, y) + d(x, x_0)}$$

$$\geq -[K_y d(x, y) + K_{x_0} d(x, x_0)],$$

so that

$$f(y) + K_y d(x, y) \geq f(x_0) - K_{x_0} d(x, x_0).$$

The last inequality is also valid for $y = x_0$ and therefore

$$F(x) = \inf_{y \in S}[f(y) + K_y d(x, y)] \geq f(x_0) - K_{x_0} d(x, x_0),$$

i.e.,

$$F(x) - F(x_0) = F(x) - f(x_0) \geq -K_{x_0} d(x, x_0).$$

Hence

$$|F(x) - F(x_0)| \leq K_{x_0} d(x, x_0),$$

for all $x \in X$.

II. $x_0 \in X \setminus S$.

For $r := d(x_0, S) > 0$ let $U := B(x_0, r/2)$.
It follows that

$$d(x, y) \geq d(x_0, y) - d(x_0, x) > r - \frac{r}{2} = \frac{r}{2}, \tag{4.1.7}$$

for every $x \in U$ and every $y \in S$.

We claim that F is bounded on U. Fix an element $z_0 \in S$. Then for any $x \in U$

$$F(x) \leq f(z_0) + K_{z_0} d(x, z_0) \leq f(z_0) + K_{z_0}(d(x, x_0) + d(x_0, z_0))$$

$$\leq f(z_0) + K_{z_0}\left(\frac{r}{2} + d(x_0, z_0)\right).$$

On the other hand, by (4.1.6), $F(x) \geq -M$, so that there exists N such that $|F(x)| \leq N$, for all $x \in U$.

Let now x be an arbitrary element of U. Then, for each $\varepsilon > 0$ there exists $y_\varepsilon \in S$ such that

$$F(x) \geq f(y_\varepsilon) + K_{y_\varepsilon} d(x, y_\varepsilon) - \varepsilon. \tag{4.1.8}$$

Since

$$F(x_0) \leq f(y_\varepsilon) + K_{y_\varepsilon} d(x_0, y_\varepsilon),$$

we get

$$F(x_0) - F(x) \le K_{y_\varepsilon}(d(x_0, y_\varepsilon) - d(x, y_\varepsilon)) + \varepsilon \le K_{y_\varepsilon} d(x_0, x) + \varepsilon. \qquad (4.1.9)$$

From (4.1.8) and (4.1.7) one obtains

$$K_{y_\varepsilon} \le \frac{F(x) - f(y_\varepsilon) + \varepsilon}{d(x, y_\varepsilon)} \le \frac{2(N + M + \varepsilon)}{r},$$

so that (4.1.9) becomes

$$F(x_0) - F(x) \le \frac{2(N + M + \varepsilon)}{r} d(x_0, x) + \varepsilon.$$

As $\varepsilon > 0$ was arbitrarily chosen, we get

$$F(x_0) - F(x) \le \frac{2(N + M)}{r} d(x_0, x).$$

Similarly

$$F(x) - F(x_0) \le \frac{2(N + M)}{r} d(x_0, x),$$

so that

$$|F(x) - F(x_0)| \le \frac{2(N + M)}{d(x_0, S)} d(x, x_0),$$

for all $x \in B(x_0, \frac{1}{2}d(x_0, S))$.

Thus the theorem is proved for f bounded.

Before proving the result in its full generality, we make the following remark.

Claim 1. *If for some $M > 0$ the Lipschitz function f satisfies the condition*

$$|f(x)| < M,$$

for all $x \in S$, then we can choose the extension F such that

$$|F(x)| < M,$$

for all $x \in X$.

Indeed, we can assume, without losing the generality, that $K_x > 1$ for each $x \in S$. For every $x \in X \setminus S$,

$$F(x) \ge -M + d(x, S) > -M,$$

so that the function $F_1 : X \to \mathbb{R}$ *given by*

$$F_1(x) = \min\{F(x), M\}, \quad x \in X,$$

is an extension of f *that is locally Lipschitz at each* $x \in X$ *and which satisfies the inequality*

$$-M < F_1(x) \leq M,$$

for each $x \in X$. *Similarly we find an extension* F_2 *of* $-f$ *that is locally Lipschitz at each* $x \in X$ *and which satisfies the inequality*

$$-M < F_2(x) \leq M \iff -M \leq -F_2(x) < M$$

for each $x \in X$. *Then the function* $H = \frac{1}{2}[F_1 - F_2]$ *is an extension of* f *that is locally Lipschitz at each* $x \in X$ *and which satisfies the inequality*

$$-M < H(x) < M$$

for each $x \in X$.

Claim 2. The proof in the general case.

   Supposing that the function $f : S \to X$ *is arbitrary (i.e., possibly unbounded), consider the function* $g = \arctan \circ f$ *which is locally Lipschitz at each* $x \in S$ *and bounded, namely*

$$|g(x)| < \frac{\pi}{2},$$

for all $x \in S$.

By *Claim 1, there exists an extension* G *of* g *that is locally Lipschitz at each* $x \in X$ *such that*

$$|G(x)| < \frac{\pi}{2}$$

for each $x \in X$.

   Then the function $F = \tan \circ G$ *is an extension of* f *that is locally Lipschitz on* X.

   It is obvious that $F(x) = f(x)$ *for* $x \in S$.

   For $x_0 \in X$ *let* $\delta > 0$ *be such that* $U := [G(x_0) - \delta, G(x_0) + \delta] \subseteq (-\pi/2, \pi/2)$. *Let* V *be a neighborhood of* x_0 *such that* $G(V) \subseteq U$ *and* G *is Lipschitz on* V. *Since, the function* $\tan$ *is Lipschitz on* U *it follows that* $F = \tan \circ G$ *is Lipschitz on* V.   □

4.2   Extension Results for Lipschitz Vector-Functions

In this section we consider the general problem of the existence of norm-preserving extensions for Lipschitz functions defined on a subset of a metric space (X, d) and with values in another metric space (X', d').

The following example shows that the extension is not always possible (see also Example 4.1.5).

Example 4.2.1 Denote by $\| \cdot \|_\infty$ the ℓ^∞-norm on $\mathbb{R}^2$,

$$\|x\|_\infty = \max\{|x_1|, |x_2|\},$$

and let $\| \cdot \|_2$ be the Euclidean norm on $\mathbb{R}^2$.

Let $S = \{a_1, a_2, a_3\}$ be the subset of $(\mathbb{R}^2, \| \cdot \|_\infty)$ formed of the points $a_1 = (1, -1)$, $a_2 = (1, 1)$, $a_3 = (-1, 1)$ and let $f : (S, \|\cdot\|_\infty) \to (\mathbb{R}^2, \|\cdot\|_2)$ be given by

$$f(1, -1) = (1, 0), \quad f(-1, 1) = (-1, 0), \quad \text{and } f(1, 1) = (0, \sqrt{3}).$$

Then f satisfies

$$\|f(u) - f(v)\|_2 = 2 = \|u - v\|_\infty,$$

for all $u, v \in S$, $u \neq v$, i.e., f is an isometry, hence $L(f) = 1$.

Put $X = S \cup \{(0, 0)\}$ and show that f does not have a Lipschitz extension F to X with $L(F) = 1$.

Indeed, taking $F(0, 0)$ to be the center of the triangle $F(a_1)$, $F(a_2)$, $F(a_3)$, i.e., $F(0, 0) = \left(0, \frac{\sqrt{3}}{3}\right)$ it follows that

$$\|F(u) - F(0, 0)\|_2 = \frac{2}{\sqrt{3}} = \frac{2}{\sqrt{3}} \|u - (0, 0)\|_\infty,$$

for all $u \in S$, implying $L(F) \geq \frac{2}{\sqrt{3}} > 1 = L(f)$. If $F(0, 0) = \eta \in \mathbb{R}^2$, with $\eta \neq (0, \frac{\sqrt{3}}{3})$, then there exists $u \in S$ such that $\|F(u) - \eta\|_2 > \frac{2}{\sqrt{3}}$ so that, as above,

$$\|F(u) - F(0, 0)\|_2 > \frac{2}{\sqrt{3}} = \frac{2}{\sqrt{3}} \|u - (0, 0)\|_\infty,$$

hence $L(F) > \frac{2}{\sqrt{3}} > 1 = L(f)$.

Remark 4.2.2 The example given in Example 4.2.1 is typical in the following sense. Let X, Y be metric spaces. Suppose that for every three point set $Z \subseteq X$ and every $x \in X \setminus Z$, every function $f : Z \to Y$ has a Lipschitz extension $F : Z \cup \{x\} \to Y$

with $L(F) = L(f)$. Then, using Zorn's lemma, it follows that for every subset S of X every Lipschitz function $f : S \to Y$ has a Lipschitz extension $F : X \to Y$ with $L(F) = L(f)$.

4.2.1 Kirszbraun and Valentine

An important case in which norm-preserving extensions of Lipschitz functions exist is presented in the following theorem.

Theorem 4.2.3 (Kirszbraun and Valentine) *Let $\mathscr{H}_1, \mathscr{H}_2$ be real Hilbert spaces and $f : S \to \mathscr{H}_2$ a Lipschitz function defined on a subset S of $\mathscr{H}_1$.*
Then there exists a Lipschitz function $F : \mathscr{H}_1 \to \mathscr{H}_2$ such that:

(a) $F|_S = f$,

and

(b) $L(F) = L(f)$.

In particular, this is true for the Euclidean spaces $\mathscr{H}_1 = \mathbb{R}^m$ and $\mathscr{H}_2 = \mathbb{R}^n$.

4.2.2 The Contraction Extension Property and the Intersection of Balls

For the proof of Theorem 4.2.3 we need some preliminary notions and results.

Consider two metric spaces (X, d) and (X', d') and a nonempty subset S of X. A function $f : S \to X'$ is called a *contraction* (*nonexpansive* sometimes) if it is 1-Lipschitz, that is,

$$d'(f(x), f(y)) \le d(x, y),$$

for all $x, y \in S$.

Definition 4.2.4 We say that a pair $((X, d), (X', d'))$ of metric spaces has the *contraction extension property* if for each subset S of X each contraction mapping $f : S \to X'$ has a contraction extension $F : X \to X'$.

A pair $((X, d), (X', d'))$ of metric spaces is said to have the *Lipschitz extension property* if for any nonempty subset S of X any Lipschitz function $f : S \to X'$ has an extension $F : X \to X'$ with the same Lipschitz constant. In particular, there exists an extension with $L(F) = L(f)$.

Consider now two families of closed balls

$$B_i = \{x \in X : d(x_i, x) \le r_i\}, \ i \in I,$$

and

$$B_i' = \{x' \in X' : d'(x_i', x') \le r_i\}, \ i \in I.$$

Definition 4.2.5 (Kirszbraun Property (K)) We say that the pair of metric spaces (X, X') has *property* (K) if for any families $\{B_i : i \in I\}$ and $\{B_i' : i \in I\}$ of closed balls, as above, satisfying the condition

$$d'(x_i', x_j') \le d(x_i, x_j),$$

for all $i, j \in I$, the following implication holds

$$\bigcap_{i \in I} B_i \ne \emptyset \ \Rightarrow \ \bigcap_{i \in I} B_i' \ne \emptyset.$$

Remark 4.2.6 Kirszbraun [359] proved the existence of norm-preserving extensions for Lipschitz functions in the case $X = X' = \mathbb{R}^n$. Valentine [666] (see also [664, 665]) remarked that the key tool in the proof of the existence of the extension is a property expressed in terms of the intersection of some families of balls (property (K) from above) and proved the existence of the extensions in the Hilbert case (for $X = X' = \mathscr{H}$). Valentine [666] also proved that property (K) is satisfied in the case $X = X' = S_{\mathbb{R}^n}$—the unit sphere of $\mathbb{R}^n$ with the Euclidean norm. We give the proof in the Hilbert case.

Theorem 4.2.7 *For any pair (X, d), (X', d') of metric spaces the following two properties are equivalent.*

1. *The pair (X, X') has the contraction extension property.*
2. *The pair (X, X') has property (K).*
   Moreover, if X and X' are normed spaces, then each of the above properties is equivalent to the following one.
3. *The pair (X, X') has the Lipschitz extension property.*

Proof $1 \Rightarrow 2$. Suppose that $x_i \in X$, $x_i' \in X'$, $r_i > 0$, $i \in I$, are such that

$$d'(x_i', x_j') \le d(x_i, x_j), \ \text{for all} \ i, j \in I, \ \text{and} \ \bigcap_{i \in I} B_i \ne \emptyset,$$

where $B_i = B[x_i, r_i]$, $i \in I$.

The function $f : \{x_i : i \in I\} \to \{x_i' : i \in I\}$ given by

$$f(x_i) = x_i', \quad i \in I,$$

satisfies the inequalities

$$d'(f(x_i), f(x_j)) \le d(x_i, x_j),$$

for all $i, j \in I$, i.e., it is a contraction. By hypothesis, there exists a contraction $F : X \to X'$ such that $F(x_i) = f(x_i) = x_i'$ for all $i \in I$.

Let $x \in \bigcap_{i \in I} B_i$. Then

$$d'(F(x), x_i') = d'(F(x), F(x_i)) \le d(x, x_i) \le r_i \,,$$

for all $i \in I$, showing that $F(x) \in \bigcap_{i \in I} B_i'$.

$2 \Rightarrow 1$. Consider the set

$$\mathscr{M} = \{(g, U) : S \subseteq U \subseteq X \text{ and } g : U \to X' \text{ is a contraction extension of } f\} \,,$$

ordered by

$$(g_1, U_1) \preceq (g_2, U_2) \iff U_1 \subseteq U_2 \text{ and } g_2|_{U_1} = g_1 \,.$$

The set $(\mathscr{M}, \preceq)$ is inductively ordered. Indeed, let (g_i, U_i), $i \in I$, be a totally ordered subset of $\mathscr{M}$. Put $U = \bigcup_{i \in I} U_i$ and define $g : U \to X'$ by $g(u) = g_i(u)$ where $i \in I$ is such that $u \in U_i$. It is easy to check that the function g is well-defined, $(g, U) \in \mathscr{M}$ and $(g_i, U_i) \preceq (g, U)$ for all $i \in I$.

Consequently, the set $\mathscr{M}$ contains a maximal element (F, Y). The proof will be done if we show that $Y = X$.

Supposing that there exists $x_0 \in X \setminus Y$ put $r_x := d(x, x_0) > 0$, $x \in Y$, and consider the balls $B[x, r_x] \subseteq X$ and $B'[F(x), r_x] \subseteq X'$, for all $x \in Y$. Then $d'(F(x), F(x')) \le d(x, x')$ for all $x, x' \in Y$ (because F is a contraction on Y). Also, the equalities $d(x_0, x) = r_x$, $x \in Y$, show that $x_0 \in \bigcap_{x \in Y} B[x, r_x]$. By hypothesis, there exists an $y_0 \in \bigcap_{x \in Y} B'[F(x), r_x]$.

Put $Z := Y \cup \{x_0\}$ and define $G : Z \to X'$ by $G(x) = F(x)$ for $x \in Y$ and $G(x_0) = y_0$. Then $G|_Y = f$ and $y_0 \in \bigcap_{x \in Y} B'[F(x), r_x]$ is equivalent to

$$d'(G(x_0), G(x)) = d'(y_0, F(x)) \le r_x = d(x_0, x) \,,$$

for all $x \in Y$, showing that G is a contraction, i.e., $(G, Z) \in \mathscr{M}$. But $(G, Z) \ne (F, Y)$ and $(F, Y) \preceq (G, Z)$, in contradiction to the maximality of (F, Y).

It is obvious that $3 \Rightarrow 1$ for arbitrary metric spaces X, X'.

Suppose now that X, X' are normed spaces and that 1 holds. Let $f : Y \to X'$ be an L-Lipschitz map, for some subset Y of X. If $L = 0$, then f is a constant function, $f(x) = c$, $x \in Y$, for some $c \in X'$, which automatically extends to $F(x) = c$, $x \in X$.

If $L > 0$, then $g = L^{-1} f$ is a contraction which, by hypothesis, has an extension to a contraction $G : X \to X'$. But then $F = LG$ is an L-Lipschitz mapping extending f. □

4.2.3   The Proof of Theorem 4.2.3

By Theorem 4.2.7 it suffices to show that the pair $(\mathscr{H}_1, \mathscr{H}_2)$ has the Kirszbraun property (K) from Definition 4.2.5

Let $B_i = B[x_i, r_i]$, $i \in I$, and $B_i' = B[y_i, r_i]$, $i \in I$, be two families of closed balls in $\mathscr{H}_1$ and $\mathscr{H}_2$, respectively, such that

$$\|y_i - y_j\| \le \|x_i - x_j\|, \quad \text{for all } i, j \in I, \quad \text{and} \quad \bigcap_{i \in I} B_i \neq \emptyset.$$

We have to prove that

$$\bigcap_{i \in I} B_i' \neq \emptyset.$$

Suppose that

$$\bigcap_{j \in J} B_j' \neq \emptyset, \tag{4.2.1}$$

for every nonempty finite subset J of I. Fix an element $i_0 \in I$ and consider the sets $C_i = B_{i_0}' \cap B_i'$, $i \in I$. Since all the balls B_i' are weakly compact, it follows that the sets C_i are weakly compact, hence weakly closed subsets of the weakly compact set B_{i_0}'. By (4.2.1) this family has the finite intersection property, so it has a nonempty intersection, hence

$$\bigcap_{i \in I} B_i' \overset{=}{=} \bigcap_{i \in I} C_i \neq \emptyset.$$

Suppose that $\xi \in \bigcap_{i \in I} B_i$ and let J be a finite subset of I. To simplify the notation suppose that $J = \{1, 2, \ldots, n\}$ for some $n \in \mathbb{N}$.

If $\xi = x_i$ for some $i \in J$, then $\|y_j - y_i\| \le \|x_j - x_i\| \le r_j$ for all $j \in J$, showing that $y_i \in \bigcap_{j \in J} B_j'$.

Suppose now that $\xi \notin \{x_1, \ldots, x_n\}$ and consider the function $g : \mathscr{H}_2 \to \mathbb{R}$ given by

$$g(y) = \max\left\{ \frac{\|y - y_i\|}{\|\xi - x_i\|} : 1 \le i \le n \right\}, \quad y \in \mathscr{H}_2.$$

Let $\lambda = \inf g\,(\mathscr{H}_2)$. Since $\lim_{\|y\| \to \infty} g(y) = \infty$, there exists $r > 0$ such that $g(y) > \lambda + 1$ for all $y \in \mathscr{H}_2$ with $\|y\| > r$. As the function g is weakly lower semicontinuous and the ball $r B_{\mathscr{H}_2}$ is weakly compact, there exists $\eta \in r B_{\mathscr{H}_2}$ such that

$$0 < g(\eta) = \inf g\left(r B_{\mathscr{H}_2}\right) = \lambda.$$

By an appropriate numbering of the points x_i, we can suppose that

> (a) $\|\eta - y_i\| = \lambda \|\xi - x_i\|$ for $1 \le i \le k$ and
>
> (b) $\|\eta - y_i\| < \lambda \|\xi - x_i\|$ for $k < i \le n$,

$$(4.2.2)$$

for some $k \in \{1, 2, \ldots, n\}$.

We show that

$$\eta \in \mathrm{co}(\{y_1, \ldots, y_k\}) \, . \tag{4.2.3}$$

If this is not true, then η can be strictly separated from $\mathrm{co}(\{y_1, \ldots, y_k\})$ (see Theorem 1.4.9), so there exists $u \in \mathscr{H}_2$, $\|u\| = 1$, such that

$$\langle \eta, u \rangle < \langle y_i, u \rangle \quad \text{for} \ 1 \le i \le k,$$

or, equivalently,

$$\langle \eta - y_i, u \rangle < 0 \quad \text{for} \ 1 \le i \le k. \tag{4.2.4}$$

For $z = \eta + tu$ we have

$$\|z - y_i\|^2 = \|\eta - y_i\|^2 + t \left(2\langle \eta - y_i, u \rangle + t\|u\|^2 \right),$$

for all $1 \le i \le k$.

Taking into account (4.2.4) and (4.2.2).(b), we can choose a sufficiently small $t > 0$ in order that $2\langle \eta - y_i, u \rangle + t\|u\|^2 < 0$ for $1 \le i \le k$ and $\|z - y_i\| < \lambda \|\xi - x_i\|$ for $k < i \le n$. But then $g(z) < g(\eta)$ in contradiction to the choice of η.

Now, by (4.2.3), there exist $\lambda_1, \lambda_2, \ldots, \lambda_k \ge 0$ with $\lambda_1 + \lambda_2 + \cdots + \lambda_k = 1$ such that

$$\eta = \lambda_1 y_1 + \lambda_2 y_2 + \cdots + \lambda_k y_k \, . \tag{4.2.5}$$

We have

$$\|y_i - y_j\|^2 = \|(y_i - \eta) + (\eta - y_j)\|^2$$
$$= \|y_i - \eta\|^2 + \|y_j - \eta\|^2 + 2\langle y_i - \eta, \eta - y_j \rangle,$$

and, similarly,

$$\|x_i - x_j\|^2 = \|x_i - \xi\|^2 + \|x_j - \xi\|^2 + 2\langle x_i - \xi, \xi - x_j \rangle.$$

Replacing these in the inequality $\|y_i - y_j\|^2 \leq \|x_i - x_j\|^2$, one obtains

$$\|y_i - \eta\|^2 + \|y_j - \eta\|^2 + 2\langle y_i - \eta, \eta - y_j \rangle \leq \|x_i - \xi\|^2 + \|x_j - \xi\|^2$$
$$+ 2\langle x_i - \xi, \xi - x_j \rangle, \qquad (4.2.6)$$

for all $1 \leq i, j \leq k$.

If $\lambda > 1$, then by (4.2.2).(a),

$$-\|y_i - \eta\|^2 < -\|x_i - \xi\|^2 \text{ and } -\|y_j - \eta\|^2 < -\|x_j - \xi\|^2.$$

Adding these inequalities to (4.2.6) one obtains

$$2\langle y_i - \eta, \eta - y_j \rangle < 2\langle x_i - \xi, \xi - x_j \rangle$$

or, equivalently,

$$\langle y_i - \eta, y_j - \eta \rangle > \langle x_i - \xi, x_j - \xi \rangle, \qquad (4.2.7)$$

for all $1 \leq i, j \leq k$.

Observe that, by (4.2.5),

$$\sum_{i,j=1}^{k} \lambda_i \lambda_j \langle y_i - \eta, y_j - \eta \rangle = \sum_{j=1}^{k} \lambda_j \langle \sum_{i=1}^{k} \lambda_i y_i - \eta, y_j - \eta \rangle = 0, \qquad (4.2.8)$$

and, by a direct calculation,

$$\sum_{i,j=1}^{k} \lambda_i \lambda_j \langle x_i - \xi, x_j - \xi \rangle = \Big\| \sum_{i=1}^{k} \lambda_i (x_i - \xi) \Big\|^2. \qquad (4.2.9)$$

Multiplying the inequality (4.2.7) by $\lambda_i \lambda_j$, summing up for $i, j = 1, \ldots, k$, and taking into account the equalities (4.2.8) and (4.2.9), one obtains the contradiction

$$0 > \Big\| \sum_{i=1}^{k} \lambda_i (x_i - \xi) \Big\|^2 \geq 0.$$

Consequently, $\lambda \leq 1$, and the relations (4.2.2) show that $\eta \in \bigcap_{i=1}^{n} B[y_i, r_i]$, i.e., (4.2.1) holds.

Theorem 4.2.3 is completely proved.

4.2.4   Flett's Theorem

As we have seen in Example 4.2.1, a Lipschitz function defined on a subset C of a normed space X and taking values in a normed space Y may not have norm-preserving extensions to X. But one can show that for some particular subsets C of X, there exists $\alpha \geq 1$, depending only on C, such that every L-Lipschitz function $f : C \to Y$ admits an extension $F : X \to Y$ satisfying

$$\|F(x) - F(y)\| \leq \alpha L \|x - y\|,$$

for all $x, y \in X$.

Definition 4.2.8 We say that a pair $((X, d), (Y, d'))$ of metric spaces has the *generalized Lipschitz extension property* if there exists a constant $\alpha \geq 1$, depending only on X and Y, such that for every subset A of X, every Lipschitz function $f : A \to Y$ admits an extension $F : X \to Y$ satisfying the condition

$$L(F) \leq \alpha L(f).$$

The following result, proved by Flett [239] emphasizes such a situation.

Theorem 4.2.9 *Let X, Y be normed spaces, $x_0 \in X$, $\delta > 0$ and C a bounded, closed, convex subset of X containing the closed ball $B[x_0, \delta]$. If the function $f : C \to Y$ is L-Lipschitz on C, then there exists an extension $F : X \to Y$ of f such that*

$$\|F(x) - F(y)\| \leq \frac{\varrho}{\delta} L \|x - y\|,$$

for all $x, y \in X$, where $\varrho = \operatorname{diam} C$.

We begin with the following result.

Proposition 4.2.10 *Consider a bounded, closed, convex subset C of a normed space X, $z \in C$, $y \in \partial C$, $\delta > 0$ such that $B := B[0, \delta] \subseteq C$. Then*

$$\|\lambda y - z\| \geq \frac{\delta}{\varrho} \|y - z\|, \tag{4.2.10}$$

for every $\lambda \geq 1$, where $\varrho = \operatorname{diam} C$.

Proof It is obvious that we can suppose $z \neq y$.

The inequality (4.2.10) is obvious if z belongs to the line passing through 0 and y since in this situation there exists $0 < \mu < 1$ or $\mu < 0$ such that $z = \mu y$ and therefore

$$\|\lambda y - z\| = (\lambda - \mu)\|y\| \geq (1 - \mu)\|y\| = \|y - z\| \geq \frac{\delta}{\varrho}\|y - z\|,$$

because $\varrho \geq 2\delta$.

Consequently, we can suppose that y does not belong to the line D passing through 0 and z.

Let P be the 2-dimensional normed subspace of X generated by y and z. In the sequel we shall use the following notation

$$(a, b) = \{ta + (1 - t)b : t \in (0, 1)\}$$
$$[a, b] = \{ta + (1 - t)b : t \in [0, 1]\} = (a, b) \cup \{a, b\},$$

for $a, b \in X$, and

$$d(x) = d(x, D),$$

for the distance from a point $x \in P$ to D.

Note that, for $x \in P$, $v \in D$ and $\alpha > 0$, we have

$$d(x + v) = d(x) \quad \text{and} \quad d(\alpha x) = \alpha d(x). \tag{4.2.11}$$

We claim that there exists $w \in B \cap P \subseteq C$ such that w and y are on the same side of D and

$$d(w) = \delta = \|w\|. \tag{4.2.12}$$

Indeed, let φ be a linear (and automatically continuous) functional φ on P such that $D = \{x \in P : \varphi(x) = 0\}$. We can further suppose that $\|\varphi\| = 1$ and $\varphi(y) > 0$. By the compactness of B there exists $w \in B$ such that

$$\varphi(w) = \sup\{\varphi(x) : x \in B[0, \delta], \varphi(x) \geq 0\}.$$

But

$$\sup\{\varphi(x) : x \in B, \varphi(x) \geq 0\} = \delta \sup\{\varphi(x) : x \in B[0, 1], \varphi(x) \geq 0\} = \delta\|\varphi\| = \delta,$$

and, by Ascoli's formula (1.4.3), $\delta = \varphi(w) = d(w, D)$. From $\varphi(w) = \delta\|\varphi\|$ it follows that $\|w\| = \delta$.

Now we prove that, for a given $\lambda \geq 1$, exactly one of the following two situations can occur:

(i) $[0, y] \cap (w, z) \neq \emptyset$,   say $[0, y] \cap (w, z) = \{u\}$,

   or

(ii) $M \cap [\lambda y, z] \neq \emptyset$,   say $M \cap [\lambda y, z] = \{v\}$,

where M is the line passing through 0 and w.

Indeed, let us suppose that (ii) is not valid and that $M = \{x : \psi(x) = 0\}$ where $\psi : P \to \mathbb{R}$ is a linear continuous functional chosen such that $\psi(z) > 0$. Since $M \cap [\lambda y, z] = \emptyset$ it follows that $\psi(x) > 0$ for all $x \in [\lambda y, z]$, in particular $\psi(\lambda y) > 0$ and so $\psi(y) > 0$. Because $w \in P$ and, by hypothesis, $w \notin [0, \lambda y]$, there exist $\alpha, \beta \in \mathbb{R}$ such that $y = \alpha w + \beta z$. Since $\varphi(y) = \alpha \varphi(w)$ and $\psi(y) = \beta \psi(z)$ we infer that $\alpha > 0$ and $\beta > 0$. Hence $\frac{1}{\alpha+\beta} y = \frac{\alpha}{\alpha+\beta} w + \frac{\beta}{\alpha+\beta} z \in (w, z) \subseteq C$, as C is convex and $w, z \in C$. Since $0 \in \text{int}(C)$ and $y \in \partial C$, it follows that $\frac{1}{\alpha+\beta} y \in [0, y]$. Hence $\frac{1}{\alpha+\beta} y \in [0, y] \cap (w, z)$, and so $[0, y] \cap (w, z) \neq \emptyset$, i.e., (i) is valid.

Let $\lambda_0 \in [1, \infty)$ be the least element in $[1, \infty)$ such that $\|\lambda_0 y - z\| = \inf\{\|\lambda y - z\| : \lambda \in [1, \infty)\}$.

If $\lambda_0 = 1$, the proof is done since $\|y - z\| \leq \|\lambda y - z\| \leq \frac{\varrho}{\delta} \|\lambda y - z\|$, so $\|\lambda y - z\| \geq \frac{\delta}{\varrho} \|y - z\|$ for every $\lambda \in [1, \infty)$, i.e., (4.2.10) is true.

Therefore we can suppose that $\lambda_0 > 1$.

Hence, $\|\lambda_0 y - z\| < \|y - z\|$ and

$$\|y - z\| \leq \|x - z\|, \tag{4.2.13}$$

for each $x \in [0, y]$.

Indeed, suppose that $\|x - z\| < \|y - z\|$ for some $x \in [0, y)$. Then, as y is a convex combination of x and $\lambda_0 y$, $\|z - y\| \leq \max\{\|z - x\|, \|z - \lambda_0 y\|\}$, in contradiction to the inequalities $\|z - y\| > \|z - x\|$ and $\|z - y\| > \|z - \lambda_0 y\|$.

Let us prove the inequality (4.2.10) in the case (i): $[0, y] \cap (w, z) = \{u\}$.

In this case there exists $\sigma \in (0, 1)$ such that $u = \sigma w + (1 - \sigma) z$ and, by (4.2.13), we have

$$\|y - z\| \leq \|u - z\| = \sigma \|w - z\| \leq \sigma \varrho \quad \text{(because } z, w \in C\text{)}. \tag{4.2.14}$$

If $\mu \geq 1$ is such that $\lambda_0 y = \mu u$, then, taking into account (4.2.11) and the fact that $z \in D$, we have

$$\|\lambda_0 y - z\| \geq d(\lambda_0 y) = d(\mu u) = \mu d(u)$$
$$\geq d(u) = d(u - z) = \sigma d(w - z) = \sigma d(w) = \sigma \delta.$$

Hence, by (4.2.14),

$$\|\lambda_0 y - z\| \ge \sigma\delta \ge \frac{\|y - z\|}{\varrho}\delta,$$

so that (4.2.10) is valid in the case (i).

Let us prove the inequality (4.2.10) in the case (ii): $M \cap [\lambda_0 y, z] = \{v\}$.

Since $\varphi(\lambda_0 y) > 0$ it follows that $\varphi(x) > 0$ for every $x \in [\lambda_0 y, z)$. As $\varphi(w) > 0$, it follows that $v = tw$, for some $t > 0$.

If $t \ge 1$, i.e., $w \in [0, v]$, then, taking into account (4.2.12), one obtains

$$\|\lambda_0 y - z\| \ge \|v - z\| \ge d(v) \ge d(w) = \delta \ge \delta \frac{\|y - z\|}{\varrho},$$

i.e., (4.2.10) is valid.

If $0 < t < 1$, then $v \in (0, w)$ and

$$\delta = \|w\| = \|v\| + \|w - v\|.$$

Consequently,

$$\|v - z\| \ge \|z\| - \|v\| = \|z\| - \delta + \|v - w\|$$

and

$$\|v - z\| \ge \|w - z\| - \|v - w\| \ge d(w) - \|v - w\| = \delta - \|v - w\|.$$

Hence, by adding the last two inequalities, we get $\|v - z\| \ge \frac{1}{2}\|z\|$. Taking $x = 0$ in (4.2.13) we get $\|y - z\| \le \|z\|$. Thus,

$$\frac{\|\lambda_0 y - z\|}{\|y - z\|} \ge \frac{\|v - z\|}{\|y - z\|} \ge \frac{1}{2}\frac{\|z\|}{\|y - z\|} \ge \frac{1}{2} \ge \frac{\delta}{\varrho},$$

so (4.2.10) is true.   □

Now we can prove Flett's theorem.

Proof of Theorem 4.2.9 Suppose first that $x_0 = 0$ and define the function $g : X \to C \subseteq X$ by

$$g(x) = \begin{cases} x', & \text{where } \{x'\} = \partial C \cap [0, x] \text{ if } x \notin C, \\ x, & \text{if } x \in C. \end{cases}$$

It is obvious that g is well-defined.

We claim that

$$\|g(x) - g(y)\| \le \frac{\varrho}{\delta}\|x - y\|,$$

for all $x, y \in X$.

In order to prove the above inequality, we have to consider the following three cases:

a) $x, y \in C$.
b) $x \in C$ and $y \in X \setminus C$.
c) $x, y \in X \setminus C$.

In the first case, we have

$$\|g(x) - g(y)\| = \|x - y\| \le \frac{\varrho}{\delta}\|x - y\|,$$

because $\varrho/\delta \ge 2$.

In the second case there exists $\lambda \ge 1$ such that $y = \lambda g(y)$ and therefore, taking into account Proposition 4.2.10,

$$\|g(x) - g(y)\| = \|x - g(y)\| \le \frac{\varrho}{\delta}\|x - \lambda g(y)\| = \frac{\varrho}{\delta}\|x - y\|.$$

In the third case, putting $x' = g(x)$ and $y' = g(y)$, it follows that $x = \lambda x'$, $y = \mu y'$ for some $\lambda, \mu \ge 1$. Without loosing the generality we can suppose that $1 \le \mu \le \lambda$, so that, by Proposition 4.2.10,

$$\|g(x) - g(y)\| = \|x' - y'\| \le \frac{\varrho}{\delta}\|\frac{\lambda}{\mu}x' - y'\|$$

$$= \frac{1}{\mu}\frac{\varrho}{\delta}\|\lambda x' - \mu y'\| \le \frac{\varrho}{\delta}\|x - y\|,$$

proving that g is ϱ/δ-Lipschitz.

Finally, the desired extension of f is given by $F = f \circ g$.

The case of an arbitrary x_0 with $B[x_0, \delta] \subseteq C$ reduces to the preceding one by considering the set $\tilde{C} := -x_0 + C$, the function $\tilde{f} : \tilde{C} \to Y$ given by $\tilde{f}(x) = f(x + x_0)$, $x \in \tilde{C}$, and replacing the norm $\| \cdot \|_X$ in X with the norm $L\| \cdot \|_X$.

If $\tilde{F}$ is the extension of $\tilde{f}$ whose existence was proved above, then $F(x) = \tilde{F}(x - x_0)$, $x \in C$, is an extension of f satisfying the requirements of the theorem. $\square$

Remark 4.2.11 In contrast to Kirszbraun's theorem which just asserts the existence of the extension, Flett's theorem gives a concrete formula for the extension.

4.3   Semi-Lipschitz Functions on Quasi-Metric Spaces

In this section we shall briefly present some results on Lipschitz functions on quasi-metric spaces and an extension result for functions in this class.

4.3.1   Quasi-Metric Spaces

A quasi-metric is a function $\varrho : X \times X \to [0, \infty)$ satisfying the conditions

(QM1) $\varrho(x, y) \geq 0$ and $\varrho(x, x) = 0$;
(QM2) $\varrho(x, y) = \varrho(y, x) = 0 \Rightarrow x = y$;
(QM3) $\varrho(x, z) \leq \varrho(x, y) + \varrho(y, z)$,

for all $x, y, z \in X$, i.e., the symmetry condition in the definition of a metric is broken. This drastically changes the properties of the metric spaces, mainly in what concerns the completeness and compactness conditions. A thorough presentation of their properties as well as of various aspects of functional analysis in spaces with asymmetric norm is given in the book [162].

To a quasi-metric ϱ one associates another one denoted by $\bar{\varrho}$ and called the conjugate of ϱ, defined by

$$\bar{\varrho}(x, y) = \varrho(y, x), \quad x, y \in X.$$

The function $\varrho^s : X \times X \to [0, \infty)$ given by

$$\varrho^s(x, y) = \max\{\varrho(x, y), \bar{\varrho}(x, y)\}, \quad x, y \in X,$$

is a metric on X.

The following inequalities hold for these quasi-metrics for all $x, y \in X$:

$$\varrho(x, y) \leq \varrho^s(x, y) \quad \text{and} \quad \bar{\varrho}(x, y) \leq \varrho^s(x, y). \tag{4.3.1}$$

An *asymmetric norm* on a real vector space X is a functional $p : X \to [0, \infty)$ satisfying the conditions:

(AN1) $p(x) = p(-x) = 0 \Rightarrow x = 0$;
(AN2) $p(\alpha x) = \alpha p(x)$;
(AN3) $p(x + y) \leq p(x) + p(y)$,

for all $x, y \in X$ and $\alpha \geq 0$.

If p satisfies only the conditions (AN2) and (AN3), then it is called an *asymmetric seminorm*. The pair (X, p) is called an *asymmetric normed* (respectively *seminormed*) *space*.

An asymmetric seminorm p defines a quasi-semimetric ϱ_p on X through the formula

$$\varrho_p(x, y) = p(y - x), \quad x, y \in X. \tag{4.3.2}$$

Defining, for any $x \in X$, the conjugate asymmetric seminorm $\bar{p}$ and the seminorm p^s by

$$\bar{p}(x) = p(-x) \quad \text{and} \quad p^s(x) = \max\{p(x), p(-x)\},$$

the inequalities (4.3.1) become

$$p(x) \le p^s(x) \quad \text{and} \quad \bar{p}(x) \le p^s(x),$$

for all $x \in X$. Obviously, p^s is a norm when p is an asymmetric norm and (X, p^s) is a normed space.

An important example is the following one.

Example 4.3.1 On the field $\mathbb{R}$ of real numbers consider the asymmetric norm $u(\alpha) = \alpha^+ := \max\{\alpha, 0\}$. Then, for $\alpha \in \mathbb{R}$, $\bar{u}(\alpha) = \alpha^- := \max\{-\alpha, 0\}$ and $u^s(\alpha) = |\alpha|$. The topology $\tau(u)$ generated by u is called the *upper topology* of $\mathbb{R}$, while the topology $\tau(\bar{u})$ generated by $\bar{u}$ is called the *lower topology* of $\mathbb{R}$. A basis of open $\tau(u)$-neighborhoods of a point $\alpha \in \mathbb{R}$ is formed of the intervals $(-\infty, \alpha + \varepsilon)$, $\varepsilon > 0$. A basis of open $\tau(\bar{u})$-neighborhoods is formed of the intervals $(\alpha - \varepsilon, \infty)$, $\varepsilon > 0$.

In this space the addition is continuous from $(\mathbb{R} \times \mathbb{R}, \tau_u \times \tau_u)$ to $(\mathbb{R}, \tau_u)$, but the multiplication is discontinuous at every point $(\alpha, \beta) \in \mathbb{R} \times \mathbb{R}$.

The balls in a quasi-metric spaces are defined as in the metric case, making distinction between those with respect to ϱ, its conjugate $\bar{\varrho}$ and to the associated metric ϱ^s:

$$B_\varrho(x, r) = \{y \in X : \varrho(x, y) < r\},$$

$$B_{\bar{\varrho}}(x, r) = \{y \in X : \bar{\varrho}(x, y) < r\} = \{y \in X : \varrho(y, x) < r\},$$

$$B_{\varrho^s}(x, r) = \{y \in X : \varrho^s(x, y) < r\} = B_\varrho(x, r) \cap B_{\bar{\varrho}}(x, r).$$

with similar definitions for the closed balls.

Correspondingly, we have three topologies $\tau(\varrho)$, $\tau(\bar{\varrho})$ and $\tau(\varrho^s)$, defining three kinds of convergence:

$$x_n \xrightarrow{\varrho} x \iff \varrho(x, x_n) \to 0;$$

$$x_n \xrightarrow{\bar{\varrho}} x \iff \bar{\varrho}(x, x_n) \to 0 \iff \varrho(x_n, x) \to 0;$$

$$x_n \xrightarrow{\varrho^s} x \iff \varrho^s(x_n, x) \to 0.$$

It is obvious that

$$x_n \xrightarrow{\varrho^s} x \iff (x_n \xrightarrow{\varrho} x \text{ and } x_n \xrightarrow{\bar{\varrho}} x).$$

Remark 4.3.2 The terminology concerning various extensions of the notion of metric is not unitary. Some authors call quasi-metric spaces (in the sense defined above) oriented metric spaces, or spaces with weak metric. Also the term "quasi-metric" is used by some authors (e.g. by Triebel [653]) to designate a function d satisfying the relaxed triangle inequality, see Remark 3.4.17.

In the following proposition we list a few of the topological properties of quasi-metric spaces.

Proposition 4.3.3 *Let (X, ϱ) be a quasi-semimetric space. The following are true.*

1. *Any ball $B_\varrho(x, r)$ is $\tau(\varrho)$-open and a ball $B_\varrho[x, r]$ is $\tau(\bar{\varrho})$-closed. The ball $B_\varrho[x, r]$ need not be $\tau(\varrho)$-closed.*
   Also, the following inclusions hold

$$B_{\varrho^s}(x, r) \subseteq B_\varrho(x, r) \text{ and } B_{\varrho^s}(x, r) \subseteq B_{\bar{\varrho}}(x, r),$$

   with similar inclusions for the closed balls.
2. *The topology $\tau(\varrho^s)$ is finer than the topologies $\tau(\varrho)$ and $\tau(\bar{\varrho})$. This means that:*
   - *any $\tau(\varrho)$-open (closed) set is $\tau(\varrho^s)$-open (closed); similar results hold for the topology $\tau(\bar{\varrho})$;*
   - *the identity mappings from $(X, \tau(\varrho^s))$ to $(X, \tau(\varrho))$ and to $(X, \tau(\bar{\varrho}))$ are continuous;*
   - *a sequence (x_n) in X is $\tau(\varrho^s)$-convergent to $x \in X$ if and only if it is $\tau(\varrho)$-convergent and $\tau(\bar{\varrho})$-convergent to x.*
3. *If ϱ is a quasi-metric, then the topologies $\tau(\varrho)$ and $\tau(\bar{\varrho})$ are T_0, but not necessarily T_1 (and so not T_2, in contrast to the case of metric spaces).*
   The topology $\tau(\varrho)$ is T_1 if and only if $\varrho(x, y) > 0$ whenever $x \neq y$. In this case, $\tau(\bar{\varrho})$ is also T_1.
4. *For every fixed $x \in X$, the mapping $\varrho(x, \cdot) : X \to (\mathbb{R}, |\cdot|)$ is $\tau(\varrho)$-usc and $\tau(\bar{\varrho})$-lsc.*
   For every fixed $y \in X$, the mapping $\varrho(\cdot, y) : X \to (\mathbb{R}, |\cdot|)$ is $\tau(\varrho)$-lsc and $\tau(\bar{\varrho})$-usc.

Similar results hold for an asymmetric seminorm p, its conjugate $\bar{p}$ and the associated seminorm p^s.

4.3.2 Semi-Lipschitz Functions

As in the case of metric spaces, their analogs in the quasi-metric case, called semi-Lipschitz functions, play an important role in the study of quasi-metric spaces.

The properties of the spaces of semi-Lipschitz functions were studied by Romaguera and Sanchis [604, 605] and Romaguera, Sánchez-Álvarez and Sanchis [606] (see also the book [162]).

Suppose that (X, ϱ) is a quasi-metric space and (Y, q) an asymmetric normed space. A function $f : X \to Y$ is called *semi-Lipschitz* provided that there exists a number $L \geq 0$ such that

$$q(f(x) - f(y)) \leq L \varrho(x, y) , \tag{4.3.3}$$

for all $x, y \in X$. A number $L \geq 0$ for which (4.3.3) holds is called a *semi-Lipschitz constant* for f and we say that f is L-*semi-Lipschitz*. We denote by $\mathrm{SLip}(X, Y)$ ($\mathrm{SLip}_{\varrho,q}(X, Y)$ if more precision is needed) the set of all semi-Lipschitz functions from X to Y.

In particular, if Y is the space $(\mathbb{R}, u)$ with $u(\alpha) = \alpha^+$ (see Example 4.3.1), the condition (4.3.3) is equivalent to

$$f(x) - f(y) \leq L \varrho(x, y) ,$$

for all $x, y \in X$. In this case one uses the notation $\mathrm{SLip}_\varrho(X) = \mathrm{SLip}_\varrho(X, \mathbb{R})$.

A function $f : X \to Y$ is called $\leq_{\varrho,q}$-*monotone* if $q(f(x) - f(y)) = 0$ whenever $\varrho(x, y) = 0$. In particular, a function $f : X \to \mathbb{R}$ is $\leq_{\varrho,u}$-monotone, called $\leq_\varrho$-*monotone*, if and only if $f(x) \leq f(y)$ whenever $\varrho(x, y) = 0$.

Obviously, a semi-Lipschitz function is $\leq_{\varrho,q}$-monotone. Since, by Proposition 4.3.3, the topology $\tau(\varrho)$ is T_1 if and only if $\varrho(x, y) = 0 \iff x = y$ for all $x, y \in X$, it follows that any function on a T_1 quasi-metric space is $\leq_{\varrho,q}$-monotone.

Remark 4.3.4 It is clear that for $\alpha, \beta \in \mathbb{R}$,

$$\alpha \leq \beta \iff \alpha - \beta \leq 0 \iff u(\alpha - \beta) = (\alpha - \beta)^+ = 0.$$

If p is an asymmetric seminorm on a vector space X, then

$$x \leq_p y \iff p(x - y) = 0$$

defines an order relation on X. Similarly, in a quasi-semimetric space (X, ϱ)

$$x \leq_\varrho y \iff \varrho(x, y) = 0$$

also defines an order relation.

Taking into account these order relations, the $\leq_{\varrho,q}$-monotonicity can be expressed by the condition

$$x \leq_\varrho y \;\Rightarrow\; f(x) \leq_q f(y),$$

justifying the term monotonicity.

Suppose now that (X, ϱ) is a quasi-metric space and (Y, q) an asymmetric normed space. For an arbitrary function $f : X \to Y$ put

$$\|f|_{\varrho,q} = \sup\left\{ \frac{q(f(x) - f(y))}{\varrho(x, y)} : x, y \in X, \, \varrho(x, y) > 0 \right\},$$

and $\|f|_\varrho = \|f|_{\varrho,u}$ when Y is $(\mathbb{R}, u)$.

Proposition 4.3.5 *Let (X, ϱ) be a quasi-metric space and (Y, q) an asymmetric normed space.*

1. *The set $\mathrm{SLip}_{\varrho,q}(X, Y)$ is a cone in the linear space $\mathrm{Lip}_{\varrho^s,q^s}(X, Y)$ of all Lip-schitz functions from the metric space (X, ϱ^s) to the normed space (Y, q^s) and $\|f|_{\varrho^s,q^s} \leq \|f|_{\varrho,q}$ for all $f \in \mathrm{SLip}_{\varrho,q}(X, Y)$.*
2. *If f is semi-Lipschitz, then $\|f|_{\varrho,q}$ is the smallest semi-Lipschitz constant for f.*
3. *A function $f : X \to Y$ is semi-Lipschitz if and only if it is $\leq_{\varrho,q}$-monotone and $\|f|_{\varrho,q} < \infty$.*

Proof

1. It is clear that $f + g, \, \alpha f \in \mathrm{SLip}_{\varrho,q}(X, Y)$ for all $f, g \in \mathrm{SLip}_{\varrho,q}(X, Y)$ and $\alpha \geq 0$.

   If $f \in \mathrm{SLip}_{\varrho,q}(X, Y)$, then there exists $L \geq 0$ such that

$$q(f(x) - f(y)) \leq L\varrho(x, y) \leq L\varrho^s(x, y), \; x, y \in X,$$

   implying

$$q^s(f(x) - f(y)) \leq L\varrho^s(x, y),$$

   for all $x, y \in X$, so that $f \in \mathrm{Lip}_{\varrho^s,q^s}(X, Y)$ and $\|f|_{\varrho^s,q^s} \leq \|f|_{\varrho,q}$.
2. The inequality $q(f(x) - f(y))/\varrho(x, y) \leq \|f|_{\varrho,q}$ implies

$$q(f(x) - f(y)) \leq \|f|_{\varrho,q}\, \varrho(x, y),$$

   for all $x, y \in X$ with $\varrho(x, y) > 0$. Since a semi-Lipschitz function is $\leq_{\varrho,q}$-monotone, $\varrho(x, y) = 0$ implies $q(f(x) - f(y)) = 0 = \|f|_{\varrho,q}\, \varrho(x, y)$. Consequently, $\|f|_{\varrho,q}$ is a semi-Lipschitz constant for f.

Suppose that L is a semi-Lipschitz constant for f. Then $q(f(x) - f(y))/\varrho(x, y) \leq L$, whenever $\varrho(x, y) > 0$, so that $\|f\|_{\varrho, q} \leq L$, showing that $\|f\|_{\varrho, q}$ is the smallest semi-Lipschitz constant for f.

3. The above reasonings also prove the validity of 3. □

The following example shows that the inequality form Proposition 4.3.5.1 can be strict.

Example 4.3.6 On a three point set $X = \{x_1, x_2, x_3\}$ consider the quasi-metric $\varrho(x_1, x_2) = 1$, $\varrho(x_2, x_1) = 2$, $\varrho(x_1, x_3) = \varrho(x_3, x_1) = 2$, $\varrho(x_2, x_3) = \varrho(x_3, x_2) = 2$, and the function $f : X \to \mathbb{R}$ given by $f(x_1) = 1$, $f(x_2) = f(x_3) = 2$. Then $\varrho^s(x_i, x_j) = 2$ for $i \neq j$ and

$$\|f\|_{\varrho, |\cdot|} = \max \left\{ \frac{|f(x_i) - f(x_j)|}{\varrho(x_i, x_j)} : 1 \leq i, j \leq 3, i \neq j \right\} = 1$$

$$> \frac{1}{2} = \max \left\{ \frac{|f(x_i) - f(x_j)|}{\varrho^s(x_i, x_j)} : 1 \leq i, j \leq 3, i \neq j \right\} = \|f\|_{\varrho^s, |\cdot|} .$$

The following proposition contains some examples of semi-Lipschitz functions.

Proposition 4.3.7 *Let (X, ϱ) be a quasi-metric space, $y \in X$ and $Y \subseteq X$ nonempty.*

1. *The functions $\varrho(\cdot, y) : X \to \mathbb{R}$ and $d(\cdot, Y) : X \to \mathbb{R}$ are semi-Lipschitz with semi-Lipschitz constant 1.*
2. *For fixed $a \in X$, the functions $f(x) = \varrho(a, x_0) - \varrho(a, x)$ and $g(x) = \varrho(x, a) - \varrho(x_0, a)$ belong to $SLip_{\varrho, 0}(X)$ and $\|f\|_\varrho, \|g\|_\varrho \leq 1$.*

Proof

1. The inequality

$$\varrho(x, y) \leq \varrho(x, x') + \varrho(x', y) , \tag{4.3.4}$$

valid for $x, x' \in X$, shows that the function $\varrho(\cdot, y)$ is semi-Lipschitz. Since the inequality (4.3.4) holds for all $y \in Y$ and fixed x, x', passing to infimum with respect to $y \in Y$ one obtains $d(x, Y) \leq \varrho(x, x') + d(x', Y)$, which means that the function $d(\cdot, Y)$ is semi-Lipschitz, too.

2. The assertions from 2 follow from 1. □

As we have seen in the previous sections of this chapter, an important problem in the study of Lipschitz functions on metric spaces is the existence of extensions for Lipschitz functions.

In the case of semi-Lipschitz functions a similar result was proved by Mustăţa [503] (see also [502, 505]). The extension problem for semi-Lipschitz functions on quasi-metric spaces was considered also by Matoušková [449]. The paper [248] discusses the existence of an extension of an asymmetric norm defined on a cone K to an asymmetric norm defined on the linear space $X = K - K$ it generates.

Theorem 4.3.8 *Let* (X, ϱ) *be a quasi-metric space,* Y *a nonempty subset of* X *and* $f : Y \to \mathbb{R}$ *an L-semi-Lipschitz function.*

1. *The functions* F, G *defined for* $x \in X$ *by*

$$F(x) = \inf\{f(y) + L\varrho(x, y) : y \in Y\} \tag{4.3.5}$$

   and

$$G(x) = \sup\{f(y') - L\varrho(y', x) : y' \in Y\} \tag{4.3.6}$$

   are L-semi-Lipschitz extensions of f.
2. *Any other L-semi-Lipschitz extension* H *of* f *satisfies the inequalities*

$$G \leq H \leq F .$$

Proof

1. Let $y, y' \in Y$ and $x \in X$. The inequalities $f(y') - f(y) \leq L\varrho(y', y) \leq L\varrho(y', x) + L\varrho(x, y)$ imply

$$f(y') - L\varrho(y', x) \leq f(y) + L\varrho(x, y) .$$

   Passing to supremum with respect to $y' \in Y$ and to infimum with respect to $y \in Y$, it follows that G, F are well-defined and $G \leq F$.

   Let $x \in Y$. Then $f(x) \leq f(y) + L\varrho(x, y)$ for every $y \in Y$, implies $f(x) \leq F(x)$. Similarly, $f(y') - L\varrho(y', x) \leq f(x)$ implies $G(x) \leq f(x)$. Taking $y = x$ in (4.3.5) and $y' = x$ in (4.3.6), it follows that $F(x) \leq f(x)$ and $G(x) \geq f(x)$, so that $G(x) = f(x) = F(x)$ for every $x \in Y$.

   To conclude, we have to show that the functions F, G are semi-Lipschitz. Let $x, x' \in X$. The inequalities

$$F(x) \leq f(y) + L\varrho(x, y) \leq f(y) + L\varrho(x, x') + L\varrho(x', y) ,$$

   valid for all $y \in Y$, yield $F(x) \leq F(x') + L\varrho(x, x')$, showing that F is semi-Lipschitz.

   Similar reasonings show that G is semi-Lipschitz too.
2. Let H be an L-semi-Lipschitz extension of f and $x \in X$. Since

$$H(x) \leq H(y) + L\varrho(x, y) = f(y) + L\varrho(x, y) ,$$

   for every $y \in Y$, passing to infimum with respect to $y \in Y$ one obtains $H(x) \leq F(x)$.

Similarly $f(y') - H(x) = H(y') - H(x) \leq L \varrho(y', x,)$ implies

$$f(y') - L \varrho(y', x) \leq H(x),$$

for every $y' \in Y$. Passing to supremum with respect to $y' \in Y$ one obtains $G(x) \leq H(x)$. □

The following corollary shows the existence of norm-preserving extensions of semi-Lipschitz functions.

Corollary 4.3.9 *Let X, Y and f be as in Theorem 4.3.8 and*

$$\|f|_\varrho = \sup\{u(f(y) - f(y'))/\varrho(y, y') : y, y' \in Y, \varrho(y, y') > 0\},$$

and let F, G be given by (4.3.5) and (4.3.6) for $L = \|f|_\varrho$. Then

1. *The functions F and G are semi-Lipschitz norm-preserving extensions of f, that is,*

$$(i) \; F|_Y = G|_Y = f \quad and \quad (ii) \; \|F|_\varrho = \|G|_\varrho = \|f|_\varrho.$$

2. *Any other semi-Lipschitz norm-preserving extension H of f satisfies the inequalities*

$$G \leq H \leq F.$$

4.4 Lipschitz Functions with Values in Quasi-Normed Spaces

It turns out that some results concerning Banach space-valued Lipschitz functions fail in the quasi-Banach case and, in some cases, the validity of some of them forces the quasi-Banach space to be locally convex, i.e., a Banach space. For some background on quasi-Banach and F-spaces, see Sect. 1.4.16.

In this section we consider only spaces over $\mathbb{R}$.

Let (Z, d) be a b-metric space (see Remark 4.3.2) and $(Y, \| \cdot \|)$ a quasi-normed space. A function $f : Z \to Y$ is called Lipschitz if there exists $L \geq 0$ (called a Lipschitz constant for f) such that

$$\|f(z) - f(z')\| \leq Ld(z, z'),$$

for all $z, z' \in Z$. One denotes by $\mathrm{Lip}(Z, Y)$ the space of all Lipschitz functions from Z to Y.

The *Lipschitz norm* $\|f\|_L$ of f is defined by

$$\|f\|_L = \sup \left\{ \frac{\|f(z) - f(z')\|}{d(z, z')} : z, z' \in Z, \ z \neq z' \right\}.$$

It turns out that $\|f\|_L$ is the smallest Lipschitz constant for f.

Since $\|f\|_L = 0$ if and only if f is constant, $\|\cdot\|$ is actually only a seminorm on $\mathrm{Lip}(Z, Y)$. To obtain a norm, one considers a fixed element $z_0 \in Z$ and the space

$$\mathrm{Lip}_0(Z, Y) = \{f \in \mathrm{Lip}(Z, Y) : f(z_0) = 0\}.$$

If $Z = X$, where X is a quasi-normed space, then one takes 0 for the fixed point z_0, and $\mathrm{Lip}_0(X, Y)$ is a quasi-normed space,

$$\|f + g\|_L \leq k(\|f\|_L + \|g\|_L),$$

for $f, g \in \mathrm{Lip}_0(X, Y)$, where $k \geq 1$ is the constant in the inequality (QN3) from Sect. 1.4.16. It is complete, provided that Y is a quasi-Banach space.

If $Y = \mathbb{R}$, then one uses the notation $\mathrm{Lip}(X)$ and $\mathrm{Lip}_0(X)$, called the Lipschitz dual of the quasi-normed space X.

It is known that the space $L^p = L^p[0, 1]$ has trivial dual for $0 < p < 1$. Albiac [15] proved that it also has a trivial Lipschitz dual, i.e., $\mathrm{Lip}_0(L^p) = \{0\}$. Later he showed that this is the case in a more general situation.

Proposition 4.4.1 (Albiac [16]) *Let $(X, \|\cdot\|)$ be a quasi-Banach space and*

$$|||x||| := \sup\{f(x) : f \in \mathrm{Lip}_0(X), \ \|f\|_L \leq 1\}, \quad x \in X.$$

Then

(i) *$|||\cdot|||$ is a seminorm on X;*
(ii) *if $\mathrm{Lip}_0(X)$ is nontrivial, then X has a nontrivial dual, i.e., $X^* \neq \{0\}$;*
(iii) *if X has a separating Lipschitz dual, then X has a separating (linear) dual and $|||\cdot|||$ is a norm on X.*

Concerning the validity of McShane's extension property (see Sect. 4.1), the following result was proved in [16].

Proposition 4.4.2 *Let $(X, \|\cdot\|)$ be a quasi-Banach space. If for every subset Z of X, every L-Lipschitz function $f : Z \to \mathbb{R}$ admits an L'-Lipschitz extension, for some $L' \geq L$, then the space X is locally convex, i.e., it is a Banach space.*

It is known that every continuous linear operator from a quasi-Banach space X to a Banach space Y admits a norm-preserving linear extension to the Banach envelope $\widehat{X}$ of X to Y (see Sect. 1.4.16). Albiac [16] showed that this is true for Lipschitz mappings too: every Lipschitz mapping $f : X \to Y$ admits a unique Lipschitz extension with the same Lipschitz constant $\widehat{f} : \widehat{X} \to Y$.

Moreover, if X, Y are normed spaces and $f : X \to Y$ is Gâteaux differentiable on the interval $[x, y] := \{x + t(y - x) : t \in [0, 1]\}$, then

$$\|f(x) - f(y)\| \leq \|x - y\| \sup\{\|f'(\xi)\| : \xi \in [x, y]\}. \tag{4.4.1}$$

Proposition 4.4.3 ([16]) *Let* $(X, \| \cdot \|)$ *be a quasi-Banach space. If every noncon-stant Gâteaux differentiable Lipschitz function* $f : [0, 1] \to X$ *satisfies the mean value inequality* (4.4.1) *for all* $x, y \in [0, 1]$, *then the space* X *is locally convex, i.e., it is a Banach space.*

From Proposition 1.3.4 it follows that a Hölder function of order $\alpha > 1$ from $[0, 1]$ to a Banach space X is constant. This fact is no longer true if X is merely quasi-Banach.

Example 4.4.4 Let $L^p = L^p[0, 1]$ for $0 < p < 1$. The function $f : [0, 1] \to L^p$ given by $f(t) = \chi_{[0,t]}$ satisfies the equality

$$\|f(s) - f(t)\|_p = |s - t|^{1/p},$$

for all $s, t \in [0, 1]$, where $\| \cdot \|_p$ is the L^p-norm (see (1.4.15)).

Indeed, for $0 \leq t < s \leq 1$,

$$\|f(s) - f(t)\|_p = \left(\int_t^s \chi_{(t,s]}^p(u) du \right)^{1/p} = |s - t|^{1/p}.$$

The Riemann integral of a function $f : [a, b] \to X$, where $[a, b]$ is an interval in $\mathbb{R}$ and X is a Banach space, can be defined as in the real case, by simply replacing the absolute value $| \cdot |$ with the norm sign $\| \cdot \|$, and has properties similar to those from the real case. For instance, the following result is true.

Proposition 4.4.5 *Let* X *be a Banach space. If* $f : [a, b] \to X$ *is continuous, then*

(i) f *is Riemann integrable, and*
(ii) *the function*

$$F(t) = \int_a^t f(s) ds, \quad t \in [a, b], \tag{4.4.2}$$

is differentiable with $F'(t) = f(t)$ *for all* $t \in [a, b]$.

Remark 4.4.6 However, there is a point where this analogy is broken, namely the Lebesgue criterion of Riemann integrability: a function $f : [a, b] \to \mathbb{R}$ is Riemann integrable if and only if it is continuous almost everywhere on $[a, b]$ (i.e., except a set of Lebesgue measure zero). In the infinite dimensional case this criterion does not hold in general, leading to the study of those Banach spaces for which it is true (or some weaker forms), see, for instance, [270, 640, 643] and the references quoted therein.

In the case of quasi-Banach spaces the situation is different. By a result attributed to Mazur and Orlicz [452] (see also [603, p. 122]) an F-space X is locally convex if and only if every continuous function $f : [0, 1] \to X$ is Riemann integrable.

Popov [578] investigated the Riemann integrability of functions defined on intervals in $\mathbb{R}$ with values in an F-space. Among other results, he proved that a Riemann integrable function $f : [a, b] \to X$ is bounded and that the function F defined by (4.4.2) is uniformly continuous, but there exists a continuous function $f : [0, 1] \to \ell^p$, where $0 < 1 < p$, such that the function F does not have a right derivative at $t = 0$. He asked whether any continuous function f from $[0, 1]$ to $L^p[0, 1]$, $0 < p < 1$, (or more general, to a quasi-Banach space X with $X^* = \{0\}$) admits a primitive. This problem was solved by Kalton [327] who proved that if X is a quasi-Banach space with $X^* = \{0\}$, then every continuous function $f : [0, 1] \to X$ has a primitive. Kalton considered the space $C^1_{Kal}(I, X)$, where $I = [0, 1]$ and X is a quasi-Banach space, of all continuously differentiable functions $f : I \to X$ such that the function $\tilde{f} : I^2 \to X$ given for $s, t \in I$ by $\tilde{f}(t, t) = f'(t)$ and $\tilde{f}(s, t) = (f(s) - f(t))/(s - t)$ if $s \neq t$, is continuous. It follows that $C^1_{Kal}(I, X)$ is a quasi-Banach space with respect to the quasi-norm

$$\|f\| = \|f(0)\| + \|f\|_L .$$

The notation $C^1_{Kal}(I, X)$ was introduced in [17]; Kalton used the notation $C^1(I; X)$.

Denote by $C(I, X)$ the Banach space (with respect to the sup-norm) of all continuous functions from I to X. The *core* of a quasi-Banach space X is the maximal subspace Z of X (denoted by $\mathrm{core}(X)$) with $Z^* = \{0\}$. One shows that such a subspace always exists, is unique and closed. Notice that $\mathrm{core}(X) = \{0\}$ implies only that X has a nontrivial dual, but not necessarily a separating one.

In [18] it is shown that if X is a quasi-Banach space with $\mathrm{core}(X) = \{0\}$, then there exists a continuous function $f : [0, 1] \to X$ failing to have a primitive.

Kalton, *op. cit.*, called a quasi-Banach X a *D-space* if the mapping

$$D : C^1_{Kal}(I, X) \to C(I, X),$$

given by $Df = f'$, is surjective and proved the following result.

Theorem 4.4.7 ([327]) *Let X be a quasi-Banach with $\mathrm{core}(X) = \{0\}$. Then X is a D-space if and only if X is locally convex (or, equivalently, a Banach space).*

It is known that every continuously differentiable function from an interval $[a, b] \subseteq \mathbb{R}$ to a Banach space X is Lipschitz with $\|f\|_L = \sup\{\|f'(t)\| : t \in [a, b]\}$ (a consequence of the Mean Value Theorem, see (4.4.1)). As it was shown in [17] this is no longer true in quasi-Banach spaces.

Theorem 4.4.8 *Let X be a non-locally convex quasi-Banach space X. Then there exists a function $F : I \to X$ such that:*

(i) *F is continuously differentiable on I;*
(ii) *F' is Riemann integrable on I and $F(t) = \int_0^t F'(s)ds$, $t \in I$;*
(iii) *F is not Lipschitz on I.*

In [19] it is proved that the usual rule of the calculation of the integral (called Barrow's rule by the authors, also known as the Leibniz rule) holds in the quasi-Banach case in the following form.

Proposition 4.4.9 *Let X be a quasi-Banach space with separating dual. If $F : [a, b] \to X$ is differentiable with Riemann integrable derivative, then*

$$\int_a^b F'(t)dt = F(b) - F(a).$$

Another pathological result concerning differentiability of quasi-Banach valued Lipschitz functions was obtained by Kalton [326].

Theorem 4.4.10 *Let X be an F-space with trivial dual. Then for every pair of distinct points $x_0, x_1 \in X$ there exists a function $f : [0, 1] \to X$ such that $f(0) = x_0$, $f(1) = x_1$ and*

$$\lim_{|s-t|\to 0} \frac{f(s) - f(t)}{s - t} = 0 \ \text{uniformly for } s, t \in [0, 1].$$

In particular $f'(t) = 0$ for all $t \in [0, 1]$.

Remark 4.4.11 Kalton [326] also remarked that if X is an F-space and $x \in X \setminus \{0\}$, then for a function $f : [0, 1] \to X$ with $f(0) = 0$ and $f(1) = x$ to exist it is necessary and sufficient that $x \in \text{core}(X)$.

If X is a Banach space and $f : [0, 1] \to X$ is continuous then it is Riemann integrable and the average function $\text{Ave}[f] : [a, b] \times [a, b] \to X$, given by

$$\text{Ave}[f](s, t) = \begin{cases} \frac{1}{t-s} \int_s^t f(u)du & \text{if } a \leq s < t \leq b, \\ f(c) & \text{if } s = t = c \in [a, b], \\ \frac{1}{s-t} \int_t^s f(u)du & \text{if } a \leq t < s \leq b, \end{cases}$$

is jointly continuous on $[a, b] \times [a, b]$, and so, separately continuous and bounded. Some pathological properties of the average function in the quasi-Banach case are examined in [17, 578] and [20].

The analog of the Radon-Nikodým Property (RNP) (see Sect. 1.6.3) for quasi-Banach spaces and its connections with the differentiability of Lipschitz mappings and martingales are discussed in [21].

4.5 Bibliographic Comments and Miscellaneous Results

Section 4.1 Theorem 4.1.1 is due to McShane [455]. The same method of extension can be found in [680, 681]. Theorem 4.1.6 can be found in [479] (see also [472]). Theorem 4.1.7 is due to Czipszer and Gehér [172].

The existence of convex extensions preserving the Lipschitz constant for convex Lipschitz functions $f : Y \to \mathbb{R}$, where Y is a nonempty convex subset of a normed space X, was proved in [164] (see also [152, 288]). The case of starshaped Lipschitz functions was treated in [496]. In both cases, the existence of minimal and maximal extensions (the analog of (4.1.2)) was proved and applications to best approximation were given.

Having a Lipschitz function $f : K \subseteq \mathbb{R}^n \to \mathbb{R}$, where K is a compact set, if $g, h : \mathbb{R}^n \to \mathbb{R}$ are Lipschitz extensions of f such that $L(f) = L(g) = L(h)$, a very simple characterization of the set $E = \{x \in \mathbb{R}^n \setminus K : f(x) = g(x)\}$ was given by Aronsson [49].

Ricceri [595] provided sufficient conditions under which a Lipschitz real-valued function defined on a subset of a Banach space X has a Lipschitz surjective open extension over the whole space X.

Mil'man [483] introduced a method of extending functions, including bounded and Lipschitz functions, which generalizes some known results and which can be used to obtain some approximation results.

Given a compact metric space (X, d), a nonempty closed subset Y of X and an L-Lipschitz function $f : X \to \mathbb{R}$, Muștăța [499] proved that the set $E_L(f|_Y)$ of all L-Lipschitz extensions of $f|_Y$ is compact in the space $C(X)$ of all continuous real-valued functions on X with the uniform topology.

For some more applications of McShane's theorem, one can also consult [500] and [505, 509, 510], Section 2.6.1 of [162] and [449], where McShane type results for semi-Lipschitz and semi-Hölder functions are presented, or [498] and [511], where the uniqueness of the extension is discussed.

If U is a bounded open subset of $\mathbb{R}^n$ and $g : \partial U \to \mathbb{R}^m$ is a Lipschitz function, then, according to Kirszbraun's theorem there exists a Lipschitz function $u : \overline{U} \to \mathbb{R}^m$ such that $u|_{\partial U} = g$ and $L(u) = L(g)$. In general, such an extension is not unique. For $m = 1$ a result due to Jensen [302] implies that there exists a unique extension u of g such that, for every open subset V of U, the functions $u : V \to \mathbb{R}$ and $u : \partial V \to \mathbb{R}$ have the same Lipschitz constant (i.e., u is absolutely minimizing Lipschitz, AML for short). For the more complicated situation $m > 1$, Sheffield and Smart [630] introduced a notion of optimality, called tightness, stronger than the AML property and obtained, in certain frameworks, existence and uniqueness results. Let us mention that an excellent study of existence, uniqueness and regularity properties of the best possible Lipschitz extensions of scalar functions is given in [49] (see also [50]). In [8], some equivalent characterizations of the AML property are given, and in [529] an explicit sub-optimal extension is built. The paper [528] deals with the problem of building absolutely minimizing Lipschitz extensions of a given function. These extensions can be characterized as being the solution of a degenerate elliptic partial differential equation having a unique viscosity solutions. The applications of optimal Lipschitz extension problems include image processing (see [139]) and brain mapping (see [457, 458]).

Section 4.2 The results from this section are due to Valentine [664–666]. The case of Euclidean spaces was considered in [359]. The proof given here to Theorem 4.2.3 is taken from [625], where the proof is given in the case when $\mathcal{H}_1$ and $\mathcal{H}_1$ are finite dimensional Euclidean spaces (Kirszbraun's theorem), but the proof can be adapted to general Hilbert spaces. A tricky and imaginative proof of this result, relying on Rockafellar's duality result, to Reich and Simons [592], and a constructive one is due to Akopyan and Tarasov [11]. For other proofs and generalizations one can consult [28, 32, 53, 236, 243, 273, 343, 379, 470], [484].

Grünbaum [272] (see also [273]) proved that a real Banach space X has the extension property (i.e., for each $f : A \to K$, where A is a subset of X and K is a closed convex subset of X, there exists $F : X \to X$ such that: (i) $F(x) = f(x)$ for each $x \in A$; (ii) $F(x) \in K$ for $x \in X$; (iii) $L(F) = L(f)$) if and only if X is an inner-product space or X is a two-dimensional space whose unit sphere is a parallelogram.

As we have seen (Theorem 4.2.3), if A is a subset of a Hilbert space $\mathcal{H}$ and $f : A \to \mathcal{H}$ is a Lipschitz function, then there exists a Lipschitz function $F : \mathcal{H} \to \mathcal{H}$ such that $F|_A = f$ and $L(F) = L(f)$. Such an extension F is not unique. Kopecká [360, 361] proved that there exists a continuous extension operator which preserves the Lipschitz constant of every function (see Sect. 5.3 for further details).

Schönbeck [626] proved that if E and F are Banach spaces such that F is strictly convex and the pair (E, F) has the contraction extension property, then E and F are Hilbert spaces. Special results have been obtained when $E = F$ (see [179, 217, 627] and [180]). Dacorogna and Gangbo [174] (see also [173], Chapter 15) proved that if E and F are Banach spaces such that the unit sphere of F is strictly convex and the dimensions of E and F are at least 2, then the following statements are equivalent:

(i)   the pair (E, F) has the contraction extension property;
(ii) the norms of both E and F are induced by an inner product;
(iii) any function with a Lipschitz constant 1 can be extended from a 3-point set to a 4-point set with the same Lipschitz constant 1.

The authors of [282] consider a wide class of generalized Lipschitz extension problems and the corresponding problem of finding absolutely minimal Lipschitz extensions. They generalize Aronsson's result on absolutely minimal Lipschitz extensions for scalar valued functions (see the survey [50]). In [382] one proves the existence of absolutely minimizing Lipschitz extensions using a method which differs from those used by Aronsson in general metrically convex compact metric spaces and by Jensen in Euclidean spaces (see [50]). Assuming Jensen's hypotheses, one provides numerical schemes for computing the viscosity solution of the equation $\Delta_\infty(u) = 0$ with Dirichlet's condition.

We include below some important contributions towards the solution of the problem of extending Lipschitz functions so that they remain Lipschitz. To state these results we recall first the following notions.

Given a metric space Y and a Banach space Z, for $X \subseteq Y$, we denote by $e(X, Y, Z)$ the infimum over all constants K such that each Lipschitz function $f : X \to Z$ can be extended to a Lipschitz function $F : Y \to Z$ with Lipschitz constant $L(F) \le KL(f)$. In case no such K exists, we set $e(X, Y, Z) = \infty$. Let us also define

$$e(Y, Z) = \sup\{e(X, Y, Z) : X \subseteq Y\},$$

$$e_n(Y, Z) = \sup\{e(X, Y, Z) : X \subseteq Y, \ X \text{ has at most } n \text{ elements}\}, \quad \text{and}$$

$$ae(X) = \sup\{e(X, Y, Z) : X \subseteq Y, Z \text{ is a Banach space}\}.$$

The number $ae(X)$ is called the absolute extendability constant of X. If $ae(X)$ is finite, then we say that X is absolutely extendable.

Marcus and Pisier [432] proved that for each $p \in (1, 2)$ there exists a constant $C(p)$ such that $e_n(L_p, L_2) \le C(p)(\log n)^{\frac{1}{p} - \frac{1}{2}}$ for every $n \in \mathbb{N}$. Johnson and Lindenstrauss [310] (see [459] for a new proof) proved that, for every metric space Y, the inequality $e_n(Y, L_2) \le 2\sqrt{\log n}$ is valid for every $n \in \mathbb{N}$ and the inequality $C(p)(\frac{\log n}{\log \log n})^{\frac{1}{p} - \frac{1}{2}} \le e_n(L_p, L_2)$ is valid for every $p \in [1, 2)$ and every $n \in \mathbb{N}$.

Ball [63] introduced the notion of Markov type and used it to prove an important extension result known as Ball's extension theorem.

Given a metric space (X, d) and $p \ge 1$, we say that X has *Markov type p* if there exists a constant $K > 0$ such that for every stationary reversible Markov chain $\{Z_t\}_{t=0}^{\infty}$ on $\{1, \dots, n\}$, every $f : \{1, \dots, n\} \to X$ and every time $t \in \mathbb{N}$, we have

$$\mathbb{E}d(f(Z_t), f(Z_0))^p \le K^p t \mathbb{E}d(f(Z_1), f(Z_0))^p.$$

The least such constant K is called the *Markov type p constant* of X and it is denoted by $M_p(X)$.

Before giving Ball's extension theorem, we introduce two geometric concepts for normed spaces. Let $(X, \|.\|)$ be a normed space and denote by δ_X and ρ_X the respective moduli of convexity and of smoothness of X (see Sect. 1.4.15).

We say that X has *modulus of convexity of power type q* if there exists a constant k such that $\delta_X(\varepsilon) \ge k\varepsilon^q$ for all $\varepsilon \in (0, 2]$. One can easily see that in this case $q \ge 2$. An equivalent condition for X to have modulus of convexity of power type q is the existence of a constant $K > 0$ such that

$$2\|x\|^q + \frac{2}{K^q}\|y\|^q \le \|x + y\|^q + \|x - y\|^q,$$

for every $x, y \in X$. The least constant K for which the previous inequality is valid is called the *q-convexity constant* of X and is denoted by $K_q(X)$.

We say that X has *modulus of smoothness of power type p* if there exists a constant s such that $\rho_X(t) \le s\,t^p$ for all $t > 0$. In this case, $p \le 2$. An equivalent condition for X to have modulus of smoothness of power type p is the existence of a constant $S > 0$ such that

$$\|x + y\|^p + \|x - y\|^p \le 2\,\|x\|^p + 2S^p\,\|y\|^p\,,$$

for every $x, y \in X$. The least constant S for which the previous inequality is valid is called the *p-smoothness constant* of X and is denoted by $S_p(X)$.

Ball, Carlen and Lieb [65] proved the inequalities $K_2(L_p) \le \frac{1}{\sqrt{p-1}}$ for $1 < p \le 2$ and $S_2(L_p) \le \sqrt{p-1}$ for $2 \le p < \infty$.

Ball's extension theorem states that $e(X, Y) \le 6M_2(X)K_2(Y)$ for every metric space X of Markov type 2 and every Banach space Y with modulus of convexity of power type 2. Because $M_2(L_2) = 1$ (see [63]), it follows that $e(L_2, L_p) \le \frac{6}{\sqrt{p-1}}$ for $1 < p \le 2$. Naor [514] showed that $e(L_2, L_p) = \infty$ for $2 < p < \infty$. Tsar'kov [654] proved that $e(X, L_2) < \infty$ for every Banach space X with modulus of smoothness of power type 2.

Naor, Peres, Schramm and Sheffied [518] showed that every normed space with modulus of smoothness of power type 2 has Markov type 2. In fact, they proved that for every normed space X with modulus of smoothness of power type q, where $1 < q \le 2$,

$$M_q(X) \le \frac{8}{(2^{q+1} - 4)^{1/q}} S_q(X).$$

In particular, $M_2(L_p) \le 4\sqrt{p-1}$ for $2 \le p < \infty$.

Therefore, using Ball's extension theorem, the following nonlinear analog of an extension theorem due to Maurey [450] is obtained: for every Banach space X with modulus of smoothness of power type 2 and every Banach space Y with modulus of convexity of power type 2, $e(X, Y) \le 24S_2(X)K_2(Y)$. In particular, we have $e(L_p, L_q) \le 24\sqrt{\frac{p-1}{q-1}}$ for $2 \le p < \infty$ and $1 < q \le 2$.

Lee and Naor [380, 381] (see also the survey [259]) proved that various classes of metric spaces are absolutely extendable. One of these classes is that of doubling metric spaces (see Sect. 3.4, the text after Definition 3.4.16).

The *doubling constant* $\lambda(X)$ of a metric space X is the infimum of the set of all numbers k such that every ball in X can be covered by k balls of half the radius. It follows that $\lambda(X)$ is finite if X is a doubling metric space.

Using a combination of the Whitney extension method and ideas related to random covers, Lee and Naor proved that there exists a universal constant $C > 0$ such that $ae(X) \le C \log \lambda(X)$. This result unifies and generalizes the Lipschitz extension theorems due to Johnson, Lindenstrauss and Schechtman [314] (see also [313]):

(i) there exists a universal constant C such that for every metric space Y, every Banach space Z and every $n \in \mathbb{N}$, the following inequality is valid: $e_n(Y, Z) \le C \log n$, and

(ii) there exists a universal constant C such that for every d-dimensional normed space Y and every Banach space Z, the following inequality is true: $e(Y, Z) \le Cd$.

Regarding (i), it is interesting to note that Lee and Naor also obtained an asymptotic improvement of the upper bound to $e_n(Y, Z) \le C \frac{\log n}{\log \log n}$. Very recently, Naor and Rabani [517] also gave a lower bound showing that there exists a universal constant C such that for every metric space Y, every Banach space Z and every $n \in \mathbb{N}$, we have $e_n(Y, Z) \ge C\sqrt{\log n}$. This result improves a corresponding one due to Johnson and Lindenstrauss [310].

Moreover, Lee and Naor, [380, 381], proved that the family of planar metrics is absolutely extendable, generalizing a result of Matoušek [446], who proved that there exists a universal constant C such that for every tree metric space Y and every Banach space Z, the following inequality is true: $e(Y, Z) \le C$.

Recall that a metric space (X, d) is called a *tree metric space* (or a *metric tree*) if for each $x, y, u, v \in X$ the following inequality is valid:

$$d(x, y) + d(u, v) \le \max\{d(x, u) + d(y, v), d(x, v) + d(y, u)\},$$

for all $x, y, u, v \in X$.

The following equivalent definition may be more illustrative: A metric tree is a metric space (X, d), satisfying the following two axioms:

(i) for every $x, y \in X$, $x \ne y$, there exists a uniquely determined isometry $\phi_{x,y} :$ $[0, d(x, y)] \to X$ with $\phi_{x,y}(0) = x$ and $\phi_{x,y}(d(x, y)) = y$;

(ii) for every one-to-one continuous mapping $f : [0, 1] \to X$ and for every $t \in [0, 1]$,

$$d(f(0), f(t)) + d(f(t), f(1)) = d(f(0), f(1)).$$

One shows that tree metric spaces are exactly subspaces of metric trees (see [129, 203], cf. [446]).

They (Lee and Naor) also proved the following results:

(i) if M is a two-dimensional Riemannian manifold of genus g, then for every subset X of M, we have $ae(X) \le C(g + 1)$ for some universal constant C;

(ii) for every $p \in (1, 2]$ there exists a constant $C(p)$ such that for every Banach space Z the inequality $e_n(L_p, Z) \le C(p)(\log n)^{\frac{1}{p}}$ is valid for each $n \in \mathbb{N}$;

(iii) for every $n \in \mathbb{N}$ and every $p \in (2, \infty)$ the inequality

$$e_n(L_2, L_p) \geq C \left(\frac{\log n}{\log \log n} \right)^{\frac{p-2}{p^2}}$$

is valid, where C is a universal constant.

Lee and Naor extended some of their results to the case when Z belongs to a certain family of metric spaces which includes Hadamard spaces. The Lipschitz extension properties of Hadamard spaces have been also studied by Lang, Pavlović and Schroeder [376]. Ohta [530], using stochastic Lipschitz extensions and gentle partitions of unity introduced by Lee and Naor, provided various classes of source and target spaces for the validity of Lipschitz extension theorems.

We define next a Lipschitz extension modulus related to $e_n(Y, Z)$ as follows: given Y, Z metric spaces and $X \subseteq Y$, denote as before by $e(X, Y, Z)$ the infimum over $K \geq 1$ such that for every Lipschitz function $f : X \rightarrow Z$ there exists $F : Y \rightarrow Z$ a Lipschitz extension of f to Y with $L(F) \leq K L(f)$. Then

$$e^n(Y, Z) = \sup\{e(X, X \cup \{y_1, \ldots, y_n\}, Z) : X \subseteq Y \text{ is closed}, \ y_1, \ldots, y_n \in Y \setminus X\}.$$

Very recently, Basso [70] derived an upper bound for this modulus proving that $e^n(Y, Z) \leq n + 1$. Subsequently, Mendel and Naor [462] gave an upper bound on $e^n(Y, Z)$ in terms of $e_n(Y, Z)$. Namely, they proved that $e^n(Y, Z) \leq e_n(Y, Z) + 2$ for every $n \in \mathbb{N}$ and every metric spaces Y, Z. In this way one can obtain upper bounds on $e^n(Y, Z)$ using known estimations on $e_n(Y, Z)$ (such as the ones given before in the special case when Z is a Banach space).

For a Banach space $(V, \|.\|)$ let us define $\mathscr{LE}(V)$ as the infimum over all $C \geq 1$ such that for each metric space (Z, d), any $X \subseteq Z$ and any $f : X \rightarrow V$ such that $\|f(x) - f(y)\| \leq K d(x, y)$ for each $x, y \in X$, there exists $F : Z \rightarrow V$ such that $F|_X = f$ and $\|F(x) - F(y)\| \leq K d(x, y)$ for each $x, y \in Z$. The constants $\mathscr{LE}_c(V)$ and $\mathscr{LE}_f(V)$ are defined in a similar way by restricting Z to be compact, respectively finite. Rieffel [596] proved that $\mathscr{LE}(V) = \mathscr{LE}_c(V) = \mathscr{LE}_f(V)$ for any finite-dimensional Banach space V. He also performed an exact calculation of $\mathscr{LE}(M_n(\mathbb{C})^{sa})$, where $M_n(\mathbb{C})^{sa}$ means the space of self-adjoint $n \times n$ matrices and gave an estimation for $\mathscr{LE}(C(K))$, where K is compact. Lindenstrauss [389] proved that for a suitable constant λ if X is a metric space and A is a closed subset of X, then for every Lipschitz function $f : A \rightarrow C(K)$ there exists a Lipschitz function $F : X \rightarrow C(K)$ such that $L(F) \leq \lambda L(f)$, where K is a compact metric space. He also proved that $\lambda \leq 20$. Lancien and Randrianantoanina [373] asked for conditions on X in order to obtain extensions with $\lambda = 1$ (isometric case) or $\lambda = 1 + \varepsilon$, $\varepsilon > 0$ (almost isometric case). If X is a finite-dimensional normed space, they showed that only $(1 + \varepsilon)$-extensions can be obtained in general and gave a four-dimensional counterexample to the isometric version. Moreover, they proved that for a finite-dimensional normed space X having the unit ball a polyhedron one always has an isometric extension. These results have been extended by Kalton

[330, 331] who provided necessary and sufficient conditions on a subset A of a metric space X for the existence of Lipschitz extensions of functions into $C(K)$ with prescribed Lipschitz constant. In this way he determined the best constant for extensions, namely he proved that $\mathscr{LE}(C(K)) = 2$.

For a metric space X let us define $\lambda(X) = \sup_{S \subseteq X} \inf_{T \in \text{Ext}(S,X)} \|T\|$, where, for $S \subseteq X$, $\text{Ext}(S, X)$ denotes the family of all linear bounded operators $T : \{f : S \to \mathbb{R} : f \text{ is Lipschitz}\} \to \{f : X \to \mathbb{R} : f \text{ is Lipschitz}\}$ satisfying $Tf|_S = f$. Brudnyi and Brudnyi (see [121–125]) verified the finiteness of $\lambda(X)$ for diverse spaces, as metric trees, Carnot groups, groups of polynomial growth, hyperbolic metric spaces of bounded geometry, doubling metric spaces, Riemannian manifolds with nonnegative Ricci curvature or with pinched negative sectional curvature, and finite direct sums of combinations of such examples.

A criterion for the validity of the generalized Lipschitz extension property for a pair $(\mathbb{R}^k, Y)$ in terms of quantitative algebraic topology was provided by Lang and Schlichenmaier [378]. The Heisenberg group $\mathbf{H}^n$ equipped with its standard Carnot-Carathéodory metric is a geodesic metric space. Balogh and Fässler [67] (see also [68]) proved that for each $k > n$, the pairs $(\mathbb{R}^k, \mathbf{H}^n)$ and $(\mathbf{H}^k, \mathbf{H}^n)$ do not have the generalized extension property, while the pair $(\mathbb{R}^2, \mathbf{H}^n)$ has it for each $n \geq 2$.

For a positive integer k and a metric space (Y, d), one denotes by $\mathscr{A}_k(Y)$ the set of all sums $\sum_{i=1}^{k} \delta_{y_i}$, where $y_1, \ldots, y_k$ are (not necessarily distinct) elements of Y. Equipped with the metric δ given by $\delta(x, y) = \min_\sigma \max_{i \in \{1, \ldots, Q\}} d(x_i, y_{\sigma(i)})$, where $x = \sum_{i=1}^{k} \delta_{x_i}$ and $y = \sum_{i=1}^{k} \delta_{y_i}$ are elements of $\mathscr{A}_k(Y)$ and σ runs over all the permutations of $\{1, \ldots, k\}$, $(\mathscr{A}_k(Y), \delta)$ is a complete metric space called the Almgren space. The functions taking values in an Almgren space are called Almgren multi-valued functions. Concerning Almgren spaces we quote from the Zentralblatt review of the book [36].

> This is the first time Almgren's fundamental regularity result for area-minimizing surfaces of codimension greater than one becomes available to the wider mathematical community. Written over a period of more than 10 years it originally appeared in 1984 as a preprint (1728 pages) consisting of three volumes each resembling a large telephone directory. Shortly after Almgren's untimely death in 1997, Vladimir Scheffer (his third doctoral student) began the monumental task of converting the typed manuscript into files which finally resulted in the publication of this impressive book (M. Grüter, Zentralblatt MATH, Zbl 0985.49001 (2000)).

A good introduction to Almgren spaces, including extension results for Lipschitz functions with values in Almgren spaces, is done in [183].

Goblet [257] (see also [256]) proved that if X is a metric space with finite Nagata dimension and Y is a complete weakly convex geodesic space, then the pair $(X, \mathscr{A}_k(Y))$ has the generalized Lipschitz extension property. In the paper [402] one exhibits an example of a $\sqrt{2/3}$-Lipschitz function f defined on a subset of $\mathbb{R}^2$ with values in the Almgren space $\mathscr{A}_2(\mathbb{R}^2)$ such that any Lipschitz extension of f has Lipschitz constant at least 1, showing that Kirszbraun's extension theorem (Theorem 4.2.3) is not valid in this case.

In [658] Tukia and Väisälä present some bi-Lipschitz extension results for functions having quasiconformal extensions with applications to the theory of

Lipschitz manifolds. MacManus [416] proved that any bi-Lipschitz function f from a subset of a line or a circle into the plane can be extended to a bi-Lipschitz function of the whole plane onto itself, with the bi-Lipschitz constant depending only on that of f. Some conditions for a subset A of $\mathbb{R}^n$ to have the following property—any bi-Lipschitz function from A to $\mathbb{R}^n$ has a bi-Lipschitz extension to the whole $\mathbb{R}^n$—were given by Alestalo and Trotsenko [30]. Huuskonen, Partanen and Väisälä [294] proved that every compact C^1-submanifold of $\mathbb{R}^n$ has the bi-Lipschitz extension property. For some other results concerning bi-Lipschitz extensions see [31, 543, 662] or [293].

Metric spaces having the generalized Lipschitz extension property (see Definition 4.2.8) were also studied in [525]. The paper [526] is concerned with the extension of Lipschitz functions with values in nuclear Fréchet spaces.

Albiac [16] proved that for each quasi-Banach space X with a separating dual, any Lipschitz function $f : X \to Z$, where Z is a Banach space, admits a unique Lipschitz extension to the Banach envelope of X. The Banach envelope of a separable quasi-Banach space X is the smallest Banach space $\widehat{X}$ containing X (see [332]).

Luukkainen and Väisälä [413] proved that if A is a closed subset of a metric space X, every locally Lipschitz function $f : A \to M$, where M is a n-dimensional locally Lipschitz manifold, has a locally Lipschitz extension to a neighborhood of A in X.

Bressan and Cortesi [116] proved that if Ω^n is the space of all nonempty compact, convex subsets of $\mathbb{R}^n$ endowed with the Pompeiu-Hausdorff metric, A is a subset of a Hilbert space $\mathscr{H}$ and $f : A \to \Omega^n$ is a Lipschitz function, then there exists a Lipschitz function $F : \mathscr{H} \to \Omega^n$ such that $F|_A = f$ and $L(F) \leq 2n\sqrt{28/3}\,L(f)$.

Brudnyi and Shvartsman [128] proved that the linear and nonlinear Lipschitz extension properties of a given metric space are not changed when the original metric d is replaced by a metric of the form $\omega \circ d$, where $\omega : [0, \infty) \to [0, \infty)$ is a concave function such that $\omega(0) = 0$.

The connection between Lipschitz extension and best approximation was studied by Muştăţa [494] (see also [165]), Mabizela [415], Park [542] and Deutsch, Li and Mazibela [190] (see Sect. 8.9). The papers [497] and [512] contain some extension results for Hölder functions.

For some more results concerning extensions of Lipschitz or locally Lipschitz functions one can also consult [383] (where a generalization of the concept of Lipschitz constant to fields of affine jets is provided), [111, 296, 297, 484], or [471].

The book [676] is an excellent survey on extension results for various classes of functions, including the Lipschitz ones. The survey papers [641] and [642] contain many interesting results concerning the extension of Lipschitz functions and of other classes of functions.

Some extension results for Lipschitz functions on geodesic metric spaces will be proved in Chap. 5.

Chapter 5
Extension Results for Lipschitz Mappings in Geodesic Spaces

Geodesic metric spaces are a natural generalization of Riemannian manifolds and provide a suitable setting for the study of problems from various areas of mathematics with important applications. In this chapter we review selected properties of Lipschitz mappings in geodesic metric spaces focusing mainly on certain extension theorems which generalize corresponding ones from linear contexts. We point out that the two-volume book by Brudnyi and Brudnyi [126, 127] vastly covers the theory of extension and trace problems ranging from classical results to recent ones and hence includes some of the aspects that we also discuss here.

In order to make the exposition self-contained, we recall first basic concepts and results from the theory of geodesic metric spaces with emphasis on the notion of curvature. In recent years, nonpositively curved spaces in the sense of Busemann and Alexandrov spaces of curvature bounded above or below have become highly relevant in geometry and geometric group theory, in the broad area of analysis in metric spaces, ergodic theory, optimal transport, convex optimization, fixed point theory, as well as in other fields. A few excellent references where these spaces and related problems are treated at length are the monographs by Alexander et al. [29], Bridson and Haefliger [117], Burago et al. [130], Jost [318], or Papadopoulos [541].

5.1 Some Definitions and Facts in Geodesic Metric Spaces

Let (X, d) be a metric space and $x, y \in X$. We say that a point $m \in X$ is a *midpoint* of x and y if

$$d(x, m) = d(m, y) = \frac{d(x, y)}{2}.$$

© Springer Nature Switzerland AG 2019
Ş. Cobzaş et al., *Lipschitz Functions*, Lecture Notes in Mathematics 2241,
https://doi.org/10.1007/978-3-030-16489-8_5

A *(unit speed) geodesic* from x to y is a mapping $c : [0, l] \subseteq \mathbb{R} \to X$ such that $c(0) = x, c(l) = y$ and

$$d(c(s), c(s')) = |s - s'|,$$

for every $s, s' \in [0, l]$. This clearly implies that $l = d(x, y)$. We say that c *starts* from x, *joins* x and y, and that x and y are its *endpoints*. One can linearly reparametrize c by the interval $[0, 1]$ to obtain a geodesic of constant speed $d(x, y)$ given by $c' : [0, 1] \to X$, $c'(t) = c(tl)$ for all $t \in [0, 1]$. Then

$$d(c'(t), c'(t')) = |t - t'| d(x, y),$$

for every $t, t' \in [0, 1]$. In the sequel, we use the term *geodesic* for such a constant speed geodesic parametrized by $[0, 1]$.

The image of a geodesic forms a *geodesic segment*. If a geodesic c joins x and y, then we say that the geodesic segment $c([0, 1])$ joins x and y, and that x and y are its endpoints. Sometimes we will also denote a geodesic segment joining x and y by $s(x, y)$. Note that geodesic segments joining two given points in a metric space may not exist and, in case they do, they are not necessarily unique.

We say that (X, d) is a *(uniquely) geodesic space* if every two points in X can be joined by a (unique) geodesic segment. If X is a geodesic space, a point $z \in X$ belongs to a geodesic segment joining x and y if and only if there exists $t \in [0, 1]$ such that $d(x, z) = td(x, y)$ and $d(z, y) = (1 - t)d(x, y)$, and we write $z = (1 - t)x \oplus ty$ if no confusion arises. In this case $z = c(t)$, where c is the unique geodesic from x to y whose image is the geodesic segment in question. In particular, any two points in a (uniquely) geodesic space have a (unique) midpoint. Conversely, complete metric spaces where every two points have a midpoint are geodesic.

A normed space $(X, \|\cdot\|)$ endowed with the natural metric $d(x, y) = \|x - y\|$ for $x, y \in X$ is a geodesic space. Given any two points $x, y \in X$, the algebraic segment determined by x and y is a geodesic segment and is the image of the geodesic $c : [0, 1] \to X$ from x to y defined by $c(t) = (1 - t)x + ty$ for $t \in [0, 1]$. A normed space is uniquely geodesic if and only if it is strictly convex.

A *geodesic triangle* $\Delta = \Delta(x_1, x_2, x_3)$ in a geodesic space (X, d) consists of three points $x_1, x_2, x_3 \in X$ (its *vertices*) and three geodesic segments (its *sides*) joining each pair of points. We denote the *perimeter* of a geodesic triangle with vertices x_1, x_2, x_3 by $P(x_1, x_2, x_3) = d(x_1, x_2) + d(x_2, x_3) + d(x_3, x_1)$. Note that its value does not depend on the choice of geodesic segments as sides of the geodesic triangle. Having fixed a geodesic triangle $\Delta(x_1, x_2, x_3)$, we denote by $[x_i, x_j]$ the side between the vertices x_i and x_j.

Geodesic spaces constitute an appropriate framework for considering various convexity concepts. We introduce next a notion due to Busemann [132, 133] which captures an important property of nonpositive sectional curvature from Riemannian geometry.

Definition 5.1.1 (Busemann Convexity) Let (X, d) be a geodesic space. We say that X is *Busemann convex* or *nonpositively curved in the sense of Busemann* if given any two geodesics $c : [0, 1] \to X$ and $c' : [0, 1] \to X$,

$$d(c(t), c'(t)) \leq (1 - t)d(c(0), c'(0)) + td(c(1), c'(1)),$$

for all $t \in [0, 1]$.

We include in the following some properties of geodesic spaces that are Busemann convex. Let (X, d) be a Busemann convex space. In particular, X satisfies the following convexity property: for every $z \in X$ and every geodesic $c : [0, 1] \to X$ we have

$$d(z, c(t)) \leq (1 - t)d(z, c(0)) + td(z, c(1)), \qquad (5.1.1)$$

for all $t \in [0, 1]$.

Another fact which can be deduced immediately is that X is uniquely geodesic. Consequently, a normed space is Busemann convex if and only if it is strictly convex.

Geodesics in X vary continuously with their endpoints in the sense that for all $x, y, x_n, y_n \in X$, where $n \in \mathbb{N}$, such that $x_n \to x$ and $y_n \to y$, if c is the geodesic from x to y, and c_n is the geodesic from x_n to y_n, then (c_n) converges uniformly to c. This follows because

$$d(c_n(t), c(t)) \leq (1 - t)d(x_n, x) + td(y_n, y),$$

for all $n \in \mathbb{N}$ and $t \in [0, 1]$.

In addition, X is also contractible. To see this, fix $z \in X$ and define $h : [0, 1] \times X \to X$ by $h(t, x) = (1 - t)z \oplus tx$. Then

$$\begin{aligned} d(h(t, x), h(t', x')) &= d((1 - t)z \oplus tx, (1 - t')z \oplus t'x') \\ &\leq d((1 - t)z \oplus tx, (1 - t)z \oplus tx') \\ &\quad + d((1 - t)z \oplus tx', (1 - t')z \oplus t'x') \\ &\leq td(x, x') + |t - t'|d(z, x'), \end{aligned}$$

for all $t, t' \in [0, 1]$ and all $x, x' \in X$. This implies that h is continuous and so X is contractible. Actually, in the above reasoning the uniqueness of geodesics in X is not used and it is enough that for any two points one can choose a geodesic joining them such that the resulting selection of geodesics (called a geodesic bicombing) satisfies the Busemann convexity condition. More explicitly, having (X, d) a metric space, we say that a mapping $\sigma : X \times X \times [0, 1] \to X$ is a *geodesic bicombing* for X if it satisfies the following properties:

(i) $\sigma_{xy} = \sigma(x, y, \cdot)$ is a geodesic from x to y;
(ii) $\sigma_{xy}(t) = \sigma_{yx}(1-t)$;
(iii) $d(\sigma_{xy}(t), \sigma_{x'y'}(t)) \le (1-t)d(x, x') + td(y, y')$,

for all $x, y, x', y' \in X$ and all $t \in [0, 1]$.

In particular, (iii) yields

$$d(z, \sigma_{xy}(t)) \le (1-t)d(z, x) + td(z, y),$$

for all $z \in X$ and $t \in [0, 1]$. Moreover, geodesics selected by σ vary continuously with their endpoints.

In [188] and [189], a geodesic bicombing is only assumed to satisfy (i), while the conditions (ii) and (iii) are called reversibility and the conical property, respectively. Thus, the geodesic bicombings that we consider are implicitly both reversible and conical.

Besides Busemann convex spaces, the class of spaces that admit a geodesic bicombing includes for instance any normed space.

Let X be a metric space with a geodesic bicombing σ. A set $A \subseteq X$ is called σ-convex if the image of σ_{xy} is contained in A for all $x, y \in A$.

Proposition 5.1.2 *Let X be a metric space with a geodesic bicombing σ and $A \subseteq X$ σ-convex. Then the closure $\overline{A}$ is σ-convex.*

Proof Let $x, y \in \overline{A}$ and $t \in [0, 1]$. Then there exist two sequences (x_n) and (y_n) in A such that $\lim_{n\to\infty} x_n = x$ and $\lim_{n\to\infty} y_n = y$. Since A is σ-convex, $\sigma_{x_n y_n}(t) \in A$ for all $n \in \mathbb{N}$. Because geodesics selected by σ vary continuously with their endpoints, we have $\lim_{n\to\infty} \sigma_{x_n y_n}(t) = \sigma_{xy}(t)$. This shows that $\sigma_{xy}(t) \in \overline{A}$ and hence $\overline{A}$ is σ-convex. □

The *(closed) σ-convex hull* of A is the smallest (closed) σ-convex set containing A. We denote the σ-convex hull and the closed σ-convex hull of A by $\mathrm{co}_\sigma(A)$ and $\overline{\mathrm{co}_\sigma}(A)$, respectively. Given $B \subseteq X$, let $G_\sigma^1(B)$ be the union of the images of σ_{xy} for all $x, y \in B$. Thus, the set A is σ-convex if and only if $G_\sigma^1(A) = A$. Setting, for $n \in \mathbb{N}$, $n \ge 2$, $G_\sigma^n(A) = G_\sigma^1(G_\sigma^{n-1}(A))$, then $\mathrm{co}_\sigma(A) = \bigcup_{n\ge 1} G_\sigma^n(A)$. As the closure of $\mathrm{co}_\sigma(A)$ is σ-convex, it coincides with $\overline{\mathrm{co}_\sigma}(A)$.

The next two properties are well-known in normed spaces. As we will use them in metric spaces with a geodesic bicombing, we show that they hold true in this context too.

Proposition 5.1.3 *Let (X, d) be a metric space with a geodesic bicombing σ and $A \subseteq X$ nonempty. Then*

$$\mathrm{diam}\, A = \mathrm{diam}\, G_\sigma^1(A) = \mathrm{diam}\, G_\sigma^2(A) = \ldots = \mathrm{diam}\, \mathrm{co}_\sigma(A) = \mathrm{diam}\, \overline{\mathrm{co}_\sigma}(A).$$

Proof Once the first equality is established, the others are straightforward. Because $A \subseteq G_\sigma^1(A)$ we only need to show that $\mathrm{diam}\, G_\sigma^1(A) \le \mathrm{diam}\, A$. To see this, take $u, u' \in G_\sigma^1(A)$. Then there exist $x, y, x', y' \in A$ and $t, t' \in [0, 1]$ such that $u = \sigma_{xy}(t)$ and $u' = \sigma_{x'y'}(t')$, and we have

$$d(u, u') = d(u, \sigma_{x'y'}(t')) \leq (1 - t')d(u, x') + t'd(u, y')$$
$$\leq \max\{d(u, x'), d(u, y')\}$$
$$= \max\{d(\sigma_{xy}(t), x'), d(\sigma_{xy}(t), y')\}$$
$$\leq \max\{(1 - t)d(x, x') + td(y, x'), (1 - t)d(x, y') + td(y, y')\}$$
$$\leq \max\{d(x, x'), d(y, x'), d(x, y'), d(y, y')\}$$
$$\leq \operatorname{diam} A.$$

Thus, $\operatorname{diam} G_\sigma^1(A) \leq \operatorname{diam} A$. $\qquad\square$

In the following result, d_H stands for the Pompeiu–Hausdorff distance defined in Sect. 1.3.4.

Proposition 5.1.4 *Let* (X, d) *be a metric space with a geodesic bicombing* σ *and* $A, B \subseteq X$ *nonempty. Then*

$$d_H(\overline{\operatorname{co}_\sigma}(A), \overline{\operatorname{co}_\sigma}(B)) \leq d_H(A, B).$$

Proof Let $z \in B$. Obviously, $d(z, \overline{\operatorname{co}_\sigma}(A)) \leq d(z, A) \leq d_H(A, B)$. Consider the set

$$C = \{x \in X : d(x, \overline{\operatorname{co}_\sigma}(A)) \leq d_H(A, B)\},$$

which is a closed. To show that C is σ-convex, let $x, y \in C$ and $t \in [0, 1]$. Given $\varepsilon > 0$, we find $u, v \in \overline{\operatorname{co}_\sigma}(A)$ such that $\max\{d(x, u), d(y, v)\} < d_H(A, B) + \varepsilon$. Then $\sigma_{uv}(t) \in \overline{\operatorname{co}_\sigma}(A)$ and

$$d(\sigma_{xy}(t), \overline{\operatorname{co}_\sigma}(A)) \leq d(\sigma_{xy}(t), \sigma_{uv}(t)) \leq (1-t)d(x, u) + td(y, v) < d_H(A, B) + \varepsilon.$$

As $\varepsilon > 0$ is arbitrary, we get $\sigma_{xy}(t) \in C$.

Since $B \subseteq C$ it follows that $\overline{\operatorname{co}_\sigma}(B) \subseteq C$, from where

$$\sup_{z \in \overline{\operatorname{co}_\sigma}(B)} d(z, \overline{\operatorname{co}_\sigma}(A)) \leq d_H(A, B).$$

In a similar way,

$$\sup_{z \in \overline{\operatorname{co}_\sigma}(A)} d(z, \overline{\operatorname{co}_\sigma}(B)) \leq d_H(A, B)$$

and we are done. $\qquad\square$

Instead of introducing the above notions using the geodesics selected by σ, one can consider all geodesics. In this way, if X is a geodesic space, we say that a set $A \subseteq X$ is *convex* if given two points in A, any geodesic segment joining them is contained in A. In particular, any convex subset of a metric space with a geodesic bicombing σ is σ-convex. As before, one defines the *(closed) convex hull* of A by replacing σ-convex sets by convex ones and we denote the convex hull and the closed convex hull of A by $\mathrm{co}(A)$ and $\overline{\mathrm{co}}(A)$, respectively.

Remark 5.1.5 In a Busemann convex space X, convexity of sets coincides with σ-convexity, where given $x, y \in X$, σ_{xy} is the unique geodesic in X from x to y.

5.1.1 Alexandrov Spaces

In this chapter we are concerned with Alexandrov spaces with lower or upper curvature bounds considered globally (or in the large). These spaces are defined in terms of comparisons with the standard model planes of constant curvature, which we introduce below. To this end, we first briefly describe the n-dimensional spaces $\mathbb{S}^n$ and $\mathbb{H}^n$ following [117].

The *n-dimensional sphere* $\mathbb{S}^n$ is the set

$$\left\{ x \in \mathbb{R}^{n+1} : (x \mid x) = 1 \right\},$$

where $(\cdot \mid \cdot)$ is the Euclidean scalar product. Endowed with the distance $d : \mathbb{S}^n \times \mathbb{S}^n \to \mathbb{R}$ that assigns to each $(x, y) \in \mathbb{S}^n \times \mathbb{S}^n$ the unique number $d(x, y) \in [0, \pi]$ such that $\cos d(x, y) = (x \mid y)$, $\mathbb{S}^n$ is a geodesic space called the *spherical n-space*. In other words, $d(x, y)$ is the length of the smallest arc of a great circle in $\mathbb{S}^n$ which joins x and y. In $\mathbb{S}^n$, every two points at distance less than π are joined by a unique geodesic segment and balls of radius smaller than $\pi/2$ are convex.

The geodesics starting from $x \in \mathbb{S}^n$ with distance between their endpoints equal to l are given by $c : [0, 1] \to \mathbb{S}^n$, $c(t) = x \cos(tl) + u \sin(tl)$, where $u \in \mathbb{R}^{n+1}$ is a unit vector satisfying $(u \mid x) = 0$. We refer to such a vector u as an initial vector of c. Any $y \in \mathbb{S}^n$ with $d(x, y) \in (0, \pi)$ determines a unique initial vector u so that $c(1) = y$.

The *spherical angle* between two geodesics starting from the same point in $\mathbb{S}^n$ with respective initial vectors u and v is the unique number $\alpha \in [0, \pi]$ such that $\cos \alpha = (u \mid v)$. The *spherical cosine law* states that in a spherical triangle with vertices $x, y, z \in \mathbb{S}^n$ and α the vertex angle at x we have

$$\cos d(y, z) = \cos d(x, y) \cos d(x, z) + \sin d(x, y) \sin d(x, z) \cos \alpha.$$

For $x = (x_1, \ldots, x_{n+1}) \in \mathbb{R}^{n+1}$ and $y = (y_1, \ldots, y_{n+1}) \in \mathbb{R}^{n+1}$, consider the following bilinear form $\langle x \mid y \rangle = -x_{n+1} y_{n+1} + \sum_{i=1}^{n} x_i y_i$. The *hyperbolic n-space*

$\mathbb{H}^n$ is a uniquely geodesic space consisting of the set

$$\{x = (x_1, \ldots, x_{n+1}) \in \mathbb{R}^{n+1} : \langle x \mid x \rangle = -1, x_{n+1} > 0\}$$

together with the hyperbolic distance $d : \mathbb{H}^n \times \mathbb{H}^n \to \mathbb{R}$ which assigns to each $(x, y) \in \mathbb{H}^n \times \mathbb{H}^n$ the unique number $d(x, y) \geq 0$ such that $\cosh d(x, y) = -\langle x \mid y \rangle$. All balls in $\mathbb{H}^n$ are convex.

The geodesics starting from $x \in \mathbb{H}^n$ with distance between their endpoints equal to l are given by $c : [0, 1] \to \mathbb{H}^n$, $c(t) = x \cosh(tl) + u \sinh(tl)$, where $u \in \mathbb{R}^{n+1}$ satisfies $\langle u \mid u \rangle = 1$ and $\langle u \mid x \rangle = 0$. We refer to such a vector u as an initial vector of c. If $y \in \mathbb{H}^n$ with $d(x, y) > 0$, one can precisely determine the unique initial vector u such that $c(1) = y$.

The *hyperbolic angle* between two geodesics starting from the same point in $\mathbb{H}^n$ with respective initial vectors u and v is the unique number $\alpha \in [0, \pi]$ such that $\cos \alpha = \langle u \mid v \rangle$. The *hyperbolic cosine law* states that in a hyperbolic triangle with vertices $x, y, z \in \mathbb{H}^n$ and α the vertex angle at x we have

$$\cosh d(y, z) = \cosh d(x, y) \cosh d(x, z) - \sinh d(x, y) \sinh d(x, z) \cos \alpha.$$

For $\kappa \in \mathbb{R}$ and $n \in \mathbb{N}$, let M_κ^n denote the following standard model spaces: if $\kappa > 0$, M_κ^n is obtained from the spherical space $\mathbb{S}^n$ by multiplying the spherical distance with $1/\sqrt{\kappa}$; if $\kappa = 0$, M_0^n is the n-dimensional Euclidean space $\mathbb{R}^n$; and if $\kappa < 0$, M_κ^n is obtained from the hyperbolic space $\mathbb{H}^n$ by multiplying the hyperbolic distance with $1/\sqrt{-\kappa}$. Based on the above, it is immediate that M_κ^n is a geodesic space. If we denote the diameter of M_κ^n by D_κ, that is, $D_\kappa = \infty$ for $\kappa \leq 0$ and $D_\kappa = \pi/\sqrt{\kappa}$ for $\kappa > 0$, then there exists a unique geodesic segment joining two points in M_κ^n if and only if the distance between them is smaller than D_κ. Note that if $\kappa \leq 0$, no restriction is imposed on the distance between the points. Furthermore, balls of radius smaller than $D_\kappa/2$ are convex. By rescaling, one obtains a cosine law in M_κ^n from the corresponding ones in $\mathbb{S}^n$ and $\mathbb{H}^n$. From a Riemannian geometry perspective, M_κ^n can be described as the complete, simply connected, n-dimensional Riemannian manifold of constant sectional curvature κ.

In the following we always assume that $\kappa \in \mathbb{R}$ if nothing else is specified.

Upper curvature bounds in the sense of Alexandrov capture the idea of upper bounds for the sectional curvature for Riemannian manifolds. We introduce next geodesic spaces that have globally curvature bounded above by κ in the sense of Alexandrov. M. Gromov named these spaces CAT(κ) after E. Cartan, A. D. Alexandrov and V. A. Toponogov (see, e.g., [117, p. 159]). We consider a comparison condition that applies for all triangles in the space independent of their size. This explains the terminology "global". The local version for sufficiently small triangles is mentioned in Sect. 5.5.

Let (X, d) be a geodesic space and $\Delta = \Delta(x_1, x_2, x_3)$ a geodesic triangle. A triangle $\overline{\Delta} = \Delta(\overline{x}_1, \overline{x}_2, \overline{x}_3)$ in M_κ^2 is said to be a *comparison triangle* for Δ if $d(x_i, x_j) = d_{M_\kappa^2}(\overline{x}_i, \overline{x}_j)$ for $i, j \in \{1, 2, 3\}$. For κ fixed, comparison triangles in M_κ^2 for geodesic triangles having perimeter less than $2D_\kappa$ always exist and are unique up to isometry. The *comparison point* in $\overline{\Delta}$ for a point $x \in [x_i, x_j]$ is the point $\overline{x} \in [\overline{x}_i, \overline{x}_j]$ such that $d(x_i, x) = d_{M_\kappa^2}(\overline{x}_i, \overline{x})$.

Definition 5.1.6 (CAT(κ) Inequality) A geodesic triangle Δ satisfies the CAT(κ) *inequality* if for every comparison triangle $\overline{\Delta}$ in M_κ^2 for Δ and every $x, y \in \Delta$ we have $d(x, y) \le d_{M_\kappa^2}(\overline{x}, \overline{y})$, where $\overline{x}, \overline{y} \in \overline{\Delta}$ are the comparison points of x and y.

Altogether, this condition says that geodesic triangles are "thin" when compared to triangles in the model plane M_κ^2.

Definition 5.1.7 (CAT(κ) Space) A geodesic space is said to be a CAT(κ) *space* if every geodesic triangle in it having perimeter less than $2D_\kappa$ satisfies the CAT(κ) inequality.

If $\kappa > 0$, a CAT(κ) space in general need not be geodesic and one only assumes that all pairs of points whose distance is less than D_κ are joined by a geodesic. However, for simplicity, we restrict our discussion to geodesic spaces.

CAT(κ) spaces inherit a rich geometric structure from the model planes M_κ^2. If x and y are two points in a CAT(κ) space, there exists a unique geodesic from x to y provided $d(x, y) < D_\kappa$. Moreover, this geodesic varies continuously with its endpoints.

Inner product spaces are always CAT(0). If a normed space is CAT(κ) for some $\kappa \in \mathbb{R}$, then it is an inner product space. Every CAT(0) space (and so every CAT(κ) space with $\kappa \le 0$) is Busemann convex, but the converse does not hold since there exist, for instance, strictly convex normed spaces whose norm does not arise from an inner product.

CAT(κ) spaces can be equivalently defined in terms of angle comparisons. Actually, Alexandrov [27] introduced the notion of angle in metric spaces and used it to define the concept of curvature bounded above. Another characterization involving triangle comparisons is the following.

Proposition 5.1.8 *A geodesic space* (X, d) *is a* CAT(κ) *space if and only if for every geodesic triangle* $\Delta(x_1, x_2, x_3)$ *in X of perimeter less than $2D_\kappa$, taking $m \in [x_2, x_3]$ a midpoint of x_2 and x_3, its comparison point $\overline{m} \in \Delta(\overline{x}_1, \overline{x}_2, \overline{x}_3) \subseteq M_\kappa^2$ satisfies*

$$d(x_1, m) \le d_{M_\kappa^2}(\overline{x}_1, \overline{m}).$$

Alternative characterizations are given in terms of explicit inequalities. For $\kappa = 0$, we include below two such conditions that will be used in this chapter.

Proposition 5.1.9 *For a geodesic space* (X, d), *the following are equivalent.*

1. *X is a* CAT(0) *space.*
2. *For every $z \in X$ and every geodesic $c : [0, 1] \to X$,*

$$d(z, c(t))^2 \le (1 - t)d(z, c(0))^2 + td(z, c(1))^2 - t(1 - t)d(c(0), c(1))^2,$$
$$(5.1.2)$$

for all $t \in [0, 1]$.
3. *For every $x, y, u, v \in X$,*

$$d(x, y)^2 + d(u, v)^2 \leq d(x, v)^2 + d(y, u)^2 + 2d(x, u)d(y, v). \qquad (5.1.3)$$

Proof To see that $1 \Rightarrow 2$, denote $u = c(0)$, $v = c(1)$ and $w = c(t)$ for some $t \in [0, 1]$. Consider $\Delta(\overline{z}, \overline{u}, \overline{v})$ a comparison triangle in $\mathbb{R}^2$ for $\Delta(z, u, v)$ and let $\overline{w}$ be the comparison point for w. Denote by α the angle at $\overline{w}$ in $\Delta(\overline{z}, \overline{w}, \overline{u})$. Applying the Euclidean cosine law in the triangles $\Delta(\overline{z}, \overline{w}, \overline{u})$ and $\Delta(\overline{z}, \overline{w}, \overline{v})$ we obtain

$$\|\overline{z} - \overline{u}\|^2 = \|\overline{z} - \overline{w}\|^2 + t^2\|\overline{u} - \overline{v}\|^2 - 2t\|\overline{u} - \overline{v}\|\|\overline{z} - \overline{w}\| \cos \alpha$$

and

$$\|\overline{z} - \overline{v}\|^2 = \|\overline{z} - \overline{w}\|^2 + (1 - t)^2\|\overline{u} - \overline{v}\|^2 - 2(1 - t)\|\overline{u} - \overline{v}\|\|\overline{z} - \overline{w}\| \cos(\pi - \alpha).$$

Multiplying the first equality by $1 - t$, the second one by t and then adding them, we obtain

$$\|\overline{z} - \overline{w}\|^2 = (1 - t)\|\overline{z} - \overline{u}\|^2 + t\|\overline{z} - \overline{v}\|^2 - t(1 - t)\|\overline{u} - \overline{v}\|^2$$
$$= (1 - t)d(z, u)^2 + td(z, v)^2 - t(1 - t)d(u, v)^2.$$

By the CAT(0) inequality, $d(z, w) \leq \|\overline{z} - \overline{w}\|$, so (5.1.2) holds.

We prove next that $2 \Rightarrow 1$. Let $z, u, v \in X$ and $\Delta(\overline{z}, \overline{u}, \overline{v})$ be a comparison triangle in $\mathbb{R}^2$ for a geodesic triangle $\Delta(z, u, v)$. Take $m \in [u, v]$ a midpoint of u and v and let $\overline{m} \in [\overline{u}, \overline{v}]$ be the comparison point for m. By (5.1.2),

$$d(z, m)^2 \leq \frac{1}{2}d(z, u)^2 + \frac{1}{2}d(z, v)^2 - \frac{1}{4}d(u, v)^2$$
$$= \frac{1}{2}\|\overline{z} - \overline{u}\|^2 + \frac{1}{2}\|\overline{z} - \overline{v}\|^2 - \frac{1}{4}\|\overline{u} - \overline{v}\|^2 = \|\overline{z} - \overline{m}\|^2,$$

from where $d(z, m) \leq \|\overline{z} - \overline{m}\|$. Using Proposition 5.1.8, we can conclude that X is a CAT(0) space.

We include the proof that $1 \Rightarrow 3$ given in [376]. The converse implication follows from [76]. If $x = u$ or $y = v$, then (5.1.3) holds. Suppose $x \neq u$ and $y \neq v$ and let $c : [0, 1] \to X$ be the geodesic from x to y. By (5.1.2), for all $t \in (0, 1)$,

$$d(u, c(t))^2 \leq (1 - t)d(u, x)^2 + td(u, y)^2 - t(1 - t)d(x, y)^2$$

and

$$d(v, c(t))^2 \leq (1 - t)d(v, x)^2 + td(v, y)^2 - t(1 - t)d(x, y)^2.$$

For any $t \in (0, 1)$ and $a, b \in \mathbb{R}$ we have $(1 - t)^2 a^2 + t^2 b^2 \geq 2t(1 - t)ab$, which yields $t^{-1}a^2 + (1 - t)^{-1}b^2 \geq (a + b)^2$. Thus,

$$d(u, v)^2 \leq (d(u, c(t)) + d(v, c(t)))^2 \leq t^{-1}d(u, c(t))^2 + (1 - t)^{-1}d(v, c(t))^2$$
$$\leq d(u, y)^2 + d(v, x)^2 + t^{-1}(1 - t)d(u, x)^2 + t(1 - t)^{-1}d(v, y)^2$$
$$- d(x, y)^2.$$

Letting $t = d(u, x)/(d(u, x) + d(v, y))$ above, we obtain (5.1.3). □

Every CAT(κ) space is also a CAT(κ') space for all $\kappa' \geq \kappa$. Thus, (5.1.2) also holds in any CAT(κ) space with $\kappa < 0$.

The following results will also be needed in the sequel.

Proposition 5.1.10 *Let* (X, d) *be a* CAT(κ) *space with* $\kappa > 0$, $z \in X$ *and* $c :$ $[0, 1] \to X$ *a geodesic satisfying* $l = d(c(0), c(1)) \in (0, D_\kappa)$, $d(z, c(0)) \leq r$ *and* $d(z, c(1)) \leq r$ *for some* $r \leq D_\kappa/2$. *Then*

$$\cos\left(\sqrt{\kappa}d(z, c(t))\right) \geq \frac{\sin\left(\sqrt{\kappa}(1 - t)l\right) + \sin\left(\sqrt{\kappa}tl\right)}{\sin\left(\sqrt{\kappa}l\right)} \cos\left(\sqrt{\kappa}r\right),$$

for all $t \in [0, 1]$. *In particular,*

$$d(z, c(t)) \leq r, \tag{5.1.4}$$

for all $t \in [0, 1]$.

Proof Reasoning as in the proof of $1 \Rightarrow 2$ from Proposition 5.1.9 and applying now the cosine law in M_κ^2 instead of the Euclidean cosine law we obtain

$$\cos\left(\sqrt{\kappa}d(z, c(t))\right) \geq \frac{\sin\left(\sqrt{\kappa}(1 - t)l\right)}{\sin\left(\sqrt{\kappa}l\right)} \cos\left(\sqrt{\kappa}d(z, c(0))\right)$$
$$+ \frac{\sin\left(\sqrt{\kappa}tl\right)}{\sin\left(\sqrt{\kappa}l\right)} \cos\left(\sqrt{\kappa}d(z, c(1))\right),$$

for all $t \in [0, 1]$. The desired inequality follows from the above one. □

This allows now to state the next property.

Proposition 5.1.11 *Let* X *be a* CAT(κ) *space and* $A \subseteq X$ *nonempty with* diam $A \leq$ $D_\kappa/2$. *Then* diam $\overline{co}(A) =$ diam A.

Proof Clearly, (5.1.4) also holds in CAT(κ) spaces with $\kappa \leq 0$. Using this inequality one obtains as in the proof of Proposition 5.1.3 that diam $A =$ diam co(A) $=$ diam $\overline{co}(A)$. The fact that geodesics in X joining points at distance less than or equal to $D_\kappa/2$ vary continuously with their endpoints shows that $\overline{\mathrm{co}(A)}$ is convex, hence $\overline{\mathrm{co}(A)} = \overline{\mathrm{co}}(A)$.   □

A characterization of CAT(κ) spaces with $\kappa > 0$ expressed in terms of a four point inequality was given very recently in [77].

Complete CAT(0) spaces are also called *Hadamard spaces* and include Hilbert spaces, the complex Hilbert ball with the hyperbolic metric, Hadamard manifolds (i.e., complete, simply connected Riemannian manifolds of nonpositive sectional curvature), Euclidean buildings of Bruhat and Tits and other complexes.

$\mathbb{R}$-trees are another special example of CAT(0) spaces which proved to be significant in different contexts (see [78]). An $\mathbb{R}$-*tree* is a uniquely geodesic space which satisfies the property that if the intersection of two geodesic segments is precisely a common endpoint of both, then their union is the unique geodesic segment that joins the other two endpoints. Let X be an $\mathbb{R}$-tree. For $x, y \in X$, denote the unique geodesic segment joining x and y by $s(x, y)$. Thus, if $x, y, z \in X$, the following implication holds

$$s(x, y) \cap s(x, z) = \{x\} \;\Rightarrow\; s(x, y) \cup s(x, z) = s(y, z).$$

It immediately follows that for all $x, y, z \in X$ there exists $w \in X$ such that

$$s(x, y) \cap s(x, z) = s(x, w),$$

and hence

$$s(y, z) = s(y, w) \cup s(z, w).$$

Moreover, it is easy to see that a metric space is an $\mathbb{R}$-tree if and only if it is a CAT(κ) space for any real κ. Triangles in an $\mathbb{R}$-tree are tripods. A standard example of a complete $\mathbb{R}$-tree is $\mathbb{R}^2$ endowed with the so-called river metric. For $x = (x_1, x_2), y = (y_1, y_2) \in \mathbb{R}^2$, the river metric is defined by

$$d(x, y) = \begin{cases} |x_2 - y_2|, & \text{if } x_1 = y_1, \\ |x_2| + |y_2| + |x_1 - y_1|, & \text{otherwise.} \end{cases}$$

The concept of lower bounds for the sectional curvature from Riemannian geometry can be similarly generalized via triangle comparison. Again we consider the global notion.

Definition 5.1.12 (CBB(κ) **Space**) A geodesic space is said to be a CBB(κ) *space* if every geodesic triangle in it having perimeter less than $2D_\kappa$ satisfies the reverse of the CAT(κ) inequality.

Such spaces are also said to have globally curvature bounded below by κ in the sense of Alexandrov. Roughly speaking, all their geodesic triangles are "fat" when compared to triangles in the model plane. Actually, in any CBB(κ) space (without taking into account some exceptional one-dimensional spaces), perimeters of geodesic triangles are at most $2D_\kappa$. If X is a CBB(κ) space for some $\kappa \in \mathbb{R}$, then it is also a CBB(κ') space for every $\kappa' \le \kappa$.

The above definition can be restated in terms of Proposition 5.1.8, where one considers the reverse inequality. If $\kappa = 0$, an equivalent characterization can be given via the reverse inequality of (5.1.2). Thus, by the parallelogram law, a Hilbert space is also CBB(0). Other examples of CBB(0) spaces include, among others, complete Riemannian manifolds of nonnegative sectional curvature, convex surfaces in $\mathbb{R}^3$ (with the intrinsic metric), and quotients of CBB(0) spaces by groups acting on them by isometries with closed orbits.

CAT(κ) and CBB(κ) spaces are also simply referred to as *Alexandrov spaces*. Although the definitions of Alexandrov spaces of curvature bounded above and of curvature bounded below seem to be alike, their properties can be very different. For instance, an important technical fact that distinguishes them is that in Alexandrov spaces of curvature bounded above, geodesics may branch (just consider the case of $\mathbb{R}$-trees), in contrast to Alexandrov spaces of curvature bounded below, where this cannot happen. The relation between these two classes of spaces is described in detail in the (currently still in progress) book by Alexander et al. [29].

Let (X, d) and (Y, d') be metric spaces and consider the direct product $X \times Y$ equipped with the metric

$$d_2((x_1, y_1), (x_2, y_2)) = \sqrt{d(x_1, x_2)^2 + d'(y_1, y_2)^2}, \qquad (5.1.5)$$

where $x_1, x_2 \in X$ and $y_1, y_2 \in Y$. If X and Y are CAT(κ) spaces, then $X \times Y$ is a CAT(κ') space with $\kappa' = \max\{0, \kappa\}$. Similarly, if X and Y are CBB(κ) spaces, then $X \times Y$ is a CBB(κ') space with $\kappa' = \min\{0, \kappa\}$.

The Metric Projection

As in Sect. 1.4.4 one can also consider the notion of metric projection in the context of metric spaces. Let (X, d) be a metric space and $A \subseteq X$. The *metric projection* P_A onto A is the mapping $P_A : X \to 2^A$ defined by

$$P_A(x) = \{y \in A : d(x, y) = d(x, A)\} \quad \text{for every } x \in X.$$

If $P_A(x)$ is a singleton for every $x \in X$, then the set A is called *Chebyshev*.

In complete CAT(κ) spaces with $\kappa > 0$ and diameter less than $D_\kappa/2$, every nonempty closed convex set is Chebyshev. Unlike in Hilbert spaces, one cannot expect in CAT(κ) spaces the metric projection onto convex Chebyshev sets to be nonexpansive. However, in this case it is a Lipschitz mapping (a Lipschitz constant is given in [46, Proposition 3.4]).

When restricting the setting to CAT(0) spaces, not only can we drop the condition on the diameter of the space, but we have further properties. Every nonempty closed convex subset of a complete CAT(0) space is Chebyshev and, as for Hilbert spaces, the convexity of a Chebyshev subset of a CAT(0) space is equivalent to the nonexpansiveness of the metric projection onto it. A systematic study of properties of Chebyshev sets in Alexandrov spaces is carried out in [46].

5.1.2 Hyperconvex Spaces

The notion of hyperconvexity was introduced by Aronszajn and Panitchpakdi [52] and captures from a metric viewpoint the concept of injectivity from Banach space theory. The fundamental idea upon which hyperconvexity was built is the following intersection property of the real line, which is essential in the proof of the Hahn-Banach theorem: a family of mutually intersecting closed and bounded intervals has a common point. For general metric spaces, if one considers closed balls instead of closed bounded intervals, this condition is known as the *binary intersection property*. If a metric space X has the binary intersection property, then it is complete. Hyperconvexity falls close to the binary intersection property and is defined as follows.

Definition 5.1.13 (Hyperconvexity) A metric space (X, d) is called *hyperconvex* if $\bigcap_{i \in I} B[x_i, r_i] \neq \emptyset$ for every family of points $(x_i)_{i \in I}$ in X and every family of positive numbers $(r_i)_{i \in I}$ such that $d(x_i, x_j) \leq r_i + r_j$ for all $i, j \in I$.

Hyperconvex metric spaces include the real line, the Banach space ℓ^∞ as well as all real finite dimensional Banach spaces with the maximum norm. However, there are Hilbert spaces that are not hyperconvex as, e.g., ℓ^2.

A metric space is hyperconvex if and only if it is geodesic and has the binary intersection property. Actually, every hyperconvex metric space admits a geodesic bicombing (see [375]). This means that hyperconvex spaces are contractible. Note that there are nonconvex subsets of $\mathbb{R}^2$ (with the maximum norm) that are hyperconvex (as metric spaces with the induced maximum metric) and there exist hyperconvex normed spaces with linear subspaces that are not hyperconvex (see [100]). If a hyperconvex metric space is uniquely geodesic, then it is an $\mathbb{R}$-tree. This property was established by Kirk [358] who also proved the converse, namely that every complete $\mathbb{R}$-tree is a hyperconvex metric space.

We give next a few more definitions needed further on.

Definition 5.1.14 (Admissible Hull) Let (X, d) be a metric space and $A \subseteq X$. The *admissible hull* cov(A) of A is the intersection of all closed balls containing A. A set is called *admissible* if it coincides with its admissible hull.

Thus, a set is admissible if and only if it is an intersection of closed balls. It is easy to see that if A is bounded, then

$$\mathrm{cov}(A) = \bigcap_{x \in X} B[x, r_x(A)],$$

where $r_x(A) = \sup_{a \in A} d(x, a)$.

Definition 5.1.15 (External Hyperconvexity) A subset E of a metric space (X, d) is *externally hyperconvex* (with respect to X) if given any family $(x_i)_{i \in I}$ of points in X and any family $(r_i)_{i \in I}$ of positive numbers satisfying

$$d(x_i, x_j) \le r_i + r_j \quad \text{and} \quad d(x_i, E) \le r_i,$$

for all $i, j \in I$, it follows that $\left(\bigcap_{i \in I} B[x_i, r_i] \right) \cap E \ne \emptyset$.

Any externally hyperconvex set is itself hyperconvex (with the induced metric). If X is a hyperconvex metric space and A is an admissible subset of it, then A is externally hyperconvex with respect to X.

5.1.3 Hyperbolic Spaces

The concept of Gromov hyperbolicity captures asymptotic properties of the standard hyperbolic space $\mathbb{H}^n$ and plays an essential role in geometric group theory. There are several ways to introduce this condition in metric spaces and we refer to Gromov's seminal work [271] and to [134] for more details. However, in the setting of geodesic spaces, all these definitions are mainly equivalent and we consider the following one.

Let (X, d) be a metric space. For $x, y, z \in X$, the number

$$(y \mid z)_x = \frac{1}{2} \left(d(y, x) + d(z, x) - d(y, z) \right) \tag{5.1.6}$$

is called the *Gromov product* of y and z with respect to x. Note that

$$0 \le (y \mid z)_x \le \min\{d(y, x), d(z, x)\}$$

and

$$(y \mid z)_x + (x \mid z)_y = d(x, y).$$

Definition 5.1.16 (δ-Hyperbolicity) Let $\delta \ge 0$. A geodesic space (X, d) is called *δ-hyperbolic* if for every geodesic triangle $\Delta(x, y, z)$ in X the following holds: if $y' \in [x, y]$ and $z' \in [x, z]$ are points with $d(x, y') = d(x, z') \le (y \mid z)_x$, then $d(y', z') \le \delta$.

In other words, triangles can be viewed as enlarged tripods because, starting from a vertex, the two sides that issue from it run together within the distance δ up to the moment when their length reaches the corresponding Gromov product and diverge after that.

If a geodesic space X is δ-hyperbolic for some $\delta \geq 0$, then it is said to be *(Gromov) hyperbolic* and δ is called a *hyperbolicity constant* for X. An important class of hyperbolic spaces is the one of Gromov hyperbolic groups. Any CAT(κ) space with $\kappa < 0$ is δ-hyperbolic, where δ depends only on κ. Furthermore, a geodesic space is an $\mathbb{R}$-tree if and only if it is 0-hyperbolic.

5.1.4 Convex Combinations

Convex combinations in geodesic spaces are difficult elements to deal with. In a uniquely geodesic space X, the convex combination of two points $x_1, x_2 \in X$ with coefficients $a_1, a_2 \in [0, 1]$ satisfying $a_1 + a_2 = 1$ is the only point, denoted by $a_1x_1 \oplus a_2x_2$, that belongs to the geodesic segment joining x_1 and x_2 with the property that $d(a_1x_1 \oplus a_2x_2, x_1) = a_2d(x_1, x_2)$ and $d(a_1x_1 \oplus a_2x_2, x_2) = a_1d(x_1, x_2)$. Clearly, $a_1x_1 \oplus a_2x_2 = a_2x_2 \oplus a_1x_1$. However, if we consider three points $x_1, x_2, x_3 \in X$ and three coefficients $a_1, a_2, a_3 \in [0, 1]$ that sum to 1, then their corresponding convex combination is not clearly defined in the same way as before since it may depend on the order of combining the points. In this case, the points

$$a_1x_1 \oplus (1 - a_1) \left(\frac{a_2}{1 - a_1}x_2 \oplus \frac{a_3}{1 - a_1}x_3 \right)$$

and

$$a_2x_2 \oplus (1 - a_2) \left(\frac{a_1}{1 - a_2}x_1 \oplus \frac{a_3}{1 - a_2}x_3 \right),$$

which may not coincide, can be, e.g., two different options for the convex combination. We detail in the following a method of constructing the convex combination of a finite set of points which goes back to [33] and was further studied in [229].

Let (X, d) be a complete Busemann convex space and $k \in \mathbb{N}$, $k \geq 2$. Recall that the standard $(k - 1)$-simplex, denoted by Δ^{k-1}, is the set of all k-tuples of nonnegative numbers that sum to 1. We describe next the construction of the convex combination for a k-tuple $(x_1, \ldots, x_k)$ of points in X and a k-tuple $(a_1, \ldots, a_k)$ of coefficients in Δ^{k-1}. We will denote this convex combination by $\dot\oplus_{i=1}^{k}a_ix_i$ or simply by $\dot\oplus a_ix_i$.

First, note that if $a_j = 1$ for some j, we set $\dot\oplus a_ix_i = x_j$. Thus, we can assume in the following that there are at least two positive coefficients. We also use the notation x_i^0 for x_i.

For $k = 2$, one has the standard definition $\oplus_{i=1}^{2} a_i x_i = a_1 x_1 \oplus a_2 x_2$. Now let $k = 3$ and consider, for $1 \leq j \leq 3$, the sequences $(x_j^n)_{n \in \mathbb{N}_0}$ given for $n \in \mathbb{N}$ by

$$
\begin{cases}
x_1^n = a_1 x_1^{n-1} \oplus (1 - a_1)\left(\frac{a_2}{1-a_1}x_2^{n-1} \oplus \frac{a_3}{1-a_1}x_3^{n-1}\right), \\
x_2^n = a_2 x_2^{n-1} \oplus (1 - a_2)\left(\frac{a_1}{1-a_2}x_1^{n-1} \oplus \frac{a_3}{1-a_2}x_3^{n-1}\right), \\
x_3^n = a_3 x_3^{n-1} \oplus (1 - a_3)\left(\frac{a_1}{1-a_3}x_1^{n-1} \oplus \frac{a_2}{1-a_3}x_2^{n-1}\right).
\end{cases}
$$

The sequences $(x_j^n)_{n \in \mathbb{N}_0}$ converge to a same point. To see this, let $1 \leq i, j \leq 3$ and $n \geq 1$. By (5.1.1),

$$
d(x_i^n, x_j^n) \leq \sum_{m=1}^{3} a_m d(x_m^{n-1}, x_j^n) \leq \sum_{m=1}^{3}\left(a_m \sum_{p=1}^{3} a_p d(x_m^{n-1}, x_p^{n-1})\right)
$$

$$
= \sum_{m=1}^{3}\sum_{p=1}^{3} a_m a_p d(x_m^{n-1}, x_p^{n-1}) = 2\sum_{\substack{m,p=1 \\ m<p}}^{3} a_m a_p d(x_m^{n-1}, x_p^{n-1})
$$

$$
\leq \left(1 - \sum_{m=1}^{3} a_m^2\right)\max_{1 \leq m, p \leq 3} d(x_m^{n-1}, x_p^{n-1}).
$$

Denote $\delta = 1 - \sum_{m=1}^{3} a_m^2$. Then $\delta \in (0, 1)$ (since we assumed that at least two coefficients are positive) and we have

$$
\max_{1 \leq i, j \leq 3} d(x_i^n, x_j^n) \leq \delta \max_{1 \leq i, j \leq 3} d(x_i^{n-1}, x_j^{n-1})
$$

and so

$$
\max_{1 \leq i, j \leq 3} d(x_i^n, x_j^n) \leq \delta^n \max_{1 \leq i, j \leq 3} d(x_i, x_j).
$$

Note that $\{x_1^n, x_2^n, x_3^n\} \subseteq \mathrm{co}\left\{x_1^{n-1}, x_2^{n-1}, x_3^{n-1}\right\}$. Therefore,

$$
\overline{\mathrm{co}}\left\{x_1^n, x_2^n, x_3^n\right\} \subseteq \overline{\mathrm{co}}\left\{x_1^{n-1}, x_2^{n-1}, x_3^{n-1}\right\}
$$

and, by Remark 5.1.5 and Proposition 5.1.3,

$$
\mathrm{diam}\,\overline{\mathrm{co}}\left\{x_1^n, x_2^n, x_3^n\right\} = \max_{1 \leq i, j \leq 3} d(x_i^n, x_j^n)
$$

$$
\leq \delta^n \max_{1 \leq i, j \leq 3} d(x_i, x_j) = \delta^n \mathrm{diam}\,\overline{\mathrm{co}}\,\{x_1, x_2, x_3\}.
$$

Applying Cantor's intersection theorem to the descending sequence of closed sets $\left(\overline{co}\left\{x_1^n, x_2^n, x_3^n\right\}\right)_{n \in \mathbb{N}_0}$, we obtain that for $1 \leq j \leq 3$, the sequences $(x_j^n)_{n \in \mathbb{N}_0}$ are Cauchy and converge to a common limit which defines the convex combination $\dot{\oplus}_{i=1}^3 a_i x_i$.

Clearly, by the construction method, this limit is independent of the order the points x_i are arranged in the 3-tuple (assuming the correspondence between points and coefficients is maintained).

Moreover, if, e.g., $a_3 = 0$, then $\dot{\oplus}_{i=1}^3 a_i x_i = \dot{\oplus}_{i=1}^2 a_i x_i$. Indeed, it is immediate that $x_1^1 = x_2^1 = x_3^1 = a_1 x_1 \oplus a_2 x_2 = \dot{\oplus}_{i=1}^2 a_i x_i$ and so $x_j^n = \dot{\oplus}_{i=1}^2 a_i x_i$ for $1 \leq j \leq 3$ and $n \geq 1$, from where $\dot{\oplus}_{i=1}^3 a_i x_i = \dot{\oplus}_{i=1}^2 a_i x_i$.

Additionally,

$$d(\dot{\oplus}_{i=1}^3 a_i x_i, \dot{\oplus}_{i=1}^3 a_i y_i) \leq \sum_{i=1}^3 a_i d(x_i, y_i),$$

for any 3-tuple (y_1, y_2, y_3) of points in X. To see this, let (y_1, y_2, y_3) be a 3-tuple of points in X. Then it is an easy consequence of Busemann convexity that

$$d(x_j^n, y_j^n) \leq \sum_{i=1}^3 a_i d(x_i^{n-1}, y_i^{n-1}),$$

for all $1 \leq j \leq 3$ and $n \geq 1$. Because $\sum_{i=1}^3 a_i = 1$, we get $\sum_{i=1}^3 a_i d(x_i^n, y_i^n) \leq \sum_{i=1}^3 a_i d(x_i^{n-1}, y_i^{n-1})$. Iterating, we finally obtain $\sum_{i=1}^3 a_i d(x_i^n, y_i^n) \leq \sum_{i=1}^3 a_i d(x_i, y_i)$ and we only need to take limit on n.

Inductively, we consider next the general case. Let $k \in \mathbb{N}$, $k \geq 4$, $(a_1, \ldots, a_k)$ be a k-tuple of coefficients in Δ^{k-1} such that at least two of them are positive, and let $(x_1, \ldots, x_k)$ be a k-tuple of points in X. Suppose that for tuples of at most $k - 1$ points in X, the convex combination with nonnegative coefficients that sum to 1 is defined and that it does not vary with respect to the order of points in the tuple. Assume also that the addition of a point with zero coefficient to any tuple of at most $k - 2$ points leaves its convex combination unchanged. Moreover, suppose that for any $(k - 1)$-tuples $(u_1, \ldots, u_{k-1})$ and $(v_1, \ldots, v_{k-1})$ of points in X and $(b_1, \ldots, b_{k-1}) \in \Delta^{k-2}$,

$$d(\dot{\oplus}_{i=1}^{k-1} b_i u_i, \dot{\oplus}_{i=1}^{k-1} b_i v_i) \leq \sum_{i=1}^{k-1} b_i d(u_i, v_i). \tag{5.1.7}$$

In particular, if for some $u \in X$, $u_i = u$ for all $1 \leq i \leq k - 1$, then

$$d(u, \dot{\oplus}_{i=1}^{k-1} b_i v_i) \leq \sum_{i=1}^{k-1} b_i d(u, v_i). \tag{5.1.8}$$

Take now, for $1 \leq j \leq k$, the sequences $(x_j^n)_{n \in \mathbb{N}_0}$ defined for $n \in \mathbb{N}$ by

$$\begin{cases} x_1^n = a_1 x_1^{n-1} \oplus (1 - a_1)\left(\dot{\oplus}_{i=2}^k \frac{a_i}{1-a_1} x_i^{n-1}\right), \\ x_2^n = a_2 x_2^{n-1} \oplus (1 - a_2)\left(\dot{\oplus}_{\substack{i=1 \\ i \neq 2}}^k \frac{a_i}{1-a_2} x_i^{n-1}\right), \\ \quad \vdots \\ x_k^n = a_k x_k^{n-1} \oplus (1 - a_k)\left(\dot{\oplus}_{i=1}^{k-1} \frac{a_i}{1-a_k} x_i^{n-1}\right). \end{cases}$$

As before (for this one needs to apply (5.1.8) instead of (5.1.1)) one can prove that the sequences $(x_j^n)_{n \in \mathbb{N}_0}$ converge to a same point which defines the convex combination $\dot{\oplus}_{i=1}^k a_i x_i$ regardless of the order the points are originally arranged.

Again, if, e.g., $a_k = 0$, then

$$\begin{cases} x_1^n = a_1 x_1^{n-1} \oplus (1 - a_1)\left(\dot{\oplus}_{i=2}^{k-1} \frac{a_i}{1-a_1} x_i^{n-1}\right), \\ x_2^n = a_2 x_2^{n-1} \oplus (1 - a_2)\left(\dot{\oplus}_{\substack{i=1 \\ i \neq 2}}^{k-1} \frac{a_i}{1-a_2} x_i^{n-1}\right), \\ \quad \vdots \\ x_k^n = \dot{\oplus}_{i=1}^{k-1} a_i x_i^{n-1}. \end{cases}$$

The sequences $(x_j^n)_{n \in \mathbb{N}_0}$, where $1 \leq j \leq k - 1$, coincide with the corresponding sequences used in the construction of $\dot{\oplus}_{i=1}^{k-1} a_i x_i$, which means that $\dot{\oplus}_{i=1}^k a_i x_i = \dot{\oplus}_{i=1}^{k-1} a_i x_i$.

Furthermore, for any k-tuple $(y_1, \ldots, y_k)$ of points in X, the inequality

$$d(\dot{\oplus}_{i=1}^k a_i x_i, \dot{\oplus}_{i=1}^k a_i y_i) \leq \sum_{i=1}^k a_i d(x_i, y_i), \tag{5.1.9}$$

follows similarly as before by applying Busemann convexity and (5.1.7). In particular, we have

$$d(x, \dot{\oplus}_{i=1}^k a_i x_i) \leq \sum_{i=1}^k a_i d(x, x_i), \tag{5.1.10}$$

for all $x \in X$. Note also that for every $n \in \mathbb{N}$,

$$\max_{1 \leq i, j \leq k} d(x_i^n, x_j^n) \leq \left(1 - \sum_{i=1}^k a_i^2\right) \max_{1 \leq i, j \leq k} d(x_i^{n-1}, x_j^{n-1}). \tag{5.1.11}$$

The above construction shows that for every k-tuple $(x_1, \ldots, x_k)$ of points in X and every k-tuple $(a_1, \ldots, a_k)$ of coefficients in Δ^{k-1}, $\dot{\oplus}_{i=1}^k a_i x_i \in \overline{\mathrm{co}}\{x_1, \ldots, x_k\}$.

Continuity Properties of Convex Combinations

We focus here on some continuity properties of convex combinations given in [229]. In the sequel we assume that (X, d) is a complete Busemann convex space and $k \in \mathbb{N}, k \geq 2$.

Proposition 5.1.17 *Let $(x_1, \ldots, x_k)$ be a k-tuple of points in X and suppose that $(a_1, \ldots, a_k)$ and $(b_1, \ldots, b_k)$ are k-tuples of coefficients in Δ^{k-1}. Then*

$$d(\dot{\oplus}_{i=1}^k a_i x_i, \dot{\oplus}_{i=1}^k b_i x_i) \leq \left(\prod_{i=3}^k i^2\right) \frac{D}{2} \sum_{i=1}^k |a_i - b_i|, \tag{5.1.12}$$

where

$$D = \max_{1 \leq i, j \leq k} d(x_i, x_j).$$

Proof If $a_j = 1$ for some $j \in \{1, \ldots, k\}$, then, by (5.1.10),

$$d(\dot{\oplus}_{i=1}^k a_i x_i, \dot{\oplus}_{i=1}^k b_i x_i) = d(x_j, \dot{\oplus}_{i=1}^k b_i x_i) \leq \sum_{i=1}^k b_i d(x_i, x_j) = \sum_{\substack{i=1 \\ i \neq j}}^k b_i d(x_i, x_j)$$

$$\leq D \sum_{\substack{i=1 \\ i \neq j}}^k b_i = \frac{D}{2} \sum_{i=1}^k |a_i - b_i|,$$

where the last equality follows because $a_i = 0$ for $i \neq j$ and

$$\sum_{\substack{i=1 \\ i \neq j}}^k b_i = 1 - b_j = a_j - b_j.$$

In a similar way one can show that (5.1.12) holds when $b_j = 1$ for some j and so we may assume henceforth that there are at least two positive coefficients in each k-tuple $(a_1, \ldots, a_k)$ and $(b_1, \ldots, b_k)$. For simplicity, we denote next $\delta = 1 - \sum_{i=1}^k b_i^2$. Note that under the previous assumption, $\delta \in (0, 1)$.

If $k = 2$,

$$d(\dot{\oplus}_{i=1}^2 a_i x_i, \dot{\oplus}_{i=1}^2 b_i x_i) = |a_1 - b_1|D = |a_2 - b_2|D = \frac{D}{2} \sum_{i=1}^2 |a_i - b_i|.$$

Let now $k = 3$ and for $1 \leq j \leq 3$ denote by $(x_j^n)_{n \in \mathbb{N}_0}$ and $(z_j^n)_{n \in \mathbb{N}_0}$ the sequences involved in the construction of the convex combinations $\oplus a_i x_i$ and $\dot{\oplus} b_i x_i$, respectively. Let also for $n \in \mathbb{N}_0$,

$$u^n = \frac{a_2}{1 - a_1} x_2^n \oplus \frac{a_3}{1 - a_1} x_3^n, \quad v^n = \frac{a_2}{1 - a_1} z_2^n \oplus \frac{a_3}{1 - a_1} z_3^n,$$

$$w^n = \frac{b_2}{1 - b_1} z_2^n \oplus \frac{b_3}{1 - b_1} z_3^n.$$

Then, for all $n \in \mathbb{N}$, $x_1^n = a_1 x_1^{n-1} \oplus (1 - a_1) u^{n-1}$, $z_1^n = b_1 z_1^{n-1} \oplus (1 - b_1) w^{n-1}$ and

$$d(x_1^n, z_1^n) \leq d(x_1^n, a_1 z_1^{n-1} \oplus (1 - a_1) v^{n-1}) + d(a_1 z_1^{n-1} \oplus (1 - a_1) v^{n-1}, z_1^n).$$

By Busemann convexity,

$$d(x_1^n, a_1 z_1^{n-1} \oplus (1 - a_1) v^{n-1}) \leq a_1 d(x_1^{n-1}, z_1^{n-1}) + (1 - a_1) d(u^{n-1}, v^{n-1})$$

$$\leq \sum_{i=1}^{3} a_i d(x_i^{n-1}, z_i^{n-1})$$

and

$$d(a_1 z_1^{n-1} \oplus (1 - a_1) v^{n-1}, z_1^n)$$

$$\leq d(a_1 z_1^{n-1} \oplus (1 - a_1) v^{n-1}, a_1 z_1^{n-1} \oplus (1 - a_1) w^{n-1})$$

$$+ d(a_1 z_1^{n-1} \oplus (1 - a_1) w^{n-1}, z_1^n)$$

$$\leq (1 - a_1) d(v^{n-1}, w^{n-1}) + |a_1 - b_1| d(z_1^{n-1}, w^{n-1}).$$

Since

$$d(v^{n-1}, w^{n-1}) = \left| \frac{a_2}{1 - a_1} - \frac{b_2}{1 - b_1} \right| d(z_2^{n-1}, z_3^{n-1})$$

$$\leq \frac{|a_1 - b_1| + |a_2 - b_2|}{1 - a_1} \max_{1 \leq i, j \leq 3} d(z_i^{n-1}, z_j^{n-1})$$

as

$$\left| \frac{a_2}{1-a_1} - \frac{b_2}{1-b_1} \right| = \left| \frac{a_2}{1-a_1} - \frac{b_2}{1-a_1} + \frac{b_2}{1-a_1} - \frac{b_2}{1-b_1} \right|$$

$$\leq \frac{|a_2-b_2|}{1-a_1} + \frac{|a_1-b_1|}{1-a_1}\frac{b_2}{1-b_1}$$

$$\leq \frac{|a_1-b_1|+|a_2-b_2|}{1-a_1} \quad \text{because} \quad \frac{b_2}{1-b_1} = \frac{b_2}{b_2+b_3} \leq 1,$$

and, by (5.1.1),

$$d(z_1^{n-1}, w^{n-1}) \leq \frac{b_2}{1-b_1}d(z_1^{n-1}, z_2^{n-1}) + \frac{b_3}{1-b_1}d(z_1^{n-1}, z_3^{n-1})$$

$$\leq \max_{1\leq i,j\leq 3} d(z_i^{n-1}, z_j^{n-1}),$$

it follows that

$$d(x_1^n, z_1^n) \leq \sum_{i=1}^{3} a_i d(x_i^{n-1}, z_i^{n-1}) + (2|a_1-b_1|+|a_2-b_2|) \max_{1\leq i,j\leq 3} d(z_i^{n-1}, z_j^{n-1}).$$

Since $2|a_1-b_1|+|a_2-b_2| \leq (3/2)\sum_{i=1}^{3}|a_i-b_i|$ and

$$\max_{1\leq i,j\leq 3} d(z_i^{n-1}, z_j^{n-1}) \leq \delta^{n-1}D \quad \text{by (5.1.11)},$$

we obtain

$$d(x_1^n, z_1^n) \leq \sum_{i=1}^{3} a_i d(x_i^{n-1}, z_i^{n-1}) + 3\delta^{n-1}\frac{D}{2}\sum_{i=1}^{3}|a_i-b_i|.$$

In fact, one can show in this way that

$$d(x_j^n, z_j^n) \leq \sum_{i=1}^{3} a_i d(x_i^{n-1}, z_i^{n-1}) + 3\delta^{n-1}\frac{D}{2}\sum_{i=1}^{3}|a_i-b_i|,$$

for all $1 \leq j \leq 3$. Thus,

$$\sum_{i=1}^{3} a_i d(x_i^n, z_i^n) \leq \sum_{i=1}^{3} a_i d(x_i^{n-1}, z_i^{n-1}) + 3\delta^{n-1}\frac{D}{2}\sum_{i=1}^{3}|a_i-b_i|.$$

Iterating we have

$$\sum_{i=1}^{3} a_i d(x_i^n, z_i^n) \leq 3(1 + \delta + \ldots + \delta^{n-1}) \frac{D}{2} \sum_{i=1}^{3} |a_i - b_i|.$$

Taking limit on n and recalling that, by the Cauchy–Schwarz inequality, $1 - \delta = \sum_{i=1}^{3} b_i^2 \geq 1/3$,

$$d(\dot{\oplus}_{i=1}^{3} a_i x_i, \dot{\oplus}_{i=1}^{3} b_i x_i) \leq 9 \frac{D}{2} \sum_{i=1}^{3} |a_i - b_i|.$$

In general, assume that (5.1.12) holds for $k - 1$, where $k \in \mathbb{N}$, $k \geq 4$. As before, denote for $1 \leq j \leq k$ by $(x_j^n)_{n \in \mathbb{N}_0}$ and $(z_j^n)_{n \in \mathbb{N}_0}$ the sequences involved in the construction of the convex combinations $\dot{\oplus} a_i x_i$ and $\dot{\oplus} b_i x_i$, respectively. For $n \in \mathbb{N}_0$, let now

$$u^n = \dot{\oplus}_{i=2}^{k} \frac{a_i}{1 - a_1} x_i^n, \quad v^n = \dot{\oplus}_{i=2}^{k} \frac{a_i}{1 - a_1} z_i^n, \quad w^n = \dot{\oplus}_{i=2}^{k} \frac{b_i}{1 - b_1} z_i^n.$$

Then, for any $n \in \mathbb{N}$, $x_1^n = a_1 x_1^{n-1} \oplus (1 - a_1) u^{n-1}$, $z_1^n = b_1 z_1^{n-1} \oplus (1 - b_1) w^{n-1}$ and again, using Busemann convexity and (5.1.9),

$$d(x_1^n, z_1^n) \leq \sum_{i=1}^{k} a_i d(x_i^{n-1}, z_i^{n-1}) + (1 - a_1) d(v^{n-1}, w^{n-1}) + |a_1 - b_1| d(z_1^{n-1}, w^{n-1}).$$

By the induction hypothesis,

$$d(v^{n-1}, w^{n-1}) \leq \left(\prod_{i=3}^{k-1} i^2 \right) \frac{\max\limits_{2 \leq i,j \leq k} d(z_i^{n-1}, z_j^{n-1})}{2} \sum_{i=2}^{k} \left| \frac{a_i}{1 - a_1} - \frac{b_i}{1 - b_1} \right|$$

$$\leq \left(\prod_{i=3}^{k-1} i^2 \right) \delta^{n-1} \frac{D}{2} \sum_{i=2}^{k} \frac{|a_1 - b_1| + |a_i - b_i|}{1 - a_1}$$

and, by (5.1.10),

$$d(z_1^{n-1}, w^{n-1}) \leq \sum_{i=2}^{k} \frac{b_i}{1 - b_1} d(z_1^{n-1}, z_i^{n-1}) \leq \max\limits_{1 \leq i,j \leq k} d(z_i^{n-1}, z_j^{n-1}) \leq \delta^{n-1} D.$$

Therefore,

$$d(x_1^n, z_1^n) \le \sum_{i=1}^{k} a_i d(x_i^{n-1}, z_i^{n-1}) + k \left(\prod_{i=3}^{k-1} i^2\right) \delta^{n-1} \frac{D}{2} \sum_{i=1}^{k} |a_i - b_i|$$

and similarly

$$d(x_j^n, z_j^n) \le \sum_{i=1}^{k} a_i d(x_i^{n-1}, z_i^{n-1}) + k \left(\prod_{i=3}^{k-1} i^2\right) \delta^{n-1} \frac{D}{2} \sum_{i=1}^{k} |a_i - b_i|,$$

for all $1 \le j \le k$. Hence, we find

$$\sum_{i=1}^{k} a_i d(x_i^n, z_i^n) \le \sum_{i=1}^{k} a_i d(x_i^{n-1}, z_i^{n-1}) + k \left(\prod_{i=3}^{k-1} i^2\right) \delta^{n-1} \frac{D}{2} \sum_{i=1}^{k} |a_i - b_i|$$

and so

$$\sum_{i=1}^{k} a_i d(x_i^n, z_i^n) \le k \left(\prod_{i=3}^{k-1} i^2\right) (1 + \delta + \ldots + \delta^{n-1}) \frac{D}{2} \sum_{i=1}^{k} |a_i - b_i|.$$

Passing to limit as before and using the fact that $1 - \delta \ge 1/k$ concludes the induction reasoning. $\qquad\square$

Theorem 5.1.18 *Let $(x_1, \ldots, x_k)$ be a k-tuple of points in X and $(a_1, \ldots, a_k)$ a k-tuple of coefficients in Δ^{k-1}. For each $n \in \mathbb{N}$, consider the k-tuples $(z_1(n), \ldots, z_k(n))$ of points in X and $(b_1(n), \ldots, b_k(n))$ of coefficients in Δ^{k-1} such that*

$$(z_1(n), \ldots, z_k(n)) \to (x_1, \ldots, x_k) \quad and \quad (b_1(n), \ldots, b_k(n)) \to (a_1, \ldots, a_k),$$

as $n \to \infty$. Then $\dot{\oplus}_{i=1}^{k} b_i(n) z_i(n) \to \dot{\oplus}_{i=1}^{k} a_i x_i$, as $n \to \infty$.

Proof By (5.1.9) and Proposition 5.1.17,

$$d(\dot{\oplus}_{i=1}^{k} b_i(n) z_i(n), \dot{\oplus}_{i=1}^{k} a_i x_i) \le d(\dot{\oplus}_{i=1}^{k} b_i(n) z_i(n), \dot{\oplus}_{i=1}^{k} b_i(n) x_i)$$

$$+ d(\dot{\oplus}_{i=1}^{k} b_i(n) x_i, \dot{\oplus}_{i=1}^{k} a_i x_i)$$

$$\le \sum_{i=1}^{k} b_i(n) d(z_i(n), x_i)$$

$$+ \left(\prod_{i=3}^{k} i^2\right) \frac{D}{2} \sum_{i=1}^{k} |a_i - b_i(n)|,$$

where D is as in Proposition 5.1.17.

Clearly, the above inequality proves the desired convergence. $\qquad\qquad\square$

Remark 5.1.19 Convex combinations can in fact be considered in the same way not only in complete Busemann convex spaces, but also in complete metric spaces with a geodesic bicombing and all continuity properties given in this subsection hold true.

Similar techniques were also considered by Navas [520] to define a notion of barycenter and finally obtain an ergodic theorem for mappings with values in Busemann convex spaces. Both approaches actually recover ideas from [230]. Recently, the barycenter construction given in [520] has been also considered in metric spaces with a geodesic bicombing in [69, 187]. In fact, one could use this barycenter map to define an alternative notion of convex combinations to the one given in this subsection. Let (X, d) be a complete Busemann convex space, $\mathscr{B}(X)$ the σ-algebra of Borel subsets of X and denote by $P_1(X)$ the set of Radon probability measures of $(X, \mathscr{B}(X))$ that have finite first moment. Then one can construct a nonexpansive mapping bar* : $(P_1(X), W) \to (X, d)$ which satisfies bar$^*(\delta_x) = x$ for all $x \in X$, where W is the 1-Wasserstein distance and δ_x is the Dirac measure at x (see [69, 187, 520]). In this way, it is possible to define the convex combination for a k-tuple $(x_1, \ldots, x_k)$ of points in X and a k-tuple $(a_1, \ldots, a_k)$ of coefficients in Δ^{k-1} as bar$^*(\sum_{i=1}^k a_i \delta_{x_i})$. Applying a general version of the Kantorovich–Rubinstein duality theorem (see [350, Theorem 1]), one can easily show that for any k-tuple $(x_1, \ldots, x_k)$ of points in X and any k-tuples $(a_1, \ldots, a_k)$ and $(b_1, \ldots, b_k)$ of coefficients in Δ^{k-1}, $W(\sum_{i=1}^k a_i \delta_{x_i}, \sum_{i=1}^k b_i \delta_{x_i}) \le (D/2) \sum_{i=1}^k |a_i - b_i|$, where $D = \max_{1 \le i, j \le k} d(x_i, x_j)$ (see also [187]). Thus, $d(\text{bar}^*(\sum_{i=1}^k a_i \delta_{x_i}), \text{bar}^*(\sum_{i=1}^k b_i \delta_{x_i})) \le (D/2) \sum_{i=1}^k |a_i - b_i|$, which is a continuity property similar to (5.1.12) with an improved bound. To see that this inequality is sharp, take $k = 2, a_1 = a_2 = 1/2$ and denote $m = \text{bar}^*(\delta_{x_1}/2 + \delta_{x_2}/2)$. For $b_1 = 1, b_2 = 0$, we get $d(m, x_1) \le d(x_1, x_2)/2$ and for $b_1 = 0, b_2 = 1$, $d(m, x_2) \le d(x_1, x_2)/2$. Thus, $m = (1/2)x_1 \oplus (1/2)x_2$ and in both cases equality is attained.

In contrast to this line, the approach that we followed here only relies on elementary properties of Busemann convex spaces. Note that one cannot use the above upper bound in (5.1.12) as the following example shows. Still, the obtained continuity properties are sufficient for our purpose.

Example 5.1.20 Consider a tripod of endpoints x_1, x_2, x_3 and center o such that $d(o, x_1) = 4$ and $d(o, x_2) = d(o, x_3) = 1$. Then, for $a_1 = a_2 = a_3 = 1/3$, $b_1 = 1/6, b_2 = 1/3$ and $b_3 = 1/2$, one can see that $\dot{\oplus}_{i=1}^3 a_i x_i \in [o, x_1]$ with $d(o, \dot{\oplus}_{i=1}^3 a_i x_i) = 8/9$ and $\dot{\oplus}_{i=1}^3 b_i x_i \in [o, x_3]$ with $d(o, \dot{\oplus}_{i=1}^3 b_i x_i) = 1/108$. Thus,

$$d(\dot{\oplus}_{i=1}^3 a_i x_i, \dot{\oplus}_{i=1}^3 b_i x_i) = \frac{97}{108} > \frac{5}{6} = \frac{D}{2} \sum_{i=1}^3 |a_i - b_i|.$$

We point out that there are many important results from the linear setting for which counterparts in the geodesic context are yet unknown. For example, as far as we know, even in the setting of CAT(0) spaces, it is still an open question whether for a finite set A, $\overline{\text{co}}(A)$ is a compact set (see also the discussion in [555]). For X a complete Busemann convex space, $k \in \mathbb{N}$, $k \geq 2$, $A = \{x_1, \ldots, x_k\} \subseteq X$ and C(A) the set of all convex combinations of $(x_1, \ldots, x_k)$ with coefficients in Δ^{k-1}, take the onto mapping $\alpha : \Delta^{k-1} \to$ C(A) defined by $\alpha((a_1, \ldots, a_k)) = \oplus_{i=1}^{k} a_i x_i$. Then, by Proposition 5.1.17, α is also continuous, so C(A) is compact and connected as a continuous image of a compact and connected set. Note that by the construction of convex combinations, C$(A) \subseteq \overline{\text{co}}(A)$, but it is not clear if these two sets are actually the same. Moreover, it is not immediate whether C(A) is contractible.

5.2 Kirszbraun and McShane Type Extension Results

Kirszbraun's theorem (Theorem 4.2.3) is a fundamental result in the theory of Lipschitz extensions and states that for any Lipschitz function $f : A \subseteq \mathbb{R}^n \to \mathbb{R}^n$ there exists a Lipschitz extension $F : \mathbb{R}^n \to \mathbb{R}^n$ with the same Lipschitz constant. The result for pairs of Hilbert spaces, pairs of hyperbolic and pairs of spherical spaces of the same dimension goes back to Valentine [664–666] (see also Remark 4.2.6). The proof method for Hilbert spaces was described in Sect. 4.2. A similar strategy proves the corresponding results for hyperbolic and spherical spaces and we also refer to [126, Theorems 1.38, 1.40] for more details.

We show in the sequel that the concept of hyperconvexity is closely related to this problem.

Definition 5.2.1 (Injectivity) A metric space X is *injective* if given any metric space Y and any set $A \subseteq Y$, every nonexpansive mapping $f : A \to X$ admits a nonexpansive extension $F : Y \to X$ (in other words, the pair (Y, X) has the contraction extension property from Definition 4.2.4).

We denote by $\ell^\infty(\Gamma)$ the Banach space of all bounded functions $h : \Gamma \to \mathbb{R}$ with the norm $\|h\|_\infty = \sup\{|h(\gamma)| : \gamma \in \Gamma\}$ and by ℓ^∞ the space $\ell^\infty(\mathbb{N})$. Let us observe that the Banach space $\ell^\infty(\Gamma)$ is injective for an arbitrary set Γ. Indeed, let Y be a metric space, $A \subseteq Y$ and $\varphi : A \to \ell^\infty(\Gamma)$ be nonexpansive. For $\gamma \in \Gamma$, consider the function $\varphi_\gamma : A \to \mathbb{R}$, $\varphi_\gamma(a) = \varphi(a)(\gamma)$. Then φ_γ is clearly nonexpansive, so, by McShane's theorem (Theorem 4.1.1), it possesses a nonexpansive extension $\Phi_\gamma : Y \to \mathbb{R}$. Setting $\Phi : Y \to \ell^\infty(\Gamma)$, $\Phi(y)(\gamma) = \Phi_\gamma(y)$, it can be shown that Φ is well-defined and nonexpansive.

Definition 5.2.2 (Absolute Nonexpansive Retract) Let Y be a metric space and $X \subseteq Y$. We say that X is a *nonexpansive retract* of Y is there exists a nonexpansive retraction from Y onto X, i.e., a nonexpansive mapping $r : Y \to X$ such that it is the identity on X. A metric space X is called an *absolute nonexpansive retract* if for every isometric embedding $e : X \to Y$ into another metric space Y, $e(X)$ is a nonexpansive retract of Y.

Theorem 5.2.3 *For a metric space (X, d), the following are equivalent.*

1. *X is hyperconvex.*
2. *Given any metric space Y, the pair (Y, X) satisfies property (K) from Definition 4.2.5.*
3. *X is injective.*
4. *X is an absolute nonexpansive retract.*

Proof We first prove the equivalence between 1 and 2. Let X be hyperconvex and (Y, d') be any metric space. To show that (Y, X) satisfies property (K), take two families of points $(x_i)_{i \in I} \subseteq X$ and $(y_i)_{i \in I} \subseteq Y$ and a family of positive numbers $(r_i)_{i \in I}$ such that $d(x_i, x_j) \leq d'(y_i, y_j)$ for all $i, j \in I$. For $i \in I$, denote the closed balls

$$B_i = \{x \in X : d(x, x_i) \leq r_i\} \quad \text{and} \quad B'_i = \{y \in Y : d'(y, y_i) \leq r_i\}.$$

Suppose that there exists a point $y \in \bigcap_{i \in I} B'_i$. Then

$$d(x_i, x_j) \leq d'(y_i, y_j) \leq d'(y_i, y) + d'(y, y_j) \leq r_i + r_j,$$

for all $i, j \in I$. Since X is hyperconvex, this shows that $\bigcap_{i \in I} B_i \neq \emptyset$.

Assume now that the pair (Y, X) satisfies property (K) for any metric space (Y, d'). Let $(x_i)_{i \in I}$ be a family of points in X and $(r_i)_{i \in I}$ a family of positive numbers such that $d(x_i, x_j) \leq r_i + r_j$ for all $i, j \in I$. We use the above notation for closed balls.

We show first that $B_i \cap B_j \neq \emptyset$ for all $i, j \in I$. To this end, fix $i, j \in I$. If $i = j$, clearly $x_i \in B_i$. Assume next $i \neq j$ and let $A = \{x_i, x_j\}$. Add a new point u to A and denote $Y = A \cup \{u\}$. One can now easily check that

$$d'(y, z) = \begin{cases} d(y, z), & \text{if } y, z \in A, \\ \frac{r_i}{r_i + r_j} d(x_i, x_j), & \text{if } y = x_i, z = u \text{ or } y = u, z = x_i, \\ \frac{r_j}{r_i + r_j} d(x_i, x_j), & \text{if } y = x_j, z = u \text{ or } y = u, z = x_j, \\ 0, & \text{if } y = u, z = u, \end{cases}$$

is a metric on Y. Using the fact that $d(x_i, x_j) \leq r_i + r_j$, we have $u \in B'_i \cap B'_j$, where $B'_i = \{y \in Y : d'(y, x_i) \leq r_i\}$. By property (K), $B_i \cap B_j \neq \emptyset$.

We prove in the following that $\bigcap_{i \in I} B_i \neq \emptyset$. For $i \in I$, define

$$a_i = \inf \{r > 0 : \text{there exists } i_r \in I \text{ such that } B_{i_r} \subseteq B[x_i, r]\}.$$

If $a_j = 0$ for some $j \in I$, then $x_j \in \bigcap_{i \in I} B_i$. Indeed, let $i \in I$. By the definition of a_j, for every $\varepsilon > 0$, there exists $j_\varepsilon \in I$ so that $B_{j_\varepsilon} \subseteq B[x_j, \varepsilon]$. Since $B_{j_\varepsilon} \cap B_i \neq \emptyset$, we have $B_i \cap B[x_j, \varepsilon] \neq \emptyset$, so $d(x_i, x_j) \leq r_i + \varepsilon$. As $\varepsilon > 0$ is arbitrary, $x_j \in B_i$.

Suppose next $a_i > 0$ for all $i \in I$. Take now $A = \{x_i : i \in I\}$, add a new point v to A and denote $Y = A \cup \{v\}$. We show that

$$
d'(y, z) = \begin{cases} d(y, z), & \text{if } y, z \in A, \\ a_i, & \text{if } y = x_i, z = v \text{ or } y = v, z = x_i, \\ 0, & \text{if } y = v, z = v, \end{cases}
$$

is a metric on Y. Let $i, j \in I$ and $\varepsilon > 0$. Then there exist $i', j' \in I$ such that $B_{i'} \subseteq B[x_i, a_i + \varepsilon]$ and $B_{j'} \subseteq B[x_j, a_j + \varepsilon]$. Because $B_{i'} \cap B_{j'} \neq \emptyset$, we get $B[x_i, a_i + \varepsilon] \cap B[x_j, a_j + \varepsilon] \neq \emptyset$, from where $d(x_i, x_j) \leq a_i + a_j + 2\varepsilon$. As $\varepsilon > 0$ is arbitrary, we obtain $d'(x_i, x_j) \leq a_i + a_j = d'(x_i, v) + d'(x_j, v)$.

Moreover, $B_{j'} \subseteq B[x_j, a_j + \varepsilon] \subseteq B[x_i, d(x_i, x_j) + a_j + \varepsilon]$. By the definition of $a_i, a_i \leq d(x_i, x_j) + a_j + \varepsilon$, hence $d'(x_i, v) \leq d'(x_i, x_j) + d'(x_j, v)$. Now it is clear that (Y, d') is a metric space.

Note that $v \in \bigcap_{i \in I} B'[x_i, a_i]$, where $B'[x_i, a_i] = \{y \in Y : d'(y, x_i) \leq a_i\}$. By property (K), $\bigcap_{i \in I} B[x_i, a_i] \neq \emptyset$. This shows that $\bigcap_{i \in I} B_i \neq \emptyset$ since $a_i \leq r_i$ for all $i \in I$.

The equivalence between 2 and 3 is established in Theorem 4.2.7.

We prove next that 3 and 4 are equivalent too. Suppose that X is injective and let $e : X \to Y$ be an isometric embedding into a metric space Y. Then $e(X)$ is injective and the identity map on $e(X)$ can be extended to a nonexpansive retraction from Y onto $e(X)$, which shows that X is an absolute nonexpansive retract.

For the converse implication, recall first that, by the Kuratowski embedding theorem (Theorem 7.3.1), every metric space X can be isometrically embedded into $\ell^\infty(X)$ by the mapping $\Phi : X \to \ell^\infty(X)$, $\Phi(x)(y) = d(x, y) - d(x_0, y)$, where x_0 is an arbitrarily fixed point in X. Assuming that X is an absolute nonexpansive retract, there exists a nonexpansive retraction $r : \ell^\infty(X) \to \Phi(X)$. Let Y be a metric space, $A \subseteq Y$ and $f : A \to \Phi(X)$ be nonexpansive. Regarding f as a function from A to $\ell^\infty(X)$ and using the fact that $\ell^\infty(X)$ is injective, we obtain a nonexpansive extension $F : Y \to \ell^\infty(X)$. Now $r \circ F : Y \to \Phi(X)$ is a nonexpansive extension of f. This shows that $\Phi(X)$ is injective, so X is as well. $\qquad\square$

Kirszbraun's theorem was extended to Alexandrov spaces by Lang and Schroeder [379]. The same problem was later approached by Alexander et al. in [28] considering a different proof method. Before stating this generalization, we give two conditions formulated in terms of the existence of extensions for isometric mappings defined on three points which ensure that a geodesic space is an Alexandrov space. Characterizations of Alexandrov spaces in this direction were obtained in [28, 379].

Proposition 5.2.4 *Let $\kappa \in \mathbb{R}$ and (X, d) be a uniquely geodesic space. If for every four points $\overline{x}_1, \ldots, \overline{x}_4 \in M_\kappa^2$ with $P(\overline{x}_1, \overline{x}_2, \overline{x}_3) < 2D_\kappa$ and every isometric* mapping $f : \{\overline{x}_1, \overline{x}_2, \overline{x}_3\} \to X$ there exists a nonexpansive extension $F :$ $\{\overline{x}_1, \ldots, \overline{x}_4\} \to X$ of f, then X is a CAT(κ) space.

Proof Let $x_1, x_2, x_3 \in X$ with $P(x_1, x_2, x_3) < 2D_\kappa$ and consider $\Delta(\overline{x}_1, \overline{x}_2, \overline{x}_3)$ a comparison triangle in M_κ^2 for $\Delta(x_1, x_2, x_3)$. Denote m the midpoint of x_2 and x_3 and let $\overline{m} \in [\overline{x}_2, \overline{x}_3]$ be the comparison point for m. The mapping $f :$ $\{\overline{x}_1, \overline{x}_2, \overline{x}_3\} \to X$ defined by $f(\overline{x}_i) = x_i$ for $i \in \{1, 2, 3\}$ is isometric, so it admits a nonexpansive extension $F : \{\overline{x}_1, \overline{x}_2, \overline{x}_3, \overline{m}\} \to X$. Then

$$d(x_2, F(\overline{m})) = d(F(\overline{x}_2), F(\overline{m})) \le d_{M_\kappa^2}(\overline{x}_2, \overline{m}) = \frac{1}{2} d_{M_\kappa^2}(\overline{x}_2, \overline{x}_3) = \frac{1}{2} d(x_2, x_3).$$

Likewise, $d(x_3, F(\overline{m})) \le d(x_2, x_3)/2$. This shows that $F(\overline{m})$ belongs to a geodesic segment joining x_2 and x_3 and, since X is uniquely geodesic, we have $F(\overline{m}) \in$ $[x_2, x_3]$. Moreover, $F(\overline{m}) = m$ and $d(x_1, m) = d(F(\overline{x}_1), F(\overline{m})) \le d_{M_\kappa^2}(\overline{x}_1, \overline{m})$. Using Proposition 5.1.8, we deduce that X is a CAT(κ) space. $\square$

Proposition 5.2.5 *Let* $\kappa \in \mathbb{R}$ *and* X *be a geodesic space. If for every four points* $x_1, \ldots, x_4 \in X$ *with* $P(x_1, x_2, x_3) < 2D_\kappa$ *and every isometric mapping* $f :$ $\{x_1, x_2, x_3\} \to M_\kappa^2$ *there exists a nonexpansive extension* $F : \{x_1, \ldots, x_4\} \to M_\kappa^2$ *of* f, *then* X *is a* CBB(κ) *space.*

Proof Let $x_1, x_2, x_3 \in X$ with $P(x_1, x_2, x_3) < 2D_\kappa$ and consider a comparison triangle $\Delta(\overline{x}_1, \overline{x}_2, \overline{x}_3)$ in M_κ^2 for a geodesic triangle $\Delta(x_1, x_2, x_3)$. Take $m \in [x_2, x_3]$ a midpoint of x_2 and x_3. The isometric mapping $f : \{x_1, x_2, x_3\} \to M_\kappa^2$ defined by $f(x_i) = \overline{x}_i$ for $i \in \{1, 2, 3\}$ admits a nonexpansive extension to $\{x_1, x_2, x_3, m\}$, which finally yields as before that X is a CBB(κ) space. $\square$

The following result generalizes Kirszbraun's theorem.

Theorem 5.2.6 *Let* $\kappa \in \mathbb{R}$, X *be a* CBB(κ) *space, and* Y *a complete* CAT(κ) *space. Suppose* $A \subseteq X$ *is nonempty and* $f : A \to Y$ *is nonexpansive with* diam $f(A) \le$ $D_\kappa/2$. *Then there exists a nonexpansive extension* $F : X \to Y$ *of* f *with* $F(X) \subseteq$ $\overline{\text{co}}(f(A))$.

We do not include a complete proof of this theorem because it relies on more subtle constructions and techniques than the ones covered in Sect. 5.1. However, we point out the key steps of the proof given in [379]. One main ingredient is the following general Helly type theorem (see Theorem 1.1.14).

Theorem 5.2.7 *Let* $\kappa \in \mathbb{R}$, (Y, d) *be a complete* CAT(κ) *space, and* A *an arbitrary index set. For* $a \in A$, *let* $y_a \in Y$, $r_a \in [0, D_\kappa/2]$ *and assume that* $r_b < D_\kappa/2$ *for some* $b \in A$. *If* $\bigcap_{a \in E} B[y_a, r_a] \ne \emptyset$ *for every finite subset* E *of* A, *then* $\bigcap_{a \in A} B[y_a, r_a] \ne \emptyset$.

Proof Let $\mathscr{E} = \{E : E$ is a finite subset of $A\}$ and denote

$$U(E) = \bigcap_{a \in E} B[y_a, r_a] \ne \emptyset \quad \text{and} \quad U^r(E) = U(E) \cap B[y_b, r],$$

for $E \in \mathcal{E}$ and $r \geq 0$. Then $U^{r_1}(E) \subseteq U^{r_2}(E)$ for all $r_1 \leq r_2$ and $E \in \mathcal{E}$. If $r \geq r_b$ and $E \in \mathcal{E}$, $\emptyset \neq U^{r_b}(E) \subseteq U^r(E)$, so $U^r(E) \neq \emptyset$.

Let $\varrho = \inf\{r \geq 0 : U^r(E) \neq \emptyset \text{ for all } E \in \mathcal{E}\}$. We show next that $U^\varrho(E) \neq \emptyset$ for all $E \in \mathcal{E}$. Let $E \in \mathcal{E}$. We have $\varrho \leq r_b < D_\kappa/2$ and the limit $\delta = \lim_{r \searrow \varrho} \operatorname{diam} U^r(E)$ exists. If $\delta = 0$, one can apply Cantor's intersection theorem to the descending sequence of nonempty and closed sets $U^{\varrho+1/n}(E)$ to get that $U^\varrho(E)$ contains exactly one point. If $\delta > 0$, then for $r \in (\varrho, D_\kappa/2)$ we choose $u_r, v_r \in U^r(E)$ with $d(u_r, v_r) \geq \delta/2$ and denote m_r the midpoint of u_r and v_r. By (5.1.2) and Proposition 5.1.10, $m_r \in U(E)$ and moreover, for r sufficiently close to ϱ, $d(y_b, m_r) \leq \varrho$, so $m_r \in U^\varrho(E)$. Thus, $U^\varrho(E) \neq \emptyset$ for all $E \in \mathcal{E}$.

Now we prove that $\inf\{\operatorname{diam} U^\varrho(E) : E \in \mathcal{E}\} = 0$. Suppose on the contrary that there exists $\varepsilon > 0$ such that $\operatorname{diam} U^\varrho(E) \geq \varepsilon$ for all $E \in \mathcal{E}$. Using a similar midpoint construction as before, it is possible to find a number $\varrho' < \varrho$ such that $U^{\varrho'}(E) \neq \emptyset$ for all $E \in \mathcal{E}$, which contradicts the definition of ϱ.

Hence, we can find a sequence (E_i) of sets in $\mathcal{E}$ such that $\lim_{i \to \infty} \operatorname{diam} U^\varrho(E_i) = 0$. Consider the set of all pairs (y_a, r_a) with $a \in E_i$ for some i and list its elements as a sequence (y_j, r_j). Denote $U_k = \bigcap_{j=1}^k B[y_j, r_j]$ and $U_k^\varrho = U_k \cap B[y_b, \varrho]$. Note that by the above, $U_k^\varrho \neq \emptyset$ and $U_{k+1}^\varrho \subseteq U_k^\varrho$ for any k. For i fixed, $U_k^\varrho \subseteq U^\varrho(E_i)$ if k is large enough. This implies that $\lim_{k \to \infty} \operatorname{diam} U_k^\varrho = 0$. By Cantor's intersection theorem, $\bigcap_{k \geq 1} U_k^\varrho = \{u\}$ for some $u \in Y$. Fix $a \in A$. Since $U^\varrho(E) \neq \emptyset$ for every $E \in \mathcal{E}$, we have that $U_k^\varrho \cap B[y_a, r_a] \neq \emptyset$ for all k. Recalling that $(U_k^\varrho)_k$ is a descending sequence, we finally obtain that $u \in B[y_a, r_a]$, which implies that $\bigcap_{a \in A} B[y_a, r_a] \neq \emptyset$. □

Proof of Theorem 5.2.6 (Sketch) First one proves that it is possible to extend a nonexpansive mapping $f : E \to Y$ defined on a finite set $E \subseteq X$ with $\operatorname{diam} f(E) \leq D_\kappa/2$ to one additional point $x \in X \setminus E$ such that the resulting extension $f_x : E \cup \{x\} \to Y$ is nonexpansive and $f_x(x) \in \overline{\operatorname{co}}(f(E))$. We omit the proof of this step which can be found in [379, Proposition 5.1].

Then we show that one can actually extend a nonexpansive mapping $f : A \to Y$ with A an arbitrary subset of X and $\operatorname{diam} f(A) \leq D_\kappa/2$ to a new point $x \in X \setminus A$ obtaining a nonexpansive mapping $f_x : A \cup \{x\} \to Y$ satisfying $f_x|_A = f$ and $f_x(x) \in \overline{\operatorname{co}}(f(A))$. For this it is enough to prove that

$$\left(\bigcap_{a \in A} B[f(a), d(a, x)] \right) \cap \overline{\operatorname{co}}(f(A)) \neq \emptyset, \tag{5.2.1}$$

as in this case we can take $f_x(x)$ any point in this intersection.

For $a \in A$, let

$$y_a = f(a) \quad \text{and} \quad r_a = \min\{d(a, x), D_\kappa/2\}.$$

If $r_a = D_\kappa/2$ for all $a \in A$, (5.2.1) obviously holds because $\operatorname{diam} f(A) \leq D_\kappa/2$. Suppose now $r_b < D_\kappa/2$ for some $b \in A$. Let E be a finite subset of A. Considering the restriction of f to E and extending it to $E \cup \{x\}$ in a nonexpansive way, we find a point y, the value of the extension at x, such that $y \in \bigcap_{a \in E} B[y_a, d(a, x)]$ and $y \in \overline{\text{co}}(f(E)) \subseteq \overline{\text{co}}(f(A))$. By Proposition 5.1.11, diam $\overline{\text{co}}(f(A)) =$ diam $f(A) \leq D_\kappa / 2$, from where $y \in (\bigcap_{a \in E} B[y_a, r_a]) \cap \overline{\text{co}}(f(A))$. Applying Theorem 5.2.7 for $\overline{\text{co}}(f(A))$ as a complete CAT(κ) space (with the induced metric), it follows that $(\bigcap_{a \in A} B[y_a, r_a]) \cap \overline{\text{co}}(f(A)) \neq \emptyset$, which yields (5.2.1).

Consider now the set $\mathscr{U}$ consisting of all pairs (D, f'), where $A \subseteq D \subseteq X$ and $f' : D \to Y$ is a nonexpansive extension of f with $f'(D) \subseteq \overline{\text{co}}(f(A))$. Clearly, $(A, f) \in \mathscr{U}$, so $\mathscr{U} \neq \emptyset$. We define a partial order on $\mathscr{U}$ by $(D_1, f_1') \preceq (D_2, f_2')$ if and only if $D_1 \subseteq D_2$ and $f_2'|_{D_1} = f_1'$. Let $\{(D_i, f_i')\}$ be a chain in $\mathscr{U}$. Then $(\bigcup_i D_i, f')$ with $f'|_{D_i} = f_i'$ is in $\mathscr{U}$ and is an upper bound of this chain, so by Zorn's lemma $\mathscr{U}$ has a maximal element (D, F). Since $F(D) \subseteq \overline{\text{co}}(f(A))$, we have diam $F(D) \leq$ diam $\overline{\text{co}}(f(A)) =$ diam $f(A) \leq D_\kappa / 2$. By the previous argument it follows that $D = X$, which ends the proof. $\square$

Remark 5.2.8 The boundedness condition for $\kappa > 0$ on the set $f(A)$ clearly implies that the generalization of Kirszbraun's theorem for pairs of spheres cannot be deduced from Theorem 5.2.6. On the other hand, when $\kappa \leq 0$, there is no restriction on the diameter of the set $f(A)$. For $\kappa = 0$, Theorem 5.2.6 can be generalized to any arbitrary Lipschitz constant by scaling the metric on either X or Y. Recalling that any Hilbert space is both a CAT(0) and a CBB(0) space, this means that Kirszbraun's theorem and its extension to arbitrary pairs of Hilbert spaces are immediate consequences of Theorem 5.2.6. For $\kappa < 0$, a similar rescaling argument can be applied for Lipschitz constants greater than 1, but for Lipschitz constants less than 1, we cannot expect Theorem 5.2.6 to hold true. Indeed, suppose one could extend all mappings $f : A \subseteq \mathbb{H}^2 \to \mathbb{H}^2$ with $L(f) < 1$ while keeping the same Lipschitz constant. Taking $\kappa \in (-1, 0)$ and recalling that M_κ^2 is obtained from $\mathbb{H}^2$ by scaling the metric with $1/\sqrt{-\kappa}$, this implies that we can extend all nonexpansive mappings defined on $A \subseteq \mathbb{H}^2$ with values in M_κ^2 to nonexpansive mappings on $\mathbb{H}^2$. By Proposition 5.2.4 this means that M_κ^2 is a CAT(-1) space, a contradiction. Thus, the generalization of Kirszbraun's theorem for pairs of hyperbolic spaces does not follow from Theorem 5.2.6.

If the target space is an $\mathbb{R}$-tree, then it was proved in [379] that we not only can extend mappings with an arbitrary Lipschitz constant, but we can also drop the curvature assumption on the source space. Because any complete $\mathbb{R}$-tree is a hyperconvex metric space, this result is a consequence of Theorem 5.2.3 (after applying a rescaling argument to get the extension for an arbitrary Lipschitz constant).

Theorem 5.2.9 *Let X be a metric space and Y a complete $\mathbb{R}$-tree. Suppose $A \subseteq X$ is nonempty and $f : A \to Y$ is a Lipschitz mapping. Then there exists a Lipschitz extension $F : X \to Y$ of f with $L(F) = L(f)$.*

If $Y = \mathbb{R}$ in the above result, one obtains as a consequence McShane's theorem (Theorem 4.1.1). Another more general version of Theorem 5.2.9 was given by

Lang [374] for mappings that take values in a Gromov hyperbolic geodesic space and satisfy the following large-scale Lipschitz condition: given two metric spaces (X, d) and (Y, d'), a mapping $f : X \to Y$ is called (L, ε)-*Lipschitz* for $L, \varepsilon \geq 0$ if

$$d'(f(x), f(y)) \leq L\, d(x, y) + \varepsilon,$$

for all $x, y \in X$.

Theorem 5.2.10 *Let A be a nonempty subset of a metric space (X, d), (Y, d') a complete δ-hyperbolic geodesic space and $f : A \to Y$ an (L, ε)-Lipschitz mapping. Then there exists an $(L, \varepsilon + 3\delta)$-Lipschitz extension $F : X \to Y$ of f to X.*

Proof Fix $y_0 \in Y$, define $\mu(x, a) = \max\{0, d'(y_0, f(a)) - L\, d(x, a) - \varepsilon/2\}$ for $x \in X$ and $a \in A$, and set $\overline{\mu}(x) = \sup_{a \in A} \mu(x, a)$. Then $\overline{\mu}(x)$ is finite since for all $a, a' \in A$,

$$\mu(x, a) \leq \max\{0, d'(y_0, f(a')) + d'(f(a'), f(a)) - L\, d(x, a) - \varepsilon/2\}$$
$$\leq \max\{0, d'(y_0, f(a')) + L\, d(a', a) + \varepsilon - L\, d(x, a) - \varepsilon/2\}$$
$$\leq d'(y_0, f(a')) + L\, d(x, a') + \varepsilon/2.$$

By the triangle inequality, $\overline{\mu}(x) + L\, d(x, x') \geq \overline{\mu}(x')$ for every $x, x' \in X$, which means that $\overline{\mu} : X \to \mathbb{R}$ is L-Lipschitz.

For every $a \in A$ choose a geodesic segment $s(a)$ joining y_0 and $f(a)$. For $x \in X$ and $a \in A$ take a point $p(x, a) \in s(a)$ such that $d'(y_0, p(x, a)) = \mu(x, a)$. If $\mu(x, a) > 0$ (i.e., $p(x, a) \neq y_0$), then $d'(p(x, a), f(a)) = L\, d(x, a) + \varepsilon/2$. We prove that

$$d'(p(x, a), p(x', a')) \leq \max\{|\mu(x, a) - \mu(x', a')| + \delta, L\, d(x, x') + 2\delta\}, \quad (5.2.2)$$

for all $x, x' \in X$ and $a, a' \in A$.

We use in the following the notation of the Gromov product given in (5.1.6). To prove (5.2.2), let $x, x' \in X$ and $a, a' \in A$. If $\mu(x, a) \leq (f(a) \mid f(a'))_{y_0}$, then $\mu(x, a) \leq d'(y_0, f(a'))$ and we take $q \in s(a')$ such that $d'(y_0, q) = \mu(x, a)$. By δ-hyperbolicity,

$$d'(p(x, a), p(x', a')) \leq d'(p(x, a), q) + d'(q, p(x', a')) \leq \delta + |\mu(x, a) - \mu(x', a')|.$$

The above inequality also holds if $\mu(x', a') \leq (f(a) \mid f(a'))_{y_0}$.

Suppose now $\mu(x, a) > (f(a) \mid f(a'))_{y_0}$ and $\mu(x', a') > (f(a) \mid f(a'))_{y_0}$. Then

$$d'(p(x, a), f(a)) = d'(y_0, f(a)) - \mu(x, a)$$
$$< d'(y_0, f(a)) - (f(a) \mid f(a'))_{y_0} = (y_0 \mid f(a'))_{f(a)}$$

and $d'(p(x', a'), f(a')) < (y_0 \mid f(a))_{f(a')}$, from where

$$d'(p(x, a), f(a)) + d'(p(x', a'), f(a')) < d'(f(a), f(a'))$$

and we choose a geodesic segment joining $f(a)$ and $f(a')$ and two points q and q' on it with $d'(q, f(a)) = d'(p(x, a), f(a))$ and $d'(q', f(a')) = d'(p(x', a'), f(a'))$. Hence

$$d'(p(x, a), p(x', a')) \leq d'(q, q') + 2\delta \leq Ld(x, x') + 2\delta,$$

where the last inequality follows because

$$d'(q, q') = d'(f(a), f(a') - d'(p(x, a), f(a)) - d'(p(x', a'), f(a'))$$

$$\leq Ld(a, a') + \varepsilon - Ld(x, a) - \varepsilon/2 - Ld(x', a') - \varepsilon/2 \leq Ld(x, x').$$

This finishes the proof of (5.2.2).

Next we construct an $(L, 2\delta)$-Lipschitz mapping $\overline{p} : X \to Y$ satisfying

$$d'(\overline{p}(a), f(a)) \leq \varepsilon/2 + \delta,$$

for all $a \in A$. We consider the cases $\delta = 0$ and $\delta > 0$ separately.

Suppose first $\delta = 0$ (i.e., Y is an $\mathbb{R}$-tree). Given $x \in X$, choose a sequence (a_n) in A with $\lim_{n\to\infty} \mu(x, a_n) = \overline{\mu}(x)$. By (5.2.2), the sequence $(p(x, a_n))$ is Cauchy and converges to some $\overline{p}(x) \in Y$. Note that this limit does not depend on the choice of the sequence (a_n). Using again (5.2.2) we see that $\overline{p}$ is L-Lipschitz since $\overline{\mu}$ is. Now let $a \in A$. If $\overline{p}(a) \in s(a)$, then

$$d'(\overline{p}(a), f(a)) = d'(y_0, f(a)) - d'(y_0, \overline{p}(a)) = d'(y_0, f(a)) - \overline{\mu}(a)$$

$$\leq d'(y_0, f(a)) - \mu(a, a) \leq \varepsilon/2.$$

Otherwise, take a sequence (a_n) in A with $\lim_{n\to\infty} p(a, a_n) = \overline{p}(a)$ and $p(a, a_n) \notin s(a)$ for all $n \in \mathbb{N}$. For $u, v \in Y$, we denote the geodesic segment joining u and v by $s(u, v)$. Let $n \in \mathbb{N}$. Since Y is an $\mathbb{R}$-tree, $s(a) \cap s(a_n) = s(y_0, z_n)$ for some $z_n \in Y$ and $s(f(a), f(a_n)) = s(f(a), z_n) \cup s(f(a_n), z_n)$. Then $p(a, a_n) \in s(f(a), f(a_n))$, which yields

$$d'(p(a, a_n), f(a)) = d'(f(a), f(a_n)) - d'(p(a, a_n), f(a_n))$$

$$\leq L d(a, a_n) + \varepsilon - L d(a, a_n) - \varepsilon/2 = \varepsilon/2,$$

hence $d'(\overline{p}(a), f(a)) \leq \varepsilon/2$.

Assume next $\delta > 0$. Given $x \in X$, choose a point $z_x \in A$ such that $\mu(x, z_x) \geq \overline{\mu}(x) - \delta$ if $x \in X \setminus A$ and $\mu(x, z_x) \geq \max\{\mu(x, x), \overline{\mu}(x) - \delta\}$ if $x \in A$. Let $\overline{p}(x) = p(x, z_x)$. Then

$$|\mu(x, z_x) - \mu(x', z_{x'})| \leq |\overline{\mu}(x) - \overline{\mu}(x')| + \delta \leq L\, d(x, x') + \delta,$$

for all $x, x' \in X$. This inequality together with (5.2.2) yields that $\overline{p}$ is $(L, 2\delta)$-Lipschitz. Now let $a \in A$. If $\mu(a, z_a) \leq (f(a) \mid f(z_a))_{y_0}$, take $q \in s(a)$ such that $d'(y_0, q) = \mu(a, z_a)$. Then

$$d'(\overline{p}(a), f(a)) \leq d'(\overline{p}(a), q) + d'(q, f(a)) \leq \delta + d'(y_0, f(a)) - \mu(a, z_a)$$

$$\leq \delta + d'(y_0, f(a)) - \mu(a, a) \leq \delta + \varepsilon/2.$$

Otherwise, if $\mu(a, z_a) > (f(a) \mid f(z_a))_{y_0}$, then

$$d'(\overline{p}(a), f(z_a)) < (f(a) \mid y_0)_{f(z_a)} \leq d(f(a), f(z_a))$$

and we take q on a geodesic segment joining $f(z_a)$ and $f(a)$ with $d'(q, f(z_a)) = d'(\overline{p}(a), f(z_a))$. Then

$$d'(\overline{p}(a), f(a)) \leq d'(\overline{p}(a), q) + d'(q, f(a)) \leq \delta + d'(f(a), f(z_a))$$

$$- d'(\overline{p}(a), f(z_a))$$

$$\leq \delta + L\, d(a, z_a) + \varepsilon - L\, d(a, z_a) - \varepsilon/2 = \delta + \varepsilon/2.$$

This finishes the construction of the mapping $\overline{p}$.

Define $F : X \to Y$ by

$$F(x) = \begin{cases} f(x) & \text{if } x \in A \\ \overline{p}(x) & \text{if } x \in X \setminus A. \end{cases}$$

Then $F|_A$ is (L, ε)-Lipschitz, while $F|_{X \setminus A}$ is $(L, 2\delta)$-Lipschitz. If $x \in X \setminus A$ and $a \in A$,

$$d'(F(x), F(a)) = d'(\overline{p}(x), f(a)) \leq d'(\overline{p}(x), \overline{p}(a)) + d'(\overline{p}(a), f(a))$$

$$\leq L\, d(x, a) + 2\delta + \varepsilon/2 + \delta \leq L\, d(x, a) + \varepsilon + 3\delta.$$

This finally shows that F is $(L, \varepsilon + 3\delta)$-Lipschitz.   □

The above proof also yields the following approximation result.

Theorem 5.2.11 *Let X be a metric space, (Y, d') a complete $\mathbb{R}$-tree and $f : X \to Y$ an (L, ε)-Lipschitz mapping. Then there exists an L-Lipschitz mapping $f' : X \to$*

Y satisfying

$$d'(f(x), f'(x)) \leq \frac{\varepsilon}{2},$$

for all $x \in X$.

5.3 Continuity of Extension Operators

The results given in Sect. 5.2 guarantee the existence of an extension for the original mapping. However, this extension is not necessarily unique and no information is given on the parameter dependence of the extensions. E. Kopecká studied the process of assigning extensions to mappings from the point of view of continuity with respect to the supremum norm providing positive answers first in Euclidean [361] and then in Hilbert spaces [360] (see also [449]). Namely, the multi-valued extension operators that assign to every nonexpansive (resp. Lipschitz) mapping all its nonexpansive extensions (resp. Lipschitz extensions with the same Lipschitz constant) are proved to be lower semicontinuous using Kirszbraun's theorem and a homotopy argument. Applying Michael's selection theorem one obtains continuous selections of these multi-valued extension operators. E. Kopecká and S. Reich further generalized these results in [362], obtaining a continuous single-valued extension operator with the additional condition that the image of the extension belongs to the closed convex hull of the image of the original mapping.

A natural question is to study this problem in geodesic spaces with curvature bounds in the sense of Alexandrov. We show in the sequel that one can indeed prove counterparts of such continuity properties in this setting too. The results included in this section are mainly taken from [228].

Let (X, d), (Y, d') be metric spaces, $A \subseteq X$ nonempty and consider $C_b(A, Y)$ the family of continuous and bounded mappings from A to Y. For each $f, g \in C_b(A, Y)$, let

$$d_\infty(f, g) = \sup_{x \in A} d'(f(x), g(x)).$$

Endowed with the supremum distance d_∞, $C_b(A, Y)$ is a metric space which is complete if Y is complete. We consider two subsets of $C_b(A, Y)$: BLip(A, Y) which includes all bounded Lipschitz mappings from A to Y and is not necessarily a closed subset of $C_b(A, Y)$ and BN(A, Y) which stands for the family of all bounded nonexpansive mappings defined from A to Y and which is closed in $C_b(A, Y)$.

Proposition 5.3.1 *If X is a metric space and (Y, d') is a Busemann convex space, then $(C_b(X, Y), d_\infty)$ admits a geodesic bicombing*

$$\sigma : C_b(X, Y) \times C_b(X, Y) \times [0, 1] \to C_b(X, Y)$$

defined by

$$\sigma(f,g,t)(x) = (1-t)f(x) \oplus tg(x),$$

for each $f, g \in C_b(X,Y)$, $t \in [0,1]$ *and* $x \in X$.

Proof For $f, g \in C_b(X,Y)$, denote $\sigma_{fg} = \sigma(f,g,\cdot)$. Note first that σ is well-defined. Indeed, if $f, g \in C_b(X,Y)$ and $t \in [0,1]$, then

$$d'(\sigma_{fg}(t)(x), \sigma_{fg}(t)(x')) \leq (1-t)d'(f(x), f(x')) + td'(g(x), g(x')),$$

for all $x, x' \in X$. It follows that $\sigma_{fg}(t)$ is continuous. Moreover, since f and g are bounded, there exist $z \in Y$ and $M > 0$ such that for all $x \in X$,

$$\max\{d'(z, f(x)), d'(z, g(x))\} \leq M,$$

which yields

$$d'(z, \sigma_{fg}(t)(x)) \leq (1-t)d'(z, f(x)) + td'(z, g(x)) \leq M.$$

Hence, $\sigma_{fg}(t) \in C_b(X,Y)$.

Let $f, g \in C_b(X,Y)$. Then $\sigma_{fg}(0) = f$, $\sigma_{fg}(1) = g$ and

$$d_\infty(\sigma_{fg}(t), \sigma_{fg}(t')) = \sup_{x\in X} d'(\sigma_{fg}(t)(x), \sigma_{fg}(t')(x))$$
$$= |t - t'| \sup_{x\in X} d'(f(x), g(x)) = |t - t'| d_\infty(f, g),$$

for all $t, t' \in [0,1]$. This shows that σ_{fg} is a geodesic.

Clearly, $\sigma_{fg}(t) = \sigma_{gf}(1-t)$ for all $t \in [0,1]$. Moreover,

$$d_\infty\big(\sigma_{fg}(t), \sigma_{f'g'}(t)\big) = \sup_{x\in X} d'\big(\sigma_{fg}(t)(x), \sigma_{f'g'}(t)(x)\big)$$
$$= \sup_{x\in X} d'\big((1-t)f(x) \oplus tg(x), (1-t)f'(x) \oplus tg'(x)\big)$$
$$\leq \sup_{x\in X} \big[(1-t)d'(f(x), f'(x)) + td'(g(x), g'(x))\big]$$
$$\leq (1-t)d_\infty(f, f') + td_\infty(g, g'),$$

for all $f, g, f', g' \in C_b(X,Y)$ and all $t \in [0,1]$. $\qquad\square$

In the above result, it is actually enough to assume that Y admits a geodesic bicombing. Moreover, one cannot conclude that $(C_b(X,Y), d_\infty)$ is Busemann convex as the following simple example shows.

Example 5.3.2 Let $X = Y = [-1, 1]$ with d' the Euclidean metric. Then (Y, d') is Busemann convex and the geodesic segments are the usual segments. Consider the functions $f, g : [-1, 1] \to [-1, 1]$ defined by

$$f(x) = \begin{cases} 1+x & \text{if } x \in [-1, 0] \\ 1-x & \text{if } x \in (0, 1] \end{cases}$$

and $g = -f$. Then $d_\infty(f, g) = 2$ and $\sigma_{fg}(1/2)(x) = 0$ for all $x \in [-1, 1]$. However, the function $h : [-1, 1] \to [-1, 1]$,

$$h(x) = \begin{cases} 1+x & \text{if } x \in [-1, -1/2] \\ -x & \text{if } x \in (-1/2, 0] \\ x & \text{if } x \in (0, 1/2] \\ 1-x & \text{if } x \in (1/2, 1] \end{cases}$$

also satisfies $d_\infty(f, h) = d_\infty(h, g) = 1$, so h and $\sigma_{fg}(1/2)$ are two distinct midpoints of f and g, which means that $C_b(X, Y)$ is not uniquely geodesic and hence not Busemann convex.

For a set B, we denote by $\mathscr{P}(B)$ the family of all its subsets. We consider two multi-valued extension mappings:

- $\Phi : \mathrm{BN}(A, Y) \to \mathscr{P}(\mathrm{BN}(X, Y))$ which assigns to each nonexpansive mapping $f \in \mathrm{BN}(A, Y)$ all its nonexpansive extensions $F \in \mathrm{BN}(X, Y)$. Note that in this case it may happen that $L(f) < L(F) \le 1$.
- $\Psi : \mathrm{BLip}(A, Y) \to \mathscr{P}(\mathrm{BLip}(X, Y))$ which assigns to each Lipschitz mapping $f \in \mathrm{BLip}(A, Y)$ all its Lipschitz extensions $F \in \mathrm{BLip}(X, Y)$ with $L(f) = L(F)$.

Both mappings Φ and Ψ have closed values in $C_b(X, Y)$.

5.3.1 Continuous Selections in Alexandrov Spaces

In this subsection we consider appropriate curvature bounds in the sense of Alexandrov on X and Y to obtain the needed continuity properties for the mappings Φ and Ψ. For this we first rely on Theorem 5.2.6. Note that we are only concerned with the case $\kappa \le 0$ (the reason why we do not consider $\kappa > 0$ will be explained further along). Moreover, according to Remark 5.2.8, for $\kappa < 0$ we will only study the mapping Φ.

Another important role in the study of the extension operators Φ and Ψ is played by Michael's selection theorem and its generalizations to nonlinear settings, which we discuss below.

Michael's Selection Theorem

Let U and Z be two topological spaces. A multi-valued mapping $\Theta : U \to \mathscr{P}(Z)$ is *lower semicontinuous* if for every open $G \subseteq Z$, the set $\{u \in U : \Theta(u) \cap G \neq \emptyset\}$ is open in U.

The classical Michael selection theorem which goes back to [466] reads as follows.

Theorem 5.3.3 *Let U be a paracompact topological space, Z a Banach space, and $\Theta : U \to \mathscr{P}(Z)$ lower semicontinuous such that for each $u \in U$, $\Theta(u)$ is a nonempty closed and convex subset of Z. Then there exists a continuous selection for Θ, i.e., a continuous mapping $\theta : U \to Z$ such that $\theta(u) \in \Theta(u)$ for every $u \in U$.*

There is no Lipschitz variant of Theorem 5.3.3. More precisely, if (U, d) is a metric space, Z is a Banach space and $\Theta : U \to \mathscr{P}(Z)$ takes nonempty, bounded, closed and convex values and is Lipschitz (i.e., for some $L \geq 0, d_H(\Theta(u), \Theta(v)) \leq L\, d(u, v)$ for every $u, v \in U$, where d_H is the Pompeiu–Hausdorff distance defined in Sect. 1.3.4), then one cannot always find a Lipschitz selection of Θ if Z is infinite-dimensional. However, for the finite-dimensional case this is indeed true. These results go back to [60, 583, 584]. In Sect. 5.3.2 we include a result in this direction in the setting of hyperconvex spaces (see Theorem 5.3.20).

We give next the proof of a result which corresponds to Michael's selection theorem when Z is a complete Busemann convex space. Taking into account the continuity properties of convex combinations given in Sect. 5.1.4, the proof method is now quite similar to the one for Banach spaces (see, e.g., [75]).

Theorem 5.3.4 *Let U be a paracompact topological space, (Z, d) a complete Busemann convex space, and $\Theta : U \to \mathscr{P}(Z)$ lower semicontinuous such that for each $u \in U$, $\Theta(u)$ is a nonempty closed and convex subset of Z. Then there exists a continuous selection for Θ.*

We prove first the following lemma.

Lemma 5.3.5 *Let U, Z and Θ be as in Theorem 5.3.4 without assuming that Θ takes closed values. Then for every $\varepsilon > 0$ there exists a continuous mapping $\theta : U \to Z$ such that $d(\theta(u), \Theta(u)) < \varepsilon$ for all $u \in U$.*

Proof Let $\varepsilon > 0$. For $z \in Z$, denote $G_z = \{u \in U : \Theta(u) \cap B(z, \varepsilon) \neq \emptyset\}$. Since Θ is lower semicontinuous, the set G_z is open and the collection $(G_z)_{z \in Z}$ is an open cover of U, so there exists a partition of unity $(p_i)_{i \in I}$ subordinated to this cover. For every $i \in I$, choose $z_i \in Z$ such that $\mathrm{spt}(p_i) \subseteq G_{z_i}$.

For $u \in U$, denote $I_u = \{i \in I : p_i(u) \neq 0\}$ and define $\theta : U \to Z$,

$$\theta(u) = \dot{\oplus}_{i \in I_u} p_i(u) z_i.$$

Because I_u is finite, θ is well-defined. To see that it is also continuous, let $u \in U$. Take a neighborhood V of u and a finite set $J \subseteq I$ such that $p_i|_V = 0$ for all

$i \in I \setminus J$. If $v \in V$, then $I_v \subseteq J$ and $\theta(v)$ actually coincides with $\dot{\oplus}_{i \in J} p_i(v) z_i$ since the two convex combinations only differ by at most a finite collection of points with zero coefficients and this does not change the value of a convex combination. Since $u \in V$, $\theta(u) = \dot{\oplus}_{i \in J} p_i(u) z_i$ and, by Theorem 5.1.18, θ is continuous at u.

Let $u \in U$. If $i \in I_u$, then $u \in G_{z_i}$ and so $d(z_i, \Theta(u)) < \varepsilon$. Let $\delta > 0$. For any $i \in I_u$, pick $y_i \in \Theta(u)$ with

$$d(y_i, z_i) \le d(z_i, \Theta(u)) + \delta.$$

Then, as $\Theta(u)$ is convex and Z is Busemann convex, $\overline{\Theta(u)}$ is closed and convex (see Remark 5.1.5 and Proposition 5.1.2), so $\dot{\oplus}_{i \in I_u} p_i(u) y_i \in \overline{\Theta(u)}$ and

$$d(\theta(u), \Theta(u)) \le d(\dot{\oplus}_{i \in I_u} p_i(u) z_i, \dot{\oplus}_{i \in I_u} p_i(u) y_i) \le \sum_{i \in I_u} p_i(u) d(y_i, z_i) \quad \text{by (5.1.9)}$$

$$\le \delta + \sum_{i \in I_u} p_i(u) d(z_i, \Theta(u)).$$

Because $\delta > 0$ is arbitrary, $d(\theta(u), \Theta(u)) \le \sum_{i \in I_u} p_i(u) d(z_i, \Theta(u)) < \varepsilon$. □

Proof of Theorem 5.3.4 Let $\Theta_0 = \Theta$ and θ_0 be the mapping obtained by applying Lemma 5.3.5 for $\varepsilon = 1$. For $u \in U$, set

$$\Theta_1(u) = \Theta_0(u) \cap B(\theta_0(u), 1).$$

Then the values of Θ_1 are convex and nonempty by the definition of θ_0. We claim that Θ_1 is lower semicontinuous. Indeed, let G be an open set in Z and suppose $v \in \{u \in U : \Theta_1(u) \cap G \ne \emptyset\}$. Then there exists $r < 1$ such that the open set

$$\{u \in U : \Theta_0(u) \cap (B(\theta_0(v), r) \cap G) \ne \emptyset\}$$

contains v. Because θ_0 is continuous, the set

$$\{u \in U : \Theta_0(u) \cap B(\theta_0(u), 1) \cap G \ne \emptyset\}$$

also contains a neighborhood of v. This proves our claim. Thus, we can apply again Lemma 5.3.5 for Θ_1 and $\varepsilon = 1/2$ to obtain a mapping θ_1.

Continuing in this way we get a sequence of continuous mappings (θ_n) with

$$d(\theta_n(u), \Theta(u)) < \frac{1}{2^n} \quad \text{and} \quad d(\theta_{n-1}(u), \theta_n(u)) < \frac{1}{2^{n-1}} + \frac{1}{2^n},$$

for all $u \in U$ and all $n \in \mathbb{N}$. Thus, (θ_n) is Cauchy and its limit is a mapping $\theta : U \to Z$ that is continuous and satisfies $\theta(u) \in \Theta(u)$ for all $u \in U$ as Θ has closed values. This shows that θ is the desired selection. □

Remark 5.3.6 Theorem 5.3.4 holds true in a more general version assuming that Z is a complete metric space with a geodesic bicombing σ and Θ takes σ-convex values instead of convex ones (see also Remark 5.1.19).

A far-reaching generalization of Michael's selection theorem to the setting of c-spaces was given by Horvath [291]. Before stating this selection result we recall the following notions: for Z a topological space, denote by $\langle Z \rangle$ the family of its nonempty and finite subsets. A mapping $F : \langle Z \rangle \to \mathscr{P}(Z)$ is a *c-structure* if firstly, for each $A \in \langle Z \rangle$, $F(A)$ is nonempty and contractible, and secondly, for every $A_1, A_2 \in \langle Z \rangle$, $A_1 \subseteq A_2$ implies $F(A_1) \subseteq F(A_2)$. In this case, the pair (Z, F) is called a *c-space* and $V \subseteq Z$ is an *F-set* if for every $A \in \langle V \rangle$ we have that $F(A) \subseteq V$. A c-space (Z, F) is called an *l.c. metric space* if (Z, d) is a metric space such that open balls are F-sets and, in addition, if $V \subseteq Z$ is an F-set, then for every $\varepsilon > 0$, $\{z \in Z : d(z, V) < \varepsilon\}$ is an F-set.

Theorem 5.3.7 *Let U be a paracompact topological space, (Z, F) an l.c. complete metric space and $\Theta : U \to \mathscr{P}(Z)$ lower semicontinuous such that for each $u \in U$, $\Theta(u)$ is a nonempty and closed F-set. Then there exists a continuous selection for Θ.*

Remark 5.3.8 We justify next why Theorem 5.3.4 is a consequence of Theorem 5.3.7. To this end we show that any complete metric space Z that admits a geodesic bicombing σ is an l.c. complete metric space together with a suitable c-structure F. Define $F : \langle Z \rangle \to \mathscr{P}(Z)$ by $F(A) = \mathrm{co}_\sigma(A)$. For any $A \in \langle Z \rangle$, $F(A)$ is nonempty and contractible. Clearly, for every $A_1, A_2 \in \langle Z \rangle$, $A_1 \subseteq A_2$ implies $F(A_1) \subseteq F(A_2)$. Thus, (Z, F) is a c-space. Note that a subset of Z is an F-set if and only if it is σ-convex and one can finally see that (Z, F) is an l.c. metric space.

Lower Semicontinuity of the Multi-Valued Extension Mappings and Continuous Selections

We study here the lower semicontinuity of the mappings Φ and Ψ, and then use the results given before to obtain continuous selections thereof.

Theorem 5.3.9 *Let $\kappa \leq 0$, (X, d) a CBB(κ) space, (Y, d') a complete CAT(κ) space and $A \subseteq X$ nonempty. Then the mapping $\Phi : \mathrm{BN}(A, Y) \to \mathscr{P}(\mathrm{BN}(X, Y))$ has nonempty values and is lower semicontinuous.*

Proof Let $F \in \mathrm{BN}(X, Y)$. We show that for every $\varepsilon > 0$ there exists $\delta > 0$ such that every $g \in \mathrm{BN}(A, Y)$ with $\sup_{a \in A} d'(F(a), g(a)) < \delta$ admits an extension $G \in \mathrm{BN}(X, Y)$ such that $d_\infty(F, G) \leq \varepsilon$.

Since F is bounded, one can take $z \in Y$ and $M \geq 1$ such that

$$\sup_{x \in X} d'(z, F(x)) \leq M.$$

Let $\varepsilon \in (0, 1)$, $\delta = \varepsilon^2 / (8M)$ and $g \in \mathrm{BN}(A, Y)$ with $\sup_{a \in A} d'(F(a), g(a)) < \delta$.

Define the mapping

$$h : (X \times \{0\}) \cup (A \times \{\varepsilon\}) \to Y$$

in the following way: for $x \in X$, $h(x, 0) = F(x)$ and for $a \in A$, $h(a, \varepsilon) = g(a)$. Recalling (5.1.5), we obtain

$$d'(h(x, 0), h(a, \varepsilon))^2 = d'(F(x), g(a))^2 \le \big(d'(F(x), F(a)) + d'(F(a), g(a))\big)^2$$

$$\le d(x, a)^2 + \delta^2 + 4\delta M < d(x, a)^2 + \varepsilon^2$$

$$= d_2((x, 0), (a, \varepsilon))^2,$$

for all $x \in X$ and $a \in A$. This shows that h is nonexpansive as both F and g are nonexpansive on X and A, respectively. Note that $X \times \mathbb{R}$ is a CBB(κ) space because $\kappa \le 0$ and X and $\mathbb{R}$ are both CBB(κ) spaces. Thus, we can apply Theorem 5.2.6 to obtain a nonexpansive extension $H : X \times \mathbb{R} \to Y$ of h. Defining now $G : X \to Y$, $G(x) = H(x, \varepsilon)$, it is straightforward that G is nonexpansive and coincides with g on A. Moreover,

$$d'(F(x), G(x)) = d'(H(x, 0), H(x, \varepsilon)) \le d_2((x, 0), (x, \varepsilon)) = \varepsilon,$$

for each $x \in X$. This also shows that G is bounded.   □

Theorem 5.3.10 *Let (X, d) be a CBB(0) space, (Y, d') a complete CAT(0) space and $A \subseteq X$ nonempty. Then the mapping $\Psi : \mathrm{BLip}(A, Y) \to \mathscr{P}(\mathrm{BLip}(X, Y))$ has nonempty values and is lower semicontinuous.*

Proof Let $F \in \mathrm{BLip}(X, Y)$ with $L(F|_A) = L(F)$. We show that for every $\varepsilon > 0$ there exists $\delta > 0$ such that every $g \in \mathrm{BLip}(A, Y)$ for which $\sup_{a \in A} d'(F(a), g(a)) < \delta$ admits an extension $G \in \mathrm{BLip}(X, Y)$ with $L(g) = L(G)$ and $d_\infty(F, G) \le \varepsilon$.

Let $\varepsilon \in (0, 1)$. Suppose first that F is constantly equal to some $y \in Y$. Let $\delta = \varepsilon$ and $g \in \mathrm{BLip}(A, Y)$ with $\sup_{a \in A} d'(y, g(a)) < \delta$. According to Theorem 5.2.6 and Remark 5.2.8, g admits an extension G_1 to X for which $L(G_1) = L(g)$. Then we can define the desired extension $G : X \to Y$ by $G(x) = P_{B[y, \varepsilon]} \circ G_1$. Note that $L(G) = L(g)$ as on the one hand G extends g and on the other hand the metric projection onto nonempty closed and convex subsets of Y is nonexpansive (see Sect. 5.1.1).

Assume now that F is not constant. Let $z \in Y$ and $M > 0$ such that

$$\sup_{x \in X} d'(z, F(x)) \le M.$$

Take $s \in (0, 1)$ for which

$$\frac{1 - s}{s^2} < \frac{\varepsilon^2}{32M(4M + 1)}.$$

Since $L(F|_A) = L(F) > 0$, there exist $x_0, y_0 \in A$ such that

$$d'(F(x_0), F(y_0)) > s\,L(F)\,d(x_0, y_0).$$

Let

$$\delta = \min \left\{ \frac{d'(F(x_0), F(y_0)) - s\,L(F)\,d(x_0, y_0)}{2}, \frac{\varepsilon^2 s^2}{32(4M + 1)} \right\}$$

and $g \in \mathrm{BLip}(A, Y)$ with $\sup_{a \in A} d'(F(a), g(a)) < \delta$.
Suppose first $L(g) \le 2L(F)$. Then,

$$d'(g(x_0), g(y_0)) \ge d'(F(x_0), F(y_0)) - d'(F(x_0), g(x_0)) - d'(F(y_0), g(y_0))$$
$$> d'(F(x_0), F(y_0)) - 2\delta \ge s\,L(F)\,d(x_0, y_0),$$

from where we obtain $L(g) \ge s\,L(F)$. Denote

$$\eta = \frac{\varepsilon}{4L(F)}$$

and define the mapping

$$h : (X \times \{0\}) \cup (A \times \{\eta\}) \to Y$$

in the following way: for $x \in X$, $h(x, 0) = (1 - s)z \oplus sF(x)$ and for $a \in A$, $h(a, \eta) = g(a)$. Thus, for all $x \in X$ and $a \in A$, we have

$$d'(h(x, 0), h(a, \eta))^2 = d'((1 - s)z \oplus sF(x), g(a))^2$$
$$\le \left(d'((1 - s)z \oplus sF(x), F(a)) + d'(F(a), g(a)) \right)^2$$
$$\le \left((1 - s)M + s\,d'(F(x), F(a)) + \delta \right)^2$$
$$\le (\delta + (1 - s)M)^2 + s^2 L(F)^2 d(x, a)^2$$
$$\quad + 4\,s\,M\,(\delta + (1 - s)M)$$
$$< s^2 L(F)^2 \left(d(x, a)^2 + \frac{(\delta + (1 - s)M)(4M + 1)}{s^2 L(F)^2} \right)$$

$$\text{since } (\delta + (1 - s)M)^2 < \delta + (1 - s)M \text{ and } s < 1$$

$$< s^2 L(F)^2 \left(d(x,a)^2 + \eta^2 \right)$$

$$\text{since } \delta + (1-s)M < \varepsilon^2 s^2 / (16(4M+1))$$

$$\leq L(g)^2 d_2((x,0),(a,\eta))^2.$$

To complete the argument that h is Lipschitz with smallest Lipschitz constant $L(g)$ one uses Busemann convexity in Y along with the fact that the mappings F and g are Lipschitz and $L(g) \geq s\, L(F)$. Since $X \times \mathbb{R}$ is a CBB(0) space, by Theorem 5.2.6 we can extend h to a Lipschitz mapping $H : X \times \mathbb{R} \to Y$ with $L(H) = L(g)$. Define $G : X \to Y$ by $G(x) = H(x,\eta)$. Clearly, G extends g and $L(G) = L(g)$. Moreover, for every $x \in X$,

$$d'(G(x),F(x)) \leq d'(G(x),(1-s)z \oplus sF(x)) + d'((1-s)z \oplus sF(x),F(x))$$

$$\leq d'(H(x,\eta),H(x,0)) + (1-s)M < L(g)\eta + \frac{\varepsilon}{2}$$

$$\leq 2L(F)\frac{\varepsilon}{4L(F)} + \frac{\varepsilon}{2} = \varepsilon.$$

If $L(g) > 2L(F)$, consider the set

$$\tilde{A} = \left\{ x \in X : d(x,A) \geq \frac{2\delta}{L(g)} \right\}$$

and define the mapping $\tilde{g} : A \cup \tilde{A} \to Y$ by setting for $a \in A$, $\tilde{g}(a) = g(a)$ and for $x \in \tilde{A}$, $\tilde{g}(x) = F(x)$. To see that $L(g) = L(\tilde{g})$ it suffices to verify that for all $a \in A$ and $x \in \tilde{A}$,

$$d'(\tilde{g}(x),\tilde{g}(a)) = d'(F(x),g(a)) \leq d'(F(x),F(a)) + d'(F(a),g(a))$$

$$< \frac{L(g)}{2} d(x,a) + \delta \leq \frac{L(g)}{2} d(x,a) + \frac{L(g)}{2} d(x,A)$$

$$\leq L(g)\, d(x,a).$$

Take G to be any extension of $\tilde{g}$ for which $L(g) = L(G)$. For $x \in \tilde{A}$, $F(x) = G(x)$. If $x \notin \tilde{A}$, there exists $a \in A$ such that $d(x,a) < 2\delta/L(g)$. Thus,

$$d'(F(x),G(x)) \leq d'(F(x),F(a)) + d'(F(a),G(a)) + d'(G(a),G(x))$$

$$< \frac{L(g)}{2}\frac{2\delta}{L(g)} + \delta + L(g)\frac{2\delta}{L(g)} = 4\delta < \varepsilon.$$

This ends the proof.   □

Using the lower semicontinuity of the mappings Φ and Ψ, we prove that they admit continuous selections. Supposing that the hypotheses of Theorem 5.3.9 are fulfilled, we observe first that Φ has nonempty and closed values in $C_b(X, Y)$. Let σ be the geodesic bicombing given in Proposition 5.3.1. Then for each $f \in \mathrm{BN}(A, Y)$, $\Phi(f)$ is σ-convex. Indeed, let $F, F' \in \Phi(f)$ and $t \in [0, 1]$. Then, for each $x, x' \in X$,

$$d'(\sigma(F, F', t)(x), \sigma(F, F', t)(x'))$$
$$= d'((1 - t)F(x) \oplus t F'(x), (1 - t)F(x') \oplus t F'(x'))$$
$$\leq (1 - t)d'(F(x), F(x')) + t d'(F'(x), F'(x')) \leq d(x, x').$$

Thus, $\sigma(F, F', t) \in \mathrm{BN}(X, Y)$. Since for all $a \in A$,

$$\sigma(F, F', t)(a) = (1 - t)F(a) \oplus t F'(a) = f(a),$$

it follows that $\Phi(f)$ is σ-convex. Similarly, when the hypotheses of Theorem 5.3.10 are satisfied, Ψ takes nonempty and closed values in $C_b(X, Y)$ that are additionally σ-convex.

We obtain next the desired single-valued continuous extension operators.

Theorem 5.3.11 *Let $\kappa \leq 0$, X a $\mathrm{CBB}(\kappa)$ space, Y a complete $\mathrm{CAT}(\kappa)$ space and $A \subseteq X$ nonempty. Then there exists a continuous mapping $\alpha : \mathrm{BN}(A, Y) \rightarrow \mathrm{BN}(X, Y)$ such that for all $g \in \mathrm{BN}(A, Y)$, $\alpha(g)(a) = g(a)$ for every $a \in A$.*

Proof We can view the mapping Φ with values in $\mathscr{P}(C_b(X, Y))$ while still preserving its lower semicontinuity. Since any metric space is a paracompact topological space we can now apply Theorem 5.3.7 taking into account Remark 5.3.8 (or the more particular Theorem 5.3.4 using Remark 5.3.6) to obtain a continuous selection mapping $\alpha : \mathrm{BN}(A, Y) \rightarrow C_b(X, Y)$. Because Φ actually takes values in $\mathscr{P}(\mathrm{BN}(X, Y))$ we obtain the conclusion. $\square$

Theorem 5.3.12 *Let X be a $\mathrm{CBB}(0)$ space, Y a complete $\mathrm{CAT}(0)$ space and $A \subseteq X$ nonempty. Then there exists a continuous mapping $\beta : \mathrm{BLip}(A, Y) \rightarrow \mathrm{BLip}(X, Y)$ such that for all $g \in \mathrm{BLip}(A, Y)$, $\beta(g)(a) = g(a)$ for every $a \in A$ and $L(\beta(g)) = L(g)$.*

Remark 5.3.13 Note that in Theorems 5.3.9 and 5.3.10 the lower curvature bound of X is only used to apply Theorem 5.2.6. However, by Theorem 5.2.9, when Y is a complete $\mathbb{R}$-tree, one can extend Lipschitz mappings (while keeping the same Lipschitz constant) when X is an arbitrary metric space. Thus, as before, one can obtain that both mappings Φ and Ψ are lower semicontinuous and admit continuous selections. This property will be improved for the mapping Φ in Sect. 5.3.2.

Remark 5.3.14 If $\kappa > 0$ the argument given in this subsection does not work in a straightforward way. Note that in this case, the product of a CBB(κ) space with $\mathbb{R}$ or even with another CBB(κ) space is not a CBB(κ) space. Moreover, the images of the mappings Φ and Ψ may fail to be F-sets when considering a c-structure F defined as in Remark 5.3.8.

A Convexity Assumption on the Images of the Extensions

In the sequel we show that one can actually choose extensions in a continuous way even when imposing the condition that the image of the extension belongs to the closure of the convex hull of the image of the original mapping. To this end, we first provide a uniform bound on the distance between the projection points from a common point onto two subsets of a CAT(0) space. A similar result in uniformly smooth Banach spaces is given in [24]. We use below the Pompeiu–Hausdorff distance defined in Sect. 1.3.4.

Lemma 5.3.15 *Let (Y, d) be a complete* CAT(0) *space, $C_1, C_2 \subseteq Y$ nonempty, closed and convex and suppose r_1 and r_2 are positive numbers. If there exists $z \in Y$ such that $C_1, C_2 \subseteq B[z, r_1]$, then for any $x \in B[z, r_2]$,*

$$d(P_{C_1}(x), P_{C_2}(x))^2 \leq 2(r_1 + r_2)d_H(C_1, C_2).$$

Proof Let $C_1, C_2 \subseteq B[z, r_1]$ and $x \in B[z, r_2]$. Denote $p_1 = P_{C_1}(x)$, $p_2 = P_{C_2}(x)$, $q_1 = P_{C_1}(p_2)$ and $q_2 = P_{C_2}(p_1)$. Clearly,

$$\max\{d(p_1, q_2), d(p_2, q_1)\} \leq d_H(C_1, C_2). \tag{5.3.1}$$

Applying the triangle inequality, $d(x, p_1) \leq d(x, z) + d(z, p_1)$ and $d(x, p_2) \leq d(x, z) + d(z, p_2)$, so

$$\max\{d(x, p_1), d(x, p_2)\} \leq r_1 + r_2. \tag{5.3.2}$$

At the same time,

$$d(x, q_1)^2 \geq d(x, p_1)^2 + d(p_1, q_1)^2. \tag{5.3.3}$$

Indeed, by convexity of C_1, we have that $d(p, x) \geq d(p_1, x)$ for every point p on the geodesic segment that joins p_1 and q_1. Hence, the angle at $\overline{p}_1$ of a comparison triangle $\Delta(\overline{p}_1, \overline{q}_1, \overline{x})$ in $\mathbb{R}^2$ is at least $\pi/2$. Similarly, $d(x, q_2)^2 \geq d(x, p_2)^2 + d(p_2, q_2)^2$.

By (5.1.3) we also have

$$d(x, q_1)^2 + d(p_1, p_2)^2 \leq d(x, p_2)^2 + d(p_1, q_1)^2 + 2d(x, p_1)d(p_2, q_1),$$

and therefore, using (5.3.3),

$$d(x, p_1)^2 + d(p_1, p_2)^2 \leq d(x, p_2)^2 + 2d(x, p_1)d(p_2, q_1).$$

Likewise,

$$d(x, p_2)^2 + d(p_1, p_2)^2 \leq d(x, p_1)^2 + 2d(x, p_2)d(p_1, q_2).$$

Adding the two inequalities from above and then applying (5.3.1) and (5.3.2) we get

$$d(p_1, p_2)^2 \leq d(x, p_1)d(p_2, q_1) + d(x, p_2)d(p_1, q_2) \leq 2\,(\bar{r}_1 + r_2)\,d_H(C_1, C_2).$$

□

Theorem 5.3.16 *Let $\kappa \leq 0$, X a $CBB(\kappa)$ space, (Y, d') a complete $CAT(\kappa)$ space and $A \subseteq X$ nonempty. Then there exists a continuous mapping $\alpha_c : BN(A, Y) \to BN(X, Y)$ such that for all $g \in BN(A, Y)$, $\alpha_c(g)(a) = g(a)$ for every $a \in A$ and $\alpha_c(g)(X) \subseteq \overline{co}\,(g(A))$.*

Proof By Theorem 5.3.11, there exists a continuous $\alpha : BN(A, Y) \to BN(X, Y)$ such that for all $g \in BN(A, Y)$, $\alpha(g)$ extends g. Define a mapping α_c on $BN(A, Y)$ by

$$\alpha_c(g)(x) = P_{\overline{co}(g(A))}\,(\alpha(g)(x)),$$

for each $g \in BN(A, Y)$ and $x \in X$.

If $g \in BN(A, Y)$, $\alpha_c(g) \in BN(X, Y)$ since the projection onto nonempty closed and convex subsets of Y is nonexpansive. Clearly, $\alpha_c(g)(X) \subseteq \overline{co}\,(g(A))$ and $\alpha_c(g)$ coincides with g on A.

Thus, we only need to prove that α_c is continuous. To this end, let $f \in BN(A, Y)$ and $\varepsilon > 0$. As α is continuous, there exists $\delta_1 < 1$ such that for every $g \in BN(A, Y)$ with $d_\infty(f, g) < \delta_1$ we have $d_\infty(\alpha(f), \alpha(g)) < \varepsilon/2$. Fix $z \in Y$ and take

$$r = \sup_{x \in X} d'(z, \alpha(f)(x)) \quad \text{and} \quad \delta = \min\left\{\delta_1, \frac{\varepsilon^2}{16(r+1)}\right\}.$$

Let $g \in BN(A, Y)$ such that $d_\infty(f, g) < \delta$ and $x \in X$. Then

$$d'(\alpha_c(f)(x), \alpha_c(g)(x)) = d'(P_{\overline{co}(f(A))}(\alpha(f)(x)), P_{\overline{co}(g(A))}(\alpha(g)(x)))$$
$$\leq d'(P_{\overline{co}(f(A))}(\alpha(f)(x)), P_{\overline{co}(g(A))}(\alpha(f)(x)))$$
$$+ d'(P_{\overline{co}(g(A))}(\alpha(f)(x)), P_{\overline{co}(g(A))}(\alpha(g)(x))).$$

For every $a \in A$, $\alpha(f)(a) = f(a)$ and

$$d'(z, g(a)) \leq d'(z, f(a)) + d'(f(a), g(a)) < r + 1,$$

so $f(A), g(A) \subseteq B[z, r+1]$. This yields $\overline{\mathrm{co}}\,(f(A)), \overline{\mathrm{co}}\,(g(A)) \subseteq B[z, r+1]$ because $B[z, r+1]$ is closed and convex. Apply now Lemma 5.3.15 with $C_1 = \overline{\mathrm{co}}\,(f(A))$, $C_2 = \overline{\mathrm{co}}\,(g(A))$ and $r_1 = r_2 = r+1$ to get

$$d'\left(P_{\overline{\mathrm{co}}(f(A))}\,(\alpha(f)(x))\,,\,P_{\overline{\mathrm{co}}(g(A))}\,(\alpha(f)(x))\right)$$

$$\leq 2\sqrt{r+1}\sqrt{d_H\,(\overline{\mathrm{co}}\,(f(A))\,,\overline{\mathrm{co}}\,(g(A)))}$$

$$\leq 2\sqrt{r+1}\sqrt{d_H\,(f(A), g(A))}$$

by Proposition 5.1.4

$$\leq 2\sqrt{r+1}\sqrt{\sup_{a \in A} d'(f(a), g(a))}.$$

At the same time,

$$d'\left(P_{\overline{\mathrm{co}}(g(A))}\,(\alpha(f)(x))\,,\,P_{\overline{\mathrm{co}}(g(A))}\,(\alpha(g)(x))\right) \leq d'\,(\alpha(f)(x), \alpha(g)(x))$$

$$\leq d_\infty\,(\alpha(f), \alpha(g))\,.$$

Hence,

$$d_\infty\,(\alpha_c(f), \alpha_c(g)) \leq 2\sqrt{r+1}\sqrt{d_\infty(f, g)} + d_\infty\,(\alpha(f), \alpha(g)) < \frac{\varepsilon}{2} + \frac{\varepsilon}{2} = \varepsilon,$$

which proves that α_c is continuous too.   □

Following the same idea of proof one can give an analogous result for bounded Lipschitz mappings.

Theorem 5.3.17 *Let X be a* CBB(0) *space, Y a complete* CAT(0) *space and $A \subseteq X$ nonempty. Then there exists a continuous mapping $\beta_c : \mathrm{BLip}(A, Y) \to \mathrm{BLip}(X, Y)$ such that for all $g \in \mathrm{BLip}(A, Y)$, $\beta_c(g)(a) = g(a)$ for every $a \in A$, $L(\beta_c(g)) = L(g)$ and $\beta_c(g)(X) \subseteq \overline{\mathrm{co}}\,(g(A))$.*

In fact, instead of constructing the continuous mappings α_c and β_c using the metric projection, one can consider the following multi-valued extension mappings:

- $\Phi_c : \mathrm{BN}(A, Y) \to \mathscr{P}\,(\mathrm{BN}(X, Y))$ which assigns to each nonexpansive mapping $f \in \mathrm{BN}(A, Y)$ all its nonexpansive extensions $F \in \mathrm{BN}(X, Y)$ satisfying $F(X) \subseteq \overline{\mathrm{co}}(f(A))$.
- $\Psi_c : \mathrm{BLip}(A, Y) \to \mathscr{P}\,(\mathrm{BLip}(X, Y))$ which assigns to each Lipschitz mapping $f \in \mathrm{BLip}(A, Y)$ all its Lipschitz extensions $F \in \mathrm{BLip}(X, Y)$ satisfying $L(f) = L(F)$ and $F(X) \subseteq \overline{\mathrm{co}}(f(A))$.

These mappings, too, are lower semicontinuous.

Theorem 5.3.18 *Let* $\kappa \leq 0$, *X a* CBB(κ) *space,* (Y, d') *a complete* CAT(κ) *space and $A \subseteq X$ nonempty. Then the mapping $\Phi_c : \mathrm{BN}(A, Y) \to \mathscr{P}(\mathrm{BN}(X, Y))$ has nonempty values and is lower semicontinuous.*

Proof We show that for every $F \in \mathrm{BN}(X, Y)$ with $F(X) \subseteq \overline{\mathrm{co}}(f(A))$ and for every $\varepsilon > 0$ there exists $\delta > 0$ such that every $g \in \mathrm{BN}(A, Y)$ with $\sup_{a \in A} d'(F(a), g(a)) < \delta$ admits an extension $G \in \mathrm{BN}(X, Y)$ with $G(X) \subseteq \overline{\mathrm{co}}(g(A))$ and $d_\infty(F, G) \leq \varepsilon$.

Let F be as above and $\varepsilon > 0$. By Theorem 5.3.9, there exists $\delta > 0$ such that every $g \in \mathrm{BN}(A, Y)$ with $\sup_{a \in A} d'(F(a), g(a)) < \delta$ admits an extension $G_1 \in \mathrm{BN}(X, Y)$ with $d_\infty(F, G_1) \leq \varepsilon/3$.

Given $y \in Y$ and $Z \subseteq Y$, we denote next $d(y, Z) = \inf_{z \in Z} d'(y, z)$. Let $g \in \mathrm{BN}(A, Y)$ with $\sup_{a \in A} d'(F(a), g(a)) < \delta$ and define $G : X \to Y$, $G(x) = P_{\overline{\mathrm{co}}(g(A))}(G_1(x))$. Clearly, G is nonexpansive, extends g and $G(X) \subseteq \overline{\mathrm{co}}(g(A))$. It remains to prove that $d_\infty(F, G) \leq \varepsilon$. Let $x \in X$. Then,

$$d'(F(x), G(x)) \leq d'(F(x), G_1(x)) + d'(G_1(x), G(x)) \leq \frac{\varepsilon}{3} + d'(G_1(x), G(x)).$$
$$(5.3.4)$$

For every $y \in \overline{\mathrm{co}}(g(A))$ we have

$$d'(G_1(x), G(x)) \leq d'(G_1(x), y) \leq d'(G_1(x), F(x)) + d'(F(x), y),$$

from where

$$d'(G_1(x), G(x)) \leq \frac{\varepsilon}{3} + d(F(x), \overline{\mathrm{co}}(g(A))). \qquad (5.3.5)$$

Consider $E = \{y \in Y : d(y, \overline{\mathrm{co}}(g(A))) \leq \varepsilon/3\}$. We know that $F(A) \subseteq E$ since for any $a \in A$,

$$d(F(a), \overline{\mathrm{co}}(g(A))) \leq d'(F(a), g(a)) \leq d_\infty(F, G_1) \leq \varepsilon \leq \frac{\varepsilon}{3}.$$

Since E is closed and convex we get $\overline{\mathrm{co}}(F(A)) \subseteq E$. But $F(X) \subseteq \overline{\mathrm{co}}(F(A))$ and so $F(x) \in E$. Thus, $d(F(x), \overline{\mathrm{co}}(g(A))) \leq \varepsilon/3$ which, in view of (5.3.5), implies that $d'(G_1(x), G(x)) \leq 2\varepsilon/3$. Using (5.3.4), we finally obtain $d'(F(x), G(x)) \leq \varepsilon$. $\qquad \square$

The same argument yields the result for bounded Lipschitz mappings.

Theorem 5.3.19 *Let X be a* CBB(0) *space, Y a complete* CAT(0) *space and $A \subseteq X$ nonempty. Then the mapping $\Psi_c : \mathrm{BLip}(A, Y) \to \mathscr{P}(\mathrm{BLip}(X, Y))$ has nonempty values and is lower semicontinuous.*

Note that one could apply, as in Sect. 5.3.1, Theorem 5.3.7 to the mappings Φ_c and Ψ_c to obtain directly Theorems 5.3.16 and 5.3.17, respectively.

5.3.2   Nonexpansive Selections in Hyperconvex Metric Spaces

Previous results for the mappings Φ and Φ_c can be strengthened if the target space is a hyperconvex metric space. More precisely, in this case extensions of nonexpansive (Lipschitz) mappings can be chosen not only continuously but in a nonexpansive (or Lipschitz) way. This fact will follow as a direct application of the next result which goes back to [352] (see also [634]). The class of externally hyperconvex subsets of a metric space X (see Definition 5.1.15) is denoted by $\mathscr{E}(X)$.

Theorem 5.3.20 *Let X be any set and (Y, d) a hyperconvex metric space. If $T :$ $X \to \mathscr{E}(Y)$ is a multi-valued mapping, then there exists a selection $f : X \to Y$ for T such that $d(f(x), f(y)) \leq d_H(T(x), T(y))$ for all $x, y \in X$.*

Proof Let $\mathscr{U}$ be the collection of all pairs (D, f'), where $D \subseteq X$ and $f' : D \to Y$ is a selection of $T|_D$ such that $d(f'(a_1), f'(a_2)) \leq d_H(T(a_1), T(a_2))$ for all $a_1, a_2 \in D$. If x_0 is an arbitrary point in X, then for any $f' : \{x_0\} \to Y$ with $f'(x_0) \in T(x_0)$, we get that $(\{x_0\}, f') \in \mathscr{U}$, so $\mathscr{U}$ is nonempty. We define a partial order on $\mathscr{U}$ by $(D_1, f_1') \preceq (D_2, f_2')$ if and only if $D_1 \subseteq D_2$ and $f_2'|_{D_1} = f_1'$. Then $(\mathscr{U}, \preceq)$ is inductively ordered, thus by Zorn's lemma we find a maximal element (D, f) of $\mathscr{U}$. Assume there exists $x \in X \setminus D$. Consider the intersection

$$Y_0 = \left(\bigcap_{a \in D} B[f(a), d_H(T(a), T(x))] \right) \cap T(x).$$

For any $a_1, a_2 \in D$,

$$d(f(a_1), f(a_2)) \leq d_H(T(a_1), T(a_2)) \leq d_H(T(a_1), T(x)) + d_H(T(a_2), T(x)).$$

Moreover, if $a \in D$, as $f(a) \in T(a)$, we have $d(f(a), T(x)) \leq d_H(T(a), T(x))$. By external hyperconvexity of $T(x)$ with respect to Y, it follows that $Y_0 \neq \emptyset$. Let f_x be the extension of f to $D \cup \{x\}$ by taking $f_x(x)$ any element in Y_0. Since $(D \cup \{x\}, f_x) \in \mathscr{U}$, (D, f) is not maximal. This contradiction shows that $D = X$.

$\square$

This theorem implies in particular that if the multi-valued mapping T is nonexpansive (Lipschitz) then f can be chosen nonexpansive (Lipschitz). The next property that we need is that the set $\mathrm{BN}(X, Y)$ endowed with the supremum distance is hyperconvex. This basically follows from [352] where the result is proved for $\mathrm{BN}(Y, Y)$ with Y hyperconvex, but the proof carries over with no modification to our case. Before proving this fact, we give the following property due to Sine [635].

Lemma 5.3.21 *Let* (Y, d) *be a hyperconvex metric space and take a nonempty subset of it of the form* $J = \bigcap_{i \in I} B[z_i, r_i]$. *Then for any* $\varrho > 0$ *we have*

$$\bigcup_{a \in J} B[a, \varrho] = \bigcap_{i \in I} B[z_i, r_i + \varrho].$$

Proof Let $y \in \bigcup_{a \in J} B[a, \varrho]$. Then there exists $a \in J$ such that $d(a, y) \le \varrho$. Thus, $d(z_i, y) \le d(z_i, a) + d(a, y) \le r_i + \varrho$ for all $i \in I$, which yields $y \in \bigcap_{i \in I} B[z_i, r_i + \varrho]$.

Suppose now $y \in \bigcap_{i \in I} B[z_i, r_i + \varrho]$. Then $d(y, z_i) \le r_i + \varrho$ for all $i \in I$. Since Y is hyperconvex, $J \cap B[y, \varrho] \ne \emptyset$, so $y \in \bigcup_{a \in J} B[a, \varrho]$. $\qquad\square$

Theorem 5.3.22 *Let* (X, d) *be a metric space and* (Y, d') *a hyperconvex metric space. Then* $\mathrm{BN}(X, Y)$ *is hyperconvex.*

Proof Let $(f_i)_{i \in I} \subseteq \mathrm{BN}(X, Y)$ and $(r_i)_{i \in I} \subseteq \mathbb{R}_+$ such that $d_\infty(f_i, f_j) \le r_i + r_j$ for all $i, j \in I$. Then for each $x \in X$ and $i, j \in I$, $d'(f_i(x), f_j(x)) \le r_i + r_j$. The set $\bigcap_{i \in I} B[f_i(x), r_i]$ is nonempty since Y is hyperconvex. Moreover, being admissible, it is also externally hyperconvex with respect to Y, and we define a mapping $J : X \to \mathscr{E}(Y)$ by setting $J(x) = \bigcap_{i \in I} B[f_i(x), r_i]$.

We claim that $d_H(J(x), J(y)) \le d(x, y)$ for all $x, y \in X$. To see this, let $x, y \in X$ and $z \in J(x)$. Then for all $i \in I$,

$$d'(z, f_i(y)) \le d'(z, f_i(x)) + d'(f_i(x), f_i(y)) \le r_i + d(x, y).$$

Given $v \in Y$ and $Z \subseteq Y$, we denote $d(v, Z) = \inf_{z \in Z} d'(v, z)$. Using Lemma 5.3.21, $z \in \bigcap_{i \in I} B[f_i(y), r_i + d(x, y)] = \bigcup_{a \in J(y)} B[a, d(x, y)]$, from where $d(z, J(y)) \le d(x, y)$. In a similar way one has that $d(w, J(x)) \le d(x, y)$ for every $w \in J(y)$ and so the claim is proved.

Applying Theorem 5.3.20 we obtain a selection $f \in \mathrm{BN}(X, Y)$ of J. Then $f \in \bigcap_{i \in I} B[f_i, r_i]$, which shows that $\mathrm{BN}(X, Y)$ is hyperconvex. $\qquad\square$

Next we show that the mapping Φ is nonexpansive.

Lemma 5.3.23 *Let* (X, d) *be a metric space,* $A \subseteq X$ *nonempty and* (Y, d') *a hyperconvex metric space. Then* Φ *is a nonexpansive multi-valued mapping.*

Proof Let $f, g \in \mathrm{BN}(A, Y)$. To conclude that $d_H(\Phi(f), \Phi(g)) \le d_\infty(f, g)$ it is enough to show that for every $F \in \Phi(f)$ there exists $G \in \Phi(g)$ such that $d_\infty(F, G) = d_\infty(f, g)$. To this end fix $F \in \Phi(f)$ and denote $r = d_\infty(f, g)$.

Let $\mathscr{U}$ be the collection of all pairs (D, g'), where $A \subseteq D \subseteq X$ and $g' \in \mathrm{BN}(D, Y)$ is an extension of g such that $\sup_{a \in D} d'(F(a), g'(a)) = r$. Since $(A, g) \in \mathscr{U}$, we have $\mathscr{U} \ne \emptyset$. We define a partial order on $\mathscr{U}$ by $(D_1, g'_1) \preceq (D_2, g'_2)$ if and only if $D_1 \subseteq D_2$ and $g'_2|_{D_1} = g'_1$. Then $(\mathscr{U}, \preceq)$ is inductively ordered, so using Zorn's lemma we find a maximal element (D, G) of $\mathcal{U}$. Suppose there exists $x \in X \setminus D$. Consider the intersection

$$Y_0 = \left(\bigcap_{a \in D} B[G(a), d(a, x)] \right) \cap B[F(x), r].$$

Since for any $a, a_1, a_2 \in D$,

$$d'(G(a_1), G(a_2)) \le d(a_1, a_2) \le d(a_1, x) + d(a_2, x)$$

and

$$d'(G(a), F(x)) \le d'(G(a), F(a)) + d'(F(a), F(x)) \le r + d(a, x),$$

by hyperconvexity of Y, it follows that $Y_0 \ne \emptyset$. Let G_x be the extension of G to $D \cup \{x\}$ by taking $G_x(x)$ any element in Y_0. Since $(D \cup \{x\}, G_x) \in \mathcal{U}$, (D, G) is not maximal. This contradiction shows that $D = X$.   □

To be able to apply Theorem 5.3.20 we show that the values of Φ are externally hyperconvex. This result improves [634, Theorem 17].

Lemma 5.3.24 *Let (X, d) be a metric space, $A \subseteq X$ nonempty and (Y, d') a hyperconvex metric space. Then $\Phi(f)$ is externally hyperconvex with respect to $\mathrm{BN}(X, Y)$ for every $f \in \mathrm{BN}(A, Y)$.*

Proof Let $f \in \mathrm{BN}(A, Y)$. According to Theorem 5.2.3, $\Phi(f)$ is nonempty. Consider $(f_i)_{i \in I} \subseteq \mathrm{BN}(X, Y)$ and $(r_i)_{i \in I} \subseteq \mathbb{R}_+$ such that $d_\infty(f_i, f_j) \le r_i + r_j$ and $d(f_i, \Phi(f)) = \inf_{F \in \Phi(f)} d_\infty(f_i, F) \le r_i$ for all $i, j \in I$. We need to prove that

$$\left(\bigcap_{i \in I} B[f_i, r_i] \right) \cap \Phi(f) \ne \emptyset.$$

We show that there exists an extension F of f in the above intersection. Let $\mathcal{U}$ be the collection of all pairs (D, f'), where $A \subseteq D \subseteq X$ and $f' \in \mathrm{BN}(D, Y)$ is an extension of f such that $\sup_{a \in D} d'(f_i(a), f'(a)) \le r_i$ for all $i \in I$. Since $d(f_i, \Phi(f)) \le r_i$ it is clear that $\sup_{a \in A} d'(f_i(a), f(a)) \le r_i$ for each $i \in I$. Thus, $(A, f) \in \mathcal{U}$ and so $\mathcal{U} \ne \emptyset$. We define a partial order on $\mathcal{U}$ by $(D_1, f_1') \preceq (D_2, f_2')$ if and only if $D_1 \subseteq D_2$ and $f_2'|_{D_1} = f_1'$. Then $(\mathcal{U}, \preceq)$ is inductively ordered, so using Zorn's lemma we find a maximal element (D, F) of $\mathcal{U}$. Suppose there exists $x \in X \setminus D$. Consider the intersection

$$Y_0 = \left(\bigcap_{a \in D} B[F(a), d(a, x)] \right) \cap \left(\bigcap_{i \in I} B[f_i(x), r_i] \right).$$

Since for any $a, a_1, a_2 \in D$ and any $i, j \in I$, $d'(f_i(x), f_j(x)) \le r_i + r_j$,

$$d'(F(a_1), F(a_2)) \le d(a_1, a_2) \le d(a_1, x) + d(a_2, x)$$

and

$$d'(F(a), f_i(x)) \le d'(F(a), f_i(a)) + d'(f_i(a), f_i(x)) \le r_i + d(a, x),$$

by hyperconvexity of Y, the set Y_0 is nonempty. Let F_x be the extension of F to $D \cup \{x\}$ by taking $F_x(x)$ any element in Y_0. Since $(D \cup \{x\}, F_x) \in \mathscr{U}$, (D, F) is not maximal. This contradiction shows that $D = X$ and we are done. $\qquad \square$

Remark 5.3.25 It does not seem that the approach applied to Φ in the hyperconvex case carries over to the mapping Ψ.

We can now give the first main result of this subsection.

Theorem 5.3.26 *Let X be a metric space, $A \subseteq X$ nonempty and Y a hyperconvex metric space. Then there exists a nonexpansive mapping α : $\mathrm{BN}(A, Y) \to \mathrm{BN}(X, Y)$ such that for all $g \in \mathrm{BN}(A, Y)$, $\alpha(g)(a) = g(a)$ for every $a \in A$.*

Proof This follows directly from Lemmas 5.3.23 and 5.3.24, and Theorems 5.3.20 and 5.3.22. $\qquad \square$

Remark 5.3.27 For $\lambda > 0$, denote by $\lambda(A, Y)$ the family of all bounded λ-Lipschitz mappings from A into Y. By applying Theorem 5.3.26 on $(X, \lambda d)$ instead of (X, d), we obtain this result also for $\lambda(A, Y)$ instead of $\mathrm{BN}(A, Y)$ and λ-Lipschitz extensions instead of nonexpansive extensions.

In particular, since complete $\mathbb{R}$-trees are hyperconvex spaces, we have the following corollary which improves the corresponding result from Sect. 5.3.1 (see Remark 5.3.13).

Corollary 5.3.28 *Let X be a metric space, $A \subseteq X$ nonempty and Y a complete $\mathbb{R}$-tree. Then the multi-valued mapping Φ admits a nonexpansive selection.*

It is now natural to wonder about results in the hyperconvex setting when considering as before a convexity condition on the images of the extensions. First, we need to clarify the notion of convex hull. A natural option in this case is to consider the admissible hull of a set (see Definition 5.1.14). Now we can define Φ_c as in Sect. 5.3.1 by replacing $\overline{\mathrm{co}}(f(A))$ with $\mathrm{cov}(f(A))$. In order to prove Theorem 5.3.26 for Φ_c in the hyperconvex setting, we only need to show that Lemmas 5.3.23 and 5.3.24 still hold true. This is indeed the case. We point out next how to modify the corresponding proofs.

Lemma 5.3.29 *Let (X, d) be a metric space, $A \subseteq X$ nonempty and (Y, d') a hyperconvex metric space. Then Φ_c is a nonexpansive multi-valued mapping with nonempty values.*

Proof First we need to show that $\Phi_c(f)$ is nonempty for $f \in \mathrm{BN}(A, Y)$ which directly follows from the fact that admissible subsets of hyperconvex spaces are hyperconvex themselves and so, by Theorem 5.2.3, an extension F of f exists such that $F \in \mathrm{BN}(X, \mathrm{cov}(f(A)))$. Given $g \in \mathrm{BN}(A, Y)$, we are looking for $G \in \Phi_c(g)$ such that $d_\infty(F, G) = d_\infty(f, g)$. We use the notation and the method from the proof of Lemma 5.3.23 taking now couples (D, g') with the additional property that $g' \in \mathrm{BN}(D, \mathrm{cov}(g(A)))$ and thus finding a maximal element (D, G) with $G \in \mathrm{BN}(D, \mathrm{cov}(g(A)))$ an extension of g such that $\sup_{a \in D} d'(F(a), G(a)) = r$. Consider now

$$Y_0 = \left(\bigcap_{a \in D} B[G(a), d(a, x)] \right) \cap \left(\bigcap_{y \in Y} B[y, r_y(g(A))] \right) \cap B[F(x), r].$$

To apply hyperconvexity, the only case which is not trivial is for pairs of balls centered at $y \in Y$ and at $F(x)$. But for this case, since $F \in \mathrm{BN}(X, \mathrm{cov}(f(A)))$, we have

$$d'(y, F(x)) \le r_y(f(A)) = \sup_{a \in A} d'(y, f(a))$$

$$\le \sup_{a \in A}(d'(y, g(a)) + d'(g(a), f(a))) \le r_y(g(A)) + r.$$

Therefore $Y_0 \ne \emptyset$. □

Lemma 5.3.30 *Let (X, d) be a metric space, $A \subseteq X$ nonempty and (Y, d') a hyperconvex metric space. Then $\Phi_c(f)$ is externally hyperconvex with respect to $\mathrm{BN}(X, Y)$ for every $f \in \mathrm{BN}(A, Y)$.*

Proof This proof follows the same patterns as the one of Lemma 5.3.24. Let $f \in \mathrm{BN}(A, Y)$, $(f_i)_{i \in I} \subseteq \mathrm{BN}(X, Y)$ and $(r_i)_{i \in I} \subseteq \mathbb{R}_+$ such that $d_\infty(f_i, f_j) \le r_i + r_j$ and $d(f_i, \Phi_c(f)) = \inf_{F \in \Phi_c(f)} d_\infty(f_i, F) \le r_i$ for all $i, j \in I$. We need to prove that

$$\left(\bigcap_{i \in I} B[f_i, r_i] \right) \cap \Phi_c(f) \ne \emptyset.$$

As in Lemma 5.3.24, one shows that there is an extension F of f in the above intersection. We briefly point out the changes in the argument. Now the maximal element (D, F) has the additional property that $F \in \mathrm{BN}(D, \mathrm{cov}(f(A)))$ and the set Y_0 to consider is given by

$$Y_0 = \left(\bigcap_{a \in D} B[F(a), d(a, x)] \right) \cap \left(\bigcap_{y \in Y} B[y, r_y(f(A))] \right) \cap \left(\bigcap_{i \in I} B[f_i(x), r_i] \right).$$

We need to check the hyperconvexity condition for $d'(f_i(x), y)$ with $i \in I$ and $y \in Y$. Recall that $d(f_i, \Phi_c(f)) \leq r_i$ and so, for $\varepsilon > 0$, there exists $z_i \in \text{cov}(f(A))$ such that $d'(f_i(x), z_i) \leq r_i + \varepsilon$. Therefore, for $y \in Y$ and $i \in I$,

$$d'(y, f_i(x)) \leq d'(y, z_i) + d'(z_i, f_i(x)) \leq r_y(f(A)) + r_i + \varepsilon.$$

The hyperconvexity condition follows now because $\varepsilon > 0$ is arbitrary.   □

Finally, we can state the second main result of this subsection.

Theorem 5.3.31 *Let X be a metric space, $A \subseteq X$ nonempty and Y a hyperconvex metric space. Then there exists a nonexpansive mapping $\alpha_c : \text{BN}(A, Y) \to \text{BN}(X, Y)$ such that for all $g \in \text{BN}(A, Y)$, $\alpha_c(g)(a) = g(a)$ for every $a \in A$ and $\alpha_c(g)(X) \subseteq \text{cov}(g(A))$.*

Remark 5.3.32 Again, by applying Theorem 5.3.31 on $(X, \lambda d)$ instead of (X, d), we have that this result holds in fact for $\lambda(A, Y)$ and λ-Lipschitz extensions.

Remark 5.3.33 Theorem 1 in [634] gives the same result as Theorem 5.3.20 but for multi-valued mappings with admissible values instead of externally hyperconvex subsets. As far as we know, Theorem 5.3.26 may be the first application of Theorem 5.3.20 where the external hyperconvexity condition plays a substantial role. In fact, values of the mapping Φ, which have been proved to be externally hyperconvex, need not be admissible. Consider, e.g., X as the real interval $[0, 2]$, $A \subseteq X$ as $[1, 2]$ and $Y = [0, 1]$. Define $f \in \text{BN}(A, Y)$ as the function constantly equal to 1. Then the functions $g(x) = 1$ for $x \in X$ and

$$h(x) = \begin{cases} x & \text{if } x \in [0, 1] \\ 1 & \text{if } x \in (1, 2] \end{cases}$$

are in $\Phi(f)$. Therefore, any ball in $\text{BN}(X, Y)$ containing $\Phi(f)$ must be of radius at least $1/2$ and, in particular, it must contain the function constantly equal to $3/4$ which is not in $\Phi(f)$.

Remark 5.3.34 In this subsection we approached the case Φ_c in a direct way and not going through the metric projection as in Sect. 5.3.1. In contrast to the case of complete $\text{CAT}(\kappa)$ spaces with $\kappa \leq 0$ where the metric projection onto nonempty closed and convex subsets is single-valued, the metric projection onto admissible subsets of hyperconvex metric spaces is in general multi-valued. However, as shown in [634] it admits a nonexpansive selection (it was later proved in [352] that the same holds for externally hyperconvex subsets). This problem was further studied and the interested reader can find more about it in [225, 227].

5.4 Dugundji Type Extension Results

In this section we give two results in connection with Dugundji's extension theorem (Theorem 1.2.12) for continuous and Lipschitz mappings defined on a closed subset of a metric space and taking values in a complete Busemann convex space. These results were obtained in [229] by applying the continuity properties of convex combinations discussed in Sect. 5.1.4. Note that, instead of complete Busemann convex spaces, we can actually consider mappings taking values in complete metric spaces with a geodesic bicombing (see also Remark 5.1.19).

5.4.1 Continuous Extensions

We prove first a counterpart of Dugundji's extension theorem for continuous mappings.

Theorem 5.4.1 *Let A be a nonempty closed subset of a metric space (X, d), (Y, d') a complete Busemann convex space and $f : A \to Y$ a continuous mapping. Then there exists a continuous extension $F : X \to Y$ of f to X such that $F(X) \subseteq \overline{co}(f(A))$.*

Proof We may assume that the open set $\Omega = X \setminus A$ is nonempty. Cover Ω by open balls $B_m = B(m, r_m)$ with $m \in \Omega$ and $r_m = d(m, A)/3$. Note that $B_m \subseteq \Omega$ for each $m \in \Omega$. Let $(p_i)_{i \in I}$ be a partition of unity subordinated to this cover. For each $i \in I$ pick two points $z_i \in \operatorname{spt}(p_i)$ and $y_i \in A$ such that $d(y_i, z_i) < 2d(z_i, A)$.

For $x \in \Omega$, denote $I_x = \{i \in I : p_i(x) \neq 0\}$ and define

$$F(x) = \begin{cases} f(x) & \text{if } x \in A \\ \dot{\oplus}_{i \in I_x} p_i(x) f(y_i) & \text{if } x \in \Omega. \end{cases}$$

Clearly, the mapping F extends f and is well-defined as I_x is always finite for each $x \in \Omega$. We claim that F is continuous. Obviously, it is continuous on the interior of A. We prove next the continuity on Ω. For $x \in \Omega$ take a neighborhood U of x and a finite set $J \subseteq I$ with $p_i|_U = 0$ for all $i \in I \setminus J$. Now, for $u \in U$, it is immediate that $I_u \subseteq J$ and so $\dot{\oplus}_{i \in J} p_i(u) f(y_i)$ is the same convex combination that defines $F(u)$ except, at most, a finite collection of points with zero coefficients which do not alter its value as explained in Sect. 5.1.4. Since $x \in U$, this also means that $F(x) = \dot{\oplus}_{i \in J} p_i(x) f(y_i)$. Therefore, Theorem 5.1.18 proves the continuity at x.

It only rests to prove that F is continuous on the boundary of A, ∂A. Let $m \in \partial A$ and U_m be an open ball in Y centered at $f(m)$. Take $\delta > 0$ such that $f(A \cap B(m, \delta)) \subseteq U_m$. We check that for $m' \in \Omega$,

$$d(m, m') < \frac{\delta}{6} \quad \text{implies} \quad F(m') \in U_m,$$

which, of course, proves the claim. Notice that $F(m')$ is a convex combination of points $f(y_i)$ with $i \in I_{m'}$, hence, by (5.1.10), it suffices to prove that $f(y_i) \in U_m$ for each $i \in I_{m'}$. To this end, given $i \in I_{m'}$, choose $m_i \in \Omega$ so that $\mathrm{spt}(p_i) \subseteq B_{m_i}$. Then $m' \in B_{m_i}$ and we have

$$d(m_i, A) \leq d(m_i, m) \leq d(m_i, m') + d(m', m) < \frac{1}{3}d(m_i, A) + \frac{\delta}{6}.$$

These inequalities imply that $d(m_i, A) < \delta/4$ and $d(m_i, m) < \delta/4$, from where

$$d(z_i, m) \leq d(z_i, m_i) + d(m_i, m) < \frac{1}{3}d(m_i, A) + \frac{\delta}{4} < \frac{\delta}{3}.$$

Finally,

$$d(y_i, m) \leq d(y_i, z_i) + d(z_i, m) < 2d(z_i, A) + \frac{\delta}{3} \leq 2d(z_i, m) + \frac{\delta}{3} < \delta.$$

This means that $y_i \in A \cap B(m, \delta)$, so $f(y_i) \in U_m$. Hence the continuity of F is proved. $\qquad\square$

Remark 5.4.2 The extension constructed in Theorem 5.4.1 acts in a nonexpansive way in the sense that for f and g continuous mappings from A to Y we have

$$d'(F(x), G(x)) = d'(\oplus_{i \in I_x} p_i(x)f(y_i), \oplus_{i \in I_x} p_i(x)g(y_i))$$

$$\leq \sum_{i \in I_x} p_i(x)d'(f(y_i), g(y_i)) \quad \text{by (5.1.9)}$$

$$\leq \sup_{x \in A} d'(f(x), g(x)),$$

for all $x \in \Omega$. Therefore,

$$\sup_{x \in X} d'(F(x), G(x)) \leq \sup_{x \in A} d'(f(x), g(x)).$$

Thus, the multi-valued extension operator which assigns to every bounded continuous mapping all its bounded continuous extensions has a selection that is nonexpansive with respect to the supremum distance.

5.4.2 Lipschitz Extensions

For a corresponding Lipschitz extension result we assume that either the domain of the mapping or its complement have finite Nagata dimension. The result that we will give is in fact a consequence of a general one proved by Lang and Schlichenmaier in

[378] for Lipschitz mappings whose target space satisfies a Lipschitz connectedness condition. Namely, a metric space Y is Lipschitz n-connected if there exists a constant C such that for every $m \in \{0, 1, \ldots, n\}$, every L-Lipschitz mapping from the unit sphere of $\mathbb{R}^{m+1}$ to Y admits a CL-Lipschitz extension to the closed unit ball of $\mathbb{R}^{m+1}$. This condition holds, in particular, in any metric space with a geodesic bicombing. Although less general, we prefer to consider here this latter assumption for the target space (dealing in fact with Busemann convexity) because in this situation one can directly define the desired extension via convex combinations, which results in a rather simple argument and shows as well that the extension acts in a nonexpansive way and its image belongs to the closed convex hull of the image of the original mapping.

The notion of Nagata dimension was introduced in [58, 513] and is defined as follows.

Definition 5.4.3 (Nagata Dimension) Let X be a metric space. The *Nagata dimension* of X is the least $n \in \mathbb{N}_0$ for which there exists a constant $c > 0$ such that for all $s > 0$, X has a cover $\mathscr{B} = (B_i)_{i \in I}$ with the property that diam $B_i \leq cs$ for every $i \in I$ and every subset of X of diameter at most s meets at most $n + 1$ members of $\mathscr{B}$.

The following technical lemma goes back to [378] and will be used in the proof of the extension result.

Lemma 5.4.4 *Let (X, d) be a metric space, $n \in \mathbb{N}_0$, $A \subseteq X$ nonempty and closed, and denote $\Omega = X \setminus A$. Suppose that Ω is nonempty and has Nagata dimension at most n with a constant c. Then one can find a cover $\mathscr{B} = (B_i)_{i \in I}$ of Ω by nonempty subsets of Ω and two numbers $\alpha, \beta > 0$ that only depend on c such that*

(i) *diam $B_i \leq \alpha d(B_i, A)$ for all $i \in I$;*
(ii) *every nonempty subset D of Ω with diam $D \leq \beta d(D, A)$ meets at most $n + 1$ members of $\mathscr{B}$.*

Proof Let $r = 3 + 2c$. For $i \in \mathbb{Z}$, consider the sets

$$R^i = \left\{ x \in \Omega : r^i \leq d(x, A) < r^{i+1} \right\},$$

which form a countable partition of Ω. Each set R^i has Nagata dimension at most n with the same constant c. Thus, we can choose a cover $\mathscr{C}^i = \left(C_k^i\right)_{k \in K_i}$ of R^i satisfying $C_k^i \subseteq R^i$, diam $C_k^i \leq cr^i$ and with the property that every subset of R^i of diameter at most r^i meets at most $n + 1$ members of $\mathscr{C}^i$.

We replace each family $\mathscr{C}^i$ by a new family $\mathscr{B}^i = \left(B_j^i\right)_{j \in J_i}$ defined as follows. The new index set J_i consists of all $k \in K_i$ for which there is no pair of points $x \in C_k^i$ and $y \in R^{i+1}$ with $d(x, y) \leq r^i$. For every $k \in K_{i-1} \setminus J_{i-1}$, we choose an index $j(i, k) \in K_i$ such that there exist $x_k^i \in C_k^{i-1}$ and $y_k^i \in C_{j(i,k)}^i$ with $d(x_k^i, y_k^i) \leq r^{i-1}$. We show next that $j(i, k) \in J_i$. To see this, let $x \in C_{j(i,k)}^i$ and $y \in R^{i+1}$.

Then

$$r^{i+1} \le d(y, A) \le d(y, x_k^i) + d(x_k^i, A) < d(y, x_k^i) + r^i,$$

so $d(y, x_k^i) > r^{i+1} - r^i$ and we have

$$d(x, y) \ge d(y, x_k^i) - d(x_k^i, y_k^i) - \operatorname{diam} C_{j(i,k)}^i > r^{i+1} - r^i - r^{i-1} - cr^i > r^i.$$

Set now for $j \in J_i$,

$$B_j^i = C_j^i \cup \left(\bigcup_{j(i,k)=j} C_k^{i-1} \right)$$

and let $\mathscr{B}^i = \left(B_j^i \right)_{j \in J_i}$ and $\mathscr{B} = \bigcup_{i \in \mathbb{Z}} \mathscr{B}^i$. Then $\mathscr{B}$ is a cover of Ω. Indeed, if $x \in \Omega$, then $x \in R^{i-1}$ for some $i \in \mathbb{Z}$ and so $x \in C_k^{i-1}$ for some $k \in K_{i-1}$. If $k \in J_{i-1}$, then $x \in B_k^{i-1}$. Otherwise, $x \in B_{j(i,k)}^i$.

We claim that every $D \subseteq R^{i-1} \cup R^i$ with $\operatorname{diam} D \le r^{i-1}$ meets at most $n + 1$ members of $\mathscr{B}^{i-1} \cup \mathscr{B}^i$. If $D \cap \left(\bigcup_{j \in J_i} C_j^i \right) = \emptyset$, then D meets at most $n + 1$ members of $\left(C_k^{i-1} \right)_{k \in K_{i-1}}$, each of which belongs to exactly one member in $\mathscr{B}^{i-1} \cup \mathscr{B}^i$. Now suppose $D \cap C_j^i \ne \emptyset$ for some $j \in J_i$. Then there exists $y \in D \cap R^i$ and if $x \in D \cap C_k^{i-1}$ for some $k \in K_{i-1}$, then $d(x, y) \le \operatorname{diam} D \le r^{i-1}$. Thus $k \in K_{i-1} \setminus J_{i-1}$ whenever $D \cap C_k^{i-1} \ne \emptyset$ and so D does not meet any member in $\mathscr{B}^{i-1}$. Consider the set

$$D' = D \cup \{y_k^i : D \cap C_k^{i-1} \ne \emptyset\}.$$

If $x \in D$ and $k \in K_{i-1} \setminus J_{i-1}$ with $D \cap C_k^{i-1} \ne \emptyset$,

$$d(x, y_k^i) \le d(y_k^i, x_k^i) + \operatorname{diam} C_k^{i-1} + \operatorname{diam} D \le (c + 2)r^{i-1} \le r^i.$$

Likewise, if $k_1, k_2 \in K_{i-1} \setminus J_{i-1}$ with $D \cap C_{k_1}^{i-1} \ne \emptyset$ and $D \cap C_{k_2}^{i-1} \ne \emptyset$,

$$d(y_{k_1}^i, y_{k_2}^i) \le d(y_{k_1}^i, x_{k_1}^i) + \operatorname{diam} C_{k_1}^{i-1} + \operatorname{diam} D + \operatorname{diam} C_{k_2}^{i-1} + d(x_{k_2}^i, y_{k_2}^i)$$

$$\le (3 + 2c)r^{i-1} = r^i.$$

Therefore, $\operatorname{diam} D' \le r^i$ and so D' meets at most $n + 1$ members of $\mathscr{C}^i$, which proves our claim.

Let $i \in \mathbb{Z}$ and $j \in J_i$. Obviously, if $x \in C_j^i$, then $d(x, A) \geq r^i$. If $x \in C_k^{i-1}$ for some $k \in K_{i-1} \setminus J_{i-1}$ with $j(i, k) = j$, then

$$d(x, A) \geq d(y_k^i, A) - d(x_k^i, y_k^i) - d(x_k^i, x) \geq r^i - (c + 1)r^{i-1}.$$

Thus, if $B_j^i \neq \emptyset$, $d(B_j^i, A) \geq r^i - (c + 1)r^{i-1}$. Using a case by case argument, one obtains

$$\text{diam } B_j^i \leq cr^i + 2(c + 1)r^{i-1} = \alpha(r^i - (c + 1)r^{i-1}) \leq \alpha d(B_j^i, A),$$

where $\alpha = 2c + 1$.

Denote $\beta = 1/r$ and take $D \subseteq \Omega$ nonempty with diam $D \leq \beta d(D, A)$. Let $i \in \mathbb{Z}$ such that $r^{i-1} \leq d(D, A) < r^i$. Then diam $D \leq d(D, A)/r < r^{i-1}$. Pick $z \in D \cap R^{i-1}$. Then for any $x \in D$,

$$d(x, A) \leq d(x, z) + d(z, A) < \text{diam } D + r^i < r^{i-1} + r^i < r^{i+1}.$$

Moreover, for $x \in D \cap R^i$, $d(x, R^{i-1}) \leq d(x, z) \leq \text{diam } D < r^{i-1}$. Hence,

$$D \subseteq R^{i-1} \cup \{x \in R^i : d(x, R^{i-1}) < r^{i-1}\}.$$

We show next that if D meets some $B \in \mathscr{B}$, then $B \in \mathscr{B}^{i-1} \cup \mathscr{B}^i$. Suppose on the contrary that there exists $x \in D \cap C_k^i$ for some $k \in K_i \setminus J_i$. Then $d(x, R^{i-1}) < r^{i-1}$ and

$$d(x, R^{i+1}) \leq d(x, x_k^{i+1}) + d(x_k^{i+1}, y_k^{i+1}) \leq \text{diam } C_k^i + r^i \leq (c + 1)r^i.$$

Thus,

$$r^{i+1} - r^i \leq d(R^{i-1}, R^{i+1}) \leq r^{i-1} + (c + 1)r^i,$$

a contradiction. This proves that $B \in \mathscr{B}^{i-1} \cup \mathscr{B}^i$ and so D meets at most $n + 1$ members of $\mathscr{B}$.

Finally, we can assume that $\mathscr{B}$ only consists of nonempty sets.   □

Remark 5.4.5 A cover of Ω with similar properties can also be constructed under the assumption that A has finite Nagata dimension instead of Ω. This case is detailed in [378].

Theorem 5.4.6 *Let $n \in \mathbb{N}_0$ and A be a nonempty and closed subset of a metric space (X, d) such that $X \setminus A$ has Nagata dimension at most n with a constant c. Suppose (Y, d') is a complete Busemann convex space and $f : A \to Y$ an L-Lipschitz mapping. Then there exists a CL-Lipschitz extension $F : X \to Y$ of f to X such that $F(X) \subseteq \overline{co}(f(A))$, where $C \geq 1$ is a constant that depends only on n and c.*

Proof We may assume that the open set $\Omega = X \setminus A$ is nonempty. Apply Lemma 5.4.4 to obtain a cover $(B_i)_{i \in I}$ of Ω by nonempty subsets of Ω and two numbers $\alpha, \beta > 0$ depending only on c such that (i) and (ii) hold. Take $\delta = \beta/(2(\beta+1)) \in (0, 1/2)$ and define the family $(\sigma_i)_{i \in I}$ of nonexpansive functions $\sigma_i : \Omega \to [0, \infty)$ by

$$\sigma_i(x) = \max\{0, \delta d(A, B_i) - d(x, B_i)\}.$$

Let $x \in \Omega$. Note first that if $x \in B_{i_0}$ for some $i_0 \in I$, then $\sigma_{i_0}(x) > 0$. To see this, suppose $d(A, B_{i_0}) = 0$. By Lemma 5.4.4.(i), diam $B_{i_0} = 0$, so $d(x, A) = 0$, which contradicts the assumption that $x \in \Omega$. This shows that $\sigma_{i_0}(x) > 0$. As $(B_i)_{i \in I}$ is a cover of Ω, $\sigma_i(x) > 0$ for at least one index $i \in I$. We claim that we have at most $n + 1$ such indices. Indeed, for every $i \in I$ with $\sigma_i(x) > 0$, $d(x, B_i) < \delta d(A, B_i)$ and we can pick $x_i \in B_i$ such that $d(x, x_i) < \delta d(A, B_i) \leq \delta d(x_i, A)$. Let D be the set of all these x_i selected before. Then

$$\text{diam } D \leq 2\delta \sup_i d(x_i, A) \leq 2\delta(\text{diam } D + d(D, A)),$$

so diam $D \leq \beta d(D, A)$. By Lemma 5.4.4.(ii), D meets at most $n + 1$ members of $(B_i)_{i \in I}$ and the claim is proved.

Denoting $\overline{\sigma} = \sum_{i \in I} \sigma_i$, one considers the family $(\varphi_i)_{i \in I}$ of functions $\varphi_i : \Omega \to [0, 1]$ given by

$$\varphi_i(x) = \frac{\sigma_i(x)}{\overline{\sigma}(x)}.$$

Clearly, $\sum_{i \in I} \varphi_i = 1$.

For each $i \in I$, let $y_i \in A$ such that $d(y_i, B_i) \leq (2 - \delta)d(A, B_i)$. This point exists since $2 - \delta > 1$. For $x \in \Omega$, denote $I_x = \{i \in I : \varphi_i(x) \neq 0\}$ and define

$$F(x) = \begin{cases} f(x) & \text{if } x \in A \\ \dot{\bigoplus}_{i \in I_x} \varphi_i(x) f(y_i) & \text{if } x \in \Omega. \end{cases}$$

The mapping F extends f and is well-defined because, by the above, I_x is always finite for each $x \in \Omega$. If $x \in \Omega$ and $i \in I_x$, then $\sigma_i(x) > 0$, so $d(x, B_i) < \delta d(A, B_i)$ and

$$d(x, y_i) \leq d(x, B_i) + \text{diam } B_i + d(y_i, B_i)$$

$$< \delta d(A, B_i) + \alpha d(A, B_i) + (2 - \delta)d(A, B_i) = (\alpha + 2)d(A, B_i).$$

$$\tag{5.4.1}$$

We show next that F is CL-Lipschitz. Note that in the sequel $C \geq 1$ stands for any constant that depends solely on n and c.

Obviously, F is L-Lipschitz on A. Now, for $b \in \Omega$ and $a \in A$, we have

$$d'(F(a), F(b)) = d'(f(a), \dot{\oplus}_{i \in I_b} \varphi_i(b) f(y_i))$$

$$\leq \sum_{i \in I_b} \varphi_i(b) d'(f(a), f(y_i)) \quad \text{by (5.1.10)}$$

$$\leq \max_{i \in I_b} d'(f(a), f(y_i)) \leq L \max_{i \in I_b} d(a, y_i).$$

If $i \in I_b$, then $d(A, B_i) \leq d(a, B_i) \leq d(a, b) + d(b, B_i) < d(a, b) + \delta d(A, B_i)$ and this yields $d(A, B_i) < (1 - \delta)^{-1} d(a, b)$. Using (5.4.1), we then get

$$d(a, y_i) \leq d(a, b) + d(b, y_i) < d(a, b) + (\alpha+2)d(A, B_i) < \left(1 + \frac{\alpha+2}{1-\delta}\right) d(a, b).$$

Thus, $d'(F(a), F(b)) \leq CL\, d(a, b)$.

Let next $a, b \in \Omega$. Suppose first that there exists $j \in I_a \cap I_b$. The set $I_a \cup I_b$ contains at most $2n + 1$ indices $i \in I$. Moreover, $F(a) = \dot{\oplus}_{i \in I_a \cup I_b} \varphi_i(a) f(y_i)$ and $F(b) = \dot{\oplus}_{i \in I_a \cup I_b} \varphi_i(b) f(y_i)$ since we are not adding more than a finite collection of points with zero coefficients which do not change a convex combination. Applying Proposition 5.1.17 we obtain

$$d'(F(a), F(b)) \leq C \max_{i,k \in I_a \cup I_b} d'(f(y_i), f(y_k)) \sum_{i \in I_a \cup I_b} |\varphi_i(a) - \varphi_i(b)|.$$

Note that

$$\max_{i,k \in I_a \cup I_b} d'(f(y_i), f(y_k)) \leq L \max_{i,k \in I_a \cup I_b} d(y_i, y_k) \leq 2L \max_{i \in I_a \cup I_b} d(y_i, y_j).$$

Let $i \in I_a$. Since

$$d(A, B_j) \leq d(y_i, B_j) \leq d(y_i, a) + d(a, B_j)$$

$$< (\alpha + 2)d(A, B_i) + \delta d(A, B_j) \quad \text{by (5.4.1),}$$

it follows that

$$d(A, B_j) \leq \frac{\alpha+2}{1-\delta} d(A, B_i).$$

In a similar way we have $d(A, B_i) \leq (\alpha + 2)(1 - \delta)^{-1} d(A, B_j)$. Then, again by (5.4.1),

$$d(y_i, y_j) \leq d(y_i, a) + d(a, y_j) \leq (\alpha + 2) \left(d(A, B_i) + d(A, B_j) \right)$$

$$\leq (\alpha + 2) \left(1 + \frac{\alpha + 2}{1 - \delta} \right) d(A, B_j).$$

Note that the above inequality can also be proved by the same argument in the case $i \in I_b$. Furthermore, for every $i \in I_a \cup I_b$,

$$|\varphi_i(a) - \varphi_i(b)| = \left| \frac{\sigma_i(a)}{\overline{\sigma}(a)} - \frac{\sigma_i(b)}{\overline{\sigma}(b)} \right| \leq \left| \frac{\sigma_i(a)}{\overline{\sigma}(a)} - \frac{\sigma_i(b)}{\overline{\sigma}(a)} \right| + \left| \frac{\sigma_i(b)}{\overline{\sigma}(a)} - \frac{\sigma_i(b)}{\overline{\sigma}(b)} \right|$$

$$= \frac{1}{\overline{\sigma}(a)} \left(|\sigma_i(a) - \sigma_i(b)| + \frac{\sigma_i(b)}{\overline{\sigma}(b)} |\overline{\sigma}(a) - \overline{\sigma}(b)| \right)$$

$$\leq \frac{1}{\overline{\sigma}(a)} \left(|\sigma_i(a) - \sigma_i(b)| + |\overline{\sigma}(a) - \overline{\sigma}(b)| \right).$$

Since there are at most $2n + 1$ indices in $I_a \cup I_b$, it follows that

$$|\varphi_i(a) - \varphi_i(b)| \leq \frac{2n + 2}{\overline{\sigma}(a)} d(a, b).$$

Thus,

$$d'(F(a), F(b)) \leq C L \, d(A, B_j) \sum_{i \in I_a \cup I_b} \frac{d(a, b)}{\overline{\sigma}(a)}.$$

Let $m \in I$ such that $a \in B_m$. Then $m \in I_a$ and

$$\overline{\sigma}(a) \geq \sigma_m(a) = \delta d(A, B_m) \geq \delta \frac{1 - \delta}{\alpha + 2} d(A, B_j).$$

This shows that $d'(F(a), F(b)) \leq C L \, d(a, b)$.

Suppose now $I_a \cap I_b = \emptyset$. Since $a \in B_m$ for some $m \in I_a$, we then have that $\sigma_m(b) = 0$, which yields $d(A, B_m) \leq \delta^{-1} d(b, B_m) \leq \delta^{-1} d(b, a)$. Applying (5.4.1),

$$d(a, y_m) \leq (\alpha + 2) d(A, B_m) \leq \frac{\alpha + 2}{\delta} d(a, b). \tag{5.4.2}$$

Hence,

$$d'(F(a), F(b)) \leq d'(F(a), F(y_m)) + d'(F(y_m), F(b))$$
$$\leq CL\,(d(a, y_m) + d(y_m, b))$$
$$\leq CL\,(2d(a, y_m) + d(a, b)) \leq CL\,d(a, b),$$

where the last inequality follows using (5.4.2) and adjusting the constant C.   □

Remark 5.4.7 A corresponding extension result can be proved in the same way when A has finite Nagata dimension (see also Remark 5.4.5).

Remark 5.4.8 As in the continuous case, the extension constructed in Theorem 5.4.6 acts in a nonexpansive way. Considering the multi-valued extension operator which assigns to every bounded L-Lipschitz mapping all its bounded CL-Lipschitz extensions, one can find a nonexpansive selection of it with respect to the supremum distance.

5.5   Bibliographic Comments and Miscellaneous Results

Section 5.1 merely covers some basic notions and properties regarding geodesic spaces that are used in the further exposition. Here we considered spaces having globally lower or upper curvature bounds, but these conditions can also be defined locally in the following way: a metric space has locally curvature bounded above (resp. below) by $\kappa \in \mathbb{R}$ if every point in it has a neighborhood that, endowed with the induced metric, is a CAT(κ) space (resp. a CBB(κ) space). However, there are very important local-to-global results which state that, under some conditions, local curvature bounds imply global ones. In addition, more general definitions can be considered in length spaces. Note that many classical results from Riemannian geometry are generalized to the setting of Alexandrov spaces. Besides the monographs mentioned at the beginning of this chapter, which contain in detail the material on Alexandrov spaces that we included here, we also refer the interested reader to the synthesis paper of Burago et al. [131], to Plaut's chapter [576], or to the lecture notes by Ballmann [66].

A more detailed discussion on hyperconvex metric spaces can be found in [226, 227]. A very interesting construction in the theory of hyperconvex metric spaces is Isbell's [296] injective hull (also known as hyperconvex hull or tight span). An injective hull for a metric space X is a pair $(e, E(X))$, where $E(X)$ is an injective metric space and $e : X \to E(X)$ is an isometric embedding such that no proper subspace of $E(X)$ containing $e(X)$ is injective. Each metric space X has an injective hull that is unique in the sense that if (e_1, E_1) and (e_2, E_2) are two injective hulls for X, then there exists an isometry $i : E_1 \to E_2$ with the property that $i \circ e_1 = e_2$. Injective hulls find applications in discrete mathematics and are a wide-spread tool in phylogenetics (see [204, 205]). Recently, injective hulls were further studied in [375] in connection to some discrete metric spaces and groups.

Although we defined the concept of Gromov hyperbolicity in geodesic spaces, one can also consider it in a metric space (X, d) using, among others, the so-called δ-inequality given for $\delta \geq 0$ by

$$(x \mid y)_u \geq \min\{(x \mid z)_u, (z \mid y)_u\} - \delta,$$

for all $x, y, z, u \in X$. Equivalently,

$$d(x, y) + d(z, u) \leq \max\{d(x, z) + d(y, u), d(x, u) + d(y, z)\} + 2\delta,$$

for all $x, y, z, u \in X$. Every geodesic space that is δ-hyperbolic (in the sense of Definition 5.1.16) satisfies the δ-inequality and, conversely, if a geodesic space satisfies the δ-inequality, then it is 4δ-hyperbolic (see [134]). Note also that any metric space that satisfies the δ-inequality can be isometrically embedded into a complete geodesic space where the δ-inequality holds, a result which was proved by Bonk and Schramm [97].

The equivalences from Theorem 5.2.3 are well-known and already appeared in part in [52] (see also [75]).

Another previous result on the generalization of Kirszbraun's theorem to metric spaces was obtained by Kuczumow and Stachura [368] in the Hilbert ball.

As shown in [374], Theorem 5.2.10 can be generalized to obtain extensions of large-scale Lipschitz mappings with values in Cartesian products of finitely many δ-hyperbolic geodesic spaces, but no such extension exists in general if the target space is an infinite-dimensional Hilbert space. Furthermore, a necessary condition for the extendability of a nonexpansive mapping from a subset of a separable metric space to a proper metric space is given in [374]. If the target space is in addition Busemann convex, then this condition is actually sufficient.

The Nagata dimension was first considered in the context of the Lipschitz extension problem by Lang and Schlichenmaier in [378], where it was proved, in particular, that the pair formed by a metric space of Nagata dimension $\leq n$ and a complete Lipschitz $(n - 1)$-connected metric space has the generalized Lipschitz extension property as given in Definition 4.2.8. Several other properties of the Nagata dimension were also investigated. A characterization of Nagata dimension in terms of the Lipschitz extension property is given in [118]. Spaces of finite Nagata dimension include important classes of metric spaces such as doubling spaces, Gromov hyperbolic spaces that are doubling in the small, $\mathbb{R}$-trees, Euclidean buildings as well as homogenous Hadamard manifolds (see [378]). Doubling metric spaces and doubling measures constitute the appropriate framework for the development of analysis in nonsmooth spaces. An excellent introduction to the needed tools and main topics in analysis in metric spaces can be found in the lecture notes by Heinonen [283, 284], or Ambrosio and Tilli [38].

The proof of Theorem 5.4.1 follows standard patterns (for a proof of a related extension result for mappings taking values in a Banach space, see, e.g., [126,

Theorem 1.8]). A corresponding result of Theorem 5.4.6 for Banach-valued Lipschitz mappings defined on subsets of doubling metric spaces can be found in [285, Theorem 4.1.21].

Other highly important and deep Lipschitz extension results are mentioned in Sect. 4.5.

Chapter 6
Approximations Involving Lipschitz Functions

In this chapter we study the problem of the uniform approximation of some classes of functions (e.g. uniformly continuous) by Lipschitz functions, based on the existence of Lipschitz partitions of unity or on some extension results for Lipschitz functions. A result due to Baire on the approximation of semi-continuous functions by continuous ones, based on McShane's extension method, is also included. The chapter ends with a study of homotopy of Lipschitz functions and a brief presentation of Lipschitz manifolds.

6.1 Uniform Approximation via the Stone–Weierstrass Theorem

The first result is of Stone–Weierstrass type.

Theorem 6.1.1 *Let* (X, d) *be a compact metric space. Then, for each continuous function* $f : X \to \mathbb{K}$, *there exists a sequence* $f_n : X \to \mathbb{K}$, $n \in \mathbb{N}$, *of Lipschitz functions such that* $f_n \xrightarrow{u} f$.

Proof Let us show that the subalgebra $\mathscr{A} := \mathrm{Lip}(X, \mathbb{R})$ of $C(X, \mathbb{R})$ satisfies the hypotheses of the Stone–Weierstrass theorem, Theorem 1.4.39.

Condition (i) of the mentioned theorem holds because for each pair of distinct points $x, y \in X$, the function $g : X \to \mathbb{R}$ given by $g(z) = d(z, x)$, $z \in X$, is 1-Lipschitz and $g(x) = 0 < d(y, x) = g(y)$. Condition (ii) is obviously satisfied.

In the complex case the condition $f \in \mathscr{A} \Rightarrow \bar{f} \in \mathscr{A}$ is obviously verified, so we can apply Theorem 1.4.40. $\qquad\qquad\square$

Remark 6.1.2 In the real case, by Proposition 2.3.9, $\mathrm{Lip}(X, \mathbb{R})$ is a sublattice of $C(X, \mathbb{R})$, so we can also apply Theorem 1.4.37.

© Springer Nature Switzerland AG 2019
Ş. Cobzaş et al., *Lipschitz Functions*, Lecture Notes in Mathematics 2241,
https://doi.org/10.1007/978-3-030-16489-8_6

6.2   Approximation via Locally Lipschitz Partitions of Unity

In this section we prove, by using the Lipschitz partition of unity, that each continuous function can be uniformly approximated by Lipschitz functions.

Theorem 6.2.1 *Let X be a metric space and B a convex subset of a real normed space Y. Then, for each continuous function $f : X \to B$, there exists a sequence $f_n : X \to B$, $n \in \mathbb{N}$, of locally Lipschitz functions such that $f_n \overset{u}{\to} f$.*

Proof Since f is continuous, for each $x \in X$ and each $n \in \mathbb{N}$ there exists an open neighborhood $V_{x,n}$ of x such that

$$\operatorname{diam} f(V_{x,n}) < \frac{1}{n}. \tag{6.2.1}$$

According to Theorem 2.6.2, for each $n \in \mathbb{N}$, there exists a Lipschitz partition of unity $f_{x,n} : X \to [0, 1]$, $x \in X$, such that the supports $\{\operatorname{spt}(f_{x,n}) : x \in X\}$ form a locally finite family and

$$\operatorname{spt}(f_{x,n}) \subseteq V_{x,n} \quad \text{for all } x \in X. \tag{6.2.2}$$

For each $x \in X$ and $n \in \mathbb{N}$, pick $y_{x,n} \in V_{x,n}$.

For each $n \in \mathbb{N}$ consider the function $f_n : X \to Y$ given by

$$f_n(y) = \sum_{x \in X} f_{x,n}(y) f(y_{x,n}), \quad y \in X.$$

As $f_{x,n}(y) \geq 0$ and $\sum_{x \in X} f_{x,n}(y) = 1$ for each $y \in X$, $f(X) \subseteq B$ and B is convex, we infer that

$$f_n(X) \subseteq B,$$

for each $n \in \mathbb{N}$.

We show that f_n is locally Lipschitz for every $n \in \mathbb{N}$. Let $n \in \mathbb{N}$ and $x_0 \in X$ be fixed.

Since the family $\{\operatorname{spt}(f_{x,n})\}_{x \in X}$ is locally finite there exists a neighborhood V of x_0 meeting only finitely many of these supports, say $\operatorname{spt}(f_{x_1,n}), \dots, \operatorname{spt}(f_{x_k,n})$. Then for every $x \in X \setminus \{x_1, \dots, x_k\}$, $f_{x,n}(y) = 0$ for all $y \in V$, hence

$$f_n(y) = \sum_{i=1}^{k} f_{x_i,n}(y) f(y_{x_i,n}) \quad \text{for all } y \in V.$$

Since the functions $f_{x_i,n}$ are locally Lipschitz there exists a neighborhood $U \subseteq V$ of x_0 such that the function $f_{x_i,n}$ is Lipschitz on U for $i = 1, \dots, k$. But then, f_n is also Lipschitz on U as a finite sum of Lipschitz functions (see Proposition 2.3.3).

Now, let $y \in X$ be arbitrary, but fixed. Let the neighborhood V of y and $x_1, \ldots, x_k$ be chosen as above. It follows that $f_{x,n}(z) = 0$ for all $z \in V$ and $x \in X \setminus \{x_1, \ldots, x_k\}$, so that

$$f_n(y) = \sum_{i=1}^{k} f_{x_i,n}(y) f(y_{x_i,n}). \tag{6.2.3}$$

By (6.2.2) and (6.2.1),

$$f_{x_i,n}(y) \neq 0 \;\Rightarrow\; y \in \mathrm{spt}(f_{x_i,n}) \subseteq V_{x_i,n}$$
$$\Rightarrow\; \|f(y) - f(y_{x_i,n})\| < 1/n. \tag{6.2.4}$$

The equality $\sum_{i=1}^{k} f_{x_i,n}(y) = 1$ with $f_{x_i,n}(y) \geq 0$, (6.2.3) and (6.2.4) yield

$$\|f_n(y) - f(y)\| = \Big\| \sum_{i=1}^{k} f_{x_i,n}(y)\big(f(y_{x_i,n}) - f(y)\big) \Big\|$$

$$\leq \sum_{i=1}^{k} f_{x_i,n}(y) \|f(y_{x_i,n}) - f(y))\| < \frac{1}{n} \cdot \sum_{i=1}^{k} f_{x_i,n}(y) = \frac{1}{n}.$$

Since y was arbitrarily chosen in X, it follows that

$$\|f_n(y) - f(y)\| < \frac{1}{n},$$

for all $y \in X$, and so $\|f_n - f\|_\infty \leq 1/n$ for all $n \in \mathbb{N}$.
Consequently, $f_n \overset{u}{\to} f$. $\qquad\qquad\qquad\qquad\qquad\qquad\qquad\qquad\qquad\qquad\quad\square$

Since, by Theorem 2.1.6, every locally Lipschitz function on a compact metric space is Lipschitz, Theorem 6.2.1 has the following corollary.

Corollary 6.2.2 *Let X be a compact metric space and B a convex subset of a real normed space Y. Then, for each continuous function $f : X \to B$, there exists a sequence $f_n : X \to B$, $n \in \mathbb{N}$, of Lipschitz functions such that $f_n \overset{u}{\to} f$.*

Remark 6.2.3 The above result was proved by Georganopoulos [250]. Taking $B = Y = \mathbb{R}$ with the usual norm, one obtains Theorem 6.1.1. Moreover, for the case $B = Y = \mathbb{R}$, the above result can be proved using Theorem 2.7.1. More precisely, for each $n \in \mathbb{N}$ there exists a locally Lipschitz function $f_n : X \to \mathbb{R}$ such that

$$f - \frac{1}{n} < f_n < f + \frac{1}{n}.$$

Therefore $f_n \overset{u}{\to} f$.

Remark 6.2.4 There exist continuous functions which are not uniform limits of sequences of Lipschitz functions. For example, the continuous function $f : \mathbb{R} \to \mathbb{R}$ given by

$$f(x) = e^x, \quad x \in \mathbb{R},$$

is not the uniform limit of any sequence of Lipschitz functions.

Indeed, let us suppose, on the contrary, that there exists a sequence of Lipschitz functions $f_n : \mathbb{R} \to \mathbb{R}$, $n \in \mathbb{N}$, such that $f_n \overset{u}{\to} f$. Then there is $n_0 \in \mathbb{N}$ such that

$$\left| f_{n_0}(x) - f(x) \right| < \frac{1}{2}$$

for all $x \in \mathbb{R}$. But

$$|f(x) - f(y)| \le \left| f_{n_0}(x) - f(x) \right| + \left| f_{n_0}(x) - f_{n_0}(y) \right| + \left| f_{n_0}(y) - f(y) \right|,$$

and so

$$\left| e^x - e^y \right| < 1 + L(f_{n_0}) \, |x - y|,$$

for all $x, y \in \mathbb{R}$. In particular, choosing $x = n + 1$ and $y = n$, for $n \in \mathbb{N}$, we get the following contradiction:

$$e^n(e - 1) < 1 + L(f_{n_0}),$$

for all $n \in \mathbb{N}$.

6.3 Approximation via Lipschitz Extensions

In this section we present a different technique to obtain approximation results for functions between pairs of metric spaces having the generalized Lipschitz extension property (see Definition 4.2.8).

Theorem 6.3.1 *Let $((X, d), (Y, d'))$ be a pair of metric spaces which has the generalized Lipschitz extension property. Then for every bounded uniformly continuous function $f : X \to Y$ and every $\varepsilon > 0$ there exists a Lipschitz function $F_\varepsilon : X \to Y$ such that*

$$\sup_{x \in X} d'(f(x), F_\varepsilon(x)) < \varepsilon.$$

Proof Let

$$\gamma := \text{diam } f(X).$$

If $\gamma = 0$, then f is a constant function and the result is trivial.
Suppose $\gamma > 0$. For $0 < \varepsilon < \gamma$ put

$$\varepsilon' = (\alpha + 1)^{-1}\varepsilon,$$

where $\alpha \geq 1$ is the number given by Definition 4.2.8.
  By the uniform continuity of the function f, there exists $\delta > 0$ such that

$$d(x, y) < \delta \implies d'(f(x), f(y)) < \varepsilon', \tag{6.3.1}$$

for all $x, y \in X$.
  Let

$$0 < \eta := \min\{\delta, \gamma^{-1}\delta\varepsilon'\}.$$

Observe that

$$\eta \leq \gamma^{-1}\delta\varepsilon' \iff \frac{\gamma}{\delta} \leq \frac{\varepsilon'}{\eta}. \tag{6.3.2}$$

Consider the family $\mathcal{M}_\eta$ of all nonempty subsets A of X such that $d(x, y) \geq \eta$ for every pair x, y of distinct elements in A. Since $\varepsilon' \leq \varepsilon < \gamma$, there exist two points $x, y \in X$ with $d'(f(x), f(y)) > \varepsilon'$, implying $d(x, y) \geq \delta \geq \eta$, i.e., $\{x, y\} \in \mathcal{M}_\eta$, that is, $\mathcal{M}_\eta \neq \emptyset$. Ordering $\mathcal{M}_\eta$ by inclusion and applying Zorn's lemma, the existence of a maximal element S of $\mathcal{M}_\eta$ follows.
  The set S satisfies the conditions

$$\begin{array}{ll} \text{(i)} \quad d(x, y) \geq \eta \text{ for all } x, y \in S \text{ with } x \neq y, \text{ and} \\ \text{(ii)} \quad \forall x \in X \setminus S, \; \exists y_x \in S \text{ such that } d(x, y_x) < \eta. \end{array} \tag{6.3.3}$$

Claim. The function f is $\frac{\varepsilon'}{\eta}$-Lipschitz on S.

Let $x, y \in S$. If $\eta \leq d(x, y) < \delta$, then, by (6.3.1),

$$d'(f(x), f(y)) < \varepsilon' = \frac{\varepsilon'}{\eta}\eta \leq \frac{\varepsilon'}{\eta}d(x, y).$$

If $d(x, y) \geq \delta$, then, by (6.3.2),

$$d'(f(x), f(y)) \leq \gamma = \frac{\gamma}{\delta}\delta \leq \frac{\gamma}{\delta}d(x, y) \leq \frac{\varepsilon'}{\eta}d(x, y).$$

Let $F : X \to Y$ be an extension of $f|_S$ such that

$$d'(F(x), F(y)) \leq \alpha \frac{\varepsilon'}{\eta} d(x, y),$$

for all $x, y \in X$, and let us show that

$$d'(f(x), F(x)) < \varepsilon,$$

for all $x \in X$.

If $x \in S$, then $F(x) = f(x)$ and so $d'(f(x), F(x)) = 0 < \varepsilon$.

For $x \in X \setminus S$ choose $y_x \in S$ according to (6.3.3) (ii). Then

$$d'(f(x), F(x)) \leq d'(f(x), f(y_x)) + d'(f(y_x), F(x))$$

$$< \varepsilon' + \alpha \frac{\varepsilon'}{\eta} d(x, y_x) < \varepsilon' + \alpha \varepsilon' = \varepsilon.$$

□

Since, for any metric space X, the pair $(X, \mathbb{K})$ has the generalized extension property, one obtains the following corollary.

Corollary 6.3.2 *Let (X, d) be a metric space. Then every bounded uniformly continuous function $f : X \to \mathbb{K}$ can be uniformly approximated by Lipschitz functions.*

6.4 Baire's Theorem on the Approximation of Semicontinuous Functions

In this section we prove an approximation theorem for semicontinuous functions on metric spaces (see Definition 1.2.14) by continuous ones. We include it here because the method of proof relies on the extension formulae (4.1.1) for Lipschitz functions.

6.4.1 Baire's Function

In the proof we shall use the following function considered by Baire (1905). This function is used to reduce the treatment of functions taking values in $\overline{\mathbb{R}} = [-\infty, \infty]$ to those with values in $[-1, 1]$.

Define $\phi : [-\infty, \infty] \to [-1, 1]$ by

$$\phi(t) = \begin{cases} -1 & t = -\infty \\ \dfrac{t}{1 + |t|} & -\infty < t < \infty \\ 1 & t = \infty. \end{cases} \tag{6.4.1}$$

It follows that ϕ is a bijection between $[-\infty, \infty]$ and $[-1, 1]$ with inverse

$$\phi^{-1}(u) = \begin{cases} -\infty & u = -1 \\ \dfrac{u}{1 - |u|} & -1 < u < 1 \\ +\infty & u = 1. \end{cases}$$

Their derivatives are

$$\phi'(t) = \frac{1}{(1 + |t|)^2}, \qquad t \in \mathbb{R},$$

and

$$(\phi^{-1})'(u) = \frac{1}{(1 - |u|)^2}, \qquad u \in (-1, 1),$$

respectively.

The functions ϕ and ϕ^{-1} are strictly increasing and continuous, and the function ϕ satisfies the Lipschitz condition on $\mathbb{R}$ with $L = 1$,

$$|\phi(t) - \phi(t')| \le |t - t'|, \quad \text{for all } t, t' \in \mathbb{R}.$$

Using the function ϕ one can define a metric on $\overline{\mathbb{R}}$ by

$$\varrho(t, t') = |\phi(t) - \phi(t')|, \quad \text{for } t, t' \in \overline{\mathbb{R}}.$$

Since ϕ is a homeomorphism between $\mathbb{R}$ and $(-1, 1)$ it follows that ϱ is topologically equivalent to $|\cdot|$ on $\mathbb{R}$. Although homeomorphic to $(\mathbb{R}, |\cdot|)$ the space $(\mathbb{R}, \varrho)$ is not complete: the sequence $t_n = n$, $n \in \mathbb{N}$, is Cauchy with respect to ϱ but has no limit in $\mathbb{R}$, so that these metrics are not uniformly equivalent (and so nor are they Lipschitz equivalent).

In fact it is easy to check that the inverse function ϕ^{-1} is not uniformly continuous. Indeed, taking $u_n = 1 - n^{-1}$ and $v_n = 1 - n^{-2}$, $n \in \mathbb{N}$, it follows that $u_n - v_n \to 0$, while $|\phi^{-1}(u_n) - \phi^{-1}(v_n)| = n^2 - n \to \infty$ for $n \to \infty$.

6.4.2   Baire's Theorem

The following result shows that the semicontinuous functions are limits of monotone sequences of continuous functions. Although the result is valid for more general topological spaces (see [223]) we restrict the presentation to metric spaces, where a simple proof is available.

Theorem 6.4.1 *Let* (X, d) *be a metric space.*

1. *Any lsc function* $f : X \to \mathbb{R} \cup \{\infty\}$ *is the pointwise limit of a nondecreasing sequence of continuous functions.*
2. *If, in addition, the function* f *is bounded from below, then the functions*

$$f_n(x) = \inf_{y \in X}[f(y) + nd(x, y)], \quad x \in X, \ n \in \mathbb{N}, \tag{6.4.2}$$

   satisfy the conditions

   (i) f_n *is n-Lipschitz;*

   (ii) $f_n \le f_{n+1}$ *and* $f_n \le f$ *for all* $n \in \mathbb{N}$; $\qquad\qquad$ (6.4.3)

   (iii) $\lim_{n \to \infty} f_n(x) = f(x)$ *for every* $x \in X$.

3. *If the function* f *is bounded and uniformly continuous, then the sequence* (f_n) *of Lipschitz functions given by* (6.4.2) *converges to* f *uniformly on* X.

Similar results hold for usc functions.

1'. *Any usc function* $g : X \to \mathbb{R} \cup \{-\infty\}$ *is the pointwise limit of a nonincreasing sequence of continuous functions.*
2'. *If, in addition, the function* g *is bounded from above, then the functions*

$$g_n(x) = \sup_{y \in X}[g(y) - nd(x, y)], \quad x \in X, \ n \in \mathbb{N}, \tag{6.4.4}$$

   satisfy the conditions

   (i') g_n *is n-Lipschitz;*

   (ii') $g_n \ge g_{n+1}$ *and* $g_n \ge g$ *for all* $n \in \mathbb{N}$; $\qquad\qquad$ (6.4.5)

   (iii') $\lim_{n \to \infty} g_n(x) = g(x)$ *for every* $x \in X$.

3'. *If the function* g *is bounded and uniformly continuous, then the sequence* (g_n) *of Lipschitz functions given by* (6.4.4) *converges to* g *uniformly on* X.

Proof We start with the proof of 2. Suppose that f is bounded below on X, say $f(x) \geq a, x \in X$, and let $f_n : X \to \mathbb{R}$ be defined by (6.4.2).

Since for every $y \in X$ the function $h_y(x) = f(y) + nd(x, y)$, $x \in X$, is n-Lipschitz, Proposition 2.3.9 implies that f_n is also n-Lipschitz.

Let us prove (6.4.3).(ii).

The inequalities

$$f_n(x) \leq f(y) + nd(x, y) \leq f(y) + (n+1)d(x, y)$$

valid for all $y \in X$, imply $f_n(x) \leq f_{n+1}(x)$.

Taking $y = x$ in the first of the above inequalities, one obtains $f_n(x) \leq f(x)$.

Let now

$$h(x) = \lim_{n \to \infty} f_n(x) = \sup_n f_n(x), \quad x \in X.$$

Since $f_n(x) \leq f(x)$, it follows that $h(x) \leq f(x)$.

Fix $x \in X$. By the definition of f_n for every $n \in \mathbb{N}$ there exists $y_n \in X$ such that

$$f(y_n) + nd(x, y_n) < f_n(x) + \frac{1}{n} \tag{6.4.6}$$

implying

$$nd(x, y_n) < f_n(x) - f(y_n) + \frac{1}{n} \leq f(x) - a + \frac{1}{n}.$$

Consequently,

$$0 \leq d(x, y_n) < \frac{1}{n}(f(x) - a) + \frac{1}{n^2} \to 0, \quad n \to \infty$$

showing that $y_n \to x$. By (6.4.6), $f(y_n) < f_n(x) + \frac{1}{n}$, so that, taking into account the lsc of the function f, one obtains

$$f(x) \leq \liminf_{n \to \infty} f(y_n) \leq \liminf_{n \to \infty} \left(\frac{1}{n} + f_n(x)\right) = h(x),$$

proving that $h(x) = f(x)$.

To show that 3 holds, let ω be a modulus of continuity for f, that is, a nondecreasing function $\omega : [0, \infty) \to [0, \infty)$ such that

$$\lim_{\delta \searrow 0} \omega(\delta) = 0 = \omega(0)$$

and

$$|f(x) - f(y)| \le \omega(d(x, y)),$$

for all $x, y \in X$.

The uniform continuity of f is equivalent to the existence of a modulus of continuity satisfying the above conditions (take, for instance, $\omega(\delta) := \sup\{|f(x) - f(y)| : x, y \in X, d(x, y) \le \delta\}$).

Let $a := \inf f(X)$ and $b := \sup\{|f(x)| : x \in X\}$. We can suppose $b > 0$ (i.e., f is not identically 0 on X).

The definition of f_n implies

$$-\infty < a \le f_n(x) \le f(x),$$

for all $x \in X$.

If $x, y \in X$ and $n \in \mathbb{N}$ are such that $nd(x, y) > 3b$, then

$$f_n(x) \le f(x) \le b \le 2b + a < nd(x, y) + f(y) - b,$$

showing that

$$f_n(x) = \inf\{f(y) + nd(x, y) : y \in X, nd(x, y) \le 3b\},$$

i.e., we can take the infimum only for those points $y \in X$ satisfying $d(x, y) \le 3b/n$.

For $\varepsilon > 0$ let $\delta_0 > 0$ be such that

$$\omega(\delta) < \frac{\varepsilon}{2},$$

for all $0 \le \delta \le \delta_0$. Let also n_0 be such that

$$\frac{3b}{n_0} \le \delta_0.$$

For $x \in X$ and $n \in \mathbb{N}$, $n \ge n_0$, let y_n be such that $d(x, y_n) \le 3b/n$ and

$$f(y_n) + nd(x, y_n) < f_n(x) + \frac{\varepsilon}{2}.$$

It follows that

$$f(y_n) - f_n(x) + nd(x, y_n) < \frac{\varepsilon}{2},$$

and so

$$f(y_n) - f_n(x) < \frac{\varepsilon}{2}.$$

Consequently,

$$0 \le f(x) - f_n(x) = f(x) - f(y_n) + f(y_n) - f_n(x) \le \omega\left(3bn^{-1}\right) + \frac{\varepsilon}{2} < \varepsilon,$$

for all $n \ge n_0$.

Since $x \in X$ was arbitrarily chosen, it follows that

$$0 \le f(x) - f_n(x) < \varepsilon,$$

for all $x \in X$ and all $n \ge n_0$, i.e., $(f_n(x))$ converges to $f(x)$ uniformly for $x \in X$.

In order to prove assertion 1, consider the function $\varphi(x) = (\phi \circ f)(x)$, where ϕ denotes the Baire function given by (6.4.1), that is,

$$\varphi(x) = \begin{cases} 1 & \text{if} \quad f(x) = \infty, \\ f(x)/(1 + |f(x)|) & \text{if} \quad f(x) \in \mathbb{R}. \end{cases}$$

It follows that φ is lsc and $-1 < \varphi(x) \le 1$ for all $x \in X$.

Let

$$\varphi_n(x) = \inf_{y \in X}[\varphi(y) + nd(x, y)], \quad x \in X, \, n \in \mathbb{N}.$$

By the first part of the proof, (φ_n) is a nondecreasing sequence of Lipschitz functions which converges pointwise to φ.

We show that $-1 < \varphi_n(x) \le 1$ for all $x \in X$ and $n \in \mathbb{N}$. Since $\varphi_n \le \varphi$, it follows that $\varphi_n \le 1$ for all $n \in \mathbb{N}$.

For $x \in X$ let (y_k) be a sequence in X such that $\lim_{k \to \infty}[\varphi(y_k) + nd(x, y_k)] = \varphi_n(x)$. If $d(x, y_k) \to 0$, then $y_k \to x$ and

$$-1 < \varphi(x) \le \liminf_{k \to \infty} \varphi(y_k) \le \liminf_{k \to \infty}[\varphi(y_k) + nd(x, y_k)] = \varphi_n(x).$$

If $d(x, y_k) \nrightarrow 0$, then, passing to a subsequence if necessary, we can suppose $d(x, y_k) \ge \alpha, \, k \in \mathbb{N}$, for some $\alpha > 0$. The inequalities $\varphi(y_k) + nd(x, y_k) > -1 + n\alpha, \, k \in \mathbb{N}$, yield for $k \to \infty$, $\varphi_n(x) \ge -1 + n\alpha > -1$.

The sequence

$$\psi_n(x) = \frac{n}{n+1}\varphi_n(x) \text{ if } \varphi_n(x) > 0 \quad \text{and} \quad \psi_n(x) = \varphi_n(x) \text{ if } \varphi_n(x) \le 0,$$

has the same properties, that is, $\psi_n \le \psi_{n+1}$, $\lim_n \psi_n(x) = \varphi(x), \, x \in X$, and, furthermore, $-1 < \psi_n(x) < 1, \, x \in X, n \in \mathbb{N}$. It follows that the function

$$f_n(x) = \phi^{-1}(\psi_n(x)) = \frac{\psi_n(x)}{1 - |\psi_n(x)|}$$

is continuous on X for each $n \in \mathbb{N}$, $f_n \leq f_{n+1}$, and $\lim\limits_{n\to\infty} f_n(x) = \phi^{-1}(\varphi(x)) = f(x)$, for all $x \in X$.

Statements $1'$, $2'$ and $3'$ for a usc function $g : X \to \mathbb{R} \cup \{-\infty\}$ reduce to the previous ones (1, 2 and 3, respectively) applied to the function $f = -g$.

Indeed,

$$f_n(x) = \inf_{y\in X}[-g(y) + nd(x, y)] = - \sup_{y\in X}[g(y) - nd(x, y)] = -g_n(x).$$

The sequence (g_n) has the required approximation properties.   □

Remark 6.4.2 The result from 2 follows from the more general Theorem 6.3.1, but in this case we can define effectively the sequence of approximating functions.

6.5   The Homotopy of Lipschitz Functions

We start with the formal definition of homotopy.

Definition 6.5.1 Two continuous functions $f, g : X \to Y$, where X and Y are topological spaces, are called *homotopic* if there exists a continuous function $F :$ $X \times [0, 1] \to Y$ such that $F(x, 0) = f(x)$ and $F(x, 1) = g(x)$, for each $x \in X$. If, in addition, F is Lipschitz, then f and g are called *Lipschitz homotopic*.

We need the following auxiliary result.

Proposition 6.5.2 *Let X be a compact metric space and Y_1 a compact subset of a normed space Y satisfying the following properties:*

i) there exists an open subset U of Y such that

$$Y_1 \subseteq U ;$$

ii) there exists a Lipschitz function $\varphi : U \to Y_1$ such that

$$\varphi|_{Y_1} = \mathrm{Id}.$$

Then, for each continuous function $F : X \to Y_1$ there exists a continuous function $\phi : X \times [0, 1] \to Y_1$ and a Lipschitz function $G : X \to Y_1$ such that

$$\phi(x, 0) = F(x) \quad and \quad \phi(x, 1) = G(x),$$

for each $x \in X$.

Moreover, if $F|_{X_0}$ is Lipschitz, where $X_0 \subseteq X$, then $\phi|_{X_0 \times [0,1]}$ is Lipschitz.

Proof As Y_1 is compact and U is open, there exists $\eta > 0$ having the property that

$$\{z \in Y : \text{there exists } y \in Y_1 \text{ such that } \|y - z\| < \eta\} \subseteq U. \tag{6.5.1}$$

For each $y \in Y_1$ there exists $r_y > 0$ such that $B(y, r_y) \subseteq U$. By the compactness of Y_1, the open cover $\{B(y, r_y/2) : y \in Y_1\}$ contains a finite subcover $B(y_k, r_k/2)$, $k = 1, \ldots, n$, where $r_k = r_{y_k}$. Then $\eta = \min\{r_1/2, \ldots, r_n/2\}$ satisfies the requirements of (6.5.1). Indeed, if $z \in Y$ is such that $\|y - z\| < \eta$ for some $y \in Y_1$, then there exists $k \in \{1, \ldots, n\}$ such that $\|y - y_k\| < r_k/2$, implying $z \in B(y_k, r_k) \subseteq U$.

Let us consider a Lipschitz function $g : X \to Y$ such that

$$\|g(x) - F(x)\| < \eta,$$

for each $x \in X$ (see Theorem 6.2.1 and Theorem 2.1.6). Then $g(X) \subseteq U$, since $F(X) \subseteq Y_1$.

As the continuous function $\phi_1 : X \times [0, 1] \to Y$, given by

$$\phi_1(x, t) = F(x) + t \cdot [g(x) - F(x)], \quad (x, t) \in X \times [0, 1],$$

has the property that $\|\phi_1(x, t) - F(x)\| = t \cdot \|g(x) - F(x)\| < \eta$ for each $x \in X$ and $t \in [0, 1]$, we get that $\phi_1(X \times [0, 1]) \subseteq U$.

Let us consider $\phi = \varphi \circ \phi_1$ and $G = \varphi \circ g$. It is clear that ϕ is continuous, G is Lipschitz and $\phi(x, 0) = \varphi(F(x)) = F(x)$, $\phi(x, 1) = \varphi(g(x)) = G(x)$, for each $x \in X$, since $\varphi|_{Y_1} = \text{Id}$.

If $F|_{X_0}$ is Lipschitz, where $X_0 \subseteq X$, we choose $M \geq 0$ such that $\|F(x_1) - F(x_2)\| \leq M \cdot d(x_1, x_2)$ and $\|g(x_1) - g(x_2)\| \leq M \cdot d(x_1, x_2)$ for each $x_1, x_2 \in X_0$. Then

$$\|\phi_1(x_1, t_1) - \phi_1(x_1, t_2)\| = |t_1 - t_2| \cdot \|g(x_1) - F(x_1)\|$$
$$\leq \eta \cdot |t_1 - t_2| \tag{6.5.2}$$

and

$$\|\phi_1(x_1, t_2) - \phi_1(x_2, t_2)\|$$
$$= \|(1 - t_2) \cdot [F(x_1) - F(x_2)] + t_2 \cdot [g(x_1) - g(x_2)]\| \tag{6.5.3}$$
$$\leq M \cdot d(x_1, x_2),$$

for all $x_1, x_2 \in X_0$ and $t_1, t_2 \in [0, 1]$.

Since

$$\|\phi_1(x_1, t_1) - \phi_1(x_2, t_2)\|$$

$$\leq \|\phi_1(x_1, t_1) - \phi_1(x_1, t_2)\| + \|\phi_1(x_1, t_2) - \phi_1(x_2, t_2)\|$$

$$\leq \max\{\eta, M\} \cdot \left(|t_1 - t_2| + d(x_1, x_2)\right) \quad \text{(by (6.5.2) and (6.5.3))}$$

$$\leq \sqrt{2} \cdot \max\{\eta, M\} \cdot \sqrt{|t_1 - t_2|^2 + d(x_1, x_2)^2}$$

$$= \sqrt{2} \cdot \max\{\eta, M\} \cdot d((x_1, t_1), (x_2, t_2)),$$

for all $x_1, x_2 \in X_0$ and $t_1, t_2 \in [0, 1]$, we infer that $\phi_1|_{X_0 \times [0,1]}$ is Lipschitz. $\square$

Theorem 6.5.3 *Let us consider X, Y and Y_1 as in the previous Proposition. If, in addition, $X \times [0, 1]$ is a quasiconvex metric space, then two Lipschitz functions $f, g : X \to Y_1$ which are homotopic are Lipschitz homotopic.*

Proof As the functions f and g are homotopic, there exists a continuous function $F : X \times [0, 1] \to Y_1$ such that $F(x, 0) = f(x)$ and $F(x, 1) = g(x)$, for each $x \in X$. Taking into account Proposition 6.5.2, we can consider a continuous function $\phi : X \times [0, 1] \times [0, 1] \to Y_1$ and a Lipschitz function $G : X \times [0, 1] \to Y_1$ such that $\phi(x, t, 0) = F(x, t)$ and $\phi(x, t, 1) = G(x, t)$, for each $x \in X$ and $t \in [0, 1]$.

Let us consider the function $\tilde{F} : X \times [0, 1] \to Y_1$ given by

$$\tilde{F}(x, t) = \begin{cases} \phi(x, 0, 3t), & \text{for } x \in X, \ t \in [0, 1/3] , \\ \phi(x, 3t - 1, 1) = G(x, 3t - 1), & \text{for } x \in X, \ t \in [1/3, 2/3] , \\ \phi(x, 1, 3 - 3t), & \text{for } x \in X, \ t \in [2/3, 1] . \end{cases}$$

The functions $F|_{X \times \{0\}} = f$ and $F|_{X \times \{1\}} = g$ are Lipschitz. Taking again into account Proposition 6.5.2, $\phi|_{X \times \{0\} \times [0,1]}$ and $\phi|_{X \times \{1\} \times [0,1]}$ are Lipschitz, so $\tilde{F}|_{X \times [0,1/3]}$ and $\tilde{F}|_{X \times [2/3,1]}$ are also Lipschitz. Obviously $\tilde{F}|_{X \times [1/3,2/3]}$ is Lipschitz. As $X \times [0, 1]$ is a quasiconvex metric space, we infer, by Theorem 2.5.6, that $\tilde{F}$ is Lipschitz.

Moreover

$$\tilde{F}(x, 0) = \phi(x, 0, 0) = F(x, 0) = f(x) \quad \text{and}$$

$$\tilde{F}(x, 1) = \phi(x, 1, 0) = F(x, 1) = g(x) ,$$

for each $x \in X$. $\square$

Remark 6.5.4 Theorem 6.5.3, which is a generalization of result due to Hu [292], shows that, in the above mentioned framework, the homotopy properties of a space of Lipschitz functions are completely determined by the homotopies determined by Lipschitz functions.

6.6 Lipschitz Manifolds

In this section we shall briefly present some results involving Lipschitz manifolds.
Given two metric spaces (X, d) and (Y, d'), a function $f : X \to Y$ is called:

- a *lipeomorphism* (or a *Lipschitz isomorphism*) if it is a bijection and both f and
  f^{-1} are Lipschitz;
- a *Lipschitz embedding* if it is a lipeomorphism between X and $f(X)$;
- a *locally Lipschitz embedding* if every $x \in X$ has a neighborhood U such that
  $f|_U$ is a Lipschitz embedding.

A Lipschitz manifold (or a manifold with corners) is a topological manifold with
an extra structure which, on the one hand, is slightly weaker than a smooth structure
but one can still do analysis with it and, on the other hand, the essential uniqueness
of it is automatic in many situations.

We present now, following [169], the definition of a Lipschitz manifold.

Definition 6.6.1 A *Lipschitz E-manifold*, where E is a Banach space, is a Hausdorff topological space X equipped with a family of charts $h_\alpha : U_\alpha \to E$, satisfying
the following conditions:

  (i) the family $(U_\alpha)_{\alpha \in A}$ is an open cover of X;
 (ii) each h_α is a homeomorphism onto the open subset $h_\alpha(U_\alpha)$ of E;
(iii) the change of coordinates $h_\beta \circ h_\alpha^{-1} : h_\alpha(U_\alpha \cap U_\beta) \to h_\beta(U_\alpha \cap U_\beta)$ is locally
       Lipschitz, for all $\alpha, \beta \in A$.

Remark 6.6.2 Let us note that actually $h_\beta \circ h_\alpha^{-1} : h_\alpha(U_\alpha \cap U_\beta) \to h_\beta(U_\alpha \cap U_\beta)$ is
a lipeomorphism, for all $\alpha, \beta \in A$.

Remark 6.6.3 A Lipschitz $\mathbb{R}^n$-manifold is called a Lipschitz manifold of dimension
n or a *Lipschitz n-manifold*. Basically the same definition as above (with the
supplementary condition that X is a second countable locally compact space), for
finite dimensional Lipschitz manifolds, can be found in [607, p. 270] or [410, pp.
97–98].

Luukkainen and Väisälä [413] prefer to use the following alternative definition
of a Lipschitz n-manifold which is basically equivalent to the above one but
conceptually simpler.

Definition 6.6.4 A *Lipschitz n-manifold* is a separable metric space X such that
every point $x \in X$ has a closed neighborhood U lipeomorphic to $[-1, 1]^n$.

Teleman [650] shows how to do analysis on finite dimensional compact connected Lipschitz manifolds so that the signature operators can be defined. Actually
Teleman generalizes the Atiyah–Singer index theorem to closed topological oriented
manifolds which admit a Lipschitz structure. The feasibility of such an approach is
based on the famous Sullivan theorem on the existence of an essentially unique
finite dimensional Lipschitz structure on every compact topological manifold of

dimension not equal to 4. According to [607], Freedman, Donaldson and others
proved that there are topological 4-manifolds with no Lipschitz structure.

Luukkainen and Väisälä [413] proved that every connected Lipschitz 1-manifold
is lipeomorphic to exactly one of the following 1-manifold: $(0, 1)$, $[0, 1)$, $[0, 1]$,
$S^1 = \{x \in \mathbb{R}^2 : \|x\| = 1\}$. They also proved that if a Lipschitz manifold is
homeomorphic to $\mathbb{R}^n$ or $S^n = \{x \in \mathbb{R}^{n+1} : \|x\| = 1\}$ for $n \neq 4$ or to $I^n = [-1, 1]^n$
for $n \neq 4, 5$, then it is lipeomorphic to it. A characterization of Lipschitz manifolds
modeled on $\mathbb{R}^n$, given by the above named authors, was extended to compact
connected Lipschitz manifolds modeled on Banach spaces by Miculescu [474].

According to the famous Whitney embedding theorem, every C^∞ manifold of
dimension n can be C^∞ embedded in $\mathbb{R}^{2n+1}$.

In 1965, Colojoară [168], at the same time with McAlpin (in his Ph.D. thesis),
gave a generalization of this result. Namely he proved that every paracompact
second countable C^∞ manifold modeled on a separable Hilbert space $\mathscr{H}$ admits
a C^∞ embedding into $\mathscr{H}$.

In 1977, Luukkainen and Väisälä [413] proved the following embedding result
for finite dimensional Lipschitz manifolds.

Theorem 6.6.5 *If X is a second countable paracompact Lipschitz n-manifold, then
there exists an injective function $f : X \to \mathbb{R}^{n(n+1)}$ such that f and $f^{-1} : f(X) \to
X$ are locally Lipschitz and $f(X)$ is closed.*

In 1995, Colojoară returned to the subject of embeddings of infinite dimensional
manifolds, providing an embedding theorem for paracompact second countable
Lipschitz manifolds modeled on separable Hilbert spaces. More precisely, he proved
the following result.

Theorem 6.6.6 ([169]) *If X is a paracompact second countable $\mathscr{H}$-Lipschitz
manifold, where $\mathscr{H}$ is a separable Hilbert space, then there exists an injective
function $h : X \to \mathscr{H}$ such that h and $h^{-1} : f(X) \to X$ are locally Lipschitz
and $h(X)$ is closed.*

Let us mention that in [475] it is proved that for each paracompact and second
countable E-Lipschitz manifold, where E could be ℓ^p, L^p, $p \in (1, \infty)$, c_0 or
$c_0 \oplus \ell^2$, there exists a continuous and injective function $f : X \to E$ having the
property that every point $x \in X$ has a neighborhood U such that $f|_U : U \to f(U)$
is a lipeomorphism. Therefore f is a locally Lipschitz embedding.

6.7 Bibliographic Comments and Miscellaneous Results

Section 6.1 Theorem 6.1.1 appears as early as 1966, see [209, Lemma 7].

Section 6.2 Theorem 6.2.1 was proved in [477] in the case $Y = \mathbb{R}$. The formulation
given here is new.

Section 6.3 Theorem 6.3.1 is from [479].

Section 6.4 is largely based on [508].

Section 6.5 is based on [476].

Luukkainen and Väisälä [413] proved that every continuous function $f : X \to M$, where X is a metric space and M is a locally Lipschitz manifold, can be approximated in the uniform norm by locally Lipschitz functions.

Boiso [92] provided a characterization of superreflexive Banach spaces as those X for which every Lipschitz function $f : X \to \mathbb{R}$ can be uniformly approximated by a difference of two convex functions which are bounded on bounded sets.

Azagra et al. [61] proved that given a separable Banach space X with a separating polynomial, there exists $C \geq 1$ (depending only on X) such that for every Lipschitz function $f : X \to \mathbb{R}$ and every $\varepsilon > 0$ one can find a Lipschitz, real analytic function $f : X \to \mathbb{R}$ such that $|f(x) - g(x)| \leq \varepsilon$ for each $x \in X$ and $L(g) \leq CL(f)$.

Bogachev and Shkarin [90] proved that if X is a separable Banach space and Y is a Banach space with RNP, then every Lipschitz function $f : X \to Y$ can be uniformly approximated by Gâteaux differentiable Lipschitz functions, i.e., for very $\varepsilon > 0$ there exists a Gâteaux differentiable Lipschitz mapping $g : X \to Y$ such that $\|f(x) - g(x)\| \leq \varepsilon$ for all $x \in X$.

The papers [289] and [290] contain some results on the approximation by polynomials or by rational functions in Lipschitz and Hölder algebras of differentiable or analytic functions. The maximal ideals in these Lipschitz function algebras are determined as well.

The paper [657] contains an exposition of Sullivan's theory on the deformation of Lipschitz and quasi-conformal embeddings in geometric topology, which is then applied to get new results on Lipschitz and quasi-conformal embeddings.

For other results concerning approximations of (or by) Lipschitz functions see [246, 467, 468, 478, 485, 658, 661].

Section 6.6 is based on [169] and [481]. Luukkainen and Väisälä [413] proved that given a Lipschitz manifold M, a metric space X and two continuous functions $f : X \to M$ and $\varepsilon : X \to (0, \infty)$, there exists a locally Lipschitz function $g : X \to M$ such that $d(f(x), g(x)) < \varepsilon(x)$ for each $x \in M$.

Tukia [656] proved that if M and N are two Lipschitz n-manifolds, where $n \leq 3$, $f : M \to N$ is a homeomorphism and $\varepsilon : M \to (0, \infty)$ is continuous, then there exists a lipeomorphism $g : M \to N$ such that $d(f(x), g(x)) < \varepsilon(x)$ for each $x \in M$.

Luukkainen [410] obtained results concerning Lipschitz approximations of homeomorphisms between two pairs of Lipschitz manifolds. A significant impact on the development of the so-called "*Lipschitz analysis*" had the Ph.D. thesis of Luukkainen, published as [408].

Miculescu [480] considered a generalized notion of Lipschitz function. Let (X, d), (Y, d') be metric spaces and $g : X \to X$ a function. A mapping $f : X \to Y$ is called g-Lipschitz if there exists $L \geq 0$ such that $d'(f(x), f(y)) \leq Ld(g(x), g(y))$ for all $x, y \in X$. The least L for which the written inequality holds is called the g-Lipschitz norm of f and is denoted by $L_g(f)$.

The author makes some connections of this notion with the Hellinger integral, proves a McShane type result (Theorem 4.1.1) for this class of generalized Lipschitz functions and a result on the approximation of bounded generalized uniformly continuous functions by generalized Lipschitz functions (the analog of Theorem 6.3.1). A function $f : X \to Y$ is called generalized uniformly continuous with respect to $g : X \to \mathbb{R}$ if for every $\varepsilon > 0$ there exists $\delta > 0$ such that $d(g(x), g(y)) < \delta$ implies $d'(f(x), f(y)) < \varepsilon$, for all $x, y \in X$.

Chapter 7
Lipschitz Isomorphisms of Metric Spaces

The main results in this chapter are Aharoni's theorem (Theorem 7.3.3) on the bi-Lipschitz embeddability of separable metric spaces in the Banach space c_0 and a result of Väisälä (Theorem 7.4.6) on the characterization of the completeness of a normed space X by the non-existence of bi-Lipschitz surjections of X onto $X \setminus \{0\}$. Other results are discussed in the final section of this chapter.

7.1 Introduction

In order to state the main result of this chapter we need the following definition.

Two metric spaces (X, d) and (Y, d') are called *Lipschitz equivalent* (or *Lipschitz isomorphic*) if there exists a bijective function $f : X \to Y$ such that both f and f^{-1} are Lipschitz (i.e., f is a Lipschitz isomorphism).

A central problem in nonlinear Banach spaces theory is to exhibit the linear properties of Banach spaces that are stable under some particular classes of non-linear functions (such as Lipschitz isomorphisms or embeddings). For instance, by a result of Mazur and Ulam [453, 454], any surjective isometry between two Banach spaces is affine (see also [75, Theorem 14.1]). Rolewicz [602] (see also [603, Section 9.3]) extended this result to quasi-Banach spaces.

Not too much is known on the Lipschitz classification of Banach spaces. It is not true that two Lipschitz isomorphic Banach spaces are always linearly isomorphic as it is shown by the following result due to Aharoni and Lindenstrauss [10]:

There exists an uncountable set Γ and a Banach space X Lipschitz isomorphic to $c_0(\Gamma)$ such that X is not linearly isomorphic to any subspace of $c_0(\Gamma)$.

Ultrapower techniques were applied by Heinrich and Mankiewicz [286] to the uniform and Lipschitz classification of Banach spaces. In [260] some canonical examples of nonseparable Banach spaces which are Lipschitz isomorphic but not linearly isomorphic are constructed. It is not known if there exist two separable

© Springer Nature Switzerland AG 2019
Ş. Cobzaş et al., *Lipschitz Functions*, Lecture Notes in Mathematics 2241,
https://doi.org/10.1007/978-3-030-16489-8_7

Banach spaces which are Lipschitz isomorphic but not linearly isomorphic. However, a Banach space which is Lipschitz isomorphic to c_0 is linearly isomorphic to c_0 (see [261, 262]). Albiac and Kalton [23] gave examples of separable quasi-Banach spaces which are Lipschitz isomorphic but not linearly isomorphic.

7.2 Schauder Bases in Banach Spaces

We are going to prove that every separable metric space is Lipschitz isomorphic to a subset of the Banach space c_0. At the very beginning we present some definitions and results that are necessary for the proof of the mentioned result. Concerning bases in Banach spaces, the exhaustive presentation from the two-volume authoritative treatise [636, 638] is highly recommended.

Let X be a Banach space. A family $(e_i)_{i\in\mathbb{N}}$ of elements in X is called a *Schauder basis* for X if each $x \in X$ admits a unique representation

$$x = \sum_{i=1}^{\infty} x_i e_i, \tag{7.2.1}$$

where x_i are scalars and the series (7.2.1) converges with respect to the norm of X.

If in addition $\|e_i\| = 1$ for all $i \in \mathbb{N}$, then the basis (e_i) is called *normalized*.

The functions $P_n : X \to X$, where $n \in \mathbb{N}\cup\{0\}$, given by

$$P_n x = \sum_{i=1}^{n} x_i e_i, \quad \text{for } n \geq 1,$$

and

$$P_0 x = 0,$$

for each $x = \sum_{i=1}^{\infty} x_i e_i \in X$, are called the *partial sum operators*.

An element $x \in X$ is called *finitely supported* if there exists $n \in \mathbb{N}_0$ such that

$$P_n x = x.$$

The minimal $n \in \mathbb{N}$ having this property is called the *length* of x and is denoted by $l(x)$.

Remark 7.2.1 By defining

$$|||x||| = \sup\{\|P_n x - P_m x\| : m, n \in \mathbb{N}_0, m \neq n\},$$

for each $x \in X$, we obtain a norm $|||\cdot|||$ on X, which is equivalent to the initial norm $\|\cdot\|$, such that in the space $(X, |||\cdot|||)$ the operators $P_n - P_m$ are of norm 1 for all $m, n \in \mathbb{N}_0$ with $m \neq n$ (in particular, the operators P_n for $n \geq 1$).

In what follows we shall assume that the norm $\|\cdot\|$ satisfies this property.

Lemma 7.2.2 *Let $(e_i)_{i \in \mathbb{N}}$ be a Schauder basis of a Banach space X and let $x = \sum_{i=1}^{\infty} x_i e_i \in X$.*

1. For $s, t \in [0, 1]$ and $m, n \in \mathbb{N}$, $m < n$, the following inequality is true

$$\|tx_m e_m + P_n x - P_m x + sx_{n+1}e_{n+1}\| \leq \|x\|. \tag{7.2.2}$$

2. For each $m \in \mathbb{N}$ and $t \in [0, 1]$ we have

$$\|tx_m e_m + x_{m+1}e_{m+1} + \ldots\| \leq \|x\|. \tag{7.2.3}$$

Proof

1. We have

$$\|x_m e_m + x_{m+1}e_{m+1} + \cdots + x_n e_n + sx_{n+1}e_{n+1}\|$$
$$= \|s(P_{n+1} - P_{m-1})x + (1-s)(P_n - P_{m-1})x\|$$
$$\leq (s\|P_{n+1} - P_{m-1}\| + (1-s)\|P_n - P_{m-1}\|)\|x\| = \|x\|,$$

for all $x \in X$. We can also apply this argument for $y = tx_m e_m + x_{m+1}e_{m+1} + \cdots + x_n e_n + x_{n+1}e_{n+1}$ to obtain

$$\|tx_m e_m + x_{m+1}e_{m+1} + \cdots + x_n e_n + sx_{n+1}e_{n+1}\|$$
$$\leq \|tx_m e_m + x_{m+1}e_{m+1} + \cdots + x_n e_n + x_{n+1}e_{n+1}\|.$$

But, as above,

$$\|tx_m e_m + x_{m+1}e_{m+1} + \cdots + x_n e_n + x_{n+1}e_{n+1}\|$$
$$= \|t(P_{n+1} - P_{m-1})x + (1-t)(P_{n+1} - P_m)x\| \leq \|x\|,$$

showing the validity of (7.2.2).
2. With the notation

$$u = tx_m e_m + x_{m+1}e_{m+1} + \ldots$$

and

$$u_k = tx_m e_m + x_{m+1}e_{m+1} + \cdots + x_{m+k}e_{m+k},$$

we have $\lim_{k\to\infty} u_k = u$. By (7.2.2) $\|u_k\| \le \|x\|$ for all $k \in \mathbb{N}$, so that $\|u\| = \lim_{k\to\infty} \|u_k\| \le \|x\|$.

$\square$

Let X be a Banach space having a Schauder basis $(e_i)_{i\in\mathbb{N}}$ such that $\|e_i\| = 1$ for each $i \in \mathbb{N}$. For $x = \sum_{i=1}^{\infty} x_i e_i \in X \setminus \{0\}$ and $\beta \in (0, \|x\|)$ let n be the first integer such that

$$\|x - P_n x\| < \beta$$

(so $\|x - P_{n-1}x\| \ge \beta$). Let $t \in (0, 1]$ be such that

$$\|x - P_n x + t x_n e_n\| = \beta.$$

We define the function Q_β by

$$Q_\beta x := P_n x - t x_n e_n'.$$

Remark 7.2.3 The existence of such a t is guaranteed by the continuity of the function $g : [0, 1] \to \mathbb{R}$ given by $g(u) = \|x - P_n x + u x_n e_n\|$, $u \in [0, 1]$ and the inequalities

$$g(0) = \|x - P_n x\| < \beta \quad \text{and} \quad g(1) = \|x - P_{n-1}x\| \ge \beta.$$

Then, for each $x \in X \setminus \{0\}$,

$$\|x - Q_\beta x\| = \beta,$$

and

$$\|Q_\beta x\| \le \|x\|. \tag{7.2.4}$$

The inequality (7.2.4) follows from the equality $Q_\beta x = (1 - t)P_n x + t P_{n-1}x$ and the fact the operators P_k are all of norm one (see Remark 7.2.1).

For every finitely supported $y \in X$ and each $x \in X \setminus \{0\}$ the following relations are valid

$$\beta = \|x - Q_\beta x\| \le \|x - y\|, \tag{7.2.5}$$

provided that

$$l(y) < l(Q_\beta x).$$

Indeed, let $m = l(y)$. Since $\|x - P_n x\| < \beta$ and $\|x - P_{n-1}x\| \ge \beta$, it follows that $x_n \ne 0$, so that $l(Q_\beta x) = n$. By hypothesis $m < n$, so that $y_k = 0$ for all $k \ge n$

and, by (7.2.3),

$$\|tx_n e_n + x_{n+1} e_{n+1} + \ldots\| = \|t(x_n - y_n)e_n + (x_{n+1} - y_{n+1})e_{n+1} + \ldots\|$$
$$\leq \|x - y\|.$$

Lemma 7.2.4 *Let X be a Banach space having a Schauder basis* $(e_i)_{i \in \mathbb{N}}$ *such that* $\|e_i\| = 1$ *for each* $i \in \mathbb{N}$, $a > 0$ *and* $\alpha > 0$. *Then there exists a sequence* $\left(z^j(a, \alpha)\right)_{j \in \mathbb{N}}$ *of finitely supported elements from X having the following properties:*

(a) $\left\|z^j(a, \alpha)\right\| \leq \alpha$ *for each* $j \in \mathbb{N}$;
(b) *the function*

$$j \mapsto l\left(z^j(a, \alpha)\right)$$

is nondecreasing and unbounded;
(c) *for each finitely supported* $z = \sum_{i=1}^{\infty} z_i e_i \in X$ *having the property that* $\|z\| \leq \alpha$ *there exists* $j \in \mathbb{N}$ *such that:*

    (i) $\left\|z - z^j(a, \alpha)\right\| < a$;
    (ii) $l\left(z^j(a, \alpha)\right) = l(z)$;

and

    (iii) $\left(z^j(a, \alpha)\right)_m \cdot z_m > 0$,

where $m = l(z)$ *and* $\left(z^j(a, \alpha)\right)_m$ *denotes the coefficient of* e_m *in the expansion of* $z^j(a, \alpha)$.

Proof For each $n \in \mathbb{N}$, the sets

$$A_n = \{z \in X : l(z) = n, z_n > 0, \|z\| \leq \alpha\}$$

and

$$B_n = \{z \in X : l(z) = n, z_n < 0, \|z\| \leq \alpha\}$$

are totally bounded, so there exist

$$A^n = \left\{z^j(a, \alpha) : j \in \{p_{2n-2} + 1, \ldots, p_{2n-1}\}\right\} \subseteq A_n$$

and

$$B^n = \left\{z^j(a, \alpha) : j \in \{p_{2n-1} + 1, \ldots, p_{2n}\}\right\} \subseteq B_n$$

such that

$$A_n \subseteq \bigcup_{y \in A^n} B(y, a),$$

and

$$B_n \subseteq \bigcup_{y \in B^n} B(y, a),$$

where $(p_n)_{n \in \mathbb{N}}$ is an increasing sequence integers and $p_0 = 0$.

The sequence $\left(z^j(a, \alpha)\right)_{j \in \mathbb{N}}$ satisfies all the requirements in the statement of the lemma. $\qquad \square$

Definition 7.2.5 Let X be a Banach space having a Schauder basis $(e_l)_{l \in \mathbb{N}}$ such that $\|e_l\| = 1$ for each $l \in \mathbb{N}$. For $n, k, j \in \mathbb{N}$, with $n \geq 6$ and $1 \leq k \leq 2n/3$, denote by i the triple (n, k, j). For $a > 0$ let

$$M_i^a = \left\{ x \in X : (n - 2)a \leq \|x\| \leq (n + 1)a \text{ and } Q_{ka}x = z^j(a, na) \right\}$$

and

$$t_i^a = (k - 1)a,$$

where the elements $z^j(a, na)$ are given by Lemma 7.2.4.

Remark 7.2.6 It is possible to have $M_i^a = \emptyset$ for some i. We shall consider only the nonempty M_i^a.

Lemma 7.2.7 *Let X be a Banach space having a Schauder basis $(e_i)_{i \in \mathbb{N}}$ such that $\|e_i\| = 1$ for each $i \in \mathbb{N}$. Given a real number $a > 0$, then, for each $x \in X$, there exists only a finite number of triples i such that*

$$d(x, M_i^a) < t_i^a.$$

Proof On the one hand, for each $n \geq 3(a^{-1} \|x\| + 1)$, $1 \leq k \leq 2n/3$ and $y \in M_i^a$ we have

$$\|y - x\| \geq \|y\| - \|x\| \geq (n - 2)a - \left(\frac{n}{3} - 1\right) a \geq ka - a = t_i^a,$$

so

$$d(x, M_i^a) \geq t_i^a,$$

for a triple i corresponding to n and k as above.

On the other hand, for each $n < 3(a^{-1} \|x\| + 1)$, we have

$$\|x\| > \frac{(n-3)a}{3} \geq a.$$

As, taking into account Lemma 7.2.4, $Q_{\frac{a}{2}} x$ is finitely supported, there exists $j_0 \in \mathbb{N}$ such that $l(z^{j_0}(a, na)) > l(Q_{\frac{a}{2}} x)$ and, consequently, for each $j \geq j_0$, we have $l(z^j(a, na)) > l(Q_{\frac{a}{2}} x)$. For such j we have

$$Q_{ka} y = z^j(a, na),$$

for each $y \in M_i^a$, where $i = (n, k, j)$, so that

$$l(Q_{ka} y) = l(z^j(a, na)) > l(Q_{\frac{a}{2}} x).$$

Using the inequality (7.2.5), we get the following inequality

$$ka = \|y - Q_{ka} y\| \leq \left\| y - Q_{\frac{a}{2}} x \right\|.$$

Hence we have

$$\|y - x\| \geq \left\| y - Q_{\frac{a}{2}} x \right\| - \left\| x - Q_{\frac{a}{2}} x \right\| \geq \|y - Q_{ka} y\| - \frac{a}{2}$$

$$= ka - \frac{a}{2} > (k-1)a = t_i^a,$$

and therefore

$$d(x, M_i^a) \geq t_i^a.$$

Thus the set $\{i : d(x, M_i^a) < t_i^a\}$ is finite. $\qquad\qquad\square$

Lemma 7.2.8 *Suppose that X is a Banach space with a Schauder basis $(e_i)_{i \in \mathbb{N}}$ such that $\|e_i\| = 1$ for all $i \in \mathbb{N}$. If $a > 0$ and $x, y \in X$ are such that*

$$\|x\| \geq \|y\| \quad and \quad \|x - y\| \geq 36a,$$

then there exists a triple $i = (n, k, j)$ having the following properties:

(a) $d(x, M_i^a) < a$ *and* $d(y, M_i^a) \geq t_i^a$;

(b) $t_i^a \leq \dfrac{\|x - y\|}{2}$;

(c) $t_i^a - a \geq \dfrac{\|x - y\|}{4}$.

Proof If

$$n = \left[a^{-1} \|x\| \right] + 1 \quad \text{and} \quad k = \left[(3a)^{-1} \|x - y\| \right],$$

where $[\alpha]$ stands for the integer part of a real number α, then

$$(n - 1)a \le \|x\| < na \tag{7.2.6}$$

and

$$3ka \le \|x - y\| < 3a(k + 1). \tag{7.2.7}$$

Hence

$$2na > 2\|x\| \ge \|x\| + \|y\| \ge \|x - y\|,$$

so that

$$\frac{2n}{3} \ge \frac{\|x - y\|}{3a} \ge k.$$

Thus

$$k \le \frac{2n}{3}$$

and, by (7.2.4) and (7.2.6),

$$\|Q_{ka}x\| \le \|x\| < na. \tag{7.2.8}$$

Taking into account (7.2.8), we can consider the element $z^j = z^j(a, na)$ provided by Lemma 7.2.4.(c), corresponding to $z = Q_{ka}x$.

We are going to show that the triple $i = (n, k, j)$ is the one that we are looking for.

Let us prove the property (a). For $u = z^j + x - Q_{ka}x$, using Lemma 7.2.4.(c), we have

$$\|u - x\| = \left\| z^j - Q_{ka}x \right\| < a, \tag{7.2.9}$$

which yields

$$\|x\| - a < \|u\| < \|x\| + a. \tag{7.2.10}$$

Moreover,

$$\left\| u - z^j \right\| = \| x - Q_{ka}x \| = ka. \tag{7.2.11}$$

Let us note that

$$ka \leq \frac{\| x - y \|}{3} \leq \frac{2}{3} \| x \| < \| x \|.$$

Hence, if $x = x_1 e_1 + \cdots + x_n e_n + \ldots$ and $t \in [0, 1]$ and m are such that

$$\| t x_m e_m + x_{m+1} e_{m+1} + \ldots \| = ka \quad (\text{i.e. } \| x - P_m x + t x_m e_m \| = ka),$$

then

$$Q_{ka}x = P_m x - t x_m e_m = x_1 e_1 + \cdots + x_{m-1} e_{m-1} + (1 - t) x_m e_m$$

and, as $l(z^j) = l(Q_{ka}x) = m$, z^j has the form $z^j = z_1 e_1 + \cdots + z_m e_m$, it follows that

$$u = z_1 e_1 + \cdots + z_{m-1} e_{m-1} + (z_m + t x_m) e_m + x_{m+1} e_{m+1} + \ldots,$$

which shows that $u_1 = z_1, \ldots, u_{m-1} = z_{m-1}, u_m = z_m + t x_m, u_{m+1} = x_{m+1}, \ldots$. With the notation $\theta = t x_m (z_m + t x_m)^{-1} \in [0, 1]$, we have $z^j = P_m u - \theta u_m e_m$, so that, using (7.2.11), we infer that $\| u - (P_m u - \theta u_m e_m) \| = ka$, which yields

$$Q_{ka}u = z^j. \tag{7.2.12}$$

By (7.2.6), (7.2.10), we also have

$$(n - 2)a = (n - 1)a - a \leq \| x \| - a \leq \| u \|$$

$$\leq \| x \| + a \leq na + a = (n + 1)a,$$

i.e.,

$$(n - 2)a \leq \| u \| \leq (n + 1)a. \tag{7.2.13}$$

Therefore, from (7.2.12) and (7.2.13), we deduce that $u \in M_i^a$ and, consequently, by (7.2.9), $d(x, M_i^a) \leq \| u - x \| < a$, which is the first inequality in (a).

On the other hand, for each $v \in M_i^a$, we have $Q_{ka}v = z^j$, so that, by (7.2.7), (7.2.9) and (7.2.11),

$$\| y - v \| \geq \| y - x \| - \| x - Q_{ka}x \| - \left\| Q_{ka}x - z^j \right\| - \| Q_{ka}v - v \|$$

$$\geq 3ka - ka - a - ka = (k - 1)a = t_i^a.$$

Thus $d(y, M_i^a) \geq t_i^a$, which is the second inequality in (a).

In order to prove the property (b), note that, by (7.2.7),

$$t_i^a = (k-1)a\frac{\|x-y\|}{3} - a < \frac{\|x-y\|}{2}.$$

Finally, (c) is valid since, by (7.2.7),

$$\frac{t_i^a - a}{\|x-y\|} \geq \frac{(k-2)a}{3(k+1)a} = \frac{k-2}{3(k+1)} \geq \frac{1}{4}.$$

The last inequality holds for $k \geq 11$ because, by (7.2.7),

$$k > \frac{\|x-y\|}{3a} - 1 \geq \frac{36a}{3a} - 1 = 12 - 1 = 11.$$

$\square$

7.3   Separable Metric Spaces Embed in c_0

In this section we prove Aharoni's result on the Lipschitz embedding of any separable metric space in c_0.

An easier result, known as the Kuratowski embedding theorem, is the following one. Recall that $\ell^\infty(\Gamma)$ denotes the Banach space of all bounded functions $h : \Gamma \to \mathbb{R}$ with the norm $\|h\|_\infty = \sup\{|h(\gamma)| : \gamma \in \Gamma\}$ and ℓ^∞ the space $\ell^\infty(\mathbb{N})$.

Theorem 7.3.1 *Every metric space (X, d) can be isometrically embedded in $\ell^\infty(X)$. If X is separable then it can be isometrically embedded in ℓ^∞.*

Proof Fix a point $x_0 \in X$ and for every $x \in X$ define the mapping $\varphi(x) : X \to \mathbb{R}$ by

$$\varphi(x)(y) = d(x, y) - d(x_0, y), \qquad y \in X.$$

Since

$$|\varphi(x)(y)| \leq d(x, x_0),$$

for all $y \in X$, it follows $\varphi(x) \in \ell^\infty(X)$ with $\|\varphi(x)\|_\infty \leq d(x, x_0)$.

Let $x, x' \in X$. The relations

$$|\varphi(x)(y) - \varphi(x')(y)| = |d(x, y) - d(x', y)| \leq d(x, x'),$$

valid for all $y \in X$, imply

$$\|\varphi(x) - \varphi(x')\|_\infty \leq d(x, x').$$

Since $|\varphi(x)(x') - \varphi(x')(x')| = d(x, x')$ it follows

$$\|\varphi(x) - \varphi(x')\|_\infty = d(x, x'),$$

for all $x, x' \in X$, showing that φ is an isometric embedding of X into $\ell^\infty(X)$.

If (X, d) is separable, take a countable dense subset $\{x_i : i \in \mathbb{N}\}$ of X and define for $x \in X$, $\varphi(x) = (\varphi(x)_i)_{i \in \mathbb{N}}$, where

$$\varphi(x)_i = d(x, x_i) - d(x_i, x_0), \quad i \in \mathbb{N}.$$

Again $\|\varphi(x)\|_\infty \leq d(x, x_0)$ so that $\varphi(x) \in \ell^\infty$.

If $x, x' \in X$, then, as above,

$$|\varphi(x)_i - \varphi(x')_i|_\infty \leq d(x, x'), \tag{7.3.1}$$

for all $i \in \mathbb{N}$.

By the density of $\{x_i\}$ in X, there exists a subsequence $(x_{i_k})_{k \in \mathbb{N}}$ such that $x_{i_k} \to x'$. But then

$$|\varphi(x)_{i_k} - \varphi(x')_{i_k}| = |d(x, x_{i_k}) - d(x', x_{i_k})| \to d(x, x') \quad \text{as } k \to \infty. \tag{7.3.2}$$

The relations (7.3.1) and (7.3.2) show that

$$\|\varphi(x) - \varphi(x')\|_\infty = \sup_{i \in \mathbb{N}} |d(x_i, x) - d(x_i, x')| = d(x, x'),$$

that is, φ is an isometric embedding of X into ℓ^∞. □

Remark 7.3.2 The general case is due to Kuratowski [369], while the separated case was considered by Fréchet [242], for which reason this theorem is also known under the name of Kuratowski–Fréchet embedding theorem.

In the proof of the main result of this section we shall work with the Banach space $c_0(E)$ of vector sequences. If E is a Banach space, then one denotes by $c_0(E)$ the space of all sequences $z = (z_n)_{n \in \mathbb{N}}$, with $z_n \in E$, $n \in \mathbb{N}$, such that $\lim_{n \to \infty} \|z_n\| = 0$. Then $c_0(E)$ is a Banach space with respect to the norm

$$\|z\| = \sup\{\|z_n\| : n \in \mathbb{N}\}.$$

Another notation for $c_0(E)$ is $(E \oplus E \oplus \dots)_{c_0}$.

In particular, $c_0(c_0)$ is isomorphic to c_0 (see [549]).

Theorem 7.3.3 *There exists a constant $k \geq 1$ having the property that for each separable metric space (X, d) there is a function $T : X \to c_0$ such that*

$$d(x, y) \leq \|T(x) - T(y)\| \leq kd(x, y),$$

for all $x, y \in X$.

Proof By the Banach-Mazur theorem (Corollary 1.4.12) every separable metric space is isometric to a subset of $C[0, 1]$ which is a Banach space having a Schauder basis $(f_i)_{i \in \mathbb{N}}$ such that $\|f_i\| = 1$ for all $i \in \mathbb{N}$.

Therefore, without loss of generality, we can assume that X is a Banach space having a Schauder basis $(e_i)_{i \in \mathbb{N}}$ satisfying the conditions $\|e_i\| = 1$, $i \in \mathbb{N}$, and

$$\|P_n - P_m\| = 1$$

for all $m, n \in \mathbb{N}$, $m \neq n$.

For $a > 0$ define the mapping $T_a : X \to c_0$ by

$$(T_a x)_i = \max\left\{0, t_i^a - d(x, M_i^a)\right\},$$

for all $x \in X$ and every triple i.

Since $T_a x$ is finitely supported fore every $x \in X$ (by Lemma 7.2.7), T_a is well-defined and $T_a x \in c_0^+$ for each $x \in X$.

We have

$$T_a 0 = 0, \tag{7.3.3}$$

for all $a > 0$.

Indeed, for any triple i and $x \in M_i^a$, we have, by (7.2.6),

$$t_i^a = (k - 1)a \leq (n - 2)a \leq \|x\|,$$

and so $t_i^a \leq d(0, M_i^a)$, for all i.

Moreover,

$$\|T_a x - T_a y\| \leq \|x - y\|, \tag{7.3.4}$$

for all $x, y \in X$. In order to prove this we consider the following four cases concerning i.

(i) $(T_a x)_i = (T_a y)_i = 0$.

Then, obviously, we have $|(T_a x)_i - (T_a y)_i| \leq \|x - y\|$.

(ii) $(T_a x)_i \neq 0$ and $(T_a y)_i = 0$.

In such a case, we have $d(x, M_i^a) < t_i^a \leq d(y, M_i^a)$ and

$$t_i^a \leq d(y, M_i^a) \leq \|x - y\| + d(x, M_i^a).$$

Therefore

$$|(T_a x)_i - (T_a y)_i| = \left| t_i^a - d(x, M_i^a) \right| \leq \|x - y\|.$$

(iii) $(T_a x)_i = 0$ and $(T_a y)_i \neq 0$.

     This case is similar to (ii).

(iv) $(T_a x)_i \neq 0$ and $(T_a y)_i \neq 0$.

     In this case

$$|(T_a x)_i - (T_a y)_i| = \left| d(x, M_i^a) - d(y, M_i^a) \right| \leq \|x - y\|.$$

Hence $|(T_a x)_i - (T_a y)_i| \leq \|x - y\|$, for every i, which implies (7.3.4).

In addition, if $\|x\| \geq \|y\|$ and $\|x - y\| \geq 36a$, then, for any triple i_0 provided by Lemma 7.2.8 we have $(T_a y)_{i_0} = 0$ and

$$\frac{\|x - y\|}{4} \leq t_{i_0}^a - a \leq t_{i_0}^a - d(x, M_{i_0}^a) = (T_a x)_{i_0} \leq t_{i_0}^a.$$

Hence

$$\|T_a x - T_a y\| \geq \left| (T_a x)_{i_0} - (T_a y)_{i_0} \right| \geq t_{i_0}^a - a \geq \frac{\|x - y\|}{4}. \qquad (7.3.5)$$

Now, for $a > 0$, consider the mapping $S_a : X \to c_0$ given by

$$(S_a x)_i = \min \{(T_a x)_i, 72a\},$$

for every $x \in X$. Let us note that $S_a x \in c_0^+$ for all $x \in X$.

Obviously, by (7.3.3), $S_a 0 = 0$, $\|S_a x\| \leq 72a$, and $S_a x$ is finitely supported for each $x \in X$.

Moreover, we have

$$\|S_a x - S_a y\| \leq \|x - y\|, \qquad (7.3.6)$$

for all $x, y \in X$.

In order to justify this statement we consider the following possible cases.

(a) $(S_a x)_i = (T_a x)_i$ and $(S_a y)_i = (T_a y)_i$.

     Then, by (7.3.4),

$$|(S_a x)_i - (S_a y)_i| = |(T_a x)_i - (T_a y)_i| \leq \|x - y\|.$$

(b) $(S_a x)_i = (S_a y)_i = 72a$.

Then

$$|(S_a x)_i - (S_a y)_i| = 0 \le \|x - y\|.$$

(c) $(S_a x)_i = (T_a x)_i$ and $(S_a y)_i = 72a$.

In this case $(T_a x)_i \le 72a \le (T_a y)_i$, so, by (7.3.4),

$$|(S_a x)_i - (S_a y)_i| = 72a - (T_a x)_i \le (T_a y)_i - (T_a x)_i \le \|x - y\|.$$

(d) $(S_a y)_i = (T_a y)_i$ and $(S_a x)_i = 72a$.

This case is similar to (c).

We conclude that

$$|(S_a x)_i - (S_a y)_i| \le \|x - y\|,$$

for every triple i, which implies (7.3.6).

For $36a \le \|x - y\| \le 72a$ we have

$$\|S_a x - S_a y\| \ge \frac{\|x - y\|}{4}. \tag{7.3.7}$$

Indeed, taking into account Lemma 7.2.8, there exists a triple i_0 such that $d(y, M_{i_0}^a) \ge t_{i_0}^a$. Then $(T_a y)_{i_0} = 0$ and, therefore, $(S_a y)_{i_0} = 0$.

If $(S_a x)_{i_0} = (T_a x)_{i_0}$, then, by (7.3.5), we have

$$\left|(S_a x)_{i_0} - (S_a y)_{i_0}\right| = (T_a x)_{i_0} = \left|(T_a x)_{i_0} - (T_a y)_{i_0}\right| \ge \frac{\|x - y\|}{4}.$$

If $(S_a x)_{i_0} = 72a$, then we have

$$\left|(S_a x)_{i_0} - (S_a y)_{i_0}\right| = 72a \ge \|x - y\| \ge \frac{\|x - y\|}{4}.$$

Consequently, as

$$\|S_a x - S_a y\| \ge \left|(S_a x)_{i_0} - (S_a y)_{i_0}\right|,$$

Equation (7.3.7) is proved.

Now we define the mapping

$$T : X \to (c_0 \oplus c_0 \oplus \ldots)_{c_0}$$

by

$$T = 4(T_1 \oplus S_{\frac{1}{2}} \oplus \cdots \oplus S_{\frac{1}{2^n}} \oplus \ldots)$$

that is,

$$Tx = 4(T_1 x, S_{\frac{1}{2}} x, \ldots, S_{\frac{1}{2^n}} x, \ldots), \quad x \in X.$$

We claim that T has all the desired properties. Indeed, on the one hand, taking into account (7.3.4) and (7.3.6), we have:

$$\|T_1 x - T_1 y\| \le \|x - y\|$$

and

$$\left\| S_{\frac{1}{2^n}} x - S_{\frac{1}{2^n}} y \right\| \le \|x - y\|,$$

for all $n \in \mathbb{N}$ and $x, y \in X$. Therefore, as

$$\|Tx - Ty\| = 4 \sup\{\|T_1 x - T_1 y\|, \left\| S_{\frac{1}{2}} x - S_{\frac{1}{2}} y \right\|, \ldots, \left\| S_{\frac{1}{2^n}} x - S_{\frac{1}{2^n}} y \right\|, \ldots\},$$

we have

$$\|Tx - Ty\| \le 4 \|x - y\|,$$

for all $x, y \in X$.

On the other hand, for any $x, y \in X$ we have the following two mutually exclusive situations:

$$\text{(i)} \ \|x - y\| \le 36 \quad \text{and} \quad \text{(ii)} \ \|x - y\| > 36.$$

In the first case, there exists $n \in \mathbb{N}$ such that $36 \cdot 2^{-n} \le \|x - y\| \le 72 \cdot 2^{-n}$ and therefore, taking into account (7.3.7), we obtain

$$\|x - y\| \le 4 \left\| S_{\frac{1}{2^n}} x - S_{\frac{1}{2^n}} y \right\| \le \|Tx - Ty\|.$$

In the case (ii), taking into account (7.3.5), we have

$$\|x - y\| \le 4 \|T_1 x - T_1 y\| \le \|Tx - Ty\|.$$

Hence

$$\|x - y\| \le \|Tx - Ty\|$$

for all $x, y \in X$.   □

350   7   Lipschitz Isomorphisms of Metric Spaces

Remark 7.3.4 It is possible to take $k > 6$ in Theorem 7.3.3. Actually, the above proof does not ensure $k = 4$ because we have to renorm $C[0, 1]$ so that, for some Schauder basis, $\|P_n - P_m\| \le 1$ for all $m, n \in \mathbb{N}_0$ with $m \ne n$.

We shall show now that $k \ge 2$.

Theorem 7.3.5 *Let $T : \ell^1 \to c_0$ be a mapping such that*

$$\|x - y\| \le \|T(x) - T(y)\| \le k \|x - y\|,$$

for all $x, y \in \ell^1$. Then $k \ge 2$.

Proof We can assume, without loss of generality, that $T0 = 0$.
Suppose, by contradiction, that $k < 2$ and let

$$e_i = (0, \ldots, 0, 1, 0, \ldots) \in \ell^1,$$

where 1 is on the ith position, and

$$\mathscr{M} = \left\{ n \in \mathbb{N} : \left|(Te_1)_n - (Te_2)_n\right| \ge 4 - 2k \right\}.$$

For $i, j \in \mathbb{N}, i \ne j, i, j \ge 3$, define

$$\mathscr{M}_{ij} = \left\{ n \in \mathbb{N} : \left|(Te_i)_n - \left(Te_j\right)_n\right| \ge 4 - 2k \right\}.$$

The sets $\mathscr{M}$ and $\mathscr{M}_{ij}$ are finite, because $Te_i \in c_0$.
We claim that

$$\mathscr{M} \cap \mathscr{M}_{ij} \ne \emptyset, \tag{7.3.8}$$

for each $i, j \in \mathbb{N}, i \ne j, i, j \ge 3$. Indeed, since

$$\|T(e_1 + e_i) - T(e_2 + e_j)\| \ge \|(e_1 + e_i) - (e_2 + e_j)\| = 4,$$

there exists $n_0 \in \mathbb{N}$ such that

$$\left|(T(e_1 + e_i))_{n_0} - \left(T(e_2 + e_j)\right)_{n_0}\right| \ge 4.$$

Therefore

$$\left|(Te_1)_{n_0} - (Te_2)_{n_0}\right| \ge \left|(T(e_1 + e_i))_{n_0} - \left(T(e_2 + e_j)\right)_{n_0}\right|$$

$$- \left|(T(e_1 + e_i))_{n_0} - (Te_1)_{n_0}\right| - \left|\left(T(e_2 + e_j)\right)_{n_0} - (Te_2)_{n_0}\right|$$

$$\ge 4 - \|T(e_1 + e_i) - Te_1\| - \|T(e_2 + e_j) - Te_2\|$$

$$\ge 4 - k - k = 4 - 2k,$$

and so $n_0 \in \mathcal{M}$. Similarly, $\left|(Te_i)_{n_0} - (Te_j)_{n_0}\right| \geq 4 - 2k$, hence $n_0 \in \mathcal{M}_{ij}$.

Since $n_0 \in \mathcal{M} \cap \mathcal{M}_{ij}$, the proof of (7.3.8) is done.

Denote by $P_{\mathcal{M}}$ the canonical projection of c_0 onto span $(\{e_n\}_{n \in \mathcal{M}})$. The set $(P_{\mathcal{M}} Te_i)_{i \geq 3}$ is bounded since

$$\|P_{\mathcal{M}} Te_i\| \leq \|Te_i\| = \|Te_i - T0\| \leq k \|e_i\| = k,$$

for all $i \geq 3$. The set $(P_{\mathcal{M}} Te_i)_{i \geq 3}$ is infinite since, according to (7.3.8), we have $P_{\mathcal{M}} Te_i \neq P_{\mathcal{M}} Te_j$, for all $i, j \in \mathbb{N}, i \neq j, i, j \geq 3$.

As $\mathcal{M}$ is finite, $(P_{\mathcal{M}} Te_i)_{i \geq 3}$ is a bounded infinite set in a finite dimensional space, so it has an accumulation point. This contradicts the definitions of the sets $\mathcal{M}$ and $\mathcal{M}_{i,j}$. ☐

7.4   A Characterization of the Completeness of Normed Spaces in Terms of bi-Lipschitz Functions

In this section we prove that a normed space X is Banach if and only if there is no bi-Lipschitz function from X onto $X \setminus \{0\}$.

Given two metric spaces (X, d) and (Y, d'), a function $f : X \to Y$ is called M-bi-Lipschitz, where $M \geq 1$, if

$$\frac{1}{M} d(x, y) \leq d'(f(x), f(y)) \leq M d(x, y), \tag{7.4.1}$$

for all $x, y \in X$.

A function $f : X \to Y$ for which there exists $M \geq 1$ such that $f : X \to Y$ is M-bi-Lipschitz is called *bi-Lipschitz*.

Two subsets $A \subseteq X$, $B \subseteq Y$ are called *bi-Lipschitz equivalent* if there exists a bijective bi-Lipschitz mapping $f : A \to B$.

We start with the following lemma.

Lemma 7.4.1 *Let $x_1 \neq x_2$ be two elements of a normed space $(X, \| \cdot \|)$ and $r \geq 2 \|x_1 - x_2\|$ a real number. Then there exists a homeomorphism $h : X \to X$ having the following properties:*

(a) $h(x_1) = x_2$;

(b) $h(x) = x$ *for each $x \in X$ such that $\|x - x_1\| \geq r$;*

(c) h *is M-bi-Lipschitz, where $M = 1 + 2r^{-1} \|x_1 - x_2\|$.*

Proof We shall consider the function $h : X \to X$ given by

$$h(x) = \begin{cases} x + \left(1 - \frac{\|x - x_1\|}{r}\right)(x_2 - x_1), & \text{if } \|x - x_1\| \leq r \\ x, & \text{if } \|x - x_1\| > r. \end{cases}$$

It is obvious that h satisfies the conditions (a) and (b).

(c) The inequality $(1 + 2t)^{-1} \le 1 - t$, valid for all $t \in [0, 2^{-1}]$, yields for $t = \|x_1 - x_2\|/r$,

$$\left(1 + 2\frac{\|x_1 - x_2\|}{r}\right)^{-1} \le 1 - \frac{\|x_1 - x_2\|}{r},$$

i.e.,

$$\frac{1}{M} \le 1 - \frac{\|x_1 - x_2\|}{r}. \tag{7.4.2}$$

For $x, y \in X$ the following four cases are possible:

(i) $\|x - x_1\| \le r$ and $\|y - x_1\| < r$;

(ii) $\|x - x_1\| \ge r$ and $\|y - x_1\| > r$;

(iii) $\|x - x_1\| \le r$ and $\|y - x_1\| > r$;

(iv) $\|x - x_1\| \ge r$ and $\|y - x_1\| < r$.

In the case (i), as $h(x) - h(y) = x - y + r^{-1}(\|y - x_1\| - \|x - x_1\|)(x_2 - x_1)$, we have

$$(\alpha) \quad \|h(x) - h(y)\| \le \|x - y\| + \frac{\|x_2 - x_1\|}{r}\|x - y\| \le M\|x - y\|$$

and

$$(\beta) \quad \|h(x) - h(y)\| \ge \|x - y\| - \frac{\|x_2 - x_1\|}{r}\left|\|x - x_1\| - \|y - x_1\|\right|$$

$$\ge \|x - y\| - \frac{\|x_2 - x_1\|}{r}\|x - y\|$$

$$= \left(1 - \frac{\|x_2 - x_1\|}{r}\right)\|x - y\| \ge \frac{1}{M}\|x - y\|,$$

(the last inequality holds by (7.4.2)).

From (α) and (β), it follows that (c) is true in the case (i).

In the case (ii), (c) obviously holds, since $h(x) - h(y) = x - y$.

In the case (iii), as $h(x) - h(y) = x - y + (1 - \frac{\|x - x_1\|}{r})(x_2 - x_1)$, we have

$$(\alpha') \quad \|h(x) - h(y)\| \le \|x - y\| + (r - \|x - x_1\|)\frac{\|x_2 - x_1\|}{r}.$$

Since $r - \|x - x_1\| < \|y - x_1\| - \|x - x_1\| \le \|x - y\|$, we get

$$\|h(x) - h(y)\| \le \left(1 + \frac{\|x_2 - x_1\|}{r}\right) \|x - y\| \le M \|x - y\|.$$

Also,

$$(\beta') \quad \|h(x) - h(y)\| = \left\| x - y - (r - \|x - x_1\|)\frac{x_1 - x_2}{r} \right\|$$

$$\ge \|x - y\| - (r - \|x - x_1\|)\frac{\|x_2 - x_1\|}{r}.$$

Since, as we have seen, $r - \|x - x_1\| < \|x - y\|$, we have

$$\|h(x) - h(y)\| \ge \left(1 - \frac{\|x_2 - x_1\|}{r}\right) \|x - y\| \ge \frac{1}{M} \|x - y\|,$$

(the last inequality holds by (7.4.2)).

From (α') and (β'), (c) is also true in the case (iii).

The case (iv) is similar to (iii).

To prove that h is onto, take an arbitrary $y \in X$. If $\|y - x_1\| \ge r$, then $h(y) = y$.
Also, if $y = x_2$, then $h(x_1) = x_2$.

Suppose now that $\|y - x_1\| < r$, $y \ne x_2$, and let $z(s) = x_2 + s(y - x_2)$, $s \ge 0$.
Since

$$\|z(s) - x_1\| \le (1 - s)\|x_2 - x_1\| + s\|y - x_1\|$$

$$< (1 - s)\frac{r}{2} + sr < r,$$

for all $s \in [0, 1]$, there exists $s > 1$ such that $\|z(s) - x_1\| = r$. Putting $z = z(s)$ and $s = t^{-1} < 1$, it follows that $y = (1 - t)x_2 + tz$.

For $x = (1 - t)x_1 + tz$ we have $\|x - x_1\| = t\|z - x_1\| = tr < r$ and

$$h(x) = (1 - t)x_1 + tz + (1 - t)(x_2 - x_1) = (1 - t)x_2 + tz = y.$$

Consequently, h is onto. From (c) it is now obvious that h is one-to-one and that h and h^{-1} are continuous. Therefore h is a homeomorphism. □

Now we present some lemmas concerning bi-Lipschitz maps.

Lemma 7.4.2 *If* $f : X \to Y$ *is M-bi-Lipschitz, then* $f^{-1} : f(X) \to X$ *is also M-bi-Lipschitz.*

Proof For $u, v \in f(X)$ put $x = f^{-1}(u)$ and $y = f^{-1}(v))$ in (7.4.1) to obtain

$$\frac{1}{M}d'(u, v) \le d(f^{-1}(u), f^{-1}(v)) \le Md'(u, v).$$

Hence f^{-1} is M-bi-Lipschitz.   □

Lemma 7.4.3 *Let* $f : X \to Y$ *be an* M-*bi-Lipschitz function and* $(x_n)_{n\in\mathbb{N}}$ *a sequence of elements from* X. *Then* $(x_n)_{n\in\mathbb{N}}$ *is a Cauchy sequence if and only if* $(f(x_n))_{n\in\mathbb{N}}$ *is a Cauchy sequence.*

Proof This follows from the inequality

$$\frac{1}{M}d(x_n, x_m) \le d'(f(x_n), f(x_m)) \le Md(x_n, x_m),$$

which is valid for all $m, n \in \mathbb{N}$.   □

Lemma 7.4.4 *If* $f : X \to Y$ *is an* M-*bi-Lipschitz function and* A *is a complete subset of* X, *then* $f(A)$ *is a complete subset of* Y.

Proof Indeed, if $(f(x_n))_{n\in\mathbb{N}}$ is a Cauchy sequence of elements from $f(A)$, then, according to Lemma 7.4.3, $(x_n)_{n\in\mathbb{N}}$ is a Cauchy sequence of elements from A. As A is a complete subset of X, there exists $x \in A$ such that $\lim_{n\to\infty} x_n = x$. Then, by continuity of f, $(f(x_n))_{n\in\mathbb{N}}$ is convergent to $f(x) \in f(A)$.   □

Lemma 7.4.5 *If* $f : X \to Y$ *is an* M-*bi-Lipschitz function, where* (X, d) *and* (Y, d') *are complete metric spaces, then* $f(X)$ *is a closed subset of* Y.

Proof This follows from Lemma 7.4.4 and the fact that any complete subset of a metric space is closed.   □

The main result of this section, which gives a characterization of Banach spaces in terms of bi-Lipschitz functions, is the following.

Theorem 7.4.6 *A normed space* $(X, \|\cdot\|)$ *is a Banach space if and only if there exists no bi-Lipschitz function applying* X *onto* $X \setminus \{0\}$.

Proof $\Rightarrow$. Assume, by contradiction, that X is complete and that there exists a bi-Lipschitz onto function $f : X \to X \setminus \{0\}$. Then, according to Lemma 7.4.5, $f(X) = X \setminus \{0\}$ is a closed subset of X, and so $\{0\}$ is an open subset of X, a contradiction.

$\Leftarrow$. Assume that the normed space $(X, \|\cdot\|)$ is not complete and show that for every $M > 1$ there exists an M-bi-Lipschitz onto function $f : X \to X \setminus \{0\}$.

For $M > 1$ let $(t_n)_{n\in\mathbb{N}}$ be a sequence of real numbers such that $t_n \in (0, 1]$ for each $n \in \mathbb{N}$ and $\prod_{n=1}^{\infty} M_n \le M$, where $M_n = 1 + t_n$, $n \in \mathbb{N}$.

As $(X, \|\cdot\|)$ is not complete, there exists a Cauchy sequence $(x_n)_{n\in\mathbb{N}}$ in X which is not convergent. By passing to a subsequence, we can assume that $\|x_{n+1} - x_n\| \le 2^{-n}t_n$ for each $n \in \mathbb{N}$ and, by replacing x_n with $x_n - x_1$, we can further assume that $x_1 = 0$.

If B_n stands for the open ball with center x_n and radius 2^{1-n}, then $B_{n+1} \subseteq B_n$ for each $n \in \mathbb{N}$. Indeed, if $x \in B_{n+1}$, then

$$\|x - x_n\| \leq \|x - x_{n+1}\| + \|x_{n+1} - x_n\| < \frac{1}{2^n} + \frac{t_n}{2^n} \leq \frac{2}{2^n} = \frac{1}{2^{n-1}},$$

i.e., $x \in B_n$.

According to Lemma 7.4.1 (applied with x_n in place of x_1, x_{n+1} in place of x_2 and $2t_n^{-1} \|x_{n+1} - x_n\|$ in place of r), for each $n \in \mathbb{N}$ there exists a homeomorphism $h_n : X \to X$ having the following properties:

(a) $h_n(x_n) = x_{n+1}$;
(b) $h_n(x) = x$ for $x \in X$ with $\|x - x_n\| \geq 2^{1-n}$ ($\Longleftrightarrow x \notin B_n$);
   (this holds because $2^{1-n} \geq r$)
(c) h_n is M_n-bi-Lipschitz,

Then the functions $g_n = h_n \circ h_{n-1} \circ \cdots \circ h_2 \circ h_1$, $n \in \mathbb{N}$, are bijections satisfying the following conditions:

(i) $g_n(0) = x_{n+1}$ for each $n \in \mathbb{N}$;
(ii) $g_n(x) = x$ for all $x \notin B_1$ and $n \in \mathbb{N}$;
(iii) g_n is M-bi-Lipschitz for each $n \in \mathbb{N}$.

Indeed,

$$g_n(0) = g_n(x_1) = (h_n \circ h_{n-1} \circ \cdots \circ h_2)(h_1(x_1))$$

$$= (h_n \circ h_{n-1} \circ \cdots \circ h_2)(x_2) = \cdots = h_n(x_n) = x_{n+1}.$$

If $x \notin B_1$, then $x \notin B_j$, and so, by (b), $h_j(x) = x$ for each $j \in \mathbb{N}$ and therefore $g_n(x) = x$ for all $n \in \mathbb{N}$, proving (ii).

Finally, since h_j is M_j-bi-Lipschitz for each $j \in \{1, 2, \ldots, n\}$, g_n is $M_1 M_2 \ldots M_n$-bi-Lipschitz, so it is M-bi-Lipschitz.

Claim. For each $x \in X \setminus \{0\}$, there exists $\lim_{n \to \infty} g_n(x)$.

Indeed, given $x \in X \setminus \{0\}$, there exists $n_0 \in \mathbb{N}$ such that $\|x\| \geq 2^{-n_0} M$. Taking into account (i),

$$\left\| g_{n_0}(x) - x_{n_0+1} \right\| = \left\| g_{n_0}(x) - g_{n_0}(0) \right\| \geq \frac{1}{M} \|x\| \geq \frac{1}{2^{n_0}},$$

so that $g_{n_0}(x) \notin B_{n_0+1}$. Then, by (b), $g_{n_0+1}(x) = h_{n_0+1}(g_{n_0}(x)) = g_{n_0}(x) \notin B_{n_0+1}$, so $g_{n_0+1}(x) \notin B_{n_0+2}$ and, by mathematical induction, $g_n(x) = g_{n_0}(x)$ for all $n \geq n_0$. Consequently, there exists $\lim_{n \to \infty} g_n(x) = g_{n_0}(x)$.

Now we can define the function $f : X \setminus \{0\} \to X$ by

$$f(x) = \lim_{n \to \infty} g_n(x), \quad \text{for all } x \in X \setminus \{0\}.$$

As each g_n is M-bi-Lipschitz, we conclude that f is also M-bi-Lipschitz.

Finally, let us prove that f is onto. To this end, let $y \in X$. As $(x_n)_{n\in\mathbb{N}}$ has no convergent subsequences, there exists a real number $s > 0$ and $n_0 \in \mathbb{N}$ such that $\|x_n - y\| \geq s$ for all $n \geq n_0$. Choose $n_1 \in \mathbb{N}$ such that $n_1 \geq n_0$ and $2^{-n_1+1} \leq s$ and put $x = g_{n_1}^{-1}(y)$. Then $y \notin B_{n_1}$, since otherwise we get the contradiction $s \leq \|x_{n_1} - y\| < 2^{-n_1+1} \leq s$. Reasoning as above it follows that $g_n(x) = y$ for all $n \geq n_1$ and, consequently, $f(x) = \lim_{n\to\infty} g_n(x) = y$. Hence f is onto. $\qquad\square$

Remark 7.4.7 The function f also satisfies the condition $f(x) = x$ for each $x \in X$ with $\|x\| \geq 1$.

Indeed, if $\|x\| \geq 1$, then $x \notin B_n$, hence, by (ii), $g_n(x) = x$ for all $n \in \mathbb{N}$ and so $f(x) = x$.

7.5 Bibliographic Comments and Miscellaneous Results

The contents of Sects. 7.2 and 7.3 of this chapter heavily rely on Aharoni's results published in 1974 (see [9]), while Sect. 7.4 is based on Väisälä's results (see [663]).

For an exhaustive treatment of the subject of Lipschitz embeddings and isomorphisms, the authoritative treatises [75] and [533] are warmly recommended.

As we have seen in Sect. 7.3, there exists a constant K having the property that for every separable metric space (X, d) there exists a function $f : X \to c_0$ (actually f takes values in c_0^+) such that

$$d(x, y) \leq \|f(x) - f(y)\| \leq Kd(x, y)$$

for each $x, y \in X$.

It can be proved that one can take any $K > 6$.

Later, in 1978, Assouad [56] refined Aharoni's result by proving:

Let (X, d) be a separable metric space. For each $\varepsilon > 0$ there exists a function $f : X \to c_0^+$ such that

$$d(x, y) \leq \|f(x) - f(y)\| \leq (3 + \varepsilon)d(x, y)$$

for each $x, y \in X$.

Further improvements were obtained by Pelant [547] in 1994 who proved the following result:

For each separable metric space (X, d) there exists a function $f : X \to c_0^+$ such that

$$\frac{1}{3}d(x, y) \leq \|f(x) - f(y)\| \leq d(x, y)$$

for each $x, y \in X$.

Moreover, he proved that:

There is no function $f : \ell^1 \to c_0^+$ such that there exists some $\varepsilon > 0$ having the property that

$$\frac{1}{3-\varepsilon} d(x, y) \le \| f(x) - f(y) \| \le d(x, y)$$

for each $x, y \in \ell^1$.

Consequently, Pelant's result is sharp.

In the paper [548], published after the death of J. Pelant, further results concerning Lipschitz and uniform equivalence of Banach spaces are obtained. It is proved that the space $C[0, \omega_1]$, where ω_1 is the first uncountable ordinal, is not uniformly homeomorphic to any subset of $c_0(\Gamma)$ for any set Γ. One also gives an example of a compact space T such that the Banach space $C(T)$ has the same property and its Cantor-Bendixson derivative $T^{(\omega_0+1)}$ is empty, where ω_0 is the first infinite ordinal.

The optimal quantitative result was obtained by Kalton and Lancien [333] who proved the following improvement of Aharoni's theorem:

For every separable metric space (X, d) there exists a function $f : X \to c_0$ such that

$$d(x, y) < \| f(x) - f(y) \| \le 2d(x, y)$$

for all $x, y \in X$ with $x \ne y$.

Taking into account Theorem 7.3.5, this result is also sharp.

Other results in this direction were obtained by Mankiewicz [430]. He proved that a complete separable metric space (called a Polish space) is Lipschitz universal for all separable metric spaces if and only if it is Lipschitz universal for all separable reflexive Banach spaces. As the space c_0 is not Lipschitz embeddable into any reflexive Banach space (see [429]), the above mentioned result is not a direct corollary of Aharoni's theorem (Theorem 7.3.3).

Let us mention that it is not known whether the metric spaces c_0 and c_0^+ are Lipschitz isomorphic.

Bi-Lipschitz functions arise in a lot of areas of mathematics, such as geometric group theory, Banach space geometry, geometric analysis, conformal geometry, quasiconformal function theory, singular Riemannian geometry (see [534]) and in theoretical computer science (for the last one see [265] and [400]).

Definition 7.5.1 Let (X, d) and (Y, d') be metric spaces. A function $f : X \to Y$ is called a *bi-Lipschitz embedding* if there exists a real number $L \ge 1$ such that f is L-bi-Lipschitz.

Now we are going to present some other results concerning the following question:

When does a metric space admit a bi-Lipschitz embedding into some classical space?

In the study of bi-Lipschitz embeddability of ultrametric spaces into some classical spaces (such as $\mathbb{R}^n$) the concept of dimension introduced Assouad [57, 59] plays a central role.

A characterization in terms of Assouad dimension $\dim_A(X, d)$ (see Definition 3.4.16) of an ultrametric space (X, d) which are bi-Lipschitz embeddable in $\mathbb{R}^n$ is given by the following result (see Corollary 4.6 from [407]):

Let n be a positive integer. Then an ultrametric space (X, d) can be bi-Lipschitz embedded in $\mathbb{R}^n$ if and only if $\dim_A(X, d) < \infty$.

The implication $\Rightarrow$ is Theorem 4.5 from [407], while the implication $\Leftarrow$ is Theorem 3.8 from [411].

This result cannot be generalized to arbitrary metric spaces since the Heisenberg group with Carnot metric gives a counter-example to the sufficiency (see Theorem 7.1 from [628]).

A weak form of this result, due to Assouad [57], states that:

An ultrametric space (X, d) can be bi-Lipschitz embedded in $\mathbb{R}^n$, for some $n \in \mathbb{N}$, if and only if $\dim_A(X, d) < \infty$.

Assouad [59] and Movahedi-Lankarani [486] proved the following result:

Let $\mathcal{H}$ be an infinite-dimensional Hilbert space $\mathcal{H}$ and let (X, d) be a compact ultrametric space. Then there exists a bi-Lipschitz function $f : X \to \mathcal{H}$.

Some results on the isomorphisms of spaces of Lipschitz functions with other concrete Banach spaces (in particular with ℓ^∞) are proved in [55].

The paper [315] treats the following problem:

Given $C \in [1, \infty)$ and $m \in \mathbb{N}$, what is the smallest k such that any metric space with m points C-embeds into a normed space of dimension k?

We say that a metric space (X, d) C-embeds into another metric space (X', d') if there exists a subset Y of X' and a bijection $f : X \to Y$ such that $L(f)L(f^{-1}) \le C$.

The paper [377] describes some basic geometric tools to construct bi-Lipschitz embeddings of metric spaces into finite-dimensional Euclidean or hyperbolic spaces.

Let us now say some words about embedding of compacta hyperspaces in specified target spaces.

Let $\mathcal{P}_k(X)$ be the family of all nonempty compact subsets of a metric space (X, d) and d_H the Pompeiu–Hausdorff metric on $\mathcal{P}_k(X)$ (see Sect. 1.3.4). We call the pair $(\mathcal{P}_k(X), d_H)$ the *compacta hyperspace* of the metric space (X, d).

A metric space (X, d) is called *uniformly disconnected* if there exists an ultrametric d' on X such that the identity function from (X, d) to (X, d') is bi-Lipschitz.

Tyson [660] studied bi-Lipschitz embeddings of the hyperspaces $\mathcal{P}_k(X)$ of special spaces X into some standards spaces (such as $\mathbb{R}^n$ or ℓ^2). For example he proved that:

If (X, d) is a separable uniformly disconnected metric space, then $\mathcal{P}_k(X)$ admits a bi-Lipschitz embedding in ℓ^2.

Let us mention a result concerning ε-isometries between Banach spaces. Let X and Y be real Banach spaces and $0 \leq \varepsilon < 1$. A function $f : X \to Y$ is called an *ε-isometry* if

$$(1 - \varepsilon) \|x - y\| \leq \|f(x) - f(y)\| \leq (1 + \varepsilon) \|x - y\|,$$

for all $x, y \in X$. Obviously, a 0-isometry is an isometry.

A classical result of Mazur and Ulam [454] states that a surjective isometry f between two real Banach spaces X and Y such that $f(0) = 0$ is linear. A generalization of this result is due to Jun and Park [321] who proved the following result:

Let $\varepsilon \in (0, 1/3)$. Then every ε-isometry f between two real Banach spaces X and Y such that $f(0) = 0$ satisfies the inequality

$$\|f(x + y) - f(x) - f(y)\| \leq c_1(3\varepsilon)^{c_2}(\|x\| + \|y\|) \quad \text{for all } x, y \in X, \qquad (7.5.1)$$

for some $c_1, c_2 > 0$.

A function f as above satisfying

$$\|f(x + y) - f(x) - f(y)\| \leq \phi(\varepsilon)(\|x\| + \|y\|) \quad \text{for all } x, y \in X, ,$$

where $\lim_{\varepsilon \searrow 0} \phi(\varepsilon) = 0$, is called *almost additive*. In [321] it is proved that an ε-isometry f also satisfies and "almost-homogeneity" condition

$$\|f(\lambda x) - \lambda f(x)\| \leq \psi_1(\varepsilon)\psi_2(\lambda)\|x\| \quad \text{for all } x \in X \text{ and } \lambda \in \mathbb{R}, \qquad (7.5.2)$$

where $\lim_{\varepsilon \searrow 0} \phi_1(\varepsilon) = 0$ and $\lim_{\lambda \to 0} \phi_2(\lambda) = 0$. The authors of [321] call a function f satisfying (7.5.1) and (7.5.2) *almost linear*.

For more results on bi-Lipschitz embeddings one can consult [487], where a sufficient condition for the existence of a bi-Lipschitz embedding of a compact metric-measure space into some $\mathbb{R}^n$ is provided. See also [629] where an example of a sub-Riemannian manifold that embeds in a bi-Lipschitz way into some Euclidean space is exhibited, or [532] and [531], where, by weakening the notion of bi-Lipschitz embeddings, it is shown that any homogeneous metric space can be embedded into an infinite-dimensional Hilbert space using an almost bi-Lipschitz function (i.e., bi-Lipschitz functions to within logarithmic corrections). Let us also mention that there are examples of homogeneous spaces that do not admit a bi-Lipschitz embedding into any $\mathbb{R}^n$ nor into an infinite dimensional Hilbert space (see [372] and [374]).

The notion of bi-Lipschitz function plays an increasingly role in fractal theory. More precisely, in the study of self-similar sets, a crucial tool is the Lipschitz equivalence property since with it many important properties of these sets are preserved and thus it is a suitable quality to decide whether two fractal sets are similar. While topology can be viewed as the study of equivalence classes of sets under homeomorphisms, fractal geometry can be considered as the study of equivalence classes under bi-Lipschitz functions which is a good compromise between isometries and continuous functions, the first ones leading to uninteresting equivalence classes and the second ones leading to pure topological classes.

Several techniques to study the Lipschitz equivalence of dust like self-similar sets (i.e., attractors of iterated function systems satisfying the strong separation condition) have been developed by Falconer and Marsh [234] who established conditions for Lipschitz equivalence based on the algebraic properties of the contraction ratios (see also [590]). These conditions imply that many self-similar subsets of Hausdorff dimension $\log 2 / \log 3$ are not Lipschitz equivalent to the classical $1/3$ Cantor set. See also [677], where self-similar sets with the same Hausdorff dimension, but which are not Lipschitz equivalent are constructed. In addition, in [406] some sufficient conditions are provided to guarantee that if two self-conformal sets E and F have Lipschitz equivalent subsets of Hausdorff positive measure, then there exists a bi-Lipschitz function f of E into, or onto, F. In [186] it is proved that each self-similar set satisfying the strong separation condition can be bi-Lipschitz embedded into each self-similar set with larger Hausdorff dimension and that a bi-Lipschitz embedding between two self-similar sets having the same Hausdorff dimension and satisfying the strong separation condition is only possible if the two sets are bi-Lipschitz equivalent.

Two dust-like self-similar sets with the same contraction ratios are always Lipschitz equivalent, but, as shown in [610], for self-similar sets with touching structures the problem of Lipschitz equivalence becomes much more challenging and intriguing. The authors establish results for the Lipschitz equivalence of self-similar sets with touching structures in $\mathbb{R}$ with arbitrarily many branches.

Another technique is due to Xi and Ruan [688] who proved that a bi-Lipschitz function between two dust-like self-similar sets has a certain measure-preserving property, generalizing a measure-preserving property due to Cooper and Pignataro [170]. Let us mention that sufficient and necessary conditions for the Lipschitz equivalence of dust-like self-similar sets in terms of graph-directed sets can be found in [686] and [688].

David and Semmes [176] posed several problems concerning the Lipschitz equivalence of non-dust-like self-similar sets which have been solved by Rao et al. [588], Xi et al. [689] and Xi and Ruan [687].

Some connections between fractal dimension and Lipschitz–Hölder parametrization are established in [240].

Related results on Lipschitz equivalence of some other fractals can be found in [185, 235, 274, 414, 438, 589, 678, 684–691].

For an up-to-date survey on the study of Lipschitz equivalence of self-similar sets one can consult [591].

We present next a notion that plays an important role in the bi-Lipschitz embeddability of a metric space into another metric space.

Given the metric spaces (X, d) and (Y, d'), a function $f : X \to Y$ is said to have *distortion at most $D \geq 1$* if there exists $s > 0$ such that

$$s\, d(x, y) \leq d'(f(x), f(y)) \leq s\, D\, d(x, y),$$

for all $x, y \in X$. The optimal such D is called the *distortion* of f. The infimum of all constants $D \geq 1$ such that that there exists $f : X \to Y$ having distortion at most D is denoted by $c_{(Y,d')}(X, d)$ (or simply $c_Y(X)$). In the case $Y = L_p$ it is denoted by $c_p(X)$. By the Euclidean distortion of X we mean $c_2(X)$.

The estimation of $c_Y(X)$ is an important problem since if we can bound it from above, then there exists a low-distortion embedding of X into Y, while if we can find a large lower bound of it, then we infer that there exists an invariant on Y which does not allow X to be embedded into Y in a nice manner.

A fundamental result on embeddings of finite metric spaces into Hilbert spaces, which is due to Bourgain (see [109]), can be stated in the following way: *for every n-point metric space X, $c_2(X) \lesssim \log n$*, where $\lesssim$ and $\gtrsim$ designate the corresponding inequalities up to universal constant factors. Moreover, Bourgain's result is asymptotically sharp as *for arbitrarily large n there exists an n-point metric space X with $c_2(X) \gtrsim \log n$* (see [401]).

Significant results containing highly non-trivial $c_1(X)$ bounds can be found in [353].

A natural question (whose motivation is also based on a classical theorem about finite-dimensional Banach spaces which is due to Dvoretzky, see [215]) is to find out if any finite metric space contains large subsets which embed into Euclidean space with low distortion. A sharp quantitative answer (due to Mendel and Naor, see [460]) can be stated as follows:

For every n-point metric space X and every $\varepsilon \in (0, 1)$, there exists $S \subseteq X$ such that $c_2(S) \lesssim \frac{1}{\varepsilon}$ and $|S| \geq n^{1-\varepsilon}$.

Assaf Naor raised the following question:

Consider a metric space X and suppose $X = A \cup B$ such that $c_2(A) < \infty$ *and* $c_2(B) < \infty$. Does this imply that $c_2(X) < \infty$?

The answer is given by the following result (due to Makarychev and Makarychev, see [425]):

Consider a metric space X and assume that X is the union of two metric subspaces A and B that embed into ℓ_a^2 and ℓ_b^2 with distortions D_A and D_B, respectively. Then X embeds into ℓ_{a+b+1}^2 with distortion at most $7D_A D_B + 2(D_A + D_B)$. If $D_A = D_B = 1$, then X embeds into ℓ_{a+b+1}^2 with distortion at most 8.93.

We mention that a and b may be finite or infinite and that we denoted the k-dimensional Euclidean space by ℓ_k^2 and ℓ_∞^2 is the Hilbert space ℓ^2.

The Ribe Program
We finish this chapter with a few comments about the Ribe program.

As Werner [679] wrote in his review of Naor's paper [515]:

Martin Ribe is a Swedish statistician who published very few, but highly influential papers on functional analysis in the 1970s. [. . .] In his landmark work [593] and [594] he proved the following rigidity result. If X and Y are uniformly homeomorphic Banach spaces, then X is (crudely) finitely representable in Y and vice versa.

(Both MathSciNet and Zentralblatt MATH mention 7 papers by Ribe. The text from Zentralblatt is slightly modified, but we preserved the basic ideas).

Let us introduce the necessary notions. A *uniform homeomorphism* between two Banach spaces X and Y is a bijection (not necessarily linear) $\varphi : X \rightarrow Y$ such that both φ and φ^{-1} are uniformly continuous. One says that a Banach space X is *crudely finitely representable* in a Banach space Y if there exists a constant $C > 0$ such that for every finite dimensional subspace F of X there exists a linear isomorphism $T : F \rightarrow T(F)$ satisfying $\|T\|\|T^{-1}\| \leq C$. Assuming that F is not trivial, by rescaling, we may assume that $\|T^{-1}\| = 1$ and the condition $\|T\| \leq C$ can be written as

$$\|x\| \leq \|Tx\| \leq C\|x\| \quad \text{for all } x \in X.$$

In other words, the linear operator T has distortion at most C in the sense defined above.

One says also that X is *C-crudely finitely representable* in Y. The space X is *finitely representable* in Y if it is $(1+\varepsilon)$-crudely finitely representable in Y for every $\varepsilon > 0$. These notions were introduced by James [301]. We mention the following examples (see [233, Ch. 6]):

- every Banach space is finitely representable in c_0;
- the space $L^p[0, 1]$ is finitely representable in ℓ^p, for very $1 \leq p < \infty$;
- if a Banach space X is finitely representable in a Hilbert space, then X is isomorphic to a Hilbert space.

Another important result is the principle of local reflexivity of Lindenstrauss and Rosenthal [394] asserting that X^{**} is always finitely representable in X, although X can be strictly contained in X^{**} (for a proof, see [233, Theorem 6.3]). Also, a Banach space X is superreflexive (see Sect. 1.4.15) if and only if every Banach space that is crudely finitely representable in X is reflexive. Some authors take this property as definition for superreflexivity (see [233, §9.2]).

Finite representability means that two Banach spaces with the property that each of them is finitely representable in the other one have "almost" the same finite dimensional subspaces. Banach space properties expressed in terms of finite dimensional subspaces are called by the Banach space specialists as "local". This opened the way to the metric characterizations of many Banach space properties under the generic name of the Ribe program. This started with Bourgain [110] who gave a metric characterization of superreflexivity and formulated the Ribe program as a search for local properties of Banach spaces which admit metric characterizations—type, cotype, etc. At the same time, as mentioned by Naor

[516], the results from the paper by Johnson and Lindenstrauss [310] *"contained inspirational (even prophetic) ideas that had major subsequent impact on the Ribe program"*. The metric analog of cotype was defined by Mendel and Naor [461].

This analogy, between properties of metric spaces and local properties of Banach space proved to be very useful for both sides, as well as in applications to some a priori unrelated domains. We quote from [515]:

> We will explain how this suggests that, despite having no a priori link to Banach spaces, general metric spaces have a hidden structure. Using this point of view, insights from Banach space theory can be harnessed to solve problems in seemingly unrelated disciplines, including group theory, algorithms, data structures, Riemannian geometry, harmonic analysis and probability theory.

Good presentations of the results and problems concerning the Ribe program are given in the surveys [64, 259, 332, 515, 516]. For results on the uniform classification of Banach spaces we recommend [74].

Chapter 8
Banach Spaces of Lipschitz Functions

In this chapter we introduce several Banach spaces of Lipschitz functions (Lipschitz functions vanishing at a fixed point, bounded Lipschitz functions, little Lipschitz functions) on a metric space and present some of their properties. A detailed study of free Lipschitz spaces is carried out, including several ways to introduce them and duality results. The study of Monge–Kantorovich and Hanin norms is tightly connected with Lipschitz spaces, mainly via the weak convergence of probability measures, a topic treated in Sects. 8.4 and 8.5. Compactness and weak compactness properties of Lipschitz operators on Banach spaces and of compositions operators on spaces of Lipschitz functions are also studied, emphasizing the key role played by the Lipschitz free Banach spaces. Another theme presented here is the Bishop–Phelps property for Lipschitz functions, meaning density results for Lipschitz functions that attain their norms. Finally, applications to best approximation in metric spaces and in metric linear spaces X are given in the last section (Sect. 8.9) of this chapter, showing how results from the linear theory can be transposed to this situation, by using as a dual space the space of Lipschitz functions defined on X.

8.1 The Basic Metric and Lipschitz Spaces

In this section we shall consider two ways to organize the space of Lipschitz functions on a metric space as a Banach space.

For a metric space X and a normed space Y denote by

$$\mathrm{Lip}(X, Y) = \{f : X \to Y : f \text{ is Lipschitz on } X\} \qquad (8.1.1)$$

the space of all Lipschitz functions from X to Y.

© Springer Nature Switzerland AG 2019
Ş. Cobzaş et al., *Lipschitz Functions*, Lecture Notes in Mathematics 2241,
https://doi.org/10.1007/978-3-030-16489-8_8

As we have remarked the functional

$$L(f) = \sup\{\|f(x) - f(y)\|/d(x, y) : x, y \in X, \ x \ne y\} \tag{8.1.2}$$

is only a seminorm on $\mathrm{Lip}(X, Y)$, as it vanishes on constant functions. In spite of this we call it the *Lipschitz norm* of f.

Remark 8.1.1 In some cases, to emphasize the fact that it is a "norm", the Lipschitz constant of $f \in \mathrm{Lip}(X, Y)$ (for X a metric space and Y a normed space) will be denoted by $\|f\|_L$.

We have two ways to norm the spaces of Lipschitz functions. The first one is to consider the space

$$\mathrm{BLip}(X, Y) = \{f \in \mathrm{Lip}(X, Y) : f \text{ is bounded on } X\}$$

of all bounded Lipschitz functions from X to Y and two equivalent norms on it given by

$$\begin{aligned}
&\text{(i)} \quad \|f\|_{\max} = \max\{L(f), \|f\|_\infty\} \quad - \quad \text{the max-norm,} \\
&\text{(ii)} \quad \|f\|_{\mathrm{sum}} = L(f) + \|f\|_\infty \quad - \quad \text{the sum-norm,}
\end{aligned} \tag{8.1.3}$$

for $f \in \mathrm{BLip}(X, Y)$, where

$$\|f\|_\infty = \sup\{\|f(x)\| : x \in X\}.$$

It is obvious that if X is of finite diameter (in particular if X is compact), then $\mathrm{BLip}(X, Y) = \mathrm{Lip}(X, Y)$. In this case we shall still use the notation $\mathrm{BLip}(X, Y)$ to indicate that we are working with one of the norms (8.1.3).

Remark 8.1.2 It is obvious that the norms (8.1.3) define the same topology on $\mathrm{BLip}(X, Y)$. The convergence of a sequence (f_n) in $\mathrm{BLip}(X, Y)$ to $f \in \mathrm{BLip}(X, Y)$ is equivalent to

$$f_n \xrightarrow{u} f \text{ and } L(f_n - f) \to 0, \text{ as } n \to \infty.$$

If X_1, X_2 are metric spaces, then the continuity of a linear operator T from $\mathrm{BLip}(X_1, Y)$ to $\mathrm{BLip}(X_2, Y)$ is equivalent to the existence of two numbers $\alpha, \beta \ge 0$ such that

$$\|Tf\|_\infty \le \alpha \|f\|_\infty \text{ and } L(Tf) \le \beta L(f),$$

for all $f \in \mathrm{BLip}(X_1, Y)$.

A *base point* (or *distinguished point*) in a metric space (X, d) is a fixed element $\theta \in X$. A metric space with a specified base point θ is called a *pointed metric space*.

If X is a normed space then one always takes $\theta = 0$. We shall mention this by saying that (X, θ) (or (X, d, θ)) is a pointed metric space.

If X, Y are pointed metric spaces with base points θ, θ', respectively, then one denotes by

$$\text{Lip}_0(X, Y) = \{f : X \to Y : f \text{ is Lipschitz on } X \text{ and } f(\theta) = \theta'\}$$

the set of all Lipschitz mappings from X to Y preserving the base points. We say that a mapping $g : X \to Y$ such that $g(\theta) = \theta'$ *preserves the base points.*

If Y is a normed space, then the functional $L(\cdot)$ given by (8.1.2) is a norm on the vector space $\text{Lip}_0(X, Y)$. It is a Banach space with respect to this norm if Y is a Banach space (see Theorem 8.1.3).

When more precision is required, one writes $\text{Lip}_\theta(X, Y)$ (or $\text{Lip}_{x_0}(X, Y)$) to specify the base point θ (resp. x_0).

If $Y = \mathbb{K}$, then

$$\text{Lip}_0(X, \mathbb{K}) = \{f : X \to \mathbb{K} : f \text{ is Lipschitz on } X \text{ and } f(\theta) = 0\}.$$

In the case $Y = \mathbb{K}$ these spaces are denoted by

$$\text{Lip}(X), \text{ BLip}(X), \text{ Lip}_0(X), \text{ Lip}_\theta(X), \text{ Lip}_{x_0}(X).$$

When necessary, it will be specified that we work with $\mathbb{R}$ or $\mathbb{C}$.

It is also possible to obtain a norm on $\text{Lip}(X)$ by considering the quotient space $\text{Lip}(X)/C$, where C denotes the one-dimensional subspace of $\text{Lip}(X)$ formed by constant functions, and the quotient norm $L(\tilde{f}) = L(f)$, for $\tilde{f} = f + C \in \text{Lip}(X)/C$. The spaces $\text{Lip}(X)/C$ and $\text{Lip}_0(X)$ are isometrically isomorphic, regardless of the base point, but in the space $\text{Lip}(X)/C$ there is no good way to define the product and the order (in the real case). For these reasons it is preferable to work with the space $\text{Lip}_0(X)$ instead of $\text{Lip}(X)/C$. It also follows that for two base points θ, θ' the mapping $f \mapsto f - f(\theta')$ is an isometric isomorphism between the spaces $\text{Lip}_\theta(X)$ and $\text{Lip}_{\theta'}(X)$, but this isomorphism is not compatible with products and does not preserve the order.

Along with these spaces, for a metric space (X, d), a normed space Y and $0 < \alpha \leq 1$ one can consider also the corresponding spaces of Hölder functions $\text{Lip}_\alpha(X, Y)$ (see Sect. 1.3.2) and

$$\text{BLip}_\alpha(X, Y) := \{f \in \text{Lip}_\alpha(X, Y) : f \text{ is bounded}\}, \cdot \qquad (8.1.4)$$

and

$$\text{Lip}_{\alpha,0}(X, Y) := \{f \in \text{Lip}_\alpha(X, Y) : f(\theta) = 0\}, \qquad (8.1.5)$$

where X is a pointed metric space with base point θ. Notice that $\mathrm{BLip}_1(X, Y) = \mathrm{BLip}(X, Y)$, $\mathrm{Lip}_{1,0}(X, Y) = \mathrm{Lip}_0(X, Y)$ and

$$\mathrm{BLip}_\alpha((X, d), Y) = \mathrm{BLip}((X, d^\alpha), Y) \quad \text{and} \quad \mathrm{Lip}_{\alpha,0}((X, d), Y) = \mathrm{Lip}_0((X, d^\alpha), Y).$$

In the case $Y = \mathbb{K}$ one uses the notation $\mathrm{Lip}_\alpha(X)$, $\mathrm{BLip}_\alpha(X)$, $\mathrm{Lip}_{\alpha,0}(X)$. Again we have

$$\mathrm{Lip}_\alpha(X) = \mathrm{Lip}(X, d^\alpha), \ \mathrm{BLip}_\alpha(X) = \mathrm{BLip}(X, d^\alpha), \ \mathrm{Lip}_{\alpha,0}(X) = \mathrm{Lip}_0(X, d^\alpha).$$

Theorem 8.1.3 *Let (X, d) be a metric space and Y a Banach space.*

1. *$\mathrm{BLip}(X, Y)$ is a Banach space with respect to any of the norms (8.1.3).*
2. *If X is a pointed metric space, then $\mathrm{Lip}_0(X, Y)$ is a Banach space with respect to the norm (8.1.2).*

The proof will be based on the following lemma.

Lemma 8.1.4 *Let (X, d) be a metric space, Y a normed space and (f_n) a sequence in $\mathrm{Lip}(X, Y)$ that is Cauchy with respect to the Lipschitz norm, and $f : X \to Y$ a function. If $f_n(x) \to f(x)$ for every $x \in X$, then f is Lipschitz on X and*

$$\lim_{n \to \infty} L(f - f_n) = 0,$$

i.e., (f_n) converges to f with respect to the Lipschitz norm $L(\cdot)$.

Proof Since a Cauchy sequence is bounded, Proposition 2.4.1 shows that the function f is Lipschitz.

Let us show that

$$\lim_{n \to \infty} L(f_n - f) = 0.$$

Given $\varepsilon > 0$ let $n_0 \in \mathbb{N}$ be such that

$$L(f_{n+k} - f_n) \leq \varepsilon,$$

for all $n \geq n_0$ and all $k \in \mathbb{N}$.

Let $x, x' \in X$. Then

$$\| f_{n+k}(x) - f_n(x) - (f_{n+k}(x') - f_n(x')) \| \leq \varepsilon d(x, x'),$$

for all $n \geq n_0$ and all $k \in \mathbb{N}$. Letting $k \to \infty$ one obtains

$$\| (f - f_n)(x) - (f - f_n)(x') \| \leq \varepsilon d(x, x'),$$

for all $n \geq n_0$. Since x, x' were arbitrarily chosen in X, it follows that

$$L(f - f_n) \leq \varepsilon,$$

for all $n \geq n_0$. This shows that the sequence (f_n) converges to f with respect to the Lipschitz norm. $\qquad\square$

Proof of Theorem 8.1.3

1. Suppose that the sequence (f_n) is Cauchy with respect to the norm $\|\cdot\|_{\max}$. Then it is Cauchy with respect to the norm $\|\cdot\|_\infty$, so it converges uniformly to a bounded function f. Since the sequence (f_n) is Cauchy with respect to the Lipschitz norm too, Lemma 8.1.4 yields its convergence to f with respect to the Lipschitz norm and the fact that f is Lipschitz. Consequently, (f_n) converges to $f \in \mathrm{BLip}(X, Y)$ with respect to the norms $\|\cdot\|_\infty$ and $L(\cdot)$, and so with respect to the norm $\|\cdot\|_{\max}$.

2. Let (X, d, θ) be a pointed metric space and (f_n) a sequence in $\mathrm{Lip}_0(X, Y)$ which is Cauchy with respect to the Lipschitz norm. Given $\varepsilon > 0$ let $n_0 \in \mathbb{N}$ be such that $L(f_{n+k} - f_n) \leq \varepsilon$, for all $n \geq n_0$ and all $k \in \mathbb{N}$. Then, for every $x \in X$,

$$\|(f_{n+k} - f_n)(x)\| = \|(f_{n+k} - f_n)(x) - (f_{n+k} - f_n)(\theta)\| \leq \varepsilon d(x, \theta),$$

for all $n \geq n_0$ and all $k \in \mathbb{N}$. This shows that, for every $x \in X$, the sequence $(f_n(x))$ is Cauchy in Y, so it converges to some $f(x)$. The function f so defined satisfies $f(\theta) = 0$ and, by Lemma 8.1.4, $f \in \mathrm{Lip}_0(X)$ and (f_n) converges to f with respect to the Lipschitz norm. $\qquad\square$

The pointwise and Lipschitz convergence of sequences in the space $\mathrm{Lip}_0(X)$ are related in the following way.

Proposition 8.1.5 *Let (X, d, θ) be a pointed metric space and Y a normed space. If (f_n) is a sequence in $\mathrm{Lip}_0(X, Y)$ such that $L(f_n - f) \to 0$, for some $f \in \mathrm{Lip}_0(X, Y)$, then (f_n) converges pointwise to f.*

If the metric space X is further bounded, then the convergence is uniform, i.e., $f_n \overset{u}{\to} f$.

Proof The first assertion follows from the inequality

$$\|f_n(x) - f(x)\| = \|(f_n - f)(x) - (f_n - f)(\theta)\| \leq L(f_n - f)d(x, \theta),$$

valid for every $x \in X$.

If X is bounded, then

$$\|f_n(x) - f(x)\| \leq L(f_n - f) \cdot \mathrm{diam}\, X,$$

for every $x \in X$, showing that $f_n \overset{u}{\to} f$. $\qquad\square$

Remark 8.1.6 A similar result does not hold in Lip(X). Taking $f_n \equiv 1$ and $f \equiv 0$, it follows that $L(f_n - f) = 0$ for all n, but for every $x \in X$, $|f_n(x) - f(x)| = 1$ for all $n \in \mathbb{N}$.

Remark 8.1.7 On the space BLip(X) one can define a product by $(f \cdot g)(x) = f(x)g(x)$, $x \in X$, and so it becomes an algebra having as unit the function $\mathbf{1}_X \equiv 1$. We have defined two norms on BLip(X): $\| \cdot \|_{max}$ and $\| \cdot \|_{sum}$ (see (8.1.3)) The norm $\| \cdot \|_{sum}$ satisfies the condition $\|fg\|_{sum} \leq \|f\|_{sum}\|g\|_{sum}$ and BLip(X) becomes a Banach algebra. The equivalent norm $\| \cdot \|_{max}$ satisfies only the inequality $\|fg\|_{max} \leq 2\|f\|_{max}\|g\|_{max}$, but as Weaver asserts in [675], this factor 2 is not a serious impediment in their study. On the other hand, as pointed out in loc. cit, the norm $\| \cdot \|_{max}$ has some advantages over $\| \cdot \|_{sum}$, being more natural in applications. A complete normed algebra where the norm satisfies the inequality $\|xy\| \leq c\|x\| \cdot \|y\|$ for some $c \geq 1$ is called a *weak Banach algebra*. Although some authors call them Banach algebras too (see [675, p. 10]).

Example 8.1.8 The space Lip$_0$[0, 1] of Lipschitz functions on [0, 1] vanishing at 0 with the Lipschitz norm is isometrically isomorphic to the Banach space L^∞[0, 1]. The isomorphism is given by the correspondence $\Phi(f) = F$, $f \in L^\infty$[0, 1], where $F(x) = \int_0^x f(t)dt$, $x \in [0, 1]$.

The inverse mapping $\Phi^{-1} : \text{Lip}_0[0, 1] \to L^\infty[0, 1]$ is given by $\Phi^{-1}(g) = g'$ a.e..

This isomorphism Φ is not compatible with products, but it preserves the order: $f_1 \leq f_2 \Rightarrow F_1 \leq F_2$. The inverse mapping Φ^{-1} does not preserve the order.

Indeed, for $f \in L^\infty[0, 1]$, the function $F(x) = \int_0^x f(t)dt$ is absolutely continuous, $F'(x) = f(x)$ a.e. on [0, 1] and

$$\left| F(x) - F(y) \right| = \left| \int_x^y f(t)dt \right| \leq \|f\|_\infty |x - y| ,$$

so that $L(F) \leq \|f\|_\infty$.

The function F remains unchanged if we modify the function f on a set of measure 0. Let E be a subset of Lebesgue measure 0 of [0, 1] such that F is differentiable at every point $x \in [0, 1] \setminus E$. Modify f on E by putting $f(x) = \|f\|_\infty$ for all $x \in E$. For every $x \in [0, 1] \setminus E$, the inequality

$$\left| \frac{F(x + y) - F(x)}{y} \right| \leq L(F) ,$$

yields for $y \to 0$, $|f(x)| = |F'(x)| \leq L(F)$, so that $\|f\|_\infty \leq L(F)$. Hence $L(f) = \|f\|_\infty$.

Remark 8.1.9 In the correspondence from Example 8.1.8 we actually work with classes of equivalences of essentially bounded measurable functions on [0, 1].

Weaver [675] considers several classes of metric spaces (X, d) adequate to the study of Lipschitz spaces:

$\mathcal{M}_2$—the class of complete metric spaces with diam $X \leq 2$;

$\mathcal{M}^k$—the class of compact metric spaces;

$\mathcal{M}_0$—the class of pointed complete metric spaces;

$\mathcal{M}_0^k$—the class of pointed compact metric spaces;

$\mathcal{M}_0^f$—the class of pointed bounded complete metric spaces $(\text{diam } X < \infty)$.

Here

$$\text{diam } Y = \sup\{d(y, y') : y, y' \in Y\},$$

for every subset Y of X.

Remark 8.1.10 Starting from a metric space $X \in \mathcal{M}_2$ one can obtain a space $X^\bullet \in \mathcal{M}_0$ by attaching an ideal point θ and defining $d(x, \theta) = 1$ for all $x \in X$.

Observe that the completeness is not a restriction in the study of spaces of Lipschitz functions.

Proposition 8.1.11 *Let (X, d) be a metric space and $\tilde{X}$ its completion. Then the space* $\text{BLip}(X)$ *can be identified with the space* $\text{BLip}(\tilde{X})$.

Similarly, if θ is a base point for both X and $\tilde{X}$, then the space $\text{Lip}_0(\tilde{X})$ can be identified with the space $\text{Lip}_0(X)$.

Proof Every Lipschitz function f on X is uniformly continuous, so it admits a unique extension $\tilde{f}$ to $\tilde{X}$. Since this extension preserves both the Lipschitz norm $L(\cdot)$ and the uniform norm $\| \cdot \|_\infty$, the proposition follows.   □

The following theorem shows that the class $\mathcal{M}_2$ is the natural one to study BLip spaces.

Theorem 8.1.12 *Let (X, d) be a metric space.*

1. Let $d' : X \times X \to \mathbb{R}_+$ be given by

$$d'(x, y) = \min\{d(x, y), 2\}, \quad x, y \in X.$$

Then d' is a metric on X satisfying the inequality $d' \leq d$. If $(Y, \tilde{d}')$ is the completion of (X, d'), then $(Y, \tilde{d}') \in \mathcal{M}_2$ and the space $(\text{BLip}(X, d), \| \cdot \|_{\max})$ is isometrically isomorphic to $(\text{BLip}(Y, \tilde{d}'), \| \cdot \|'_{\max})$, where $\| \cdot \|'_{\max}$ is the norm (8.1.3).(i) corresponding to the metric $\tilde{d}'$.
2. Let $X \in \mathcal{M}_2$ and $(X^\bullet, \theta) \in \mathcal{M}_0$ as in Remark 8.1.10. Defining, for $f \in \text{BLip}(X)$, $\tilde{f}(x) = f(x)$, for $x \in X$ and $\tilde{f}(\theta) = 0$, it follows that the mapping

$f \mapsto \tilde{f}$ is an isometric isomorphism between the spaces $(\mathrm{BLip}(X), \| \cdot \|_{\max})$ and $(\mathrm{Lip}_0(X^{\bullet}), L(\cdot))$.

3. Let $(X, d, \theta) \in \mathcal{M}_0^f$. Then the identity mapping defines an isomorphism of $\mathrm{Lip}_0(X)$ onto a codimension-one subspace of $\mathrm{BLip}(X)$. If $d(x, \theta) \leq 1$ for all $x \in X$, then this isomorphism is an isometry.

Proof (Sketch)

1. It is easily seen that d' is a metric on X and that $d' \leq d$. . One checks also that the identity mapping is an isometric isomorphism between $(\mathrm{BLip}(X, d'), \| \cdot \|'_{\max})$ and $(\mathrm{BLip}(X, d), \| \cdot \|_{\max})$. Indeed, for every $f \in \mathrm{BLip}(X, d')$,

$$ |f(x) - f(y)| \leq L'(f)d'(x, y) \leq L'(f)d(x, y), $$

for all $x, y \in X$, so that $L(f) \leq L'(f)$. It follows that $\mathrm{BLip}(X, d') \subseteq \mathrm{BLip}(X, d)$ and $\|f\|_{\max} \leq \|f\|'_{\max}$.

If $f \in \mathrm{BLip}(X, d)$, then

$$ \sup_{0 < d(x,y) \leq 2} \frac{|f(x) - f(y)|}{d'(x, y)} = \sup_{0 < d(x,y) \leq 2} \frac{|f(x) - f(y)|}{d(x, y)} \leq L(f) \leq \|f\|_{\max}, $$

and

$$ \sup_{d(x,y) > 2} \frac{|f(x) - f(y)|}{d'(x, y)} = \sup_{d(x,y) > 2} \frac{|f(x) - f(y)|}{2} \leq \|f\|_{\infty} \leq \|f\|_{\max}. $$

Consequently, $L'(f) \leq \|f\|_{\max}$, so that $\mathrm{BLip}(X, d) \subseteq \mathrm{BLip}(X, d')$ and $\|f\|'_{\max} \leq \|f\|_{\max}$.

But, by Proposition 8.1.11, $\mathrm{BLip}(X, d')$ can be identified with $\mathrm{BLip}(Y, \tilde{d}')$.

2. It is clear that the correspondence $f \mapsto \tilde{f}$ is an algebraic isomorphism between $\mathrm{BLip}(X)$ and $\mathrm{Lip}_0(X^{\bullet})$. The relations

$$ \frac{|\tilde{f}(x) - \tilde{f}(y)|}{d(x, y)} = \frac{|f(x) - f(y)|}{d(x, y)} \leq L(f) \leq \|f\|_{\max}, $$

and

$$ \frac{|\tilde{f}(x) - \tilde{f}(\theta)|}{d(x, \theta)} = |f(x)| \leq \|f\|_{\infty} \leq \|f\|_{\max}, $$

valid for all $x, y \in X$, show that $L(\tilde{f}) \leq \|f\|_{\max}$.

On the other hand, it is obvious that $L(f) \leq L(\tilde{f})$. Since

$$|f(x)| = |\tilde{f}(x) - \tilde{f}(\theta)| \leq L(\tilde{f}),$$

for all $x \in X$, it follows $\|f\|_\infty \leq L(\tilde{f})$, hence $\|f\|_{\max} \leq L(\tilde{f})$.
Consequently, $\|f\|_{\max} = L(\tilde{f})$.

3. Let $r(X) = \sup\{d(x, \theta) : x \in X\}$. Then, for any $f \in \mathrm{Lip}_0(X)$,

$$|f(x)| = |f(x) - f(\theta)| \leq L(f)d(x, \theta) \leq L(f)r(X),$$

implying $\|f\|_\infty \leq r(X)L(f)$, and so

$$L(f) \leq \|f\|_{\max} \leq L(f)\max\{r(X), 1\},$$

showing that the identity map is an isomorphism. The space $\mathrm{Lip}_0(X)$ is complemented in $\mathrm{BLip}(X)$ by the one-dimensional space C of constant functions, so it has codimension one.

If $d(x, \theta) \leq 1$, for all $x \in X$, then $\max\{r(X), 1\} = 1$, and so $L(f) = \|f\|_{\max}$.
□

Remark 8.1.13 The isomorphism from Theorem 8.1.12.1 is not only a Banach space isometric isomorphism between $\mathrm{BLip}(X)$ and $\mathrm{BLip}(Y)$, but it also identifies these spaces in what concerns products and order. Consequently, the study of spaces $\mathrm{BLip}(X)$ reduces to the case $X \in \mathcal{M}_2$.

Remark 8.1.14 The metrics d and d' from Theorem 8.1.12.1 are topologically equivalent, but not Lipschitz equivalent.

Indeed, as $d' \leq d$, $x_n \xrightarrow{d} x$ implies $x_n \xrightarrow{d'} x$. If $x_n \xrightarrow{d'} x$, then, for some $n_0 \in \mathbb{N}$, $d'(x_n, x) < 2$ for all $n \geq n_0$. It follows that $d(x_n, x) = d'(x_n, x)$, $n \geq n_0$, implying $x_n \xrightarrow{d} x$.

On $\mathbb{R}$ consider the usual metric $d(x, y) = |x - y|$. For $x_n = n + 1$, $n \in \mathbb{N}$, $d(x_n, 0) = n + 1 \geq 2$, so that $d'(x_n, 0) = 2$ for all n. It follows that there is no $L > 0$ such that $d(x_n, 0) = n + 1 \leq 2L = Ld'(x_n, 0)$ for all $n \in \mathbb{N}$.

8.2 Lipschitz Free Banach Spaces

The use of Lipschitz free Banach spaces allows to extend a lot of results from the linear case to the Lipschitz case, by replacing linear functionals and operators by Lipschitz functions and Lipschitz mappings. Although the notion of Lipschitz free Banach space was known since 1956 (see [44]) the interest in the study of these spaces and their connections with the uniform and Lipschitz classification of Banach spaces was revived by the papers [260] and [329]. Using the technique of Lipschitz free Banach spaces one proves in [260] that if a separable Banach space X embeds isometrically into a Banach space Y, then Y contains an isometric linear copy of X, but this is false for every nonseparable weakly compactly generated Banach space X. Canonical examples of nonseparable Banach spaces which are Lipschitz isomorphic but not linearly isomorphic are constructed. In [329] the structure of Lipschitz and Hölder-type spaces and their preduals on general metric spaces is studied. Applications to the uniform structure of Banach spaces are provided. The main problem treated in the paper is Weaver's question (raised in [675]) whether the "little" Lipschitz space $\text{lip}_\alpha(K)$ over a compact metric space is necessarily isomorphic to c_0. This was known to be true for compact subsets in $\mathbb{R}^n$, but Kalton, *loc cit*, proved that if K is a compact convex subset of ℓ^2, then $\text{lip}_\alpha(K)$ is isomorphic to c_0 if and only if K is finite-dimensional.

In this section we shall show that $\text{Lip}_0(X)$ is always a dual Banach space. Three constructions of the predual will be presented along with their basic properties.

A *Lipschitz free Banach space* over a pointed metric space (X, d) is a Banach space $\widetilde{X}$ together with an isometric embedding $i_X : X \to \widetilde{X}$ satisfying the conditions:

(i) $\widetilde{X} = \overline{\text{span}}(i_X(X))$, and
(ii) for every Banach space Y and every mapping $F \in \text{Lip}_0(X, Y)$ there exists a unique continuous linear operator $\widetilde{F} : \widetilde{X} \to Y$ such that $\widetilde{F} \circ i_X = F$ and $\|\widetilde{F}\| = L(F)$.

8.2.1　The Arens-Eells Space

This is a constructive approach to obtain a predual of $\text{Lip}_0(X)$, proposed by Arens and Eells in [44], see also [675].

Let (X, d) be a metric. A *molecule* on X is a function $m : X \to \mathbb{R}$ with finite support $\text{spt}(m) := \{x \in X : m(x) \neq 0\}$, and such that $\sum_{x \in X} m(x) = 0$. Denote by $\text{Mol}(X)$ the space of molecules on X. For $x, y \in X$ put $m_{x,y} = \chi_x - \chi_y$, where χ_x denotes the characteristic function of the set $\{x\}$. One can show that every $m \in \text{Mol}(X)$ can be written, in at least one way, in the form $m = \sum_{i=1}^n a_i m_{x_i, y_i}$. For instance, if $\text{spt}(m) = \{x_1, x_2, \ldots, x_n\}$ and $m(x_i) = a_i$, $i = 1, 2, \ldots, n$, with $a_1 + a_2 + \cdots + a_n = 0$, then

$$m = a_1 m_{x_1, x_2} + (a_1 + a_2) m_{x_2, x_3} + \cdots + (a_1 + \cdots + a_{n-1}) m_{x_{n-1}, x_n}.$$

Put

$$\|m\|_{Æ} = \inf\left\{\sum |a_i| d(x_i, y_i) : m = \sum a_i m_{x_i, y_i}\right\}, \tag{8.2.1}$$

meaning that the infimum is taken over all representations of m in the form $m = \sum_{i=1}^{n} a_i m_{x_i, y_i}$, $n \in \mathbb{N}$. For the moment it follows that $\| \cdot \|_{\text{Æ}}$ is a seminorm on the vector space $\text{Mol}(X)$, but soon we shall see that it is in fact a norm (see Theorem 8.2.4.1). From (8.2.1) it is obvious that

$$\|m_{x,y}\|_{\text{Æ}} \le d(x, y), \tag{8.2.2}$$

for all $x, y \in X$.

Remark 8.2.1 It is easy to check that the seminorm $\| \cdot \|_{\text{Æ}}$ is the biggest seminorm on $\text{Mol}(X)$ satisfying this inequality for all $x, y \in X$.

Suppose now that X has a base point θ. Then the functions $m_{x,\theta}$, $x \in X \setminus \{\theta\}$, form an algebraic base of $\text{Mol}(X)$—every $m \in \text{Mol}(X)$ with $\text{spt}(m) = \{x_1, \ldots, x_n\}$, $m(x_i) = a_i$, $i = 1, \ldots, n$, can be written as $m = \sum_{i=1}^{n} a_i m_{x_i, \theta}$. For every function $f : X \to \mathbb{R}$ vanishing at θ the formula $\varphi(m) = \sum_{i=1}^{n} a_i f(x_i)$ defines a unique linear functional φ on $\text{Mol}(X)$. Conversely, if $\varphi : \text{Mol}(X) \to \mathbb{R}$ is linear, then $f(x) = \varphi(m_{x,\theta})$, $x \in X$, defines a function $f : X \to \mathbb{R}$ vanishing at θ. In particular

$$f(x) - f(y) = \varphi(m_{x,\theta}) - \varphi(m_{y,\theta}) = \varphi(m_{x,\theta} - m_{y,\theta}) = \varphi(m_{x,y}),$$

for all $x, y \in X$. If $f \in \text{Lip}_0(X)$, then for every representation $m = \sum_{i=1}^{k} a_i m_{x_i, y_i}$,

$$|\varphi(m)| \le \sum_{i=1}^{k} |a_i| \, |\varphi(m_{x_i, y_i})| = \sum_{i=1}^{k} |a_i| \, |f(x_i) - f(y_i)| \le L(f) \sum_{i=1}^{k} |a_i| d(x_i, y_i),$$

implying $|\varphi(m)| \le L(f) \|m\|_{\text{Æ}}$. The continuity of φ follows as well as the inequality $\|\varphi\| \le L(f)$.

If $\varphi \in \text{Mol}(X)^*$, and $f : X \to \mathbb{R}$ is defined by $f(x) = \varphi(m_{x,\theta})$, $x \in X$, then $m_{x,y} = m_{x,\theta} - m_{y,\theta}$ and, by (8.2.2),

$$|f(x) - f(y)| = |\varphi(m_{x,y})| \le \|\varphi\| \|m_{x,y}\|_{\text{Æ}} \le \|\varphi\| d(x, y),$$

for all $x, y \in X$. Consequently, $f \in \text{Lip}_0(X)$ and $L(f) \le \|\varphi\|$.

Denoting by $\text{Æ}(X)$ the completion of the normed space $(\text{Mol}(X), \| \cdot \|_{\text{Æ}})$, it follows that every functional $\varphi \in \text{Mol}(X)^*$ automatically extends to a linear continuous functional $\tilde{\varphi}$ on $\text{Æ}(X)$ with the same norm as φ. For convenience, we shall denote this functional by the same symbol φ.

Taking in account the above considerations it follows.

Theorem 8.2.2 *Let (X, d) be a metric space with base point θ and $\text{Mol}(X)$ and $\text{Æ}(X)$ as above. The application $\Lambda : \text{Æ}(X)^* \to \text{Lip}_0(X)$ given for $\varphi \in \text{Æ}(X)^*$ by*

$$(\Lambda\varphi)(x) = \varphi(m_{x,\theta}), \quad x \in X,$$

is an isometric isomorphism of $\text{Æ}(X)^*$ onto $\text{Lip}_0(X)$.

Based on this theorem we shall consider a function $f \in \mathrm{Lip}_0(X)$ acting on $\mathrm{Mol}(X)$ by the rule

$$\langle f, m \rangle = \left(\Lambda^{-1} f \right)(m) = \sum a_i (f(x_i) - f(y_i)), \qquad (8.2.3)$$

for $m = \sum a_i m_{x_i, y_i} \in \mathrm{Mol}(X)$.

Remark 8.2.3 The considerations preceding Theorem 8.2.2 show that for any fixed point $x_0 \in X$, the space $\mathcal{E}(X)^*$ is isometrically isomorphic to $\mathrm{Lip}_{x_0}(X) := \{f \in \mathrm{Lip}(X) : f(x_0) = 0\}$. The mapping realizing this isomorphism is given by

$$\Lambda_{x_0} \varphi(x) = \varphi(m_{x,x_0}), \quad x \in X, \; \varphi \in \mathcal{E}(X)^*.$$

Other properties are mentioned in the following theorem.

Theorem 8.2.4 *Let* (X, d, θ) *be a pointed metric space.*

1. *The functional* $\| \cdot \|_{\mathcal{E}}$ *given by* (8.2.1) *is a norm on* $\mathcal{E}(X)$, *and for every molecule* $m \in \mathrm{Mol}(X)$

$$\|m\|_{\mathcal{E}} = \sup\{|\langle f, m \rangle| : f \in \mathrm{Lip}_0(X), \, L(f) \le 1\},$$

and there exists $f \in \mathrm{Lip}_0(X)$, $L(f) = 1$ *such that* $\langle f, m \rangle = \|m\|_{\mathcal{E}}$.
2. *The mapping* $i_X : X \to \mathcal{E}(X)$ *given by*

$$i_X(x) = m_{x,\theta}, \quad x \in X, \qquad (8.2.4)$$

is an isometry, that is,

$$d(x, y) = \|m_{x,y}\|_{\mathcal{E}} = \|i_X(x) - i_X(y)\|_{\mathcal{E}}, \qquad (8.2.5)$$

for all $x, y \in X$.
3. *On bounded subsets of* $\mathrm{Lip}_0(X)$ *the weak* topology agrees with the topology of pointwise convergence. This means that for any bounded net* $(f_i : i \in I)$ *in* $\mathrm{Lip}_0(X)$

$$f_i \xrightarrow{w^*} f \iff \forall x \in X, \; f_i(x) \to f(x).$$

4. *Let* E *be a Banach space. Then for every* $F \in \mathrm{Lip}_0(X, E)$ *there exists a unique continuous linear mapping* $\Psi(F) : \mathcal{E}(X) \to E$ *such that* $\Psi(F) \circ i_X = F$. *Furthermore* $\|\Psi(F)\| = L(F)$ *and the spaces* $\mathrm{Lip}_0(X, E)$ *and* $\mathscr{L}(\mathcal{E}(X), E)$ *are isometrically isomorphic.*

5. *If (Y, θ') is a pointed metric space, then for every $F \in \mathrm{Lip}_0(X, Y)$ there exists a unique continuous linear operator $\Phi(F) : Æ(X) \to Æ(Y)$ such that $i_Y \circ F = \Phi(F) \circ i_X$ and $\|\Phi(F)\| = L(F)$.*

Proof (Sketch)

1. It suffices to show that $\|m\|_Æ > 0$ for every molecule $m \neq 0$. Let $m = \sum_{i=1}^{n} a_i m_{x_i, \theta}$ with $a_i \neq 0$, $i = 1, \ldots, n$, x_i pairwise distinct and different from θ. By Proposition 2.1.1 there exists a Lipschitz function f on X such that $f(x_1) = 1$ and $f(x) = 0$ for all $x \in \{x_2, \ldots, x_n\} \cup \{\theta\}$, implying

$$L(f)\|m\|_Æ \geq |\langle f, m\rangle| = |a_1| > 0 \,.$$

2. For a fixed $y \in X$ consider the isomorphism $\Lambda_y : Æ(X)^* \to \mathrm{Lip}_y(X)$ from Remark 8.2.3. Let $f_y \in \mathrm{Lip}_y(X)$ be given by $f_y(x) = d(x, y)$, $x \in X$. Then $L(f_y) = 1$ so that

$$\|m_{x,y}\|_Æ \geq \langle f_y, m_{x,y}\rangle = f_y(x) - f_y(y) = d(x, y) \,,$$

which, in conjunction with (8.2.2) yields the equality (8.2.5).

The equalities

$$d(x, y) = \|m_{x,y}\|_Æ = \|m_{x,\theta} - m_{y,\theta}\|_Æ = \|i_X(x) - i_X(y)\|_Æ \,,$$

show that the mapping given by (8.2.4) is an isometry.

3. The proof (see [675, Theorem 2.2.2, p. 39]) is based on a result of Shmulian (see [456, Corollary 2.7.12]) asserting that a subspace W of the dual E^* of a Banach space E is w^*-closed provided that $W \cap B_{E^*}$ is w^*-closed.

4. For $F \in \mathrm{Lip}_0(X, E)$ define $A(m_{x,\theta}) = F(x)$, $x \in X$, and extend it by linearity to $\mathrm{Mol}(X)$, i.e., put $A(m) = \sum_{i=1}^{n} a_i F(x_i)$ for every molecule $m = \sum_{i=1}^{n} a_i m_{x_i, \theta}$.

Then for every representation $m = \sum_{j=1}^{k} b_j m_{y_j, z_j}$ of a molecule m we have $A(m) = \sum_{j=1}^{k} b_j (F(y_j) - F(z_j))$ so that

$$\|A(m)\| \leq \sum_{j=1}^{k} |b_j| \|F(y_j) - F(z_j)\| \leq L(F) \sum_{j=1}^{k} |b_j| d(y_j, z_j) \,.$$

Consequently,

$$\|A(m)\| \leq L(F)\|m\|_Æ \,,$$

for all $m \in \mathrm{Mol}(X)$, yielding the continuity of A and the inequality

$$\|A\| \leq L(F) \,.$$

On the other hand, taking into account (8.2.5), we obtain

$$\|F(x) - F(y)\| = \|A(m_{x,y})\| \le \|A\| \cdot \|m_{x,y}\|_{\mathscr{E}} = \|A\| \, d(x, y),$$

for all $x, y \in X$, which yields $L(F) \le \|A\|$ and so $\|A\| = L(F)$.

Since $\mathrm{Mol}(X)$ is dense in $\mathscr{E}(X)$, the operator A has a unique continuous linear extension $\Psi(F) : \mathscr{E}(X) \to E$ satisfying also the equality $\|\Psi(F)\| = L(F)$.

5. Apply 4 to $E = \mathscr{E}(Y)$ and $G = i_Y \circ F : X \to \mathscr{E}(Y)$. The required operator is $\Phi(F) = \Psi(G)$.   □

8.2.2 Lipschitz Free Banach Spaces Generated by Evaluation Functionals

The first who remarked the possibility to use evaluation functionals in the construction of free Lipschitz spaces was Michael [469]. Later the method was rediscovered by Kadets [322], see also [206]. Let (X, d) be a metric space with a base point θ.

We start with a consequence of McShane's extension theorem.

Proposition 8.2.5 *Let (X, d, θ) be a pointed metric space. For every pair x, y of distinct points in X there exists a Lipschitz function $f \in \mathrm{Lip}_0(X)$ with $L(f) = 1$ such that $f(x) - f(y) = d(x, y)$.*

In particular, for every $x \ne \theta$ in X there exists a Lipschitz function $f \in \mathrm{Lip}_0(X)$ with $L(f) = 1$ such that $f(x) = d(\theta, x)$.

Proof Suppose first that $x \ne y$, $x, y \in X \setminus \{\theta\}$. Let

$$g(x) = d(\theta, x), \quad g(y) = d(\theta, x) - d(x, y), \quad \text{and } g(\theta) = 0.$$

Then

$$|g(x) - g(\theta)| = g(x) = d(\theta, x),$$
$$|g(y) - g(\theta)| = |d(\theta, x) - d(x, y)| \le d(\theta, y) \text{ and}$$
$$g(x) - g(y) = d(x, y).$$

It follows that $L(g) = 1$. Extending g to a function $f \in \mathrm{Lip}_0(X)$ with $L(f) = L(g)$ one obtains the desired function.

The second assertion follows from the first one taking $y = \theta$.   □

For $x \in X$ define $e_x : \mathrm{Lip}_0(X) \to \mathbb{R}$ by

$$e_x(f) = f(x), \quad f \in \mathrm{Lip}_0(X). \tag{8.2.6}$$

The functionals e_x are called *evaluation functionals*.

Proposition 8.2.6 *The functional e_x is linear and continuous with $\|e_x\| = d(\theta, x)$ for all $x \in X$. The mapping $i_X(x) = e_x$, $x \in X$, is an isometric embedding of X into $\left(\mathrm{Lip}_0(X)\right)^*$, i.e.,*

$$\|e_x - e_y\| = d(x, y),\tag{8.2.7}$$

for all $x, y \in X$.

If X is a Banach space, then i_X is nowhere Gâteaux differentiable.

Proof The linearity of e_x is obvious. Suppose $x \neq \theta$. The inequality $|e_x(f)| = |f(x)| \leq L(f)d(\theta, x)$ valid for all $f \in \mathrm{Lip}_0(X)$ yields the continuity of e_x and the inequality $\|e_x\| \leq d(x, \theta)$. By Proposition 8.2.5, there exists $f \in \mathrm{Lip}_0(X)$ with $L(f) = 1$ such that $f(x) = d(\theta, x)$, implying $\|e_x\| \geq |e_x(f)| = d(x, \theta)$, and so $\|e_x\| = d(\theta, x)$.

Let now $x, y \in X$. Observe first that

$$|e_x(f) - e_y(f)| = |f(x) - f(y)| \leq L(f)d(x, y),$$

for all $f \in \mathrm{Lip}_0(X)$, implying $\|e_x - e_y\| \leq d(x, y)$. Apply again Proposition 8.2.5 to obtain a function $f \in \mathrm{Lip}_0(X)$ such that $L(f) = 1$ and $f(x) - f(y) = d(x, y)$. Then

$$\|e_x - e_y\| \geq |(e_x - e_y)(f)| = |f(x) - f(y)| = d(x, y),$$

so that $\|e_x - e_y\| = d(x, y)$.

Suppose that X is Banach and that, for some $x, h \in X$ with $h \neq 0$, the limit

$$\lim_{t \to 0} \frac{i_X(x + th) - i_X(x)}{t} = \psi(x, h)$$

exists.

Taking into account (8.2.7), one obtains the contradiction

$$\lim_{t \to 0} \frac{|t|}{t}\|h\| = \lim_{t \to 0} \frac{\|e_{x+th} - e_x\|}{t} = \lim_{t \to 0} \frac{\|i_X(x + th) - i_X(x)\|}{t} = \|\psi(x, h)\|.$$

$\square$

Let

$$X_e = \{e_x : x \in X\} \subseteq \left(\mathrm{Lip}_0(X)\right)^* \quad \text{and} \quad \mathbb{F}(X) = \overline{\mathrm{span}}(X_e),\tag{8.2.8}$$

where the closure is the norm-closure in $\left(\mathrm{Lip}_0(X)\right)^*$. By Proposition 8.2.6, the set X_e is an isometric copy of X.

The following theorem shows that $\mathbb{F}(X)$ is a predual of $\mathrm{Lip}_0(X)$.

Theorem 8.2.7 *The space* $\mathrm{Lip}_0(X)$ *is isometrically isomorphic to* $\mathbb{F}(X)^*$. *An isometric isomorphism* $\Lambda : \mathrm{Lip}_0(X) \to \mathbb{F}(X)^*$ *is determined by the condition*

$$(\Lambda f)(\sum_i t_i e_{x_i}) = \sum_i t_i f(x_i),$$

for $\sum_i t_i e_{x_i} \in \mathrm{span}(X_e)$.

Proof For $f \in \mathrm{Lip}_0(X)$ define ψ on $\mathrm{span}(X_e)$ by

$$\psi\left(\sum_{i=1}^n t_i e_{x_i}\right) = \sum_{i=1}^n t_i f(x_i). \tag{8.2.9}$$

It is clear that ψ is a linear functional and the relations

$$\left\|\sum_{i=1}^n t_i e_{x_i}\right\| = \sup_{L(g)\leq 1} \left|\langle\sum_{i=1}^n t_i e_{x_i}, g\rangle\right|$$

$$\geq \frac{1}{L(f)}\left|\langle\sum_{i=1}^n t_i e_{x_i}, f\rangle\right| = \frac{1}{L(f)}\left|\sum_{i=1}^n t_i f(x_i)\right|,$$

imply

$$\left|\psi\left(\sum_{i=1}^n t_i e_{x_i}\right)\right| \leq L(f)\cdot\left\|\sum_{i=1}^n t_i e_{x_i}\right\|.$$

Consequently, ψ is a continuous linear functional on $\mathrm{span}(X_e)$ with $\|\psi\| \leq L(f)$, so it admits a unique extension Λf to $\mathbb{F}(X) = \overline{\mathrm{span}}(X_e)$. This extension satisfies $\|\Lambda f\| \leq L(f)$, $f \in \mathrm{Lip}_0(X)$, so that $\|\Lambda\| \leq 1$, i.e., Λ is nonexpansive.

Define now $\Gamma : \mathbb{F}(X)^* \to \mathrm{Lip}_0(X)$ by $\Gamma\varphi := f : X \to \mathbb{R}$, where

$$f(x) = \varphi(e_x), \quad \text{for } x \in X \text{ and } \varphi \in \mathbb{F}(X)^*.$$

It is clear that $f(\theta) = 0$ and, by (8.2.7),

$$|f(x) - f(x')| = |\varphi(e_x - e_{x'})| \leq \|\varphi\|\|e_x - e_{x'}\| = \|\varphi\|d(x, x'),$$

showing that $f \in \mathrm{Lip}_0(X)$ and $L(\Gamma\varphi) \leq \|\varphi\|$, for all $\varphi \in \mathbb{F}(X)^*$, so that $\|\Gamma\| \leq 1$.
Observe now that, for $\varphi \in \mathbb{F}(X)^*$,

$$(\Lambda \circ \Gamma)(\varphi)(\sum_i t_i e_{x_i}) = \Lambda(\Gamma\varphi)(\sum_i t_i e_{x_i}) = \sum_i t_i(\Gamma\varphi)(x_i)$$

$$= \sum_i t_i \varphi(e_{x_i}) = \varphi(\sum_i t_i e_{x_i}),$$

for every $\sum_i t_i e_{x_i} \in \operatorname{span}(X_e)$, which implies $(\Lambda \circ \Gamma)(\varphi) = \varphi$.

On the other hand, for $f \in \operatorname{Lip}_0(X)$,

$$(\Gamma \circ \Lambda)(f)(x) = \Gamma(\Lambda f)(x) = (\Lambda f)(e_x) = f(x),$$

for all $x \in X$.

Consequently,

$$\forall \varphi \in \mathbb{F}(X)^*, \quad (\Lambda \circ \Gamma)(\varphi) = \varphi, \quad \text{and}$$
$$\forall f \in \operatorname{Lip}_0(X), \quad (\Gamma \circ \Lambda)(f) = f.$$

Also

$$L(f) = \|\Gamma(\Lambda f)\| \le \|\Lambda f\| \le L(f),$$

so that $\|\Lambda f\| = L(f)$, for all $f \in \operatorname{Lip}_0(X)$. Similarly $L(\Gamma \varphi) = \|\varphi\|$ for all $\varphi \in \mathbb{F}(X)^*$, showing that the operator Λ is an isometric isomorphism of $\operatorname{Lip}_0(X)$ onto $\mathbb{F}(X)^*$. $\qquad\square$

Remark 8.2.8 It follows that every $f \in \operatorname{Lip}_0(X)$ may be viewed as a continuous linear functional $\varphi_f \in \mathbb{F}(X)^*$ acting on $\operatorname{span}(X_e)$ by the rule ·

$$\left\langle \sum_i t_i e_{x_i}, f \right\rangle = \sum_i t_i f(x_i),$$

and, further, $\|\varphi_f\| = L(f)$.

In this case there is also a correspondence between Lipschitz mapping from X to Y and linear operators from $\mathbb{F}(X)$ to $\mathbb{F}(Y)$.

Theorem 8.2.9 *Let (X, d, θ), (Y, d', θ') be pointed metric spaces, $F \in \operatorname{Lip}_0(X, Y)$ and let i_X, i_Y be the isometric mappings of X, Y onto X_e, Y_e, respectively. Then there exists a unique linear continuous operator $\widetilde{F} : \mathbb{F}(X) \to \mathbb{F}(Y)$ such that $\widetilde{F} \circ i_X = i_Y \circ F$ and $\|\widetilde{F}\| = L(F)$. The operator $\widetilde{F}$ is determined by the condition*

$$\widetilde{F}\left(\sum_i t_i e_{x_i} \right) = \sum_i t_i e_{F(x_i)}, \tag{8.2.10}$$

for every $\sum_i t_i e_{x_i} \in \operatorname{span}(X_e)$.

Proof Define first $\tilde{F} : \mathrm{span}(X_e) \to \mathrm{span}(Y_e)$ by (8.2.10). Then

$$\left\| \sum_i t_i e_{x_i} \right\| = \sup \left\{ \left| \sum_i t_i f(x_i) \right| : f \in \mathrm{Lip}_0(X),\ L(f) \le 1 \right\}$$

$$\ge \frac{1}{L(F)} \sup \left\{ \left| \sum_i t_i (g \circ F)(x_i) \right| : g \in \mathrm{Lip}_0(Y),\ L(g) \le 1 \right\}$$

$$= \frac{1}{L(F)} \left\| \sum_i t_i e_{F(x_i)} \right\|,$$

proving the continuity of $\tilde{F}$ and the inequality

$$\left\| \tilde{F} \left(\sum_i t_i e_{x_i} \right) \right\| \le L(F) \left\| \sum_i t_i e_{x_i} \right\|.$$

It follows that the operator $\tilde{F}$ has a unique extension to $\mathbb{F}(X) = \mathrm{cl}(\mathrm{span}(X_e))$, denoted by the same symbol $\tilde{F}$, satisfying $\|\tilde{F}\| \le L(F)$.

Since

$$L(F) = \sup_{x \ne x'} \frac{d'(F(x), F(x'))}{d(x, x')} = \sup_{x \ne x'} \frac{\|e_{F(x)} - e_{F(x')}\|}{\|e_x - e_{x'}\|}$$

$$= \sup_{x \ne x'} \frac{\|\tilde{F}(e_x - e_{x'})\|}{\|e_x - e_{x'}\|} \le \|\tilde{F}\|,$$

it follows that $\|\tilde{F}\| = L(F)$.   □

Other properties of $\mathbb{F}(X)$ are collected in the following theorem

Theorem 8.2.10 *Let X, Y, Z be pointed metric spaces with base points $\theta, \theta', \theta''$, respectively.*

1. *If $F \in \mathrm{Lip}_0(X, Y)$ and $G \in \mathrm{Lip}_0(Y, Z)$, then*

$$\widetilde{G \circ F} = \tilde{G} \circ \tilde{F} \quad \text{and} \quad \widetilde{\mathrm{Id}}_X = \mathrm{Id}_{\mathbb{F}(X)}.$$

   If $F : X \to Y$ is a Lipschitz isomorphism, then $\tilde{F} : \mathbb{F}(X) \to \mathbb{F}(Y)$ is a linear isomorphism and $(\tilde{F})^{-1} = \widetilde{F^{-1}}$.
2. *The weak and strong topologies of $\mathbb{F}(X)$ agree on X_e.*
3. *If the metric space X is separable, then the space $\mathbb{F}(X)$ is also separable.*
4. *Let (X, d) be a metric space with base point θ, Y a subspace of X containing θ, and $j : Y \to X$ the embedding mapping. Then:*

   (i) *$\tilde{j} : \mathbb{F}(Y) \to \mathbb{F}(X)$ is an isometric embedding, and*
   (ii) *for any $y \in X$, $e_y \in \mathbb{F}(Y) \iff y \in \overline{Y}$.*

5. *If the metric space (X, d) is complete and X_1, X_2 are closed subsets of X such that $X_1 \cap X_2 = \{\theta\}$, then $\mathbb{F}(X_1 \cup X_2) = \mathbb{F}(X_1) \oplus \mathbb{F}(X_2)$.*
6. *Suppose that X be a Banach space. Then:*

   (i) *there exists a norm-one surjective continuous linear operator $T : \mathbb{F}(X) \to X$, determined by the condition*

$$T\left(\sum_i t_i e_{x_i}\right) = \sum_i t_i x_i, \qquad (8.2.11)$$

   for $\sum_i t_i e_{x_i} \in \mathrm{span}(X_e)$;
   (ii) *the space X is isometrically isomorphic to a quotient space of $\mathbb{F}(X)$;*
   (iii) *the operator $i_X \circ T$ is a norm-one projection of $\mathbb{F}(X)$ onto $i_X(X)$.*

7. *Suppose again that X is a Banach space and Y a closed subspace of X. If $R : X \to Y$ is a Lipschitz retraction of X onto Y, then $\widetilde{R}$ is a continuous linear projection of $\mathbb{F}(X)$ onto $\mathbb{F}(Y)$ with $\|\widetilde{R}\| = L(R)$, where $\widetilde{R}$ is the linear operator associated to R by Theorem 8.2.9.*
   Conversely, if S is a continuous linear projection of $\mathbb{F}(X)$ onto $\mathbb{F}(Y)$, then $\overline{S} = T \circ S \circ i_X$ is a continuous linear projection of X onto Y with $\|\overline{S}\| \leq \|S\|$, where $T : \mathbb{F}(Y) \to Y$ is the operator from the statement 6.(i) of the theorem.

Proof The equalities from 1 follow directly from the definitions.

2. Supposing that the net $(e_{x_i} : i \in I)$ converges weakly to e_x, the following equivalences hold true

$$e_{x_i} \xrightarrow{w} e_x \iff \forall \varphi \in \mathbb{F}(X)^*, \ \varphi(e_{x_i}) \to \varphi(e_x)$$

$$\iff \forall f \in \mathrm{Lip}_0(X), \ f(x_i) \to f(x).$$

Consider the function $f(y) = d(x, y) - d(x, \theta)$, $y \in X$. Then $f \in \mathrm{Lip}_0(X)$ and

$$f(x_i) \to f(x) \iff d(x, x_i) \to 0 \iff \|e_{x_i} - e_x\| \to 0.$$

3. If $\{y_i : i \in \mathbb{N}\}$ is a countable dense subset of X, then $Z := \{\sum_{i=1}^{n} q_i e_{y_i} : q_i \in \mathbb{Q}, n \in \mathbb{N}\}$ is a countable subset of $\mathrm{span}(X_e) \subseteq \mathbb{F}(X)$. Let us show that Z is also dense in $\mathbb{F}(X)$. For $u \in \mathbb{F}(X) = \overline{\mathrm{span}}(X_e)$ and $\varepsilon > 0$, let $\sum_{i=1}^{n} t_i e_{x_i} \in \mathrm{span}(X_e)$ such that $\|u - \sum_{i=1}^{n} t_i e_{x_i}\| \leq \varepsilon$. Since

$$\left\| \sum_{i=1}^{n} t_i e_{x_i} - \sum_{i=1}^{n} q_i e_{y_i} \right\| \leq \sum_{i=1}^{n} |t_i - q_i| \|e_{x_i} - e_{y_i}\|$$

$$= \sum_{i=1}^{n} |t_i - q_i| d(x_i, y_i),$$

it is clear that we can choose $q_i \in \mathbb{Q}$ and the elements y_i such that this difference is less than or equal to ε, implying

$$\left\| u - \sum_{i=1}^{n} q_i e_{y_i} \right\| \le \left\| u - \sum_{i=1}^{n} t_i e_{x_i} \right\| + \left\| \sum_{i=1}^{n} t_i e_{x_i} - \sum_{i=1}^{n} q_i e_{y_i} \right\| \le 2\varepsilon \,.$$

4. (i) Observe that for finite sets of elements $y_i \in Y$ and $t_i \in \mathbb{R}$, $\sum_i t_i e_{y_i}$ can be considered as acting on both $\mathrm{Lip}_0(Y)$ and $\mathrm{Lip}_0(X)$ by the same formula $\sum_i t_i f(y_i)$. To make distinction we shall denote the corresponding norms by $\| \cdot \|_Y$ and $\| \cdot \|_X$, respectively, and similarly for the Lipschitz norms $L_Y(\cdot)$ and $L_X(\cdot)$.

   Appealing to (8.2.10), we have

$$\tilde{j}\left(\sum_i t_i e_{y_i} \right)(f) = \left(\sum_i t_i e_{j(y_i)} \right)(f) = \sum_i t_i f(y_i) \,,$$

   for all $f \in \mathrm{Lip}_0(X)$.

   Since every $g \in \mathrm{Lip}_0(Y)$ has an extension $f \in \mathrm{Lip}_0(X)$ with $L_X(f) = L_Y(g)$, it follows that $\{g \in \mathrm{Lip}_0(Y) : L_Y(g) \le 1\} = \{f|_Y : f \in \mathrm{Lip}_0(X), L_X(f) \le 1\}$ and

$$\left\| \sum_i t_i e_{y_i} \right\|_Y = \sup \left\{ \left| \sum_i t_i g(y_i) \right| : g \in \mathrm{Lip}_0(Y), L_Y(g) \le 1 \right\}$$

$$= \sup \left\{ \left| \sum_i t_i f(y_i) \right| : f \in \mathrm{Lip}_0(X), L_X(f) \le 1 \right\}$$

$$= \left\| \tilde{j}\left(\sum_i t_i e_{y_i} \right) \right\|_X \,.$$

   Consequently, $\left\| \tilde{j}\left(\sum_i t_i e_{y_i} \right) \right\|_X = \left\| \sum_i t_i e_{y_i} \right\|_Y$, which implies $\| \tilde{j}(u) \| = \| u \|$ for all $u \in \mathbb{F}(Y)$.

   (ii) If $y \in \overline{Y}$, then there exists a sequence (y_n) in Y such that $y_n \to y$. It follows that $e_{y_n} \to e_y$ in $\mathbb{F}(X)$. Since $\mathbb{F}(Y)$ is a Banach space (and so closed in $\mathbb{F}(X)$) and $e_{y_n} \in \mathbb{F}(Y)$, this implies $e_y \in \mathbb{F}(Y)$.

   To prove the converse, suppose that $y \notin \overline{Y}$. Then $d = d(y, Y) > 0$, so that, by Proposition 2.1.1, there exists $f \in \mathrm{Lip}_0(X)$ such that $f(Y) = \{0\}$ and $f(y) = 1$. To f one associates a continuous linear functional ψ on $\mathbb{F}(X)$, given by the formula (8.2.9). Let $w_n = \sum_i t_i^n e_{y_i^n}$ be a sequence in $\mathrm{span}(Y_e)$ converging to e_y. Then $\psi(w_n) = 0$ for all n, but $\psi(e_y) = f(y) = 1$, a contradiction.

5. The equality $\mathbb{F}(X_1 \cup X_2) = \mathbb{F}(X_1) + \mathbb{F}(X_2)$ is obvious (follows from $\mathrm{span}((X_1 \cup X_2)_e) = \mathrm{span}((X_1)_e) + \mathrm{span}((X_2)_e)$. If $e_x \in \mathbb{F}(X_1) \cap \mathbb{F}(X_2)$, then, by 4.(ii), $x \in \overline{X}_1 \cap \overline{X}_2 = X_1 \cap X_2 = \{\theta\}$ and $e_x = e_\theta = 0$, showing that $\mathbb{F}(X_1 \cup X_2) = \mathbb{F}(X_1) + \mathbb{F}(X_2)$ (direct algebraic sum). Since $\mathbb{F}(X_1)$, $\mathbb{F}(X_2)$ are closed

(as complete) in the Banach space $\mathbb{F}(X_1 \cup X_2)$, it follows that $\mathbb{F}(X_1 \cup X_2) = \mathbb{F}(X_1) \oplus \mathbb{F}(X_2)$ (direct topological and algebraic sum).

6. Let $T : \text{span}(X_e) \to X$ be given by (8.2.11). The relations

$$\left\| T\left(\sum_i t_i e_{x_i} \right) \right\| = \left\| \sum_i t_i x_i \right\| = \sup \left\{ \left| \sum_i t_i x^*(x_i) \right| : x^* \in X^*, \ \|x^*\| \le 1 \right\}$$

$$\le \sup \left\{ \left| \sum_i t_i f(x_i) \right| : f \in \text{Lip}_0(X), \ L(f) \le 1 \right\} = \left\| \sum_i t_i e_{x_i} \right\|$$

show that T is linear continuous and $\|T\| \le 1$, so that it can be uniquely extended to a continuous linear operator $T : \mathbb{F}(X) \to X$ with $\|T\| \le 1$.

Since T is obviously surjective, it follows that X is isomorphic to $\mathbb{F}(X)/\ker(T)$. Let us show that the factor mapping $\widehat{T}(e_x + Z) = T(e_x) = x$, $x \in X$, where $Z = \ker T$, is in fact an isometry. Indeed, $\|\widehat{T}\| = \|T\|$ and since, by Proposition 8.2.6, $\|e_x\| = \|x\|$ for $x \in X$, we have

$$\|e_x + Z\| \ge \left\| \widehat{T}(e_x + Z) \right\| = \|T(e_x)\| = \|x\|$$

$$= \|e_x\| \ge \inf\{\|e_x + u\| : u \in Z\} = \|e_x + Z\|.$$

Consequently, $\|x\| = \left\| \widehat{T}(e_x + Z) \right\| = \|e_x + Z\|$ for all $x \in X$ and $\|T\| = \|\widehat{T}\| = 1$.

We have proved the assertions (i) and (ii).

To prove (iii), let $P = i_X \circ T$. Since i_X is an isometry of X onto $i_X(X)$ it follows that $\|P\| = 1$ and $P(e_x) = i_X(T(e_x)) = i_X(x) = e_x$ for all $x \in X$, showing that P is a projection of $\mathbb{F}(X)$ onto $i_X(X)$.

7. By Theorem 8.2.9,

$$\widetilde{R}\left(\sum_i t_i e_{y_i} \right) = \sum_i t_i e_{Ry_i} = \sum_i t_i e_{y_i},$$

for every $\sum_i t_i e_{y_i} \in \text{span}(Y_e)$, implying $\widetilde{R}u = u$ for all $u \in \mathbb{F}(Y)$. By the same theorem $\|\widetilde{R}\| = L(R)$.

Conversely, if $\overline{S} = T \circ S \circ i_X$, then $L(\overline{S}) \le L(T) \cdot \|S\| \cdot \|i_X\| \le \|S\|$ and

$$(T \circ S \circ i_X)(y) = T(S(e_y)) = T(e_y) = y,$$

for all $y \in Y$, showing that $\overline{S}$ is a Lipschitz retraction of X onto Y with $L(\overline{S}) \le \|S\|$. $\square$

As a consequence of Theorem 8.2.10 one obtains another important property of the space $\mathbb{F}(X)$.

Theorem 8.2.11 *If (X, d, θ) is a pointed metric space and Y a Banach space, then for every $F \in \text{Lip}_0(X, Y)$ there exists a unique continuous linear mapping $\widehat{F}$:*

$\mathbb{F}(X) \to Y$ *such that* $\widehat{F} \circ i_X = F$ *and* $\|\widehat{F}\| = L(F)$. *The map* $\widehat{F}$ *is determined by the condition*

$$\widehat{F}(\sum_i t_i e_{x_i}) = \sum_i t_i F(x_i) \quad \text{for} \quad \sum_i t_i e_{x_i} \in \operatorname{span}(X_e), \tag{8.2.12}$$

and $\widetilde{F} = \widehat{i_Y \circ F}$. *The correspondence* $F \mapsto \widehat{F}$ *is an isometric isomorphism between the spaces* $\operatorname{Lip}_0(X, Y)$ *and* $\mathscr{L}(\mathbb{F}(X), Y)$.

Proof Define $\widehat{F} : \operatorname{span}(X_e) \to Y$ by $\widehat{F}(\sum_i t_i e_{x_i}) = \sum_i t_i F(x_i)$. If $T : \operatorname{span}(Y_e) \to Y$ is the application $T(\sum_i t_i e_{y_i}) = \sum_i t_i y_i$ considered in the proof of the assertion 6 of Theorem 8.2.10, then $\widehat{F} = \widetilde{F} \circ T$, where $\widetilde{F} : \mathbb{F}(X) \to \mathbb{F}(Y)$ is the mapping associated to F by Theorem 8.2.9. Since $\|T\| \leq 1$ and $\|\widetilde{F}\| = L(F)$, it follows that $\|\widehat{F}\| \leq \|T\| \cdot \|\widetilde{F}\| \leq L(F)$, implying the continuity of $\widehat{F}$ and the inequality $\|\widehat{F}\| \leq L(F)$. The relations

$$L(F) = \sup_{x \neq x'} \frac{\|F(x) - F(x')\|}{d(x, x')} = \sup_{x \neq x'} \frac{\|\widehat{F}(e_x - e_{x'})\|}{\|e_x - e_{x'}\|} \leq \|\widehat{F}\|,$$

yield the reverse inequality $\|\widehat{F}\| \geq L(F)$ and so $\|\widehat{F}\| = L(F)$. Denoting by the same symbol the unique extension of $\widehat{F}$ to $\mathbb{F}(X) = \overline{\operatorname{span}}(X_e)$ one obtains the result.

For $A \in \mathscr{L}(\mathbb{F}(X), Y)$ the mapping $F := A \circ i_X$ satisfies $F(\theta) = 0$ and $L(F) \leq \|A\|$, so that it belongs to $\operatorname{Lip}_0(X, Y)$. Since, for any $\sum_i t_i e_{x_i} \in \operatorname{span}(X_e)$,

$$\widehat{F}(\sum_i t_i e_{x_i}) = \sum_i t_i A(i_X(x_i)) = \sum_i t_i A(e_{x_i}) = A(\sum_i t_i e_{x_i}),$$

it follows that $\widehat{A \circ i_X} = A$. This shows that the mapping $F \mapsto \widehat{F}$ is surjective, and so it is an isometric isomorphism between $\operatorname{Lip}_0(X, Y)$ and $\mathscr{L}(\mathbb{F}(X), Y)$.

Since

$$\widehat{i_Y \circ F}(\sum_i t_i e_{x_i}) = \sum_i t_i (i_Y \circ F)(x_i) = \sum_i t_i e_{F(x_i)} = \widetilde{F}(\sum_i t_i e_{x_i}),$$

it follows that $\widetilde{F} = \widehat{i_Y \circ F}$.   □

The following two examples of spaces $\mathbb{F}(X)$ are given in Kadets [322].

Example 8.2.12

1. If (X, d, θ) is a pointed metric space such that $d(x, y) = 1$ for all $x, y \in X$, $x \neq y$, then the space $\mathbb{F}(X)$ is isomorphic to $\ell^1(X \setminus \{\theta\})$.
2. Let X be an interval in $\mathbb{R}$ of the form $[0, b]$ or $[0, \infty)$ and define $\Phi : X \to \mathbb{R}$ by $\Phi(e_x) = \chi_{[0,x]}$, $x \in X$. Extending this function by linearity and continuity to $\mathbb{F}(X)$ one obtains an isomorphism between $\mathbb{F}(X)$ and $L^1(X)$.

For the first example put $X' = X \setminus \{\theta\}$ and denote by $\alpha : X' \to \mathbb{R}$ a generic element of $\ell^1(X')$. Then

$$\sum_{x \in X'} |\alpha(x)| \|e_x\| = \sum_{x \in X'} |\alpha(x)| = \|\alpha\|_1 < \infty,$$

so that the element $\Phi(\alpha) = \sum_{x \in X'} \alpha(x)e_x \in \mathbb{F}(X)$ is well-defined and $\|\Phi(\alpha)\| \leq \|\alpha\|_1$.

The function $f(x) = \operatorname{sign} \alpha(x)$, $x \in X$, is Lipschitz with $L(f) = 2$, so that

$$\Big\| \sum_{x \in X'} \alpha(x)e_x \Big\| \geq \frac{1}{2} \Big| \Big\langle \sum_{x \in X'} \alpha(x)e_x, f \Big\rangle \Big|$$

$$= \frac{1}{2} \sum_{x \in X'} \alpha(x)f(x) = \frac{1}{2} \sum_{x \in X'} |\alpha(x)| = \frac{1}{2} \|\alpha\|_1 .$$

Consequently, the linear operator $\Phi : \ell^1(X') \to \mathbb{F}(X)$ defined above satisfies

$$\frac{1}{2} \|\alpha\|_1 \leq \|\Phi(\alpha)\| \leq \|\alpha\|_1 , \tag{8.2.13}$$

for all $\alpha \in \ell^1(X')$.

Every element $u = \sum_{i=1}^{n} t_i e_{x_i} \in \operatorname{span}(X_e)$ corresponds by Φ to $\alpha \in \ell^1(X')$, where $\alpha(x_i) = t_i$, $i = 1, \ldots, n$, and $\alpha(x) = 0$ otherwise. If $u \in \mathbb{F}(X)$, then there exists a sequence (u_n) in $\operatorname{span}(X_e)$ converging to u. If $\alpha_n \in \ell^1(X')$ are such that $\Phi(\alpha_n) = u_n$, then by (8.2.13)

$$\frac{1}{2} \|\alpha_n - \alpha_m\|_1 \leq \|\Phi(\alpha_n - \alpha_m)\| = \|u_n - u_m\| ,$$

for all $m, n \in \mathbb{N}$. It follows that (α_n) is a Cauchy sequence, so it converges to some $\alpha \in \ell^1(X')$. But then $\Phi(\alpha) = \lim_{n \to \infty} \Phi(\alpha_n) = \lim_{n \to \infty} u_n = u$, proving the surjectivity of Φ and the fact that Φ is an isomorphism of $\ell^1(X')$ onto $\mathbb{F}(X)$.

The second example can be treated in a similar way defining $\Phi : X_e \to L^1(X)$ by $\Phi(e_x) = \chi_{[0,x]}$, $x \in X$, and taking into account that the space generated by the characteristic functions of subintervals of X (the space of step functions) is dense in $L^1(X)$.

We sketch the details following [206]. Write $w = \sum_{i=1}^{n} \lambda_i e_{x_i} \in \operatorname{span}([0, \infty)_e)$ with $0 < x_1 < \cdots < x_n$ and $\lambda_i \neq 0$, $i = 1, \ldots, n$, as $w = \sum_{i=1}^{n} t_i (e_{x_i} - e_{x_{i-1}})$, where $x_0 = 0$ and $t_1 = \lambda_1 + \cdots + \lambda_n$, $t_2 = \lambda_2 + \cdots + \lambda_n, \ldots, t_n = \lambda_n$ and assign it the function $f = \Phi(w) = \sum_{i=1}^{n} t_i \chi_{[x_{i-1}, x_i]}$. Then $f \in L^1(\mathbb{R}_+)$ and

$$\|f\|_{L^1} = \sum_{i=1}^{n} |t_i|(x_i - x_{i-1}) .$$

We have

$$\sum_{i=1}^{n} |t_i|(x_i - x_{i-1}) = \sum_{i=1}^{n} |t_i| \, \|e_{x_i} - e_{x_{i-1}}\| \geq \|w\| . \qquad (8.2.14)$$

Let now $h = \sum_{i=1}^{n} \operatorname{sign}(t_i) \chi_{[x_{i-1}, x_i]}$ and $g(t) = \int_0^t h(s) ds$, $t \in \mathbb{R}_+$. Then $g \in \operatorname{Lip}_0(\mathbb{R}_+)$ and $L(g) = \|h\|_\infty = 1$.

The relations

$$\sum_{i=1}^{n} |t_i|(x_i - x_{i-1}) = \sum_{i=1}^{n} t_i(g(x_i) - g(x_{i-1})) = \langle w, g \rangle \leq L(g)\|w\| = \|w\| ,$$

combined with (8.2.14) yield

$$\|w\| = \sum_{i=1}^{n} |t_i|(x_i - x_{i-1}) = \|f\|_{L^1} .$$

Since the step functions are dense in $L^1(\mathbb{R}_+)$ and $\operatorname{span}([0, \infty)_e)$ is dense in $\mathbb{F}(\mathbb{R}_+)$, it follows that $\mathbb{F}(\mathbb{R}_+) \cong L^1(\mathbb{R}_+)$.

Another Interpretation of $\mathbb{F}(X)$

Let (X, d, θ) be a pointed metric space. A functional $\sum_{i=1}^{n} t_i e_{x_i}$ from $\operatorname{span}(X_e)$ can be viewed as a Borel measure μ on X with finite support $\{x_1, \ldots, x_n\}$ acting by the rule $\mu(\{x_i\}) = t_i$ and $\mu(Y) = 0$ for any Borel subset Y of X with $Y \cap \{x_1, \ldots, x_n\} = \emptyset$. Then

$$\int_X f d\mu = \sum_{i=1}^{n} t_i f(x_i) = \Big(\sum_{i=1}^{n} t_i e_{x_i}\Big)(f) ,$$

for every $f \in \operatorname{Lip}_0(X)$. The space $\mathbb{F}(X)$ will be the completion of this space with respect to the norm

$$\|\mu\| = \sup\{ |\int_X f d\mu| : f \in \operatorname{Lip}_0(X), \ L(f) \leq 1 \} .$$

In this case every Borel measure μ supported on a compact subset K of X can be viewed as a member of $\mathbb{F}(X)$ acting by the formula

$$\mu(f) = \int_K f d\mu, \quad f \in \operatorname{Lip}_0(X) .$$

8.2.3 Pestov's Approach

A similar construction of the Lipschitz free Banach space was proposed by Pestov [553]. Let (X, d) be a metric space with a base point θ. Let $X' = X \setminus \{\theta\}$ and let $\mathscr{F}(X)$ be the free vector space having X' as a Hamel basis and θ as the null element. Define a functional $\|\cdot\| : \mathscr{F}(X) \to \mathbb{R}$ by

$$\left\| \sum_i t_i x_i \right\| = \sup\left\{ \left| \sum_i t_i f(x_i) \right| : f \in \mathrm{Lip}_0(X),\ L(f) \leq 1 \right\}, \qquad (8.2.15)$$

for $\sum_i t_i x_i \in \mathscr{F}(X)$.

Lemma 8.2.13 *The functional* $\|\cdot\|$ *defined by* (8.2.15) *is a norm on* $\mathscr{F}(X)$ *and the metric space* X *embeds isometrically in* $(\mathscr{F}(X), \|\cdot\|)$, *i.e.,*

$$\|x - x'\| = d(x, x'), \qquad (8.2.16)$$

for all $x, x' \in X$.

Proof It is obvious that the functional $\|\cdot\|$ given by (8.2.15) is a seminorm on $\mathscr{F}(X)$. Let $u = \sum_{i=1}^n t_i x_i \in \mathscr{F}(X)$, with $t_i \neq 0$, $x_i \neq \theta$ and x_i pairwise distinct. The function $g(x) = d(x, \{\theta, x_2, \ldots, x_n\})$, $x \in X$, is in $\mathrm{Lip}_0(X)$ and $L(g) \leq 1$, so that

$$\left\| \sum_{i=1}^n t_i x_i \right\| \geq \left| \sum_{i=1}^n t_i d(x_i, \{\theta, x_2, \ldots, x_n\}) \right| = |t_1| d(x_1, \{\theta, x_2, \ldots, x_n\})) > 0,$$

showing that $\|\cdot\|$ is a norm on $\mathscr{F}(X)$.

Now, for $x, x' \in X$ and $f \in \mathrm{Lip}_0(X)$ with $L(f) \leq 1$,

$$|f(x) - f(x')| \leq d(x, x'),$$

so that $\|x - x'\| \leq d(x, x')$.

By Proposition 8.2.5, there exists a function $g \in \mathrm{Lip}_0(X)$ with $L(g) = 1$ such that $|g(x) - g(x')| = d(x, x')$, implying

$$\|x - x'\| \geq |g(x) - g(x')| = d(x, x'),$$

and so $\|x - x'\| = d(x, x')$. $\qquad\square$

Take now $\overline{X}$ to be the completion of the normed space $(\mathscr{F}(X), \|\cdot\|)$ and show that it satisfies the requirements of a Lipschitz free Banach space. Denote by i_X the embedding of X in $\overline{X}$.

Theorem 8.2.14 *Let* (X, d, θ) *be a pointed metric space.*

1. *If* Y *a Banach space, then for every* $F \in \mathrm{Lip}_0(X, Y)$ *there exists a unique continuous linear operator* $\overline{F} : \overline{X} \to Y$ *with* $\overline{F} \circ i_X = F$ *and* $\|\overline{F}\| = L(F)$.
2. *If* (Y, d', θ') *is another pointed metric space, then for every* $F \in \mathrm{Lip}_0(X, Y)$ *there exists a unique continuous linear operator* $\overline{F} : \overline{X} \to \overline{Y}$ *such that* $\overline{F} \circ i_X = i_Y \circ F$ *and* $\|\overline{F}\| = L(F)$.

Proof

1. Define first $\tilde{F} : \mathscr{F}(X) \to Y$ by

$$\tilde{F}\left(\sum t_i x_i\right) = \sum t_i F(x_i) \, .$$

Then for every $y^* \in B_{Y^*}$, $f := \frac{1}{L(F)}(y^* \circ F) \in \mathrm{Lip}_0(X)$ and $L(f) \leq 1$, so that for $u = \sum_i t_i x_i \in \mathscr{F}(X)$,

$$\|u\| \geq \frac{1}{L(F)} \cdot \sup_{y^* \in B_{Y^*}} \left| y^*\left(\sum_i t_i F(x_i)\right) \right| = \frac{1}{L(F)} \left\| \sum_i t_i F(x_i) \right\| = \frac{1}{L(F)} \|\tilde{F}(u)\| \, .$$

It follows that $\|\tilde{F}(u)\| \leq L(F)\|u\|$, $u \in \mathscr{F}(X)$, hence $\|\tilde{F}\| \leq L(F)$.

On the other hand, taking into account (8.2.16), one obtains

$$L(F) = \sup_{x \neq x'} \frac{\|F(x) - F(x')\|}{d(x, x')} = \sup_{x \neq x'} \frac{\|\tilde{F}(x - x')\|}{\|x - x'\|} \leq \|\tilde{F}\| \, ,$$

hence $\|\tilde{F}\| = L(F)$.

Now, the operator $\tilde{F}$ can be uniquely extended to a continuous linear operator $\overline{F} : \overline{X} \to Y$ and $\|\overline{F}\| = \|\tilde{F}\| = L(F)$.

The assertions from 2 can be obtained from 1 as 5 was obtained from 4 in the proof of Theorem 8.2.4. Indeed, applying 1 to the operator $i_Y \circ F : X \to \overline{Y}$ we obtain the existence of an operator $\overline{F} : \overline{X} \to \overline{Y}$ such that $\overline{F} \circ i_X = i_y \circ F$ and $\|\overline{F}\| = L(i_X \circ F) = L(F)$ (the last equality holds because i_X is an isometric embedding). $\qquad\square$

8.2.4 A Result of Dixmier and Ng

This is a general method, obtained by Ng [524], to recognize a dual Banach space which, in fact, extends a result of Dixmier from 1948 [197]. Let X be a normed space with dual X^*. For $Y \subseteq X$ and $Z \subseteq X^*$ one defines the *absolute polars* of these sets by

$$Y^\circ = \{x^* \in X^* : |x^*(y)| \leq 1, \ \forall y \in Y\} \quad \text{and} \quad Z_\circ = \{x \in X : |x^*(x)| \leq 1, \ \forall x^* \in Z\}.$$

If Y, Z are subspaces, then

$$Y^\circ = Y^\perp := \{x^* \in X^* : x^*|_Y = 0\} \quad \text{and} \quad Z_\circ = Z_\perp := \{x \in X : x^*(x) = 0, \forall x^* \in Z\}.$$

The following results, called bipolar theorems, hold true:

$$(Y^\circ)_\circ = \overline{\text{aco}}(Y) \quad \text{and} \quad (Z_\circ)^\circ = w^*\text{-cl}(\text{aco}(Z)),$$

where aco(W) denotes the absolutely convex hull of a subset W of a vector space. It follows that

$$(Y^\circ)_\circ = Y \iff Y \subseteq X \text{ is absolutely convex and closed, and}$$
$$(Z_\circ)^\circ = Z \iff Z \subseteq X^* \text{ is absolutely convex and } w^*\text{-closed}.$$
(8.2.17)

Theorem 8.2.15 (Dixmier-Ng) *Let E be a normed space and τ a Hausdorff locally convex topology on E such that the closed unit ball B_E of E is τ-compact.*

1. The set

$$V := \{v : E \to \mathbb{K} : v \text{ linear and } \tau\text{-continuous on } B_E\}$$

is a linear space satisfying

$$(E, \tau)^* \subseteq V \subseteq (E, \|\cdot\|)^*.$$
(8.2.18)

2. The mapping $Q : E \to V^*$, given for $x \in E$ by

$$Q(x)(v) = v(x), \quad v \in V,$$
(8.2.19)

is an isometric isomorphism between E and V^*. In particular, E itself is a Banach space.

Proof

1. It is clear that V is a linear subspace of the algebraic dual of E.

   The first inclusion in (8.2.18) is obvious. Let $v : E \to \mathbb{K}$ be linear and τ-continuous on B_E. Since the ball B_E is τ-compact, $v(B_E)$ will be a compact subset of $\mathbb{K}$, hence bounded, implying the $\|\cdot\|$-continuity of the functional v and the validity of the second inclusion.

   Based on the second inclusion from (8.2.18), we equip the space V with the norm induced by the norm of E^*.

2. The functional $Q(x)$ given by (8.2.19) is linear with respect to v and

$$|Q(x)(v)| = |v(x)| \leq \|v\| \|x\|,$$

so that $Q(x) \in V^*$ and $\|Q(x)\| \le \|x\|$. Since Q is also linear with respect to x, the continuity of Q and the inequality $\|Q\| \le 1$ follow.

Notice that Q is injective. Indeed, taking into account (8.2.18) and the fact that τ is a Hausdorff locally convex topology, we have

$$Q(x) = Q(x') \iff \forall v \in V, \ v(x) = v(x')$$
$$\Rightarrow \forall x^* \in (E, \tau)^*, \ x^*(x) = x^*(x')$$
$$\iff x = x'.$$

Observe now that the restriction of Q to B_E is $(\tau, \sigma(V^*, V))$-continuous. Indeed, let (x_i) be a net in B_E τ-convergent to $x \in B_E$. Taking into account the definition of the space V, $v(x_i) \to v(x)$ for every $v \in V$, hence

$$x_i \xrightarrow{\tau} x \ \Rightarrow \forall v \in V, \ v(x_i) \to v(x)$$
$$\iff \forall v \in V, \ Q(x_i)(v) \to Q(x)(v)$$
$$\iff Q(x_i) \xrightarrow{\sigma(V^*, V)} Q(x).$$

It follows that $D = Q(B_E)$ is $\sigma(V^*, V)$-compact, so that $(D_\circ)^\circ = D$ (by (8.2.17)).
But

$$D_\circ = \{v \in V : |v(x)| \le 1, \ \forall x \in B_E\} = \{v \in V : \|v\| \le 1\} = B_V, \qquad (8.2.20)$$

hence

$$D = (D_\circ)^\circ = (B_V)^\circ = B_{V^*}.$$

Consequently, $Q(B_E) = B_{V^*}$ and, taking into account the fact that Q is injective, this implies that Q is a linear isometry of E onto V^*.   □

Let (X, d, θ) be a pointed metric space, We show now that, if $E = \mathrm{Lip}_0(X)$ and τ is the topology τ_p of pointwise convergence, then the space V from Theorem 8.2.15 is exactly the free Lipschitz space $\mathbb{F}(X)$ generated by the evaluation functionals (see (8.2.8)).

The topology τ_p of pointwise convergence is the topology induced by the product topology of $\mathbb{K}^X$ (see Definition 1.2.22) and is determined by the condition

$$f_i \xrightarrow{\tau_p} f \iff \forall x \in X, \ f_i(x) \to f(x),$$

for any net $(f_i : i \in I)$ in $\mathbb{K}^X$ and $f \in \mathbb{K}^X$.

For convenience, we shall use the notation $X^{\#}$ for $\mathrm{Lip}_0(X)$. The following result is an analog of the Alaoglu–Bourbaki theorem (Theorem 1.4.18) for the unit ball of the Lipschitz dual.

Proposition 8.2.16 *The closed unit ball $B_{X^{\#}}$ of $X^{\#}$ is τ_p-compact.*

Proof Observe first that $B_{X^{\#}}$ is closed in $\mathbb{K}^X$ with respect to the product topology τ_p.

Indeed, let $(f_i : i \in I)$ be a net in $B_{X^{\#}}$ τ_p-convergent to $f \in \mathbb{K}^X$. For $x, y \in X$, the inequality

$$|f_i(x) - f_i(y)| \le d(x, y),$$

valid for all $i \in I$, yields at limit $|f(x) - f(y)| \le d(x, y)$, showing that $f \in B_{X^{\#}}$.

Let $f \in B_{X^{\#}}$. The inequality

$$|f(x)| = |f(x) - f(\theta)| \le d(x, \theta), \quad x \in X,$$

shows that $f \in \prod_{x \in X} B_{\mathbb{K}}[0, d(x, \theta)]$. Each ball $B_{\mathbb{K}}[0, d(x, \theta)]$ is compact in $\mathbb{K}$, so that $\prod_{x \in X} B_{\mathbb{K}}[0, d(x, \theta)]$ is a τ_p-compact subset of $\mathbb{K}^X$, hence its τ_p-closed subset $B_{X^{\#}}$ will be τ_p-compact too. $\qquad\square$

Consequently, we can apply the procedure from Theorem 8.2.15 to the space $E = X^{\#}$ and the topology τ_p.

Let

$$\widehat{X} = \{(x, y) \in X^2 : x \ne y\},$$

and, for $(x, y) \in \widehat{X}$, let

$$e_{x,y} = (e_x - e_y)/d(x, y).$$

Theorem 8.2.17 *Let (X, d, θ) be a pointed metric space and let V be the space given by Theorem 8.2.15 for $E = X^{\#}$ and $\tau = \tau_p$ — the pointwise topology. Then*

1. *the closed unit ball of V is the closed absolutely convex hull in the space $(X^{\#})^*$ of the set $\{e_{x,y} : (x, y) \in \widehat{X}\}$;*
2. *the space V is the closed linear hull in the space $(X^{\#})^*$ of the set $\{e_x : x \in X\}$. Consequently, $V = \mathbb{F}(X)$.*

Proof

1. For every $x \in X$, the evaluation functional e_x, given by $e_x(f) = f(x)$, $f \in X^{\#}$, is τ_p-continuous, so it belongs to the linear space V, as well as the functionals

$e_{x,y}$, $(x, y) \in \widehat{X}$. Since Q maps $X^{\#}$ onto V^* we have

$$Q(B_{X^{\#}}) = \{Q(f) : f \in X^{\#}, \ |e_{x,y}(f)| \leq 1, \ \forall (x, y) \in \widehat{X}\}$$
$$= \{Q(f) : f \in X^{\#}, \ |Q(f)(e_{x,y})| \leq 1, \ \forall (x, y) \in \widehat{X}\}$$
$$= \{F \in V^* : |F(e_{x,y})| \leq 1, \ \forall (x, y) \in \widehat{X}\}$$
$$= \{e_{x,y} : (x, y) \in \widehat{X}\}^{\circ}.$$

The equalities (8.2.20) show that $Q(B_{X^{\#}})_{\circ} = B_V$, so that, by (8.2.17),

$$B_V = (\{e_{x,y} : (x, y) \in \widehat{X}\}^{\circ})_{\circ} = \overline{\mathrm{aco}}(\{e_{x,y}; (x, y) \in \widehat{X}\}).$$

2. Let $W = \mathrm{span}(\{e_x : x \in X\})$ and $Z = \mathrm{span}(\{e_{x,y} : (x, y) \in \widehat{X}\})$. By 1, $\mathrm{cl}(Z) = V$. Since $e_{x,y} \in W$, it follows that $Z \subseteq W$. The relations

$$e_x = e_x - e_\theta = d(x, \theta) e_{x,\theta} \in Z$$

imply $W \subseteq Z$, so that $W = Z$ and $\mathbb{F}(X) = \mathrm{cl}(W) = \mathrm{cl}(Z) = V$.   □

8.2.5 The Lipschitz Conjugate Operator

In this subsection we shall present, following [160], the Lipschitz analog of the conjugate of a linear mapping. Compactness properties for Lipschitz operators and their conjugates will be presented in Sect. 8.6.

We consider now that X, Y are Banach spaces with the origins as base points. Then the space $\mathscr{L}(X, Y)$ of all continuous linear operators from X to Y is contained in the space $\mathrm{Lip}_0(X, Y)$ of all Lipschitz mappings $F : X \to Y$ satisfying $F(0) = 0$ and the operator norm

$$\|A\| := \sup\{\|Ax\| : x \in X, \ \|x\| \leq 1\}$$

of $A \in \mathscr{L}(X, Y)$ agrees with its Lipschitz norm $L(A)$:

$$L(A) = \sup_{x \neq x'} \frac{\|Ax - Ax'\|}{\|x - x'\|} = \sup_{x \neq x'} \left\| A\left(\frac{x - x'}{\|x - x'\|}\right)\right\| = \sup\{\|Au\| : \|u\| = 1\} = \|A\|,$$

provided that $X \neq \{0\}$. If $X = \{0\}$, then $A = O$ (the null operator) and $L(A) = 0 = \|A\|$.

In particular, $X^* \subseteq \mathrm{Lip}_0(X)$ and $\|x^*\| = L(x^*)$ for $x^* \in X^*$.

For $F \in \mathrm{Lip}_0(X, Y)$ define $F^{\#} : \mathrm{Lip}_0(Y) \to \mathrm{Lip}_0(X)$ by

$$F^{\#}g = g \circ F, \quad g \in \mathrm{Lip}_0(Y).$$

Proposition 8.2.18 *Under the above hypotheses, $F^{\#}$ is a continuous linear operator from $\mathrm{Lip}_0(Y)$ to $\mathrm{Lip}_0(X)$ with norm $\|F^{\#}\| = L(F)$.*

Proof From

$$L(g \circ F) \leq L(g)L(F)$$

it follows that $F^{\#}$ is well-defined. Since the linearity is obvious, the above inequality shows its continuity and the inequality

$$\|F^{\#}\| \leq L(F).$$

For $0 < \varepsilon < L(F)$ let x_1, x_2 be distinct points in X such that

$$\|F(x_1) - F(x_2)\|/\|x_1 - x_2\| > L(F) - \varepsilon.$$

If $y^* \in X^*$, $\|y^*\| = 1$, is such that

$$y^*(F(x_1) - F(x_2)) = \|F(x_1) - F(x_2)\|,$$

then

$$\|F^{\#}\| = \sup_{g \neq 0} \frac{L(F^{\#}g)}{L(g)} \geq \frac{L(y^* \circ F)}{L(y*)} = L(y^* \circ F)$$

$$\geq \frac{|y^*(F(x_1)) - y^*(F(x_2))|}{\|x_1 - x_2\|} = \frac{\|F(x_1) - F(x_2)\|}{\|x_1 - x_2\|} > L(F) - \varepsilon.$$

Letting $\varepsilon \searrow 0$, one obtains $\|F^{\#}\| \geq L(F)$, and so $\|F^{\#}\| = L(F)$. $\qquad\square$

For $F \in \mathrm{Lip}_0(X, Y)$ let $\Phi(F) : Æ(X) \to Æ(Y)$ be the continuous linear operator attached to F by Theorem 8.2.4.5. We shall show that the Banach space conjugate $\Phi(F)^* : Æ(Y)^* \to Æ(X)^*$ of $\Phi(F)$ corresponds in a canonical way to the Lipschitz conjugate $F^{\#}$. Denote by Λ_1, Λ_2 the isometrical isomorphisms between $Æ(X)^*$ and $\mathrm{Lip}_0(X)$, and $Æ(Y)^*$ and $\mathrm{Lip}_0(Y)$, respectively, defined in Theorem 8.2.2.

Theorem 8.2.19 *We have*

$$\Lambda_1 \circ \Phi(F)^* = F^{\#} \circ \Lambda_2, \tag{8.2.21}$$

that is, the following diagram is commutative

$$
\begin{array}{ccc}
Æ(Y)^* & \xrightarrow{\ \Phi(F)^*\ } & Æ(X)^* \\[2pt]
{\scriptstyle \Lambda_2}\downarrow & & \downarrow{\scriptstyle \Lambda_1} \\[6pt]
\mathrm{Lip}_0(Y) & \xrightarrow{\ F^{\#}\ } & \mathrm{Lip}_0(X).
\end{array}
$$

Proof Observe that

$$\Phi(F)(m_{x,0}) = m_{F(x),0}, \quad \text{for all } x \in X,$$

and, for all $\psi \in \mathcal{E}(X)^*$,

$$\Phi(F)^*(\psi)(m_{x,0}) = \psi(m_{F(x),0}), \quad \text{for all } x \in X.$$

Consequently,

$$\left(\Lambda_1 \circ \Phi(F)^*\right)(\psi)(x) = \Lambda_1\left(\Phi(F)^*\psi\right)(x) = \left(\Phi(F)^*\psi\right)(m_{x,0}) = \psi(m_{F(x),0}),$$

and

$$\left(F^\# \circ \Lambda_2\right)(\psi)(x) = F^\#(\Lambda_2\psi)(x) = ((\Lambda_2\psi) \circ F)(x)$$
$$= (\Lambda_2\psi)(F(x)) = \psi(m_{F(x),0}),$$

for all $x \in X$ and all $\psi \in \mathcal{E}(X)^*$, proving the equality (8.2.21). $\square$

Similar results hold for the free Lipschitz spaces $\mathbb{F}(X)$ generated by evaluation functionals and for the free space $\overline{X}$ considered by Pestov (see Sect. 8.2.3). We shall discuss the case of $\mathbb{F}(X)$.

Theorem 8.2.20 *Let* X, Y *be Banach spaces and* $F \in \mathrm{Lip}_0(X, Y)$ *and* $F^\#$: $\mathrm{Lip}_0(Y) \to \mathrm{Lip}_0(X)$ *the Lipschitz conjugate operator associated to* F.
Let $\Lambda_1 : \mathrm{Lip}_0(X) \to \mathbb{F}(X)$ *and* $\Lambda_2 : \mathrm{Lip}_0(Y) \to \mathbb{F}(Y)$ *be the isomorphisms given by Theorem 8.2.7 and* $\widetilde{F} : \mathbb{F}(X) \to \mathbb{F}(Y)$ *the linear continuous operator corresponding to* F *by Theorem 8.2.9.*
Then

$$\Lambda_1 \circ F^\# = \widetilde{F}^* \circ \Lambda_2,$$

where $\widetilde{F}^*$ *denotes the Banach conjugate of the continuous linear operator* $\widetilde{F}$.
This means that the following diagram is commutative

$$\mathrm{Lip}_0(Y) \xrightarrow{\ F^\# \ } \mathrm{Lip}_0(X)$$
$$\Lambda_2 \downarrow \qquad\qquad \Lambda_1 \downarrow$$
$$\mathbb{F}(Y)^* \xrightarrow{\ \widetilde{F}^* \ } \mathbb{F}(X)^*.$$

Proof Indeed, for any $g \in \mathrm{Lip}_0(Y)$,

$$\Lambda_1(F^\# g)(\sum_i t_i e_{x_i}) = \Lambda_1(g \circ F)(\sum_i t_i e_{x_i}) = \sum_i t_i g(F(x_i)),$$

and

$$\tilde{F}^*(\Lambda_2 g)(\sum_i t_i e_{x_i}) = ((\Lambda_2 g) \circ \tilde{F})(\sum_i t_i e_{x_i}) = (\Lambda_2 g)(\sum_i t_i e_{F(x_i)}) = \sum_i t_i g(F(x_i)),$$

for all $\sum_i t_i e_{x_i} \in \text{span}(X_e)$. □

The Lipschitz conjugate operator is related to the Banach conjugate in the following way.

Proposition 8.2.21 *Let X, Y be Banach spaces and $A : X \to Y$ a continuous linear operator. Then*

$$A^{\#}|_{Y^*} = A^*.$$

Proof Indeed, for any $y^* \in Y^*$,

$$A^{\#} y^* = y^* \circ A = A^* y^*.$$

□

8.3 Little Lipschitz Functions

This is an important subclass of Lipschitz functions. Let (X, d, θ) be a pointed metric space. A Lipschitz function $f : X \to \mathbb{K}$ is called *little Lipschitz* if

$$\lim_{\delta \to 0+} \sup\{|f(x) - f(y)|/d(x, y) : 0 < d(x, y) \le \delta\} = 0. \tag{8.3.1}$$

This means that for every $\varepsilon > 0$ there exists $\delta > 0$ such that for all $x, y \in X$,

$$d(x, y) \le \delta \implies |f(x) - f(y)| \le \varepsilon d(x, y).$$

Condition (8.3.1) can be written also in the equivalent form

$$\lim_{d(x,y) \to 0} \frac{|f(x) - f(y)|}{d(x, y)} = 0. \tag{8.3.2}$$

The space of little Lipschitz functions is denoted by $\text{lip}(X)$, that of little Lipschitz functions vanishing at θ by $\text{lip}_0(X)$ and that of bounded little Lipschitz functions by $\text{blip}(X)$.

Starting from Hölder spaces (see (8.1.4) and (8.1.5)), one can define, in an analogous way, the little Hölder spaces

$$\text{lip}_\alpha(X), \text{blip}_\alpha(X), \text{lip}_{\alpha,0}(X),$$

as formed of all functions f in $\mathrm{Lip}_\alpha(X), \mathrm{BLip}_\alpha(X)$ and $\mathrm{Lip}_{\alpha,0}(X)$, respectively, satisfying the condition

$$\lim_{d(x,y)\to 0} \frac{|f(x) - f(y)|}{(d(x,y))^\alpha} = 0\,.$$

Again we adopt the conventions:

$$\mathrm{lip}_1(X) = \mathrm{lip}(X),\ \ \mathrm{blip}_1(X) = \mathrm{blip}(X),$$

$$\mathrm{lip}_{1,0}(X) = \mathrm{lip}_0(X)\,.$$

As in the case of big Hölder spaces, the following equalities hold

$$\mathrm{lip}_\alpha(X, d) = \mathrm{lip}(X, d^\alpha),\ \ \mathrm{blip}_\alpha(X, d) = \mathrm{blip}(X, d^\alpha),$$

$$\mathrm{lip}_{\alpha,0}(X, d) = \mathrm{lip}_0(X, d^\alpha)\,.$$

8.3.1 De Leeuw's Map

This is a map considered first by de Leeuw [182] in the study of spaces of Lipschitz functions.

For a pointed metric space (X, d, θ) let $\Delta(X) := \{(x, x) : x \in X\}$ be the diagonal of the Cartesian product $X^2 = X \times X$ and $\widehat{X} = X^2 \setminus \Delta(X) = \{(x, y) \in X^2 : x \neq y\}$. For a function $f : X \to \mathbb{K}$ let $\Phi f : \widehat{X} \to \mathbb{K}$ be given by

$$(\Phi f)(x, y) = \frac{f(x) - f(y)}{d(x, y)}, \quad (x, y) \in \widehat{X}\,. \tag{8.3.3}$$

It is obvious that f is Lipschitz if and only if Φf is bounded and in this case

$$L(f) = \|\Phi f\|_\infty\,.$$

The following property is a direct consequence of the definitions.

Proposition 8.3.1 *Let X be a pointed metric space, $(f_i : i \in I)$ a net in $\mathrm{Lip}_0(X)$ and $f \in \mathrm{Lip}_0(X)$. Then*

$$f_i \to f\ \text{pointwise} \iff \Phi f_i \to \Phi f\ \text{pointwise}\,.$$

The mapping Φ has the following properties.

Theorem 8.3.2 *Let* (X, d, θ) *be a pointed metric space.*

1. Φ *is a linear isometry of* $\mathrm{Lip}_0(X)$ *into* $\ell^\infty(\widehat{X})$.
2. $\Phi(f \cdot g) = f \cdot \Phi(g) + \Phi(f) \cdot g,$ *for all* $f, g \in \mathrm{Lip}_0(X) \cap \ell^\infty(X)$.
3. $\Phi\big(\mathrm{Lip}_0(X)\big)$ *is* w^*-*closed in* $\ell^\infty(\widehat{X})$.

The proofs of 1 and 2 are straightforward. The proof of 3 is based on the Krein-Shmulian theorem which states that a subspace F of E^*, E a Banach space, is w^*-closed if and only if $F \cap B_{E^*}$ is w^*-closed.

The assertion 3 of the above theorem implies that $\mathrm{Lip}_0(X)$ is a dual space. This follows from the following result.

Theorem 8.3.3 *Let* E *be a Banach space and* F *a* w^*-*closed subspace of* E^*. *Then*

$$F \cong (E/F_\perp)^* \,,$$

where $F_\perp = \{x \in E : x^*(x) = 0, \ \forall x^* \in F\}$.

In our case $\mathrm{Lip}_0(X)$ is isometrically isomorphic to the w^*-closed subspace $\Phi\big(\mathrm{Lip}_0(X)\big)$ of $\ell^\infty(\widehat{X})$, so that we can apply the above theorem.

Theorem 8.3.4 *Let* (X, d, θ) *be a pointed metric space. Then*

$$\mathrm{Lip}_0(X) \cong \Big(\ell^1(\widehat{X}) / \big(\mathrm{Lip}_0(X)\big)_\perp\Big)^* .$$

Considering the w^*-topology of $\mathrm{Lip}_0(X)$ given by the duality from Theorem 8.2.7, the following result holds.

Proposition 8.3.5 *Let* (X, d, θ) *be a pointed metric space. The* w^*-*topology of* $\mathrm{Lip}_0(X)$ *agrees on bounded sets with the topology of pointwise convergence, that is, for every bounded net* $(f_i : i \in I)$ *in* $\mathrm{Lip}_0(X)$ *and* $f \in \mathrm{Lip}_0(X)$

$$f_i \xrightarrow{w^*} f \iff \forall x \in X, \ f_i(x) \to f(x).$$

If a bounded net $(f_i : i \in I)$ *converges pointwise to* f, *then the convergence is uniform on every totally bounded subset of* X.

Consequently, if the space X *is compact, then the topology of pointwise convergence agrees with the topology of uniform convergence.*

Proof The proof of Proposition 2.4.1 for sequences of Lipschitz functions can be adapted to nets.   □

8.3.2 Properties of the Space $\mathrm{lip}_0(X)$

We shall present the basic properties of the spaces of little Lipschitz functions following the treatise [675].

In this subsection (X, d, θ) will be a pointed compact metric space.

To see some difficulties that arise in the study of these spaces we start with some examples.

Example 8.3.6 Let $X = [0, 1]$. It follows that every function $f \in \mathrm{lip}(X)$ is differentiable and $f' \equiv 0$, so that $\mathrm{lip}(X)$ is formed only of constant functions.

Example 8.3.7 Suppose that X is totally disconnected (such as the Cantor ternary set). Then for every clopen subset Y of X the characteristic function χ_Y is in $\mathrm{lip}(X)$. It follows that $\mathrm{lip}(X)$ contains the linear subspace generated by these functions.

Example 8.3.8 Let (X, d, θ) be a pointed compact metric space and $0 < \alpha < 1$. Then d^α is a metric on X, topologically equivalent to d, and $\mathrm{Lip}_0(X, d) \subseteq \mathrm{Lip}_0(X, d^\alpha)$. In fact, the stronger inclusion $\mathrm{Lip}_0(X, d) \subseteq \mathrm{lip}_0(X, d^\alpha)$ holds.

The first inclusion follows from the inequality

$$|f(x) - f(y)| \le L(f)d(x, y) \le L(f)(\mathrm{diam}\ X)^{1-\alpha}d(x, y)^{\alpha}.$$

Given $\varepsilon > 0$, let $\delta = (\varepsilon/L(f))^{1/(1-\alpha)}$. Then

$$|f(x) - f(y)| \le L(f)\delta^{1-\alpha}d(x, y)^{\alpha} = \varepsilon d(x, y)^{\alpha},$$

for all $x, y \in X$ with $d(x, y) \le \delta$, proving the validity of the second inclusion.

The space $\widehat{X} = X^2 \setminus \Delta(X)$ and de Leeuw's mapping $\Phi : \mathrm{lip}_0(X) \to \ell^\infty(\widehat{X})$, given by (8.3.3), play a key role in the study of the space $\mathrm{lip}_0(X)$.

Proposition 8.3.9 *Let (X, d, θ) be a pointed compact metric space and $f \in \mathrm{Lip}_0(X)$. Then*

$$f \in \mathrm{lip}_0(X) \iff \Phi f \in C_0(\widehat{X}).$$

The subspace $\mathrm{lip}_0(X) = \Phi^{-1}(C_0(\widehat{X}))$ is closed in $\mathrm{Lip}_0(X)$, so it is a Banach space (with respect to the Lipschitz norm).

Proof The proof is based on the remark that a function $F \in C(\widehat{X})$ belongs to $C_0(\widehat{X})$ if and only if the function $\widetilde{F} : X^2 \to \mathbb{K}$, given by $\widetilde{F}(x, y) = F(x, y)$ for $(x, y) \in \widehat{X}$ and $\widetilde{F}(x, x) = 0$ for $x \in X$, belongs to $C(X^2)$.

Since the space $C_0(\widehat{X})$ is closed in $\ell^\infty(\widehat{X})$ it follows that $\mathrm{lip}_0(X) = \Phi^{-1}(C_0(\widehat{X}))$ is closed in $\mathrm{Lip}_0(X)$.   □

Remark 8.3.10 The space $\mathrm{lip}_0(X)$ is a Banach algebra and a Banach lattice, that is,

(i) $f, g \in \mathrm{lip}_0(X) \Rightarrow f \cdot g \in \mathrm{lip}_0(X)$;

(ii) $f, g \in \mathrm{lip}_0(X) \Rightarrow f \vee g, f \wedge g \in \mathrm{lip}_0(X)$.

As in the case of Lipschitz functions, a natural question is that of norm-preserving extensions of little Lipschitz functions from subsets to the whole space.

Example 8.3.6 shows that such an extension is not always possible. Indeed, if $Y \subseteq X$ is finite, then every function $f : Y \to \mathbb{R}$ is in $\mathrm{lip}_0(Y)$, but $\mathrm{lip}_0(X)$ contains only constant functions which, restricted to Y, do not agree with f if f is not constant.

Therefore an extra condition is needed to guarantee the possibility of the extension, a condition that was found by Weaver (see [675, Definition 3.2.1]).

Definition 8.3.11 Let (X, d, θ) be a pointed compact metric space. One says that $\mathrm{lip}_0(X)$ *separates points uniformly* if there exists $a > 1$ such that for every $x, y \in X$ there exists $f \in \mathrm{lip}_0(X)$ with

$$L(f) \le a \quad \text{and} \quad |f(x) - f(y)| = d(x, y).$$

If the condition $L(f) \le a$ is dropped, then one says simply that $\mathrm{lip}_0(X)$ *separates points*.

By Proposition 8.2.5 this property holds in the space $\mathrm{Lip}_0(X)$. Some examples of spaces with this property are given in [675, Proposition 3.2.2].

Proposition 8.3.12

1. *If $K \subseteq [0, 1]$ is the Cantor ternary set, then $\mathrm{lip}_0(X)$ separates points uniformly.*
2. *If (X, d, θ) is a pointed compact metric space, then $\mathrm{lip}_0(X, d^\alpha)$ separates points uniformly, for every $0 < \alpha < 1$.*

The following theorem shows that this is the right condition for the existence of the extensions.

Theorem 8.3.13 *Suppose that (X, d, θ) is a pointed compact metric space such that $\mathrm{lip}_0(X, d^\alpha)$ separates points uniformly and let Y be a subset of X containing θ.*

1. *Every function $f \in \mathrm{lip}_0(Y, \mathbb{R})$ with $L(f) < 1$ admits an extension $\tilde{f} \in \mathrm{lip}_0(X)$ with $L(\tilde{f}) < 1$ and $\|\tilde{f}\|_\infty = \|f\|_\infty$.*
2. *Every function $f \in \mathrm{lip}_0(Y, \mathbb{C})$ with $L(f) < 1$ admits an extension $\tilde{f} \in \mathrm{lip}_0(X)$ with $L(\tilde{f}) < \sqrt{2}$ and $\|\tilde{f}\|_\infty = \|f\|_\infty$.*

The proof of this theorem is long (based on three lemmas) and difficult, see [675, pp. 78–81].

Another important result concerning little Lipschitz spaces is that, under some hypotheses on the metric space X, the space $\mathrm{Lip}_0(X)$ is in fact the double dual space of $\mathrm{lip}_0(X)$.

Theorem 8.3.14 *Let (X, d, θ) be a pointed compact metric space. If $\mathrm{lip}_0(X)$ separates points uniformly, then $\mathrm{lip}_0(X)^* \cong \mathcal{E}(X)$ and $\mathrm{lip}_0(X)^{**} \cong \mathrm{Lip}_0(X)$.*
   The same is true in the case of the space $\mathbb{F}(X)$.

Proof (Sketch) One defines $\Gamma : \mathcal{E}(X) \to \mathrm{lip}_0(X)^*$ by

$$(\Gamma m)(f) = \langle f, m \rangle, \quad f \in \mathrm{lip}_0(X),$$

for every molecule m, where $\langle f, m \rangle$ is given by (8.2.3). One shows that Γ is a linear isometry that extends to an isometric isomorphism of $\mathcal{Æ}(X)$ onto $\mathrm{lip}_0(X)^*$. But then the conjugate operator Γ^* will be an isometric isomorphism of $\mathrm{lip}_0(X)^{**}$ onto $\mathcal{Æ}(X)^* \cong \mathrm{Lip}_0(X)$. In proving the surjectivity of Γ one appeals to the isometric isomorphism Φ between $\mathrm{lip}_0(X)$ and $C_0(\widehat{X})$ from Proposition 8.3.9 and to the representation of continuous linear functionals on $C_0(\widehat{X})$ as Radon measures.

In the case of the space $\mathbb{F}(X)$ one can consider the evaluation functionals $\tilde{e}_x$, $x \in X$, acting on $\mathrm{lip}_0(X)$ by the rule $\tilde{e}_x(f) = f(x)$, $f \in \mathrm{lip}_0(X)$. In fact these are the restrictions of the evaluation functionals $e_x \in \mathrm{Lip}_0(X)^*$ considered in Sect. 8.2.2, to the closed subspace $\mathrm{lip}_0(X)$ of $\mathrm{Lip}_0(X)$. Put $X_e := \{e_x : x \in X\} \subseteq \mathrm{Lip}_0(X)^*$ and $X_{\tilde{e}} := \{\tilde{e}_x : x \in X\} \subseteq \mathrm{lip}_0(X)^*$ and consider the mapping $\Gamma : \mathrm{span}(X_e) \to \mathrm{span}(X_{\tilde{e}}) \subseteq \mathrm{lip}_0(X)^*$ given by

$$\Gamma\left(\sum_i t_i e_{x_i}\right) = \sum_i t_i \tilde{e}_{x_i}\,,$$

for $\sum_i t_i e_{x_i} \in \mathrm{span}(X_e)$. This correspondence is linear and isometric, so that it extends to an isometric linear mapping from $\mathbb{F}(X)$ to $\mathrm{lip}_0(X)^*$. One shows that this mapping is also onto, and so it is an isometric isomorphism.

Consequently, $\Gamma(\mathbb{F}(X)) = \overline{\mathrm{span}}(X_{\tilde{e}}) = \mathrm{lip}_0(X)^*$.

The space $\mathrm{Lip}_0(X)$ is isometrically isomorphic to the space $\mathrm{lip}_0(X)^{**} \cong \mathbb{F}(X)^*$, the isomorphism $\Psi : \mathrm{Lip}_0(X) \to \mathrm{lip}_0(X)^{**}$ being determined by the condition

$$\Psi(f)(\tilde{e}_x) = f(x), \quad x \in X, \ f \in \mathrm{Lip}_0(X)\,.$$

Its inverse Ψ^{-1} satisfies

$$\Psi^{-1}(\varphi)(x) = \varphi(\tilde{e}_x), \quad x \in X, \ \varphi \in \mathrm{lip}_0(X)^{**}\,.$$

$\square$

Taking into account the examples of spaces satisfying the uniform separation condition, given in Proposition 8.3.12, one obtains the following result.

Corollary 8.3.15

1. If $K \subseteq [0, 1]$ is the Cantor ternary set, then $\mathrm{lip}_0(K)^{**} \cong \mathrm{Lip}_0(K)$.
2. If (X, d, θ) is a pointed compact metric space, then $\mathrm{lip}_0(X, d^\alpha)^{**} \cong \mathrm{Lip}_0(X, d^\alpha)$.

In fact the uniform separability condition is, besides sufficient, also necessary for the validity of Theorem 8.3.14.

Proposition 8.3.16 *For a pointed compact metric space (X, d, θ) the following conditions are equivalent.*

1. *There exists $a > 1$ such that for every $x, y \in X$ there exists $f \in \mathrm{lip}_0(X)$ satisfying the conditions $L(f) \le a$ and $|f(x) - f(y)| = d(x, y)$.*

2. *For all* $a > 1$ *and* $x, y \in X$ *there exists* $f \in \mathrm{lip}_0(X)$ *satisfying the conditions* $L(f) \le a$ *and* $|f(x) - f(y)| = d(x, y)$.
3. *There exists* $b > 1$ *such that for every* $f \in \mathrm{Lip}_0(X)$ *and every* $A \subseteq X$, *A finite, there exists* $g \in \mathrm{lip}(X)$ *satisfying the conditions* $L(g) \le b L(f)$ *and* $|g(x) - g(y)| = d(x, y)$.
4. *For all* $b > 1$, $f \in \mathrm{Lip}_0(X)$ *and* $A \subseteq X$, *A finite, there exists* $g \in \mathrm{lip}(X)$ *satisfying the conditions* $L(g) \le b L(f)$ *and* $|g(x) - g(y)| = d(x, y)$.
5. $\mathrm{lip}_0(X)^{**} \cong \mathrm{Lip}_0(X)$.

8.4 The Kantorovich–Rubinstein Metric

In this section we shall introduce some metrics on spaces of Borel measures on a compact metric space (X, d) (usually called Kantorovich–Rubinstein metrics), derived from some norms on these spaces. In this way one obtains a new representation of the space $\mathrm{Lip}_0(X)$ as a dual space. These metrics are essential tools in the treatment of the Monge–Kantorovich mass transport problem and at the same time are related with the weak convergence of sequences of measures. The use of a sesquilinear integral for functions and measures taking values in a Hilbert space, developed in [146], allows us to extend some results to the more general context of measures with values in a Hilbert space.

8.4.1 A Sesquilinear Integral

We have seen in Sect. 1.6 that one could define reasonable integrals for vector functions with respect to scalar measures and for scalar functions with respect to vector measures. Taking advantage of the scalar product properties in a Hilbert space $\mathscr{H}$, one can define an integral for functions and measures both taking values in $\mathscr{H}$.

In this subsection we shall briefly present, following [146] (where further details can be found), the construction and some properties of this sesquilinear integral. In order to distinguish it from the usual integral we shall use the notation $\int \langle f, d\mu \rangle$. This integral will be the key ingredient in the definition of the Kantorovich–Rubinstein metric for measures with values in a Hilbert space.

Let (X, d) be a metric space and $\mathscr{H}$ a Hilbert space with scalar product $\langle \cdot, \cdot \rangle$ and corresponding norm $\|x\| = \sqrt{\langle x, x \rangle}$. The σ-algebra of Borel subsets of X is denoted by $\mathscr{B}(X)$. Let $C(X, \mathscr{H}) = \{f : X \to \mathscr{H} : f \text{ is continuous}\}$. The vector space $C_b(X, \mathscr{H}) = \{f : X \to \mathscr{H} : f \text{ is continuous and bounded}\}$ is a Banach space with norm $\|f\|_\infty := \sup\{\|f(x)\| : x \in X\}$. Actually, $C_b(X, \mathscr{H})$ is a closed subspace of the Banach space $B(X, \mathscr{H}) = \{f : X \to \mathscr{H} : f \text{ is bounded}\}$ equipped with the norm $\|\cdot\|_\infty$ and $C_b(X, \mathscr{H}) = C(X, \mathscr{H}) \cap B(X, \mathscr{H})$.

If X is compact, then $C(X, \mathcal{H}) = C_b(X, \mathcal{H})$.

For $f : X \to \mathbb{K}$ and $\xi \in \mathcal{H}$ define $f\xi : X \to \mathcal{H}$ by

$$(f\xi)(x) = f(x)\xi, \quad x \in X.$$

If f is bounded then $f\xi$ is also bounded and $\|f\xi\|_\infty = \|f\|_\infty \|\xi\|$.

A function $f : X \to \mathcal{H}$ will be called *simple* if it has the form $f = \sum_{i=1}^m \chi_{A_i} \xi_i$, with $A_i \in \mathcal{B}(X)$ and $\xi_i \in \mathcal{H}$, where χ_A denotes the characteristic function of the set A, i.e., $\chi_A(x) = 1$ if $x \in A$ and $\chi_A(x) = 0$ otherwise. We shall suppose that the sets A_i are mutually disjoint and $\bigcup_{i=1}^m A_i = X$. The set $S(X, \mathcal{H})$ of all simple functions is a vector subspace of $B(X, \mathcal{H})$.

A function $\mu : \mathcal{B}(X) \to \mathcal{H}$ is called a σ-additive vector measure if

$$\mu\left(\bigcup_{i=1}^\infty A_i\right) = \sum_{i=1}^\infty \mu(A_i),$$

for every family $\{A_i : i \in \mathbb{N}\}$ of pairwise disjoints elements of $\mathcal{B}(X)$ (see Sect. 1.6). Recall that the variation of μ over $A \in \mathcal{B}(X)$, denoted by $|\mu|(A)$, is defined by

$$|\mu|(A) = \sup\{\sum_{i=1}^m \|\mu(A_i)\| : (A_i)_{i \in \{1,2,\dots,m\}} \text{ is a partition of } A, \ m \in \mathbb{N}\},$$

i.e., the supremum is taken over all possible partitions of A. We say that μ *is of bounded variation* if $|\mu|(X) < \infty$. The number

$$\|\mu\| := |\mu|(X) \tag{8.4.1}$$

is called the *total variation* of the measure μ (in some cases it will be denoted by $var(\mu)$). Any σ-additive measure $\mu : \mathcal{B}(X) \to \mathbb{K}$ is of bounded variation.

The set $cabv(X, \mathcal{H})$ of all σ-additive vector measures $\mu : \mathcal{B}(X) \to \mathcal{H}$ of bounded variation is a vector space (with respect to the natural operations) and (8.4.1) is a complete norm on $cabv(X, \mathcal{H})$, i.e., $cabv(X, \mathcal{H})$ is a Banach with respect to this norm (*cabv* comes from "countable additive with bounded variation").

The closed unit ball of the space $(cabv(X, \mathcal{H}), \|\cdot\|)$ will be denoted by

$$B_{var}(cabv(X, \mathcal{H})) := \{\mu \in cabv(X, \mathcal{H}) : \|\mu\| \le 1\}.$$

Then a ball of center 0 and radius $r > 0$ will be given by $r B_{var}(cabv(X, \mathcal{H}))$.

The topology on $cabv(X, \mathcal{H})$ generated by this norm will be called the *variational topology* and will be denoted by τ_{var}.

For a sequence $(\mu_n)_n \subseteq cabv(X, \mathcal{H})$ and $\mu \in cabv(X, \mathcal{H})$, $\mu_n \xrightarrow{var} \mu$ means that $(\mu_n)_n$ converges to μ in τ_{var}.

Remark 8.4.1 Notice that $\mu_n \xrightarrow{var} \mu$ implies $\mu_n(A) \to \mu(A)$, uniformly with respect to $A \in \mathcal{B}(X)$.

Indeed, for $\varepsilon > 0$ there exists n_0 such that $|\mu_n - \mu|(X) = \|\mu_n - \mu\| \le \varepsilon$ for all $n \ge n_0$. But then

$$\|(\mu_n - \mu)(A)\| \le |\mu_n - \mu|(A) \le |\mu_n - \mu|(X) \le \varepsilon,$$

for all $A \in \mathcal{B}(X)$ and all $n \ge n_0$.

For $\mu \in cabv(X, \mathbb{K})$ and $\xi \in \mathcal{H}$, define $\mu\xi : \mathcal{B}(X) \to \mathcal{H}$ by

$$\mu\xi(A) = \mu(A)\xi, \quad A \in \mathcal{B}(X).$$

Then $\mu\xi$ belongs to $cabv(X, \mathcal{H})$ and $\|\mu\xi\| = \|\mu\| \|\xi\|$. This is due to the fact that

$$\sum_{i=1}^{m} \|\mu(A_i)\xi\| = \left(\sum_{i=1}^{m} |\mu(A_i)| \right) \|\xi\|,$$

for every partition A_i, $1 \le i \le n$, of a set $A \in \mathcal{B}(X)$.

For $(f_n)_n \subseteq B(X, \mathcal{H})$ and $f \in B(X, \mathcal{H})$, we denote by $f_n \xrightarrow{u} f$ the fact that $(f_n)_n$ converges uniformly to f (i.e., $(f_n)_n$ converges to f in the Banach space $B(X, \mathcal{H})$). The closure of $S(X, \mathcal{H})$ in $(B(X, \mathcal{H}), \|\cdot\|_\infty)$ is the space of *totally measurable functions*, denoted by $TM(X, \mathcal{H})$, i.e., $TM(X, \mathcal{H}) = \overline{S(X, \mathcal{H})}$. If the metric space X is compact, then $C(X, \mathcal{H}) \subseteq TM(X, \mathcal{H})$.

Now, let $\mu \in cabv(X, \mathcal{H})$. For any $f = \sum_{i=1}^{m} \chi_{A_i}\xi_i \in S(X, \mathcal{H})$, *the integral of f with respect to μ* is defined by

$$\int \langle f, d\mu \rangle = \sum_{i=1}^{m} \langle \xi_i, \mu(A_i) \rangle.$$

It is easily seen that the definition does not depend on the representation of the simple function f.

Because $\left| \int \langle f, d\mu \rangle \right| \le \|f\|_\infty \|\mu\|$, the continuous linear functional $u :$ $S(X, \mathcal{H}) \to \mathbb{K}$, given by $u(f) = \int \langle f, d\mu \rangle$, can be extended by uniform continuity to a continuous linear functional (with the same norm) $v : \overline{S(X, \mathcal{H})} = TM(X, \mathcal{H}) \to \mathbb{K}$. For any $f \in TM(X, \mathcal{H})$, we define *the integral of f with respect to μ* by

$$\int \langle f, d\mu \rangle = v(f).$$

Hence, for any $f \in TM(X, \mathscr{H})$, one has

$$\int \langle f, d\mu \rangle = \lim_{n \to \infty} \int \langle f_n, d\mu \rangle,$$

where $(f_n)_n \subseteq S(X, \mathscr{H})$ is such that $f_n \overset{u}{\to} f$. One shows that the result does not depend on the sequence $(f_n)_n$ used, so that this integral is well-defined. It is sesquilinear, i.e.,

$$\int \langle \alpha f + \beta g, d\mu \rangle = \alpha \int \langle f, d\mu \rangle + \beta \int \langle g, d\mu \rangle \quad \text{and}$$

$$\int \langle f, d(\alpha\mu + \beta v) \rangle = \overline{\alpha} \int \langle f, d\mu \rangle + \overline{\beta} \int \langle g, d\mu \rangle,$$

for $\alpha, \beta \in \mathbb{K}$, $f, g \in TM(X, \mathscr{H})$ and $\mu, v \in cabv(X, \mathscr{H})$.

Notice that, for any $f \in TM(X, \mathscr{H})$ and any $\mu \in cabv(X, \mathscr{H})$, one has

$$\left| \int \langle f, d\mu \rangle \right| \le \|f\|_\infty \|\mu\|, \tag{8.4.2}$$

and $\int \langle f, d\mu \rangle$ makes sense for any $f \in C(X, \mathscr{H}) \subseteq TM(X, \mathscr{H})$.

In the particular case when $\mathscr{H} = \mathbb{C}$, due to the fact that the scalar product in $\mathbb{C}$ is given by the formula $\langle \alpha, \beta \rangle = \alpha\overline{\beta}$, the integral in the sense defined here is related with the usual integral by the equality

$$\int \langle f, d\mu \rangle = \int f d\overline{\mu},$$

for $f \in TM(X, \mathbb{C})$ and $\mu \in cabv(X, \mathbb{C})$, where $\overline{\mu} \in cabv(X, \mathbb{C})$ acts via $\overline{\mu}(A) = \overline{\mu(A)}$, $A \in \mathscr{B}(X)$.

Consider now the general case. Let $(e_i)_{i \in I}$ be an orthonormal basis of $\mathscr{H}$ and let $\mu \in cabv(X, \mathscr{H})$ and $f \in TM(X, \mathscr{H})$.

For any $i \in I$ let

$$f_i : X \to \mathbb{K} \quad \text{be given by} \quad f_i(x) = \langle f(x), e_i \rangle, \quad x \in X, \quad \text{and}$$

$$\mu_i : \mathscr{B}(X) \to \mathbb{K} \quad \text{be given by} \quad \mu_i(A) = \langle e_i, \mu(A) \rangle, \quad A \in \mathscr{B}(X).$$

It follows that $f_i \in TM(X, \mathbb{K})$, $\mu_i \in cabv(X, \mathbb{K})$ and

$$\int \langle f, d\mu \rangle = \sum_i \int f_i d\mu_i. \tag{8.4.3}$$

It is clear that it is sufficient to prove (8.4.3) for simple functions. If $f = \sum_{k=1}^{m} \chi_{A_k} \xi_k$, then $f_i = \sum_{k=1}^{m} \langle \xi_k, e_i \rangle \chi_{A_k}$ and

$$\int f_i d\mu_i = \sum_{k=1}^{m} \langle \xi_k, e_i \rangle \langle e_i, \mu(A_k) \rangle \,.$$

Taking into account the completeness and the orthonormality of the system $\{e_i : i \in I\}$, it follows that

$$\int \langle f, d\mu \rangle = \sum_{k=1}^{m} \langle \xi_k, \mu(A_k) \rangle = \sum_{k=1}^{m} \left\langle \sum_{i \in I} \langle \xi_k, e_i \rangle e_i, \sum_{j \in I} \langle \mu(A_k), e_j \rangle e_j \right\rangle$$

$$= \sum_{k=1}^{m} \sum_{i \in I} \langle \xi_k, e_i \rangle \overline{\langle \mu(A_k), e_i \rangle} = \sum_{i \in I} \left(\sum_{k=1}^{m} \langle \xi_k, e_i \rangle \langle e_i, \mu(A_k) \rangle \right)$$

$$= \sum_{i \in I} \int f_i d\mu_i \,.$$

Observe that for $f \in TM(X, \mathbb{K})$, $\mu \in cabv(X, \mathbb{K})$ and $\xi, \eta \in \mathcal{H}$ one has

$$\int \langle f\xi, d(\mu\eta) \rangle = \left(\int \langle f, d\mu \rangle \right) \cdot \langle \xi, \eta \rangle \,,$$

which, in the case $\xi = \eta$, $\|\xi\| = 1$, becomes $\int \langle f\xi, d(\mu\xi) \rangle = \int \langle f, d\mu \rangle$.

For $x \in X$, let $\delta_x \in cabv(X, \mathbb{K})$ be the Dirac measure concentrated at x, defined by $\delta_x(A) = 1$ if $x \in A$ and $\delta_x(A) = 0$ if $x \in X \setminus A$, for all $A \in \mathscr{B}(X)$. Then, for any $\xi \in \mathcal{H}$, $\delta_x \xi \in cabv(X, \mathcal{H})$ and $\|\delta_x \xi\| = \|\xi\|$. Also

$$\int \langle f, d(\delta_x \xi) \rangle = \langle f(x), \xi \rangle \,,$$

for every $f \in TM(X, \mathcal{H})$.

Remark 8.4.2 Suppose that the metric space X is compact. By the Riesz representation theorem (Theorem 1.4.29) there exists a conjugate linear isometric mapping $\Phi : \mathcal{H} \to \mathcal{H}^*$, given by $\Phi(\eta)(\xi) = \langle \xi, \eta \rangle$ for all $\xi, \eta \in \mathcal{H}$. By a result of Dinculeanu [196, Corollary 2, p. 387] $C(X, \mathcal{H})^* \cong cabv(X, \mathcal{H}^*)$, so that one obtains a conjugate linear isometric isomorphism $F : cabv(X, \mathcal{H}) \to C(X, \mathcal{H})^*$ given by $F(\mu) = v_\mu$, where $v_\mu(f) = \int \langle f, d\mu \rangle$ for any $\mu \in cabv(X, \mathcal{H})$ and any $f \in C(X, \mathcal{H})$.

8.4.2 Lipschitz Functions

Let (X, d) be a metric space having at least two elements and let $\mathcal{H}$ be a non null Hilbert space with scalar product $\langle \cdot, \cdot \rangle$ and the corresponding norm $\|\cdot\|$.

Recall that a function $f : X \to \mathcal{H}$ is a *Lipschitz function* if there exists a number $M \in (0, \infty)$ such that $\|f(x) - f(y)\| \leq Md(x, y)$ for all $x, y \in X$. For such an f, we define the *Lipschitz norm* $\|f\|_L$ of f by

$$\|f\|_L = \sup \left\{ \frac{\|f(x) - f(y)\|}{d(x, y)} : x, y \in X, \ x \neq y \right\}. \qquad (8.4.4)$$

The set

$$\mathrm{Lip}(X, \mathcal{H}) = \{f : X \to \mathcal{H} : f \text{ is a Lipschitz function}\}$$

is a vector space (with respect to the usual operation of addition and multiplication by scalars) and (8.4.4) is a seminorm on $\mathrm{Lip}(X, \mathcal{H})$ (by an abuse of language we called it "Lipschitz norm").

Notice that $\|f\|_L$ is the least Lipschitz constant for f and $\|f\|_L = 0$ if and only if f is constant.

Fixing an element $x_0 \in X$ put

$$\mathrm{Lip}_{x_0}(X, \mathcal{H}) := \{f \in \mathrm{Lip}(X, \mathcal{H}) : f(x_0) = 0\}.$$

Then $\|\cdot\|_L$ is a norm on $\mathrm{Lip}_{x_0}(X, \mathcal{H})$ and $\mathrm{Lip}_{x_0}(X, \mathcal{H})$ is a Banach space with respect to this norm.

Along with $\mathrm{Lip}(X, \mathcal{H})$ we consider also the vector space $\mathrm{BLip}(X, \mathcal{H})$ of all bounded Lipschitz functions $f : X \to \mathcal{H}$. Put

$$\|f\|_\infty = \sup\{\|f(x)\| : x \in X\},$$

and on $\mathrm{BLip}(X, \mathcal{H})$ consider the following equivalent norms:

$$\|f\|_{\mathrm{sum}} = \|f\|_L + \|f\|_\infty,$$

and

$$\|f\|_{\mathrm{max}} = \max\{\|f\|_L, \|f\|_\infty\}.$$

Then $\mathrm{BLip}(X, \mathcal{H})$ is a Banach space when equipped with any of these two norms.

The closed unit ball in $\mathrm{Lip}(X, \mathscr{H})$ with respect to the seminorm $\|\cdot\|_L$ is denoted by $B_\ell(\mathrm{Lip}(X, \mathscr{H}))$,

$$B_\ell(\mathrm{Lip}(X, \mathscr{H})) := \{f \in \mathrm{Lip}(X, \mathscr{H}) : \|f\|_L \le 1\}. \tag{8.4.5}$$

The closed unit ball in $\mathrm{BLip}(X, \mathscr{H})$ with respect to the norm $\|\cdot\|_{\mathrm{sum}}$ is denoted by $B_{\mathrm{sum}}(\mathrm{BLip}(X, \mathscr{H}))$,

$$B_{\mathrm{sum}}(\mathrm{BLip}(X, \mathscr{H})) := \{f \in \mathrm{BLip}(X, \mathscr{H}) : \|f\|_{\mathrm{sum}} \le 1\},$$

while that with respect to the norm $\|\cdot\|_{\mathrm{max}}$ by

$$B_{\mathrm{max}}(\mathrm{BLip}(X, \mathscr{H})) := \{f \in \mathrm{BLip}(X, \mathscr{H}) : \|f\|_{\mathrm{max}} \le 1\}.$$

The closed balls of center 0 and radii $r > 0$ are obtained by multiplying the corresponding closed unit balls by r.

If the metric space X is bounded (i.e., diam $X < \infty$), in particular if X is compact, then $\mathrm{BLip}(X, \mathscr{H}) = \mathrm{Lip}(X, \mathscr{H})$, and we shall denote by $B_{\mathrm{sum}}(\mathrm{Lip}(X, \mathscr{H}))$ and $B_{\mathrm{max}}(\mathrm{Lip}(X, \mathscr{H}))$ the closed unit balls with respect to the norms $\|\cdot\|_{\mathrm{sum}}$ and $\|\cdot\|_{\mathrm{max}}$, respectively.

Clearly $\mathrm{BLip}(X, \mathscr{H}) \subseteq \mathrm{Lip}(X, \mathscr{H}) \subseteq C(X, \mathscr{H})$,

$$B_{\mathrm{sum}}(\mathrm{BLip}(X, \mathscr{H})) \subseteq B_\ell(\mathrm{Lip}(X, \mathscr{H})) \quad \text{and} \quad B_{\mathrm{max}}(\mathrm{BLip}(X, \mathscr{H})) \subseteq B_\ell(\mathrm{Lip}(X, \mathscr{H})).$$

If the metric space X is compact, then $\mathrm{Lip}(X, \mathscr{H})$ is dense in $C(X, \mathscr{H})$ with respect to the uniform norm $\|\cdot\|_\infty$ (see Corollary 6.2.2).

8.4.3 The Kantorovich–Rubinstein Norm

In this subsection (X, d) is a compact metric space and $\mathscr{H}$ a Hilbert space.

In order to define the Kantorovich–Rubinstein metric we need some auxiliary results concerning measures.

Lemma 8.4.3 Let $\mu_1, \mu_2 : \mathscr{B}(X) \to \mathbb{R}_+$ be two finite positive σ-additive measures. Then

$$\mu_1 = \mu_2 \iff \int \langle f, d\mu_1 \rangle = \int \langle f, d\mu_2 \rangle \quad \forall f \in \mathrm{Lip}(X, \mathbb{R}).$$

Proof One must prove only the implication $\Leftarrow$.

By the regularity of μ_1, μ_2 it is sufficient to show that

$$\mu_1(K) = \mu_2(K),$$

for every compact subset K of X.

Let $\varepsilon > 0$ and $K \subseteq X$ compact. Again, by the regularity of the measures μ_1, μ_2, there exist the open sets $D_1 \supset K$, $D_2 \supset K$ such that

$$\mu_1(D_1) \le \mu_1(K) + \varepsilon \quad \text{and} \quad \mu_2(D_2) \le \mu_2(K) + \varepsilon.$$

By Proposition 2.1.1 there exists a Lipschitz function $f : X \to [0, 1]$ such that $f(x) = 1$ on K and $f(x) = 0$ on $X \setminus D$, where $D = D_1 \cap D_2$.
But then

$$\mu_1(K) \le \int_X \langle f, d\mu_1 \rangle = \int_X \langle f, d\mu_2 \rangle \le \mu_2(K) + \mu_2(D \setminus K) \le \mu_2(K) + \varepsilon.$$

Since $\varepsilon > 0$ is arbitrary, it follows that $\mu_1(K) \le \mu_2(K)$. By symmetry, $\mu_2(K) \le \mu_1(K)$, and so $\mu_1(K) = \mu_2(K)$.   □

Theorem 8.4.4 *Let (X, d) be a compact metric space and $\mathscr{H}$ a Hilbert space. For any $\mu \in cabv(X, \mathscr{H})$, one has the equivalence:*

$$\mu = 0 \iff \int \langle f, d\mu \rangle = 0 \quad \text{for all} \ f \in \mathrm{Lip}(X, \mathscr{H}).$$

Proof We have to prove only the implication $\Leftarrow$.

The case $\mathscr{H} = \mathbb{R}$ follows from Lemma 8.4.3 applied to μ_+ and μ_-, where $\mu = \mu_+ - \mu_-$ is the Jordan decomposition of the measure μ. The case $\mathscr{H} = \mathbb{C}$ follows from the case $\mathscr{H} = \mathbb{R}$ applied to μ_1, μ_2 where $\mu = \mu_1 + i\mu_2$.

Consider now the general case. Let $(e_i)_{i \in I}$ be an orthonormal basis of $\mathscr{H}$. Then, as it was remarked (see (8.4.3)),

$$\int \langle f, d\mu \rangle = \sum_i \int f_i d\mu_i,$$

for all $\mu \in cabv(X, \mathscr{H})$ and $f \in TM(X, \mathscr{H})$. Here, for any $i \in I$,

$$f_i : X \to \mathbb{K} \quad \text{is given by} \quad f_i(x) = \langle f(x), e_i \rangle, \quad x \in X,$$

and

$$\mu_i : \mathscr{B}(X) \to \mathbb{K} \quad \text{is given by} \quad \mu_i(A) = \langle e_i, \mu(A) \rangle, \quad A \in \mathscr{B}(X).$$

It follows that $\mu_i \in cabv(X, \mathbb{K})$ for every $i \in I$, and, if $f \in \mathrm{Lip}(X, \mathscr{H})$, then $f_i \in \mathrm{Lip}(X, \mathbb{K})$ for all $i \in I$ (with the same Lipschitz constant as f).

Fix now a $j \in I$ and for $\varphi \in \mathrm{Lip}(X, \mathbb{K})$ let $f = \varphi e_j$ be given by $f(x) = \varphi(x)e_j$, $x \in X$. Then $f \in \mathrm{Lip}(X, \mathscr{H})$, $f_j = \varphi$ and, for $i \ne j$, $f_i(x) = \langle f(x), e_i \rangle = 0$ for all $x \in X$, i.e., $f_i = 0$ for all $i \ne j$.

But then

$$\int \varphi d\mu_j = \int f_j d\mu_j = \int \langle f, d\mu \rangle = 0,$$

for all $\varphi \in \text{Lip}(X, \mathbb{K})$, implying $\mu_j = 0$ (i.e., $\mu_j(A) = 0$ for all $A \in \mathscr{B}(X)$). Consequently, $\mu_i = 0$ for all $i \in I$, and so

$$\mu(A) = \sum_i \overline{\mu_i(A)}\, e_i = 0,$$

for all $A \in \mathscr{B}(X)$, i.e., $\mu = 0$. $\qquad\qquad\square$

Theorem 8.4.5 *Let (X, d) be a compact metric space and $\mathscr{H}$ a Hilbert space. For any $\mu \in cabv(X, \mathscr{H})$, define*

$$\|\mu\|_{KR} = \sup\{\left|\int \langle f, d\mu \rangle\right| : f \in \text{Lip}(X, \mathscr{H}),\ \|f\|_{\text{sum}} \le 1\}. \tag{8.4.6}$$

Then the functional $\mu \mapsto \|\mu\|_{KR}$ is a norm on $cabv(X, \mathscr{H})$ and one has

$$\|\mu\|_{KR} \le \|\mu\| \tag{8.4.7}$$

for all $\mu \in cabv(X, \mathscr{H})$.

Proof Let $\mu \in cabv(X, \mathscr{H})$. Then, in view of the conjugate linear identification $C(X, \mathscr{H})^* \cong cabv(X, \mathscr{H})$ (see Remark 8.4.2), one has

$$\|\mu\| = \sup\{\left|\int \langle f, d\mu \rangle\right| : f \in B(C(X, \mathscr{H}))\},$$

where $B(C(X, \mathscr{H})) = \{f \in C(X, \mathscr{H}) : \|f\|_\infty \le 1\}$. The inequality $\|f\|_\infty \le \|f\|_{\text{sum}}$ implies $B_{\text{sum}}(\text{Lip}(X, \mathscr{H})) \subseteq B(C(X, \mathscr{H}))$, hence $\|\mu\|_{KR} \le \|\mu\|$.

It is obvious that $\mu \mapsto \|\mu\|_{KR}$ is a seminorm. To finish the proof, one must show that $\mu = 0$ if $\|\mu\|_{KR} = 0$.

But $\|\mu\|_{KR} = 0$ implies $\int \langle f, d\mu \rangle = 0$ for any $f \in \text{Lip}(X, \mathscr{H})$, so that, by Theorem 8.4.4, $\mu = 0$. $\qquad\qquad\square$

The norm $\|\cdot\|_{KR}$ given by (8.4.6) is called the *Kantorovich-Rubinstein norm* (KR-norm for short).

In the following proposition we collect some simple properties of this norm.

Proposition 8.4.6 *Let (X, d) be a compact metric space and $\mathscr{H}$ a Hilbert space.*

1. For any $\mu \in cabv(X, \mathscr{H})$ and any $f \in \text{Lip}(X, \mathscr{H})$

$$\left|\int \langle f, d\mu \rangle\right| \le \|f\|_{\text{sum}} \|\mu\|_{KR}. \tag{8.4.8}$$

2. *Let $\xi \in \mathscr{H}$, $\|\xi\| = 1$, and $x, y \in X$, $x \neq y$. Then*

$$\text{(i)} \qquad \|\delta_x \xi\|_{KR} = 1;$$

$$\text{(ii)} \qquad \|(\delta_x - \delta_y)\xi\|_{KR} \leq d(x, y);$$

$$\text{(iii)} \qquad \|(\delta_x - \delta_y)\xi\| = 2.$$

$(8.4.9)$

Proof

1. Obviously, (8.4.8) holds for $f = 0$. If $f \neq 0$, then $g := f/\|f\|_{\text{sum}}$ satisfies $\|g\|_{\text{sum}} = 1$, and so

$$\left| \int \langle f, d\mu \rangle \right| = \|f\|_{\text{sum}} \left| \int \langle g, d\mu \rangle \right| \leq \|f\|_{\text{sum}} \|\mu\|_{KR} .$$

2. (i) Put $\mu = \delta_x \xi$ and let $f \in B_{\text{sum}}(\text{Lip}(X, \mathscr{H}))$. Then $\int \langle f, d\mu \rangle = \langle f(x), \xi \rangle$, hence $\left| \int \langle f, d\mu \rangle \right| \leq \|f(x)\| \|\xi\| = \|f(x)\| \leq 1$, and so $\|\mu\|_{KR} \leq 1$.

   To prove the converse inequality, take $f(x) = \xi$ for all $x \in X$, so that $\|f\|_{\text{sum}} = \|f\|_\infty = 1$, hence

$$\|\mu\|_{KR} \geq \left| \int \langle f, d\mu \rangle \right| = \langle \xi, \xi \rangle = 1 .$$

   (ii) Putting $\mu = (\delta_x - \delta_y)\xi$, we have $\int \langle f, d\mu \rangle = \langle f(x) - f(y), \xi \rangle$ for any $f \in \text{Lip}(X, \mathscr{H})$. Hence, if $f \in B_{\text{sum}}(\text{Lip}(X, \mathscr{H}))$, then

$$\left| \int \langle f, d\mu \rangle \right| \leq \|f(x) - f(y)\| \|\xi\| = \|f(x) - f(y)\| \leq d(x, y)$$

   and passing to supremum, we get (8.4.9).(ii).

   (iii) Put again $\mu = (\delta_x - \delta_y)\xi$. Let A_i, $1 \leq i \leq m$, be a partition of X. If there exists $i \in I$ such that $x, y \in A_i$, then $\sum_{i=1}^{m} \|\mu(A_i)\| = 0$, otherwise $\sum_{i=1}^{m} \|\mu(A_i)\| = 2$, implying $\|\mu\| \leq 2$.

   For $r := d(x, y)/2 > 0$ consider the partition $A_1 = B_X(x, r)$, $A_2 = B_X(y, r)$, $A_3 = X \setminus (A_1 \cup A_2)$ of X. Then $x \in A_1$ and $y \in A_2$, so that $\sum_{i=1}^{3} \|\mu(A_i)\| = 2$, implying $\|\mu\| \geq 2$.

   It follows that $\|\mu\| = |\mu|(X) = 2$.   □

The topology generated by $\| \cdot \|_{KR}$ on $cabv(X, \mathscr{H})$ will be denoted by τ_{KR} and called the *Kantorovich–Rubinstein topology*. The convergence of sequences with respect to this topology is denoted by $\mu_n \xrightarrow{KR} \mu$.

In the sequel, we shall make some considerations concerning the comparison between the variational topology τ_{var} and the Kantorovich–Rubinstein topology τ_{KR}.

Due to the inequality $\|\mu\|_{KR} \le \|\mu\|$, we have $\tau_{KR} \subseteq \tau_{var}$. Of course, if X is finite, one has $\tau_{KR} = \tau_{var}$. The case when X is infinite is discussed in the following theorem.

Theorem 8.4.7 *Assume X is infinite. Then the inclusion $\tau_{KR} \subseteq \tau_{var}$ is strict. Also, in this case, the normed space $(cabv(X, \mathscr{H}), \|\cdot\|_{KR})$ is not complete.*

Proof Because X is infinite, it has an accumulation point x_0. Let $(x_n)_{n\in\mathbb{N}}$ be a sequence in $X \setminus \{x_0\}$ converging to x_0. According to (8.4.9).(ii), for any $\xi \in \mathscr{H}$ with $\|\xi\| = 1$, one has $\delta_{x_n}\xi \xrightarrow{KR} \delta_{x_0}\xi$, whereas, according to (8.4.9).(iii), the assertion $\delta_{x_n}\xi \xrightarrow{var} \delta_{x_0}\xi$ is false. Hence the inclusion $\tau_{KR} \subseteq \tau_{var}$ must be strict.

The fact that $(cabv(X, \mathscr{H}), \|\cdot\|_{KR})$ is not complete follows from the inequality $\|\cdot\|_{KR} \le \|\cdot\|$ and the fact that the norms $\|\cdot\|_{KR}$ and $\|\cdot\|$ are not equivalent. Indeed, if $cabv(X, \mathscr{H})$ would be complete with respect to $\|\cdot\|_{KR}$, then, by the Banach isomorphism theorem, the identity operator I would be an isomorphism between $(cabv(X, \mathscr{H}), \|\cdot\|)$ and $(cabv(X, \mathscr{H}), \|\cdot\|_{KR})$, which would imply the equivalence of the norms $\|\cdot\|$ and $\|\cdot\|_{KR}$. $\qquad\square$

Remark 8.4.8

(a) If X is infinite, one can find a sequence $(\mu_n)_{n\in\mathbb{N}} \subseteq cabv(X, \mathscr{H})$ such that $\|\mu_n\|_{KR} = 1$ and $\|\mu_n\| > n$ for all n.
(b) By the inequality (8.4.7),

$$\mu_n \xrightarrow{var} \mu \;\Rightarrow\; \mu_n \xrightarrow{KR} \mu,$$

for every sequence $(\mu_n)_n$ in $cabv(X, \mathscr{H})$.

The converse implication is not true for infinite X.

Example 8.4.9 Take $X = [0, 1]$, $\mathscr{H} = \mathbb{R}$. Then, by (8.4.9).(ii), $\delta_{\frac{1}{n}} \xrightarrow{KR} \delta_0$. Because $\delta_{\frac{1}{n}}((0, 1]) = 1$ and $\delta_0((0, 1]) = 0$ it follows that the sequence $(\delta_{\frac{1}{n}}((0, 1]))_n$ does not converge to $\delta_0((0, 1])$.

Consequently, the convergence of a sequence in $cabv(X, \mathscr{H})$ in the KR-norm does not imply pointwise convergence (meaning the convergence $\mu_n(A) \to \mu(A)$ for every $A \in \mathscr{B}(X)$), and so, by Remark 8.4.1, nor the convergence with respect to the variational topology.

We shall see later (Theorem 8.4.13) that, for bounded sequences, the convergence in the KR-norm agrees with the weak* convergence.

8.4.4   The Weak* Topology on $cabv(X, \mathscr{H})$

Let again (X, d) be a compact metric space and $\mathscr{H}$ a Hilbert space.

We shall introduce a new topology on $cabv(X, \mathscr{H})$, defined on the basis of the fact that $cabv(X, \mathscr{H})$ can be identified with the dual of $C(X, \mathscr{H})$.

The *weak* topology* on $cabv(X, \mathscr{H})$ is the Hausdorff locally convex topology on $cabv(X, \mathscr{H})$ generated by the family of seminorms $(p_f)_{f \in C(X, \mathscr{H})}$, where, for any $f \in C(X, \mathscr{H})$, $p_f : cabv(X, \mathscr{H}) \to \mathbb{R}_+$ is given by

$$p_f(\mu) = \Big| \int \langle f, d\mu \rangle \Big|, \quad \mu \in cabv(X, \mathscr{H}).$$

The weak* topology will be denoted by τ_{w^*}. Sometimes the notions referring to this topology will be appealed as w^*-compact, w^*-convergent, etc.

For any $\mu \in cabv(X, \mathscr{H})$, a neighborhood basis at μ is formed of the sets

$$V(\mu; g_1, g_2, \ldots, g_n; \varepsilon) := \{\nu \in cabv(X, \mathscr{H}) : \Big| \int \langle g_i, d(\mu - \nu) \rangle \Big| < \varepsilon, \ 1 \le i \le n\},$$

for all $\varepsilon > 0$ and $g_i \in C(X, \mathscr{H})$, $i = 1, \ldots, n$, $n \in \mathbb{N}$.

For a sequence $(\mu_k)_k \subseteq cabv(X, \mathscr{H})$ and for $\mu \in cabv(X, \mathscr{H})$, we shall write $\mu_k \overset{w^*}{\longrightarrow} \mu$ to denote the fact that $(\mu_k)_k$ converges to μ in τ_{w^*}. We have

$$\mu_k \overset{w^*}{\longrightarrow} \mu \iff \lim_{k \to \infty} \int \langle f, d\mu_k \rangle = \int \langle f, d\mu \rangle, \tag{8.4.10}$$

for every $f \in C(X, \mathscr{H})$.

Notice that the Alaoglu–Bourbaki theorem (Theorem 1.4.18) implies that, for any $r > 0$, the ball $r B_{var}(cabv(X, \mathscr{H})) = \{\mu \in cabv(X, \mathscr{H}) : \|\mu\| \le r\}$ is weak* compact (i.e., compact in τ_{w^*}).

In the sequel, we shall fix $m \in \mathbb{N}$ and we shall work with $\mathscr{H} = \mathbb{K}^m$.

Theorem 8.4.10 *For any $r > 0$, $r B_{var}(cabv(X, \mathscr{H}))$ is w^*-compact.*

On $r B_{var}(cabv(X, \mathbb{K}^m))$ the topology τ_{w^*} is metrizable, hence complete with respect to any metric generating τ_{w^*}.

Proof The w^*-compactness of $r B_{var}(cabv(X, \mathscr{H}))$ follows from the Alaoglu–Bourbaki Theorem (Theorem 1.4.18). The metrizability of τ_{w^*} follows from the separability of $C(X, \mathbb{K}^m)$ (see Theorem 2.7.3 and Remark 2.7.4), viewing $cabv(X, \mathbb{K}^m)$ as the dual of $C(X, \mathbb{K}^m)$ (see [214, V, 5.1, p. 426]). □

We shall need also the following result.

Theorem 8.4.11 (Arzela–Ascoli-Type Theorem) *Let (X, d) be a compact metric space. For any $m \in \mathbb{N}$, the closed unit ball $B_{sum}(Lip(X, \mathbb{K}^m))$ is relatively compact in $C(X, \mathbb{K}^m)$.*

Proof By the Arzela–Ascoli theorem (Corollary 1.4.44) it is sufficient to show that the ball $B_{sum}(Lip(X, \mathbb{K}^m))$ is bounded and equicontinuous in $C(X, \mathbb{K}^m)$.

The boundedness follows from the inequalities

$$\|f\|_\infty \le \|f\|_{\text{sum}} \le 1 \,,$$

valid for all $f \in B_{\text{sum}}(\text{Lip}(X, \mathbb{K}^m))$.

For $x, x' \in X$ with $d(x, x') < \varepsilon$ we have

$$\|f(x) - f(x')\| \le \|f\|_L \, d(x, x') < \varepsilon \text{ for all } f \in B_{\text{sum}}(\text{Lip}(X, \mathbb{K}^m)) \,,$$

proving the (uniform) equicontinuity of $B_{\text{sum}}(\text{Lip}(X, \mathbb{K}^m))$. □

Remark 8.4.12 It is natural to ask whether the previous result remains valid for an arbitrary Hilbert space $\mathscr{H}$ instead of $\mathbb{K}^m$, i.e., whether $B_{\text{sum}}(\text{Lip}(X, \mathscr{H}))$ is relatively compact in $C(X, \mathscr{H})$ also for infinite dimensional $\mathscr{H}$. The answer is negative, as we shall see later (Sect. 8.4.6).

We begin the investigation of the connection between the topologies τ_{w^*} and τ_{KR}.

Theorem 8.4.13 *Let (X, d) be a compact metric space, $\mathscr{H}$ a Hilbert space, $r > 0$, (μ_k) a sequence in $r B_{var}(cabv(X, \mathscr{H}))$ and $\mu \in r B_{var}(cabv(X, \mathscr{H}))$. Then:*

1.

$$\mu_k \xrightarrow{KR} \mu \ \Rightarrow \ \mu_k \xrightarrow{w^*} \mu \,; \tag{8.4.11}$$

2. if $\mathscr{H} = \mathbb{K}^m$, then the converse is also true, that is, in this case,

$$\mu_k \xrightarrow{KR} \mu \ \Longleftrightarrow \ \mu_k \xrightarrow{w^*} \mu \,. \tag{8.4.12}$$

Proof Obviously, it suffices to give the proof for $r = 1$.

1. To prove (8.4.11), suppose that (μ_k) is a sequence in $B_{var}(cabv(X, \mathscr{H}))$ which is KR-convergent to some $\mu \in B_{var}(cabv(X, \mathscr{H}))$.

    Since the sequence (μ_k) is bounded with respect to the total variation norm in $cabv(X, \mathscr{H})$ and the set $\text{Lip}(X, \mathscr{H})$ is dense in $C(X, \mathscr{H})$ with respect to the uniform norm (see Corollary 6.2.2), it follows that it is sufficient to show that

$$\lim_{k \to \infty} \int \langle f, d\mu_k \rangle = \int \langle f, d\mu \rangle \,,$$

for every $f \in \text{Lip}(X, \mathscr{H})$,

    Let $f \in \text{Lip}(X, \mathscr{H})$, $f \ne 0$. By the definition (8.4.6) of the norm $\| \cdot \|_{KR}$ and the hypothesis, for every $\varepsilon > 0$ there exists $k_0 \in \mathbb{N}$ such that

$$\sup\{|\int \langle g, d(\mu_k - \mu) \rangle| : g \in B_{\text{sum}}(\text{Lip}(X, \mathscr{H}))\} = \|\mu_k - \mu\|_{KR} \le \frac{\varepsilon}{\|f\|_{\text{sum}}} \,,$$

for all $k \geq k_0$. Then $g := f/\|f\|_{\text{sum}}$ is of sum-norm 1, and

$$\left| \int \langle f, d(\mu_k - \mu) \rangle \right| = \|f\|_{\text{sum}} \left| \int \langle g, d(\mu_k - \mu) \rangle \right| \leq \varepsilon \,,$$

for all $k \geq k_0$.

2. We have to prove only the implication $\mu_k \xrightarrow{w^*} \mu \Rightarrow \mu_k \xrightarrow{KR} \mu$ from (8.4.12). Suppose, by contradiction, that there exists a sequence (μ_k) in $B_{var}(cabv(X, \mathbb{K}^m))$ and $\mu \in B_{var}(cabv(X, \mathbb{K}^m))$ such that $\mu_k \xrightarrow{w^*} \mu$ but (μ_k) is not KR-convergent to μ.

Then there exist $\varepsilon > 0$ and a subsequence $(\mu_{k_i})_{i \in \mathbb{N}}$ of (μ_k) such that

$$\|\mu_{k_i} - \mu\|_{KR} > \varepsilon \quad \text{for all } i.$$

Passing to this subsequence, we can suppose that

(i) $\mu_k \xrightarrow{w^*} \mu$, and

(ii) $\|\mu_k - \mu\|_{KR} > \varepsilon \quad \text{for all} \quad k \in \mathbb{N}.$

(8.4.13)

Taking into account the definition (8.4.6) of the KR-norm, for every $k \in \mathbb{N}$ there exists $f_k \in B_{\text{sum}}(\text{Lip}(X, \mathbb{K}^m))$ such that

$$\left| \int \langle f_k, d(\mu_k - \mu) \rangle \right| > \varepsilon \,.$$

By Theorem 8.4.11 there exist a subsequence (f_{k_i}) of (f_k) and $f \in C(X, \mathbb{K}^m)$ such that

$$\lim_{i \to \infty} \|f_{k_i} - f\|_\infty = 0 \,.$$

Let $i_1 \in \mathbb{N}$ be such that

$$\|f_{k_i} - f\|_\infty < \frac{\varepsilon}{4} \,,$$

for all $i \geq i_1$.

By (8.4.13).(i) and (8.4.10), there exists $i_2 \in \mathbb{N}$ such that

$$\left| \int \langle f, d(\mu_{k_i} - \mu) \rangle \right| < \frac{\varepsilon}{2} \,,$$

for all $i \geq i_2$. But then, taking into account (8.4.2),

$$\varepsilon < \left| \int \langle f_{k_i}, d(\mu_{k_i} - \mu) \rangle \right| \leq \left| \int \langle f, d(\mu_{k_i} - \mu) \rangle \right| + \left| \int \langle f_{k_i} - f, d(\mu_{k_i} - \mu) \rangle \right|$$

$$< \frac{\varepsilon}{2} + \| f_{k_i} - f \|_\infty \| \mu_{k_i} - \mu \| < \varepsilon,$$

for all $i \geq i_0 := \max\{i_1, i_2\}$, a contradiction. $\qquad\square$

Remark 8.4.14 The last results admit the following interpretation. Let $m \in \mathbb{N}$ and $r > 0$. On $r B_{var}(cabv(X, \mathbb{K}^m))$ we have two metrizable topologies—τ_{KR} and τ_{w^*}. Theorem 8.4.13 says that the convergent sequences are the same for these topologies, hence they are equal.

Again Theorem 8.4.10 says that $r B_{var}(cabv(X, \mathbb{K}^m))$ is compact with respect to τ_{w^*}, hence for τ_{KR} too. Consequently, $r B_{var}(cabv(X, \mathbb{K}^m))$ is a compact (hence complete) metric space for the metric given by $\| \cdot \|_{KR}$.

We get the following result (see also Theorem 8.4.7).

Theorem 8.4.15 *Let $m \in \mathbb{N}$. For any $r > 0$ the set $r B_{var}(cabv(X, \mathbb{K}^m))$, equipped with the metric generated by the KR-norm $\| \cdot \|_{KR}$, is a compact, hence complete, metric space, its topology being exactly the topology induced by τ_{w^*} (in spite of the fact that the normed space $(cabv(X, \mathbb{K}^m), \| \cdot \|_{KR})$ is not complete if X is infinite).*

Remark 8.4.16 This coincidence is no longer valid in general, i.e., with an arbitrary Hilbert space $\mathscr{H}$ instead of $\mathbb{K}^m$ (see Sect. 8.4.6).

8.4.5 The Modified Kantorovich–Rubinstein Norm

In this subsection, we shall be concerned with the so-called "modified Kantorovich–Rubinstein norm", defined only on a subspace of $cabv(X, \mathscr{H})$. This new norm is strongly related to the Kantorovich–Rubinstein norm and generates a very important distance on some subsets of $cabv(X, \mathscr{H})$, which generalizes the classical Kantorovich–Rubinstein metric on the space of probabilities (see, e.g. [211], or [210]).

Again (X, d) is a compact metric space and $\mathscr{H}$ a Hilbert space.

For any $\xi \in \mathscr{H}$, let us define

$$cabv(X, \mathscr{H}, \xi) = \{\mu \in cabv(X, \mathscr{H}) : \mu(X) = \xi\}.$$

Clearly $\delta_x \xi \in cabv(X, \mathscr{H}, \xi)$ for every $x \in X$. It is easily seen that $cabv(X, \mathscr{H}, 0)$ is a vector subspace of $cabv(X, \mathscr{H})$.

Lemma 8.4.17 *For any $\xi \in \mathscr{H}$, $cabv(X, \mathscr{H}, \xi)$ is weak* closed in $cabv(X, \mathscr{H})$.*

Proof For an arbitrary $\eta \in \mathscr{H}$ consider the constant function $\varphi_\eta \in C(X, \mathscr{H})$ given by $\varphi_\eta(x) = \eta$, $x \in X$.

Let $\mu \in cabv(X, \mathscr{H})$ be a w^*-adherent point of $cabv(X, \mathscr{H}, \xi)$ and let $(\mu_i)_i$ be a net in $cabv(X, \mathscr{H}, \xi)$ such that $\mu_i \xrightarrow{w^*} \mu$, i.e., $\int \langle f, d\mu_i \rangle \to \int \langle f, d\mu \rangle$ for every $f \in C(X, \mathscr{H})$. Then, for $f = \varphi_\eta$,

$$\langle \eta, \xi \rangle = \langle \eta, \mu_i(X) \rangle = \int \langle \varphi_\eta, d\mu_i \rangle \to \int \langle \varphi_\eta, d\mu \rangle = \langle \eta, \mu(X) \rangle \text{ as } i \to \infty.$$

Consequently, $\langle \eta, \mu(X) \rangle = \langle \eta, \xi \rangle$ for every $\eta \in \mathscr{H}$, hence $\mu(X) = \xi$, that is, $\mu \in cabv(X, \mathscr{H}, \xi)$.   □

For any $\mu \in cabv(X, \mathscr{H}, 0)$, let us define

$$\|\mu\|_{KR}^\circ = \sup\{|\int \langle f, d\mu \rangle| : f \in B_\ell(\mathrm{Lip}(X, \mathscr{H}))\}, \qquad (8.4.14)$$

where $B_\ell(\mathrm{Lip}(X, \mathscr{H}))$ is the closed unit ball in $\mathrm{Lip}(X, \mathscr{H})$ with respect to the Lipschitz seminorm $\| \cdot \|_L$ (see (8.4.5)).

Denote by $\mathbf{1}_X : X \to \mathbb{K}$ the function identically equal to 1 on X.

Theorem 8.4.18 *For any $\mu \in cabv(X, \mathscr{H}, 0)$, the following inequalities hold true*

(i) $\|\mu\|_{KR}^\circ \le \|\mu\| \operatorname{diam} X$,

(ii) $\|\mu\|_{KR} \le \|\mu\|_{KR}^\circ \le \|\mu\|_{KR}(1 + \operatorname{diam} X)$. $(8.4.15)$

Proof Observe that for $f \in B_\ell(\mathrm{Lip}(X, \mathscr{H}))$ and $x_0 \in X$, one has

$$\left| \int \langle f, d\mu \rangle \right| = \left| \int \langle f - \mathbf{1}_X f(x_0), d\mu \rangle + \int \langle \mathbf{1}_X f(x_0), d\mu \rangle \right|$$

$$= \left| \int \langle f - \mathbf{1}_X f(x_0), d\mu \rangle + \langle f(x_0), \mu(X) \rangle \right|$$

$$= \left| \int \langle f - \mathbf{1}_X f(x_0), d\mu \rangle \right|$$

$$\le \| f - \mathbf{1}_X f(x_0) \|_\infty \|\mu\| \le \|\mu\| \operatorname{diam} X.$$

Passing to supremum with respect to $f \in B_\ell(\mathrm{Lip}(X, \mathscr{H}))$ one obtains the inequality (i).

The first inequality in (ii) follows from the inclusion

$$B_{\mathrm{sum}}(\mathrm{Lip}(X, \mathscr{H})) \subseteq B_\ell(\mathrm{Lip}(X, \mathscr{H}))$$

and the definitions (8.4.6) and (8.4.14) of KR-norm and of the modified KR-norm, respectively.

For $f \in \mathrm{Lip}(X, \mathscr{H})$ with $\|f\|_L \leq 1$ and $x_0 \in X$, $\|f - \mathbf{1}_X f(x_0)\|_L = \|f\|_L \leq 1$ and $\|f(x) - f(x_0)\| \leq \|f\|_L d(x, x_0) \leq \mathrm{diam}\, X$ for all $x \in X$, so that

$$\|f - \mathbf{1}_X f(x_0)\|_{\mathrm{sum}} = \|f - \mathbf{1}_X f(x_0)\|_\infty + \|f - \mathbf{1}_X f(x_0)\|_L \leq 1 + \mathrm{diam}\, X.$$

Taking into account (8.4.8) and the above calculations, one obtains

$$\left| \int \langle f, d\mu \rangle \right| = \left| \int \langle f - \mathbf{1}_X f(x_0), d\mu \rangle \right| \leq \|f - \mathbf{1}_X f(x_0)\|_{\mathrm{sum}} \|\mu\|_{KR}$$

$$\leq (1 + \mathrm{diam}\, X) \|\mu\|_{KR}.$$

Passing again to supremum with respect to $f \in B_\ell(\mathrm{Lip}(X, \mathscr{H}))$ one obtains the second inequality in (ii). $\qquad\square$

Remark 8.4.19 According to the definition (8.4.14), one has

$$\left| \int \langle f, d\mu \rangle \right| \leq \|\mu\|_{KR}^\circ \|f\|_L, \tag{8.4.16}$$

for every $\mu \in cabv(X, \mathscr{H}, 0)$ and every $f \in \mathrm{Lip}(X, \mathscr{H})$.

Indeed, if $\|f\|_L = 0$, i.e., $f \equiv \xi \in \mathscr{H}$ (f is constant), then $\int \langle f, d\mu \rangle = \langle \xi, \mu(X) \rangle = 0$. If $\|f\|_L > 0$, then the function $g := f/\|f\|_L$ belongs to $B_\ell(\mathrm{Lip}(X, \mathscr{H}))$, so that

$$\left| \int \langle f, d\mu \rangle \right| = \|f\|_L \cdot \left| \int \langle g, d\mu \rangle \right| \leq \|f\|_L \|\mu\|_{KR}^\circ.$$

Theorem 8.4.20 *The function* $\| \cdot \|_{KR}^\circ$ *defined by (8.4.14) is a norm on the subspace* $cabv(X, \mathscr{H}, 0)$ *of* $cabv(X, \mathscr{H})$. *The norms* $\| \cdot \|_{KR}$ *and* $\| \cdot \|_{KR}^\circ$ *are equivalent on* $cabv(X, \mathscr{H}, 0)$.

Proof By (8.4.15).(i), $\|\mu\|_{KR}^\circ$ takes finite values and $\|\mu\|_{KR}^\circ = 0$ if and only if $\mu = 0$. The fact that $\| \cdot \|_{KR}^\circ$ is a seminorm is obvious.

The equivalence of the norms $\| \cdot \|_{KR}$ and $\| \cdot \|_{KR}^\circ$ on $cabv(X, \mathscr{H}, 0)$ follows from the inequalities (8.4.15).(ii). $\qquad\square$

The norm $\| \cdot \|_{KR}^\circ$ defined above on $cabv(X, \mathscr{H}, 0)$ is called *the modified Kantorovich-Rubinstein norm* (the modified KR-norm or the KR°-norm).

Theorem 8.4.20 says that on $cabv(X, \mathscr{H}, 0)$ the topology τ_{KR° generated by $\| \cdot \|_{KR}^\circ$ and τ_{KR} coincide. For a sequence $(\mu_n)_{n \in \mathbb{N}} \subseteq cabv(X, \mathscr{H}, 0)$ and for $\mu \in cabv(X, \mathscr{H}, 0)$, one has the equivalence

$$\mu_n \xrightarrow{KR^\circ} \mu \iff \mu_n \xrightarrow{KR} \mu.$$

Theorem 8.4.21 *Let (X, d) be a compact metric space and $\mathcal{H}$ a Hilbert space.*

1. Let $x, y \in X$, $x \neq y$, and $\xi \in \mathcal{H}$, $\|\xi\| = 1$. Then $(\delta_x - \delta_y)\xi \in cabv(X, \mathcal{H}, 0)$ and

$$\left\|(\delta_x - \delta_y)\xi\right\|_{KR}^{\circ} = d(x, y).$$

2. The set

$$Y := \Big\{ \sum_{k=1}^{m} a_k(\delta_{x_k} - \delta_{y_k}) : a_k \in \mathbb{K}^m, \ x_k, y_k \in X, \ k = 1, \ldots, m, \ m \in \mathbb{N} \Big\},$$

is a subspace of $cabv(X, \mathbb{K}^m, 0)$, dense in $cabv(X, \mathbb{K}^m, 0)$ with respect to the modified KR-norm.

Proof

1. Let $f \in B_\ell(\text{Lip}(X, \mathcal{H}))$. Putting $\mu = (\delta_x - \delta_y)\xi$, one has

$$\left| \int \langle f, d\mu \rangle \right| = \left| \langle f(x) - f(y), \xi \rangle \right| \leq \|f(x) - f(y)\| \|\xi\|$$
$$= \|f(x) - f(y)\| \leq d(x, y),$$

hence

$$\|\mu\|_{KR}^{\circ} \leq d(x, y).$$

Define $f : X \to \mathcal{H}$ by $f(z) = d(z, x)\xi$, $z \in X$. Then $f \in B_\ell(\text{Lip}(X, \mathcal{H}))$, because, for $z, z' \in X$, one has

$$\|f(z) - f(z')\| = \|(d(z, x) - d(z', x))\xi\|$$
$$= |d(z, x) - d(z', x)| \leq d(z, z').$$

Consequently,

$$\|\mu\|_{KR}^{\circ} \geq \left| \int \langle f, d\mu \rangle \right| = |\langle f(x) - f(y), \xi \rangle|$$
$$= |\langle -d(y, x)\xi, \xi \rangle| = d(x, y).$$

2. Observe that any measure ν of the form $\nu = \sum_{k=1}^{m} a_k \delta_{x_k}$, where $a_1, \ldots, a_m \in \mathbb{K}^m$ satisfy the condition $\sum_{k=1}^{m} a_k = 0$, belongs to Y. Indeed, such a measure can be represented as

$$\nu = a_1(\delta_{x_1} - \delta_{x_2}) + (a_1 + a_2)(\delta_{x_2} - \delta_{x_3}) + \cdots + (a_1 + \cdots + a_{n-1})(\delta_{x_{n-1}} - \delta_{x_n}) \in Y.$$

Let $\mu \in cabv(X, \mathbb{K}^m, 0)$, $\mu \neq 0$, and $\varepsilon > 0$. Put $\varepsilon' = \varepsilon / |\mu|(X)$ and let $U_k = B_X[x_k, \varepsilon'/2]$, $k = 1, 2, \ldots, m$, be a cover of the compact metric space X. The sets $A_1 := U_1$ and $A_k := U_k \setminus (U_1 \cup U_2 \cup \cdots \cup U_{k-1})$ for $k = 2, 3, \ldots, m$, form a partition of X. We can suppose $A_k \neq \emptyset$ for all $k \in \{1, 2, \ldots, m\}$, so that $x_k \in A_k$ and diam $A_k \leq \varepsilon'$, for all $k = 1, \ldots, m$. It follows that

$$|f(x) - f(x_k)| \leq d(x, x_k) \leq \varepsilon', \quad \forall x \in A_k,$$

for all $k \in \{1, 2, \ldots, m\}$ and any $f \in B_\ell(\mathrm{Lip}(X, \mathbb{K}^m))$.

Let

$$\nu = \sum_{k=1}^{m} \mu(A_k)\delta_{x_k}.$$

Since $\sum_{k=1}^{m} \mu(A_k) = \mu(X) = 0$, it follows that ν belongs to Y and

$$\int_{A_k} \langle f, d\nu \rangle = \langle f(x_k), \nu(A_k) \rangle = \langle f(x_k), \mu(A_k) \rangle = \int_{A_k} \langle \mathbf{1}_X f(x_k), d\mu \rangle$$

for every $f \in B_\ell(\mathrm{Lip}(X, \mathbb{K}^m))$. But then

$$\left| \int_{A_k} \langle f, d(\mu - \nu) \rangle \right| = \left| \int_{A_k} \langle f - \mathbf{1}_X f(x_k), d\mu \rangle \right| \leq \varepsilon' |\mu|(A_k), \quad k = 1, 2, \ldots, m,$$

implying

$$\left| \int \langle f, d(\mu - \nu) \rangle \right| = \left| \sum_{k=1}^{m} \int_{A_k} \langle f, d(\mu - \nu) \rangle \right| \leq \varepsilon' \sum_{k=1}^{m} |\mu|(A_k) = \varepsilon' |\mu|(X) = \varepsilon.$$

Since this holds for all $f \in B_\ell(\mathrm{Lip}(X, \mathbb{K}^m))$, it follows that $\|\mu - \nu\|_{KR}^\circ \leq \varepsilon$. $\square$

Remark 8.4.22 For convenience, we agreed to write $a(\delta_x - \delta_y)$ instead of $(\delta_x - \delta_y)a$, for $a \in \mathbb{K}^m$ and $x, y \in X$.

We shall show that the dual space $(cabv(X, \mathbb{K}, 0), \| \cdot \|_{KR}^\circ)^*$ is isometrically isomorphic to $\mathrm{Lip}_{x_0}(X, \mathbb{K}) = \{f \in \mathrm{Lip}_{x_0}(X, \mathbb{K}) : f(x_0) = 0\}$ for any base point $x_0 \in X$. In one direction the result holds for an arbitrary Hilbert space.

Theorem 8.4.23 *Let (X, d) be a compact metric space, $x_0 \in X$ and $\mathscr{H}$ a Hilbert space. Suppose that $cabv(X, \mathscr{H}, 0)$ is equipped with the modified KR-norm $\| \cdot \|_{KR}^\circ$ and $\mathrm{Lip}_{x_0}(X, \mathscr{H})$ with the Lipschitz norm.*

1. For every $f \in \mathrm{Lip}_{x_0}(X, \mathscr{H})$ the functional $u_f : cabv(X, \mathscr{H}, 0) \to \mathbb{K}$ defined by

$$u_f(\mu) = \overline{\int \langle f, d\mu \rangle}, \quad \mu \in cabv(X, \mathscr{H}, 0), \tag{8.4.17}$$

is linear and continuous with norm

$$\|u_f\| = \|f\|_L \, . \tag{8.4.18}$$

   The mapping $f \mapsto u_f$ is an isometric linear embedding of $(\mathrm{Lip}_{x_0}(X, \mathcal{H}), \| \cdot \|_L)$ *into* $(cabv(X, \mathcal{H}, 0), \| \cdot \|_{KR}^{\circ})^*$.

2. *Every continuous linear functional v on* $(cabv(X, \mathbb{K}, 0), \| \cdot \|_{KR}^{\circ})$ *is of the form*

$$v(\mu) = u_f(\mu) := \int f d\mu, \quad \mu \in cabv(X, \mathbb{K}, 0) \, , \tag{8.4.19}$$

for a uniquely determined function $f \in \mathrm{Lip}_{x_0}(X, \mathbb{K})$ with $\|f\|_L = \|v\|$. Consequently, the mapping $f \mapsto u_f$ is an isometric isomorphism of $(\mathrm{Lip}_{x_0}(X, \mathbb{K}), \| \cdot \|_L)$ *onto* $(cabv(X, \mathbb{K}, 0), \| \cdot \|_{KR}^{\circ})^*$, *that is,*

$$(\mathrm{Lip}_{x_0}(X, \mathbb{K}), \| \cdot \|_L) \cong (cabv(X, \mathbb{K}, 0), \| \cdot \|_{KR}^{\circ})^* \, .$$

Proof

1. Since the integral is conjugate linear with respect to μ, it follows that u_f is linear. By (8.4.16),

$$|u_f(\mu)| \leq \|f\|_L \|\mu\|_{KR}^{\circ} \, ,$$

for every $\mu \in cabv(X, \mathcal{H}, 0)$, showing that u_f is also continuous and

$$\|u_f\| \leq \|f\|_L \, .$$

Suppose $f \neq 0$. For $x, y \in X$ with $f(x) \neq f(y)$, take

$$\xi = (f(x) - f(y))/\|f(x) - f(y)\|$$

and let

$$\mu_{x,y} = \frac{1}{d(x, y)} (\delta_x - \delta_y)\xi \, .$$

Then, by Theorem 8.4.21.1, $\|\mu_{xy}\|_{KR}^{\circ} = 1$, so that

$$\|u_f\| \geq \left| \int \langle f, d\mu_{x,y} \rangle \right| = \frac{\|f(x) - f(y)\|}{d(x, y)} \, .$$

Since this inequality holds for all $x, y \in X$ with $f(x) \neq f(y)$, it follows that

$$\|u_f\| \geq \sup \left\{ \frac{\|f(x) - f(y)\|}{d(x, y)} : x, y \in X, \ f(x) \neq f(y) \right\} = \|f\|_L \, ,$$

proving the equality (8.4.18).

The linearity of the mapping $f \mapsto u_f$ follows from the properties of the bilinear integral $\int \langle f, d\mu \rangle$.

2. In this case $\overline{\int \langle \overline{f}, d\mu \rangle} = \int f d\mu$, so that the definitions (8.4.19) and (8.4.17) are equivalent, hence we have to prove that every continuous linear functional on $(cabv(X, \mathbb{K}, 0), \|\cdot\|_{KR}^\circ)$ is of the form

$$u_f(\mu) = \int f d\mu, \quad \mu \in cabv(X, \mathbb{K}, 0), \qquad (8.4.20)$$

for some $f \in \mathrm{Lip}_{x_0}(X, \mathbb{K})$.

For $v \in (cabv(X, \mathbb{K}, 0), \|\cdot\|_{KR}^\circ)^*$ define $f : X \to \mathbb{K}$ by

$$f(x) = v(\delta_{x,x_0}), \quad x \in X,$$

where, for $x, y \in X$, $\delta_{x,y} = \delta_x - \delta_y$.

Then, by the first assertion of Theorem 8.4.21,

$$|f(x) - f(y)| = |v(\delta_x - \delta_y)| \le \|v\| \|\delta_x - \delta_y\|_{KR}^\circ = \|v\| d(x, y),$$

for all $x, y \in X$, showing that f is Lipschitz and $\|f\|_L \le \|v\|$.

For $x, y \in X$,

$$v(\delta_{x,y}) = f(x) - f(y) = \int f d\delta_{x,y} = u_f(\delta_{x,y}),$$

where u_f is given by (8.4.20). By the linearity of the functionals v and u_f it follows that $v(\nu) = u_f(\nu)$ for every $\nu \in Y := \mathrm{span}(\{\delta_{x,y} : x, y \in X\})$. Since, by Theorem 8.4.21.2, Y is dense in $(cabv(X, \mathbb{K}, 0), \|\cdot\|_{KR^\circ})$, the continuity of the functionals v and u_f implies their equality on $cabv(X, \mathbb{K}, 0)$.

To finish the proof, observe that, by the first part of the theorem,

$$\|v\| = \|u_f\| = \|f\|_L.$$

$\square$

Remark 8.4.24 From the proof it is clear that the choice of the base point x_0 does not matter, that is, for any $x_0 \in X$ we have an isometric isomorphism $(\mathrm{Lip}_{x_0}(X, \mathbb{K}), \|\cdot\|_L) \cong (cabv(X, \mathbb{K}, 0), \|\cdot\|_{KR}^\circ)^*$ given by (8.4.19).

For $r > 0$ and $\xi \in \mathscr{H}$ with $\|\xi\| \le r$, put

$$B_{r,var}(cabv(X, \mathscr{H}, \xi)) := r B_{var}(cabv(X, \mathscr{H})) \cap cabv(X, \mathscr{H}, \xi). \qquad (8.4.21)$$

Then $B_{r,var}(cabv(X, \mathscr{H}, \xi)) \ne \emptyset$ because $\|\delta_x \xi\| = \|\xi\| \le r$ for every $x \in X$.

On a nonempty set $A \subseteq cabv(X, \mathcal{H})$, one can consider the following distances:

- *the variational distance* given by $d_{\|\cdot\|}(\mu, \nu) = \|\mu - \nu\|$;
- *the Kantorovich–Rubinstein distance* given by $d_{KR}(\mu, \nu) = \|\mu - \nu\|_{KR}$;
- assuming that $A - A \subseteq cabv(X, \mathcal{H}, 0)$, *the modified Kantorovich–Rubinstein distance* given by $d_{KR}^{\circ}(\mu, \nu) = \|\mu - \nu\|_{KR}^{\circ}$.

We shall mainly work in the particular case when $A = B_{r,var}(cabv(X, \mathcal{H}, \xi))$ with $\|\xi\| \leq r$. On such $B_{r,var}(cabv(X, \mathcal{H}, \xi))$ the last two distances are equivalent (Theorem 8.4.20).

In Sect. 8.5 we shall consider another distance, namely the distance generated by the Hanin norm (8.5.30).

Theorem 8.4.25 *Let $r > 0$ and $\xi \in \mathcal{H}$ be such that $\|\xi\| \leq r$.*

1. *The (nonempty) set $B_{r,var}(cabv(X, \mathcal{H}, \xi))$ is weak* closed in $r B_{var}(cabv(X, \mathcal{H}))$, hence it is weak* compact.*
   The metrics d_{KR} and d_{KR}° are equivalent on $B_{r,var}(cabv(X, \mathcal{H}, \xi))$.
2. *The set $B_{r,var}(cabv(X, \mathbb{K}^m, \xi))$ has in addition the property that the topology generated by one of the equivalent metrics d_{KR} or d_{KR}° is equal to the restriction of the weak* topology to $B_{r,var}(cabv(X, \mathcal{H}, \xi))$, hence it is compact with respect to the topology generated by any of this metrics.*
3. *Let $\mathcal{H} = \mathbb{K} = \mathbb{R}$ and $\xi \geq 0$. The set*

$$B_{r,var}^{+}(cabv(X, \mathbb{R}, \xi)) := B_{r,var}(cabv(X, \mathbb{R}, \xi)) \cap cabv_{+}(X, \mathbb{R}),$$

where $cabv_{+}(X, \mathbb{R}) = \{\mu \in cabv(X, \mathbb{R}) : \mu \geq 0\}$, equipped with one of the equivalent metrics d_{KR} or d_{KR}°, is a compact, hence complete, metric space, and its topology is equal to the restriction of the weak* topology to $B_{r,var}^{+}(cabv(X, \mathbb{R}, \xi))$.
   For $r = 1$ and $\xi = 1$, $B_{1,var}^{+}(cabv(X, \mathbb{R}, 1))$ is exactly the set $P_1(X)$ of all probability Borel measures on $\mathcal{B}(X)$.

Proof

1. Because $r B_{var}(cabv(X, \mathcal{H}))$ is w^*-compact (by Theorem 8.4.10) and, by Lemma 8.4.17, $cabv(X, \mathcal{H}, \xi)$ is w^*-closed in $cabv(X, \mathcal{H})$, it follows that

$$B_{r,var}(cabv(X, \mathcal{H}, \xi)) = r B_{var}(cabv(X, \mathcal{H})) \cap cabv(X, \mathcal{H}, \xi)$$

is w^*-compact. The equivalence of the considered metrics follows from Theorem 8.4.20.
2. This follows from Theorem 8.4.15.
3. All that remains to prove in this case is the fact that $cabv_{+}(X, \mathbb{R})$ is weak* closed in $cabv(X, \mathbb{R})$. Let $(\mu_i)_{i \in I}$ be a net in $cabv_{+}(X, \mathbb{R})$ which is τ_{w^*}-convergent to some $\mu \in cabv(X, \mathbb{R})$. It follows that $\lim_i \int \langle f, d\mu \rangle_i = \int \langle f, d\mu \rangle$ for every

$f \geq 0$ in $C(X, \mathbb{R})$. As $\int \langle f, d\mu \rangle_i \geq 0$ for every $i \in I$, this implies $\int \langle f, d\mu \rangle \geq 0$ for every $f \geq 0$ in $C(X, \mathbb{R})$. By the Riesz-Kakutani representation theorem (Theorem 1.5.26) it follows that $\mu \geq 0$, i.e., $\mu \in cabv_+(X, \mathbb{R})$. $\square$

8.4.6   Infinite Dimensional Extensions Do Not Work

In this subsection we present a counterexample showing that Theorem 8.4.11 does not hold for an infinite dimensional Hilbert space $\mathcal{H}$. Also the equivalence of the topologies τ_{KR} and τ_{w*} on $cabv(X, \mathbb{K}^m)$, stated in Theorem 8.4.13, does not hold for an arbitrary Hilbert space $\mathcal{H}$ instead of $\mathbb{K}^m$.

The Counterexample
Let $X = \{1, 2\}$ with $d(1, 2) = 1$, $d(1, 1) = d(2, 2) = 0$, and $\mathcal{H} = \ell^2$.

Note that in this case $C(X, \ell^2) = \mathrm{Lip}(X, \ell^2) = (\ell^2)^X$, i.e., any function on X with values in ℓ^2 is Lipschitz (and so continuous).

For any $f : X \to \ell^2$ we have

$$\|f\|_L = \|f(1) - f(2)\| \quad \text{and} \quad \|f\|_\infty = \max\{\|f(1)\|, \|f(2)\|\}.$$

For $\mu \in cabv(X, \ell^2)$ we agree to write $\mu(i)$ instead of $\mu(\{i\})$, $i = 1, 2$. Then

$$\int \langle f, d\mu \rangle = \langle f(1), \mu(1) \rangle + \langle f(2), \mu(2) \rangle.$$

Also

$$\int \langle f, d\mu \rangle = \langle f(1) - f(2), \mu(1) \rangle,$$

because

$$\mu \in cabv(X, \ell^2, 0) \iff \mu(2) = -\mu(1).$$

We have also

$$\|\mu\|_{KR}^\circ = \|\mu(1)\|,$$

for every $\mu \in cabv(X, \ell^2, 0)$.
Indeed,

$$\left| \int \langle f, d\mu \rangle \right| = |\langle f(1) - f(2), \mu(1) \rangle| \leq \|\mu(1)\|,$$

since

$$f \in B_\ell(\text{Lip}(X, \ell^2)) \iff \|f(1) - f(2)\| = \|f\|_L \le 1,$$

so that $\|\mu\|_{KR}^\circ \le \|\mu(1)\|$.

For $f \in B_\ell(\text{Lip}(X, \ell^2))$, given by $f(1) = \mu(1)/\|\mu(1)\|$, $f(2) = 0$, we have

$$\|\mu\|_{KR}^\circ \ge \left| \int \langle f, d\mu \rangle \right| = \frac{1}{\|\mu(1)\|} |\langle \mu(1), \mu(1) \rangle| = \|\mu(1)\|.$$

It follows that the mapping $\mu \mapsto \mu(1)$, $\mu \in cabv(X, \ell^2, 0)$, is an isometric isomorphism of $(cabv(X, \ell^2, 0), \| \cdot \|_{KR}^\circ)$ onto ℓ^2.

Let $e_k = (\delta_{ik})_{i\in\mathbb{N}}$, $k \in \mathbb{N}$, be the canonical orthonormal basis of ℓ^2 and let $\mu_k \in cabv(X, \ell^2, 0)$ be defined by

$$\mu_k(1) = e_k = -\mu_k(2), \quad k \in \mathbb{N}.$$

Then, for every $f \in C(X, \ell^2)$,

$$\int \langle f, d\mu_k \rangle = \langle f(1) - f(2), e_k \rangle \to 0 \quad \text{as} \quad k \to \infty,$$

showing that

$$\mu_k \xrightarrow{w^*} 0.$$

Since

$$\|\mu_k\|_{KR}^\circ = \|\mu(1)\| = \|e_k\| = 1,$$

the sequence (μ_k) is not τ_{KR°-convergent to 0. As, by Theorem 8.4.20), the norms $\| \cdot \|_{KR}$ and $\| \cdot \|_{KR}^\circ$ are equivalent on $cabv(X, \ell^2, 0)$ it follows that (μ_k) is not τ_{KR}-convergent to 0. Consequently, the implication from (8.4.11) in Theorem 8.4.13 cannot be reversed in general.

Taking $f_k(1) = e_k = f_k(2)$, then $f_k \in B_{\text{sum}}(\text{Lip}(X, \ell^2))$ because $\|f_k\|_{\text{sum}} = \|f_k\|_\infty = 1$ for all $k \in \mathbb{N}$. But $\|f_i - f_j\|_\infty = \sqrt{2}$ for $i \ne j$, $i, j \in \mathbb{N}$, showing that the sequence $(f_k) \subseteq C(X, \ell^2)$ does not admit $\| \cdot \|_\infty$-convergent subsequences. Consequently, the ball $B_{\text{sum}}(\text{Lip}(X, \ell^2))$ is not compact in $(C(X, \ell^2), \| \cdot \|_\infty)$, showing that the result from Theorem 8.4.11 cannot be extended beyond the finite dimensional case.

8.4.7 The Mass Transfer Problem

Let (X, d) be a compact metric space, $cabv(X)$ the space of finite real-valued Borel measures on X and $cabv°(X)$ its subspace formed of all measures $\mu \in cabv(X)$ with $\mu(X) = 0$, i.e., $cabv°(X) = cabv(X, \mathbb{R}, 0)$ (with the notations from the preceding subsections). Throughout this section we shall also use the notation: $cabv_+(X) = cabv_+(X, \mathbb{R})$, $\text{Lip}(X) = \text{Lip}(X, \mathbb{R})$ and $C(X) = C(X, \mathbb{R})$.

For $\mu \in cabv°(X)$, denote by Λ_μ the set of all measures $\lambda \in cabv_+(X^2)$ satisfying the condition

$$\lambda(X \times A) - \lambda(A \times X) = \mu(A) \quad \text{for all } A \in \mathscr{B}(X). \tag{8.4.22}$$

Along with Λ_μ consider also the narrow class $\tilde{\Lambda}_\mu \subseteq \Lambda_\mu$ of all $\lambda \in cabv_+(X^2)$ satisfying the condition

$$\lambda(X \times A) = \mu_+(A) \quad \text{and} \quad \lambda(A \times X) = \mu_-(A) \text{ for all } A \in \mathscr{B}(X), \tag{8.4.23}$$

where $\mu = \mu_+ - \mu_-$ is the Jordan decomposition of the measure μ.

For $\mu \in cabv°(X)$, $\mu \neq 0$, the measure λ_μ defined by

$$\lambda_\mu(A \times B) = \frac{\mu_-(A)\mu_+(B)}{\mu_-(X)} \quad A, B \in \mathscr{B}(X), \tag{8.4.24}$$

satisfies the condition (8.4.23), so $\lambda_\mu \in \tilde{\Lambda}_\mu \subseteq \Lambda_\mu$ (recall that $\mu(X) = 0$ and so $\mu_+(X) = \mu_-(X)$).

It is easy to check that Λ_μ satisfies the relations:

$$0 \in \Lambda_0; \quad \Lambda_\mu + \Lambda_\nu \subseteq \Lambda_{\mu+\nu}; \quad \alpha\Lambda_\mu = \Lambda_{\alpha\mu}, \tag{8.4.25}$$

for all $\mu, \nu \in cabv°(X)$ and $\alpha > 0$.

We have also

$$\Lambda_{-\mu} = \{\lambda^* : \lambda \in \Lambda_\mu\},$$

where for a Borel measure γ on $X \times X$, γ^* is the "transposed" measure given by

$$\gamma^*(A \times B) = \gamma(B \times A) \quad \text{for all } A \times B \in \mathscr{B}(X \times X).$$

For every $\lambda \in \Lambda_\mu$,

$$\lambda \in \tilde{\Lambda}_\mu \iff var(\lambda) = \frac{1}{2} var(\mu).$$

Define the functional $\| \cdot \|_d^\circ : cabv^\circ(X) \to \mathbb{R}_+$ by

$$\|\mu\|_d^\circ = \inf \left\{ \int d(x,y) d\lambda(x,y) : \lambda \in \Lambda_\mu \right\}, \quad \mu \in cabv^\circ(X). \quad (8.4.26)$$

One shows that

$$\|\mu\|_d^\circ = \inf \left\{ \int d(x,y) d\lambda(x,y) : \lambda \in \tilde{\Lambda}_\mu \right\}, \quad \mu \in cabv^\circ(X). \quad (8.4.27)$$

Remark 8.4.26 Observe that for a Borel measure μ on $\mathscr{B}(X)$, $\tilde{\Lambda}(\mu) = M(\mu_+, \mu_-)$, where $M(\mu_+, \mu_-)$ is the set defined at the beginning of Sect. 8.4.8. Also $\|\mu\|_d^\circ = W(\mu_+, \mu_-)$ (see (8.4.29)). The fact that μ_+, μ_- are not necessarily probability measures is inessential, because we can replace $\mu \in \Lambda_\mu$ by $\mu/\mu_-(X)$ (provided that $\mu \neq 0$).

Here $d(x,y)$ can be interpreted as the cost of the transportation of a unit mass for x to y. For $\mu \in cabv^\circ(X)$, $\mu_+(A)$ is the mass to be transported and $\mu_-(A)$ the added mass, so that the condition $\mu_+(A) = \mu_-(A)$ says that the transported mass must be equal with the added mass. A measure $\lambda \in \Lambda_\mu$ can be interpreted as a translocation plan and $\int d(x,y)d\lambda(x,y)$ the cost of this translocation. The mass transfer problem consists in finding the plan λ_0 minimizing the cost. The problem of mass transfer was posed by the French mathematician Gaspar Monge a long time ago (in 1871)—how to fill up a hole with the material from a given pile of sand, in an optimal way, i.e., with a minimal cost. The problem was actually very difficult and Monge proposed a complicated geometrical solution. More than 100 years after, namely in 1887, Appell approached the problem with complicated variational methods (in fact, the solution given by Appel was not quite complete—he did not prove the existence of a minimal plan). Kantorovich, the inventor of linear programming, attacked the problem in a totally different way. First he considered the discreet variant of the problem, interpreting it as a problem of linear programming and totally solving it (see [338]) in a way suitable for the successful use of computers (these results constitute a major part of the reasons for the 1975 Nobel prize for economy, shared by Kantorovich with Koopmans). Initially Kantorovich was not aware of the Monge problem, but he mentioned the connections in the note [339]. Afterwards, he transformed the problem in an abstract way, working with a compact metric space instead of a finite set and with a measure instead of a vector in $\mathbb{R}^m$. Alone, or jointly with his student Rubinstein (see [341] and [342]), he succeeded in completely solving the new abstract problem. The necessary mathematical tools were the theory of normed spaces and different metrics on spaces of measures, usually called now the Kantorovich–Rubinstein metrics. For these reasons the problem is called now the Monge–Kantorovich mass transport problem. In fact the Kantorovich problem is a generalization of the Monge problem, but they are not equivalent (see [87]).

After that the field attracted the attention of many renowned mathematicians who attacked the problem using methods from various domains of mathematics— functional analysis, optimization, calculus of variations and partial differential equations, differential geometry, etc. The actual stage of the research is masterly presented in the books [585, 617] and [672]. Interesting historical information and comments can be found in [88] and [668–670] (see also [671]). In 2004 a conference was organized in Sankt Petersburg to celebrate the 90th anniversary of Kantorovich (actually the anniversary year was 2002 but the celebration was postponed to 2004 due to some local organizational problems). Some of the talks delivered at this conference, containing further information on the remarkable contributions of Kantorovich to applied mathematics and on his life, were published as vol. 312 (2004) of Zapiski Nauchnyh Seminarov POMI, and translated cover-to-cover as volume 133 (2006), no. 4, of J. Math. Sci (New York).

We do not resist the temptation to include here the nice presentation of Gaspard Monge given by Bogachev and Kolesnikov in [88].

Gaspard Monge was a very universal scholar who made significant contributions to descriptive geometry, differential geometry, engineering, organization of science, and higher education. He was an active participant in the French Revolution (as the Naval Minister he signed the decision of the Court to execute Louis XVI), a comrade-in-arms of Napoleon, a participant in his Egyptian expedition, and the person who ensured Napoleon's election to the Academy of Sciences (in those years, as now, membership in the Academy did not necessarily require scientific achievements). Among Monge's students there were such renowned scientists as Cauchy, Poisson, Meusnier, Carnot (the Carnot cycle was an invention of his son S. Carnot), Poncelet, and Coriolis. Later, after the fall of Napoleon and the French "perestroika", he was expelled from the Academy (by adroit colleagues who were quick in changing course), but much later the Academy of Sciences announced prizes for a solution of the optimal mass transfer problem posed by him for the concrete function $h(x, y) = |x - y|$ (though the very existence of a solution was tacitly assumed, and the problem was concerned with the study of some of its geometric properties). The problem was solved by Appel in 1887 who received such a prize. However, also his solution did not give a proof of the existence of an optimal map, but only established certain properties of such a map.

We shall present now the results of Kantorovich and Rubinstein, following the book [340].

Theorem 8.4.27 Let (X, d) be a compact metric space. The functional $\| \cdot \|_d^\circ$ given by (8.4.26) is a norm on the space $cabv^\circ(X)$. The mapping $f \mapsto u_f$, where

$$u_f(\mu) = \int f \, d\mu, \quad \mu \in cabv^\circ(X),$$

is an isometric isomorphism of the space $(\mathrm{Lip}_{x_0}(X), \| \cdot \|_L)$ onto $(cabv^\circ(X), \| \cdot \|_d^\circ)^*$.

Taking into account this duality one obtains the following equality.

Corollary 8.4.28 *The norms* $\| \cdot \|_{KR}^{\circ}$ *given by (8.4.14) and* $\| \cdot \|_d^{\circ}$ *are equal*

$$\|\mu\|_d^{\circ} = \sup \left\{ \int f d\mu : f \in B_{\ell}(\mathrm{Lip}(X)) \right\}$$

$$= \sup \left\{ \left| \int f d\mu \right| : f \in B_{\ell}(\mathrm{Lip}(X)) \right\} = \|\mu\|_{KR}^{\circ} .$$

Remark 8.4.29 The above results are consequences of those from Sects. 8.4.5 and 8.4.8.

A characterization of optimal transportation plans is given in the following theorem.

Theorem 8.4.30 *Let* $\mu \in cabv^{\circ}(X)$. *A measure* $\lambda_0 \in \Lambda_{\mu}$ *satisfies the condition*

$$\int d(x, y) d\lambda_0(x, y) = \|\mu\|_d^{\circ}$$

if and only if there exists a function $h : X \to \mathbb{R}$ *such that*

$$h(x) - h(y) \leq d(x, y) \quad \text{for all } x, y \in X,$$
$$h(x) - h(y) = d(x, y) \quad \text{for all } (x, y) \in \mathrm{spt}(\lambda_0) .$$

This characterization has the following consequence.

Corollary 8.4.31 *If the measure* $\mu \in cabv^{\circ}(X)$ *has finite support, then there exists* $\lambda_0 \in \Lambda_{\mu}$ *such that*

$$\int d(x, y) d\lambda_0(x, y) = \|\mu\|_d^{\circ} .$$

Furthermore, $\lambda_0(X \times A) = \mu_+(A)$ *and* $\lambda_0(A \times X) = \mu_-(A)$ *for all* $A \in \mathscr{B}(X)$, *i.e.,* $\lambda_0 \in \tilde{\Lambda}_{\mu}$.

Since the norms $\| \cdot \|_{KR}^{\circ}$ and $\| \cdot \|_d^{\circ}$ are equivalent, Theorem 8.4.25 yields the compactness of the ball $B_{\|\cdot\|_d^{\circ}}(cabv(X))$. Based on this result one can prove the following general result.

Theorem 8.4.32 *For every measure* $\mu \in cabv^{\circ}(X)$ *the set* Λ_{μ} *contains a measure* λ_0 *such that*

$$\int d(x, y) d\lambda_0(x, y) = \|\mu\|_d^{\circ} .$$

In fact, the measure λ_0 *belongs to* $\tilde{\Lambda}_{\mu}$ *(i.e., it satisfies (8.4.23)) and*

$$\lambda_0(X \times X) = \mu_+(X) = \mu_-(X) = \frac{1}{2} \|\mu\| ,$$

where $\|\mu\| = |\mu|(X)$ is the total variation of the measure μ.

Remark 8.4.33 In fact, the existence of an optimal plan is given by the Kantorovich–Rubinstein duality theorem (Theorem 8.4.35).

8.4.8 The Kantorovich–Rubinstein Duality

In this subsection we shall work with probability measures. Let (X, d) be a compact metric space and let $P_1(Y)$ denote the set of probability measures on a metric space Y, meaning positive Borel measures ν with $\nu(Y) = 1$.

For $\mu, \nu \in P_1(X)$ denote by $M(\mu, \nu)$ the set of all measures $\lambda \in cabv_+(X^2)$ such that

$$\lambda(A \times X) = \mu(A) \quad \text{and} \quad \lambda(X \times A) = \nu(A) \quad \text{for all } A \in \mathscr{B}(X).$$

One says that λ is a measure with *marginals* μ and ν.
This property is equivalent to

$$\int_X f \, d\mu + \int_X g \, d\nu = \int_{X^2} (f(x) + g(y)) d\lambda(x, y), \tag{8.4.28}$$

for all $f, g \in C(X)$.
Indeed, if $\lambda \in M(\mu, \nu)$, then

$$\int_{X^2} f(x) d\lambda(x, y) = \int_X f(x) d\lambda(x, X) = \int_X f(x) d\mu(x),$$

and

$$\int_{X^2} g(y) d\lambda(x, y) = \int_X g(y) d\lambda(X, y) = \int_X g(y) d\nu(y).$$

Conversely, if (8.4.28) holds, then if $f = \chi_A$, for some $A \in \mathscr{B}(X)$, and $g = 0$, the equality $\chi_A(x) = \chi_{A \times X}(x, y)$, $(x, y) \in X^2$, implies

$$\mu(A) = \int_X f \, d\mu = \int_{X^2} f(x) d\lambda(x, y) = \int_{X^2} \chi_{A \times X} d\lambda(x, y) = \lambda(A \times X).$$

Similarly $\lambda(X \times A) = \nu(A)$.
Consider the quantities

$$W(\mu, \nu) = \inf \left\{ \int d(x, y) d\lambda(x, y) : \lambda \in M(\mu, \nu) \right\}, \tag{8.4.29}$$

called the *Wasserstein distance* between μ and v, and

$$\gamma(\mu, v) = \sup\left\{\left|\int f d\mu - \int f dv\right| : f \in \mathrm{Lip}(X), \|f\|_L \leq 1\right\},$$

the modified Kantorovich–Rubinstein distance between μ and v.

Remark 8.4.34 The term Wasserstein distance comes from Vasershtein [667]. In fact Dobrušin [198] attributed it to Vasershtein (see the comments in [88, p. 796]). This term is used in many papers, as Rachev and Shortt [586], for instance.

The aim of this subsection is to prove the following result.

Theorem 8.4.35 (Kantorovich–Rubinstein Duality) *Let (X, d) be a compact metric space. For every $\mu, v \in P_1(X)$ the following equality holds true*

$$W(\mu, v) = \gamma(\mu, v).$$

Moreover, there exists a measure $\lambda_0 \in M(\mu, v)$ such that

$$W(\mu, v) = \int d(x, y) d\lambda_0(x, y),$$

i.e., the infimum in the definition of the Wasserstein distance is attained.

The proof is adapted from [536]. Let us consider another quantity associated to $\mu, v \in P_1(X)$, namely

$$m_d(\mu, v) = \sup\left\{\int f d\mu + \int g dv\right\},$$

where the supremum is taken over all $f, g \in C(X)$ such that $f(x) + g(y) < d(x, y)$ for all $x, y \in X$.

Lemma 8.4.36 *The equality*

$$\gamma(\mu, v) = m_d(\mu, v) \tag{8.4.30}$$

holds for all $\mu, v \in P_1(X)$.

Proof For $f : X \to \mathbb{R}$ with $\|f\|_L \leq 1$ and $\varepsilon > 0$ let $g(y) = -f(y) - \varepsilon$, $y \in X$. Then

$$f(x) + g(y) = f(x) - f(y) - \varepsilon \leq d(x, y) - \varepsilon < d(x, y),$$

for all $x, y \in X$, so that

$$m_d(\mu, v) \geq \int f d\mu + \int g dv = \int f d\mu - \int f dv - \varepsilon.$$

Since $f \in B_\ell(\mathrm{Lip}(X))$ was arbitrarily chosen, it follows that

$$m_d(\mu, \nu) \geq \gamma(\mu, \nu) - \varepsilon,$$

and so, by the arbitrariness of $\varepsilon > 0$,

$$m_d(\mu, \nu) \geq \gamma(\mu, \nu). \tag{8.4.31}$$

For $f, g \in C(X)$ such that

$$f(x) + g(y) < d(x, y), \quad \text{for all } x, y \in X, \tag{8.4.32}$$

put

$$h(x) = \inf\{d(x, y) - g(y) : y \in X\}, \quad x \in X.$$

Let $x, x' \in X$ and $\varepsilon > 0$. Take $y_0 \in X$ such that $d(x', y_0) - g(y_0) < h(x') + \varepsilon$. Then

$$h(x) - h(x') < d(x, y_0) - g(y_0) - d(x', y_0) + g(y_0) + \varepsilon \leq d(x, x') + \varepsilon.$$

This shows that $h \in B_1(\mathrm{Lip}(X))$. The inequality (8.4.32) implies $f(x) \leq h(x)$ for all $x \in X$. Since, by the definition of the function h,

$$h(x) \leq d(x, x) - g(x) = -g(x) \implies g(x) \leq -h(x),$$

for all $x \in X$, it follows that

$$\int f d\mu + \int g d\nu \leq \int h d\mu - \int h d\nu \leq \gamma(\mu, \nu),$$

for all $f, g \in C(X)$ satisfying (8.4.32).

Consequently,

$$m_d(\mu, \nu) \leq \gamma(\mu, \nu),$$

which, combined with (8.4.31), yields (8.4.30). □

We shall need also the following consequence of Hahn–Banach extension theorem.

Lemma 8.4.37 *Let E be a topological vector space, U a nonempty open convex subset of E and Z a linear subspace of E with $U \cap Z \neq \emptyset$. If $\varphi : Z \to \mathbb{R}$ is a linear functional such that $\sup \varphi(U \cap Z) < \infty$, then there exists a linear functional $\psi : E \to \mathbb{R}$ such that*

$$\psi|_Z = \varphi \quad \text{and} \quad \sup \psi(U) = \sup \varphi(U \cap Z).$$

Proof Suppose first that $0 \in U$ and let p_U be the Minkowski functional corresponding to U (see Sect. 1.1.4). Then p_U is sublinear (i.e., subadditive and positively homogeneous) and

$$\{x \in E : p_U(x) < 1\} \subseteq U \subseteq \{x \in E : p_U(x) \le 1\}. \tag{8.4.33}$$

If $\beta := \sup \varphi(U \cap Z)$, then $\varphi(z) \le \beta p_U(z)$ for all $z \in Z$. Indeed, if $p_U(z) > 0$, then, for every $\varepsilon > 0$, $z' := z/(\varepsilon + p_U(z))$ satisfies $p_U(z') < 1$, so that $z' \in U$ and $\varphi(z') \le \beta \iff \varphi(z) \le \beta(\varepsilon + p_U(z))$. Consequently, $\varphi(z) \le \beta p_U(z)$.

If $p_U(z) = 0$, then, for every $t > 0$, $p_U(tz) = tp_U(z) = 0 < 1$, so that $tz \in U$ and $t\varphi(z) = \varphi(tz) \le \beta$, implying $\varphi(z) \le 0 = \beta p_U(z)$.

By the Hahn–Banach theorem, there exists a linear functional $\psi : E \to \mathbb{R}$ such that

$$\psi|_Z = \varphi \quad \text{and} \quad \psi(x) \le \beta p_U(x) \quad \text{for all} \ x \in E.$$

But then, by the second inclusion in (8.4.33), $\psi(x) \le \beta p_U(x) \le \beta$ for every $x \in U$. Since $\sup \psi(U \cap Z) = \sup \varphi(U \cap Z) = \beta$, it follows that $\sup \psi(U) = \beta$.

The general case can be reduced to that considered above. Let $x_0 \in Z \cap U$ be fixed. The set $\tilde{U} := U - x_0$ is open convex, $0 \in \tilde{U}$ and $\sup \varphi(\tilde{U}) = \sup \varphi(U) - \varphi(x_0)$, so that, by the first part of the proof, there exists a linear extension ψ of φ such that $\sup \psi(\tilde{U}) = \sup \varphi(\tilde{U} \cap Z)$. But then

$$\sup \varphi(U \cap Z) - \varphi(x_0) = \sup \varphi(\tilde{U} \cap Z) = \sup \psi(\tilde{U})$$
$$= \sup \psi(U) - \psi(x_0) = \sup \psi(U) - \varphi(x_0),$$

implying $\sup \psi(U) = \sup \varphi(U \cap Z)$. $\square$

Proof of Theorem 8.4.35 Consider the Banach space $E = C(X^2)$ with the sup-norm $\| \cdot \|_\infty$ and let

$$U = \{f \in E : f(x, y) < d(x, y), \ \text{for all} \ x, y \in X\}.$$

The set U is obviously convex and nonempty $(d(\cdot, \cdot) - \varepsilon \mathbf{1}_{X^2} \in U$ for every $\varepsilon > 0)$. It is also open. Indeed, for any $f \in C(X^2)$ the compactness of X^2 implies the existence $(x_0, y_0) \in X^2$ such that

$$\alpha := \inf\{d(x, y) - f(x, y) : (x, y) \in X^2\} = d(x_0, y_0) - f(x_0, y_0) > 0.$$

If $g \in E$ satisfies $\|f - g\|_\infty < \alpha/2$, then

$$d(x, y) - g(x, y) = d(x, y) - f(x, y) + f(x, y) - g(x, y)$$
$$\ge \alpha - \|f - g\|_\infty > \alpha/2 > 0,$$

showing that $g \in U$.

Consider the subspace Z of E formed of all functions $h \in E$ of the form $h(x, y) = f(x) + g(y)$, $x, y \in X$, for some $f, g \in C(X)$, and the linear functional $\varphi : Z \to \mathbb{R}$ given by

$$\varphi(h) = \int f d\mu + \int g dv,$$

for every $h \in Z$, $h(x, y) = f(x) + g(y)$, $x, y \in X$. For a fixed $\lambda \in M(\mu, v)$, the equality (8.4.28) shows that $\varphi(h) = \int h(x, y) d\lambda(x, y)$ for all $h \in Z$, so that φ is well-defined (i.e., it does not depend on the particular representation of h as a sum of functions $f, g \in C(X)$) and linear. Let ψ be a linear functional on E such that

$$\sup \psi(U) = \sup \varphi(U \cap Z)$$

$$= \sup\{\int f d\mu + \int g dv : f, g \in C(X), \ f(x) + g(y) < d(x, y), \ \forall(x, y) \in X^2\}$$

$$= m_d(\mu, v).$$

Let us show that

$$\psi(h) \geq 0,$$

for every $h \in C(X^2)$ with $h \geq 0$. Suppose that $\psi(h) < 0$ for some $h \geq 0$ in $C(X^2)$. Then, for a fixed $\varepsilon > 0$ and every $t > 0$, the function

$$h_t(x, y) = d(x, y) - t h(x, y) - \varepsilon \mathbf{1}_{X^2}, \ (x, y) \in X^2,$$

satisfies $h_t(x, y) < d(x, y)$ for all $(x, y) \in X^2$, so it belongs to U. But then

$$\psi(d) - t\psi(h) - \varepsilon\psi(\mathbf{1}_{X^2}) = \psi(h_t) \leq \sup \psi(U) = m_d(\mu, v) < \infty.$$

Letting $t \to \infty$ one obtains a contradiction.

Since ψ is bounded from above on the open convex set U, it is continuous. Since it is also positive, the Riesz representation theorem (Theorem 1.5.26) yields a measure $\lambda_0 \in cabv_+(X^2)$ such that

$$\psi(h) = \int h(x, y) d\lambda_0(x, y),$$

for all $h \in C(X^2)$.

The equality $\psi|_Z = \varphi$ is equivalent to

$$\int (f(x) + g(y)) d\lambda_0(x, y) = \int f d\mu + \int g dv,$$

for every $f, g \in C(X)$, so that, by (8.4.28), $\lambda_0 \in M(\mu, v)$.

For every $\varepsilon > 0$ the function $d(\cdot,\cdot) - \varepsilon \mathbf{1}_{X^2}$ belongs to U, so that

$$-\varepsilon\lambda_0(X^2) + \int d(x,y)d\lambda_0(x,y) \le \sup \psi(U) \le \int d(x,y)d\lambda_0(x,y),$$

implying

$$m_d(\mu,v) = \sup \psi(U) = \int d(x,y)d\lambda_0(x,y) \ge W(\mu,v). \qquad (8.4.34)$$

We have

$$\int f d\mu + \int g dv = \int (f(x)+g(y))d\lambda(x,y) \le \int d(x,y)d\lambda(x,y),$$

for all $f,g \in C(X)$ satisfying (8.4.32) and all $\lambda \in M(\mu,v)$.

Keeping λ fixed and taking the supremum with respect to all these f,g, one obtains

$$m_d(\mu,v) \le \int d(x,y)\lambda(x,y),$$

for all $\lambda \in M(\mu,v)$, so that

$$m_d(\mu,v) \le W(\mu,v),$$

which, combined with (8.4.34), yields

$$m_d(\mu,v) = W(\mu,v).$$

Taking into account Lemma 8.4.36, one obtains $\gamma(\mu,v) = m_d(\mu,v) = W(\mu,v)$. Furthermore, the infimum in the definition (8.4.29) of $W(\mu,v)$ is attained at the measure λ_0 constructed above. $\qquad \square$

Remark 8.4.38 In [536] the proof of the duality is given for a separable metric space X. There are numerous extensions of this duality, both concerning the space X as well as results where the distance $d(x,y)$ is replaced by a function $c(x,y)$ satisfying some appropriate semicontinuity conditions, see, for instance, [88, 218, 219, 386, 387, 586] and the books [585] and [672].

8.5 Hanin's Norm and Applications

Hanin succeeded to extend the modified Kantorovich–Rubinstein norm from $cabv(X,\mathbb{R},0)$ to the whole space $cabv(X,\mathbb{R})$, the extended norm being equivalent to $\|\cdot\|_{KR}^\circ$ on $cabv(X,\mathbb{R},0)$. The norm was introduced in [277] and subsequently developed in [278–280]. We present this norm and its basic properties, following essentially [278] and [280]. In Sect. 8.5.6 we shall show, using approximately the same line of reasoning as in the scalar case, that $\| \cdot \|_{KR}^{\circ}$ can be extended from $cabv(X, \mathscr{H}, 0)$ to a norm $\| \cdot \|_H$ on $cabv(X, \mathscr{H})$, for an arbitrary Hilbert space $\mathscr{H}$.

8.5.1 Definition and First Properties

We shall consider here only the real case $\mathscr{H} = \mathbb{R}$ and the Hanin norm (8.5.7).

For an arbitrary metric space (X, d) let $C_b(X)$ be the Banach space of real-valued bounded continuous functions on X with the sup-norm

$$\|f\|_{\infty} = \sup\{|f(x)| : x \in X\}.$$

Consider also the Banach space $\mathrm{BLip}(X)$ of all bounded Lipschitz functions on X with the norm

$$\|f\|_{\max} = \max\{\|f\|_L, \|f\|_{\infty}\}.$$

Let $cabv(X)$ be the space of all real-valued Borel measures of finite variation on the σ-algebra $\mathscr{B}(X)$ of Borel subsets of X and let

$$cabv^{\circ}(X) = \{\mu \in cabv(X) : \mu(X) = 0\}.$$

For $\mu \in cabv^{\circ}(X)$ consider the sets Λ_{μ} and $\tilde{\Lambda}_{\mu}$ of all $\lambda \in cabv_+(X \times X)$ satisfying (8.4.22) and (8.4.23), respectively. The function λ_{μ} given by (8.4.24) belongs to $\tilde{\Lambda}_{\mu}$.

The condition (8.4.22) is equivalent to

$$\int_X f d\mu = \int_{X \times X} [f(x) - f(y)] d\lambda(x, y), \quad \text{for all } f \in C_b(X). \qquad (8.5.1)$$

On $cabv^{\circ}(X)$ consider the functional

$$\|\mu\|_d^{\circ} = \inf\left\{ \int_{X \times X} d(x, y) d\lambda(x, y) : \lambda \in \Lambda_{\mu} \right\}$$

$$= \inf\left\{ \int_{X \times X} d(x, y) d\lambda(x, y) : \lambda \in \tilde{\Lambda}_{\mu} \right\}, \qquad (8.5.2)$$

(see (8.4.26), (8.4.27)).

Kantorovich and Rubinstein (see Sect. 8.4.7) proved that the mapping $f \mapsto u_f$ where

$$u_f(\mu) = \int_X f d\mu, \quad \mu \in cabv^\circ(X)$$

is an isometric isomorphism of $(\mathrm{Lip}_{x_0}(X), \|\cdot\|_L)$ onto $(cabv^\circ(X), \|\cdot\|_d^\circ)^*$.

This isomorphism implies that

$$\|\mu\|_d^\circ = \sup\left\{ \int_X f d\mu : f \in \mathrm{Lip}(X), \|f\|_L \le 1 \right\} = \|\mu\|_{KR}^\circ, \quad \mu \in cabv^\circ(X).$$

Rachev and Shortt [586] extended this duality to a separable metric space (X, d), with the metric $d(x, y)$ replaced by a function $c(x, y)$ having some appropriate properties related to the metric d, and $cabv^\circ(X)$ replaced with the space of measures μ in $cabv^\circ(X)$ satisfying the finite moment condition

$$\int_X c(x_0, y)d|\mu|(y) < \infty, \tag{8.5.3}$$

for some $x_0 \in X$.

Let now

$$cabv_d^\circ(X) = \{\mu \in cabv^\circ(X) : \|\mu\|_d^\circ < \infty\}.$$

Because

$$\|\mu\|_d^\circ \le \int_X d(x_0, y)d|\mu|(y),$$

a sufficient condition for a measure $\mu \in cabv^\circ(X)$ to belong to $cabv_d^\circ(X)$ is to satisfy (8.5.3) with $c(x, y) = d(x, y)$.

Notice that, if (8.5.3) holds for some $x_0 \in X$, then, by the triangle inequality,

$$\int_X d(x, y)d|\mu|(y) < \infty,$$

for every $x \in X$.

It follows that $cabv_d^\circ(X)$ is a subspace of $cabv^\circ(X)$ and the functional $\|\cdot\|_d^\circ$ is a norm on $cabv_d^\circ(X)$ (to prove this one can use the relations (8.4.25)).

From (8.5.1) one obtains

$$\left|\int_X f d\mu\right| \le \int_{X\times X} |f(x) - f(y)|d\lambda(x, y) \le \|f\|_L \int_{X\times X} d(x, y)d\lambda(x, y),$$

for every $\lambda \in \Lambda_\mu$, implying

$$\left| \int_X f d\mu \right| \leq \|f\|_L \|\mu\|_d^\circ, \tag{8.5.4}$$

for every $f \in \text{Lip}(X)$ and $\mu \in cabv_d^\circ(X)$.

Since

$$\text{spt}(\lambda_\mu) \subseteq \text{spt}(\mu_-) \times \text{spt}(\mu_+) \quad \text{for every } \lambda \in \tilde{\Lambda}_\mu,$$

$$\text{spt}(\mu) = \text{spt}(\mu_-) \cup \text{spt}(\mu_+),$$

and

$$\sup\{d(x, y) : (x, y) \in \text{spt}(\mu_-) \times \text{spt}(\mu_+)\}$$

$$\leq \sup\{d(x, y) : x, y \in \text{spt}(\mu_-) \cup \text{spt}(\mu_+)\}$$

$$= \text{diam}(\text{spt}(\mu)),$$

it follows that

$$\int_{X \times X} d(x, y) d\lambda_\mu(x, y) = \int_{\text{spt}(\mu_-) \times \text{spt}(\mu_+)} d(x, y) d\lambda_\mu(x, y)$$

$$\leq \text{diam}(\text{spt}(\mu))\mu_+(X) = \frac{1}{2} \text{diam}(\text{spt}(\mu))\|\mu\|,$$

where $\lambda_\mu \in \tilde{\Lambda}_\mu$ is given by (8.4.24).

Taking into account the definition (8.5.2) of $\|\mu\|_d^\circ$, one obtains

$$\|\mu\|_d^\circ \leq \frac{1}{2} \|\mu\| \cdot \text{diam}(\text{spt}(\mu)), \tag{8.5.5}$$

for all $\mu \in cabv_d^\circ(X)$.

Recall that the support $\text{spt}(\mu)$ of a Borel measure μ is the complement of the biggest open set G on which μ vanishes (i.e., $\mu(B) = 0$ for all Borel subsets B of G).

Remark 8.5.1 Notice that, by (8.5.5), $\mu \in cabv_d^\circ(X)$ for every $\mu \in cabv^\circ(X)$ with bounded support.

Let $x, y \in X$, $x \neq y$. Then $\text{spt}(\delta_x - \delta_y) = \{x, y\}$ so that, by (8.5.5),

$$\|\delta_x - \delta_y\|_d^\circ \leq d(x, y).$$

Let $f(z) = d(z, y)$, $z \in X$. Then $\|f\|_L = 1$ and, by (8.5.4),

$$d(x, y) = |f(x) - f(y)| = \left| \int_X f d(\delta_x - \delta_y) \right| \leq \|f\|_L \|\delta_x - \delta_y\|_d^\circ.$$

Consequently,

$$\|\delta_x - \delta_y\|_d^\circ = d(x, y), \tag{8.5.6}$$

for all $x, y \in X$ with $x \neq y$ (compare with Theorem 8.4.21).

Consider now the *Hanin norm* $\|\cdot\|_H$ defined on $cabv(X)$ by

$$\|\mu\|_H = \inf\{\|v\|_d^\circ + \|\mu - v\| : v \in cabv_d^\circ(X)\}, \quad \mu \in cabv(X). \tag{8.5.7}$$

Proposition 8.5.2 *Let (X, d) be an arbitrary metric space. The functional $\|\cdot\|_H$ defined by (8.5.7) is a norm on the space $cabv(X)$ of all Borel measures with finite variation on X and the following relations hold true:*

(i) $\left|\int_X f d\mu\right| \leq \|f\|_{\max} \|\mu\|_H$ *for all $f \in \mathrm{BLip}(X)$ and $\mu \in cabv(X)$;*

(ii) $|\mu(X)| \leq \|\mu\|_H \leq \|\mu\|$ *for all $\mu \in cabv(X)$;*

(iii) $\|\mu\|_H = \|\mu\|$ *for every $\mu \in cabv_+(X)$;   in particular*

$\|\delta_x\|_H = \|\delta_x\| = 1$;

(iv) $\|\mu\|_H \leq \|\mu\|_d^\circ \leq \dfrac{1}{2} \max\{2, \mathrm{diam}\,(\mathrm{spt}(\mu))\}\|\mu\|_H$ *for all $\mu \in cabv_d^\circ(X)$;*

(v) $\|\delta_x - \delta_y\|_H = \min\{2, d(x, y)\}$.

Proof Taking into account (8.5.4),

$$\left|\int_X f d\mu\right| \leq \left|\int_X f dv\right| + \left|\int_X f d(\mu - v)\right|$$
$$\leq \|f\|_L \|v\|_d^\circ + \|f\|_\infty \|\mu - v\| \leq \|f\|_{\max}(\|v\|_d^\circ + \|\mu - v\|),$$

for every $v \in cabv_d^\circ(X)$, implying (i).

The first inequality in (ii) follows from the evaluations

$$|\mu(X)| = |(\mu - v)(X)| \leq \|\mu - v\| \leq \|v\|_d^\circ + \|\mu - v\|,$$

valid for every $v \in cabv_d^\circ(X)$.

The second one follows taking $v = 0$ in the definition (8.5.7) of the Hanin norm.

If $\mu \geq 0$, then $\|\mu\| = \mu(X)$, so that (iii) follows from (ii).

Let us prove (iv).

If $\mu \in cabv_d^\circ(X)$, then we can take $v = \mu$ in (8.5.7) to obtain $\|\mu\|_H \leq \|\mu\|_d^\circ$.

By (8.5.5),

$$\|\mu\|_d^\circ \leq \|v\|_d^\circ + \|\mu - v\|_d^\circ \leq \|v\|_d^\circ + \|\mu - v\| \cdot \frac{1}{2} \mathrm{diam}\,(\mathrm{spt}(\mu)).$$

If diam $(\text{spt}(\mu)) \le 2$, then

$$\|\mu\|_d^\circ \le \|\nu\|_d^\circ + \|\mu - \nu\|,$$

for all $\nu \in cabv_d^\circ(X)$, implying

$$\|\mu\|_d^\circ \le \|\mu\|_H. \tag{8.5.8}$$

If diam $(\text{spt}(\mu)) > 2$, then $1 - \frac{1}{2} \operatorname{diam}(\text{spt}(\mu)) < 0$ and

$$\|\nu\|_d^\circ + \|\mu - \nu\| \cdot \frac{1}{2} \operatorname{diam}(\text{spt}(\mu))$$

$$= (\|\nu\|_d^\circ + \|\mu - \nu\|) \cdot \frac{1}{2} \operatorname{diam}(\text{spt}(\mu)) + \|\nu\|_d^\circ \cdot \left(1 - \frac{1}{2} \operatorname{diam}(\text{spt}(\mu))\right)$$

$$\le (\|\nu\|_d^\circ + \|\mu - \nu\|) \cdot \frac{1}{2} \operatorname{diam}(\text{spt}(\mu)),$$

for all $\nu \in cabv_d^\circ(X)$, implying

$$\|\mu\|_d^\circ \le \frac{1}{2} \operatorname{diam}(\text{spt}(\mu)) \cdot \|\mu\|_H. \tag{8.5.9}$$

The inequalities (8.5.8) and (8.5.9) show that the second inequality in (iv) holds too.

It remains to show that

$$\|\delta_x - \delta_y\|_H = \min\{2, d(x, y)\}.$$

We have diam $(\text{spt}(\delta_x - \delta_y)) = \operatorname{diam}(\{x, y\}) = d(x, y)$ and by (8.5.6), $\|\delta_x - \delta_y\|_d^\circ = d(x, y)$.

Then, by (iv) applied to $\mu = \delta_x - \delta_y$ one obtains

$$\|\delta_x - \delta_y\|_H \le d(x, y) \le \frac{1}{2} \max\{d(x, y), 2\}\|\delta_x - \delta_y\|_H.$$

If $d(x, y) \le 2$, then

$$\|\delta_x - \delta_y\|_H \le d(x, y) \le \|\delta_x - \delta_y\|_H,$$

implying

$$\|\delta_x - \delta_y\|_H = d(x, y). \tag{8.5.10}$$

Suppose now that $d(x, y) > 2$. Observe first that, by (iii),

$$\|\delta_x - \delta_y\|_H \le \|\delta_x\|_H + \|\delta_y\|_H = 2.$$

By (iv) and (8.5.6)

$$d(x, y) = \|\delta_x - \delta_y\|_d^\circ \le \frac{1}{2}\max\{d(x, y), 2\}\|\delta_x - \delta_y\|_H$$

$$= \frac{1}{2}d(x, y)\|\delta_x - \delta_y\|_H,$$

implying $\|\delta_x - \delta_y\|_H \ge 2$, so that

$$\|\delta_x - \delta_y\|_H = 2. \qquad (8.5.11)$$

The equalities (8.5.10) and (8.5.11) prove the validity of (v).   □

8.5.2   The Density of Measures with Finite Support

The following density result is the analog of Theorem 8.4.21.2.

Proposition 8.5.3 *Let (X, d) be a separable complete metric space. Then the set of all Borel measures on X with finite support is dense in $(cabv(X), \|\cdot\|_H)$.*

The proof is based on the following lemma.

Lemma 8.5.4 *Let (X, d) be a separable complete metric space. Then the set of all measures in $cabv^\circ(X)$ with finite support is dense in $(cabv_d^\circ(X), \|\cdot\|_d^\circ)$*

Proof The proof is similar to that of Theorem 8.4.21.2, taking advantage on the separability of X instead of compactness. Recall that we are working with the definition (8.5.2) of the norm $\|\cdot\|_d^\circ$.

Let $\mu \in cabv_d^\circ(X)$ and $\varepsilon > 0$. Let also $\{x_n : n \in \mathbb{N}\}$ be a countable dense subset of X. Then the balls $B[x_n, \varepsilon/2]$, $n \in \mathbb{N}$, cover X and so do the sets given by

$$X_1 = B[x_1, \varepsilon/2], \quad X_n = B[x_n, \varepsilon/2] \setminus (X_1 \cup \cdots \cup X_{n-1}), n \ge 2.$$

In addition, the sets X_n are pairwise disjoint and each of them of diameter $\le \varepsilon$. We can suppose further that each set X_n is nonempty so that $x_n \in X_n$, $n \in \mathbb{N}$.

By (8.5.2) there exists $\lambda \in \Lambda_\mu$ such that

$$\int_{X \times X} d(x, y)d\lambda(x, y) < \infty.$$

There exists also $m \in \mathbb{N}$ such that the set $Y_m := X_1 \cup \cdots \cup X_m$ satisfies the condition

$$\int_{X \times X} d(x, y) d\lambda(x, y) \leq \int_{Y_m \times Y_m} d(x, y) d\lambda(x, y) + \varepsilon . \qquad (8.5.12)$$

Define λ_m and μ_m by

(i) $\lambda_m(A \times B) = \lambda((A \times B) \cap (Y_m \times Y_m))$, $A \times B \in \mathscr{B}(X \times X)$, and

(ii) $\quad \mu_m(A) = \lambda_m(X \times A) - \lambda_m(A \times X) \qquad (8.5.13)$

$$= \lambda(X \times (A \cap Y_m)) - \lambda((A \cap Y_m) \times X), \quad A \in \mathscr{B}(X).$$

It is easy to check that

$$\mu_m \in cabv^\circ(X), \quad \mathrm{spt}(\mu_m) \subseteq Y_m, \quad \|\mu_m\| \leq 2\,var(\lambda) \quad \text{and} \quad \lambda - \lambda_m \in \Lambda_{\mu - \mu_m} . \qquad (8.5.14)$$

Also, by (8.5.12),

$$\|\mu - \mu_m\|_d^\circ \leq \int_{X \times X} d(x, y) d(\lambda - \lambda_m)(x, y) \leq \varepsilon . \qquad (8.5.15)$$

Define now the measure ν_k by

$$\nu_k(A) = \mu_m(A \cap X_k) - \mu_m(X_k)\,\delta_{x_k}(A), \quad A \in \mathscr{B}(X) ,$$

for $k = 1, 2, \ldots, m$.

Then $\nu_k \in cabv^\circ(X)$, $\|\nu_k\| \leq 2\|\mu_m\|$ and $\mathrm{spt}(\nu_k) \subseteq X_k$ so that, by (8.5.5),

$$\|\nu_k\|_d^\circ \leq \frac{1}{2}\,\mathrm{diam}\,X_k \cdot \|\nu_k\| \leq \varepsilon \cdot \|\mu_m\| = \varepsilon \cdot |\mu_m|(X_k) . \qquad (8.5.16)$$

Finally, put

$$\bar{\nu}_m = \sum_{k=1}^m \nu_k \quad \text{and} \quad \sigma_m = \sum_{k=1}^m \mu_m(X_k)\delta_{x_k} .$$

Then σ_m is a measure with finite support $\{x_1, \ldots, x_m\}$ and $\sigma_m(X) = \mu_m(Y_m) = 0$, i.e., $\sigma_m \in cabv^\circ(X)$.

Taking into account (8.5.16) and (8.5.14), one obtains

$$\|\mu_m - \sigma_m\|_d^\circ = \|\bar{\nu}_m\|_d^\circ \leq \sum_{k=1}^m \|\nu_k\|_d^\circ \leq \varepsilon\,|\mu_m|(Y_m) = \varepsilon\|\mu_m\| \leq 2\varepsilon\,var(\lambda) ,$$

which, combined with (8.5.15), yields

$$\|\mu - \sigma_m\|_d^\circ \le \|\mu - \mu_m\|_d^\circ + \|\mu_m - \sigma_m\|_d^\circ \le \varepsilon(1 + 2var(\lambda)).$$

Since $\varepsilon > 0$ was arbitrarily chosen, this proves the required density result. □

Based on this lemma we can prove now Proposition 8.5.3.

Proof of Proposition 8.5.3 Let $\mu \in cabv(X)$ and $\varepsilon > 0$. By the regularity of the measure $|\mu|$ there exists a compact subset K of X such that

$$|\mu|(X) \le |\mu|(K) + \varepsilon.$$

Since $|\mu|(X) = |\mu|(K) + |\mu|(\complement(K))$ this implies

$$|\mu|(\complement(K)) \le \varepsilon.$$

Let the measure ν be defined by

$$\nu(A) := \mu(A \cap K), \quad A \in \mathscr{B}(X).$$

For any partition $A_1, \ldots, A_n$ of X,

$$\sum_{i=1}^n |\mu(A_i) - \nu(A_i)| = \sum_{i=1}^n |\mu(A_i \cap \complement(K))| \le |\mu|(\complement(K)) \le \varepsilon,$$

so that

$$\|\mu - \nu\| \le \varepsilon. \tag{8.5.17}$$

For $x_0 \in K$ the measure $\gamma := \nu - \nu(K)\delta_{x_0}$ has the support contained in K and $\gamma(X) = 0$, that is, $\gamma \in cabv_d^\circ(X)$ (see Remark 8.5.1). By Lemma 8.5.4 there exists a measure $\sigma \in cabv_d^\circ(X)$ with finite support such that

$$\|\gamma - \sigma\|_d^\circ \le \varepsilon. \tag{8.5.18}$$

The measure $\varrho := \sigma + \nu(K)\delta_{x_0}$ has finite support and, by the definition (8.5.7) of $\|\cdot\|_H$ and (8.5.17), (8.5.18),

$$\|\mu - \varrho\|_H \le \|\gamma - \sigma\|_d^\circ + \|\mu - \varrho - (\gamma - \sigma)\|$$
$$= \|\gamma - \sigma\|_d^\circ + \|\mu - \nu\| \le 2\varepsilon.$$

□

8.5.3 The Dual of $(cabv(X), \| \cdot \|_H)$

Now we are able to determine the dual of the space $(cabv(X), \| \cdot \|_H)$.
For $f \in \mathrm{BLip}(X)$ define the functional $u_f : cabv(X) \to \mathbb{R}$ by

$$u_f(\mu) = \int_X f d\mu, \quad \mu \in cabv(X). \tag{8.5.19}$$

Then u_f is obviously linear and, by Proposition 8.5.2.(i),

$$|u_f(\mu)| \le \|f\|_{\max} \cdot \|\mu\|_H, \quad \mu \in cabv(X).$$

Consequently, u_f is a continuous linear functional on $(cabv(X), \| \cdot \|_H)$ with norm satisfying the inequality

$$\|u_f\| \le \|f\|_{\max}. \tag{8.5.20}$$

Theorem 8.5.5 *Let (X, d) be a separable complete metric space. Then*

$$(\mathrm{BLip}(X), \| \cdot \|_{\max}) \cong (cabv(X), \| \cdot \|_H)^*,$$

an isometric linear isomorphism of $(\mathrm{BLip}(X), \| \cdot \|_{\max})$ onto $(cabv(X), \| \cdot \|_H)^*$ being given by the mapping $f \mapsto u_f$, $f \in \mathrm{BLip}(X)$.

Proof By Proposition 8.5.2.(iii), for every $x \in X$, $\|\delta_x\|_H = 1$ and

$$|f(x)| = \Big| \int_X f d\delta_x \Big| = |u_f(\delta_x)| \le \|u_f\| \|\delta_x\|_H = \|u_f\|,$$

implying

$$\|f\|_\infty \le \|u_f\|. \tag{8.5.21}$$

By Proposition 8.5.2.(iv), for all $x, y \in X$, $\|\delta_x - \delta_y\|_H \le d(x, y)$, so that

$$|f(x) - f(y)| = \Big| \int_X f d(\delta_x - \delta_y) \Big| = |u_f(\delta_x - \delta_y)|$$

$$\le \|u_f\| \|\delta_x - \delta_y\|_H \le \|u_f\| d(x, y),$$

implying

$$\|f\|_L \le \|u_f\|. \tag{8.5.22}$$

The inequalities (8.5.20), (8.5.21) and (8.5.22) yield

$$\|u_f\| = \|f\|_{\max}.$$

Consequently, the mapping $f \mapsto u_f$ is a linear isometry of $(\mathrm{BLip}(X), \|\cdot\|_{\max})$ into $(cabv(X), \|\cdot\|_H)^*$.

It remains to show that this mapping is also onto (i.e., surjective). For $v \in (cabv(X), \|\cdot\|_H)^*$ define $f : X \to \mathbb{R}$ by

$$f(x) = v(\delta_x), \quad x \in X.$$

Then, for all $x \in X$, $|f(x)| \leq \|v\|\|\delta_x\|_H = \|v\|$, so that $\|f\|_\infty \leq \|v\|$. Also

$$|f(x) - f(y)| = |v(\delta_x - \delta_y)|$$
$$\leq \|v\|\|\delta_x - \delta_y\|_H \leq \|v\|d(x, y),$$

for all $x, y \in X$, implying $\|f\|_L \leq \|v\|$. It follows that $f \in \mathrm{BLip}(X)$ and $\|f\|_{\max} \leq \|v\|$.

Now, the equalities

$$v(\delta_x) = f(x) = \int_X f d\delta_x = u_f(\delta_x), \quad x \in X,$$

and the linearity of v and of the integral, imply that $v(\sigma) = u_f(\sigma)$ for every σ in the linear space generated by the Dirac measures δ_x, $x \in X$, i.e., for every Borel measure with finite support. Since, by Proposition 8.5.3, these measures are dense in the space $(cabv(X), \|\cdot\|_H)$ it follows that $v(\mu) = u_f(\mu)$ for all $\mu \in cabv(X)$.

By the first part of the proof, $\|v\| = \|u_f\| = \|f\|_{\max}$.   □

It is known that for any normed space E, for every $x \in E$,

$$\|x\| = \sup\{|x^*(x)| : x^* \in B_{E^*}\},$$

and, in the real case,

$$\|x\| = \sup\{x^*(x) : x^* \in B_{E^*}\}.$$

Based on these results one obtains the following representation for the Hanin norm.

Corollary 8.5.6 *Let (X, d) be a separable metric space. Then the norm $\|\cdot\|_H$ on $cabv(X)$ can be calculated by the formula*

$$\|\mu\|_H = \sup\left\{\int_X f d\mu : f \in \mathrm{BLip}(X), \|f\|_{\max} \leq 1\right\}, \quad \mu \in cabv(X).$$
$$(8.5.23)$$

Proof Indeed,

$$\|\mu\| = \sup \left\{ u_f(\mu) : f \in \mathrm{BLip}(X),\ \|f\|_{\max} \le 1 \right\}$$

$$= \sup \left\{ \int_X f\, d\mu : f \in \mathrm{BLip}(X),\ \|f\|_{\max} \le 1 \right\},$$

for all $\mu \in cabv(X)$. $\square$

8.5.4 The Weak* Convergence of Borel Measures

We have seen in Theorem 8.4.13 that for bounded sequences in $cabv(X, \mathbb{K}^m)$ weak* convergence agrees with the convergence in the Kantorovich–Rubinstein norm. We show that a similar result holds for the Hanin norm.

For every $\mu \in cabv(X)$ let $x_\mu^* : C_b(X) \to \mathbb{R}$ be defined by

$$x_\mu^*(f) = \int_X f\, d\mu, \quad f \in C_b(X). \tag{8.5.24}$$

The functional x_μ^* is obviously linear and the inequality

$$\left| \int_X f\, d\mu \right| \le \|\mu\|\, \|f\|_\infty,$$

shows that x_μ^* is also continuous, with norm $\|x_\mu^*\| \le \|\mu\| (= var(\mu))$. Consequently, the space $cabv(X)$ equipped with the total variation norm can be viewed as a subspace of the dual space of the Banach space $(C_b(X), \|\cdot\|_\infty)$.

Let us define the weak* convergence (w^*-convergence) of a sequence (μ_n) in $cabv(X)$ to $\mu \in cabv(X)$ by the condition

$$\mu_n \xrightarrow{w^*} \mu \iff \lim_{n \to \infty} \int_X f\, d\mu_n = \int_X f\, d\mu \quad \text{for all } f \in C_b(X). \tag{8.5.25}$$

Theorem 8.5.7 *Let (X, d) be a separable metric space, (μ_n) a bounded (with respect to the total variation norm) sequence in $cabv(X)$ and $\mu \in cabv(X)$. Then*

$$\mu_n \xrightarrow{w^*} \mu \iff \mu_n \xrightarrow{\|\cdot\|_H} \mu.$$

Proof Suppose that $\mu_n \xrightarrow{w^*} \mu$. By Theorem 6.8 in [544] this is equivalent to

$$\sup_{f \in \mathscr{F}} \left| \int_X f\, d(\mu_n - \mu) \right| \to 0 \quad \text{as } n \to \infty, \tag{8.5.26}$$

8 Banach Spaces of Lipschitz Functions

for every equicontinuous norm-bounded family $\mathscr{F}$ of functions in $C_b(X)$. Taking $\mathscr{F}$ to be the closed unit ball of $(cabv(X), \|\cdot\|_H)$ and taking into account formula (8.5.23), the relation (8.5.26) is equivalent to $\|\mu_n - \mu\|_H \to 0$, that is, $\mu_n \xrightarrow{\|\cdot\|_H} \mu$.

By Theorem 6.1 in [544], $\mu_n \xrightarrow{w^*} \mu$ if and only if (8.5.25) holds for every uniformly continuous function $f \in C_b(X)$.

Suppose now that $\mu_n \xrightarrow{\|\cdot\|_H} \mu$, with $\|\mu_n\| \le \beta$, $n \in \mathbb{N}$, and $\|\mu\| \le \beta$, for some $\beta > 0$. Let f be a uniformly continuous function in $C_b(X)$ and $0 < \varepsilon < 1$.

Since, by Corollary 6.3.2, every uniformly continuous bounded function can be uniformly approximated by Lipschitz functions, there exists a Lipschitz function g such that $\|f - g\|_\infty \le \varepsilon/(4\beta)$. It follows that $\|g\|_\infty \le \frac{\varepsilon}{4\beta} + \|f\|_\infty$, so that $g \in \mathrm{BLip}(X)$.

Let $n_0 \in \mathbb{N}$ be such that

$$\|\mu_n - \mu\|_H \le \frac{\varepsilon}{2\|g\|_{\max}}$$

for all $n \ge n_0$.

Then, taking into account Proposition 8.5.2, one obtains

$$\left| \int_X f d(\mu_n - \mu) \right| \le \left| \int_X g d(\mu_n - \mu) \right| + \left| \int_X (f - g) d(\mu_n - \mu) \right|$$

$$\le \|g\|_{\max} \|\mu_n - \mu\|_H + \|f - g\|_\infty \|\mu_n - \mu\| \le \varepsilon,$$

for all $n \ge n_0$. □

Remark 8.5.8 In [544] (and in [81] as well) one considers only probability measures, but as it is remarked in [280], the proofs given there can be adapted to the situation considered here.

Remark 8.5.9 If the metric space X is compact, then

$$(C(X), \|\cdot\|_\infty)^* \cong (cabv(X), \|\cdot\|),$$

where $\|\cdot\|$ is the total variation norm, so that, by the Alaoglu–Bourbaki theorem, the unit ball $B_{var}(cabv(X))$ of $cabv(X)$ is w^*-compact. By Theorem 8.5.7 it follows that it is also compact with respect to the topology generated by the Hanin norm.

8.5.5 Double Duality

For a metric space (X, d) denote by $\mathrm{blip}(X)$ the space of bounded functions in $\mathrm{lip}(X)$,

$$\mathrm{blip}(X) = \{f \in \mathrm{lip}(X) : f \text{ is bounded}\},$$

(see Sect. 8.3).

We shall present Hanin's results on the double duality $(\text{blip}(X))^{**} \cong \text{BLip}(X)$. The idea is to show first the duality $(\text{blip}(X))^{*} \cong (cabv(X), \|\cdot\|_H)$ and apply then Theorem 8.5.5. Recall that we are working with the norm

$$\|f\|_{\max} = \max\{\|f\|_{\infty}, \|f\|_L\}$$

on $\text{BLip}(X)$ and on its subspace $\text{blip}(X)$. If X is of finite diameter (in particular, if X is compact), then $\text{Lip}(X) = \text{BLip}(X)$ and $\text{lip}(X) = \text{blip}(X)$, but we keep the notation $\text{blip}(X)$ to indicate that one works with the norm $\|\cdot\|_{\max}$.

If (X, d) is a separable complete metric space, then, by Proposition 8.5.2.(i), the mapping

$$\iota : (cabv(X), \|\cdot\|_H) \to (\text{blip}(X)^{*}, \|\cdot\|),$$

given for $\mu \in cabv(X)$ by

$$\iota(\mu)(f) = \int_X f d\mu, \quad \text{for } f \in \text{blip}(X), \tag{8.5.27}$$

is linear and of norm not greater than 1.

Lemma 8.5.10 *For any compact metric space (X, d), $\iota(cabv(X))$ is norm-dense in* $\text{blip}(X)^{*}$.

Proof Let $\Delta(X) = \{(x, x) : x \in X\}$ be the diagonal of $X \times X$ and $Y = X \sqcup \widehat{X}$ be the disjoint union of the topological spaces X and $\widehat{X} = X \times X \setminus \Delta(X) = \{(x, y) \in X \times X : x \neq y\}$. For $f \in \text{blip}(X)$ let

$$\hat{f}(x, x) = f(x) \qquad\qquad \text{for } x \in X, \text{ and}$$

$$\hat{f}(x, y) = \frac{f(x) - f(y)}{d(x, y)} \qquad \text{for } (x, y) \in \widehat{X},$$

(see Sect. 8.3.1 on de Leeuw's map).

Then Y is a locally compact space and $\hat{f} \in C_0(Y)$ for every $f \in \text{blip}(X)$. Indeed, consider on $X \times X$ the metric $\varrho((x, y), (x', y')) = \max\{d(x, x'), d(y, y')\}$ generating the product topology. For $(x, y) \in \widehat{X}$, $r := \text{dist}_\varrho((x, y), \Delta(X)) > 0$ and the closed ball $B_\varrho[(x, y), r/2]$ is a compact neighborhood of (x, y) contained in $\widehat{X}$. For $x \in X$ one can take X as a compact neighborhood of x.

Let us show now that $\hat{f} \in C_0(Y)$ for $f \in \text{blip}(X)$. Indeed, for $\varepsilon > 0$ there exists $\delta > 0$ such that

$$0 < d(x, y) < \delta \implies \frac{|f(x) - f(y)|}{d(x, y)} < \varepsilon.$$

The set $K := \{(x, y) \in X \times X : d(x, y) \geq \delta\}$ is closed in $X \times X$ and so compact, as well as $\tilde{K} = X \sqcup K$. For every $(x, y) \in Y \setminus \tilde{K}$, $|\hat{f}(x, y)| < \varepsilon$, showing that $\hat{f} \in C_0(Y)$.

Furthermore, for every $f \in \mathrm{blip}(X)$,

$$\|\hat{f}\|_\infty = \max\{\|f\|_X, \|f\|_L\} = \|f\|_{\max},$$

where

$$\|f\|_X = \sup\{|f(x)| : x \in X\}.$$

Consequently, the mapping $\gamma : \mathrm{blip}(X) \to C_0(Y)$ given by $\gamma(f) = \hat{f}$ is a linear isometry.

Let $\varphi \in \mathrm{blip}(X)^*$. Then φ induces a continuous linear functional $\hat{\varphi}$ on the subspace $\gamma(\mathrm{blip}(X))$ of $C_0(Y)$, which can be extended to a continuous linear functional ψ on $C_0(Y)$. By Riesz' representation theorem (Theorem 1.5.29) there exist the measures $\nu \in cabv(X)$ and $\lambda \in cabv(\widehat{X})$ such that

$$\varphi(f) = \int_X f(x)d\nu(x) + \int_{\widehat{X}} \frac{f(x) - f(y)}{d(x, y)} d\lambda(x, y),$$

for all $f \in \mathrm{blip}(X)$.

For $n \in \mathbb{N}$ consider the sets

$$A_n := \{(x, y) \in \widehat{X} : d(x, y) \geq 1/n\} \text{ and}$$

$$B_n := \{(x, y) \in \widehat{X} : d(x, y) < 1/n\} = \widehat{X} \setminus A_n,$$

and the linear functionals

$$\varphi_n(f) = \int_X f(x)d\nu(x) + \int_{A_n} \frac{f(x) - f(y)}{d(x, y)} d\lambda(x, y), \quad \text{for } f \in \mathrm{blip}(X).$$

Then

$$|\varphi_n(f)| \leq \|f\|_X(\|\nu\| + 2n\|\lambda\|).$$

It follows that φ_n is a continuous linear functional on the subspace $\mathrm{blip}(X)$ of $(C(X), \|\cdot\|_X)$. Extending it to a continuous linear functional on $(C(X), \|\cdot\|_X)$ and appealing again to Riesz' representation theorem, there exists a measure $\mu_n \in cabv(X)$ such that

$$\varphi_n(f) = \int_X f d\mu_n = \iota(\mu_n)(f),$$

for all $f \in \mathrm{blip}(X)$. But then, for every $f \in \mathrm{blip}(X)$ with $\|f\|_{\max} \le 1$,

$$|(\varphi - \iota(\mu_n))(f)| \le \int_{B_n} \frac{|f(x) - f(y)|}{d(x, y)} d\lambda(x, y) \le |\lambda|(B_n).$$

It follows that $\|\varphi - \iota(\mu_n)\| \le |\lambda|(B_n) \to 0$ as $n \to \infty$, because $B_n \supset B_{n+1}$ and $\bigcap_{n=1}^{\infty} B_n = \emptyset$. $\qquad\square$

The characterization of metric spaces for which the double duality holds will be given in terms of the following property satisfied by a metric space (X, d):

Property (A): for every pair x, y of distinct points in X and every $\varepsilon > 0$ there exists a function $g \in \mathrm{blip}(X)$ such that

$$g(x) = 1, \quad g(y) = 0 \quad \text{and} \quad \|g\|_L \le \frac{1 + \varepsilon}{\min\{d(x, y), 2\}}.$$

This property is related to the following one:

Property (B): for every finite subset Z of X, every $\varepsilon > 0$ and every function $f : Z \to \mathbb{R}$ there exists a function $g \in \mathrm{blip}(X)$ such that

$$g|_Z = f \quad \text{and} \quad \|g\|_{\max} \le (1 + \varepsilon)\|f\|_{Z,\max}.$$

Here

$$\|f\|_Z = \sup_{z \in Z} |f(z)|, \quad \|f\|_{Z,L} = \sup_{z,z' \in Z,\, z \ne z'} \frac{|f(z) - f(z')|}{d(z, z')} \quad \text{and}$$

$$\|f\|_{Z,\max} = \max\{\|f\|_Z, \|f\|_{Z,L}\}.$$

Lemma 8.5.11 *If the metric space (X, d) satisfies the condition (A), then it satisfies the condition (B).*

Proof Let Z, f and ε be as in the hypotheses of the condition (B). For every pair x, y of distinct points in Z let $g_{x,y}$ be the function given by (A) and let

$$h_{x,y} = f(y) + [f(x) - f(y)]g_{x,y}.$$

Then $h_{x,y}(x) = f(x)$, $h_{x,y}(y) = f(y)$ and

$$\|h_{x,y}\|_L = |f(x) - f(y)| \cdot \|g_{x,y}\|_L \le (1 + \varepsilon)\frac{|f(x) - f(y)|}{\min\{d(x, y), 2\}} \le (1 + \varepsilon)\|f\|_{Z,\max}.$$

Indeed, if $d(x, y) \le 2$, then

$$\frac{|f(x) - f(y)|}{\min\{d(x, y), 2\}} = \frac{|f(x) - f(y)|}{d(x, y)} \le \|f\|_{Z,L} \le \|f\|_{Z,\max}.$$

If $d(x, y) > 2$, then

$$\frac{|f(x) - f(y)|}{\min\{d(x, y), 2\}} = \frac{|f(x) - f(y)|}{2} \leq \|f\|_Z \leq \|f\|_{Z,\max} .$$

Let

$$g_0 = \max_{x \in Z} \min_{y \in Z} h_{x,y} .$$

Let $z \in Z$. Then $g_0(z) \geq \min_{y \in Z} h_{z,y}(z) = f(z)$. If $w \in Z$ is such that $g_0(z) = \min_{y \in Z} h_{w,y}(z)$, then $g_0(z) \leq h_{w,z}(z) = f(z)$, so that $g_0(z) = f(z)$.

Since every function $h_{x,y}$ satisfies $\|h_{x,y}\|_L \leq (1+\varepsilon)\|f\|_{Z,\max}$, Proposition 2.3.9 implies that $\|g_0\|_L \leq (1+\varepsilon)\|f\|_{Z,\max}$.

Putting $\beta := (1+\varepsilon)\|f\|_{Z,\max}$, it follows that the function

$$g(x) = \begin{cases} \beta & \text{if } g_0(x) > \beta, \\ g_0(x) & \text{if } |g_0(x)| \leq \beta, \\ -\beta & \text{if } g_0(x) < -\beta, \end{cases}$$

satisfies all the requirements from (B). Indeed, if $z \in Z$, then

$$|g_0(z)| = |f(z)| \leq \|f\|_Z < (1+\varepsilon)\|f\|_{Z,\max} ,$$

so that $g(z) = g_0(z) = f(z)$.

On the other hand, the inequalities

$$\frac{|g(x) - g(y)|}{d(x, y)} \leq \frac{|g_0(x) - g_0(y)|}{d(x, y)} \leq \|g_0\|_L \leq \beta,$$

and

$$\|g\|_\infty = \min\{\|g_0\|_\infty, \beta\} \leq \|g_0\|_\infty ,$$

imply $\|g\|_{\max} \leq \beta = (1+\varepsilon)\|f\|_{Z,\max}.$ □

Let $cabv(X)^c$ be the completion of the space $cabv(X)$ with respect to the Hanin norm $\|\cdot\|_H$ and denote the extension of this norm to $cabv(X)^c$ by the same symbol $\|\cdot\|_H$.

The mapping $\iota : cabv(X) \to \mathrm{blip}(X)^*$ given by (8.5.27) extends to a linear mapping of norm not greater than 1 from $cabv(X)^c$ to $\mathrm{blip}(X)^*$, denoted by the same letter ι.

Theorem 8.5.12 *Let (X, d) be a compact metric space. Then the mapping $\iota : cabv(X)^c \to \mathrm{blip}(X)^*$ is an isometric isomorphism if and only if the metric space X satisfies the condition* (A).

Proof Suppose that the metric space X satisfies the condition (A). By Lemma 8.5.10, it suffices to show that $\|\iota(\mu)\| = \|\mu\|_H$ for every $\mu \in cabv(X)$.

Let $\mu \in cabv(X)$ and $\varepsilon > 0$. By Proposition 8.5.3, there exists a measure $\nu \in cabv(X)$ with finite support Z such that $\|\mu - \nu\|_H \le \varepsilon$. For $f \in BLip(X)$ with $\|f\|_{\max} \le 1$ choose $g \in blip(X)$ according to the condition (B). Then $\int_X (f - g)d\nu = 0$, so that, taking into account the inequality (i) from Proposition 8.5.2, one obtains

$$\int_X f d\mu = \int_X (f - g)d\nu + \int_X f d(\mu - \nu) + \int_X g d(\nu - \mu) + \int_X g d\mu$$

$$\le \varepsilon + \varepsilon(1 + \varepsilon) + \int_X g d\mu.$$

$$(8.5.28)$$

Since, by Corollary 8.5.6,

$$\|\mu\|_H = \sup\{\int_X f d\mu : f \in BLip(X), \|f\|_{\max} \le 1\},$$

and

$$\|\iota(\mu)\| = \sup\{\int_X g d\mu : g \in blip(X), \|g\|_{\max} \le 1\},$$

it follows that $\|\mu\|_H \ge \|\iota(\mu)\|$. The inequalities (8.5.28) yield

$$\|\mu\|_H \le \varepsilon(2 + \varepsilon) + (1 + \varepsilon)\|\iota(\mu)\|.$$

As $\varepsilon > 0$ is arbitrary, it follows that $\|\mu\|_H \le \|\iota(\mu)\|$, so that $\|\mu\|_H = \|\iota(\mu)\|$.

Suppose now that the mapping ι is an isometry, that is, $\|\iota(\mu)\| = \|\mu\|_H$ for every $\mu \in cabv(X)$.

Then, for $x, y \in X$, $x \ne y$,

$$\sup\{h(x) - h(y) : h \in blip(X), \|h\|_{\max} \le 1\}$$

$$= \sup\{\int_X h d(\delta_x - \delta_y) : h \in blip(X), \|h\|_{\max} \le 1\}$$

$$= \|\iota(\delta_x - \delta_y)\| = \|\delta_x - \delta_y\|_H = \min\{d(x, y), 2\}.$$

It follows that, for every $\varepsilon > 0$, there exists $h \in blip(X)$, $\|h\|_{\max} \le 1$, such that $h(x) - h(y) \ge (1 + \varepsilon)^{-1} \min\{d(x, y), 2\}$. But then the function

$$g(t) := (h(t) - h(y))/(h(x) - h(y))$$

satisfies the conditions $g(x) = 1$, $g(y) = 0$ and

$$\|g\|_L = \frac{\|h\|_L}{|h(x) - h(y)|} \leq \frac{1 + \varepsilon}{\min\{d(x, y), 2\}},$$

that is the condition (A) is satisfied.   □

Consider the conjugate ι^* of ι, $\iota^*: (\mathrm{blip}(X), \|\cdot\|_{\max})^{**} \to (cabv(X)^c, \|\cdot\|_H)^*$.

Supposing the metric space X compact, denote by $j : (\mathrm{BLip}(X), \|\cdot\|_{\max}) \to (cabv(X), \|\cdot\|_H)^*$ the isometric isomorphism from Theorem 8.5.5. As the spaces $(cabv(X), \|\cdot\|_H)^*$ and $(cabv(X)^c, \|\cdot\|_H)^*$ are isometrically isomorphic, we can suppose that j is an isometric isomorphism between the spaces $(\mathrm{BLip}(X), \|\cdot\|_{\max})$ and $(cabv(X)^c, \|\cdot\|_H)^*$. Putting

$$J := j^{-1} \circ \iota^* : (\mathrm{blip}(X), \|\cdot\|_{\max})^{**} \to (\mathrm{BLip}(X), \|\cdot\|_{\max}),$$

one obtains the following result.

Theorem 8.5.13 *Let (X, d) be a compact metric space. The mapping J considered above is an isometric isomorphism if and only if the metric space X satisfies the condition (A).*

Examples of metric spaces satisfying the condition (A) can be obtained in the following way. Denote by Γ the set of all nondecreasing functions $\omega : \mathbb{R}_+ \to \mathbb{R}_+$ satisfying the conditions:

$$\text{(i)} \quad \omega(0) = 0 = \lim_{t \searrow 0} \omega(t); \qquad \text{(ii)} \quad \lim_{t \to \infty} \frac{\omega(t)}{t} = \infty;$$

$$\text{(iii)} \quad \frac{\omega(t)}{t} \quad \text{is nonincreasing for } t > 0.$$

Notice that the functions in Γ are subadditive:

$$\omega(s + t) \leq \omega(s) + \omega(t),$$

for all $s, t \geq 0$.

Proposition 8.5.14 ([277, 278]) *If (X, d) is a compact metric space and $\omega \in \Gamma$, then $(X, \omega(d))$ is a compact metric space satisfying the condition (A). Consequently,*

$$(\mathrm{blip}(X, \omega(d)), \|\cdot\|_{\max})^{**} \cong (\mathrm{BLip}(X, \omega(d)), \|\cdot\|_{\max}).$$

Taking $\omega \in \Gamma$ given by $\omega(t) = t^\alpha$ for $0 < \alpha < 1$, one obtains isomorphism results for spaces of Hölder functions. The case $X = [0, 1]$ was treated by de Leeuw [182] and by Johnson [309] for an arbitrary compact metric space X, rediscovered in [62].

8.5.6 Hanin's Norm in the Hilbert Case

In this subsection we show, following [147], that an analog of Hanin's norm can be defined on the space $cabv(X, \mathcal{H})$ of measures with values in a Hilbert space $\mathcal{H}$. The construction rests on the properties of the modified Kantorovich–Rubinstein norm $\|\cdot\|_{KR}^{\circ}$ on $cabv(X, \mathcal{H}, 0)$ (see Definition 8.4.14), as presented in Sect. 8.4.5.

Suppose now that (X, d) is a compact metric space and $\mathcal{H}$ is a Hilbert space. The problem with the modified Kantorovich–Rubinstein norm is the fact that it cannot be defined on the whole space $cabv(X, \mathcal{H})$. To be more precise, it is difficult to extend $\|\cdot\|_{KR}^{\circ}$ beyond $cabv(X, \mathcal{H}, 0)$ using the definition (8.4.14), because one can obtain infinite values, as the following result shows.

Proposition 8.5.15 *Define* $p\colon cabv(X, \mathcal{H}) \to \overline{\mathbb{R}}_+$ *by*

$$p(\mu) = \sup\Big\{\Big|\int \langle f, d\mu\rangle\Big| : f \in B_\ell(\mathrm{Lip}(X, \mathcal{H}))\Big\}.$$

Then p *is an extended seminorm, i.e.,* $p(\mu + \nu) \le p(\mu) + p(\nu)$ *and* $p(\alpha\mu) = |\alpha|\, p(\mu)$ *(with the convention* $0 \cdot \infty = 0$*), for all* $\mu, \nu \in cabv(X, \mathcal{H})$ *and all* $\alpha \in \mathbb{K}$. *Also*

$$p(\mu) < \infty \iff \mu \in cabv(X, \mathcal{H}, 0), \tag{8.5.29}$$

for every $\mu \in cabv(X, \mathcal{H})$.

Proof The only fact which must be proved is the implication $\Rightarrow$ in the equivalence (8.5.29).

Let $\mu \in cabv(X, \mathcal{H})$ with $\mu(X) \ne 0$. If $\xi \in \mathcal{H}$, $\|\xi\| = 1$, is such that $\langle \xi, \mu(X)\rangle = \|\mu(X)\| > 0$, then, for any $n \in \mathbb{N}$, the function $f_n \in C(X, \mathcal{H})$ given by $f_n(x) = n\xi$, $x \in X$, satisfies $\|f_n\|_L = 0$ and $\int f_n d\mu = n\|\mu(X)\| \to \infty$ as $n \to \infty$. Hence $p(\mu) = \infty$, because all f_n are in $B_\ell(\mathrm{Lip}(X, \mathcal{H}))$. $\qquad\square$

Define $\|\cdot\|_H : cabv(X, \mathcal{H}) \to \mathbb{R}_+$ by

$$\|\mu\|_H = \inf\{\|\nu\|_{KR}^{\circ} + \|\mu - \nu\| : \nu \in cabv(X, \mathcal{H}, 0)\}, \tag{8.5.30}$$

for all $\mu \in cabv(X, \mathcal{H})$.

Proposition 8.5.16 *The following inequalities hold:*

(i) $\|\mu\|_H \le \|\mu\|$ *for all* $\mu \in cabv(X, \mathcal{H})$;

(ii) $\|\mu\|_H \le \|\mu\|_{KR}^{\circ} \le \|\mu\|_H \max\{1, \mathrm{diam}\, X\}$,

   for all $\mu \in cabv(X, \mathcal{H}, 0)$;

(iii) $\Big|\int \langle f, d\mu\rangle\Big| \le \|\mu\|_H \|f\|_{\max}$,

   for all $f \in \mathrm{BLip}(X, \mathcal{H})$ *and* $\mu \in cabv(X, \mathcal{H})$.

$$\tag{8.5.31}$$

Proof Taking $\nu = 0$, we get $\|\mu\|_H \le \|\mu\|$. If $\mu \in cabv(X, \mathscr{H}, 0)$, then, taking $\nu = \mu$, one obtains $\|\mu\|_H \le \|\mu\|_{KR}^\circ$.

By (8.4.15).(i),

$$\|\mu\|_{KR}^\circ \le \|\nu\|_{KR}^\circ + \|\mu - \nu\|_{KR}^\circ$$
$$\le \|\nu\|_{KR}^\circ + \|\mu - \nu\| \, \text{diam} \, X \,,$$

for every $\nu \in cabv(X, \mathscr{H}, 0)$.

If diam $X \le 1$, then

$$\|\mu\|_{KR}^\circ \le \|\nu\|_{KR}^\circ + \|\mu - \nu\| \,.$$

If diam $X > 1$, then

$$\|\mu\|_{KR}^\circ \le (\|\nu\|_{KR}^\circ + \|\mu - \nu\|) \, \text{diam} \, X + \|\nu\|_{KR}^\circ (1 - \text{diam} \, X)$$
$$\le (\|\nu\|_{KR}^\circ + \|\mu - \nu\|) \, \text{diam} \, X \,.$$

Passing to infimum with respect to $\nu \in cabv(X, \mathscr{H}, 0)$ one obtains

$$\|\mu\|_{KR}^\circ \le \|\mu\|_H \max\{1, \text{diam} \, X\} \,.$$

To prove (8.5.31).(iii), observe that, by (8.4.16),

$$\left| \int \langle f, d\mu \rangle \right| \le \left| \int \langle f, d\nu \rangle \right| + \left| \int \langle f, d(\mu - \nu) \rangle \right|$$
$$\le \|\nu\|_{KR}^\circ \|f\|_L + \|\mu - \nu\| \, \|f\|_\infty$$
$$\le (\|\nu\|_{KR}^\circ + \|\mu - \nu\|) \, \|f\|_{\max} \,,$$

for all $\nu \in cabv(X, \mathscr{H}, 0)$. Due to the arbitrariness of ν, (8.5.31).(iii) follows. $\square$

Theorem 8.5.17

1. *The functional* $\| \cdot \|_H : cabv(X, \mathscr{H}) \to \mathbb{R}_+$ *is a norm on* $cabv(X, \mathscr{H})$ *satisfying the inequalities*

$$|\mu(X)| \le \|\mu\|_H \le \|\mu\| \,, \tag{8.5.32}$$

for all $\mu \in cabv(X, \mathscr{H})$. *It follows that the topology generated by* $\| \cdot \|_H$ *is weaker than the variational topology generated by* $\| \cdot \|$.
2. *On* $cabv(X, \mathscr{H}, 0)$,

$$\|\mu\|_H \le \|\mu\|_{KR}^\circ \le \|\mu\|_H \max\{1, \text{diam} \, X\} \,, \tag{8.5.33}$$

so that the modified KR-norm $\| \cdot \|_{KR}^{\circ}$ and the restriction of $\| \cdot \|_H$ to this subspace are equivalent. If diam $X \leq 1$, *then* $\|\mu\|_H = \|\mu\|_{KR}^{\circ}$ *for every* $\mu \in cabv(X, \mathcal{H})$.

Proof Observe that the inequalities (8.5.32) are contained in Proposition 8.5.16. The first can be obtained by taking $f \equiv 1$ in (8.5.31).(iii) and the second one is (i).

It is routine to check that $\| \cdot \|_H$ is a seminorm. Let us show that $\| \cdot \|_H$ is a norm, i.e., for any $\mu \in cabv(X, \mathcal{H})$, $\mu = 0$ if $\|\mu\|_H = 0$. Indeed, if $\|\mu\|_H = 0$, then, by (8.5.31).(iii), $\int \langle f, d\mu \rangle = 0$ for all $f \in \mathrm{BLip}(X, \mathcal{H})$, so that, by the definition (8.4.6) of the Kantorovich–Rubinstein norm, $\|\mu\|_{KR} = 0$, and thus $\mu = 0$. $\square$

Corollary 8.5.18 *For any $\mu \in cabv(X, \mathcal{H}, 0)$ one has*

$$\frac{1}{\max\{1, \mathrm{diam}\ X\}} \|\mu\|_{KR} \leq \|\mu\|_H \leq (1 + \mathrm{diam}\ X) \|\mu\|_{KR}, \quad (8.5.34)$$

hence, the restrictions of $\| \cdot \|_H$ and $\| \cdot \|_{KR}$ to $cabv(X, \mathcal{H}, 0)$ are equivalent. If diam $X \leq 1$, *then*

$$\|\mu\|_{KR} \leq \|\mu\|_H \leq 2 \|\mu\|_{KR}.$$

Proof The inequalities (8.5.34) follow from (8.4.15).(ii) and (8.5.31).(ii). $\square$

Definition 8.5.19 The norm $\| \cdot \|_H$ given by (8.5.30) is called the *Hanin norm* on $cabv(X, \mathcal{H})$.

As usual, we present the associated notations. The topology on $cabv(X, \mathcal{H})$, generated by $\| \cdot \|_H$, will be called the *Hanin topology* and it will be denoted by τ_{Han}. The convergence of a sequence $(\mu_n)_n \subseteq cabv(X, \mathcal{H})$ to $\mu \in cabv(X, \mathcal{H})$ with respect to this topology will be denoted by $\mu_n \xrightarrow{\mathrm{Han}} \mu$. Finally, on any $\emptyset \neq A \subseteq cabv(X, \mathcal{H})$, $\| \cdot \|_H$ generates the *Hanin metric* d_{Han} given by $d_{\mathrm{Han}}(\mu, \nu) = \|\mu - \nu\|_H$ for any μ, ν in A.

Theorem 8.5.17 is very important. Due to the equivalence of $\| \cdot \|_H$ and $\| \cdot \|_{KR}^{\circ}$ on $cabv(X, \mathcal{H}, 0)$, it follows that the metrics d_{Han} and d_{KR}° are equivalent on any $B_{r,var}(cabv(X, \mathcal{H}, \xi))$ for every $\|\xi\| \leq r$ (for the definition of $B_{r,var}$ see (8.4.21)). Hence, in the statement of Theorem 8.4.25, one can add d_{Han} to the previous equivalent metrics d_{KR} and d_{KR}°. Consequently, the following result holds

Corollary 8.5.20 *For $r > 0$ and $\xi \in \mathcal{H}$ with $\|\xi\| \leq r$, the metrics d_{KR}, d_{KR}° and d_H are equivalent on $B_{r,var}(cabv(X, \mathcal{H}, \xi))$.*

Theorem 8.5.21 *Let $x, y \in X$, $x \neq y$, and $\xi \in \mathcal{H}$, $\|\xi\| = 1$. Then*

$$\|\delta_x \xi\|_H = 1$$

and

$$\frac{1}{\max\{1, \operatorname{diam} X\}} d(x, y) \le \left\| (\delta_x - \delta_y)\xi \right\|_H \le d(x, y). \qquad (8.5.35)$$

Hence, if diam $X \le 1$, *then* $\left\| (\delta_x - \delta_y)\xi \right\|_H = d(x, y)$.
For $\mathcal{H} = \mathbb{K}$ *and* $\xi = 1$, *we have*

$$\frac{1}{\max\{1, \operatorname{diam} X\}} d(x, y) \le \left\| \delta_x - \delta_y \right\|_H \le d(x, y).$$

Proof For $\xi \in \mathcal{H}$, $\|\xi\| = 1$, define again $\varphi_\xi : X \to \mathcal{H}$ by $\varphi_\xi(x) = \xi$, $x \in X$, and notice that $\left\| \varphi_\xi \right\|_{\max} = \left\| \varphi_\xi \right\|_\infty = 1$.
For $\mu = \delta_x \xi$ use (8.5.31) to get

$$1 = \langle \xi, \xi \rangle = \langle \xi, \mu(X) \rangle = \left| \int \langle \varphi_\xi, d\mu \rangle \right| \le \|\mu\|_H \le \|\mu\| = 1.$$

Put $\nu = (\delta_x - \delta_y)\xi \in cabv(X, \mathcal{H}, 0)$. By Theorem 8.4.21

$$\|\nu\|_{KR}^\circ = d(x, y),$$

so that the inequalities (8.5.33) yield

$$\|\nu\|_H \le \|\nu\|_{KR}^\circ \le \|\nu\|_H \max\{1, \operatorname{diam} X\}.$$

These inequalities are equivalent to (8.5.35).   □

8.6 Compactness Properties of Lipschitz Operators

The importance of compact and weakly compact operators in functional analysis is well-known. In this section we show, following [308], that it is possible to define these properties for Lipschitz operators in a way that is in concordance with the compactness of associated linear operators on the corresponding Lipschitz free Banach spaces. Some compactness properties in spaces of Lipschitz functions were studied in [158].

8.6.1 Compact and Weakly Compact Linear Operators

For reader's convenience we recall here some properties of compact and weakly compact linear operators. A good presentation of the properties of these operators is given in Megginson [456, Sections 3.4 and 3.5] (see also [683]).

Let E_1, E_2 be Banach spaces with closed unit balls B_{E_1}, B_{E_2} and duals E_1^*, E_2^*. A linear operator $A : E_1 \to E_2$ is called

- *compact* if the set $A(B_{E_1})$ is relatively compact in E_2;
- *weakly compact* if the set $A(B_{E_1})$ is relatively weakly compact in E_2.

It is immediate that any compact (weakly compact) linear operator is continuous.

We shall use the notations $\mathscr{K}(E_1, E_2)$ and $\mathscr{W}\mathscr{K}(E_1, E_2)$ to denote the spaces of these operators. They are contained in $\mathscr{L}(E_1, E_2)$.

Compact Operators

The following properties are almost immediate, but very useful.

Proposition 8.6.1 *The following are equivalent.*

1. *The linear operator $A : E_1 \to E_2$ is compact.*
2. *For any bounded subset Z of E_1 the set $A(Z)$ is relatively compact in E_2.*
3. *For every bounded sequence (x_n) in E_1 the sequence $(A(x_n))$ contains a convergent subsequence.*

The following theorem emphasizes the so-called ideal properties of spaces of compact operators.

Theorem 8.6.2 *Let E_1, E_2 be Banach spaces.*

1. *If (A_n) is a sequence of compact operators converging in the norm of the space $\mathscr{L}(E_1, E_2)$ to the operator A, then A is compact. Consequently, the space $\mathscr{K}(E_1, E_2)$ is norm-closed in $\mathscr{L}(E_1, E_2)$.*
2. *Let E_0, E_3 be other Banach spaces and*

$$A \in \mathscr{K}(E_1, E_2), B \in \mathscr{L}(E_0, E_1), C \in \mathscr{L}(E_2, E_3).$$

   Then AB and CA are compact operators. It follows that, for any Banach space E, $\mathscr{K}(E)$ is a closed bilateral ideal in the Banach algebra $\mathscr{L}(E)$.

A fundamental result in the theory of linear operators is Schauder's theorem on the compactness of the conjugate operator.

Theorem 8.6.3 (Schauder's Theorem) *Let E_1, E_2 be Banach spaces. An operator $A \in \mathscr{L}(E_1, E_2)$ is compact if and only if its adjoint $A^* \in \mathscr{L}(E_2^*, E_1^*)$ is compact.*

There is another characterization of the compactness of an operator in terms of the continuity of its adjoint with respect to a topology that we briefly describe now.

The Bounded Weak* (bw^*) Topology

Let E be a normed space. The bounded weak* (bw^*) topology is the topology of E^* characterized by the property that a subset Y of E^* is bw^*-closed if and only if for every bounded subset Z of E^*, the set $Y \cap Z$ is relatively w^*-closed in Z.

There are several characterizations of this topology, see, for instance, [456, Section 2.7] or [178, p. 48]. We present only a few.

Proposition 8.6.4 *Each of the following conditions characterizes the bw^*-topology of E^*.*

1. *A net in E^* is bw^*-convergent to 0 if and only if it converges to 0 uniformly on every totally bounded subset of E.*
2. *A net in E^* is bw^*-convergent to 0 if and only if it converges to 0 uniformly on every compact subset of E.*
3. *A subset Y of E^* is bw^*-closed if and only if it contains the limit of each bounded w^*-convergent net in Y.*

Remark 8.6.5 The topology bw^* is stronger than w^* and weaker than the norm topology, but a linear functional on E is w^*-continuous if and only if it is bw^*-continuous.

Now we can present the promised characterization.

Theorem 8.6.6 *Let E_1, E_2 be Banach spaces. A linear operator $A \in \mathscr{L}(E_1, E_2)$ is compact if and only if $A^* \in \mathscr{L}((E_2^*, bw^*), (E_1^*, \|\cdot\|))$.*

Weakly Compact Operators

They have properties similar to those of compact operators as for instance, those contained in Proposition 8.6.1.

Proposition 8.6.7 *The following are equivalent.*

1. *The linear operator $A : E_1 \to E_2$ is weakly compact.*
2. *For any bounded subset Z of E_1 the set $A(Z)$ is relatively weakly compact in E_2.*
3. *For every bounded sequence (x_n) in E_1 the sequence $(A(x_n))$ contains a weakly convergent subsequence.*

Weakly compact operators have also the ideal property.

Theorem 8.6.8 *Let E_1, E_2 be Banach spaces.*

1. *If (A_n) is a sequence of weakly compact operators converging in the norm of the space $\mathscr{L}(E_1, E_2)$ to the operator A, then A is weakly compact. Consequently, the space $\mathscr{W}\mathscr{K}(E_1, E_2)$ is norm-closed in $\mathscr{L}(E_1, E_2)$.*
2. *Let E_0, E_3 be other Banach spaces, $A \in \mathscr{W}\mathscr{K}(E_1, E_2)$, $B \in \mathscr{L}(E_0, E_1)$, $C \in \mathscr{L}(E_2, E_3)$. Then AB and CA are weakly compact operators. It follows that, for any Banach space E, $\mathscr{W}\mathscr{K}(E)$ is a closed bilateral ideal in the Banach algebra $\mathscr{L}(E)$.*

Reflexivity implies that all continuous linear operator are weakly compact.

Proposition 8.6.9 *If one of the Banach spaces E_1, E_2 is reflexive, then every operator $A \in \mathscr{L}(E_1, E_2)$ is weakly compact.*

A deep result in the theory of weakly compact operators, obtained by Davis et al. [177] (see also [683, Theorem II.C.5]), is the factorization through reflexive Banach spaces.

Theorem 8.6.10 *Let E_1, E_2 be Banach spaces. An operator $A \in \mathscr{L}(E_1, E_2)$ is weakly compact if and only if there exist a reflexive Banach space E and the operators $C \in \mathscr{L}(E_1, E)$ and $B \in \mathscr{L}(E, E_2)$ such that $A = BC$.*

For a Banach space denote by $j_E : E \to E^{**}$ its canonical isometric embedding in the bidual given for $x \in E$ by

$$j_E(x)(x^*) = x^*(x), \quad x^* \in E^*.$$

Theorem 8.6.11 *Let E_1, E_2 be Banach spaces and $A \in \mathscr{L}(E_1, E_2)$. Then the following are equivalent.*

1. *The operator A is weakly compact.*
2. $A^{**}(E_1^{**}) \subseteq j_{E_2}(E_2)$.
3. *The adjoint operator A^* belongs to $\mathscr{L}\big((E_2^*, \sigma(E_2^*, E_2)), (E_1^*, \sigma(E_1^*, E_1^{**}))\big)$.*

The analog of Schauder's theorem for weakly compact operators was proved by Gantmacher [247] (see also [456, Theorem 3.5.8]).

Theorem 8.6.12 (Gantmacher's Theorem) *Let E_1, E_2 be Banach spaces. An operator $A \in \mathscr{L}(E_1, E_2)$ is weakly compact if and only if its adjoint $A^* \in \mathscr{L}(E_2^*, E_1^*)$ is weakly compact.*

8.6.2 Lipschitz Compact and Weakly Compact Operators

In this subsection and in the next one we shall present, following [308], the Lipschitz versions of the above results for compact and weakly compact operators.

Let (X, d, θ) be a pointed metric space and E a Banach space. By the *Lipschitz image* of a mapping $F \in \mathrm{Lip}_0(X, E)$ we mean the set

$$\mathrm{Lipim}(F) := \{(F(x) - F(y))/d(x, y) : (x, y) \in \widehat{X}\},$$

where $\widehat{X} = \{(x, y) \in X^2 : x \neq y\}$.

Definition 8.6.13 A mapping $F \in \mathrm{Lip}_0(X, E)$ is called *Lipschitz compact (Lipschitz weakly compact)* if its Lipschitz image is relatively compact (respectively, relatively weakly compact) in E.

We shall denote by $\mathrm{Lip}_{0k}(X, E)$ and $\mathrm{Lip}_{0w}(X, E)$ the spaces of these operators.

The following three propositions show that the definition is the right one. We show that, in the case of linear operators, Lipschitz compactness and compactness agree. To this end we need the following simple lemma.

Lemma 8.6.14 *Let E_1, E_2 be Banach spaces. A linear operator $A : E_1 \to E_2$ is compact if and only if the set $A(S_{E_1})$ is relatively compact in E_2.*

Proof Suppose that the set $A(S_{E_1})$ is relatively compact in E_2. Let (x_n) be a sequence in $B_X \setminus \{0\}$. By hypothesis, the sequence $\big(A(x_n/\|x_n\|)\big)$ contains a convergent subsequence $A(x_{n_k}/\|x_{n_k}\|) \to y$. Since the sequence $(\|x_{n_k}\|)$ is bounded, it contains a convergent subsequence $\|x_{n_{k_i}}\| \to \alpha$. But then

$$A(x_{n_{k_i}}) = \|x_{n_{k_i}}\| \, A\left(\frac{x_{n_{k_i}}}{\|x_{n_{k_i}}\|}\right) \to \alpha y \,,$$

proving the compactness of A. The converse implication is trivial.   □

Proposition 8.6.15 *Let E_1, E_2 be Banach spaces. A linear operator $A : E_1 \to E_2$ is Lipschitz compact if and only if it is compact.*

Proof The proposition follows from the following equivalences

$$A \text{ is Lipschitz compact} \iff \left\{A\left(\frac{x-y}{\|x-y\|}\right) : x \neq y\right\} \text{ is relatively compact}$$

$$\iff A(S_{E_1}) \text{ is relatively compact} \iff A \text{ compact}.$$

     □

The second result is the equivalence between Lipschitz compactness of F and the compactness of the associated linear operator $\widehat{F}$.

Proposition 8.6.16 *Let X be a pointed metric space and E a Banach space. A mapping $F \in \mathrm{Lip}_0(X, E)$ is Lipschitz compact if and only if the linear operator $\widehat{F}$ associated to F by Theorem 8.2.11 is compact.*

Proof Let $W = \{e_{x,y} : (x, y) \in \widehat{X}\}$, where $e_{x,y} = (e_x - e_y)/d(x, y)$ for $x \neq y$. Then, by (8.2.12),

$$\widehat{F}(W) = \{(F(x) - F(y))/d(x, y) : (x, y) \in \widehat{X}\} = \mathrm{Lipim}(F) \,.$$

By Theorem 8.2.17, $B_{\mathbb{F}(X)} = \overline{\mathrm{aco}}(W)$. Consequently, the inclusions

$$\widehat{F}(W) \subseteq \widehat{F}(\overline{\mathrm{aco}}(W)) \subseteq \overline{\mathrm{aco}}(\widehat{F}(W))$$

are equivalent to

$$\mathrm{Lipim}(F) \subseteq \widehat{F}(B_{\mathbb{F}(X)}) \subseteq \overline{\mathrm{aco}}(\mathrm{Lipim}(F)) \,. \qquad\qquad (8.6.1)$$

Since, for every relatively compact subset Z of E, the set $\overline{aco}(Z)$ is compact, the above inclusions prove the equivalence between the Lipschitz compactness of F and the compactness of the linear operator $\widehat{F}$. □

Proposition 8.6.17 *Let X be a pointed metric space, E a Banach space and $F \in \mathrm{Lip}_0(X, E)$. Then the following are equivalent.*

1. *The Lipschitz operator F is Lipschitz weakly compact.*
2. *The associated continuous linear operator $\widehat{F} : \mathbb{F}(X) \to E$ is weakly compact.*
3. *There exist a reflexive Banach space E_1, a bounded linear operator $A \in \mathcal{L}(E_1, E)$ and a Lipschitz operator $G \in \mathrm{Lip}_0(X, E_1)$ such that $F = A \circ G$.*

Proof The inclusions from (8.6.1) and the fact that the closed absolutely convex hull of a relatively weakly compact subset of a Banach space is weakly compact prove the equivalence $1 \Longleftrightarrow 2$.

$2 \Rightarrow 3$. By the mentioned factorization result (Theorem 8.6.10), there exist a reflexive Banach space E_1 and the operators $B \in \mathcal{L}(\mathbb{F}(X), E_1)$, $A \in \mathcal{L}(E_1, E)$ such that $\widehat{F} = A \circ B$. Let $G = B \circ i_X$, where $i_X : X \to \mathbb{F}(X)$ is the isometric embedding from Proposition 8.2.6. It follows that $G \in \mathrm{Lip}_0(X, E_1)$ and, by Theorem 8.2.11, $F = \widehat{F} \circ i_X = A \circ B \circ i_X = A \circ G$.

$3 \Rightarrow 2$. Let $F = A \circ G$, where A and G are as in 3. If we show that $\widehat{F} = A \circ \widehat{G}$, then, by the factorization theorem for weakly compact linear operators, it follows that $\widehat{F}$ is weakly compact.

For $\sum_i t_i e_{x_i} \in \mathrm{span}(X_e)$,

$$\widehat{A \circ G}\Big(\sum_i t_i e_{x_i}\Big) = \sum_i t_i A(G(x_i)) = A\Big(\sum_i t_i G(x_i)\Big)$$

$$= A\Big(\widehat{G}\Big(\sum_i t_i e_{x_i}\Big)\Big) = (A \circ \widehat{G})\Big(\sum_i t_i e_{x_i}\Big),$$

proving the equality $\widehat{F} = \widehat{A \circ G} = A \circ \widehat{G}$. □

Another important property of compact (weakly compact) linear operators is the ideal property: the composition of a compact (weakly compact) linear operator with a continuous linear operator is a compact (weakly compact) linear operator. This implies that the spaces $\mathcal{K}(E)$ ($\mathcal{WK}(E)$) of compact (weakly compact) linear operators on the Banach space E is a bilateral ideal in the Banach algebra $\mathcal{L}(E)$ of all continuous linear operators on E. By Theorems 8.6.2 and 8.6.8, these ideals are also closed.

In the case of Lipschitz compact operators this result takes the following form.

Proposition 8.6.18

1. *Let X, Y be pointed metric spaces, E_1, E_2 Banach spaces, $g \in \mathrm{Lip}_0(Y, X)$ and $A \in \mathcal{L}(E_1, E_2)$. If $F \in \mathrm{Lip}_0(X, E_1)$ is a Lipschitz compact (Lipschitz weakly compact) operator then the operator $A \circ F \circ g$ is also Lipschitz compact (Lipschitz weakly compact).*

2. *Let X be a pointed metric space and E a Banach space. If (F_n) is sequence of Lipschitz compact (Lipschitz weakly compact) operators converging in Lipschitz norm to $F \in \mathrm{Lip}_0(X, E)$, then F is a Lipschitz compact (Lipschitz weakly compact) operator.*

Proof

1. We keep the notations from Theorems 8.2.9 and 8.2.11. The proof will be done if we show that

$$A \circ \widehat{F \circ g} = A \circ \widehat{F} \circ \tilde{g}. \qquad (8.6.2)$$

For any $\sum_i t_i e_{y_i} \in \mathrm{span}(Y_e)$ we have

$$(A \circ \widehat{F \circ g})\Big(\sum_i t_i e_{y_i}\Big) = \sum_i t_i A(F(g(y_i))),$$

and

$$(A \circ \widehat{F} \circ \tilde{g})\Big(\sum_i t_i e_{y_i}\Big) = (A \circ \widehat{F})\Big(\sum_i t_i e_{g(y_i)}\Big)$$

$$= A\Big(\sum_i t_i F(g(y_i))\Big) = \sum_i t_i A(F(g(y_i))),$$

proving the equality (8.6.2).

2. Taking into account Theorem 8.2.11,

$$\|\widehat{F_n} - \widehat{F}\| = \|\widehat{F_n - F}\| = L(F_n - F) \to 0 \quad \text{as} \quad n \to \infty,$$

and the result follows from the corresponding result for linear operators.   □

8.6.3 The Analogs of the Schauder and Gantmacher Theorems for Lipschitz Operators

Let E_1, E_2 be two Banach spaces.

In this subsection we shall try to extend these results to Lipschitz operators. We shall need the analog of the following equality

$$\{A^* : A \in \mathscr{L}(E_1, E_2)\} = \mathscr{L}((E_2^*, \sigma(E_2^*, E_2)), (E_1^*, \sigma(E_1^*, E_1))), \qquad (8.6.3)$$

that is, the space of adjoint operators agrees with the space of all linear operators $B : E_2^* \to E_1^*$ continuous with respect to the w^*-topologies of the spaces E_2^* and E_1^*. To this end we have to define an appropriate w^*-topology on $X^{\#} = \mathrm{Lip}_0(X)$.

For a pointed metric space X let w^* be the topology determined by the isometric isomorphism $\Lambda : X^\# \to \mathbb{F}(X)^*$ from Theorem 8.2.7. This means that

$$U \subseteq X^\# \text{ is } w^*\text{-open} \iff \Lambda^{-1}(U) \text{ is } w^*\text{-open in } \mathbb{F}(X)^*,$$

or, in terms of nets,

$$f_i \xrightarrow{w^*} f \text{ in } X^\# \iff \Lambda(f_i) \xrightarrow{w^*} \Lambda(f) \text{ in } \mathbb{F}(X)^*,$$

for every net (f_i) in $X^\#$ and $f \in X^\#$. In the next proposition we discuss the relations of this topology with the pointwise topology τ_p of $X^\#$.

Proposition 8.6.19 *Let X be a pointed metric space, (f_i) a net in $X^\#$ and $f \in X^\#$.*

1. *If $f_i \xrightarrow{w^*} f$, then $f_i \xrightarrow{\tau_p} f$.*
2. *If (f_i) is bounded, then $f_i \xrightarrow{w^*} f \iff f_i \xrightarrow{\tau_p} f$. This means that the topologies w^* and τ_p agree on bounded subsets of $X^\#$.*

Proof

1. Suppose that $f_i \xrightarrow{w^*} f$. Then $\Lambda(f_i) \xrightarrow{w^*} \Lambda(f)$ (in $\mathbb{F}(X)^*$) and, since for every $x \in X$, $e_x \in \mathbb{F}(X)$, we have

$$\Lambda(f_i)(e_x) \to \Lambda(f)(e_x) \iff f_i(x) \to f(x),$$

   showing that $f_i \xrightarrow{\tau_p} f$.
2. Let (f_i) be a bounded net in $X^\#$, τ_p-convergent to $f \in X^\#$. Then $(\Lambda(f_i))$ is a bounded net in $\mathbb{F}(X)^*$ such that $\Lambda(f_i)(w) \to \Lambda(f)(w)$ for every $w \in \mathrm{span}(X_e)$. The boundedness of the net $(\Lambda(f_i))$ and the density of $\mathrm{span}(X_e)$ in $\mathbb{F}(X)$ imply that $(\Lambda(f_i))$ is w^*-convergent to $\Lambda(f)$. $\qquad\square$

We have to change a bit the notion of Lipschitz conjugate operator of a mapping $F \in \mathrm{Lip}_0(X, E)$: instead of $F^\# : E^\# \to X^\#$ given by $F^\#(g) = g \circ F$ we shall work with its restriction F^t to E^*

$$F^t := F^\#|_{E^*} .$$

With these conventions the following result holds.

Theorem 8.6.20 *Let X be a pointed metric space and E a Banach space. The correspondence $F \mapsto F^t$ is an isometric isomorphism of $\mathrm{Lip}_0(X, E)$ onto $\mathscr{L}((E^*, w^*), (X^\#, w^*))$.*

Proof We show first that

$$\Lambda \circ F^t = \widehat{F}^* . \tag{8.6.4}$$

For $x^* \in E^*$ and $x \in X$ we have

$$\Lambda(F^t(x^*))(e_x) = \Lambda(x^* \circ F)(e_x) = x^*(F(x)),$$

and

$$\widehat{F}^*(x^*)(e_x) = (x^* \circ \widehat{F})(e_x) = x^*(F(x)).$$

By linearity it follows that the equality (8.6.4) holds on $\mathrm{span}(X_e)$ and, by continuity and density, on $\mathbb{F}(X)$.

Taking into account Theorem 8.2.11, it follows that the mapping $F \mapsto \widehat{F}$ is an isometric isomorphism of $\mathrm{Lip}_0(X, E)$ onto $\mathscr{L}(\mathbb{F}(X), E)$.

By (8.6.3), the mapping $\widehat{F} \mapsto \widehat{F}^*$ is an isometric isomorphism of $\mathscr{L}(\mathbb{F}(X), E)$ onto $\mathscr{L}((E^*, w^*), (\mathbb{F}(X)^*, w^*))$.

Finally, taking into account the fact that Λ^{-1} is an isometric isomorphism of $\mathbb{F}(X)^*$ onto $X^\#$ (Theorem 8.2.7) which is also (w^*, w^*)-continuous (by the definition of the topology w^* on $X^\#$), it follows that the mapping $\widehat{F}^* \mapsto \Lambda^{-1} \circ \widehat{F}^*$ is an isometric isomorphism of $\mathscr{L}((E^*, w^*), (\mathbb{F}(X)^*, w^*))$ onto $\mathscr{L}((E^*, w^*), (X^\#, w^*))$.

But, by (8.6.4), $\Lambda^{-1} \circ \widehat{F}^* = F^t$. Combining the above isomorphisms, it follows that the mapping $F \mapsto F^t$ is an isometric isomorphism of $\mathrm{Lip}_0(X, E)$ onto $\mathscr{L}((E^*, w^*), (X^\#, w^*))$.   □

We are now in position to state and prove the Lipschitz analog of the Schauder theorem.

Theorem 8.6.21 *Let X be a pointed metric space, E a Banach space and $F \in \mathrm{Lip}_0(X, E)$. Then the following are equivalent.*

1. *The operator F is Lipschitz compact.*
2. *The Lipschitz adjoint F^t is a compact linear operator from E^* to $X^\#$.*
3. *The operator F^t is continuous from (E^*, bw^*) to $X^\#$.*

Proof The equivalence $1 \iff 2$ will be a consequence of the following equivalences. First, by Proposition 8.6.16,

$$F \text{ is Lipschitz compact} \iff \widehat{F} \in \mathscr{K}(\mathbb{F}(X), E).$$

By Schauder's theorem for linear operators,

$$\widehat{F} \in \mathscr{K}(\mathbb{F}(X), E) \iff \widehat{F}^* \in \mathscr{K}(E^*, \mathbb{F}(X)^*).$$

Since $\Lambda : X^\# \to \mathbb{F}(X)^*$ is an isometric isomorphism (Theorem 8.2.7),

$$\widehat{F}^* \in \mathscr{K}(E^*, \mathbb{F}(X)^*) \iff \Lambda^{-1} \circ \widehat{F}^* \in \mathscr{K}(E^*, X^\#).$$

But, by (8.6.4), $\Lambda^{-1} \circ \widehat{F}^* = F^t$.

$1 \Longleftrightarrow 3$. By Proposition 8.6.16 and Theorem 8.6.6, we have

F is Lipschitz compact $\Longleftrightarrow$ $\widehat{F}$ is Lipschitz compact

$$\Longleftrightarrow \widehat{F}^* \in \mathscr{L}((E^*, bw^*), \mathbb{F}(X)^*) \Longleftrightarrow F^t = \Lambda^{-1} \circ \widehat{F}^* \in \mathscr{L}((E^*, bw^*), X^{\#}).$$

$\square$

The analog of Gantmacher's theorem has the following form.

Theorem 8.6.22 *Let X be a pointed metric space, E a Banach space and $F \in \mathrm{Lip}_0(X, E)$. Then the following are equivalent.*

1. The operator F is Lipschitz weakly compact.
2. The Lipschitz adjoint F^t is a weakly compact linear operator from E^* to $X^{\#}$.
3. The operator F^t is continuous from (E^*, w^*) to $(X^{\#}, w)$.

Proof $1 \Longleftrightarrow 2$. By Proposition 8.6.17,

$$F \text{ is Lipschitz weakly compact} \Longleftrightarrow \widehat{F} \in \mathscr{W}\mathscr{K}(\mathbb{F}(X), E).$$

By Gantmacher's theorem (Theorem 8.6.12),

$$\widehat{F} \in \mathscr{W}\mathscr{K}(\mathbb{F}(X), E) \Longleftrightarrow \widehat{F}^* \in \mathscr{W}\mathscr{K}(E^*, \mathbb{F}(X)^*).$$

Since $\Lambda : X^{\#} \to \mathbb{F}(X)^*$ is an isometric isomorphism (Theorem 8.2.7),

$$\widehat{F}^* \in \mathscr{K}(E^*, \mathbb{F}(X)^*) \Longleftrightarrow \Lambda^{-1} \circ \widehat{F}^* \in \mathscr{K}(E^*, X^{\#}).$$

Finally, the equivalence follows from the fact that $\Lambda^{-1} \circ \widehat{F}^* = F^t$ (by (8.6.4)).
$1 \Longleftrightarrow 3$. By Proposition 8.6.17 and Theorem 8.6.11, we have

F is Lipschitz weakly compact $\Longleftrightarrow$ $\widehat{F}$ is Lipschitz weakly compact $\Longleftrightarrow$

$$\widehat{F}^* \in \mathscr{L}((E^*, w^*), (\mathbb{F}(X)^*, w)) \Longleftrightarrow F^t = \Lambda^{-1} \circ \widehat{F}^* \in \mathscr{L}((E^*, w^*), (X^{\#}, w)).$$

The last equivalence holds because Λ is also an isomorphism between the spaces $(X^{\#}, w)$ and $(F(X)^*, w)$.
$\square$

8.7 Composition Operators

In this section we shall present this important class of operators, which play a key role in the study of function spaces and in applications, with emphasis on their compactness properties.

8.7.1 Definition and Basic Properties

We start with the definition of composition operators. For convenience, when X, Y are metric spaces we shall denote their metrics by the same symbol d (as it is customary in the case of normed spaces).

Let X, Y be metric spaces, $g : Y \to X$ a Lipschitz mapping, $0 < \alpha \le 1$. The *composition operator* C_g is defined by

$$C_g(f) = f \circ g, \quad f \in \mathrm{Lip}_\alpha(X). \tag{8.7.1}$$

The following two propositions are concerned with the properties of C_g.

Proposition 8.7.1 *Let X, Y be metric spaces, $g : Y \to X$ a Lipschitz mapping, $0 < \alpha \le 1$ and C_g the composition operator defined by (8.7.1)*

1. *C_g is a continuous linear operator from $\mathrm{BLip}_\alpha(X)$ to $\mathrm{BLip}_\alpha(Y)$.*
2. *If X and Y are pointed metric spaces with base points θ and θ', respectively, and g preserves the base points $(g(\theta') = \theta)$, then C_g is a continuous linear operator from $\mathrm{Lip}_0(X, d^\alpha)$ to $\mathrm{Lip}_0(Y, d^\alpha)$ with norm $\|C_g\| = L(g)^\alpha$.*

The proof will be based on the following simple fact.

Lemma 8.7.2 *Let $A \subseteq \mathbb{R}_+$. For any $\beta > 0$ put $A^\beta = \{t^\beta : t \in A\}$. Then*

$$\sup(A^\beta) = (\sup A)^\beta. \tag{8.7.2}$$

If X, Y are metric spaces and $g : Y \to X$ is Lipschitz, then

$$L(g)^\beta = \sup \left\{ \frac{|g(y) - g(y')|^\beta}{d^\beta(y, y')} : y, y' \in Y, x \ne y \right\}. \tag{8.7.3}$$

Proof Since $t \le \sup A$ implies $t^\beta \le (\sup A)^\beta$, for all $t \in A$, it follows that $\sup(A^\beta) \le (\sup A)^\beta$.

Applying the above inequality with β, A replaced with $1/\beta$ and A^β, respectively, one obtains

$$\sup\left((A^\beta)^{1/\beta}\right) \le (\sup A^\beta)^{1/\beta} \iff (\sup A)^\beta \le \sup(A^\beta).$$

The equality (8.7.3) follows from (8.7.2) taking into account the definition (1.3.4) of the Lipschitz norm $L(g)$.   □

Proof of Proposition 8.7.1 We start with the proof of the second statement.

1. Recall that by $L_\alpha(f)$ we denote the Hölder norm of a Hölder function f (see (1.3.6)). We have

$$|f(g(y)) - f(g(y'))| \le L_\alpha(f) d^\alpha(g(y), g(y')) \le L_\alpha(f) L(g)^\alpha d^\alpha(y, y'),$$

for all $y, y' \in Y$, so that

$$L_\alpha(C_g(f)) \leq L(g)^\alpha L_\alpha(f),$$

for all $f \in \mathrm{Lip}_0(X, d^\alpha)$. Hence $C_g(f) \in \mathrm{Lip}_0(Y, d^\alpha)$, C_g is linear and continuous and

$$\|C_g\| \leq L(g)^\alpha.$$

For fixed $y \in Y$ define $f_y : X \to \mathbb{R}$ by $f_y(x) = d^\alpha(g(y), x) - d^\alpha(g(y), \theta)$, $x \in X$. Then $f_y \in \mathrm{Lip}_0(X, d^\alpha)$ with $L_\alpha(f_y) = 1$. For any $y' \in Y \setminus \{y\}$ we have

$$\|C_g\| \geq L_\alpha(C_g(f_y)) = L_\alpha(f_y \circ g) \geq \frac{|f_y(g(y')) - f_y(g(y))|}{d^\alpha(y, y')} = \frac{d^\alpha(g(y), g(y'))}{d^\alpha(y, y')}.$$

Since the inequality between the extreme terms holds for all distinct $y, y' \in Y$, taking into account the equality (8.7.3), one obtains $L(g)^\alpha \leq \|C_g\|$, and so $\|C_g\| = L(g)^\alpha$.
2. For any $f \in \mathrm{blip}(X)$, $\|f \circ g\|_\infty \leq \|f\|_\infty$, and, by the proof of the second point of the proposition, $L_\alpha(f \circ g) \leq L(g)^\alpha L_\alpha(f)$. It follows that $f \circ g \in \mathrm{BLip}_\alpha(Y)$ as well as the continuity of the linear operator $C_g : \mathrm{BLip}_\alpha(X) \to \mathrm{BLip}_\alpha(Y)$. □

As it was shown in [306] (see also [305]), a similar result holds in the case of the little Hölder spaces.

Proposition 8.7.3 *Let X, Y be metric spaces, $g : Y \to X$ a Lipschitz mapping, C_g the composition operator defined by (8.7.1) and $0 < \alpha \leq 1$. Then the following hold.*

1. *$f \circ g \in \mathrm{lip}_\alpha(Y)$ for every $f \in \mathrm{lip}_\alpha(X)$.*
2. *C_g is a continuous linear operator from $\mathrm{blip}_\alpha(X)$ to $\mathrm{blip}_\alpha(Y)$.*
3. *If X, Y are pointed metric spaces and g preserves the base points, then C_g is a continuous linear operator from $\mathrm{lip}_0(X, d^\alpha)$ to $\mathrm{lip}_0(Y, d^\alpha)$ with norm $\|C_g\| = L(g)^\alpha$.*

Proof

1. For $\varepsilon > 0$ let $\delta > 0$ be such that

$$|f(x) - f(x')| \leq \frac{\varepsilon}{L(g)^\alpha} d^\alpha(x, x'),$$

for all $x, x' \in X$ with $d(x, x') \leq \delta$. Then $d(g(y), g(y')) \leq \delta$, for $y, y' \in Y$ with $d(y, y') \leq \delta/L(g)$, so that

$$|f(g(y)) - f(g(y'))| \leq \frac{\varepsilon}{L(g)^\alpha} d^\alpha(g(y), g(y')) \leq \varepsilon d^\alpha(y, y'),$$

showing that $f \circ g \in \mathrm{lip}_\alpha(Y)$.

2. Let $f \in \mathrm{blip}_\alpha(X)$. By 1, $C_g f \in \mathrm{lip}_\alpha(Y)$ and the inequality $|f(g(y))| \le \|f\|_\infty$ shows that $C_g f$ is also bounded with $\|C_g f\|_\infty \le \|f\|_\infty$.

As in the proof of Proposition 8.7.1, one proves the inequality $L_\alpha(C_g(f)) \le L(g)^\alpha L_\alpha(f)$. Consequently, $C_g f \in \mathrm{blip}_\alpha(Y)$ and the operator $C_g : \mathrm{blip}_\alpha(X) \to \mathrm{blip}_\alpha(Y)$ is linear and continuous with $\|C_g\| \le (1 + L(g)^\alpha)\|f\|_\alpha$.

3. Let $f \in \mathrm{lip}_0(X, d^\alpha)$. By 1 and the fact that g preserves the base points, it follows that $f \circ g \in \mathrm{lip}_0(Y, d^\alpha)$. The inequality $L_\alpha(C_g f)) \le L(g)^\alpha L_\alpha(f)$ shows that C_g is continuous with

$$\|C_g\| \le L(g)^\alpha. \tag{8.7.4}$$

Suppose now that $0 < \alpha < 1$ and prove that the opposite inequality to (8.7.4) holds too. The function used in the proof of Proposition 8.7.1 to obtain the converse inequality does not belong to $\mathrm{lip}_0(X, d^\alpha)$, so we have to modify it adequately.

If g is constant, then $g \equiv \theta$ and C_g is the null operator. Supposing g not constant, let $y, y' \in Y$ be such that $g(y) \ne g(y')$. For a fixed β, $\alpha < \beta < 1$, define the function $h : X \to \mathbb{K}$ by

$$h(x) = \frac{d^\beta(x, g(y)) - d^\beta(x, g(y'))}{2d^{\beta-\alpha}(g(y), g(y'))}, \quad x \in X.$$

Let us show first that h is Hölder of order α with $L_\alpha(h) = 1$.
For $x \ne x'$ in X,

$$|h(x) - h(x')| = \frac{|d^\beta(x, g(y)) - d^\beta(x, g(y')) - d^\beta(x', g(y)) + d^\beta(x', g(y'))|}{2d^{\beta-\alpha}(g(y), g(y'))}.$$

But

$$|d^\beta(x, g(y)) - d^\beta(x, g(y')) - d^\beta(x', g(y)) + d^\beta(x', g(y'))|$$
$$\le |d^\beta(x, g(y)) - d^\beta(x', g(y))| + |d^\beta(x, g(y')) - d^\beta(x', g(y'))|$$
$$\le 2d^\beta(x, x'),$$

and, with another arrangement of the terms,

$$|d^\beta(x, g(y)) - d^\beta(x, g(y')) - d^\beta(x', g(y)) + d^\beta(x', g(y'))|$$
$$\le |d^\beta(x, g(y)) - d^\beta(x, g(y'))| + |d^\beta(x', g(y)) - d^\beta(x', g(y'))|$$
$$\le 2d^\beta(g(y), g(y')),$$

so that

$$|h(x) - h(x')| \le \frac{\min\{d^\beta(x, x'), d^\beta(g(y), g(y'))\}}{d^{\beta-\alpha}(g(y), g(y'))}.$$

If $d(x, x') \leq d(g(y), g(y'))$, then

$$\frac{|h(x) - h(x')|}{d^\alpha(x, x')} \leq \frac{\cdot d^\beta(x, x')}{d^\alpha(x, x')d^{\beta-\alpha}(g(y), g(y'))} = \frac{d^{\beta-\alpha}(x, x')}{d^{\beta-\alpha}(g(y), g(y'))} \leq 1.$$

$$(8.7.5)$$

If $d(x, x') \geq d(g(y), g(y'))$, then

$$\frac{|h(x) - h(x')|}{d^\alpha(x, x')} \leq \frac{d^\beta(g(y), g(y'))}{d^\alpha(x, x')d^{\beta-\alpha}(g(y), g(y'))} = \frac{d^\alpha(g(y), g(y'))}{d^\alpha(x, x')} \leq 1.$$

For $x = g(y)$ and $x' = g(y')$ one obtains

$$\frac{|h(g(y)) - h(g(y'))|}{d^\alpha(g(y), g(y'))} = \frac{2d^\beta(g(y), g(y'))}{2d^\alpha(g(y), g(y'))d^{\beta-\alpha}(g(y), g(y'))} = 1.$$

so that $L_\alpha(h) = 1$.
Since

$$\lim_{d(x,x')\to 0} \frac{d^{\beta-\alpha}(x, x')}{d^{\beta-\alpha}(g(y), g(y'))} = 0,$$

the inequalities (8.7.5) show that h belongs to lip(X, d^α).
But then the function $f(x) = h(x) - h(\theta)$, $x \in X$, belongs to lip$_0(X, d^\alpha)$ and $L_\alpha(f) = 1$. It follows that

$$\|C_g\| \geq L_\alpha(C_g(f)) = L_\alpha(f \circ g) \geq \frac{|f(g(y)) - f(g(y'))|}{d^\alpha(y, y')} = \frac{d^\alpha(g(y), g(y'))}{d^\alpha(y, y')}.$$

Taking the supremum, one obtains

$$\|C_g\| \geq \sup\left\{\frac{d^\alpha(g(y), g(y'))}{d^\alpha(y, y')} : y, y' \in Y, \ g(y) \neq g(y')\right\} = L(g)^\alpha.$$

$\square$

Other properties of the operator C_g are collected in the following proposition.

Proposition 8.7.4 ([675], Proposition 1.8.4) *Let X, Y be pointed metric spaces, $g : Y \to X$ a Lipschitz mapping preserving the base points and C_g the composition operator from* Lip$_0(X)$ *to* Lip$_0(Y)$. *Then the following assertions hold true.*

1. *C_g is surjective $\iff$ $g : Y \to g(Y)$ is bi-Lipschitz.*
2. *C_g is injective $\iff$ $g(Y)$ is dense in X.*
3. *C_g is bijective $\iff$ g is a bi-Lipschitz map from Y onto X.*

By an *algebra homomorphism* between two Banach algebras A, B we mean a continuous linear mapping $T : A \to B$ such that $T(ab) = T(a)T(b)$ for all $a, b \in A$. We shall work with the sum-norm $\|f\|_{\text{sum}} = L(f) + \|f\|_\infty$ (see (8.1.3)). The composition operator acts on the Banach algebras BLip and is an algebra homomorphism.

Theorem 8.7.5 *Let X, Y be metric spaces and $g : Y \to X$ a Lipschitz mapping.*

1. *The operator C_g given by (8.7.1) is an algebra homomorphism between the Banach algebras $A = \text{BLip}(X)$ and $B = \text{BLip}(Y)$ endowed with the norms $\|\cdot\|_{\text{sum}}$.*

   Conversely, if the metric spaces X, Y are compact, then any algebra homomorphism $T : A \to B$ is of the form $T = C_g$ for some Lipschitz mapping $g : Y \to X$.

2. *If Y is compact, then the operator $C_g : A \to B$ is compact if and only if g further satisfies the condition*

$$\lim_{d(y,y')\to 0} \frac{d(g(y), g(y'))}{d(y, y')} = 0. \qquad (8.7.6)$$

Proof The inequalities

$$\|f \circ g\|_\infty \le \|f\|_\infty \quad \text{and} \quad L(f \circ g) \le L(f)L(g) \qquad (8.7.7)$$

show that $C_g(f) \in \text{BLip}(Y)$ for every $f \in \text{BLip}(X)$, that is, C_g is well-defined.

It is easy to check that the operator C_g is an algebra homomorphism from A to B. Since

$$\|C_g(f)\|_{\text{sum}} = \|f \circ g\|_\infty + L(f \circ g) \le \|f\|_\infty + L(f)L(g)$$
$$\le (1 + L(g))\|f\|_{\text{sum}},$$

for all $f \in \text{BLip}(X)$, it follows that C_g is continuous and $\|C_g\| \le 1 + L(g)$.

For of the proof of the rest of the assertions from 1 we refer to Sherbert [631] (see also [632]) and for the proof of 2 to Kamowitz and Scheinberg [336] (in the case $Y = X$), but an extension of Kamowitz and Scheinberg's result on the compactness of the composition operator will be given in Theorem 8.7.12. □

A result analogous to that contained in Theorem 8.7.5 holds in the case of little Lipschitz spaces too.

Theorem 8.7.6 ([675], Corollary 4.5.6) *Let X, Y be compact pointed metric spaces such that $\text{lip}_0(X)$ separates points uniformly. Then any algebra homomorphism from $\text{lip}_0(X)$ to $\text{lip}_0(Y)$ is the restriction of C_g to $\text{lip}_0(X)$, for some base point preserving Lipschitz map $g : Y \to X$.*

For the uniform separation property, see Definition 8.3.11.

8.7.2 Compactness of the Composition Operators

In this subsection we prove, following [307], a compactness criterion for the composition operator, improving the result of Kamowitz and Scheinberg [336] (see Theorem 8.7.5). Compact endomorphisms on some Lipschitz algebras of analytic functions were studied in [72, 423, 424] and on Lipschitz algebras of differentiable functions in [421, 422].

If X, Y are metric spaces with base points θ and θ', respectively, we agree to denote by $\mathrm{Lip}_0(X, Y)$ the set of all Lipschitz functions $g : X \to Y$ such that $g(\theta) = \theta'$. Then $\mathrm{Lip}_0(X, X)$ is the set of all Lipschitz functions $g : X \to X$ such that $g(\theta) = \theta$.

Let X be a pointed metric space. Recall that, for a Lipschitz mapping $g \in \mathrm{Lip}_0(Y, X)$, the composition operator $C_g : \mathrm{Lip}_0(X) \to \mathrm{Lip}_0(Y)$ is given by

$$C_g(f) = f \circ g, \quad f \in \mathrm{Lip}_0(X).$$

The following simple result will be used several times in the proofs. The conjugate $C_g^* : \mathrm{Lip}_0(Y)^* \to \mathrm{Lip}_0(X)^*$ of the composition operator satisfies the following equality

$$C_g^*(e_y) = e_{g(y)}, \quad \text{for all} \quad y \in Y, \tag{8.7.8}$$

where e_y is the evaluation functional given by (8.2.6) (see also Proposition 8.2.6).

Indeed, for every $f \in \mathrm{Lip}_0(X)$,

$$C_g^*(e_y)(f) = e_y(C_g f) = e_y(f \circ g) = f(g(y)) = e_{g(y)}(f).$$

The mapping $g : Y \to X$ is called *supercontractive* if it satisfies the condition (8.7.6) or, equivalently, for every $\varepsilon > 0$ there exists $\delta > 0$ such that, for all $y, y' \in Y$,

$$d(y, y') \leq \delta \implies d(g(y), g(y')) \leq \varepsilon d(y, y'). \tag{8.7.9}$$

We say also that g is a *supercontraction*.

Remark 8.7.7 A Lipschitz function $g : X \to \mathbb{K}$ is supercontractive if and only if it is little Lipschitz (see (8.3.2)).

Theorem 8.7.8 *Let X, Y be pointed metric spaces, with X separable and Y bounded, and $g \in \mathrm{Lip}_0(Y, X)$. Then the composition operator $C_g : \mathrm{Lip}_0(X) \to \mathrm{Lip}_0(Y)$ is compact if and only if the mapping g is supercontractive and $g(Y)$ is totally bounded.*

The proof will be based on the following result.

Proposition 8.7.9 *Let X, Y be pointed metric spaces with X separable and $g \in \mathrm{Lip}_0(Y, X)$. The composition operator $C_g : \mathrm{Lip}_0(X) \to \mathrm{Lip}_0(Y)$ is compact if and* *only if every bounded sequence* (f_n) *in* $\mathrm{Lip}_0(X)$, *converging pointwise to* 0, *contains a subsequence* (f_{n_k}) *such that* $L(f_{n_k} \circ g) \to 0$.

Proof Suppose that the operator C_g is compact. Let (f_n) be a bounded sequence in $\mathrm{Lip}_0(X)$ such that $f_n(x) \to 0$ for every $x \in X$. The compactness of C_g implies the existence of a subsequence (f_{n_k}) of (f_n) and of $\varphi \in \mathrm{Lip}_0(Y)$ such that

$$L(f_{n_k} \circ g - \varphi) \to 0 \ \text{ as } \ k \to \infty.$$

It follows that $f_{n_k}(g(y)) \to \varphi(y)$ for every $y \in Y$. By hypothesis, $f_{n_k}(g(y)) \to 0$ for every $y \in Y$, so that $\varphi \equiv 0$ and $L(f_{n_k} \circ g) \to 0$.

Conversely, let (f_n) be a sequence in the closed unit ball of $\mathrm{Lip}_0(X) = \mathbb{F}(X)^*$. By Theorem 8.2.10.3, the space $\mathbb{F}(X)$ is separable, so that the restriction of the w^*-topology to the closed unit ball of $\mathbb{F}(X)^*$ is metrizable. Since this ball is w^*-compact, the sequence (f_n) contains a subsequence (f_{n_k}) w^*-convergent to some $f \in \mathrm{Lip}_0(X)$. But the w^*-topology of $\mathrm{Lip}_0(X)$ agrees on bounded sets with the topology of pointwise convergence (see Theorem 8.2.4.3), hence $f_{n_k}(x) - f(x) \to 0$, for every $x \in X$. By hypothesis, there exists a further subsequence $(f_{n_{k_i}})$ such that

$$L((f_{n_{k_i}} - f) \circ g) \to 0 \ \text{ as } \ i \to \infty.$$

Since

$$(f_{n_{k_i}} - f) \circ g = f_{n_{k_i}} \circ g - f \circ g = C_g f_{n_{k_i}} - C_g f,$$

this proves the compactness of the operator C_g. $\square$

Proof of Theorem 8.7.8 Supposing that $g \in \mathrm{Lip}_0(Y, X)$ is supercontractive and that $g(Y)$ is totally bounded, let us prove that the operator C_g is compact.

Let (f_n) be a sequence in the closed unit ball of $\mathrm{Lip}_0(X)$ that converges pointwise to 0. If we show that $L(f_n \circ g) \to 0$, then, by Proposition 8.7.9, the operator C_g will be compact.

Given $\varepsilon > 0$ choose $\delta > 0$ according to (8.7.9). The total boundedness of $g(Y)$ implies the existence of $y_1, \ldots, y_m \in X$ such that

$$\forall y \in Y, \ \exists i \in \{1, \ldots, m\} \ \text{ such that } \ d(g(y), g(y_i)) \le \delta\varepsilon. \tag{8.7.10}$$

The pointwise convergence of the sequence (f_n) implies the existence of $n_0 \in \mathbb{N}$ such that

$$|f_n(g(y_i))| \le \delta\varepsilon, \quad i = 1, 2, \ldots, m, \tag{8.7.11}$$

for all $n \ge n_0$.

Let $n \geq n_0$ and $y, y' \in X$. If $0 < d(y, y') < \delta$, then

$$\frac{|f_n(g(y)) - f_n(g(y'))|}{d(y, y')} \leq \frac{d(g(y), g(y'))}{d(y, y')} \leq \varepsilon. \tag{8.7.12}$$

If $d(y, y') \geq \delta$, choose $1 \leq i, j \leq m$ satisfying (8.7.10) for y and y', respectively. Taking into account (8.7.11), one obtains

$$\frac{|f_n(g(y)) - f_n(g(y'))|}{d(y, y')}$$

$$\leq \frac{|f_n(g(y)) - f_n(g(y_i))| + |f_n(g(y_i)) - f_n(g(y_j))| + |f_n(g(y_j)) - f_n(g(y'))|}{\delta}$$

$$\tag{8.7.13}$$

$$\leq \frac{d(g(y), g(y_i)) + 2\delta\varepsilon + d(g(y_j), g(y'))}{\delta}$$

$$\leq \frac{4\delta\varepsilon}{\delta} = 4\varepsilon.$$

The inequalities (8.7.12), (8.7.13) imply $L(f_n \circ g) \leq 4\varepsilon$ for all $n \geq n_0$, proving the convergence of the sequence $(L(f_n \circ g))$ to 0.

The reverse implication will be proved by contradiction.

Suppose first that the set $g(Y)$ is not totally bounded. Then there exist $\varepsilon > 0$ and a sequence (y_n) in Y such that $d(g(y_n), g(y_m)) \geq \varepsilon$ for all $n, m \in \mathbb{N}$ with $n \neq m$. Since

$$d(g(y_n), g(y_m)) = \|e_{g(y_n)} - e_{g(y_m)}\|, \ e_{g(y_n)} = C_g^*(e_{y_n}) \quad \text{and}$$

$$\|e_{y_n}\| = d(\theta', y_n) \leq \text{diam } Y,$$

(see Proposition 8.2.6), it follows that (e_{y_n}) is a bounded sequence in $\text{Lip}_0(Y)^*$ such that $(C_g^*(e_{y_n}))$ does not contain convergent subsequences. This implies that the conjugate operator $C_g^* : \text{Lip}_0(Y)^* \to \text{Lip}_0(X)^*$ is not compact, and so, by Schauder's theorem (Theorem 8.6.3), nor is the operator C_g.

Suppose now that the function g is not supercontractive. Then there exist $\varepsilon > 0$ and two sequences $(y_n), (y_n')$ in Y such that

$$0 < d(y_n, y_n') < \frac{1}{n} \quad \text{and} \quad \frac{d(g(y_n), g(y_n'))}{d(y_n, y_n')} \geq \varepsilon,$$

for all $n \in \mathbb{N}$.

Consider the functions

$$f_n(x) = \frac{1}{n} \left[\exp(-nd(\theta, g(y_n))) - \exp(-nd(x, g(y_n))) \right], \quad x \in X, \ n \in \mathbb{N}.$$

Let $x, x' \in X$, $x \neq x'$. By the Mean Value Theorem (MVT) applied to the function $h(t) = \exp(-nt)$, there exists a number ξ_n between $d(x, g(y_n))$ and $d(x', g(y_n))$ such that

$$\frac{|f_n(x) - f_n(x')|}{d(x, x')} = \frac{|\exp(-nd(x', g(y_n))) - \exp(-nd(x, g(y_n)))|}{nd(x, x')}$$

$$= \exp(-n\xi_n) \cdot \frac{|d(x', g(y_n)) - d(x, g(y_n))|}{d(x, x')} < 1,$$

because $|d(x', g(y_n)) - d(x, g(y_n))| \leq d(x, x')$ and $\exp(-n\xi_n) < 1$. It follows that $L(f_n) \leq 1$ for all $n \in \mathbb{N}$.

By the definition of f_n and the inequality $\exp(-nt) \leq 1$ for $t \geq 0$ we have $|f_n(x)| \leq 2/n$ for all $x \in X$, which implies that the sequence (f_n) converges to 0, uniformly on X.

Again, by MVT, there exists η_n between 0 and $d(g(y_n), g(y'_n))$ such that

$$L(f_n \circ g) \geq \frac{|f_n(g(y_n)) - f_n(g(y'_n))|}{d(y_n, y'_n)} = \frac{|1 - \exp(-nd(g(y_n), g(y'_n)))|}{nd(y_n, y'_n)}$$

$$= \exp(-n\eta_n) \cdot \frac{d(g(y_n), g(y'_n))}{d(y_n, y'_n)}.$$

The inequalities

$$\eta_n < d(g(y_n), g(y'_n)) \leq L(g)d(y_n, y'_n) < \frac{1}{n}L(g),$$

imply $\exp(-n\eta_n) \geq \exp(-L(g))$ and so

$$L(f_n \circ g) \geq \exp(-L(g))\,\varepsilon$$

for all $n \in \mathbb{N}$. It follows that the sequence $(L(f_n \circ g))$ does not contain any subsequence converging to 0, so that, by Proposition 8.7.9, the operator C_g is not compact.   □

Remark 8.7.10 In [307] the compactness criterion for a composition operator acting on $\mathrm{Lip}_0(X)$ is formulated for an arbitrary metric space X. But in the proof (see the second paragraph of the proof of Proposition 8.7.9) one uses the fact that a sequence (f_n) in the closed unit ball of $\mathrm{Lip}_0(X) = \mathbb{F}(X)^*$ admits a w^*-convergent subsequence, a fact that is not true in general. For a study of Banach spaces having this property, see, for instance, Chapter XIII in [193].

The case of the Banach space $\mathrm{BLip}(X)$ of bounded Lipschitz functions with the norm $\|f\|_{\max} = \max\{L(f), \|f\|_\infty\}$ can be reduced to the Lip_0 case by appealing to Theorem 8.1.12 and using the following result.

Proposition 8.7.11 *Let X, Y be metric spaces with diameters at most 2 and $X^\bullet, Y^\bullet$ the spaces obtained from X, Y by attaching the ideal points θ, θ' and defining $d(\theta, x) = 1 = d(\theta', y)$ for all $x \in X$ and $y \in Y$. For $g \in \mathrm{Lip}(Y, X)$ let $\overline{g} : Y^\bullet \to X^\bullet$ be given by $\overline{g}(y) = g(y)$ for $y \in Y$ and $g(\theta') = \theta$. Then the following results hold.*

1. The mapping $\overline{g}$ is Lipschitz, i.e., $\overline{g} \in \mathrm{Lip}_0(Y^\bullet, X^\bullet)$.
2. $\overline{g}$ is supercontractive if and only if g is supercontractive.
3. $\overline{g}(Y^\bullet)$ is totally bounded if and only if $g(Y)$ is totally bounded.

Proof

1. Since, for $y, y' \in Y$,

$$d(\overline{g}(y), \theta) = 1 = d(y, \theta'),$$

and

$$d(\overline{g}(y), \overline{g}(y')) = d(g(y), g(y')) \leq L(g)d(y, y'),$$

it follows that $L(\overline{g}) \leq \max\{L(g), 1\}$.
2. In the definition (8.7.9) we can suppose $0 < \delta < 1$. Since $d(y, \theta) = 1 > \delta$, it follows that there is no requirement on $\overline{g}$ for pairs of points of the form (y, θ'), $y \in Y$, proving the equivalence from 2.
3. If, for some $\varepsilon > 0$, $g(y_1), \ldots, g(y_m)$ is an ε-net for $g(Y)$, then $g(y_1), \ldots, g(y_m), \theta$ is an ε-net for $\overline{g}(Y^\bullet)$.
   Conversely, suppose that $g(y_1), \ldots, g(y_m), \theta$ is an ε-net for $\overline{g}(Y^\bullet)$, where $0 < \varepsilon < 1$. Since $B(\theta, \varepsilon) = \{\theta\}$, it follows that $g(y_1), \ldots, g(y_m)$ is an ε-net for $g(Y)$. □

The compactness result for the composition operator on the space $\mathrm{BLip}(X)$ is the following.

Theorem 8.7.12 *Let X, Y be metric spaces with diameters at most 2, with X separable, and $g \in \mathrm{Lip}(Y, X)$. Then the composition operator $C_g : \mathrm{BLip}(X) \to \mathrm{BLip}(Y)$ is compact if and only if g is supercontractive and $g(Y)$ is totally bounded.*

Proof Let $X^\bullet, Y^\bullet$ be the spaces obtained from X, Y by attaching the ideal points θ, θ', respectively, and let $T_1 : \mathrm{BLip}(X) \to \mathrm{Lip}_0(X^\bullet)$ and $T_2 : \mathrm{BLip}(Y) \to \mathrm{Lip}_0(Y^\bullet)$ be the isometric isomorphisms given by Theorem 8.1.12.2. Let $\overline{g} \in \mathrm{Lip}_0(Y^\bullet, X^\bullet)$ be the function defined in Proposition 8.7.11 and $C_{\overline{g}} : \mathrm{Lip}_0(X^\bullet) \to \mathrm{Lip}_0(Y^\bullet)$ the corresponding composition operator.

Since

$$T_2 C_g = C_{\overline{g}} T_1,$$

i.e., $C_g = T_2^{-1} C_{\bar{g}} T_1$, it follows that the compactness of C_g is equivalent to the compactness of $C_{\bar{g}}$. The characterization result follows now by Theorem 8.7.8 and Proposition 8.7.11.   □

The composition operator acts on little Lipschitz spaces too. We start by the following simple result. Let X, Y be metric spaces. We shall denote by $\mathscr{C}_g$ the restriction of composition operator C_g to little Lipschitz spaces:

$$\mathscr{C}_g = C_g|_{\mathrm{lip}_0(X)} \quad \text{or} \quad \mathscr{C}_g = C_g|_{\mathrm{blip}(X)}.$$

By Proposition 8.7.3, $\mathscr{C}_g$ is a continuous linear operator from $\mathrm{blip}(X)$ to $\mathrm{blip}(Y)$ and from $\mathrm{lip}_0(X)$ to $\mathrm{lip}_0(Y)$ as well.

The compactness criteria proved in the case of the Lip spaces hold in the case of little Lipschitz spaces too.

Theorem 8.7.13 *Let X, Y be pointed metric spaces.*

1. *If X is separable and Y is bounded, $g \in \mathrm{Lip}_0(Y, X)$ is supercontractive and $g(Y)$ is totally bounded, then the composition operator $\mathscr{C}_g : \mathrm{lip}_0(X) \to \mathrm{lip}_0(Y)$ is compact.*
2. *Conversely, if Y is bounded, $\mathrm{lip}_0(X)$ separates the points uniformly and $g \in \mathrm{Lip}_0(Y, X)$ is such that the composition operator $\mathscr{C}_g$ is compact, then the set $g(Y)$ is totally bounded and the mapping g is supercontractive.*

Similar results hold for the spaces $\mathrm{blip}(X)$ and $\mathrm{blip}(Y)$.

Proof

1. By Proposition 8.3.9, $\mathrm{lip}_0(X)$ is a closed subspace of $\mathrm{Lip}_0(X)$. By Theorem 8.7.8, the operator C_g is compact, and so will be $C_g \circ J : \mathrm{lip}_0(X) \to \mathrm{Lip}_0(Y)$, where J is the embedding of $\mathrm{lip}_0(X)$ into $\mathrm{Lip}_0(X)$. Since $(C_g \circ J)(\mathrm{lip}_0(X)) \subseteq \mathrm{lip}_0(Y)$, it follows that the operator $\mathscr{C}_g$ is compact too.
2. We prove now the converse. The separation hypothesis yields a number $a > 1$ such that for every $y, y' \in X$ with $g(y) \neq g(y')$ there exists $f_{y,y'} \in \mathrm{lip}_0(X)$ satisfying the conditions

$$L(f_{y,y'}) \leq a \quad \text{and} \quad |f_{y,y'}(g(y)) - f_{y,y'}(g(y'))| = d(g(y), g(y')). \tag{8.7.14}$$

In what follows, for $x \in X$ and $y \in Y$ we shall denote by $\tilde{e}_x$ and $\tilde{e}_y$ the restrictions of the evaluation functionals e_x and e_y to $\mathrm{lip}_0(X)$ and $\mathrm{lip}_0(Y)$, respectively. Supposing that $g(Y)$ is not totally bounded there exist $\varepsilon > 0$ and a sequence (y_n) in Y such that $d(g(y_n), g(y_m)) \geq \varepsilon$ for all $n, m \in \mathbb{N}$ with $n \neq m$. For $n \neq m$ let $f_{n,m} \in \mathrm{lip}_0(X)$ satisfying (8.7.14) for $y = y_n$ and $y' = y_m$. Then

$$\varepsilon \leq d(g(y_n), g(y_m)) = |f_{n,m}(g(y_n)) - f_{n,m}(g(y_m))| \leq L(f_{n,m}) \|\tilde{e}_{g(y_n)} - \tilde{e}_{g(y_m)}\|$$

$$\leq a \|\tilde{e}_{g(y_n)} - \tilde{e}_{g(y_m)}\| = a \|(\mathscr{C}_g^*)(\tilde{e}_{y_n}) - (\mathscr{C}_g^*)(\tilde{e}_{y_m})\|,$$

where $\mathscr{C}_g^* : \mathrm{lip}_0(Y)^* \to \mathrm{lip}_0(X)^*$ is the operator conjugate to $\mathscr{C}_g$.

This implies that the sequence $\left(\left(\mathscr{C}_g^*\right)(\tilde{e}_{y_n})\right)_{n\in\mathbb{N}}$ has no convergent subsequences (with respect to the conjugate norm). Since, by Proposition 8.2.6, $\|\tilde{e}_{y_n}\| \le \|e_{y_n}\| = d(y_n, \theta') \le \text{diam } Y$, it follows that the conjugate operator $\mathscr{C}_g^*$ is not compact and so, by Schauder's theorem (Theorem 8.6.3), nor is the operator $\mathscr{C}_g$.

To prove that g is supercontractive let $\varepsilon > 0$ be given and denote by B the closed unit ball of $\text{lip}_0(X)$. Since the set $\mathscr{C}_g(B)$ is relatively compact it is totally bounded, so there exist $f_1, \ldots, f_m \in B$ such that

$$\forall f \in B, \ \exists k \in \{1, 2\ldots, m\} \text{ such that } L(f \circ g - f_k \circ g) \le \varepsilon. \quad (8.7.15)$$

Since $f_k \circ g \in \text{lip}_0(Y)$, there exists $\delta > 0$ such that

$$d(y, y') \le \delta \ \Rightarrow \ |f_k(g(y)) - f_k(g(y'))| \le \varepsilon d(y, y'), \quad k = 1, 2, \ldots, m, \quad (8.7.16)$$

for all $y, y' \in Y$.

For $y, y' \in Y$ such that $d(y, y') \le \delta$ and $g(y) \ne g(y')$ choose $f_{y,y'} \in \text{lip}_0(X)$ according to (8.7.14). For $a^{-1}f_{y,y'} \in B$ pick $k \in \{1, \ldots, m\}$ according to (8.7.15). Taking into account (8.7.16), one obtains

$$a^{-1}d(g(y), g(y')) = a^{-1}|f_{y,y'}(g(y)) - f_{y,y'}(g(y'))|$$
$$\le \left|(a^{-1}f_{y,y'} - f_k)(g(y)) - (a^{-1}f_{y,y'} - f_k)(g(y'))\right|$$
$$+ |f_k(g(y)) - f_k(g(y'))| \le 2\varepsilon d(y, y').$$

It follows that $d(g(y), g(y')) \le 2a\varepsilon d(y, y')$, for all $y, y' \in Y$ with $d(y, y') \le \delta$, showing that g is supercontractive.

If $X, Y \in \mathscr{M}_2$ (i.e., they have diameters at most 2), then the case of the spaces blip can be reduced to that of the spaces lip_0, as in the proof of Theorem 8.7.12. Indeed, consider the pointed metric spaces $X^\bullet, Y^\bullet$ corresponding to X, Y as in Theorem 8.1.12. By Remark 8.7.7, a function h from a metric space Z to $\mathbb{K}$ is little Lipschitz if and only if it is supercontractive, so that, by Proposition 8.7.11.2, $\tilde{h} \in \text{lip}_0(Z^\bullet)$ for every $h \in \text{blip}(Z)$. It follows that the correspondences given by Theorem 8.1.12.2 are isometric isomorphisms between the spaces $\text{blip}(X)$ and $\text{lip}_0(X^\bullet)$ and $\text{blip}(Y)$ and $\text{lip}_0(Y^\bullet)$. Reasoning as in the proof of Theorem 8.7.12, the case of the spaces blip reduces to that of the spaces lip_0. $\quad\square$

8.7.3 Weakly Compact Composition Operators

In this subsection we prove, following [304], that every weakly compact composition operator on a Lipschitz space is compact.

Let X be a metric space with base point θ. For a base point-preserving Lipschitz map $g : X \to X$ we shall denote by C_g the composition operator on $\mathrm{Lip}_0(X)$ (see (8.7.1)) and by $\mathscr{C}_g$ its restriction to the closed subspace $\mathrm{lip}_0(X)$.

Theorem 8.7.14 *Let X be pointed compact metric space such that $\mathrm{lip}_0(X)$ separates points uniformly and $g : X \to X$ a base point-preserving Lipschitz map. If the composition operator $\mathscr{C}_g : \mathrm{lip}_0(X) \to \mathrm{lip}_0(X)$ is weakly compact, then it is compact.*

A similar result holds in the case of Lip spaces.

Theorem 8.7.15 *Let X be a pointed compact metric space such that $\mathrm{lip}_0(X)$ separates points uniformly and $g : X \to X$ a base point-preserving Lipschitz map. If the composition operator $C_g : \mathrm{Lip}_0(X) \to \mathrm{Lip}_0(X)$ is weakly compact, then it is compact.*

Proof We prove Theorem 8.7.15 assuming that Theorem 8.7.14 holds.

The space $\mathrm{Lip}_0(X)$ is isometrically isomorphic to the space $\mathrm{lip}_0(X)^{**}$, the isomorphism $\Psi : \mathrm{Lip}_0(X) \to \mathrm{lip}_0(X)^{**}$ being determined by the condition

$$\Psi(f)(e_x) = f(x), \quad x \in X, \ f \in \mathrm{Lip}_0(X), \tag{8.7.17}$$

where e_x denotes the evaluation functional acting on $\mathrm{lip}_0(X)$.

Its inverse Ψ^{-1} satisfies

$$\Psi^{-1}(\varphi)(x) = \varphi(e_x), \quad x \in X, \ \varphi \in \mathrm{lip}_0(X)^{**}, \tag{8.7.18}$$

(see Theorem 8.3.14).

Also $\mathscr{C}_g^* : \mathrm{lip}_0(X)^* \to \mathrm{lip}_0(X)^*$ satisfies

$$\mathscr{C}_g^*(e_x) = e_{g(x)}, \quad x \in X. \tag{8.7.19}$$

The proof is similar to the proof of the corresponding equality in the case of the space $\mathrm{Lip}_0(X)$, see (8.7.8).

If we show that

$$C_g = \Psi^{-1}\mathscr{C}_g^{**}\Psi \quad (\Longleftrightarrow \ \mathscr{C}_g^{**} = \Psi C_g \Psi^{-1}), \tag{8.7.20}$$

then the weak compactness of C_g is equivalent to the weak compactness of $\mathscr{C}_g^{**}$. Applying twice Gantmacher's theorem (Theorem 8.6.12), one obtains, successively, the weak compactness of $\mathscr{C}_g^*$ and of $\mathscr{C}_g$. By Theorem 8.7.14 this implies the compactness of $\mathscr{C}_g$, which, by Theorem 8.7.13, is equivalent to the fact that g is supercontractive. By Theorem 8.7.8 this yields the compactness of the operator C_g.

Let us prove (8.7.20). For $f \in \mathrm{Lip}_0(X)$,

$$\Psi^{-1}\mathscr{C}_g^{**}\Psi(f)(x) = \mathscr{C}_g^{**}\Psi(f)(e_x) = \Psi(f)(\mathscr{C}_g^* e_x)$$

$$= \Psi(f)(e_{g(x)}) = f(g(x)) = C_g(f)(x),$$

for all $x \in X$.

Since the linear combinations of evaluation functionals are dense in $\mathbb{F}(X) \cong \mathrm{lip}_0(X)^*$, the equality (8.7.20) follows. □

The proof of Theorem 8.7.14 is more involved and needs some auxiliary results.

Lemma 8.7.16 *Let X be a pointed compact metric space and $g : X \to X$ a continuous mapping. If g is not supercontractive, then there exist a real number $\varepsilon > 0$, two sequences $(x_n), (y_n)$ in X converging to a point $x_0 \in X$ such that $d(g(x_n), g(y_n)) > \varepsilon d(x_n, y_n) > 0$, for all $n \in \mathbb{N}$, and a function $f \in \mathrm{Lip}_0(X)$ such that $f(g(x_n)) = d(g(x_n), g(y_n))$ and $f(g(y_n)) = 0$ for all $n \in \mathbb{N}$.*

Proof Supposing the mapping g not supercontractive (see (8.7.9)) we prove first the existence of $\varepsilon > 0$, of two sequences $(x_n), (y_n)$ in X and of numbers $r_n > 0$ such that

(i) $x_n \to x_0$ and $y_n \to x_0$ for some $x_0 \in X$;

(ii) $d(g(x_n), g(y_n)) > \varepsilon d(x_n, y_n) > 0$ for all $n \in \mathbb{N}$;

(iii) $B(g(x_n), r_n) \cap B(g(x_{n'}), r_{n'}) = \emptyset$ for $n \neq n'$,   and   (8.7.21)

$$g(y_n) \notin \bigcup_{j=1}^{\infty} B(g(x_j), r_j) \quad \text{for all } n \in \mathbb{N}.$$

Since the mapping g is not supercontractive, there exist $\varepsilon > 0$ and two sequences $(u_n), (v_n)$ in X such that

$$0 < d(u_n, v_n) < \frac{1}{n} \quad \text{and} \quad d(g(u_n), g(v_n)) > \varepsilon d(u_n, v_n), \quad (8.7.22)$$

for all $n \in \mathbb{N}$.

The compactness of X implies the existence of a subsequence (u_{n_k}) of (u_n) converging to some $x_0 \in X$. Since $d(u_n, v_n) \to 0$ as $n \to \infty$, it follows that $v_{n_k} \to x_0$. Hence, we can suppose, without restricting the generality, that the sequences $(u_n), (v_n)$ satisfy (8.7.22) and further

$$u_n \to x_0 \quad \text{and} \quad v_n \to x_0,$$

for some $x_0 \in X$.

The continuity of the mapping g implies

$$g(u_n) \to g(x_0) \quad \text{and} \quad g(v_n) \to g(x_0). \tag{8.7.23}$$

To obtain the existence of the sequence of disjoint open balls satisfying the conditions (iii) from (8.7.21), consider the sets

$$A = \{n \in \mathbb{N} : g(u_n) = g(x_0)\} \quad \text{and} \quad B = \{n \in \mathbb{N} : g(v_n) = g(x_0)\}.$$

We have to distinguish two cases.

Case I.   *One of the sets A, B is infinite.*

Suppose that this set is B and, for simplicity, suppose that $g(v_n) = g(x_0)$ for all $n \in \mathbb{N}$, i.e., $B = \mathbb{N}$. Since $g(u_n) \neq g(v_n)$, it follows $g(u_n) \neq g(x_0)$ for all $n \in \mathbb{N}$. Put $m_1 = 1, x_1 = u_{m_1}$ and

$$r_1 = \frac{1}{3} d(g(u_{m_1}), g(x_0)).$$

By (8.7.23) there exists $m_2 > m_1$ such that $x_2 = u_{m_2}$ satisfies the condition

$$d(g(x_2), g(x_0)) < r_1.$$

Continuing by induction one obtains the natural numbers $1 = m_1 < m_2 < \dots$ such that the sequence $x_k = u_{m_k}$ satisfies the condition

$$d(g(x_{k+1}), g(x_0)) < r_k,$$

where

$$r_k = \frac{1}{3} d(g(x_k), g(x_0)),$$

for all $k \in \mathbb{N}$.

Put $y_k = v_{m_k}$, $k \in \mathbb{N}$, and show that the sequences (x_k), (y_k) and the numbers $r_k > 0$ satisfy the conditions (8.7.21).

Indeed, let $z_k = g(x_k)$, $k = 0, 1, 2, \dots$. We have to show that

$$B(z_k, r_k) \cap B(z_{k+j}, r_{k+j}) = \emptyset,$$

for all $k, j \in \mathbb{N}$.

If some w belongs to this intersection, then

$$d(z_k, z_{k+j}) \le d(z_k, w) + d(z_{k+j}, w)$$

$$< r_k + r_{k+j} = \frac{1}{3}\left(d(z_k, z_0) + d(z_{k+j}, z_0)\right)$$

$$< \left(\frac{1}{3} + \frac{1}{3^{j+1}}\right) d(z_k, z_0).$$

On the other hand,

$$d(z_k, z_{k+j}) \ge d(z_k, z_0) - d(z_{k+j}, z_0)$$

$$> \left(1 - \frac{1}{3^j}\right) d(z_k, z_0),$$

leading to the contradiction

$$1 - \frac{1}{3^j} < \frac{1}{3} + \frac{1}{3^{j+1}} \iff 1 < \frac{2}{3^j}.$$

It is obvious that $g(y_k) = g(x_0) \notin B(x_k, r_k)$ for all $k \in \mathbb{N}$. Furthermore,

$$d(g(x_k), g(y_k)) = d(g(x_k), g(x_0)) = 3r_k \quad \text{for all } k \in \mathbb{N}.$$

Case II. *The sets A, B are both finite.*

Again, for simplicity, suppose that $A = B = \emptyset$, i.e., $g(u_n) \ne g(x_0) \ne g(v_n)$ for all $n \in \mathbb{N}$. Let $z_n = g(u_n)$, $w_n = g(v_n)$, $n \in \mathbb{N}$, and $z_0 = g(x_0)$.

Take $m_1 = 1$ and let

$$r_1 = \frac{1}{3} \min\{d(z_{m_1}, w_{m_1}), d(z_{m_1}, z_0), d(w_{m_1}, z_0)\}.$$

Let $m_2 > m_1$ be such that

$$\max\{d(z_{m_2}, w_{m_2}), d(z_{m_2}, z_0), d(w_{m_2}, z_0)\} < r_1$$

and put

$$r_2 = \frac{1}{3} \min\{d(z_{m_2}, w_{m_2}), d(z_{m_2}, z_0), d(w_{m_2}, z_0)\}.$$

By induction, choose $m_{k+1} > m_k$ such that

$$\max\{d(z_{m_{k+1}}, w_{m_{k+1}}), d(z_{m_{k+1}}, z_0), d(w_{m_{k+1}}, z_0)\} < r_k,$$

where

$$r_k = \frac{1}{3} \min\{d(z_{m_k}, w_{m_k}), d(z_{m_k}, z_0), d(w_{m_k}, z_0)\},$$

and put

$$r_{k+1} = \frac{1}{3} \min\{d(z_{m_{k+1}}, w_{m_{k+1}}), d(z_{m_{k+1}}, z_0), d(w_{m_{k+1}}, z_0)\}.$$

Note that $r_{k+1} < r_k/3$. Now we define for $k \in \mathbb{N}$, $x_k = u_{m_k}$ and $y_k = v_{m_k}$. The sequences (x_k), (y_k) and the numbers $r_k > 0$ satisfy the conditions (8.7.21).

To see this we only need to check that (8.7.21).(iii). The condition $B(z_{m_k}, r_k) \cap B(z_{m_{k+j}}, r_{k+j}) = \emptyset$ holds because these balls are smaller (in the sense of inclusion) than those from Case I.

Fix now $k \in \mathbb{N}$. We show that $w_{m_k} \notin \bigcup_{j=1}^{\infty} B(z_{m_j}, r_j)$ by considering the following three cases.

1. If $j = k$, then $w_{m_k} \notin B(z_{m_k}, r_k)$ as $d(w_{m_k}, z_{m_k}) \geq 3r_k$.
2. If $j \leq k - 1$ and $w_{m_k} \in B(z_{m_j}, r_j)$, then

$$3r_j \leq d(z_{m_j}, z_0) \leq d(z_{m_j}, w_{m_k}) + d(w_{m_k}, z_0) < r_j + r_{k-1} \leq 2r_j,$$

a contradiction.
3. If $j \geq k + 1$ and $w_{m_k} \in B(z_{m_j}, r_j)$, then

$$3r_k \leq d(w_{m_k}, z_0) \leq d(z_{m_j}, w_{m_k}) + d(z_{m_j}, z_0) < r_j + r_{j-1} \leq 2r_k,$$

a contradiction.

Notice that, by the construction of the sequences (x_n), (y_n) and of the numbers r_n, we have

$$3r_{n+1} \leq d(g(x_{n+1}), g(y_{n+1})) < r_n, \tag{8.7.24}$$

for all $n \in \mathbb{N}$.

Suppose now that the sequences (x_n), (y_n) and the numbers $r_n > 0$ satisfy (8.7.21) and (8.7.24). Then the function

$$h_n(x) = \max\left\{0, 1 - \frac{d(x, g(x_n))}{r_n}\right\}, \quad x \in X,$$

is Lipschitz with

$$L(h_n) \leq 1/r_n, \quad h_n(g(x_n)) = 1 \quad \text{and} \quad h_n(x) = 0 \text{ for } x \in X \setminus B(g(x_n), r_n), \tag{8.7.25}$$

for all $n \in \mathbb{N}$.

Consider the function

$$\tilde{f}(x) = \sum_{n=1}^{\infty} d(g(x_{n+1}), g(y_{n+1}))h_n(x), \quad x \in X.$$

Observe that $\tilde{f}(x) = 0$ for $x \in X \setminus \bigcup_{n=1}^{\infty} B(g(x_n), r_n)$ and so $\tilde{f}(g(y_n)) = 0$ for all $n \in \mathbb{N}$. If $x \in \bigcup_{n=1}^{\infty} B(g(x_n), r_n)$ then, since the balls $B(g(x_n), r_n)$ are pairwise disjoint, there is exactly one $m \in \mathbb{N}$ such that $x \in B(g(x_m), r_m)$, implying $\tilde{f}(x) = d(g(x_m), g(y_m))h_m(x)$. In particular, $\tilde{f}(g(x_n)) = d(g(x_n), g(y_n))$ for all $n \in \mathbb{N}$.

By (8.7.24) and (8.7.25),

$$L(d(g(x_{n+1}), g(y_{n+1}))h_n) = d(g(x_{n+1}), g(y_{n+1})) L(h_n) < r_n \frac{1}{r_n} = 1,$$

for all $n \in \mathbb{N}$.

We want to show that the function $\tilde{f}$ is Lipschitz.

If $x, x' \in X \setminus \bigcup_{k=1}^{\infty} B(g(x_k), r_k)$, then $\tilde{f}(x) = \tilde{f}(x') = 0$.

If $x \in B(g(x_n), r_n)$ and $x' \in X \setminus \bigcup_{k=1}^{\infty} B(g(x_k), r_k)$, then $\tilde{f}(x') = h_n(x') = 0$, so that

$$|\tilde{f}(x) - \tilde{f}(x')| = d(g(x_{n+1}), g(y_{n+1})) |h_n(x) - h_n(x')| \le d(x, x').$$

The same inequality holds if $x, x' \in B(g(x_n), r_n)$.

Finally, suppose that $x \in B(g(x_n), r_n)$ and $x' \in B(g(x_{n'}), r_{n'})$ for some $n \neq n'$. Then

$$\tilde{f}(x) = d(g(x_{n+1}), g(y_{n+1}))h_n(x) + d(g(x_{n'+1}), g(y_{n'+1}))h_{n'}(x) \quad \text{and}$$

$$\tilde{f}(x') = d(g(x_{n+1}), g(y_{n+1}))h_n(x') + d(g(x_{n'+1}), g(y_{n'+1}))h_{n'}(x'),$$

because $h_{n'}(x) = 0 = h_n(x')$. Hence,

$$|\tilde{f}(x) - \tilde{f}(x')| \le \big(L(d(g(x_{n+1}), g(y_{n+1}))h_n)$$
$$+ L(d(g(x_{n'+1}), g(y_{n'+1}))h_{n'}) \big) d(x, x') \le 2d(x, x').$$

Consequently, $\tilde{f}$ is Lipschitz with $L(\tilde{f}) \le 2$.

To finish the proof we shall modify the function $\tilde{f}$ in order to obtain a function $f \in \mathrm{Lip}_0(X)$ and such that the requirements of the lemma are still satisfied.

If $g(x_0) = \theta$ (the base point of X), then

$$0 = \tilde{f}(g(y_n)) \to \tilde{f}(g(x_0)) = \tilde{f}(\theta)$$

implies $\tilde{f}(\theta) = 0$, so we can take $f = \tilde{f} \in \mathrm{Lip}_0(X)$.

If $g(x_0) \neq \theta$, let $\beta := d(g(x_0), \theta)/2 > 0$. Since $g(x_n) \to g(x_0)$, there exists $n_0 \in \mathbb{N}$ such that $d(g(x_n), \theta) \geq \beta$ for all $n \geq n_0$.

It follows that the function

$$f(x) = \left(1 - \max\{0, 1 - \beta^{-1}d(x, \theta)\}\right)\tilde{f}(x), \quad x \in X,$$

is Lipschitz (see Sect. 2.3) and satisfies the conditions $f(\theta) = 0$, $f(g(x_n)) = \tilde{f}(g(x_n)) = d(g(x_n), g(y_n))$ for all $n \geq n_0$ and $f(g(y_n)) = 0$ for all $n \in \mathbb{N}$. Consequently, the function $f \in \mathrm{Lip}_0(X)$ and the sequences $(x_{n+n_0})_{n\in\mathbb{N}}$, $(y_{n+n_0})_{n\in\mathbb{N}}$ satisfy the requirements of the lemma. $\square$

A second lemma that will be used in the proof is the following one.

Lemma 8.7.17 *The operator* $\mathscr{C}_g : \mathrm{lip}_0(X) \to \mathrm{lip}_0(X)$ *is weakly compact if and only if*

$$f \circ g \in \mathrm{lip}_0(X) \text{ for all } f \in \mathrm{Lip}_0(X).$$

Proof By Theorem 8.6.11, the operator $\mathscr{C}_g$ is weakly compact if and only if

$$\mathscr{C}_g^{**}\left(\mathrm{lip}_0(X)^{**}\right) \subseteq j_X\left(\mathrm{lip}_0(X)\right), \tag{8.7.26}$$

where j_X denotes the canonical embedding of $\mathrm{lip}_0(X)$ into its bidual.

Let $\Psi : \mathrm{Lip}_0(X) \to \mathrm{lip}_0(X)^{**}$ be the isometric isomorphism given by (8.7.17). Its inverse is determined by the condition (8.7.18).

Suppose first that $\mathscr{C}_g$ is weakly compact. Then the inclusion (8.7.26) holds, so that, for $f \in \mathrm{Lip}_0(X)$,

$$\mathscr{C}_g^{**}\Psi(f) = j_X(h),$$

for some $h \in \mathrm{lip}_0(X)$.

Taking into account (8.7.19),

$$\mathscr{C}_g^{**}\Psi(f)(e_x) = \Psi(f)(\mathscr{C}_g^* e_x) = \Psi(f)(e_{g(x)}) = f(g(x)) = (f \circ g)(x),$$

for all $x \in X$.

On the other hand

$$j_X(h)(e_x) = e_x(h) = h(x),$$

for all $x \in X$.

Consequently, $f \circ g = h \in \mathrm{lip}_0(X)$.

Suppose now that $f \circ g \in \mathrm{lip}_0(X)$ for all $f \in \mathrm{Lip}_0(X)$. Let $\tilde{C}_g : \mathrm{Lip}_0(X) \to \mathrm{lip}_0(X)$ be the composition operator corresponding to g, that is, $\tilde{C}_g(f) = f \circ g$, $f \in$

$\text{Lip}_0(X)$. We show that

$$\mathscr{C}_g^{**} = j_X \widetilde{C}_g \Psi^{-1},\tag{8.7.27}$$

which will imply (8.7.26).

We have

$$j_X \widetilde{C}_g \Psi^{-1}(\varphi)(e_x) = j_X \left(\Psi^{-1}(\varphi) \circ g \right)(e_x)$$

$$= e_x \left(\Psi^{-1}(\varphi) \circ g \right) = \Psi^{-1}(\varphi)(g(x)) = \varphi(e_{g(x)}),$$

for all $\varphi \in \text{lip}_0(X)^{**}$ and $x \in X$.

On the other hand, by (8.7.19),

$$\mathscr{C}_g^{**}(\varphi)(e_x) = \varphi(\mathscr{C}_g^* e_x) = \varphi(e_{g(x)}),$$

proving the equality (8.7.27). $\qquad\qquad\qquad\qquad\qquad\qquad\qquad\qquad\square$

We can prove now Theorem 8.7.14.

Proof of Theorem 8.7.14 Suppose that the composition operator $\mathscr{C}_g : \text{lip}_0(X) \to \text{lip}_0(X)$ is not compact. Then, by Theorem 8.7.13, the function g is not supercontractive. By Lemma 8.7.16, there exist a function $f \in \text{Lip}_0(X)$, a number $\varepsilon > 0$ and two sequences $(x_n), (y_n)$ in X such that

$$d(x_n, y_n) \to 0, \quad d(g(x_n), g(y_n)) > \varepsilon d(x_n, y_n) > 0,$$

and

$$f(g(x_n)) = d(g(x_n), g(y_n)), \quad f(g(y_n)) = 0,$$

for all $n \in \mathbb{N}$.

Then

$$\frac{|f(g(x_n)) - f(g(y_n))|}{d(x_n, y_n)} = \frac{d(g(x_n), g(y_n))}{d(x_n, y_n)} > \varepsilon,$$

for all $n \in \mathbb{N}$, which implies that the function $f \circ g$ is not in $\text{lip}_0(X)$ and so, by Lemma 8.7.17, the operator $\mathscr{C}_g$ is not weakly compact. $\qquad\qquad\square$

Remark 8.7.18 Let (X, d) be a compact metric space and $\text{Lip}(X)$ the Banach algebra of Lipschitz functions on X with the sum-norm (8.1.3).(ii). Golbaharan and Mahyar [267] introduced an operator generalizing both the multiplication and the composition operator. Given $u : X \to \mathbb{K}$ and $g : X \to X$ define

$$(uC_g)(f) = u(f \circ g),\tag{8.7.28}$$

for every $f : X \to \mathbb{K}$.

Obviously, for $u \equiv 1$, $uC_g = C_g$ is the composition operator corresponding to g, and for $g(x) \equiv x$, $uC_g = M_u$ is the multiplication operator corresponding to u. They showed that uC_g is a continuous linear operator on $\text{Lip}(X)$ if and only if u is Lipschitz on X and

$$\sup\left\{|u(x)|\frac{d(g(x), g(y))}{d(x, y)} : x, y \in X, \ x \neq y\right\} < \infty.$$

They also gave characterizations of some properties of uC_g—injectivity, surjectivity, compactness—in terms of the properties of the mappings u and g, similar to those given in Proposition 8.7.4 and Theorem 8.7.12.

Golbaharan [266] proved that every weakly compact operator uC_g on $\text{Lip}(X)$ is compact (compare with Theorem 8.7.15).

8.7.4 Composition Operators on Spaces of Vector Lipschitz Functions

In this subsection we shall present some extensions to the vector case of the Kamowitz-Scheinberg criterion [336] (see Theorem 8.7.5) on the compactness of the composition operators.

The first results in this direction were obtained in Botelho and Jamison [106] who extended the method of Kamowitz and Scheinberg to the vector case. Esmaeli and Mahyar [224], apparently unaware of [106], obtained similar results with different methods, reducing the vector case to the scalar one by using some appropriate Lipschitz functions. Also they work in the slightly more general context of operators acting between two different Hölder spaces, while Botelho and Jamison considered only endomorphisms of Lipschitz spaces.

We start by presenting the results of Esmaeli and Mahyar [224]. Let X, Y be compact Hausdorff spaces, E, F Banach spaces and $S(X, E)$, $S(Y, F)$ subspaces of $C(X, E)$, $C(Y, F)$ (Banach spaces of continuous functions with the sup-norm), respectively. Consider a function $g : Y \to X$ and a mapping $w : Y \to L(E, F)$ (the space of linear operators, not necessarily continuous), whose value at $y \in Y$ is denoted by w_y or by $w(y)$. A weighted composition operator is an operator $T : S(X, E) \to S(Y, F)$ whose value at $f \in S(X, E)$ is calculated by the formula

$$(Tf)(y) = C_{g,w}(f)(y) = w_y(f(g(y))), \quad y \in Y.\tag{8.7.29}$$

Observe that such an operator is always linear.

Proposition 8.7.19 *Let X, Y be compact Hausdorff spaces, E, F Banach spaces and $S(X, E)$, $S(Y, F)$ Banach subspaces of $C(X, E)$, $C(Y, F)$, respectively, such that the topology of pointwise convergence is weaker than their norm topology. If w applies Y into $\mathscr{L}(E, F)$, then any composition operator $T : S(X, E) \to S(Y, F)$ of the form (8.7.29) is a continuous linear operator.*

Proof The proof follows from the Closed Graph Theorem. Suppose $f_n \to 0$ and $Tf_n \to h \in S(Y, F)$. Then $f_n(x) \to 0$ for all $x \in X$, so that $f_n(g(y)) \to 0$ for all $y \in Y$. But then

$$h(y) = \lim_{n \to \infty} Tf_n(y) = \lim_{n \to \infty} w_y(f_n(g(y))) = 0,$$

for every $y \in Y$, showing that $h = 0 = T0$. □

Note 8.7.20 In this subsection we shall be interested in spaces of Hölder functions $\mathrm{BLip}_\alpha(X, E)$ and $\mathrm{blip}_\alpha(X, E)$, for $0 < \alpha \leq 1$ and X a metric space (see (8.1.4) and Sect. 8.3), with the convention that $\mathrm{Lip}_1 = \mathrm{Lip}$, $\mathrm{BLip}_1 = \mathrm{BLip}$, $\mathrm{lip}_1 = \mathrm{lip}$, $\mathrm{blip}_1 = \mathrm{blip}$, and $L_1(f) = L(f)$ for f in one of these spaces (see (1.3.3), (1.3.4) and (1.3.6)). The definitions of Hölder functions f with values in a normed space E can be obtained from the corresponding definitions of scalar Hölder functions by replacing $|f(x) - f(y)|$ with $\|f(x) - f(y)\|$.

Since the convergence in the Lipschitz and Hölder norms implies pointwise convergence, the conclusion of Proposition 8.7.19 holds for the spaces BLip_α and lip_α.

Although, for compact X, $\mathrm{BLip}_\alpha(X, E)$ and $\mathrm{Lip}_\alpha(X, E)$ coincide as sets, we use the notation $\mathrm{BLip}_\alpha(X, E)$ to indicate that we are working with the norm

$$\|f\|_\alpha = L_\alpha(f) + \|f\|_\infty,$$

with the convention that $\|f\|_1 = \|f\|_{\mathrm{sum}}$ (see (8.1.3)).

A similar convention is adopted for the spaces lip_α and blip_α.

For $f \in \mathrm{BLip}_\alpha(X)$ and $e \in E$ consider the function $f_e \in \mathrm{BLip}_\alpha(X, E)$ given by

$$f_e(x) = f(x)e, \quad x \in X.$$

Then

$$\|f_e\|_\infty = \|f\|_\infty \cdot \|e\| \quad \text{and} \quad L_\alpha(f_e) = L_\alpha(f) \cdot \|e\|,$$

so that

$$\|f_e\|_\alpha = \|f\|_\alpha \cdot \|e\|.$$

Also, we shall denote by 1_e the mapping $1_e : X \to E$ given by $1_e(x) = e$, $x \in X$. It follows that $L_\alpha(1_e) = 0$ and $\|1_e\|_\infty = \|e\|$, so that $\|1_e\|_\alpha = \|e\|$.

Let now X, Y be compact metric spaces. In order to simplify the notation we agree, as in the case of normed spaces, to denote the metrics on X and Y by the same symbol d.

A mapping $g : Y \to X$ is called *supercontractive on a subset Z of Y* if

$$\lim_{d(y,y')\to_Z 0} \frac{d(g(y), g(y'))}{d(y, y')} = 0,\qquad (8.7.30)$$

where $d(y, y') \to_Z 0$ means that $d(y, y') \to 0$ with $y, y' \in Z$, $y \neq y'$.

Explicitly, this means that for every $\varepsilon > 0$ there exists $\delta > 0$ such that

$$d(g(y), g(y')) \leq \varepsilon d(y, y'),$$

for all $y, y' \in Z$ with $d(y, y') < \delta$.

For $w : Y \to \mathscr{L}(E, F)$ and $\beta > 0$ let

$$D_\beta(w) = \{y \in Y : \|w_y\| \geq \beta\} \quad \text{and}$$

$$D(w) = \{y \in Y : \|w_y\| > 0\} = \bigcup_{\beta>0} D_\beta(w).\qquad (8.7.31)$$

Remark 8.7.21 If w is continuous, then

$$\|w\|_\infty := \sup\{\|w_y\| : y \in Y\} < \infty, \quad D(w) = \bigcup_{0<\beta\leq\|w\|_\infty} D_\beta(w),$$

and $D_\beta(w)$ is a nonempty closed, and so compact, subset of Y, for every $0 < \beta \leq \|w\|_\infty$. Also, every compact subset Z of Y is contained in $D_\beta(w)$, where $\beta = \sup\{\|w_y\| : y \in Z\}$.

The following theorem contains some necessary conditions on w and g ensuring that $C_{w,g}$ is properly defined.

Theorem 8.7.22 *Let X, Y be compact metric spaces, $0 < \alpha \leq 1$ and T a composition operator of the form (8.7.29) from $\mathrm{BLip}_\alpha(X, E)$ to $\mathrm{BLip}_\alpha(Y, F)$. If T is continuous, then $w \in \mathrm{BLip}_\alpha(Y, \mathscr{L}(E, F))$ and g is continuous on $D(w)$ and Lipschitz on $D_\beta(w)$ for every $0 < \beta \leq \|w\|_\infty$.*

Proof

I. $w \in \mathrm{BLip}_\alpha(Y, \mathscr{L}(E, F))$.

For every $e \in E$ and $y \in Y$,

$$\|w_y(e)\| = \|T1_e(y)\| \leq \|T1_e\|_\alpha \leq \|T\|\|1_e\|_\alpha = \|T\|\|e\|,$$

so that $w_y \in \mathcal{L}(E, F)$ and $\|w_y\| \leq \|T\|$. For $y_1 \neq y_2$ in Y,

$$\frac{\|w_{y_1}(e) - w_{y_2}(e)\|}{d^\alpha(y_1, y_2)} = \frac{\|T1_e(y_1) - T1_e(y_2)\|}{d^\alpha(y_1, y_2)}$$

$$\leq L_\alpha(T1_e) \leq \|T1_e\|_\alpha \leq \|T\| \|1_e\|_\alpha = \|T\| \|e\|,$$

which implies $\|w_{y_1} - w_{y_2}\| \leq \|T\| d^\alpha(y_1, y_2)$.

Consequently,

$$w \in \text{Lip}_\alpha(Y, \mathcal{L}(E, F)) \quad \text{and} \quad L_\alpha(w) \leq \|T\|.$$

Thus $w : Y \to (\mathcal{L}(E, F), \| \cdot \|)$ is continuous so that, by the compactness of Y, $\sup\{\|w_y\| : y \in Y\} < \infty$ and, since we have seen that $L_\alpha(w) \leq \|T\|$, it follows that $w \in \text{BLip}_\alpha(Y, \mathcal{L}(E, F))$.

II. *The continuity of g on $D(w)$.*

Let (y_n) be a sequence in $D(w)$ converging to some $y_0 \in D(w)$. Pick $e \in E$ such that $w_{y_0}(e) \neq 0$ and let $f : X \to \mathbb{R}$ be given by $f(x) = d(x, g(y_0))$, $x \in X$. The inequalities $|d(x, g(y_0)) - d(x', g(y_0))| \leq d(x, x') = d^{1-\alpha}(x, x') d^\alpha(x, x') \leq (\text{diam } X)^{1-\alpha} d^\alpha(x, x')$ show that $f \in \text{BLip}_\alpha(X)$ with $L_\alpha(f) \leq (\text{diam } X)^{1-\alpha}$ and $\|f\|_\alpha \leq \text{diam } X + (\text{diam } X)^{1-\alpha}$. But then $f_e(x) = d(x, g(y_0)) e$ satisfies the equalities

$$Tf_e(y_n) = w_{y_n}(f_e(g(y_n))) = d(g(y_n), g(y_0)) w_{y_n}(e),$$

and

$$Tf_e(y_0) = w_{y_n}(f_e(g(y_0))) = d(g(y_0), g(y_0)) w_{y_0}(e) = 0.$$

Since w_y is continuous $\lim_{n \to \infty} \|w_{y_n} - w_{y_0}\| = 0$, so that $\lim_{n \to \infty} w_{y_n}(e) = w_{y_0}(e)$, hence there exists $n_0 \in \mathbb{N}$ such that

$$\|w_{y_n}(e)\| \geq \frac{1}{2} \|w_{y_0}(e)\| > 0,$$

for all $n \geq n_0$. Then $Tf_e(y_n) \to Tf_e(y_0)$ (because $Tf_e \in \text{BLip}_\alpha(Y, F)$) and, for all $n \geq n_0$,

$$d(g(y_n), g(y_0)) \cdot \frac{1}{2} \|w_{y_0}(e)\| \leq d(g(y_n), g(y_0)) \|w_{y_n}(e)\|$$

$$= \|Tf_e(y_n)\| \to \|Tf_e(y_0)\| = 0.$$

It follows that $d(g(y_n), g(y_0)) \to 0$, that is, $g(y_n) \to g(y_0)$ as $n \to \infty$.

III. *For every* $0 < \beta \le \|w\|_\infty$ *the function* g *is Lipschitz on the set* $D_\beta(w)$.

For $y \in D_\beta(w)$ define $f_y : X \to \mathbb{R}$ by $f_y(x) = d^\alpha(x, g(y))$, $x \in X$. Then

$$|d^\alpha(x, g(y)) - d^\alpha(x', g(y))| \le d^\alpha(x, x'),$$

for all $x, x' \in X$, so that $L_\alpha(f_y) \le 1$. It follows that $f_y \in \mathrm{BLip}_\alpha(X)$, with

$$\|f_y\|_\alpha \le 1 + (\mathrm{diam}\ X)^\alpha.$$

Let $e \in E$ with $\|e\| = 1$. The function $f_{y,e}(x) = d^\alpha(x, g(y))e$, $x \in X$, belongs to $\mathrm{BLip}_\alpha(X, E)$ (with $\|f_{y,e}\|_\alpha = \|f_y\|_\alpha \le 1 + (\mathrm{diam}\ X)^\alpha$) and satisfies the equality

$$Tf_{y,e}(y') = d^\alpha(g(y'), g(y)) \cdot w_{y'}(e),$$

for all $y' \in D_\beta(w)$, so that $Tf_{y,e}(y) = 0$.

It follows that

$$d^\alpha(g(y'), g(y)) \cdot \|w_{y'}(e)\| = \|Tf_{y,e}(y') - Tf_{y,e}(y)\|$$
$$\le L_\alpha(Tf_{y,e})\, d^\alpha(y', y) \le \|Tf_{y,e}\|_\alpha\, d^\alpha(y', y)$$
$$\le \|T\|\, \|f_{y,e}\|_\alpha d^\alpha(y', y) \le c\|T\| d^\alpha(y', y),$$

where $c = 1 + (\mathrm{diam}\ X)^\alpha$. Taking the supremum with respect to $e \in S_E$ it follows that

$$\beta\, d^\alpha(g(y'), g(y)) \le d^\alpha(g(y), g(y')) \cdot \|w_{y'}\| \le c\|T\| d^\alpha(y, y'),$$

which implies

$$d(g(y), g(y')) \le \lambda d(y, y'),$$

for all $y, y' \in D_\beta(w)$, where $\lambda = (c\|T\|/\beta)^{1/\alpha}$. $\square$

Remark 8.7.23 By Remark 8.7.21 the statement "*g is Lipschitz on* $D_\beta(w)$ *for every* $0 < \beta \le \|w\|_\infty$" is equivalent to "*g is Lipschitz on every compact subset of* Y".

A similar result holds for little Lipschitz spaces.

Theorem 8.7.24 *Let* X, Y *be compact metric spaces,* $\alpha \in (0, 1)$ *and* T *a composition operator of the form* (8.7.29) *from* $\mathrm{blip}_\alpha(X, E)$ *to* $\mathrm{blip}_\alpha(Y, F)$. *If* T *is continuous, then* $w \in \mathrm{BLip}_\alpha(Y, \mathscr{L}(E, F))$, g *is continuous on* $D(w)$ *and Lipschitz on* $D_\beta(w)$ *for every* $0 < \beta \le \|w\|_\infty$.

In the proof we shall need the following elementary result.

Lemma 8.7.25 *Let $0 < \alpha < 1$ and $a > 0$. The function $h(t) = (t+a)^\alpha - a^\alpha$, $t \geq 0$, satisfies the inequality*

$$|h(t) - h(t')| = |(t + a)^\alpha - (t' + a)^\alpha| \leq |t - t'|^\alpha, \tag{8.7.32}$$

for all $t, t' \geq 0$. Also h is Lipschitz with $L(h) = \alpha\, a^{\alpha-1}$.

Proof One can prove by elementary methods that

$$|s^\alpha - 1| \leq |s - 1|^\alpha,$$

for all $s \geq 0$. The equivalence

$$|u^\alpha - v^\alpha| \leq |u - v|^\alpha \iff \left|\left(\frac{u}{v}\right)^\alpha - 1\right| \leq \left|\frac{u}{v} - 1\right|^\alpha,$$

shows the validity of $|u^\alpha - v^\alpha| \leq |u - v|^\alpha$ for all $u, v > 0$. Taking $u = t + a$ and $v = t' + a$ one obtains (8.7.32). By (8.7.32) $L_\alpha(h) \leq 1$. Also h is strictly increasing on $[0, \infty)$, because $h'(t) = \alpha(t + a)^{\alpha-1} > 0$ for all $t \geq 0$. The derivative h' is strictly decreasing on $[0, \infty)$ and $0 < h'(t) \leq h'(0) = \alpha a^{\alpha-1}$ for all $t \geq 0$. By Proposition 2.2.1, this implies that h is Lipschitz with $L(h) = \alpha\, a^{\alpha-1}$. □

Proof of Theorem 8.7.24 The proof is similar to that given to Theorem 8.7.22, exploiting the fact that for $\alpha \in (0, 1)$ $\mathrm{Lip}(Z, E) \subseteq \mathrm{lip}_\alpha(Z, E)$ for every metric space Z. Steps I and II are treated in the same way. Only Step III needs some additional explanation.

III. *The function g is Lipschitz on every set $D_\beta(w)$, $0 < \beta \leq \|w\|_\infty$.*

For $y_1 \neq y_2$ in Y fixed, consider the function $f = f_{y_1,y_2} : X \to \mathbb{R}$ given, for $x \in X$, by

$$f(x) = (d(x, g(y_2)) + d(g(y_1), g(y_2)))^\alpha - d^\alpha(g(y_1), g(y_2)) = h(d(x, g(y_2))),$$

where h is as in Lemma 8.7.25 for $a = d(g(y_1), g(y_2))$.

Taking into account the properties of the function h given in the mentioned lemma, it follows that

$$|f(x)| = h(d(x, g(y_2))) \leq h(\delta) < (\delta + a)^\alpha \leq (2\delta)^\alpha,$$

where $\delta = \mathrm{diam}\, X$, and

$$|f(x) - f(x')| = |h(d(x, g(y_2))) - h(d(x', g(y_2)))|$$
$$\leq |d(x, g(y_2)) - d(x', g(y_2))|^\alpha \leq d^\alpha(x, x'),$$

for all $x, x' \in X$.

Consequently,

$$f \in \mathrm{BLip}_\alpha(X) \quad \text{and} \quad \|f\|_\alpha \leq 1 + (2\,\mathrm{diam}\,X)^\alpha .$$

Also, the relations

$$|f(x) - f(x')| = |h(d(x, g(y_2))) - h(d(x', g(y_2)))|$$
$$\leq L(h)|d(x, g(y_2)) - d(x', g(y_2))| \leq L(h)d(x, x'),$$

show that $f \in \mathrm{Lip}(X) \subseteq \mathrm{lip}_\alpha(X)$, that is, $f \in \mathrm{blip}_\alpha(X)$.

By the definition of f

$$f(g(y_2)) = 0 \quad \text{and} \quad f(g(y_1)) = (2^\alpha - 1)\,d^\alpha(g(y_1), g(y_2)).$$

For every $e \in E$, $\|e\| = 1$, the function $f_e(x) = f(x)\,e$ satisfies the equality

$$T(f_e)(y) = f(g(y))\,w_y(e),$$

so that

$$T(f_e)(y_1) = (2^\alpha - 1)\,d^\alpha(g(y_1), g(y_2))\,w_{y_1}(e) \quad \text{and} \quad T(f_e)(y_2) = 0.$$

Consequently,

$$\frac{d^\alpha(g(y_1), g(y_2))}{d^\alpha(y_1, y_2)}\,\|w_{y_1}(e)\| = \frac{1}{2^\alpha - 1}\,\frac{\|T(f_e)(y_1) - T(f_e)(y_2)\|}{d^\alpha(y_1, y_2)}$$

$$\leq \frac{1}{2^\alpha - 1}\,L_\alpha(T(f_e)) \leq \frac{1}{2^\alpha - 1}\,\|T(f_e)\|_\alpha$$

$$\leq \frac{1}{2^\alpha - 1}\,\|T\|\,\|f_e\|_\alpha \leq \frac{1}{2^\alpha - 1}\,\|T\|\,\left(1 + (2(\mathrm{diam}\,X)^\alpha\right).$$

Since these inequalities hold for every $e \in E$ with $\|e\| = 1$, it follows that

$$\frac{d^\alpha(g(y_1), g(y_2))}{d^\alpha(y_1, y_2)}\,\beta \leq \frac{d^\alpha(g(y_1), g(y_2))}{d^\alpha(y_1, y_2)}\,\|w_{y_1}\| \leq \frac{1}{2^\alpha - 1}\,\|T\|\,\left(1 + (2(\mathrm{diam}\,X)^\alpha\right).$$

Consequently, $d(g(y_1), g(y_2)) \leq \lambda\,d(y_1, y_2)$, where $\lambda > 0$ is given by

$$\lambda^\alpha = \frac{1}{\beta(2^\alpha - 1)}\,\|T\|\,\left(1 + (2(\mathrm{diam}\,X)^\alpha\right).$$

$\square$

The compactness result is the following one.

Theorem 8.7.26 *Let* X, Y *be compact metric spaces,* $0 < \alpha \leq 1$ *and* T *a composition operator of the form* (8.7.29) *from* $\mathrm{BLip}_\alpha(X, E)$ *to* $\mathrm{BLip}_\alpha(Y, F)$.

1. *If T is compact, then $w \in \mathrm{BLip}_\alpha(Y, \mathcal{K}(E, F))$ and g is continuous on $D(w)$ and a supercontraction on $D_\beta(w)$, for every $0 < \beta \leq \|w\|_\infty$.*
2. *If $w \in \mathrm{blip}_\alpha(Y, \mathcal{K}(E, F))$ and g is supercontraction on $D(w)$, then the composition operator T is compact.*

Proof

1. By Theorem 8.7.22 $w \in \mathrm{BLip}_\alpha(Y, \mathcal{L}(E, F))$ and g is continuous on $D(w)$.

   Let us show that $w_y \in \mathcal{K}(E, F)$ for every $y \in Y$. To this end fix an element $y \in Y$.

   Let (e_n) be a sequence in the closed unit ball of E. Then $\|1_{e_n}\|_\alpha = \|e_n\| \leq 1$, so that, by the compactness of T, there exists a subsequence $(T1_{e_{n_k}})_{k \in \mathbb{N}}$ convergent in the sum-norm of $\mathrm{BLip}_\alpha(Y, F)$ to some $f \in \mathrm{BLip}_\alpha(Y, F)$. In particular,

   $$w_y(e_{n_k}) = T1_{e_{n_k}}(y) \to f(y) \quad \text{as} \quad k \to \infty,$$

   proving the compactness of w_y.

   So it remains to show that g is a supercontraction on the set $D_\beta(w)$ for every $0 < \beta \leq \|w\|_\infty$.

   Let $0 < \gamma < \beta$. For $y_0 \in D_\beta(w)$ there exists $e \in E$, $\|e\| = 1$, such that $\|w_{y_0}(e)\| > \gamma$. By the continuity of w there exists $\delta > 0$ such that $\|w_y(e)\| > \gamma$ for all y in the compact set $B_\delta := B[y_0, \delta]$.

   Define the operator $S : \mathrm{BLip}_\alpha(X) \to \mathrm{BLip}_\alpha(B_\delta)$ by $Sf = f \circ g|_{B_\delta}$ and let $f_e(x) = f(x)e$, $x \in X$. Then

   $$Tf_e(y) = f(g(y)) \cdot w_y(e) = Sf(y) \cdot w_y(e), \tag{8.7.33}$$

   for all $y \in B_\delta$.

   Let (f_n) be a sequence in $\mathrm{BLip}_\alpha(X)$ and $f_{n,e}(x) = f_n(x)e$, $x \in X$, $n \in \mathbb{N}$.

   Claim I. *If $(Tf_{n,e})_{n \in \mathbb{N}}$ is a Cauchy sequence in $\mathrm{BLip}_\alpha(Y, F)$, then (Sf_n) is a Cauchy sequence in $\mathrm{BLip}_\alpha(B_\delta)$.*

   For a function $\varphi : Z \to G$, where Z is a set and G a normed space, put

   $$\|\varphi\|_W = \sup\{\|\varphi(z)\| : z \in W\},$$

   for $\emptyset \neq W \subseteq Z$.

   Let $g_n = f_{n,e}$, $n \in \mathbb{N}$. By (8.7.33) and the definition of B_δ,

   $$|(Sf_n - Sf_{n'})(y)| = \frac{\|(Tg_n - Tg_{n'})(y)\|}{\|w_y(e)\|} \leq \frac{1}{\gamma}\|Tg_n - Tg_{n'}\|_Y,$$

   for all $y \in B_\delta$.

Consequently,

$$\|Sf_n - Sf_{n'}\|_{B_\delta} \leq \frac{1}{\gamma} \|Tg_n - Tg_{n'}\|_Y\,, \qquad\qquad (8.7.34)$$

which shows that (Sf_n) is Cauchy with respect to the sup-norm on B_δ.

By (8.7.33),

$$(Tg_n - Tg_{n'})(y') = (Sf_n - Sf_{n'})(y')\, w_y(e) + (Sf_n - Sf_{n'})(y')\, (w_{y'}(e) - w_y(e))\,,$$

so that

$$\frac{|(Sf_n - Sf_{n'})(y) - (Sf_n - Sf_{n'})(y')|}{d^\alpha(y, y')}$$

$$\leq \frac{\|(Tg_n - Tg_{n'})(y) - (Tg_n - Tg_{n'})(y')\|}{d^\alpha(y, y')\|w_y(e)\|} + \frac{|(Sf_n - Sf_{n'})(y')| \cdot \|w_{y'}(e) - w_y(e)\|}{d^\alpha(y, y')\|w_y(e)\|}$$

$$\leq \frac{1}{\gamma} \cdot L_\alpha(Tg_n - Tg_{n'}) + \frac{\|(Tg_n - Tg_{n'})(y')\|}{\|w_y(e)\|\,\|w_{y'}(e)\|} \cdot \frac{\|w_{y'} - w_y\|}{d^\alpha(y, y')}$$

$$\leq \frac{1}{\gamma} \cdot L_\alpha(Tg_n - Tg_{n'}) + \frac{1}{\gamma^2}\|Tg_n - Tg_{n'}\|_Y\, L_\alpha(w)\,,$$

for all $y, y' \in B_\delta$ with $y \neq y'$.

It follows that

$$L_\alpha(Sf_n - Sf_{n'}) \leq \frac{1}{\gamma} \cdot L_\alpha(Tg_n - Tg_{n'}) + \frac{1}{\gamma^2}\|Tg_n - Tg_{n'}\|_Y\, L_\alpha(w)\,. \qquad (8.7.35)$$

The inequalities (8.7.34) and (8.7.35) show that (Sf_n) is a Cauchy sequence in $\mathrm{BLip}_\alpha(B_\delta)$.

Claim II. The operator S is compact.

It is sufficient to show that for every bounded sequence (f_n) in $\mathrm{BLip}_\alpha(X)$, the sequence (Sf_n) contains a Cauchy subsequence. Indeed, if (f_n) is a bounded sequence in $\mathrm{BLip}_\alpha(X)$, then, by the compactness of T, the sequence $(Tf_{n,e})_{n\in\mathbb{N}}$ contains a Cauchy subsequence $(Tf_{n_k,e})_{k\in\mathbb{N}}$. By Claim I., the sequence (Sf_{n_k}) is Cauchy in $\mathrm{BLip}_\alpha(B_\delta)$.

We can prove now that g is supercontractive on $D_\beta(w)$ for every $0 < \beta \leq \|w\|_\infty$.

By Claim II., for every $z \in D_\beta(w)$ there exists $\delta_z > 0$ such that $B[z, \delta_z] \subseteq D_\beta(w)$ and the operator $S_z : \mathrm{BLip}_\alpha(X) \to \mathrm{BLip}_\alpha(B[z, \delta_z])$ given by $S_z f = f \circ g|_{B[z,\delta_z]}$ is compact. By the compactness of $D_\beta(w)$ there exists $z_1, \ldots, z_m \in D_\beta(w)$ such that $D_\beta(w) \subseteq \bigcup_{i=1}^m B(z_i, \delta_i)$, where $\delta_i = \delta_{z_i}$. By Theorem 8.7.12, the function g is supercontractive on every ball $B[z_i, \delta_i]$, $i = 1, \ldots, m$. Let us show that this implies the supercontractiveness of g on $D_\beta(w)$.

Indeed, assuming the contrary, then there exist $\varepsilon > 0$ and two sequences $(y_n), (y_n')$ in $D_\beta(w)$ such that

$$y_n \neq y_{n'}, \quad \lim_{n \to \infty} d(y_n, y_n') = 0 \quad \text{and} \quad \frac{d(g(y_n), g(y_n'))}{d(y_n, y_n')} \geq \varepsilon, \text{ for all } n, n' \in \mathbb{N}.$$

$$(8.7.36)$$

By the compactness of $D_\beta(w)$, the sequence (y_n) contains a subsequence (y_{n_k}) convergent to some $y \in D_\beta(w)$. The second relation in (8.7.36) shows that $\lim_{k \to \infty} y_{n_k}' = y$, too. If $i \in \{1, 2, \ldots, m\}$ is such that $y \in B(z_i, \delta_i)$, then $y_{n_k}, y_{n_k}' \in B(z_i, \delta_i)$ for sufficiently large k. By (8.7.36) this implies that g is not supercontractive on $B[z_i, \delta_i]$, a contradiction.

This finishes the proof of the first assertion.

2. We start with a claim similar to Claim $I.$.

Claim III. *Suppose that the hypotheses of 2 are fulfilled. If (f_n) is a sequence in the unit ball of* $\mathrm{BLip}_\alpha(X, E)$ *such that $(T f_n)$ is Cauchy with respect to the sup-norm* $\| \cdot \|_Y$, *then $(T f_n)$ is Cauchy with respect to the Lipschitz norm L_α too.*

Put $\|w\|_\infty := \sup\{\|w_y\| : y \in Y\}$. For $\varepsilon > 0$ let $\delta > 0$ be such that

$$\frac{d(g(y), g(y'))}{d(y, y')} < \varepsilon^{1/\alpha},$$

$$(8.7.37)$$

for all $y, y' \in D(w)$ with $0 < d(y, y') < \delta$, and

$$\frac{\|w_y - w_{y'}\|}{d^\alpha(y, y')} < \varepsilon,$$

$$(8.7.38)$$

for all $y, y' \in Y$ with $0 < d(y, y') < \delta$.

Choose now $n_0 \in \mathbb{N}$ such that

$$\|T f_n - T f_{n'}\|_Y < \frac{\delta^\alpha}{2} \varepsilon,$$

$$(8.7.39)$$

for all $n, n' \geq n_0$.

We shall evaluate

$$\frac{\|(T f_n - T f_{n'})(y) - (T f_n - T f_{n'})(y')\|}{d^\alpha(y, y')}$$

$$= \frac{\|w_y(f_n - f_{n'})(g(y)) - w_{y'}(f_n - f_{n'})(g(y'))\|}{d^\alpha(y, y')},$$

$$(8.7.40)$$

for $y \neq y'$ in Y and $n, n' \geq n_0$.

We distinguish three cases.

Case 1. $d(y, y') \geq \delta$.

In this case, by (8.7.39),

$$\frac{\|(Tf_n - Tf_{n'})(y) - (Tf_n - Tf_{n'})(y')\|}{d^\alpha(y, y')} \leq \frac{2}{\delta^\alpha} \|Tf_n - Tf_{n'}\|_Y < \varepsilon. \qquad (8.7.41)$$

Case 2. $y, y' \in D(w)$ with $0 < d(y, y') < \delta$.

Taking into account (8.7.40), (8.7.37) and (8.7.38) one obtains

$$\frac{\|(Tf_n - Tf_{n'})(y) - (Tf_n - Tf_{n'})(y')\|}{d^\alpha(y, y')} \leq \frac{\|(w_y - w_{y'})(f_n - f_{n'})(g(y))\|}{d^\alpha(y, y')}$$

$$+ \frac{\|w_{y'}((f_n - f_{n'})(g(y)) - (f_n - f_{n'})(g(y')))\|}{d^\alpha(y, y')} \leq \frac{\|w_y - w_{y'}\|}{d^\alpha(y, y')} \|f_n - f_{n'}\|_X$$

$$(8.7.42)$$

$$+ \|w_{y'}\| \frac{\|(f_n - f_{n'})(g(y)) - (f_n - f_{n'})(g(y'))|}{d^\alpha(y, y')}$$

$$\leq 2\varepsilon + \|w\|_\infty L_\alpha(f_n - f_{n'}) \frac{d^\alpha(g(y), g(y'))}{d^\alpha(y, y')} < (2 + 2\|w\|_\infty)\varepsilon.$$

Case 3. $y \in D(w)$ and $y' \notin D(w)$ are such that $0 < d(y, y') < \delta$.

In this case $w_{y'} = O$ (the null operator in $\mathscr{L}(E, F)$), so that, calculating as in (8.7.42), one obtains

$$\frac{\|(Tf_n - Tf_{n'})(y) - (Tf_n - Tf_{n'})(y')\|}{d^\alpha(y, y')}$$

$$(8.7.43)$$

$$= \frac{\|(w_y - w_{y'})(f_n - f_{n'})(g(y))\|}{d^\alpha(y, y')} \leq \frac{\|w_y - w_{y'}\|}{d^\alpha(y, y')} \|f_n - f_{n'}\|_X < 2\varepsilon.$$

Since the case $y, y' \notin D(w)$ is trivial, the inequalities (8.7.41), (8.7.42) and (8.7.43) show that

$$L_\alpha(Tf_n - Tf_{n'}) < (2 + 2\|w\|_\infty)\varepsilon,$$

for all $n, n' \geq n_0$.

We can prove now the compactness of T. As in the proof of the first assertion it is sufficient to show that for every sequence (f_n) in the unit ball of $\mathrm{BLip}_\alpha(X, E)$, the sequence (Tf_n) contains a subsequence which is Cauchy in $\mathrm{BLip}_\alpha(Y, F)$.

We shall consider first $\{Tf_n : n \in \mathbb{N}\}$ as a subset of $C(Y, F)$. The inequalities

$$\|Tf_n(y) - Tf_n(y')\| \leq \|T\| \|f_n(y) - f_n(y')\| \leq \|T\| L_\alpha(f_n)d^\alpha(y, y') \leq \|T\| d^\alpha(y, y')$$

show that the set $\{Tf_n : n \in \mathbb{N}\}$ is equicontinuous in $C(Y, F)$. The compactness of w_y implies that the set $\{Tf_n(y) : n \in \mathbb{N}\} = \{w_y(f_n(g(y))) : n \in \mathbb{N}\}$ is relatively compact in F. By the Arzela–Ascoli theorem (Theorem 1.4.42) it follows that the set $\{Tf_n : n \in \mathbb{N}\}$ is relatively compact in $C(Y, F)$. Consequently, the sequence (Tf_n) contains a subsequence (Tf_{n_k}) convergent with respect to the sup-norm to some $f \in C(Y, F)$. It follows that (Tf_{n_k}) is a Cauchy sequence with respect to the sup-norm, so that, by Claim *III.*, it is Cauchy with respect to the Lipschitz norm L_α, too, and so it is Cauchy with respect to the norm $\|\cdot\|_\alpha$ on $\mathrm{BLip}_\alpha(Y, F)$. $\qquad\square$

Using Theorem 8.7.24 and the characterization of compact composition operators on little Lipschitz spaces (Theorem 8.7.13) one can obtain a characterization of compact composition operators on little Hölder spaces.

Theorem 8.7.27 *Let* X, Y *be compact metric spaces,* $\alpha \in (0, 1)$ *and* T *a composition operator of the form* (8.7.29) *from* $\mathrm{blip}_\alpha(X, E)$ *to* $\mathrm{blip}_\alpha(Y, F)$.

1. *If* T *is compact, then* $w \in \mathrm{BLip}_\alpha(Y, \mathscr{K}(E, F))$ *and* g *is continuous on* $D(w)$ *and a supercontraction on* $D_\beta(w)$ *for every* $0 < \beta \leq \|w\|_\infty$.
2. *If* $w \in \mathrm{blip}_\alpha(Y, \mathscr{K}(E, F))$ *and* g *is a supercontraction on* $D(w)$, *then the composition operator* T *is compact.*

We shall present now some results obtained by Botelho and Jamison [106]. We consider a property of a function $g : X \to X$ related to supercontractiveness (see (8.7.30)).

Let (X, d) be a metric space. For $Y \subseteq X$ put

$$h_{g,Y}(x) = \limsup_{y \to_Y x} \frac{d(g(x), g(y))}{d(x, y)}, \quad x \in Y,$$

and let

$$\varrho_Y(g) = \sup\{h_{g,Y}(x) : x \in Y\}.$$

Here $y \to_Y x$ means that y tends to x and $y \in Y \setminus \{x\}$. We put $h_g = h_{g,X}$ and $\varrho(g) = \varrho_X(g)$.

We say that the function g is *weakly supercontractive* on Y if $\varrho_Y(g) = 0$, that is,

$$\lim_{y \to_Y x} \frac{d(g(x), g(y))}{d(x, y)} = 0,$$

for every $x \in Y$.

Lemma 8.7.28 *Let* (X, d) *be a metric space and* $g : X \to X$.

1. *If* $h_g(x) < \infty$, *then* g *is continuous at* x.
2. *If* g *is supercontractive on a subset* Y *of* X, *then it is weakly supercontractive on* Y.

3. (Federer [236, p. 64]) *If X is a convex subset of a normed space E and $\varrho(g) < \infty$, then g is Lipschitz on X with $L(g) = \varrho(g)$. Consequently, if g is weakly supercontractive on X, then g is constant on X.*

Proof

1. If g is not continuous at x, then there exist $\varepsilon > 0$ and a sequence $x_n \to x$ such that

$$d(g(x_n), g(x)) \geq \varepsilon , \qquad (8.7.44)$$

for all $n \in \mathbb{N}$. Denoting $\beta := \limsup_{n\to\infty} d(g(x_n), g(x))/d(x_n, x)$, it follows that $\beta \leq h_g(x) < \infty$. If (x_{n_k}) is such that $\lim_{k\to\infty} d(g(x_{n_k}), g(x))/d(x_{n_k}, x) = \beta$, then

$$\lim_{k\to\infty} d(g(x_{n_k}), g(x)) = 0 ,$$

in contradiction to (8.7.44).
2. This assertion is obvious (see (8.7.9)).
3. For $x_0, x_1 \in X$ let $x_t = x_0 + t(x_1 - x_0)$, $t \in [0, 1]$. For $\beta > \varrho(g)$ let

$$S = \{t \in [0, 1] : \|g(x_t) - g(x_0)\| \leq t\beta\|x_1 - x_0\|\} \quad \text{and} \quad \tau = \sup S .$$

The continuity of g at x_τ implies $\tau \in S$.
Suppose $\tau < 1$. If for some $t > \tau$, $\|g(x_t) - g(x_\tau)\| \leq (t - \tau)\beta\|x_1 - x_0\|$, then

$$\|g(x_t) - g(x_0)\| \leq \|g(x_t) - g(x_\tau)\| + \|g(x_\tau) - g(x_0)\| \leq t\beta\|x_1 - x_0\| ,$$

implying $t \in S$, in contradiction to $\tau = \sup S$.
Consequently, $\|g(x_t) - g(x_\tau)\| > (t - \tau)\beta\|x_1 - x_0\| = \beta\|x_t - x_\tau\|$ for all $\tau < t \leq 1$, yielding the contradiction

$$h_g(x_\tau) \geq \limsup_{t\searrow\tau} \frac{\|g(x_t) - g(x_\tau)\|}{\|x_t - x_\tau\|} \geq \beta > \varrho(g) .$$

It follows that $\tau = 1$ and $\|g(x_1) - g(x_0)\| \leq \beta\|x_1 - x_0\|$.
Since $x_0, x_1 \in X$ were arbitrarily chosen, it follows that $L(g) \leq \beta$ for every $\beta > \varrho(g)$, and so $L(g) \leq \varrho(g)$.
It follows that g is Lipschitz, and the inequality $\|g(x) - g(y)\|/d(x, y) \leq L(g)$ implies $h_g(x) \leq L(g)$, $x \in X$, and so $\varrho(g) \leq L(g)$. Thus, $\varrho(g) = L(g)$. $\qquad \square$

As we shall work only with the weak supercontractiveness of g on $D(w)$ we mention the following auxiliary result.

Lemma 8.7.29 *The mapping g is weakly supercontractive on $D(w)$ if and only if it is weakly supercontractive on $D_\beta(w)$, for every $\beta > 0$, where $D(w)$ and $D_\beta(w)$ are given by (8.7.31).*

Proof We show that if g is not weakly supercontractive on $D(w)$, then there exists $\beta > 0$ such that g is not weakly supercontractive on $D_\beta(w)$.

Suppose that $h_g(x_0) > 0$ for some $x_0 \in D(w)$. Then there exist $\varepsilon > 0$ and a sequence (x_n) in X such that

$$d(x_n, x_0) \to 0 \quad \text{and} \quad d(g(x_n), g(x_0)) > \varepsilon \, d(x_n, x_0) > 0 \text{ for all } n \in \mathbb{N}.$$
$$(8.7.45)$$

By the continuity of the mapping w there exists $n_0 \in \mathbb{N}$ such that

$$\|w_{x_n}\| \geq \frac{1}{2}\|w_{x_0}\|,$$

for all $n \geq n_0$. Putting $\beta := (1/2)\|w_{x_0}\| > 0$, it follows that the sequence $(x_n)_{n \geq n_0}$ is in $D_\beta(w)$. By (8.7.45) it follows that

$$\limsup_{y \to_{D_\beta(w)} x} \frac{d(g(x), g(y))}{d(x, y)} \geq \varepsilon > 0,$$

hence g is not weakly supercontractive on $D_\beta(w)$.

The reverse implication is obvious. $\qquad\qquad\qquad\qquad\qquad\qquad\qquad\qquad\square$

We present now another sufficient condition for the compactness of the composition operator.

Theorem 8.7.30 *Let X be a compact convex subset of a normed space, E a finite dimensional Banach space, and $w : X \to \mathscr{L}(E)$ and $g : X \to X$ Lipschitz mappings. If the mapping g is weakly supercontractive on $D(w) = \{x \in X : w_x \neq O_E\}$, then the operator $C_{g,w} : \mathrm{BLip}(X, E) \to \mathrm{BLip}(X, E)$ given by (8.7.29) is compact.*

Proof Let (f_n) be a sequence in the closed unit ball of $\mathrm{BLip}(X, E)$. Since for every $x \in X$, $\|f_n(x)\| \leq \|f_n\|_\infty \leq 1$, it follows that the set $\{f_n(x) : n \in \mathbb{N}\}$ is bounded in E, hence totally bounded, because E is finite dimensional.

The inequalities $\|f_n(x) - f_n(y)\| \leq L(f_n)d(x, y) \leq d(x, y)$, $x, y \in X$, imply the equicontinuity of the family $\{f_n : n \in \mathbb{N}\}$, so that, by the Arzela–Ascoli theorem, there exist a continuous function $f : X \to E$ and a subsequence (f_{n_k}) of (f_n) such that $\|f_{n_k} - f\|_\infty \to 0$ as $k \to \infty$. By Proposition 2.4.1, the function f is Lipschitz and $L(f) \leq 1$.

The relations

$$\|C_{g,w}(f_{n_k})(x) - C_{g,w}(f)(x)\| = \|w_x(f_{n_k}(g(x))) - w_x(f(g(x)))\|$$
$$\leq \|w_x\| \cdot \|f_{n_k}(g(x)) - f(g(x))\|$$
$$\leq \|w\|_\infty \cdot \|f_{n_k} - f\|_\infty,$$

valid for all $x \in X$, imply

$$\|C_{g,w}(f_{n_k}) - C_{g,w}(f)\|_\infty \to 0 \quad \text{as} \quad k \to \infty.$$

The proof will be done if we show that

$$L(C_{g,w}(f_{n_k}) - C_{g,w}(f)) \to 0 \quad \text{as} \quad k \to \infty. \tag{8.7.46}$$

For $x, y \in X$, let

$$R_{x,y}^k := \|C_{g,w}(f_{n_k} - f)(x) - C_{g,w}(f_{n_k} - f)(y)\|$$
$$= \|w_x(f_{n_k}(g(x))) - w_x(f(g(x))) + w_y(f(g(y))) - w_y(f_{n_k}(g(y)))\|.$$

Condition (8.7.46) is equivalent to

$$\lim_{k \to \infty} \sup_{x \neq y} \frac{R_{x,y}^k}{d(x, y)} = 0. \tag{8.7.47}$$

Observe that

$$R_{x,y}^k \leq Q_{x,y}^k + H_{x,y}^k, \tag{8.7.48}$$

where

$$Q_{x,y}^k := \|w_x(f_{n_k}(g(x)) - f_{n_k}(g(y)))\| + \|w_x(f(g(x)) - f(g(y)))\|,$$

and

$$H_{x,y}^k := \|(w_x - w_y)(f_{n_k}(g(y)) - f(g(y)))\|$$
$$\leq \|w_x - w_y\| \|f_{n_k}(g(y)) - f(g(y))\| \tag{8.7.49}$$
$$\leq L(w)\|f_{n_k} - f\|_\infty d(x, y).$$

We shall show that $Q^k_{x,y} = 0$ for all $x, y \in X$, so that, by (8.7.48) and (8.7.49)

$$0 \le \frac{R^k_{x,y}}{d(x,y)} \le \frac{H^k_{x,y}}{d(x,y)} \le L(w)\|f_{n_k} - f\|_\infty \to 0 \text{ as } k \to \infty,$$

for all $x, y \in X$, $x \ne y$, showing that (8.7.47) holds.

Let us prove that $Q^k_{x,y} = 0$. For two distinct points $x, y \in X$ consider the segment $[x, y] = \{z(t) : t \in [0,1]\}$, where $z(t) = x + t(y - x)$, $t \in [0,1]$. We have to examine several cases.

If $x, y \in X \setminus D(w)$, then $w_x = w_y = O_E$, so that $Q^k_{x,y} = 0$.

If $[x, y] \subseteq D(w)$, then, by Lemma 8.7.28.3 applied to the convex set $[x, y]$, $g|_{[x,y]}$ is a constant function, hence $g(x) = g(y)$ and so $Q^k_{x,y} = 0$.

Suppose now that $x \in D(w)$ and $[x, y] \cap (X \setminus D(w)) \ne \emptyset$. If

$$t_1 := \min\{t \in [0,1] : \|w_{z(t)}\| = 0\},$$

then $0 < t_1 < 1$, $w_{z(t_1)} = O_E$ and $[x, z(t)] \subseteq D(w)$, for every $0 \le t < t_1$. It follows that the function g is constant on $[x, z(t)]$, so that $Q^k_{x,z(t)} = 0$ for every $0 \le t < t_1$, and, by continuity,

$$Q^k_{x,z(t_1)} = \lim_{t \nearrow t_1} Q^k_{x,z(t)} = 0.$$

We have also $Q^k_{z(t_1),y} = 0$ because $w_{z(t_1)} = O_E$. The inequality

$$Q^k_{x,y} \le Q^k_{x,z(t_1)} + Q^k_{z(t_1),y}$$

yields $Q^k_{x,y} = 0$.

Consequently, $Q^k_{x,y} = 0$ in all cases.   □

8.7.5 The Arens Product

The proof given in [304] to Theorem 8.7.14 uses the Arens product on $\mathrm{lip}_0(X)^{**}$. The proof given here avoids the use of this tool, but, for completeness, we shall present it briefly. It is used in [62] to study the amenability of the Banach algebras of Lipschitz functions.

The *Arens product* is a binary operation defined by R. Arens (see [41, 42]) on the bidual A^{**} of a Banach algebra A. Denoting by a, b the elements of A, by x^*, y^* those of A^* and by φ, ψ elements in A^{**}, the Arens product $\diamond$ is defined gradually by

$$x^* \diamond a \in A^*, \quad (x^* \diamond a)(b) = x^*(ab), \quad b \in B;$$

$$\varphi \diamond x^* \in A^*, \quad (\varphi \diamond x^*)(a) = \varphi(x^* \diamond a), \quad a \in A, \text{ and}$$

$$\varphi \diamond \psi \in A^{**}, \quad (\varphi \diamond \psi)(x^*) = \varphi(\psi \diamond x^*), \quad x^* \in A^*.$$

The Banach algebra A is said to be *Arens regular* if the algebra $(A^{**}, \diamond)$ is commutative.

Lemma 8.7.31 *Let X be compact pointed metric space such that $\mathrm{lip}_0(X)$ separates points uniformly. Then the Arens product on $\mathrm{lip}_0(X)^{**}$ coincides with the pointwise product,*

$$(\varphi \diamond \psi)(u) = \varphi(u)\psi(u), \quad u \in \mathrm{lip}_0(X)^*, \tag{8.7.50}$$

and so the Banach algebra $\mathrm{lip}_0(X)$ is Arens regular.

Proof Since $\mathrm{lip}_0(X)^*$ agrees with the closed linear space generated by the evaluation functionals $\{e_x : x \in X\} \subseteq \mathrm{lip}_0(X)^*$, it is sufficient to verify that (8.7.50) holds only for these functionals.

We shall denote by f, g elements in $\mathrm{lip}_0(X)$, by u, v elements in $\mathrm{lip}_0(X)^*$ and by φ, ψ elements in $\mathrm{lip}_0(X)^{**}$.

Observe that

$$(e_x \diamond f)(g) = e_x(fg) = f(x)g(x), \quad \text{and so} \quad e_x \diamond f = f(x)e_x \in \mathrm{lip}_0(X)^*.$$

Also

$$(\varphi \diamond e_x)(f) = \varphi(e_x \diamond f) = \varphi(f(x)e_x) = f(x)\,\varphi(e_x) = \varphi(e_x)e_x(f),$$

for all $f \in \mathrm{lip}_0(X)$, that is,

$$\varphi \diamond e_x = \varphi(e_x)e_x \in \mathrm{lip}_0(X)^*.$$

Finally,

$$(\varphi \diamond \psi)(e_x) = \varphi(\psi \diamond e_x) = \varphi(\psi(e_x)e_x) = \psi(e_x)\varphi(e_x).$$

Consequently,

$$(\varphi \diamond \psi)(u) = \varphi(u)\psi(u),$$

for all $u \in \mathrm{lip}_0(X)^*$. $\square$

8.7.6 The Nemytskii Superposition Operator

Another important operator acting of function spaces is the Nemytskii superposition operator, see the book [40] for a good presentation.

Let I, J be intervals in $\mathbb{R}$ and $h : I \times J \to \mathbb{R}$. Consider two spaces $F(I, J)$, $G(I)$ of functions $f : I \to J$ and $g : I \to \mathbb{R}$, respectively, and let $H : F(I, J) \to G(I)$ be defined by

$$(Hf)(x) = h(x, f(x)), \quad \text{for } x \in I \text{ and } f \in F(I, J). \tag{8.7.51}$$

The operator H defined by (8.7.51) is called the *Nemytskii superposition operator*.

In this subsection we shall present some results of Matkowski on the characterization of this operator in the case of Lipschitz spaces. The space $\mathrm{Lip}(I)$ of Lipschitz functions on the interval I is considered equipped with the norm

$$\|f\| = |f(x_0)| + L(f), \tag{8.7.52}$$

where x_0 is a fixed point in I.

The first result was obtained in [439].

Theorem 8.7.32 *Let $h : [a, b] \times \mathbb{R} \to \mathbb{R}$ and suppose that the operator H defined by (8.7.51) applies $\mathrm{Lip}[a, b]$ into $\mathrm{Lip}[a, b]$. If H is Lipschitz with respect to the norm (8.7.52), then there exist two functions $\varphi, \psi \in \mathrm{Lip}[a, b]$ such that*

$$h(x, y) = y\varphi(x) + \psi(x), \quad (x, y) \in [a, b] \times \mathbb{R}. \tag{8.7.53}$$

A local version of this result was proved in [440].

Extensions of this result to spaces of Hölder functions were given in [442]. Let I, J be intervals in $\mathbb{R}$. For $0 < \alpha \leq 1$ denote by $\mathrm{Lip}^\alpha(I, J)$ the set of functions $f : I \to J$ such that

$$\|f\|_\alpha := |f(x_0)| + L_\alpha(f) < \infty, \tag{8.7.54}$$

where

$$L_\alpha(f) = \sup\{|f(x) - f(x')|/|x - x'|^\alpha : x, x' \in I, \ x \neq x'\}.$$

Put $\mathrm{Lip}^\alpha(I) = \mathrm{Lip}^\alpha(I, \mathbb{R})$.

Theorem 8.7.33 *Let $h : I \times J \to \mathbb{R}$ be such that the operator H defined by* (8.7.51) *applies* $\mathrm{Lip}^\alpha(I \times J)$ *into* $\mathrm{Lip}^\alpha(I)$.

1. *If there exists a continuous function $\gamma : [0, \infty) \to [0, \infty)$ with $\gamma(0) = 0$ such that*

$$\|Hf - Hg\|_\alpha \le \gamma(\|f - g\|_\alpha), \quad \text{for all}\ \ f, g \in \mathrm{Lip}^\alpha(I \times J),$$

   then h is of the form (8.7.53) *for some functions $\varphi, \psi \in \mathrm{Lip}^\alpha(I)$.*
2. *The same result holds if H is supposed to be uniformly continuous.*

In [441, 443] the vector case is treated. One considers a set X equipped with two metrics, ϱ and d, two normed spaces Y, Z and a convex subset W of Y with $\mathrm{int}(W) \ne \emptyset$. The Lipschitz norms (of the form (8.7.52)) on the spaces $\mathrm{Lip}((X, d), Y)$ and $\mathrm{Lip}((X, \varrho), Y)$ are denoted by $\| \cdot \|_d$ and $\| \cdot \|_\varrho$, respectively.

Theorem 8.7.34 *Suppose that the function $h : X \times W \to Z$ is continuous with respect to the second variable (i.e., $h(x, \cdot) : W \to Z$ is continuous for every $x \in X$). If the operator H defined by* (8.7.51) *applies* $\mathrm{Lip}((X, d), W)$ *into* $\mathrm{Lip}((X, \varrho), Z)$ *and there exists a function $\gamma : [0, \infty) \to [0, \infty)$, bounded from above in a right neighborhood of 0, such that*

$$\|Hf - Hg\|_\varrho \le \gamma(\|f - g\|_d) \ \text{for all}\ \ f, g \in \mathrm{Lip}((X, d), W),$$

then there exist the mappings $A \in \mathrm{Lip}((X, \varrho), \mathscr{L}(Y, Z))$ and $B \in \mathrm{Lip}((X, \varrho), Z)$ such that

$$h(x, y) = A(x)(y) + B(x) \ \text{for all}\ \ (x, y) \in X \times W.$$

A mapping T between two metric spaces (X_1, d_1) and (X_2, d_2) is called *uniformly bounded* if for every $\varepsilon > 0$ there exists $\delta(\varepsilon) > 0$ such that

$$\mathrm{diam}_1\ S \le \delta(\varepsilon) \ \Rightarrow\ \mathrm{diam}_2\ T(S) \le \varepsilon,$$

for all $S \subseteq X_1$.

With this convention, the conclusion of Theorem 8.7.34 holds if H is supposed to be merely uniformly bounded.

The papers [444, 445] are concerned with set-valued versions of the superposition operator.

8.8 The Bishop–Phelps–Bollobás Property

The Bishop–Phelps theorem on the denseness of the support points and support functionals of a bounded convex set is one of the most important results in functional analysis with many applications in optimization theory and in other areas of mathematics. In this section we shall present, following [324] and [258], some variants of this property in the space $\text{Lip}_0(X)$. For completeness, we survey some results obtained in the linear case.

8.8.1   The Bishop–Phelps–Bollobás Theorem in Banach Spaces

Let X be a real Banach space and C a nonempty subset of X. One says that a functional $x^* \in X^*$ *supports* the set C (or that x^* is a *support functional* of the set C) if there exists $x_0 \in C$ such that $x^*(x_0) = \sup x^*(C)$. A point $x_0 \in C$ for which there exists a non-zero functional $x^* \in X^*$ such that $x^*(x_0) = \sup x^*(C)$ is called a *support point* of the set C.

If C is a closed convex subset with nonempty interior of a topological vector space, then every boundary point of C is a support point (see Theorem 1.4.10). Note also that a support point x_0 of the set C belongs to the boundary of C. Indeed, suppose that for some $x_0 \in \text{int } C$ and $x^* \in X^*$, $x^* \neq 0$, $x^*(x_0) = \sup x^*(C)$. Let $r > 0$ be such that $B(x_0, r) = x_0 + rB_X \subseteq C$. Since $x^* \neq 0$ there exists $u \in rB_X$ such that $x^*(u) > 0$, leading to the contradiction $x^*(x_0 + u) = x^*(x_0) + x^*(u) = \sup x^*(C) + x^*(u) > \sup x^*(C)$.

If $C = B_X$ and $x^* \in X^*$ supports B_X at $x_0 \in B_X$, then $x^*(x_0) = \sup x^*(B_X) = \|x^*\|$. In this case we say that x^* *attains its norm* at x_0. Notice that, in this case, $\|x_0\| = 1$ provided that $x^* \neq 0$.

Theorem 8.8.1 (Bishop–Phelps Theorem) *Let X be a real Banach space and C a nonempty closed convex subset of X.*

1. *The support points of the set C are dense in the boundary of C.*
2. *If the set C is nonempty closed convex and bounded, then the support functionals of the set C are dense in X^*. Also the norm-one support functionals of the set C are dense in S_{X^*}.*

Remark 8.8.2 By James' theorem (see Theorem 1.4.20) a Banach space X is reflexive if and only if every continuous linear functional supports the unit ball of X. A Banach space X is called *subreflexive* if the support functionals of its unit ball are dense in X^*. In [556–558] Phelps proved the subreflexivity of many classical Banach spaces by giving characterizations of the support functionals of their unit balls. The general result was obtained later in [83] and extended to bounded closed convex subsets in [84]. A proof is given also in the well-known book of Phelps (see [565] and [567]), and in many other books on functional analysis or optimization. The story of the discovery of this result is nicely described in Phelps' paper [569] (see also [568]). A more general approach is proposed in [561] and [563].

Analyzing the proof given in [83], Bollobás [93] found a quantitative version of the Bishop–Phelps result, allowing to approximate simultaneously a point and a functional by a support point and a support functional, respectively. He applied this result to the study of numerical ranges of continuous linear operators (see [98]).

Theorem 8.8.3 (Bollobás [93]) *Suppose that $x \in S_X$ and $x^* \in S_{X^*}$ are such that $|1 - x^*(x)| \leq \varepsilon^2/2$, where $0 < \varepsilon < 1/2$. Then there exist $y \in S_X$ and $y^* \in S_{X^*}$ such that $y^*(y) = 1$, $\|y - x\| \leq \varepsilon$ and $\|y^* - x^*\| < \varepsilon + \varepsilon^2$.*

Remark 8.8.4 Bollobás [93] showed that his result is optimal in the following sense: For any $0 < \varepsilon < 1$ there exist a Banach space X (a renorming of $\mathbb{R}^2$), $x \in S_X$ and $x^* \in S_{X^*}$ with $x^*(x) = 1 - \varepsilon^2/2$, but either $\|y - x\| \geq \varepsilon$ or $\|y^* - x^*\| \geq \varepsilon + \varepsilon^2$, for every $y \in S_X$ and $y^* \in S_{X^*}$ such that $y^*(y) = 1$.

Remark 8.8.5 For $x \in B_X$ and $x^* \in B_{X^*}$, the condition $|1 - x^*(x)| \leq \delta$ is equivalent to $x^*(x).1 - \delta$.

One obtains a better approximation asking $x^*(x)$ to be closer to $\|x^*\|$.

Theorem 8.8.6 ([3]) *Let $\varepsilon > 0$. Suppose that $x \in B_X$ and $x^* \in S_{X^*}$ are such that $|1 - x^*(x)| \leq \varepsilon^2/4$. Then there exist $y \in S_X$ and $y^* \in S_{X^*}$ such that $y^*(y) = 1$, $\|y - x\| \leq \varepsilon$ and $\|y^* - x^*\| \leq \varepsilon$.*

The following versions of the Bishop–Phelps–Bollobás theorem were proved in [144, Corollary 2.4] by appealing to a result of Phelps [563].

Theorem 8.8.7 *Let X be a Banach space (over $\mathbb{R}$ or $\mathbb{C}$).*

1. *If $x \in B_X$ and $x^* \in B_{X^*}$ are such that $\operatorname{Re} x^*(x) > 1 - \varepsilon^2/2$, where $0 < \varepsilon < \sqrt{2}$, then there exist $y \in S_X$ and $y^* \in S_{X^*}$ such that*

$$y^*(y) = 1, \ \ \|y - x\| < \varepsilon \ \ and \ \ \|y^* - x^*\| < \varepsilon.$$

2. *If $x \in B_X$ and $x^* \in B_{X^*}$ are such that $\operatorname{Re} x^*(x) > 1 - \delta$, where $0 < \delta < 2$, then there exist $y \in S_X$ and $y^* \in S_{X^*}$ such that*

$$y^*(y) = 1, \ \ \|y - x\| < \sqrt{2\delta} \ \ and \ \ \|y^* - x^*\| < \sqrt{2\delta}..$$

As it is remarked in [138, Corollary 3.2], similar results can be obtained by a direct application of the following Brøndsted–Rockafellar result.

Theorem 8.8.8 (Brøndsted–Rockafellar [119], see [567], Theorem 3.17) *Suppose that f is a convex proper lsc function on a Banach space X. Then for every $x \in \operatorname{dom}(f)$, $\varepsilon > 0$, $\lambda > 0$ and every $x^* \in \partial_\varepsilon f(x)$ there exists $y \in \operatorname{dom}(f)$ and $y^* \in X^*$ such that*

$$y^* \in \partial f(y), \ \ \ \|y - x\| \leq \varepsilon/\lambda \ \ and \ \ \|y^* - x^*\| \leq \lambda.$$

In particular, the domain of ∂f is dense in $\operatorname{dom}(f)$.

Here, for a proper convex lsc function $f : X \to \mathbb{R} \cup \{\infty\}$ and $\varepsilon > 0$, the ε-*subdifferential* of f at a point $x \in \mathrm{dom}(f)$ is given by

$$\partial_\varepsilon f(x) = \{x^* \in X^* : x^*(y) - x^*(x) \le f(y) - f(x) + \varepsilon \text{ for all } y \in X\},$$

while the *subdifferential* of f is obtained taking $\varepsilon = 0$ in the above definition, that is,

$$\partial f(x) = \{x^* \in X^* : x^*(y) - x^*(x) \le f(y) - f(x) \text{ for all } y \in X\}.$$

Optimal results were obtained by Chica et al. [144] (see also [145]) who defined some moduli to measure the best possible Bishop–Phelps–Bollobás-type result in a given Banach space. In order to state these results we first need to introduce some notation.

For a real Banach space X with dual X^* put

$$\Pi(X) = \{(x, x^*) \in S_X \times S_{X^*} : x^*(x) = 1\}.$$

Definition 8.8.9 The *Bishop–Phelps–Bollobás* (BPB)-*modulus* of X is the function $\Phi_X : (0, 2) \to \mathbb{R}_+$ whose value at $\delta \in (0, 2)$ is the infimum of all $\varepsilon > 0$ such that for every $(x, x^*) \in B_X \times B_{X^*}$ with $x^*(x) > 1 - \delta$, there exists $(y, y^*) \in \Pi(X)$ such that $\|y - x\| < \varepsilon$ and $\|y^* - x^*\| < \varepsilon$.

The *spherical Bishop–Phelps–Bollobás* (BPB)-*modulus* of X is the function $\Phi_X^s : (0, 2) \to \mathbb{R}_+$ whose value at $\delta \in (0, 2)$ is calculated as above, with the modification that the conclusion holds only for those $(x, x^*) \in S_X \times S_{X^*}$ satisfying $x^*(x) > 1 - \delta$.

Remark 8.8.10 It is shown in [144] that it does not matter if in the above definitions some (or all) of the strict inequality signs are replaced by $\le$ or by $\ge$, accordingly.

Remark 8.8.11 In [144] one works with complex scalars and the condition $\mathrm{Re}\, x^*(x) > 1 - \delta$. As it is remarked in the same paper, the dual of a complex Banach space X is isometric (taking real parts of functionals) to the dual of the subjacent real space $X_\mathbb{R}$ and $\Pi(X)$ does not change if we consider X as a real Banach space. Consequently, only the real structure of the space is playing a role in the above definitions. For simplicity, we restrict the presentation to the real case.

Endowing $X \times X^*$ with the d_∞ metric

$$d_\infty((x, x^*), (y, y^*)) = \max\{\|x - y\|, \|x^* - y^*\|\},$$

the corresponding Pompeiu–Hausdorff distance between two subsets A, B of $X \times X^*$ is given by

$$d_H(A, B) = \max\{\sup_{a \in A} d_\infty(a, B), \sup_{b \in B} d_\infty(b, A)\},$$

where, for $u \in X \times X^*$ and $V \subseteq X \times X^*$

$$d_\infty(u, V) = \inf\{d_\infty(u, v) : v \in V\}.$$

For $0 < \delta < 2$ put

$$A_X(\delta) = \{(x, x^*) \in B_X \times B_{X^*} : x^*(x) > 1 - \delta\},$$
$$A_X^s(\delta) = \{(x, x^*) \in S_X \times S_{X^*} : x^*(x) > 1 - \delta\},$$

and

$$D_X(\delta) = \{\varepsilon > 0 : \forall (x, x^*) \in A_X(\delta), \exists (y, y^*) \in \Pi(X), \|y - x\| < \varepsilon, \|y^* - x^*\| < \varepsilon\},$$
$$D_X^s(\delta) = \{\varepsilon > 0 : \forall (x, x^*) \in A_X^s(\delta), \exists (y, y^*) \in \Pi(X), \|y - x\| < \varepsilon, \|y^* - x^*\| < \varepsilon\}.$$

It follows that $\Pi(X) \subseteq A_X^s(\delta) \subseteq A_X(\delta)$ and

$$\Phi_X(\delta) = d_H(A_X(\delta), \Pi(X)) \quad \text{and} \quad \Phi_X^s(\delta) = d_H(A_X^s(\delta), \Pi(X)).$$

Also $D_X(\delta) \subseteq D_X^s(\delta)$ implies

$$\Phi_X(\delta) = \inf D_X(\delta) \geq \inf D_X^s(\delta) = \Phi_X^s(\delta).$$

For $0 < \delta_1 < \delta_2 < 2$, the inclusions $A_X(\delta_1) \subseteq A_X(\delta_2)$ and $A_X^s(\delta_1) \subseteq A_X^s(\delta_2)$ yield $D_X(\delta_1) \supset D_X(\delta_2)$ and $D_X^s(\delta_1) \supset D_X^s(\delta_2)$, respectively, so that $\Phi_X(\delta_1) \leq \Phi_X(\delta_2)$ and $\Phi_X^s(\delta_1) \leq \Phi_X^s(\delta_2)$.

The result from [144] is the following.

Theorem 8.8.12 *For every Banach space X and every $\delta \in (0, 2)$,*

$$\Phi_X^s(\delta) \leq \Phi_X(\delta) \leq \sqrt{2\delta}.$$

The proof is based on a result of Phelps on support properties of convex sets.

Theorem 8.8.13 ([563], Corollary 2.2) *Let C be a closed convex set of a Banach space X, $x^* \in S_{X^*}$, $\varepsilon > 0$ and $x \in C$ satisfy*

$$x^*(x) \leq \sup x^*(C) + \varepsilon.$$

Then for every $0 < \lambda < 1$ there exist $y^* \in X^*$ and $y \in C$ such that

$$y^*(y) = \sup y^*(C), \quad \|y - x\| \leq \varepsilon/\lambda \quad \text{and} \quad \|y^* - x^*\| \leq \lambda.$$

Remark 8.8.14 In [144] one uses the above result in the case $C = B_X$ in the following form:

> Let X be a Banach space and let $x^* \in S_{X^*}$, $x \in B_X$ and $\eta > 0$ be such that $x^*(x) > 1 - \eta$. Then for every $\lambda \in (0, 1)$ there exist $y^* \in X^*$ and $y \in S_X$ such that
>
> $$y^*(y) = \|y^*\|, \quad \|y - x\| \le \eta/\lambda \quad \text{and} \quad \|y^* - x^*\| \le \lambda.$$

8.8.2 The Bishop–Phelps Theorem for Weak*-Closed Convex Subsets of the Dual Space

Let X be a real Banach space and C a convex subset of the dual space X^*. A point $x^* \in C$ is called a w^*-*support point* of C if there exists $x_0 \in X \setminus \{0\}$ such that

$$x^*(x_0) = \sup C(x_0), \tag{8.8.1}$$

where, for $x \in X$, $C(x) = \{x^*(x) : x^* \in C\}$. An element $x_0 \in X \setminus \{0\}$ satisfying (8.8.1) is called a w^*-*support functional* of C.

The following results were obtained by Phelps in [562].

Theorem 8.8.15 *Let X be a real Banach space and C a w^*-closed convex subset of the dual space X^*.*

1. *The w^*-support points of C are norm-dense in the norm-boundary of C.*
2. *The w^*-support functionals of C are dense among those for which $\sup C(x) < \infty$.*

The theorem has the following corollary.

Corollary 8.8.16 *Let X be a real Banach space and C a w^*-closed convex subset of the dual space X^*. Then C is the intersection of its w^*-closed supporting half-spaces.*

8.8.3 The Bishop–Phelps Theorem Fails in the Complex Case and in Locally Convex Spaces

If X is a complex Banach space, then $x^* \in X^*$ is called a support functional of a subset C of X if there exists $x_0 \in C$ such that

$$|x^*(x_0)| = \sup\{|x^*(x)| : x \in C\}.$$

Phelps [566] asked whether the Bishop–Phelps theorem holds in the complex case. Lomonosov [403] gave a negative answer to this question (a nice presentation of this result along with some further remarks is given in the talk [568]). Let $H^\infty = H^\infty(D)$ be the Banach, space with respect to the sup-norm, of all bounded functions, analytic in the unit open disc $D \subseteq \mathbb{C}$. Then the evaluation functionals $e_z(f) = f(z)$, $f \in H^\infty$, $z \in D$, belong to $(H^\infty)^*$. By some results in complex analysis there exists a Banach space X such that $X^* = H^\infty(D)$ and $e_z \in X$ for every $z \in D$. If C is the closed convex hull of the set $\{e_z : z \in D\}$, then every support functional $g \in H^\infty$ of C is of the form $g = \alpha \operatorname{id}_D$, for some $\alpha \in \mathbb{C}$, that is, the set of complex support functionals of C is contained in the one-dimensional subspace $\mathbb{C} \cdot \operatorname{id}_D$, where id_D is the identity function on D, $\operatorname{id}_D(z) = z$, $z \in D$.

Consider now the one-dimensional subspace L of X generated by e_0 (the evaluation functional at $0 \in D$) and let $\pi : X \to X/L$ be the corresponding projection. Then $(X/L)^* \cong L^\perp$, where

$$L^\perp := \{x^* \in X^* : x^*(u) = 0, \ \forall u \in L\} = \{g \in H^\infty : g(0) = 0\}.$$

Lomonosov [403, Theorem 2] showed that the set $C_1 = \pi(C)$ has no non-zero support functionals.

In [404] he interpreted these results in the language of the algebra $\mathscr{L}(\mathscr{H})$ of operators on a Hilbert space $\mathscr{H}$ and proved in [405] that if the Bishop–Phelps theorem holds for a uniform dual algebra A of operators on a Hilbert space $\mathscr{H}$, then the algebra A is selfadjoint. Consequently, the Bishop–Phelps theorem fails for uniform non-selfadjoint subalgebras of $\mathscr{L}(\mathscr{H})$ (see also the survey paper [47]).

Another question concerns the validity of the Bishop–Phelps theorem in complete locally convex spaces. Peck [545] announced that the locally convex space $\prod_{n=1}^\infty X_n$, where X_n are non-reflexive real Banach spaces, contains a closed bounded convex set without support points. The details appeared in [546], where some situations in which a Bishop–Phelps-type theorem holds in the locally convex setting are also discussed.

For other results and some open problems in this area see, for instance, the paper [2].

8.8.4   Norm-Attaining Operators

Let X, Y be Banach spaces. An operator $T \in \mathscr{L}(X, Y)$ is said to be *norm-attaining* if there exists $x_0 \in S_X$ such that $\|T x_0\| = \|T\| (= \sup\{\|T x\| : x \in B_X\})$. Denote by $\mathrm{NA}(X, Y)$ the set of all norm-attaining operators. In the seminal paper [83] Bishop and Phelps asked if their result can be extended to spaces of operators, i.e., whether the set of norm-attaining operators is dense or not in $\mathscr{L}(X, Y)$.

Definition 8.8.17 A pair (X, Y) of Banach spaces is said to have:

- *the Bishop–Phelps property* if the set of norm-attaining operators is dense in $\mathscr{L}(X, Y)$;
- *the Bishop-Phelps-Bollobas property* if for every $\varepsilon > 0$ there exists $\delta(\varepsilon) > 0$ such that for every $T \in \mathscr{L}(X, Y)$ and $x_0 \in S_X$ such that $\|Tx_0\| > 1 - \delta(\varepsilon)$ there exists a norm-one operator $S \in \mathscr{L}(X, Y)$ and $u_0 \in S_X$ satisfying the conditions

$$\|Su_0\| = 1, \quad \|x_0 - u_0\| < \varepsilon \quad \text{and} \quad \|S - T\| < \varepsilon.$$

As the study of these properties is strongly connected with some geometric properties of the involved Banach spaces, we start by presenting some of these properties.

The Radon–Nikodým Property (RNP) and Exposed Points

The Radon–Nikodým Property (RNP) for Banach spaces was defined in Sect. 1.6.3. Here we extend the definition by considering sets with the RNP.

Definition 8.8.18 A bounded closed convex subset C of a Banach space X is said to have the *Radon–Nikodým property* (RNP) if for every finite positive measure space $(\Omega, \mathscr{A}, \mu)$ and every vector measure $\nu : \mathscr{A} \to X$, of bounded variation and absolutely continuous with respect to μ and such that the average range $\{\nu(A)/\mu(A) : A \in \mathscr{A}, \mu(A) > 0\}$ is contained in C, there exists a Bochner integrable function $f : \Omega \to X$ such that $\nu(A) = \int_A f d\mu$ for every $A \in \mathscr{A}$.

It follows that the space X has the RNP if and only if its closed unit ball B_X has the RNP, or, equivalently, if every closed bounded convex subset of X has the RNP.

A point $x_0 \in C$ is called an *exposed point* of C if there exists $x^* \in X^* \setminus \{0\}$ such that $x^*(x_0) = \sup x^*(C)$ and $x^*(x) < x^*(x_0)$ for all $x \in C \setminus \{x_0\}$. The functional x^* is called an *exposing functional* for C. A point $x_0 \in C$ is called a *strongly exposed point* if there exists $x^* \in X^* \setminus \{0\}$ such that $x^*(x_0) = \sup x^*(C)$ and $\|x_n - x_0\| \to 0$ for every maximizing sequence (x_n) (i.e., a sequence (x_n) in C such that $x^*(x_n) \to \sup x^*(C)$). In this case, one says that x^* is a *strongly exposing functional* (or that x^* *strongly exposes* C at x_0).

A strongly exposed point is exposed, and an exposed point is an extreme point.

For a bounded subset A of X, $x^* \in X^*$ and $\alpha > 0$ the *slice* $S(A, x^*, \alpha)$ is defined by

$$S(A, x^*, \alpha) = \{x \in A : x^*(x) > \sup x^*(A) - \alpha\}, \tag{8.8.2}$$

and a *w^*-slice* of a subset D of X^* is defined by

$$S(D, x, \alpha) = \{x^* \in D : x^*(x) > \sup D(x) - \alpha\}, \tag{8.8.3}$$

for $x \in X$ and $\alpha > 0$.

In the case $A = B_X$ the equality (8.8.2) becomes

$$S(B_X, x^*, \alpha) = \{x \in B_X : x^*(x) > \|x^*\| - \alpha\},$$

while the equality (8.8.3) becomes for $D = B_{X^*}$

$$S(B_{X^*}, x, \alpha) = \{x^* \in B_{X^*} : x^*(x) > \|x\| - \alpha\}.$$

A subset C of X is called *dentable* if it admits slices of arbitrarily small diameter, that is, for every $\varepsilon > 0$ there exists $x^* \neq 0$ in X and $\alpha > 0$ such that diam $S(C, x^*, \alpha) < \varepsilon$. The definition of a w^*-dentable subset of X^* is obtained by analogy, working with w^*-slices.

We mention the following results concerning dentability, slices and exposed points.

Proposition 8.8.19 *Let X be a Banach space.*

1. *A bounded subset C of X is dentable if and only if for every $\varepsilon > 0$ there exists a point $x_\varepsilon \in C$ such that*

$$x_\varepsilon \notin \overline{\mathrm{co}}\, (C \setminus B[x_\varepsilon, \varepsilon]).$$

2. *A point $x_0 \in C$ is a strongly exposed point of C if and only if there exists $x^* \in X^* \setminus \{0\}$ such that $x_0 \in S(C, x^*, \alpha)$ for all $\alpha > 0$ and*

$$\lim_{\alpha \searrow 0} \mathrm{diam}\, S(C, x^*, \alpha) = 0.$$

We shall need the following result [5, Lemma 2.1].

Lemma 8.8.20 *Let X be a uniformly convex Banach space with modulus of convexity δ_X. Then for $x^* \in S_{X^*}$ and $\varepsilon > 0$,*

$$\mathrm{diam}\, S(B_X, x^*, \delta_X(\varepsilon)) \leq \varepsilon.$$

Proof Let $x, y \in S(B_X, x^*, \delta_X(\varepsilon))$. Then $x^*(x) > 1 - \delta_X(\varepsilon)$ and $x^*(y) > 1 - \delta_X(\varepsilon)$, so that

$$\|x + y\| \geq x^*(x + y) > 2(1 - \delta_X(\varepsilon)),$$

which implies $\|x - y\| < \varepsilon$.   □

For us the following characterization of the RNP will be of interest (see [567]).

Proposition 8.8.21 *For a closed bounded convex subset C of a Banach space X the following are equivalent.*

1. *The set C has the RNP.*
2. *Every norm closed bounded convex subset of C has a dense (in X^*) G_δ set of support functionals.*
3. *Every norm closed bounded convex subset of C has a dense (in X^*) G_δ set of strongly exposing functionals.*
4. *Every closed convex subset of C is dentable.*

Stegall's Variational Principle

The following remarkable result was proved by Stegall, see [645, 646].

Theorem 8.8.22 (Stegall's Variational Principle) *Let X be a Banach space, C a subset of X with the RNP and $f : X \to \mathbb{R}$ usc and bounded from above. Then for every $\varepsilon > 0$ there exists $x^* \in X^*$ with $\|x^*\| < \varepsilon$ such that $f + x^*$ and $f + |x^*|$ strongly expose C.*

There is a tight connection between spaces with the RNP and Asplund spaces: a Banach space is an Asplund space if and only if its conjugate has the RNP. Recall that Asplund spaces can be defined as those Banach spaces on which every continuous convex function is Fréchet differentiable on a dense set (see Definition 3.4.14 and the comments following it).

Density Results

We start by presenting some density results related to the Bishop–Phelps property (see Definition 8.8.17). The first who attacked this problem was Lindenstrauss [388]. We mention first the following result from Lindenstrauss' paper.

Proposition 8.8.23 ([388], Proposition 4) *If Y is a strictly convex Banach space such that $\mathcal{K}(c_0, Y) \neq \mathcal{L}(c_0, Y)$, then $\overline{NA(c_0, Y)} \neq \mathcal{L}(c_0, Y)$. In particular this holds if Y is a strictly convex renorming of c_0.*

Proof Let $(e_k)_{k \in \mathbb{N}}$ be the canonical basis of c_0. Suppose that $T \neq O$ in $\mathcal{L}(c_0, Y)$ and $x = (x_k)_{k \in \mathbb{N}} \in S_{c_0}$ are such that $\|Tx\| = \|T\|$. If $n \in \mathbb{N}$ is such that $|x_k| < 1/2$ for all $k > n$, then $|x_k \pm \frac{1}{2}| < 1$, so that $\|x \pm \frac{1}{2} e_k\| = 1$, for all $k > n$ and

$$\|T\| = \|Tx\| \leq \frac{1}{2} \left(\|Tx - \frac{1}{2}Te_k\| + \|Tx + \frac{1}{2}Te_k\| \right) \leq \|T\|.$$

The strict convexity of Y implies $Te_k = 0$ for all $k > n$, that is, T has finite dimensional range, denoted as $T \in \mathcal{F}(c_0, Y)$. But then

$$\overline{NA(c_0, Y)} \subseteq \overline{\mathcal{F}(c_0, Y)} \subseteq \mathcal{K}(c_0, Y) \neq \mathcal{L}(c_0, Y).$$

$\square$

He also showed that if Y is a strictly convex Banach space isomorphic to c_0, then the space $X = c_0 \oplus Y$ equipped with the norm $\|(y, z)\| = \max\{\|y\|, \|z\|\}$ has the property that $NA(X, X)$ is not dense in $\mathscr{L}(X, X)$.

In the same paper Lindenstrauss proved that for every Banach spaces X, Y the set $\{T \in \mathscr{L}(X, Y) : T^{**} \in NA(X^{**}, Y^{**})\}$ is norm-dense in $\mathscr{L}(X, Y)$, hence every reflexive Banach space has property A (see below). The result was improved by Zizler [696] who used a variational principle to prove that the set of operators $T \in \mathscr{L}(X, Y)$ such that $T^* \in NA(Y^*, X^*)$ is norm-dense in $\mathscr{L}(X, Y)$. Iwanik [300] (see also [299]) showed that Lindenstrauss' proof can be adapted to yield this result. It is obvious that if T attains its norm, then T^* also attains its norm. Indeed, if $\|Tx\| = \|T\|$ for some $x \in S_X$, then there exists $y^* \in S_{Y^*}$ such that $y^*(Tx) = \|Tx\|$, so that $(T^*y^*)(x) = y^*(Tx) = \|Tx\| = \|T\| = \|T^*\|$, which implies $\|T^*y^*\| = \|T^*\|$. This shows that Zizler's result extends Lindenstrauss' result. Using again a variational principle, Poliquin and Zizler [577] proved that Zizler's density results can be achieved by summing a given operator with a rank-one operator of arbitrarily small norm. A variational principle with applications to norm-attaining operators is also proved in [4].

Lindenstrauss [388] considered two properties of Banach spaces. One says that a Banach space X has:

• *property* A if $NA(X, Y)$ is dense in $\mathscr{L}(X, Y)$ for every Banach space Y;
• *property* B if $NA(Y, X)$ is dense in $\mathscr{L}(Y, X)$ for every Banach space Y.

Property B means, in fact, that in the Bishop–Phelps theorem the scalar field $\mathbb{R}$ can be replaced by the Banach space X. Bourgain [108] showed that a Banach space X has property A if and only if it has the Radon–Nikodým property.

Lindenstrauss also proved that a Banach space Y has property B if it has the so-called *property* β:

There exist the pairs $(y_i, y_i^*) \in S_Y \times S_{Y^*}$, $i \in I$, and $0 \le \lambda < 1$ such that

(i) $y_i^*(y_i) = 1$, for all $i \in I$, (ii) $|y_i^*(y_j)| \le \lambda$, for all $i \ne j$,

(iii) $\|y\| = \sup_i |y_i^*(y)|$, for all $y \in Y$.

Clearly, the spaces $c_0(I)$ and $\ell^\infty(I)$ have this property with $y_i = e_i$, $y_i^* = \delta_i$, $i \in I$, where $e_i(j) = 1$ for $j = i$ and $e_i(j) = 0$ otherwise, and $\delta_i(x) = x(i)$ for $x : I \to \mathbb{R}$.

An analogous condition for property A was introduced by Schachermayer [620] (see also [619]). A Banach space X has property A if it has the so-called *property* α:

There exist the pairs $(x_i, x_i^*) \in S_X \times S_{X^*}$, $i \in I$, and $0 \le \lambda < 1$ such that

(i) $x_i^*(x_i) = 1$, for all $i \in I$, (ii) $|x_i^*(x_j)| \le \lambda$, for all $i \ne j$,

(iii) $B_X = \overline{aco}\{x_i : i \in I\}$,

where aco(A) denotes the absolutely convex hull of a set A and $\overline{\text{aco}}$ its closed absolutely convex hull.

It is clear that the space $\ell^1(I)$ has property α with the same families of vectors and functionals as in the case of property β. Lindentrauss, *op. cit.*, proved that:

- the space $L^1(\mu)$ has property A $\iff$ μ is purely atomic (i.e., the space $L^1(\mu)$ agrees with $\ell^1(I)$);
- the space $C(K)$, K a compact metric space, has property A $\iff$ K is finite;
- if the unit ball B_X of X is the closed absolutely convex hull of a family of uniformly exposed points, then X has property A;
- if X has property A and admits a strictly convex renorming, then the unit ball B_X of X is the closed convex hull of its exposed points;
- if X has property A and admits a locally uniformly convex renorming, then the unit ball B_X of X is the closed convex hull of its strongly exposed points.

We mention also the following results related to the RNP:

- if $C \subseteq X$ has the RNP, then for every Banach space Y, the subset of operators $T \in \mathscr{L}(X, Y)$ such that $\sup\{\|Tx\| : x \in C\}$ is attained contains a G_δ-set dense set in $\mathscr{L}(X, Y)$ (J. Bourgain, Israel J. Math. 28 (1977), 265–271);
- the Banach space X has the RNP if and only if for every Banach space Z isomorphic to X and every Banach space Y, the set NA(Z, Y) is dense in $\mathscr{L}(Z, Y)$ (R. E. Huff, Rev. Roumaine Math. Pures Appl. 25 (1980), 239–241).

We shall present now some density and non-density results for some concrete Banach spaces. A good presentation of these results, along with some density results for norm-attaining multilinear mappings and polynomials, is given in the survey paper by Acosta [1]. For the density of compact norm-attaining operators in spaces of compact operators, see the survey [437]. Notice that, for two Banach spaces X, Y, the correspondence $B \mapsto A$, where, for $B \in \mathscr{L}_2(X \times Y)$ and $x \in X$, $Ax \in Y^*$ is given by the equality $(Ax)(y) = B(x, y)$, $y \in Y$, establishes an isometric isomorphism between the space $\mathscr{L}_2(X \times Y)$ of continuous bilinear forms on $X \times Y$ and the space $\mathscr{L}(X, Y^*)$.

Some Positive Results

The set NA(X, Y) is norm-dense in $\mathscr{L}(X, Y)$ in the following cases:

- $X = L^1(\mu)$ and $Y = L^1(\nu)$, where μ, ν are arbitrary measures (A. Iwanik, Pacific J. Math. 83 (1979), 381–386);
- $X = L^1(\mu)$ and $Y = L^\infty[0, 1]$ (C. Finet and R. Payá, Israel J. Math. 108 (1998), 139–143);
- $X = L^1(\mu)$ and $Y = L^\infty(\nu)$ for μ arbitrary and ν localizable (R. Payá, and Y. Saleh, Arch. Math. 75 (2000), 380–388);
- $X = C(K)$ and $Y = C(S)$, K, S compact Hausdorff spaces (J. Johnson and J. Wolfe, Studia Math. 65 (1979), 7–19);
- X Asplund and $Y = C(K)$, K a compact Hausdorff space (J. Johnson and J. Wolfe, Studia Math. 65 (1979), 7–19);

- $X = L^1(\mu)$, μ a finite positive measure, and Y has the RNP (J. J. Uhl, Pacific J. Math. 63 (1976), 293–300). In the same paper a partial converse is proved: if Y is strictly convex and $NA(L^1[0, 1], Y)$ is dense in $\mathscr{L}(L^1[0, 1], Y)$, then Y has the RNP.

Some Negative Results
- $NA(L^1[0, 1], C[0, 1])$ is not dense in $\mathscr{L}(L^1[0, 1], C[0, 1])$ (Schachermayer [620]); another example with $C(S)$, S a compact metric space, instead of $C[0, 1]$, was given by J. Johnson and J. Wolfe, Proc. Amer. Math. Soc. 86 (1982), 609–612;
- for every infinite dimensional Banach space X there exists a norm-one operator $T \in \mathscr{L}(X, c_0)$ which does not attain its norm (M. Martín, J. Merí and R. Payá, J. Math. Anal. Appl. 318 (2006), 175–189, Lemma 2.2);
- there exists a compact linear operator between two Banach spaces which cannot be approximated by norm-attaining operators (M. Martín, J. Funct. Anal. 267 (2014), 1585–1592; see also the survey on this topic by the same author, Rev. R. Acad. Cienc. Exactas Fís. Nat., Ser. A Mat., RACSAM 110 (2016), No. 1, 269–284).

The following spaces do not have property B:

- $Y = \ell^p$—there exists a Banach space X such that for all $p, 1 < p < \infty$, $NA(X, \ell^p)$ is not dense in $\mathscr{L}(X, \ell^p)$ (W. T. Gowers, Israel J. Math. 69 (1990), 129–151);
- any strictly convex (or $\mathbb{C}$-strictly convex in the case of complex scalars) Banach space Y (M. D. Acosta, Proc. Roy. Soc. Edinburgh, 129A (1999), 1107–1114).

Quantitative Results of Bishop–Phelps–Bollobás Type
The study of the Bishop–Phelps–Bollobás (BPB) property for operators (see Definition 8.8.17) was initiated in [3].

In the same paper it is shown that if the space Y has property β, then the pair (X, Y) has the BPB property, for any Banach space X. In the case of an arbitrary pair (X, Y) of Banach spaces the Bishop–Phelps and the Bishop–Phelps–Bollobás property can be distinct notions: the pair (ℓ^1, Y) has the Bishop–Phelps property for every Banach space Y (i.e., ℓ^1 has property A in the sense of [388]), but there exist Banach spaces Y for which (ℓ^1, Y) does not have the BPB property, see [3, Theorem 4.1].

For some recent results on this topic, see the papers [6, 7, 48, 148, 354, 355, 367], and the references quoted therein.

8.8.5   Support Functionals in Spaces of Lipschitz Functions

We shall present now, following [324] and [258], some results on norm-attaining functionals in spaces of Lipschitz functions.

Let (X, d) be a metric space with a base point x_0. Let $\mathrm{Lip}_0(X)$ be the space of Lipschitz functions vanishing at x_0 equipped with the Lipschitz norm

$$L(f) = \sup\left\{ \frac{|f(x) - f(y)|}{d(x, y)} : x, y \in X, \, x \neq y \right\}.$$

If X is a real Banach space, then we take $x_0 = 0$, denote $\mathrm{Lip}_0(X)$ by $X^{\#}$ and call it the *Lipschitz dual* of X.

Definition 8.8.24 One says that a function $f \in \mathrm{Lip}_0(X)$ *attains its norm in the strong sense*, if there exist $x, y \in X$, $x \neq y$, such that

$$L(f) = \frac{|f(x) - f(y)|}{d(x, y)}.$$

In this case we also say that f strongly attains its norm at the pair x, y. The set of functions in $\mathrm{Lip}_0(X)$ that strongly attain their norm is denoted by $\mathrm{SA}(\mathrm{Lip}_0(X))$ or by $\mathrm{SA}(X^{\#})$.

Notice that if f is linear, $f = x^*$ for some $x^* \in X^*$, then

$$\|x^*\| = L(x^*) = \left| x^*\left(\frac{y - x}{\|y - x\|} \right) \right|,$$

that is, the functional x^* attains its norm at $(y - x)/\|y - x\| \in S_X$.

Since some results hold only for some special classes of metric spaces, we briefly recall the definitions and some properties of these spaces, referring to Chap. 5 for further details (one can consult also [85, 264] or [541]).

A metric space (X, d) is called *convex* (or *Menger convex*) if for every pair x, y of distinct points in X there exists $z \in X \setminus \{x, y\}$ such that

$$d(x, y) = d(x, z) + d(z, y).$$

A *geodesic path* in X is an isometric mapping from an interval $[a, b] \subseteq \mathbb{R}$ into X, i.e., a mapping $\sigma : [a, b] \to X$ such that

$$d(\sigma(t), \sigma(t')) = |t - t'|,$$

for all $t, t' \in [a, b]$. A metric space is called a *geodesic space* if for every pair x, y of distinct points in X there exists an isometric mapping $\sigma : [0, d(x, y)] \to X$ such that $\sigma(0) = x$ and $\sigma(d(x, y)) = y$, i.e., any points $x, y \in X$ can be connected by a geodesic path. It is obvious that a geodesic space is Menger convex.

As it was shown by Menger [463], the converse holds under the hypothesis of completeness.

Theorem 8.8.25 *A Menger convex complete metric space is a geodesic metric space.*

This result in a slightly different form, using instead of Menger convexity the existence of midpoints between every two points in the space, is mentioned in Sect. 5.1 of Chap. 5. Thus, in complete metric spaces, the property of being geodesic is equivalent to Menger convexity, which is in its turn equivalent to the existence of midpoints between every two points in the space.

Remark 8.8.26 If X is a geodesic metric space, then in the calculation of the Lipschitz norm of a function $f \in \mathrm{Lip}_0(X)$ it is sufficient to take only points in X where f is different from 0, that is,

$$L(f) = \sup \left\{ \frac{|f(x) - f(y)|}{d(x,y)} : x, y \in X, \ x \neq y, \ f(x) \neq 0 \text{ and } f(y) \neq 0 \right\}.$$

Indeed, denoting by $\tilde{L}(f)$ the above supremum, it is obvious that $\tilde{L}(f) \leq L(f)$. We have to examine only the situation when $f(z_0) \neq 0$ and $f(z_1) = 0$, for a pair z_0, z_1 of points in X. Let $\sigma : [0, \alpha] \to X$ be an isometric mapping such that $\sigma(0) = z_0$ and $\sigma(\alpha) = z_1$, where $\alpha = d(z_0, z_1)$.
Let

$$t_0 = \inf\{t \in [0, \alpha] : f(\sigma(t)) = 0\}.$$

By the continuity of f, $0 < t_0 \leq 1$, $f(\sigma(t_0)) = 0$ and $f(\sigma(t)) \neq 0$ for all t, $0 \leq t < t_0$. Let $0 < t_n < t_0$ with $t_n \to t_0$ and let $y_n = \sigma(t_n)$. Then

$$\frac{|f(z_0) - f(z_1)|}{d(z_0, z_1)} = \frac{|f(z_0)|}{d(z_0, z_1)} = \frac{|f(z_0) - f(\sigma(t_0))|}{d(z_0, z_1)}$$

$$\leq \frac{|f(z_0) - f(\sigma(t_0))|}{d(z_0, \sigma(t_0))} = \lim_{n \to \infty} \frac{|f(z_0) - f(y_n)|}{d(z_0, y_n)} \leq \tilde{L}(f),$$

implying $L(f) \leq \tilde{L}(f)$.
Note the following simple result.

Proposition 8.8.27 *Let (X, d) be a metric space. If $f \in \mathrm{Lip}_0(X)$ strongly attains its norm at the pair x, y of distinct points in X, then it strongly attains the norm at x, z and y, z, for every $z \in X \setminus \{x, y\}$ such that $d(x, y) = d(x, z) + d(z, y)$. In this case,*

$$f(z) = \frac{d(z, y) f(x) + d(x, z) f(y)}{d(x, y)}.$$

In particular, if f is defined on a convex subset X of a normed space Z, then $f((1 - t)x + ty) = (1 - t)f(x) + tf(y)$ for all $t \in [0, 1]$, that is, f is affine on the algebraic segment $[x, y] = \{x + t(y - x) : t \in [0, 1]\}$.

Proof We can suppose $f(x) - f(y) \geq 0$. Then

$$f(x) - f(y) = L(f)d(x, y) = L(f)d(x, z) + L(f)d(z, y)$$
$$\geq (f(x) - f(z)) + (f(z) - f(y)) = f(x) - f(y),$$

which yield the equalities

$$f(x) - f(z) = L(f)d(x, z) \quad \text{and} \quad f(z) - f(y) = L(f)d(z, y),$$

showing that f strongly attains its norm at the pairs x, z and y, z.

From the first equality from above

$$f(z) = f(x) - L(f)d(x, z) = f(x) - \frac{f(x) - f(y)}{d(x, y)} \cdot d(x, z)$$
$$= \frac{f(x)[d(x, y) - d(x, z)] + f(y)d(x, z)}{d(x, y)} = \frac{f(x)d(y, z) + f(y)d(x, z)}{d(x, y)}.$$

$\square$

This notion is too strong to obtain density results as shown by the following simple example.

Example 8.8.28 The set $SA(\mathrm{Lip}_0[0, 1])$ is not dense in $\mathrm{Lip}_0[0, 1]$.

It is known that $\mathrm{Lip}_0[0, 1]$ is isometrically isomorphic to the space $L^\infty[0, 1]$ through the mapping $\Phi(f) = f'$ (see Example 8.1.8). Let A be a nowhere dense closed subset of $[0, 1]$ of positive Lebesgue measure. Then the characteristic function χ_A of A satisfies

$$L(h_A - f) = \|\chi_A - f'\|_\infty \geq \frac{1}{2}, \tag{8.8.4}$$

for every $f \in SA(\mathrm{Lip}_0[0, 1])$, where $h_A = \Phi^{-1}(\chi_A)$.

Indeed, by Proposition 8.8.27, every $f \in SA(\mathrm{Lip}_0[0, 1])$ is applied through Φ to a function which has the derivative $\pm L(f)$ on some non-degenerate interval (a, b) contained in $[0, 1]$. Since $\|\chi_A\|_\infty = 1$, the inequality (8.8.4) holds if $\|f'\|_\infty \leq 1/2$. If $L(f) = \|f'\|_\infty > 1/2$, then $|f'(t)| > 1/2$ for all t in the interval (a, b). Since A is nowhere dense, there exists a smaller interval $(c, d) \subseteq (a, b)$ such that $A \cap (c, d) = \emptyset$. But then, for every $t \in (c, d)$,

$$\|\chi_A - f'\|_\infty \geq |\chi_A(t) - f'(t)| = |f'(t)| > 1/2.$$

Based on this example one can prove a more general result.

Proposition 8.8.29 *Let (X, d) be a geodesic metric space. Then $SA(\mathrm{Lip}_0(X))$ is not dense in $\mathrm{Lip}_0(X)$.*

Proof Let $x_1 \in X \setminus \{x_0\}$ and let $\gamma : [x_0, x_1] \to [0, d(x_0, x_1)]$ be an isometric mapping. Without restricting the generality we can suppose $d(x_0, x_1) = 1$, that is, $\gamma : [x_0, x_1] \to [0, 1]$ is an isometric mapping with $\gamma(x_0) = 0$ and $\gamma(x_1) = 1$. Extend it to a 1-Lipschitz function $\varphi : X \to [0, 1]$ (see Remark 4.1.2) and let $g = h_A \circ \varphi$, where h_A is the function constructed in Example 8.8.28.

We shall show that

$$L(f - g) \geq \frac{1}{2},$$

for every function $f \in SA(\text{Lip}_0(X))$.

Supposing, on the contrary, that

$$L(f - g) < \frac{1}{2},$$

for some $f \in SA(\text{Lip}_0(X))$, it follows that $L(f) \geq L(g) - L(f - g) > 1/2$. Let $x, y \in X$, $x \neq y$, be such that

$$|f(x) - f(y)| = L(f)d(x, y).$$

Let $\sigma : [0, d(x, y)] \to X$ be an isometric mapping such that

$$\sigma(0) = x \quad \text{and} \quad \sigma(d(x, y)) = y.$$

Then, by Proposition 8.8.27, $|f(z) - f(z')| = L(f)d(z, z')$ for all z, z' in the set $\Delta = \sigma([0, d(x, y)])$. If $z \neq z'$ are in Δ, then

$$|f(z) - f(z')| = L(f)d(z, z') > \frac{1}{2}d(z, z') \quad \text{and}$$

$$|(f - g)(z) - (f - g)(z')| \leq L(f - g)d(z, z') < \frac{1}{2}d(z, z'),$$

so that

$$|g(z) - g(z')| \geq |f(z) - f(z')| - |(f - g)(z) - (f - g)(z')| > 0.$$

It follows that the function $g = h_A \circ \varphi$ is injective and so φ must be injective as well, hence φ applies the (connected) set Δ on a nondegenerate interval $[a, b] \subseteq [0, 1]$. Since A is nowhere dense, there exists a nondegenerate interval (α, β) contained in $[a, b]$ such that $(\alpha, \beta) \cap A = \emptyset$. By definition, the function h_A is constant on (α, β), in contradiction to the injectivity of g.   □

Remark 8.8.30 In the same way one can show that $SA(X^\#)$ is not dense in $X^\#$ for every Banach space X. Indeed, for $x^* \in X^*$, $\|x^*\| = 1$, $h_A \circ x^* \in X^\# \setminus \overline{SA(X^\#)}$, where h_A is the function used in Example 8.8.28.

Although the norm density of Lipschitz functions strongly attaining their norm does not hold, a weak density result can be proven.

Theorem 8.8.31 *For every Banach space X the set of Lipschitz functions strongly attaining their norm is weakly sequentially dense in $X^{\#}$, that is, for every $f \in X^{\#}$ there exists a sequence (f_n) in $SA(X^{\#})$ which converges weakly to f.*

For the proof we need an auxiliary result.

Lemma 8.8.32 *Let $f_n \in X^{\#}$ with $L(f_n) = 1$, $n \in \mathbb{N}$, and $U_n = \{x \in X : f_n(x) \neq 0\}$. If the sets U_n are pairwise disjoint, then the sequence (f_n) is isometrically equivalent to the unit basis of c_0, i.e., for every $a = (a_k) \in c_0$ the series $\sum_{k=1}^{\infty} a_k f_k$ converges and*

$$L\left(\sum_{k=1}^{\infty} a_k f_k\right) = \max\{|a_k| : k \in \mathbb{N}\} = \|a\|_{c_0}.$$

In particular, the sequence (f_n) converges weakly to 0.

Proof We prove first that

$$L\left(\sum_{k=1}^{n} a_k f_k\right) = \max\{|a_k| : 1 \leq k \leq n\}, \tag{8.8.5}$$

for every finite set $a_1, \ldots, a_n$ of real numbers.

Denote $F_n = \sum_{k=1}^{n} a_k f_k$ and observe that

$$
\begin{aligned}
L(F_n) &= \sup\left\{\frac{|\sum_{k=1}^{n} a_k(f_k(x) - f_k(y))|}{\|x - y\|} : x, y \in X, \; x \neq y\right\} \\
&\geq \sup\left\{\frac{|\sum_{k=1}^{n} a_k(f_k(x) - f_k(y))|}{\|x - y\|} : x, y \in U_i, \; x \neq y\right\} \\
&= \sup\left\{\frac{|a_i||f_i(x) - f_i(y)|}{\|x - y\|} : x, y \in U_i, \; x \neq y\right\} = |a_i| \, L(f_i) = |a_i|,
\end{aligned}
$$

for every $i \in \{1, 2, \ldots, n\}$. (The equality before the last one holds by Remark 8.8.26).

It follows that $L(F_n) \geq \max\{|a_i| : 1 \leq i \leq n\}$.

To prove the converse we have to show that

$$\frac{|F_n(x) - F_n(y)|}{\|x - y\|} \leq \max\{|a_k| : 1 \leq k \leq n\},$$

for all $x, y \in X$, $x \neq y$, with $F_n(x) \neq 0 \neq F_n(y)$.

If $x, y \in U_i$ for some $i \in \{1, 2, \ldots, n\}$, then

$$\frac{|F_n(x) - F_n(y)|}{\|x - y\|} = \frac{|a_i|\,|f_i(x) - f_i(y)|}{\|x - y\|} \le |a_i| L(f_i) = |a_i| \le \max_{1 \le k \le n} |a_k|\,.$$

Suppose now that and $x \in U_i$ and $y \in U_j$, where i, j are distinct elements in $\{1, 2, \ldots, n\}$. Put $z_t = x + t(y - x), t \in [0, 1]$. If $t \in [0, 1]$ is such that $z_t \in U_k$, where $k \in \{i, j\}$, then, by the continuity of the function f_k and the openness of the set U_k, there exists $\delta > 0$ such that

$$z_{t'} \in U_k \quad \text{for all} \quad t' \in (t - \delta, t + \delta) \cap [0, 1]\,. \tag{8.8.6}$$

Let now $t_0 = \inf\{t \in [0, 1] : f_i(z_t) = 0\}$. By (8.8.6), $t_0 > 0$ and, by the definition of t_0, $z_t \in U_i$ for all $0 \le t < t_0$. If $f_j(z_{t_0}) \ne 0$, then there exists $\gamma > 0$ such that $z_t \in U_j$ for all $t \in (t_0 - \gamma, t_0 + \gamma) \cap [0, 1]$, leading to the contradiction $z_t \in U_i \cap U_j = \emptyset$, for all $t \in (t_0 - \gamma, t_0) \cap [0, 1]$. Hence $f_j(z_{t_0}) = 0$ and $0 < t_0 < 1$.

If $f_i(z_{t_0}) \ne 0$, then there exists $\delta > 0$ such that $(t_0 - \delta, t_0 + \delta) \subseteq (0, 1)$ and $z_t \in U_i$ for all $t \in (t_0 - \delta, t_0 + \delta)$ It follows, $f(z_t) \ne 0$ for all $0 \le t < t_0 + \delta$ yielding the contradiction $t_0 \ge t_0 + \delta$.

Consequently, $f_i(z_{t_0}) = 0 = f_j(z_{t_0})$ and $0 < t_0 < 1$.

With the notation $z = z_{t_0}$, one obtains

$$\frac{|F_n(x) - F_n(y)|}{\|x - y\|} = \frac{|a_i f_i(x) - a_j f_j(y)|}{\|x - y\|}$$

$$\le \frac{|a_i|\,|f_i(x) - f_i(z)|}{\|x - z\|} \cdot \frac{\|x - z\|}{\|x - y\|} + \frac{|a_j|\,|f_j(y) - f_j(z)|}{\|y - z\|} \cdot \frac{\|y - z\|}{\|x - y\|}$$

$$\le \max_{1 \le k \le n} |a_k| L(f_i) \cdot \frac{\|x - z\|}{\|x - y\|} + \max_{1 \le k \le n} |a_k| L(f_j) \cdot \frac{\|z - y\|}{\|x - y\|}$$

$$= \max_{1 \le k \le n} |a_k| \cdot \frac{\|x - z\| + \|z - y\|}{\|x - y\|} = \max_{1 \le k \le n} |a_k|\,,$$

showing that $L(F_n) \le \max_{1 \le k \le n} |a_k|$.

Let now $a = (a_k) \in c_0$. Then, by (8.8.5),

$$L(F_{n+p} - F_{n-1}) = L\left(\sum_{k=n}^{n+p} a_k f_k\right) = \max_{n \le k \le n+p} |a_k| \to 0\,,$$

as $n \to \infty$, uniformly with respect to $p \in \mathbb{N}$, showing that the sequence (F_n) is Cauchy in $X^{\#}$. Consequently, it has a limit $\sum_{k=1}^{\infty} a_k f_k \in X^{\#}$. Also

$$L\left(\sum_{k=1}^{\infty} a_k f_k\right) = \lim_{n \to \infty} L\left(\sum_{k=1}^{n} a_k f_k\right) = \lim_{n \to \infty} \max_{1 \le k \le n} |a_k| = \|a\|_{c_0}\,.$$

Finally, take a sequence $a = (a_k) \in c_0$ with $a_k \neq 0$ for all k (e.g. $a_k = 1/k$). Then the series $\sum_{k=1}^{\infty} a_k f_k$ converges, so that, for every $x^* \in (X^{\#})^*$, the series

$$\sum_{k=1}^{\infty} a_k x^*(f_k) = x^* \left(\sum_{k=1}^{\infty} a_k f_k \right)$$

converges, implying $\lim_{n \to \infty} x^*(f_n) = 0$.

This shows that the sequence (f_n) converges weakly to 0. $\square$

Proof of Theorem 8.8.31 Let $f \in X^{\#}$ with $L(f) = 1$. Consider a family of pairwise disjoint open balls U_n with centers x_n and radii r_n. For each n choose $\varepsilon_n \in (0, \frac{1}{2})$ and $y_n \in U_n$ such that $0 < \|x_n - y_n\| = \varepsilon_n r_n$. Consider also the set $X_n = (X \setminus U_n) \cup \{x_n, y_n\}$ and define $h_n : X_n \to \mathbb{R}$ by

$$h_n(x) = \begin{cases} f(x) & \text{for } x \in X_n \setminus \{x_n\} \\ f(y_n) - s_n(1 + 2\varepsilon_n)\|x_n - y_n\| & \text{for } x = x_n, \end{cases}$$

where $s_n = \text{sign}(f(y_n) - f(x_n))$. We check now that

$$L(h_n) = 1 + 2\varepsilon_n = \frac{|h_n(x_n) - h_n(y_n)|}{\|x_n - y_n\|}.$$

The second equality from above is obvious. Also, if $x, y \in X_n \setminus \{x_n\}$, then

$$|h_n(x) - h_n(y)| = |f(x) - f(y)| \leq \|x - y\|.$$

It remains to examine the case $x = x_n$ and $y \in X_n \setminus \{x_n, y_n\}$. In this case

$$\frac{|h_n(x_n) - h_n(y)|}{\|x_n - y\|} = \frac{|f(y_n) - s_n(1 + 2\varepsilon_n)\|x_n - y_n\| - f(y)|}{\|x_n - y\|}$$

$$\qquad\qquad\qquad\qquad\qquad\qquad\qquad\qquad (8.8.7)$$

$$\leq \frac{|f(x_n) - f(y)|}{\|x_n - y\|} + \frac{\left| f(y_n) - f(x_n) - s_n(1 + 2\varepsilon_n)\|x_n - y_n\| \right|}{\|x_n - y\|}.$$

But

$$|s_n(1 + 2\varepsilon_n)\|x_n - y_n\| - (f(y_n) - f(x_n))|$$

$$= |s_n((1 + 2\varepsilon_n)\|x_n - y_n\| - |f(y_n) - f(x_n)|)|$$

$$= (1 + 2\varepsilon_n)\|x_n - y_n\| - |f(y_n) - f(x_n)|$$

$$\leq (1 + 2\varepsilon_n)\|x_n - y_n\| = (1 + 2\varepsilon_n)\varepsilon_n r_n,$$

because $|f(y_n) - f(x_n)| \leq \|x_n - y_n\| \leq (1 + 2\varepsilon_n)\|x_n - y_n\|$.

Taking into account these facts and that $\|x_n - y\| \geq r_n$, from (8.8.7) one obtains

$$\frac{|h_n(x_n) - h_n(y)|}{\|x_n - y\|} \leq 1 + \frac{(1 + 2\varepsilon_n)\varepsilon_n r_n}{r_n} = 1 + \varepsilon_n + 2\varepsilon_n^2 < 1 + 2\varepsilon_n,$$

because $0 < \varepsilon_n < 1/2$.

Let f_n be a norm-preserving extension of h_n to X. Then

$$L(f_n) = 1 + 2\varepsilon_n = \frac{|f_n(x_n) - f_n(y_n)|}{\|x_n - y_n\|},$$

that is, $f_n \in \mathrm{SA}(X^\#)$ for all n.

By the definitions of the functions h_n and f_n, $\{x \in X : f_n(x) - f(x) \neq 0\} \subseteq U_n$. Since the sets U_n are pairwise disjoint, we can apply Lemma 8.8.32 to conclude that the sequence $(f_n - f)$ converges weakly to 0. Consequently, (f_n) is a sequence in $\mathrm{SA}(X^\#)$ converging weakly to f. Since the condition $L(f) = 1$ does not restrict the generality, it follows that $\mathrm{SA}(X^\#)$ is sequentially weakly dense in $X^\#$. $\square$

Remark 8.8.33 Taking ε_n and r_n tending to 0, one obtains further convergence properties of the sequence (f_n). Namely, $L(f_n) \to L(f)$ and $(f_n(x))$ converges to $f(x)$ uniformly for $x \in X$. Note that, although this yields norm-convergence with respect to the sup-norm, it is not necessarily the case when considering the Lipschitz norm, as pointed out in Remark 8.8.30.

Indeed,

$$L(f_n) = 1 + 2\varepsilon_n \to 1 = L(f),$$

as $n \to \infty$.

Observe that $f_n(x) - f(x) = 0$ for $x \in (X \setminus U_n) \cup \{y_n\}$.

Let $x \in U_n \setminus \{y_n\}$. Since $f_n(y_n) = f(y_n)$, we have

$$|f_n(x) - f(x)| \leq |f_n(x) - f_n(y_n)| + |f(y_n) - f(x)|$$

$$\leq 2(1 + \varepsilon_n)\|x - y_n\| \leq 4(1 + \varepsilon_n)r_n \to 0 \quad \text{as } n \to \infty.$$

It follows that the sequence $(f_n(x))_{n \in \mathbb{N}}$ converges to $f(x)$ uniformly with respect to $x \in X$.

Actually $\|x - y_n\|$ satisfies an inequality stronger than $\|x - y_n\| \leq 2r_n$, namely

$$\|x - y_n\| \leq \|x - x_n\| + \|x_n - y_n\| \leq (1 + \varepsilon_n)r_n,$$

but this is not needed for the uniform convergence.

Remark 8.8.34 In [324] Theorem 8.8.31 is proved (following similar ideas) for a class of metric spaces, called local metric spaces, introduced in [298] in the study of the Daugavet property for spaces of Lipschitz functions (see also [323]). One says that a metric space (X, d) is a local metric space if for every $f \in \mathrm{Lip}_0(X)$ and every $\varepsilon > 0$ there exist two distinct points x, y in X such that $d(x, y) < \varepsilon$ and

$$\frac{f(x) - f(y)}{d(x, y)} > L(f) - \varepsilon .$$

Every local metric space is infinite (in fact, it has no isolated points) and every geodesic metric space is local (see [298]).

Since the strong attaining condition is too restrictive, the authors of [324] proposed another version of this property.

Definition 8.8.35 Let X be a Banach space. One says that $f \in X^{\#}$ *attains its norm in the direction* $u \in S_X$ if there exists a sequence $(x_n, y_n) \in X \times X \setminus \Delta(X)$ such that

$$\lim_{n \to \infty} \frac{x_n - y_n}{\|x_n - y_n\|} = u \quad \text{and} \quad \lim_{n \to \infty} \frac{f(x_n) - f(y_n)}{\|x_n - y_n\|} = L(f) .$$

The set of all functions in $X^{\#}$ that attain the norm in some direction in S_X is denoted by $\mathrm{DA}(X^{\#})$.

The following remark shows that this is also an appropriate generalization of the linear case.

Remark 8.8.36

1. A linear continuous functional $x^* \in X^*$ attains its norm in the directions $u \in S_X$ if and only if $x^*(u) = \|x^*\|$.
2. If the Banach space X is finite dimensional, then $\mathrm{DA}(X^{\#}) = X^{\#}$.

The assertion 1 follows from the equalities

$$\|x^*\| = \lim_{n \to \infty} \frac{x^*(x_n) - x^*(y_n)}{\|x_n - y_n\|} = x^* \left(\lim_{n \to \infty} \frac{x_n - x_n}{\|x_n - y_n\|} \right) = x^*(u) .$$

To prove 2, for $f \in X^{\#}$ let $(x_n, y_n) \in X \times X$, $x_n \neq y_n$, $n \in \mathbb{N}$ be such that

$$\lim_{n \to \infty} \frac{f(x_n) - f(y_n)}{\|x_n - y_n\|} = L(f) .$$

Extracting a convergent subsequence $(x_{n_k} - y_{n_k})/\|x_{n_k} - y_{n_k}\| \to u \in S_X$, it follows that u and this subsequence satisfy the requirements of Definition 8.8.35.

We consider also the Bishop–Phelps–Bollobás property for Lipschitz functions.

Definition 8.8.37 A Banach space X has the *directional Bishop–Phelps–Bollobás property for Lipschitz functions* (Lip-BPB for short) if for every $\varepsilon > 0$ there exists $\delta > 0$ such that for every $f \in X^{\#}$ with $L(f) = 1$ and every $x \neq y$ in X such that

$(f(x) - f(y))/\|x - y\| > 1 - \delta$, there exist $g \in X^{\#}$ with $L(g) = 1$ and $u \in S_X$ such that g attains its norm in the direction u, $L(f - g) < \varepsilon$ and $\left\| \frac{x-y}{\|x-y\|} - u \right\| < \varepsilon$.

8.8.6 Norm-Attaining Seminorms

In this subsection we shall consider the special case of continuous seminorms defined on a Banach space X. We denote by $\mathrm{Sem}(X)$ the set of all seminorms on X.

Consider the set $K \subseteq \ell^{\infty}(S_X, \mathbb{R})$ formed of all $\varphi \in \ell^{\infty}(S_X, \mathbb{R})$ satisfying the conditions

(i) $\varphi(z) \geq 0$;   (ii) $\varphi(z) = \varphi(-z)$, for all $z \in S_X$;

(iii) $\varphi\left(\dfrac{x + y}{\|x + y\|} \right) \leq \dfrac{\|x\|}{\|x + y\|} \varphi\left(\dfrac{x}{\|x\|} \right) + \dfrac{\|y\|}{\|x + y\|} \varphi\left(\dfrac{y}{\|y\|} \right)$,

for all $x, y \in X \setminus \{0\}$ with $x + y \neq 0$.

It is clear that K is a closed convex cone in $\ell^{\infty}(S_X, \mathbb{R})$. For $\varphi \in K$ let

$$p_{\varphi}(x) = \|x\| \varphi\left(\frac{x}{\|x\|} \right) \quad \text{for } x \neq 0 \text{ and } p_{\varphi}(0) = 0.$$

Proposition 8.8.38 *Let X be a Banach space.*

1. *The correspondence $\varphi \mapsto p_{\varphi}$ is a bijection between K and $\mathrm{Sem}(X)$ satisfying the condition*

$$L(p_{\varphi}) = \|\varphi\|_{\infty}. \tag{8.8.8}$$

2. *For every $p \in \mathrm{Sem}(X)$ there exist a Banach space Y and an operator $T \in \mathscr{L}(X, Y)$ such that, $p(x) = \|Tx\|$ for all $x \in X$.*
3. *Let Y be a Banach space and $T \in \mathscr{L}(X, Y)$ such that, for some $p \in \mathrm{Sem}(X)$, $\|Tx\| = p(x)$ for all $x \in X$. Then*

$$\|T\| = \|p\|_{\infty},$$

where $\|p\|_{\infty} = \sup\{p(x) : x \in S_X\}$, so that

$$p(z) = \|p\|_{\infty} \iff \|Tz\| = \|T\|,$$

for every $z \in S_X$.

Proof

1. It is easy to check that for $\varphi \in K$, the mapping p_φ is a seminorm on X. The relations

$$p_\varphi(x) = \|x\| \, \varphi \left(\frac{x}{\|x\|} \right) \leq \|\varphi\|_\infty \|x\|,$$

valid for all $x \in X$, $x \neq 0$, prove the continuity of p_φ.
   For all $x \neq y$ in X

$$\frac{|p_\varphi(x) - p_\varphi(y)|}{\|x - y\|} \leq \frac{p_\varphi(x - y)}{\|x - y\|} = p_\varphi \left(\frac{x - y}{\|x - y\|} \right) = \varphi \left(\frac{x - y}{\|x - y\|} \right) \leq \|\varphi\|_\infty,$$

   implying $L(p_\varphi) \leq \|\varphi\|_\infty$.
   For all $z \in S_X$,

$$L(p_\varphi) \geq |p_\varphi(z) - p_\varphi(0)| = p_\varphi(z) = \varphi(z),$$

   so that $L(p_\varphi) \geq \|\varphi\|_\infty$.
   It is obvious that for $p \in \mathrm{Sem}(X)$, $\varphi := p|_{S_X}$ belongs to K and $p = p_\varphi$.

2. Take Y to be the completion of the quotient space $X/\ker p$ with respect to the quotient seminorm $\tilde{p}(x + \ker p) = \inf\{p(x + z) : z \in \ker p\}$ and T be the composition of the quotient map $j : X \to X/\ker p$, given by $j(x) = x + \ker p$, with the embedding of $X/\ker p$ into Y.
   Since $p(x + z) \leq p(x) + p(z) = p(x)$ and $p(x) \leq p(x + z) + p(-z) = p(x + z)$ for all $z \in \ker p$, it follows that $p(x + z) = p(x)$ for all $z \in \ker p$, so that $\tilde{p}(x + \ker p) = p(x)$.
   But then, by the definition of the operator T, $\|Tx\| = \tilde{p}(x + \ker p) = p(x)$ for all $x \in X$.

3. The proof follows from the equalities $p(x) = \|Tx\|$ and

$$\|p\|_\infty = \sup\{p(x) : x \in S_X\} = \sup\{\|Tx\| : x \in S_X\} = \|T\|.$$

$\square$

Remark 8.8.39 The density character of a Banach space X is the smallest of the cardinal numbers β for which there exists a dense subset of X of cardinality β. Denoting it by $\mathrm{dens}(X)$, it is obvious that $\mathrm{dens}(X) = \aleph_0$ is equivalent to the separability of X. By Fabian et al. [233, Exercise 5.30, p. 273] any Banach space X is linearly isometric to a subspace of $\ell^\infty(\Gamma)$, where $\Gamma = \mathrm{dens}(Y)$.

One can show also that in Proposition 8.8.38.2 one can take Y to be $\ell^\infty(\Gamma, \mathbb{R})$, where Γ is the density character of the Banach space X. This follows from the fact that the density character of Y is smaller than or equal to the density character Γ of X, so Y embeds isometrically into $\ell^\infty(\Gamma, \mathbb{R})$ and one can extend T to be an element of $\mathscr{L}(X, \ell^\infty(\Gamma, \mathbb{R}))$.

Based on the equality (8.8.8), we shall use the notation

$$\|p\|_\infty = \sup\{p(z) : z \in S_X\},$$

for $p \in \mathrm{Sem}(X)$.

Seminorms have good norm-attaining properties.

Proposition 8.8.40 *Let X be a Banach space. Then for every $p \in \mathrm{Sem}(X)$ the following are equivalent.*

1. *p attains its norm as an element of $\ell^\infty(S_X, \mathbb{R})$, i.e., there exists $z \in S_X$ such that $p(z) = \|p\|_\infty$.*
2. *$p \in \mathrm{SA}(X^\#)$.*
3. *$p \in \mathrm{DA}(X^\#)$.*
4. *There exists a Banach space Y and a norm-attaining operator $T \in \mathscr{L}(X, Y)$ such that $\|Tx\| = p(x)$ for all $x \in X$.*

Proof $1 \Rightarrow 2$. If p attains its sup-norm at some $z \in S_X$, then

$$|p(z) - p(0)|/\|z - 0\| = p(z) = \|p\|_\infty = L(p).$$

The implication $2 \Rightarrow 3$ is true for an arbitrary Lipschitz function.

$3 \Rightarrow 4$. By Proposition 8.8.38 there exist a Banach space Y and $T \in \mathscr{L}(X, Y)$ such that $p(x) = \|Tx\|$ for all $x \in X$. We have to show that T attains its norm. Let $x_n \neq y_n$, $n \in \mathbb{N}$, be such that

$$\lim_{n\to\infty} \frac{x_n - y_n}{\|x_n - y_n\|} = u \in S_X \quad \text{and} \quad \lim_{n\to\infty} \frac{p(x_n) - p(y_n)}{\|x_n - y_n\|} = L(p).$$

Then

$$\|Tu\| = \lim_{n\to\infty} \frac{\|Tx_n - Ty_n\|}{\|x_n - y_n\|} \geq \lim_{n\to\infty} \frac{\big|\|Tx_n\| - \|Ty_n\|\big|}{\|x_n - y_n\|} = \lim_{n\to\infty} \frac{|p(x_n) - p(y_n)|}{\|x_n - y_n\|}$$

$$= L(p) = \|p\|_\infty = \|T\|,$$

implying $\|Tu\| = \|T\|$, that is, T attains its norm at u.

The implication $4 \Rightarrow 1$ follows from Proposition 8.8.38.3. $\square$

First we show that the Bishop–Phelps–Bollobás property holds for the uniform norm.

Proposition 8.8.41 *Let X be a Banach space. Then for every $\varepsilon > 0$ there exists $\delta > 0$ such that for every $p \in \mathrm{Sem}(X)$ with $\|p\|_\infty = 1$ and every $x \in S_X$ such that $p(x) > 1 - \delta$, there exist $p_0 \in \mathrm{Sem}(X)$ with $\|p_0\|_\infty = 1$ and $x_0 \in S_X$ such that $p_0(x_0) = 1$, $\|x - x_0\| < \varepsilon$ and $\|p - p_0\|_\infty < \varepsilon$.*

Proof The proof will be based on the Bishop–Phelps–Bollobás property for operators.

Let $\varepsilon > 0$ and $p \in \mathrm{Sem}(X)$ with $\|p\|_\infty = 1$. By Proposition 8.8.38 there exists an operator $T \in \mathscr{L}(X, \ell^\infty(\Gamma, \mathbb{R}))$ such that $p(z) = \|Tz\|$ for all $z \in X$, where Γ is the density character of X. Take δ according to the BPB property for $\mathscr{L}(X, \ell^\infty(\Gamma, \mathbb{R}))$ and let $x \in S_X$ be such that $p(x) > 1 - \delta$. It follows that $\|T\| = \|p\|_\infty = 1$ and $\|Tx\| = p(x) > 1 - \delta$, so that, by Theorem 2.2 in [3], there exist $T_0 \in \mathscr{L}(X, \ell^\infty(\Gamma, \mathbb{R}))$ with $\|T_0\| = 1$ and $x_0 \in S_X$ such that $\|T_0 x_0\| = 1$, $\|x - x_0\| < \varepsilon$ and $\|T - T_0\| < \varepsilon$. Let $p_0(z) = \|T_0 z\|$, $z \in X$. Then

$$\|p_0\|_\infty = \|T_0\| = \|T_0 x_0\| = p_0(x_0),$$

and

$$|p(z) - p_0(z)| = \big|\|Tz\| - \|T_0 z\|\big| \leq \|Tz - T_0 z\| \leq \|T - T_0\| < \varepsilon,$$

for all $z \in S_X$, implying $\|p - p_0\|_\infty < \varepsilon$. $\qquad\square$

For the Lipschitz norm we can prove only a Bishop–Phelps-type result.

Proposition 8.8.42 *Let X be a Banach space with the RNP. Then the set of Lipschitz norm-attaining seminorms is dense in $\mathrm{Sem}(X)$ with respect to the Lipschitz norm.*

Proof Let $p \in \mathrm{Sem}(X)$. By Stegall's Variational Principle (Theorem 8.8.22), for every $\varepsilon > 0$ there exists $x^* \in X^*$ with $\|x^*\| < \varepsilon$ such that $p + |x^*|$ attains its supremum on B_X. It follows that the seminorm $q := p + |x^*|$ attains its norm at some $x_0 \in S_X$ and

$$L(q - p) = L(|x^*|) = \||x^*|\|_\infty = \|x^*\| < \varepsilon.$$

$\qquad\square$

8.8.7 The Lip-BPB Property

In this subsection we shall prove some results on the Lip-BPB property for Lipschitz functions. The main result will be its validity for uniformly convex Banach spaces.

Theorem 8.8.43 *Any uniformly convex Banach space has the directional Lip-BPB property for Lipschitz functions.*

We start with some preliminary results.

Lemma 8.8.44 *Let X be a Banach space, $f \in X^\#$, $L(f) = 1$, $\delta \in (0, 2)$ and let $x \neq y$ in X be such that*

$$\frac{f(x) - f(y)}{\|x - y\|} > 1 - \delta.$$

Then there exists $g \in X^\#$ with $L(g) = 1$ and $L(f - g) < \sqrt{2\delta}$ such that for every $h \in X^\#$ with $L(h) = 1$ and $(h(x) - h(y))/\|x - y\| = 1$ there exist the points $u_n \neq v_n$, $n \in \mathbb{N}$, in X, such that

$$\lim_{n \to \infty} \frac{g(u_n) - g(v_n)}{\|u_n - v_n\|} = 1 \quad and \quad \frac{h(u_n) - h(v_n)}{\|u_n - v_n\|} > 1 - \sqrt{2\delta}, \quad for\ all\ n \in \mathbb{N}.$$
$$(8.8.9)$$

Proof The proof relies on some results on free Lipschitz spaces, see Sect. 8.2.2. Let $\mathbb{F}(X)$ be the free Lipschitz space corresponding to X. Then $\mathbb{F}(X)^* = X^\#$ and every $f \in X^\#$ can be viewed as a continuous linear functional on $\mathbb{F}(X)$, of norm $L(f)$, acting by the rule

$$\left\langle \sum t_i e_{x_i}, f \right\rangle = \sum t_i f(x_i),$$

for $\sum t_i e_{x_i} \in X_e$, where X_e denotes the space generated by the evaluation functionals $e_x \in (X^\#)^*$, $x \in X$, given by $e_x(f) = f(x)$ for $f \in X^\#$. Then, $\|e_x - e_y\| = \|x - y\|$ and for every $f \in X^\#$,

$$L(f) = \sup \left\{ \frac{f(x) - f(y)}{\|x - y\|} : x, y \in X, x \neq y \right\}$$

$$= \sup \left\{ \left\langle \frac{e_x - e_y}{\|x - y\|}, f \right\rangle : x, y \in X, x \neq y \right\}.$$

Putting $W := \{(e_x - e_y)/\|x - y\| : x, y \in X, x \neq y\} \subseteq S_{\mathbb{F}(X)}$, it follows that

$$\overline{co}(W) = B_{\mathbb{F}(X)}.$$
$$(8.8.10)$$

Returning to the proof of Lemma 8.8.44, let $f \in X^\#$ satisfy the hypotheses of this lemma. If $\xi := (e_x - e_y)/\|x - y\| \in S_{\mathbb{F}(X)}$, then, by hypothesis, $\langle \xi, f \rangle > 1 - \delta$, so that by the Bishop–Phelps–Bollobás theorem (Theorem 8.8.7), there exist $g \in X^\#$, $L(g) = 1$, and $\zeta \in \mathbb{F}(X)$, $\|\zeta\| = 1$, such that

$$\langle \zeta, g \rangle = 1, \quad \|\xi - \zeta\| < \sqrt{2\delta} \quad and \quad L(f - g) < \sqrt{2\delta}.$$

Let ν be such that $\|\xi - \zeta\| < \nu < \sqrt{2\delta}$ and let (δ_n) be a sequence of positive numbers converging to 0. By (8.8.10) there exists a sequence $\zeta_n \in co(W)$, $n \in \mathbb{N}$, converging to ζ, such that

$$\|\xi - \zeta_n\| < \nu \quad and \quad \langle \zeta_n, g \rangle > 1 - \delta_n,$$

for all $n \in \mathbb{N}$.

Let $h \in X^\#$ with $L(h) = 1$ be such that $\langle \xi, h \rangle = 1$. Then

$$\langle \zeta_n, h \rangle = \langle \xi, h \rangle - \langle \xi - \zeta_n, h \rangle > 1 - \nu,$$

for all $n \in \mathbb{N}$.

Let now $\alpha_n \in (0, 1)$, $n \in \mathbb{N}$, be such that

$$\lim_{n \to \infty} \alpha_n = 0 \quad \text{and} \quad \lim_{n \to \infty} \frac{\delta_n}{\alpha_n} = 0. \tag{8.8.11}$$

Then

$$\alpha_n \langle \zeta_n, h \rangle + (1 - \alpha_n) \langle \zeta_n, g \rangle > \alpha_n (1 - \nu) + (1 - \alpha_n)(1 - \delta_n) = 1 - \alpha_n \nu - (1 - \alpha_n)\delta_n .$$

Since $\zeta_n \in \mathrm{co}(W)$, there exists $w_n \in W$ such that

$$A(w_n) := \alpha_n \langle w_n, h \rangle + (1 - \alpha_n) \langle w_n, g \rangle > 1 - \alpha_n \nu - (1 - \alpha_n)\delta_n . \tag{8.8.12}$$

This last inequality implies

$$\langle w_n, g \rangle > 1 - \delta_n - \frac{\alpha_n}{1 - \alpha_n} \sqrt{2\delta} \quad \text{and} \quad \langle w_n, h \rangle > 1 - \nu - \frac{\delta_n}{\alpha_n}(1 - \alpha_n) . \tag{8.8.13}$$

Indeed, if the first inequality from above does not hold then, taking into account that $\langle w_n, h \rangle \leq 1$, one obtains

$$A(w_n) \leq \alpha_n + (1 - \alpha_n)(1 - \delta_n) - \alpha_n \sqrt{2\delta} < 1 - \alpha_n \nu - (1 - \alpha_n)\delta_n ,$$

in contradiction to (8.8.12). (The last inequality is equivalent to $\nu < \sqrt{2\delta}$, so it is true by the choice of ν).

If the second inequality from (8.8.13) does not hold then, taking into account that $\langle w_n, g \rangle \leq 1$, one obtains

$$A(w_n) \leq \alpha_n - \alpha_n \nu - (1 - \alpha_n)\delta_n + 1 - \alpha_n = 1 - \alpha_n \nu - (1 - \alpha_n)\delta_n ,$$

in contradiction to (8.8.12).

Now, from (8.8.13) and (8.8.11) it follows that

$$\lim_{n \to \infty} \langle w_n, g \rangle = 1 \quad \text{and} \quad \langle w_n, h \rangle > 1 - \sqrt{2\delta}, \quad \text{for } n \in \mathbb{N} \text{ large enough.} \tag{8.8.14}$$

By the definition of the set W, for every $n \in \mathbb{N}$ there exists $u_n \neq v_n$ in X such that

$$w_n = \frac{e_{u_n} - e_{v_n}}{\|u_n - v_n\|},$$

and the relations (8.8.14) are equivalent to the corresponding ones in (8.8.9), which ends the proof of Lemma 8.8.44. $\qquad\square$

For completeness we present a proof of the equality (8.8.10), which follows from the next result.

Lemma 8.8.45 *If V is a closed convex subset of the unit ball B_X of a Banach space X such that $\sup x^*(V) = \|x^*\|$ for every $x^* \in X^*$, then $V = B_X$.*

Proof Observe that for any ball $B[x_0, R] = x_0 + R\, B_X$ and $x^* \in X^*$,

$$\inf x^*(B[x_0, R]) = x^*(x_0) + R \inf x^*(B_X) = x^*(x_0) - R\,\|x^*\|\,.$$

If there exists $x_0 \in S_X \setminus V$, then $d := d(x_0, V) > 0$ and $U := B[x_0, r] = x_0 + r B_X$ is disjoint from V for every $0 < r < d$. Fix $0 < r < \min\{d, 1\}$. By the separation theorem there exists $x^* \in X^* \setminus \{0\}$ such that $\|x^*\| = \sup x^*(V) \le \inf x^*(U)$.

Since $B[x_0, r/n] \subseteq B[x_0, r]$ for every $n \in \mathbb{N}$, it follows that

$$\|x^*\| \le \inf x^*(B[x_0, r/n]) = x^*(x_0) - \frac{r}{n}\,\|x^*\|\,,$$

which yields for $n \to \infty$, $\|x^*\| \le x^*(x_0)$, so that $x^*(x_0) = \|x^*\|$. If $1 - r < \alpha < 1$, then $\alpha x_0 \in U$ and

$$\|x^*\| \le \alpha x^*(x_0) = \alpha \|x^*\| < \|x^*\|\,,$$

a contradiction.

Consequently, $V = B_X$. $\square$

The following result differs from the directional Lip-BPB property (see Definition 8.8.37) by the fact that we do not ask the convergence of the approximate support sequence.

Lemma 8.8.46 *Let X be a Banach space. Suppose that for every $\varepsilon > 0$ there exists $\delta > 0$ such that for every $f \in X^\#$ with $L(f) = 1$ and every pair x, y of distinct points in X such that*

$$\frac{f(x) - f(y)}{\|x - y\|} > 1 - \delta\,,$$

there exist $g \in S_{X^\#}$ and the sequences u_n, v_n, $u_n \ne v_n$, $n \in \mathbb{N}$, in X such that

$$\lim_{n\to\infty} \frac{g(u_n) - g(v_n)}{\|u_n - v_n\|} = 1, \quad L(f - g) < \varepsilon \ \text{ and } \ \left\| \frac{x - y}{\|x - y\|} - \frac{u_n - v_n}{\|u_n - v_n\|} \right\| < \varepsilon\,.$$

Then X has the directional Lip-BPB *property.*

Proof Let $f \in X^\#$ with $L(f) = 1$, and $\varepsilon > 0$ be given. Consider a sequence $\varepsilon_n > 0$, $n \in \mathbb{N}$, such that $\sum_{n=1}^{\infty} \varepsilon_n < \varepsilon$.

Let $\delta_1 > 0$ corresponding to ε_1 in accordance with the hypotheses of the theorem. We shall show that the requirements of Definition 8.8.37 are satisfied for $\delta = \delta_1$.

Suppose that $x, y \in X$, $x \neq y$, are such that

$$\frac{f(x) - f(y)}{\|x - y\|} > 1 - \delta_1,$$

and put $f_1 = f$ and $(x_1, y_1) = (x, y)$.

Then there exist $g_1 \in X^\#$ with $L(g_1) = 1$ and the sequences $u_n^1, v_n^1 \in X$, $u_n^1 \neq v_n^1$, $n \in \mathbb{N}$, such that

$$\lim_{n \to \infty} \frac{g_1(u_n^1) - g_1(v_n^1)}{\|u_n^1 - v_n^1\|} = 1, \quad L(f_1 - g_1) < \varepsilon_1 \text{ and}$$

$$\left\| \frac{x_1 - y_1}{\|x_1 - y_1\|} - \frac{u_n^1 - v_n^1}{\|u_n^1 - v_n^1\|} \right\| < \varepsilon_1,$$

for all $n \in \mathbb{N}$.

The first equality from above implies the existence of $n_1 \in \mathbb{N}$ such that

$$\frac{g_1(u_{n_1}^1) - g_1(v_{n_1}^1)}{\|u_{n_1}^1 - v_{n_1}^1\|} > 1 - \delta_2,$$

where δ_2 corresponds to ε_2. We can suppose further that $\delta_2 < 1/2$.

Putting $f_2 = g_1$ and $(x_2, y_2) = (u_{n_1}^1, v_{n_1}^1)$, the existence of a function $g_2 \in X^\#$ with $L(g_2) = 1$ follows, as well as of the sequences $u_n^2, v_n^2 \in X$, $u_n^2 \neq v_n^2$, $n \in \mathbb{N}$, such that

$$\lim_{n \to \infty} \frac{g_2(u_n^2) - g_2(v_n^2)}{\|u_n^2 - v_n^2\|} = 1, \quad L(f_2 - g_2) < \varepsilon_2 \text{ and}$$

$$\left\| \frac{x_2 - y_2}{\|x_2 - y_2\|} - \frac{u_n^2 - v_n^2}{\|u_n^2 - v_n^2\|} \right\| < \varepsilon_2,$$

for all $n \in \mathbb{N}$.

Again, there exists $n_2 \in \mathbb{N}$ such that

$$\frac{g_2(u_{n_2}^2) - g_1(v_{n_2}^2)}{\|u_{n_2}^2 - v_{n_2}^2\|} > 1 - \delta_3,$$

where $\delta_3 < 1/3$ corresponds to ε_3.

Put $f_3 = g_2$, $(x_3, y_3) = (u_{n_2}^2, v_{n_2}^2)$ and apply the hypotheses of the lemma for ε_3.

In this way we obtain the numbers $\delta_k < 1/k$, the functions f_k, g_k in $S_{X^\#}$ with $f_{k+1} = g_k$, and the points $x_k, y_k \in X$, $x_k \neq y_k$, $k \in \mathbb{N}$. Putting $w_k := (x_k - y_k)/$

$\|x_k - y_k\|$ it follows that

$$\text{(i)} \qquad L(f_k - f_{k+1}) < \varepsilon_k;$$

$$\text{(ii)} \qquad \|w_k - w_{k+1}\| < \varepsilon_k; \tag{8.8.15}$$

$$\text{(iii)} \qquad \frac{f_k(x_k) - f_k(y_k)}{\|x_k - y_k\|} > 1 - \delta_k,$$

for all $k \in \mathbb{N}$.

The first condition implies that (f_k) is a Cauchy sequence with respect to the Lipschitz norm, so there exists $g \in S_{X^\#}$ such that $\lim_{k\to\infty} L(f_k - g) = 0$.

The second one implies that (w_k) is a Cauchy sequence in X, so there exists $u \in S_X$ such that $\lim_{k\to\infty} \|w_k - u\| = 0$.

Let us show now that g, u and the sequences $(x_n), (y_n)$ satisfy the requirements of Definition 8.8.37 (see also Definition 8.8.35).

Writing $f_n = f_1 + (f_2 - f_1) + \cdots + (f_n - f_{n-1})$ and taking into account the condition (i) from (8.8.15), it follows that

$$L(f_n - f) = L(f_n - f_1) \leq \sum_{k=2}^{n} L(f_k - f_{k-1}) < \sum_{k=1}^{n-1} \varepsilon_k.$$

Letting $n \to \infty$ one obtains $L(g - f) \leq \sum_{k=1}^{\infty} \varepsilon_k < \varepsilon$.

Similarly, the condition (ii) from (8.8.15) yields $\|w_1 - u\| < \varepsilon$, where

$$w_1 = \frac{x_1 - y_1}{\|x_1 - y_1\|} = \frac{x - y}{\|x - y\|}.$$

Finally, by (8.8.15).(iii),

$$\frac{g(x_k) - g(y_k)}{\|x_k - y_k\|} = \frac{f_k(x_k) - f_k(y_k)}{\|x_k - y_k\|} + \frac{(g - f_k)(x_k) - (g - f_k)(y_k)}{\|x_k - y_k\|}$$

$$> 1 - \delta_k - L(g - f_k) \to 1 \quad \text{as } k \to \infty.$$

Since $g \in S_{X^\#}$, this shows that $\lim_{k\to\infty} \frac{g(x_k)-g(y_k)}{\|x_k-y_k\|} = 1 = L(g)$ and so Lemma 8.8.46 is proved. $\qquad\square$

We can prove now Theorem 8.8.43.

Proof of Theorem 8.8.43 For $\varepsilon > 0$ let $\delta > 0$ be such that $0 < \sqrt{2\delta} < \delta_X(\varepsilon)$. Let $f \in S_{X^\#}$ and $x \neq y$ in X be such that

$$\frac{f(x) - f(y)}{\|x - y\|} > 1 - \delta.$$

Let $x^* \in S_{X^*}$ be a functional that supports B_X at $(x - y)/\|x - y\|$ (see Theorem 1.4.10). Then

$$\frac{x^*(x) - x^*(y)}{\|x - y\|} = x^* \left(\frac{x - y}{\|x - y\|} \right) = 1 \,,$$

so that we can apply Lemma 8.8.44 to $h = x^*$ to find $g \in S_{X^\#}$ and sequences $u_n, v_n \in X$, $u_n \neq v_n$, $n \in \mathbb{N}$, such that

$$L(f - g) < \sqrt{2\delta}, \quad \lim_{n \to \infty} \frac{g(u_n) - g(v_n)}{\|u_n - v_n\|} = 1 \,,$$

and

$$x^* \left(\frac{u_n - v_n}{\|u_n - v_n\|} \right) = \frac{x^*(u_n) - x^*(v_n)}{\|u_n - v_n\|} > 1 - \sqrt{2\delta} > 1 - \delta_X(\varepsilon) \,,$$

for all $n \in \mathbb{N}$.

This last inequality means that the elements $(u_n - v_n)/\|u_n - v_n\|$ belong to the slice $S(B_X, x^*, \delta_X(\varepsilon))$ (see (8.8.2)), for all $n \in \mathbb{N}$. Since $(x - y)/\|x - y\|$ also belongs to $S(B_X, x^*, \delta_X(\varepsilon))$, Lemma 8.8.20 implies

$$\left\| \frac{x - y}{\|x - y\|} - \frac{u_n - v_n}{\|u_n - v_n\|} \right\| \leq \varepsilon \,.$$

Appealing to Lemma 8.8.46 it follows that the space X has the directional Lip-BPB property. $\qquad\square$

8.8.8 Asymptotically Uniformly Smooth Banach Spaces and Norm-Attaining Lipschitz Operators

In this subsection we shall present following [258] some results on norm-attaining Lipschitz mappings between Banach spaces.

Let X, Y be two real Banach spaces and let $(X, Y)^\# = \mathrm{Lip}_0(X, Y)$ the Banach space of all Lipschitz mappings from X to Y vanishing at $0 \in X$.

Definition 8.8.47 Let $f \in (X, Y)^\#$. One says that f attains its norm:

- *strongly* at the pair $x, x' \in X$ of distinct points in X if

$$\|f(x) - f(x')\| = L(f)\|x - x'\|;$$

- *in the direction $u \in S_X$ if there exists $x \in X$ such that $\|D_u^+ f(x)\| = L(f)$,* where

$$D_u^+ f(x) = \lim_{t \searrow 0} \frac{f(x+tu) - f(x)}{t}$$

denotes the derivative of f in the direction u;
- *in the direction $y \in Y$ if $\|y\| = L(f)$ and there exist a sequence $((u_n, v_n))$ with $u_n \neq v_n$, in X^2 such that*

$$\lim_{n \to \infty} \frac{f(u_n) - f(v_n)}{\|u_n - v_n\|} = y .$$

Remark 8.8.48 Observe that the definitions given here to the directionally norm-attaining Lipschitz functions differ from that given in Definition 8.8.35. For instance, if $Y = \mathbb{R}$, then there always exists a sequence $((u_n, v_n))$, $u_n \neq v_n$, in X^2 such that $\lim_{n \to \infty} [f(u_n) - f(v_n)]/\|u_n - v_n\| = L(f)$, meaning that f attains its norm in the direction $L(f) \in \mathbb{R}$.

Also it is obvious that if there exists $u \in S_X$ and $x \in X$ such that $\|D_u^+ f(x)\| = L(f)$, then f attains its norm in the direction $y = D_u^+ f(x) \in Y$.

Godefroy [258] considered a class of Banach spaces, called asymptotically uniformly smooth, and emphasized their relevance in the study of norm-attaining Lipschitz functions.

Definition 8.8.49 For a Banach space X, $x \in S_X, \tau > 0$ and a closed finite-codimensional subspace Z of X put

$$\varrho(\tau, x, Z) = \sup_{z \in S_Z} \|x + \tau z\| - 1 ,$$

and let

$$\varrho(\tau, x) = \inf_Z \varrho(\tau, x, Z) ,$$

where the infimum is taken over all closed finite-codimensional subspaces Z of X.
Finally, put

$$\varrho(\tau) = \sup_{x \in S_X} \varrho(\tau, x) .$$

The function ϱ (or ϱ_X) is called the *modulus of asymptotic uniform smoothness* of X and the space X is called *asymptotically uniformly smooth* (AUSm for short) if

$$\lim_{\tau \searrow 0} \frac{\varrho_X(\tau)}{\tau} = 0 .$$

The space X is called *asymptotically uniformly flat* if there exists $\tau_0 > 0$ such that $\varrho_X(\tau_0) = 0$. Since the modulus of asymptotic uniform smoothness is a non-decreasing function of τ, it follows that an asymptotically uniformly flat Banach space is asymptotically uniformly smooth.

An example of AUSm space is the space c_0. In [262] it is shown that a separable Banach space has an equivalent asymptotically uniformly flat norm if and only if it is isomorphic to a subspace of c_0.

In the particular case of a Banach space with separable dual, the modulus admits the following practical way of computing it.

Proposition 8.8.50 *Let X be a Banach space with separable dual, $\tau \in (0, 1]$ and $x \in S_X$. Put*

$$\eta(\tau, x) = \sup[\limsup_{n\to\infty} \|x + x_n\| - 1],$$

where the supremum is taken over all sequences (x_n) contained in τB_X which converge weakly to 0, and let $\eta(\tau) = \sup_{x \in S_X} \eta(\tau, x)$. Then

$$\varrho_X(\tau, x) = \eta(\tau, x) \quad and \quad \varrho_X(\tau) = \eta(\tau).$$

We mention the following renorming result.

Theorem 8.8.51 ([258]) *Let X, Y be separable Banach spaces with X AUSm. If there exists a Lipschitz isomorphism $f : X \to Y$, then there exists an AUSm equivalent norm on Y whose modulus ϱ_Y satisfies the inequality*

$$\varrho_Y(\tau/4M) \le 2\varrho_X(\tau),$$

for every $\tau \in (0, 1]$, where $M = L(f) \cdot L(f^{-1})$.

This theorem appears (without proof) as Theorem 5.4 in [263], where it was a consequence of the previous Theorem 5.3. For this reason, the author supplies in [258] the non trivial calculations, based on a concrete expression of the new norm.

Remark 8.8.52 As the author remarks in [258], the renorming $\| \cdot \|_1$ from Theorem 8.8.51 can be also obtained by appealing to the Lipschitz free Banach space $\mathbb{F}(X)$ (see Sect. 8.2.1). Indeed if $f : X \to Y$ is a Lipschitz isomorphism with $f(0) = 0$ and $\tilde{f} \in \mathscr{L}(\mathbb{F}(X), Y)$ is its linear correspondent (see Theorem 8.2.4), then the norm $\| \cdot \|_1$ is the quotient norm of the canonical norm on $\mathbb{F}(X)$ obtained through $\tilde{f}$. Since the mapping $f \mapsto \tilde{f}$ is a linear isometry between the spaces $\mathrm{Lip}_0(X, Y)$ and $\mathscr{L}(\mathbb{F}(X), Y)$, the norm attainment statements have linear translations. For instance, f attains its norm on a couple x, x' of distinct points in X if and only if $\tilde{f}$ attains its norm on the molecule $m_{x,x'}/\|x - x'\| = (\chi_x - \chi_{x'})/\|x - x'\|$.

We shall present now some results on norm attainment.

Proposition 8.8.53 *Let X, Y be separable Banach spaces such that X is AUSm. If there exists a Lipschitz isomorphism f from X onto Y which attains its Lipschitz norm in some direction $y \in Y$, then there exists a constant $c > 0$ such that $\varrho_Y(\tau/c, y) \leq 2\varrho_X(\tau)$ for all $\tau \in (0, 1]$.*

Proof We can assume $L(f) = 1$ and so $\|y\| = 1$. Let $\|\cdot\|$ be the original norm on Y and $\|\cdot\|_1$ the equivalent AUSm norm constructed in Theorem 8.8.51. As it is shown in [258], the new norm satisfies the inequality $\|\cdot\| \leq \|\cdot\|_1$. Let

$$y = \lim_{n\to\infty} [f(u_n) - f(v_n)]/\|u_n - v_n\|,$$

for a sequence $((u_n, v_n))_{n\in\mathbb{N}}$ in X^2 with $u_n \neq v_n$, $n \in \mathbb{N}$. Then

$$\frac{|f(u_n) - f(v_n)|}{\|u_n - v_n\|_1} \leq \frac{|f(u_n) - f(v_n)|}{\|u_n - v_n\|} \leq 1, \quad n \in \mathbb{N},$$

implies $\|y\|_1 \leq 1$, and so and so $1 = \|y\| = \|y\|_1$.

Theorem 8.8.51 and the fact that $\|\cdot\| \leq \|\cdot\|_1$ imply $\varrho_Y(\tau/4M, y) \leq 2\varrho_X(\tau)$ for all $\tau \in (0, 1]$, where $M = L(f^{-1})$. $\qquad\square$

Remark 8.8.54 The assumptions of Proposition 8.8.53 are satisfied also in the cases when f attains its norm at a point $x \in X$ in the direction $u \in S_X$ (i.e., $\|D_u^+ f(x)\| = L(f)$), or strongly attains its norm at a pair x, x' of distinct points in X.

Recall that one says that a Banach space X has the Kadets-Klee property if weak and strong convergence of sequences in S_X agree.

Corollary 8.8.55 *Let X be a separable asymptotically uniformly flat Banach space and Y a Banach space having the Kadets-Klee property. If f is a Lipschitz isomorphism between X and Y, then there is no $y \in Y$ such that f attains its Lipschitz norm in the direction y.*

Proof Assume, on the contrary, that some f with $L(f) = 1$ attains its Lipschitz norm in the direction $y \in S_Y$. Since Y is Lipschitz isomorphic to X, its dual Y^* is separable (see [262, Theorem 2.1]). If $\tau_0 \in (0, 1]$ is such that $\rho_X(\tau_0) = 0$, then, by Proposition 8.8.53, $\rho_Y(\tau_1, y) = 0$, where $\tau_1 = \tau_0/c$. By Proposition 8.8.50, there exists a weakly-null sequence (y_n) with $\|y_n\| = \tau_1$, $n \in \mathbb{N}$, such that

$$\lim_{n\to\infty} \|y + y_n\| = 1.$$

Putting $z_n = (y+y_n)/\|y+y_n\|$, $n \in \mathbb{N}$, it follows that the sequence (z_n) is contained in S_Y, converges weakly but not in norm to y, in contradiction to the hypothesis. $\qquad\square$

Remark 8.8.56 Corollary 8.8.55 fails to hold if X is assumed to be only AUSm. Indeed, the space ℓ^p, $1 < p < \infty$, is AUSm and uniformly convex, but $f = \mathrm{Id}_X$ serves as a counterexample.

Observe that Corollary 8.8.55 holds if X is asymptotically uniformly flat and Y is locally uniformly convex, but fails if Y is supposed only strictly convex. To see this, take a continuous injective linear map $T : c_0 \to \ell^2$. It follows that T takes weakly-null sequences in c_0 to norm-null sequences, implying that $\|x\|_1 = \|x\|_\infty + \|Tx\|_{\ell^2}$ is an equivalent norm on c_0 which is asymptotically uniformly flat and strictly convex, but again, $f = \mathrm{Id}_{c_0}$ trivially attains its norm.

Concerning c_0 the following result holds.

Proposition 8.8.57 ([258]) *Let X be an infinite dimensional subspace of c_0 equipped with the restriction of the canonical norm, and Y a strictly convex normed space. Then no injective Lipschitz map from X to Y strongly attains its Lipschitz norm.*

Proof Assume that an injective Lipschitz map $f : X \to Y$ attains its norm on a pair x, x' of distinct points in X. It follows that a geodesic between x and x' is applied by f to a geodesic from $f(x)$ to $f(x')$. Since Y is strictly convex any geodesic between two points y, y' in Y agrees with the algebraic interval $[y, y']$. So the proposition will be proved if we show that there are at least two geodesics connecting the points x and x'. We can assume $\|x\| = 1$ and $x' = 0$. The set $F = \{n \in \mathbb{N} : |x(n)| > 1/2\}$ is finite, so that, since X is infinite dimensional, there exists $y \in X$, $y \neq 0$, with $\|y\| \leq 1/2$ such that $y(n) = 0$ for all $n \in F$. Then $\|x - y\| = \|x + y\| = 1$ and, for any point $z \in [(x - y)/2, (x + y)/2]$, the mapping $g : [0, 1] \to X$ given by $g(0) = 0$, $g(1/2) = z$ and $g(1) = x$ which is affine on $[0, 1/2]$ and on $[1/2, 1]$ is a geodesic between 0 and x.   □

8.9 Applications to Best Approximation in Metric Spaces

In the study of best approximation in a normed space X a special role is played by the dual space X^*, used to obtain duality results and characterizations of the best approximation elements in terms of some continuous linear functionals (see [637]). We show that some of these results can be extended to metric spaces, or to metric linear spaces, by using Lipschitz duals instead of the linear duals.

8.9.1 Best Approximation in Arbitrary Metric Spaces

Phelps [560] obtained some results relating the approximation properties of the annihilator of a subspace Y of a normed space X with the extension properties of continuous linear functionals from Y to X (Hahn–Banach theorem). Various situations when such a duality occurs are presented in [159].

For a subspace Y of X let

$$Y^\perp = \left\{ x^* \in X^* : x^*|_Y = 0 \right\}$$

be the *annihilator* space of Y in X^*. Phelps [560] proved that the subspace $Y^\perp$ is always proximinal and that it is Chebyshevian if and only if every $y^* \in Y^*$ has a unique norm-preserving extension $x^* \in X^*$.

Based on the extension results for Lipschitz functions one can show that similar results hold in the case of Lipschitz functions. The research in this direction started in [490] and continued in [164, 492, 495, 500–502, 604]. Some of these results were rediscovered in [415]. For a survey see the paper [165].

Let (X, d) be a metric space, θ a fixed point in X and Y a subset of X containing θ. Denote by $\mathrm{Lip}_0(X)$ $\left(\mathrm{Lip}_0(Y)\right)$ the spaces of all Lipschitz functions on X (respectively on Y) vanishing at θ. They are Banach spaces with respect to the corresponding Lipschitz norms.

We shall denote these norm by the symbols $\|\cdot\|_X$ and $\|\cdot\|_Y$, respectively.

Put

$$Y^\perp = \left\{ F \in \mathrm{Lip}_0(X) : F|_Y = 0 \right\} , \qquad (8.9.1)$$

and, for $f \in \mathrm{Lip}_0(Y)$, let

$$*E\,(f) = \left\{ F \in \mathrm{Lip}_0(X) : F \text{ is a norm-preserving extension of } f \right\} ,$$

i.e.,

$$F \in E\,(f) \iff F|_Y = f \quad \text{and} \quad \|F\|_X = \|f\|_Y .$$

By McShane's extension theorem (Theorem 4.1.1) the set $E\,(f)$ is nonempty for every $f \in \mathrm{Lip}_0(Y)$.

Theorem 8.9.1 *Let* (X, d), θ, Y *be as above and* $Y^\perp$ *be defined by (8.9.1)*

1. The space $Y^\perp$ *is always proximinal in* $\mathrm{Lip}_0(X)$ *and for every* $F \in \mathrm{Lip}_0(X)$

$$d\left(F, Y^\perp\right) = \|F|_Y\| , \qquad (8.9.2)$$

and

$$P_{Y^\perp}\,(F) = F - E\,(F|_Y) . \qquad (8.9.3)$$

2. The space $Y^\perp$ *is Chebyshevian in* $\mathrm{Lip}_0(X)$ *if and only if every* $f \in \mathrm{Lip}_0(Y)$ *has a unique norm-preserving extension* $F \in \mathrm{Lip}_0(X)$.

Proof For $F \in \mathrm{Lip}_0(X)$ and arbitrary $G \in Y^{\perp}$ we have

$$\|F|_Y\| = \sup\left\{\frac{|F(x) - F(y)|}{d(x, y)} : x, y \in Y, x \neq y\right\}$$

$$= \sup\left\{\frac{|(F - G)(x) - (F - G)(y)|}{d(x, y)} : x, y \in Y; x \neq y\right\} \leq$$

$$\leq \sup\left\{\frac{|(F - G)(x) - (F - G)(y)|}{d(x, y)} : x, y \in X; x \neq y\right\} = \|F - G\|,$$

implying

$$\|F|_Y\| \leq \inf\left\{\|F - G\| : G \in Y^{\perp}\right\} = d(F, Y^{\perp}).$$

By Theorem 4.1.1 there exists $G \in \mathrm{Lip}_0(X)$ such that $G|_Y = F|_Y$ and $\|G\| = \|F|_Y\|$. It follows that $F - G \in Y^{\perp}$ and

$$d(F, Y^{\perp}) \leq \|F - (F - G)\| = \|G\| = \|F|_Y\|,$$

showing that (8.9.2) holds.

Also $G \in Y^{\perp}$ is a nearest point to F in $Y^{\perp}$ if and only if

$$\|F - G\| = d(F, Y^{\perp}) = \|F|_Y\|.$$

Since $(F - G)|_Y = F|_Y$ we have $F - G \in E(F|_Y)$ which is equivalent to $G \in F - E(F|_Y)$.

Therefore

$$G \in P_{Y^{\perp}}(F) \iff G \in F - E(F|_Y),$$

proving the formula (8.9.3).

The second assertion of the theorem is an immediate consequence of the formula (8.9.3).   □

Since, by Theorem 4.1.1, any norm-preserving extension of $F|_Y$ is contained between two extremal extensions F_1, F_2 (the functions F, G from (4.1.1)), it follows that $E(F|_Y)$ is a singleton if and only if

$$\sup\{F|_Y(y) - \|F|_Y\|d(x, y) : y \in Y\} = \inf\{F|_Y(y) + \|F|_Y\|d(x, y) : y \in Y\},$$

for all $x \in X$.

Since

$$\inf F|_Y(Y) + \|F|_Y\|d(x, Y) \leq \inf_{y \in Y}\left(F|_Y(y) + \|F|_Y\|d(x, y)\right),$$

and

$$\sup F|_Y (Y) - \|F|_Y\| d (x, Y) \geq \sup_{y \in Y} \left(F|_Y (y) - \|F|_Y\| d (x, y) \right),$$

we have

$$d (x, Y) \leq \frac{\sup (F|_Y) (Y) - \inf (F|_Y) (Y)}{2 \|F|_Y\|}, \tag{8.9.4}$$

for every $x \in X$ and every $F \in \mathrm{Lip}_0(X) \setminus Y^\perp$.

Using this inequality one can give conditions on Y so that its annihilator $Y^\perp$ is Chebyshevian. A metric space (X, d) is called *uniformly discrete* if there exists $\delta > 0$ such that $d(x, y) \geq \delta$ for all pairs of distinct points x, y in X.

Proposition 8.9.2 *Let (X, d) be a pointed metric space and Y a subset of X containing the base point θ.*

1. *If $\overline{Y} = X$, then $Y^\perp$ is Chebyshevian in $\mathrm{Lip}_0(X)$.*
2. *If $Y^\perp$ is Chebyshevian and Y is not uniformly discrete, then $\overline{Y} = X$.*

Proof

1. It is obvious that if Y is dense in X, then every Lipschitz function on Y admits a unique Lipschitz extension to X, and so $Y^\perp$ is Chebyshevian. Actually, in this case $Y^\perp = \{0\}$ so it is trivially Chebyshevian.
2. If Y is not uniformly discrete there exist $x_n \neq y_n$ in Y with $d(x_n, y_n) \to 0$ as $n \to \infty$. Consider the functions

$$f_n(x) = d(x, x_n) - d(x, y_n) - (d(\theta, x_n) - d(\theta, y_n)), \quad x \in X, \ n \in \mathbb{N}.$$

It is obvious that $f_n \in \mathrm{Lip}_0(X)$ for all $n \in \mathbb{N}$. Suppose that the set $\{n \in \mathbb{N} : d(\theta, x_n) - d(\theta, y_n) \geq 0\}$ is infinite. Without loosing the generality we can suppose that this holds for all $n \in \mathbb{N}$. It follows that

$$f_n(x_n) < 0 \ \text{ and } \ f_n(y_n) \geq 0,$$

for all $n \in \mathbb{N}$.

Define the functions $\psi_n : \mathbb{R} \to [0, 1]$ by

$$\psi_n(t) = \begin{cases} 1 & \text{for } t < f_n(x_n) \\ \frac{t}{f_n(x_n)} & \text{for } f_n(x_n) \leq t \leq 0 \\ 0 & \text{for } t > 0. \end{cases}$$

The function ψ is Lipschitz with $L(\psi) = 1/|f_n(x_n)|$. Let $F_n = \psi_n \circ f_n$, $n \in \mathbb{N}$. Then $F_n \in \mathrm{Lip}_0(X)$ and

$$\|F_n|_Y\| \geq \frac{|\psi_n(f_n(x_n)) - \psi_n(f_n(y_n))|}{d(x_n, y_n)} = \frac{1}{d(x_n, y_n)}.$$

By (8.9.4),

$$d(x, Y) \leq \frac{\sup F_n(Y) - \inf F_n(Y)}{2\|F_n|_Y\|} \leq \frac{1}{2} d(x_n, y_n).$$

Since $d(x_n, y_n) \to 0$ as $n \to \infty$, it follows that $d(x, Y) = 0$, and so $x \in \overline{Y}$. The case when the set $\{n \in \mathbb{N} : d(x_0, x_n) - d(x_0, y_n) \leq 0\}$ is infinite can be treated similarly. $\square$

Similar results hold in the case of the extension of convex Lipschitz functions.

Let Y be a convex subset of a normed space X. It was proved in [164] that every convex function $f \in \mathrm{Lip}(Y)$ admits a norm-preserving convex extension $F \in \mathrm{Lip}(X)$ and that, as in the case of Lipschitz extensions, all these extensions are contained between two extremal extensions F_1, F_2.

Now, if $0 \in Y$ is the fixed point, then put

$$K_Y = \{f \in \mathrm{Lip}_0(Y) : f \text{ is convex on } Y\}.$$

It follows that K_Y is a convex cone and let

$$X_c = K_X - K_X$$

be the linear space generated by the cone

$$K_X = \{F \in \mathrm{Lip}_0(X) : F \text{ is convex on } X\}.$$

Let also

$$Y_c = K_Y - K_Y,$$

and

$$Y_c^\perp = \{F \in X_c : F|_Y = 0\}.$$

Theorem 8.9.3 ([164]) *Let Y be a convex subset of a normed space X.*

1. If $F \in K_X$, then

$$\|F|_Y\| = d\left(F, Y_c^\perp\right).$$

2. *The space $Y_c^\perp$ is K_X-proximinal and, for $F \in K_X$, a function $G \in Y_c^\perp$ is a nearest point to F in $Y_c^\perp$ if and only if $G = F - H$, where H is a norm-preserving convex extension of $F|_Y$.*
3. *The space $Y_c^\perp$ is K_X-Chebyshevian if and only if every $f \in K_Y$ has a unique convex norm-preserving extension to X.*

Similar results hold for starshaped Lipschitz functions (see [493, 496]).

Characterizations of Best Approximation

As it was shown in [491, 494], one can obtain characterizations of best approximation elements in metric spaces, similar to those known in normed spaces (see [637, Ch. 1, §1]), by replacing the linear functionals with Lipschitz functions.

Let (X, d) be a metric space with a base point θ and Y a nonempty subset of X containing θ. Denote by $\mathrm{Lip}_0(X), \mathrm{Lip}_0(Y)$ the spaces of real-valued Lipschitz functions vanishing at θ with the Lipschitz norms $\|\cdot\|_X$ and $\|\cdot\|_Y$, respectively.

For $x \in X$ put

$$d(x, Y) = \inf\{d(x, y) : y \in Y\} \quad \text{and} \quad P_Y(x) = \{y \in Y : d(x, y) = d(x, Y)\}.$$

The elements in $P_Y(x)$ are called elements of best approximation of x by elements in Y (or nearest points to x in Y).

Proposition 8.9.4 *Let (X, d, θ) be a pointed metric space, $\theta \in Y \subseteq X$ and $x_0 \in X$. If $\delta := d(x_0, Y) > 0$, then there exists a Lipschitz mapping $f \in \mathrm{Lip}_0(X)$ such that*

$$(i) \quad f|_Y = 0, \quad (ii) \quad f(x_0) = 1 \quad and \quad (iii) \quad \|f\|_X = 1/\delta.$$

Proof Take $f(x) = \delta^{-1}d(x, Y)$. Then, by Proposition 2.1.1, f is Lipschitz with $\|f\|_X = \delta^{-1}$, $f(x_0) = 1$ and $f(y) = 0$ for all $y \in Y$. $\square$

The following result is a partial analogue of the Ascoli's formula (1.4.3) for the distance to a closed hyperplane in a normed space.

Proposition 8.9.5 *Let $f \in \mathrm{Lip}_0(X)$, $f \neq 0$, and $Z = f^{-1}(0)$. Then, for every $x \in X$,*

$$d(x, f^{-1}(0)) \geq \frac{|f(x)|}{\|f\|_X}. \tag{8.9.5}$$

If there exists $z_0 \in Z$ such that $|f(x) - f(z_0)| = \|f\|_X d(x, z_0)$, then $z_0 \in P_Z(x)$.

Proof For every $z \in Z$

$$\frac{|f(x)|}{\|f\|_X} = \frac{|f(x) - f(z)|}{\|f\|_X} \leq d(x, z),$$

implying $|f(x)|/\|f\|_X \leq d(x, Z)$.

If, for some $z_0 \in Z$, $|f(x) - f(z_0)| = \|f\|_X d(x, z_0)$, then

$$d(x, Z) \geq \frac{|f(x)|}{\|f\|_X} = \frac{|f(x) - f(z_0)|}{\|f\|_X} = d(x, z_0) \geq d(x, Z),$$

showing that $d(x, z_0) = d(x, Z)$.   □

The characterization result is the following one.

Proposition 8.9.6 *Let Y be a subset of the pointed metric space (X, d, θ) containing the base point θ, $x_0 \in X \setminus Y$ and $y_0 \in Y$. Then $y_0 \in P_Y(x_0)$ if and only if there exists $f \in \mathrm{Lip}_0(X)$ such that*

(i) $\|f\|_X = 1$,   (ii) $f|_Y = 0$ *and*

(iii) $|f(x_0) - f(y_0)| = d(x_0, y_0)$.

Proof Suppose that $f \in \mathrm{Lip}_0(X)$ satisfies the conditions (i)–(iii). By (ii), $Y \subseteq f^{-1}(0)$, so that by (8.9.5),

$$d(x_0, Y) \geq d(x_0, f^{-1}(0)) \geq \frac{|f(x_0)|}{\|f\|_X} = |f(x_0) - f(y_0)| = d(x_0, y_0) \geq d(x_0, Y),$$

showing that $d(x_0, y_0) = d(x_0, Y)$.

Suppose now that $\delta := d(x_0, Y) = d(x_0, y_0) > 0$. Let $g \in \mathrm{Lip}_0(X)$ be the functional given by Proposition 8.9.4 and $f = \delta g$. Then $\|f\|_X = 1$ and $f|_Y = 0$, that is, (i) and (ii) hold. Also

$$|f(x_0) - f(y_0)| = \delta |g(x_0)| = \delta = d(x_0, y_0),$$

showing that (iii) holds too.   □

Let

$$Y^\perp = \left\{ f \in \mathrm{Lip}_0(X) : f|_Y = 0 \right\},$$

and, for $x \in X$, put

$$s_{Y^\perp}(x) = \sup \left\{ \frac{|f(x)|}{\|f\|_X} : f \in Y^\perp, \ f \neq 0 \right\}.$$

The following proposition contains a duality result for best approximation.

Proposition 8.9.7 *Let (X, d) be a metric space with a base point θ, Y a subset of X containing θ and $x_0 \in X \setminus Y$. Then*

$$s_{Y^\perp}(x_0) = d(x_0, Y) .$$

Consequently, $y_0 \in Y$ belongs to $P_Y(x_0)$ if and only if $d(x_0, y_0) = s_{Y^\perp}(x_0)$.

Proof Since

$$\frac{|f(x_0)|}{\|f\|_X} = \frac{|f(x_0) - f(y)|}{\|f\|_X} \leq d(x_0, y),$$

for all $f \in Y^\perp \setminus \{0\}$ and $y \in Y$, it follows that $s_{Y^\perp}(x_0) \leq d(x_0, Y)$. If $d(x_0, Y) = 0$, then we have equality. If $\delta := d(x_0, Y) > 0$, then, choosing $f \in Y^\perp$ according to Proposition 8.9.4, one obtains the reverse inequality

$$s_{Y^\perp}(x_0) \geq \frac{|f(x_0)|}{\|f\|_X} = \delta = d(x_0, Y).$$

$\square$

Remark 8.9.8 In [637, Ch.I, Theorem 1.12] the author gives characterization of the elements of best approximation by a subspace Y of a Banach space X in terms of the extreme points of the unit ball B_{X^*} of the dual space X^*. The key tool is the existence of norm-preserving extensions of extreme functionals of the unit ball B_{Y^*} which are extreme points of B_{X^*}. Similar results in the case of Lipschitz functions are obtained in [492]. Characterizations of the extreme points of the unit ball of the space $\mathrm{Lip}_0(X)$ are given in [156].

Extensions of these results to characterizations of best approximation elements in quasi-metric spaces in terms of semi-Lipschitz or semi-Hölder functions were given by Mustăţa in [502–504, 506, 507, 510, 511].

8.9.2   Lipschitz Duals of Metric Linear Spaces and Best Approximation

A metric linear space is a topological vector space (X, τ) whose topology is defined by a translation invariant metric, i.e., a metric d on X satisfying the condition

$$d(x + z, y + z) = d(x, y),$$

for all $x, y, z \in X$. Then $q(x) := d(x, 0)$, $x \in X$, is an F-norm on X (see Definition 1.4.55). Any F-norm q on a vector space X generates a vector topology and $d(x, y) := q(x - y)$, $x, y \in X$, is a translation invariant metric on X generating the same topology as q.

Schnatz [624] proposes as a nonlinear dual of a metric linear space the space $\mathrm{Lip}_{od}(X)$ formed of all odd Lipschitz functions, i.e., those Lipschitz functions f satisfying

$$|f(x) - f(y)| \leq Lq(x - y) \quad \text{and} \quad f(-x) = -f(-x), \tag{8.9.6}$$

for all $x, y \in X$. As usual, the number L in the above inequality is called a Lipschitz constant for f and the smallest Lipschitz constant is denoted by $L(f)$ and called the Lipschitz norm of f.

It follows that any function $f : X \to \mathbb{R}$ satisfying (8.9.6) vanishes at 0, $f(0) = 0$, so that $\mathrm{Lip}_{od}(X)$ is a linear subspace of $\mathrm{Lip}_0(X)$.

For a function $f : X \to \mathbb{R}$ put

$$\|f\| := \sup\left\{ \frac{|f(x)|}{q(x)} : x \in X, \, x \neq 0 \right\}. \tag{8.9.7}$$

If f is Lipschitz and odd, then

$$\|f\| \leq L(f).$$

It follows that (8.9.7) defines a norm on the space $\mathrm{Lip}_{od}(X)$. For $r > 0$ let

$$B_r(\mathrm{Lip}_{od}(X)) := \{f \in \mathrm{Lip}_{od}(X) : L(f) \leq r\}.$$

The role of "duals" in characterizing the best approximation elements in metric linear spaces will be played by these spaces. On $\mathrm{Lip}_{od}(X)$ and $B_r(\mathrm{Lip}_{od}(X))$ one introduce the "weak" topologies, denoted by w_{od}, as the topologies induced by the product topology on $\mathbb{R}^X$.

Proposition 8.9.9 *Let X be a metric linear space.*

1. *A net $(f_i : i \in I)$ in $\mathrm{Lip}_{od}(X)$ converges to $f \in \mathrm{Lip}_{od}(X)$ with respect to w_{od} if and only if*

$$f_i(x) \to f(x) \text{ for every } x \in X.$$

The same is true in $B_r(\mathrm{Lip}_{od}(X))$.
2. *If a net $(f_i : i \in I)$ in $B_r(\mathrm{Lip}_{od}(X))$ converges pointwise to some $f \in \mathbb{R}^X$, then $f \in B_r(\mathrm{Lip}_{od}(X))$.*
3. *If a net $(f_i : i \in I)$ in $\mathrm{Lip}_{od}(X)$ converges to some $f \in \mathrm{Lip}_{od}(X)$ with respect to the norm (8.9.7), then it also converges pointwise.*
4. *The set $B_r(\mathrm{Lip}_{od}(X))$ is closed in $\mathrm{Lip}_{od}(X)$ with respect to both of the topologies w_{od} and that generated by the norm (8.9.7).*

Proof The equivalence from 1 is a general property of the product topology (see Sect. 1.2.7).

2. Since, for every $x, y \in X$ and every $i \in I$, $|f_i(x) - f_i(y)| \leq rq(x - y)$, passing to limit with respect to $i \in I$ one obtains $|f(x) - f(y)| \leq rq(x - y)$. Similarly, the equality $f_i(-x) = -f_i(x)$, valid for every $x \in X$ and $i \in I$, yields $f(-x) = -f(x)$.
3. This follows from the inequality $|f_i(x) - f(x)| \leq \|f_i - f\|q(x)$, $i \in I$.

4. The closedness with respect to w_{od} follows from 2. Taking into account 3, this implies the closedness with respect to $\|\cdot\|$.   □

Remark 8.9.10 In [624] one shows that actually $B_r(\mathrm{Lip}_{od}(X))$ is w_{od}-compact, for every $r > 0$. The proof follows the lines of the proof of Proposition 8.2.16, since the w_{od}-topology is the topology induced by the product topology on $\mathbb{R}^X$, as the topology τ_p.

As in the linear case, the key tool in characterizing the nearest points is the existence of some special functions obtained via some extension results.

We start with a general result.

Proposition 8.9.11 *Let X be a vector space, Y a subspace of X and $p : X \to \mathbb{R}$ an even subadditive functional with $p(0) = 0$. Then every odd function $f : Y \to \mathbb{R}$ satisfying*

$$f(y) - f(y') \le p(y - y'),\tag{8.9.8}$$

for all $y, y' \in Y$, admits an extension $F : X \to \mathbb{R}$ satisfying

$$F(x) - F(x') \le p(x - x'),\tag{8.9.9}$$

for all $x, x' \in Y$.

Proof As in the proof of McShane's theorem (Theorem 4.1.1) one shows that the functions defined by

$$F_1(x) = \sup\{f(y) - p(x - y) : y \in Y\} \text{ and}$$
$$F_2(x) = \inf\{f(y) + p(x - y) : y \in Y\},$$

for every $x \in X$, are extensions of f satisfying (8.9.9). Also $F_1 \le F_2$.

Consider the function $F : X \to \mathbb{R}$ defined by

$$F(x) = \begin{cases} F_1(x) & \text{if } F_1(x) > 0 \\ F_2(x) & \text{if } F_2(x) < 0 \\ 0 & \text{if } F_1(x) \le 0 \le F_2(x). \end{cases}\tag{8.9.10}$$

It is easy to check that $F(y) = f(y)$ for $y \in Y$. Since f is odd we have

$$F_1(-x) = -F_2(x),$$

and so $F(-x) = -F(x)$.

Analyzing various situations that can occur, according to the definition of F, one shows that F satisfies also (8.9.9), so it is the required extension of f.   □

Remark 8.9.12 Taking into account the facts that p is even, the condition (8.9.8) is equivalent to

$$|f(y) - f(y')| \le p(y - y'),$$

for all $y, y' \in Y$. A similar remark holds for (8.9.9).

Based on these result one obtains the norm-preserving extension (with respect to the norm (8.9.7)) of odd Lipschitz functions.

Proposition 8.9.13 *Let* (X, d) *be a metric linear space and* Y *a subspace of* X. *Then every* $f \in \text{Lip}_{\text{od}}^r(Y)$ *admits an extension* $F \in \text{Lip}_{\text{od}}^r(X)$, *preserving the norm* (8.9.7).

Proof We shall denote by $\| \cdot \|_Y, \| \cdot \|_X$ the norms in $\text{Lip}_{\text{od}}(Y)$ and $\text{Lip}_{\text{od}}(X)$ given by (8.9.7), respectively.

Let $f \in \text{Lip}_{\text{od}}(X)$. We shall apply Proposition 8.9.11 with $p = rq$. Then for $f \in \text{Lip}_{\text{od}}(Y)$ the function F defined by (8.9.10) is an odd Lipschitz function extending f and satisfying

$$F(x) - F(x') \le rq(x - x') \quad \text{for all} \quad x, x' \in X,$$

which, by Remark 8.9.12, implies $L(F) \le r$, that is, $F \in B_r(\text{Lip}_{\text{od}}(X))$.

It remains to show that F and f have the same norm. Since f is a restriction of F, it follows that

$$\|F\|_X = \sup_{x \in X \setminus \{0\}} \frac{|F(x)|}{q(x)} \ge \sup_{y \in Y \setminus \{0\}} \frac{|F(y)|}{q(y)} = \|f\|_Y.$$

To prove the reverse inequality suppose first that $F(x) > 0$. In this case

$$F(x) = \sup\{f(y) - rq(x - y) : y \in Y\}$$

$$= \sup\{f(y) - rq(x - y) : y \in Y, \ f(y) - rq(x - y) > 0\}.$$

Let $y \in Y$ be such that $f(y) - rq(x - y) > 0$. If $q(x) \ge q(y)$, then

$$\frac{f(y) - rq(x - y)}{q(x)} \le \frac{f(y) - rq(x - y)}{q(y)} \le \frac{f(y)}{q(y)} \le \|f\|_Y.$$

If $q(x) < q(y)$, then

$$\frac{f(y)}{q(y)}(q(y) - q(x)) \le rq(x - y),$$

which is equivalent to

$$\frac{f(y) - rq(x-y)}{q(x)} \le \frac{f(y)}{q(y)} \le \|f\|_Y .$$

It follows that

$$\frac{|F(x)|}{q(x)} = \frac{F(x)}{q(x)} \le \|f\|_Y . \tag{8.9.11}$$

If $F(x) < 0$, then $F(-x) = -F(x) > 0$ and

$$\frac{|F(x)|}{q(x)} = \frac{-F(x)}{q(x)} = \frac{F(-x)}{q(-x)} \le \|f\|_Y . \tag{8.9.12}$$

Since, for $F(x) = 0$, $|F(x)|/q(x) = 0 \le \|f\|_Y$, the inequalities (8.9.11) and (8.9.12) show that

$$\frac{|F(x)|}{q(x)} \le \|f\|_Y ,$$

for all $x \in X$, so that

$$\|F\|_X = \sup_{x \in X} \frac{|F(x)|}{q(x)} \le \|f\|_Y .$$

$\square$

As in the linear case the characterization of best approximation elements is based on the existence of some special functionals related to the distance function. We mention first some properties of the distance function with respect to an F-norm.

Let X be a metric linear space with the topology generated by an F-norm q. The F-norm q is called *monotone* if

$$|s| \le |t| \implies q(sx) \le q(tx) ,$$

for all $x \in X$ and $s, t \in \mathbb{R}$.

Examples of monotone norms are $q(x) = \sum_{k=1}^{\infty} |x_k|^p$ and $q(f) = \int_0^1 |f(t)|^p dt$ for $x = (x_k)_{k \in \mathbb{N}} \in \ell^p$ and $f \in L^p[0, 1]$, for $0 < p < 1$ (see (1.4.18)).

Proposition 8.9.14 *Let q be an F-norm on a vector space X and Y a subspace of X. The distance*

$$d(x, Y) = \inf\{q(x-y) : y \in Y\}$$

has the following properties:

(i) $d(-x, Y) = d(x, Y)$,   (ii) $d(x + y, Y) = d(x, Y)$,

(iii) $|d(x, Y) - d(x', Y)| \leq q(x - x')$,
$$\text{(8.9.13)}$$

for all $x, x' \in X$ and $y \in Y$.
   If the F-norm q is monotone, then

$$d(\alpha x, Y) \leq d(\beta x, Y),$$
$$\text{(8.9.14)}$$

for all $x \in X$ and $\alpha, \beta \in \mathbb{R}$ with $0 < |\alpha| \leq |\beta|$.

Proof The equalities (i) and (ii) from (8.9.13) are obvious. The proof of the inequality (iii) is similar to that of (1.3.8). The proof of (8.9.14) follows from the following relations:

$$d(\alpha x, Y) = \inf\{q(\alpha x - y) : y \in Y\} = \inf\{q(\alpha x - \alpha y) : y \in Y\}$$

$$\leq \inf\{q(\beta x - \beta y) : y \in Y\} = \inf\{q(\beta x - y) : y \in Y\} = d(\beta x, Y).$$

$$\square$$

The existence result for these special functionals is the following one.

Proposition 8.9.15 *Let q be a monotone F-norm on a vector space X and Y a subspace of X. Then, for every $x \in X \setminus \overline{Y}$, there exists a function $f \in \mathrm{Lip}_{od}(X)$ with $L(f) \leq 2$, satisfying the following conditions:*

(i) $\|f\|_X = 1$,   (ii) $f(x) = d(x, Y)$ *and*

(iii) $f(y) = 0$ *and* $f(x + y) = f(x)$ *for all* $y \in Y$.
$$\text{(8.9.15)}$$

Proof On $Z := \mathbb{R}x \dotplus Y$ define the functional φ by

$$\varphi(\alpha x + y) = s(\alpha)d(\alpha x + y, Y) \text{ for } \alpha \in \mathbb{R} \text{ and } y \in Y,$$

where $s(\alpha) = \mathrm{sign}(\alpha)$.
   Then φ is odd, $\varphi(\alpha x + y) = s(\alpha)d(\alpha x, Y)$ and $\varphi(x) = d(x, Y)$. Also $\varphi(y) = d(y, Y) = 0$ and $\varphi(x + y) = d(x, Y) = \varphi(x)$ for all $y \in Y$, that is, φ satisfies (ii) and (iii) from (8.9.15) with $f = \varphi$.
   Since

$$\frac{|\varphi(\alpha x + y)|}{q(\alpha x + y)} = \frac{d(\alpha x + y, Y)}{q(\alpha x + y)} \leq 1,$$

for all $\alpha x + y \neq 0$ in Z, it follows that $\|\varphi\|_Z \leq 1.$

If (y_n) is a sequence in Y such that $\lim_n q(x + y_n) = d(x, Y)$, then

$$\|\varphi\|_Z \geq \frac{\varphi(x + y_n)}{q(x + y_n)} = \frac{d(x, Y)}{q(x + y_n)} \to 1 \text{ as } n \to \infty,$$

hence $\|\varphi\|_Z \geq 1$, and so $\|\varphi\|_Z = 1$, that is, (i) holds too with $f = \varphi$.

We show that $L(f) \leq 2$.

Let $y, y' \in Y$. If $\alpha, \alpha' \in \mathbb{R}$ have the same sign or one of them is null, then, by (8.9.13).(iii),

$$|\varphi(\alpha x + y) - \varphi(\alpha' x + y')| = |d(\alpha x + y, Y) - d(\alpha' x + y', Y)|$$
$$\leq q(\alpha x + y - (\alpha' x + y')).$$

If $\alpha, \alpha' \in \mathbb{R} \setminus \{0\}$ have different signs, then

$$|\varphi(\alpha x + y) - \varphi(\alpha' x + y')| = d(\alpha x + y, Y) + d(\alpha' x + y', Y).$$

Suppose, for instance, that $\alpha > 0$ and $\alpha' < 0$. Then, by (8.9.14),

$$d(\alpha x + y, Y) = d(\alpha x, Y) \leq d((\alpha + |\alpha'|)x, Y)$$
$$\leq q((\alpha + |\alpha'|)x + y - y') = q((\alpha x + y) - (\alpha' x + y')).$$

A similar inequality holds for $d(\alpha' x + y', Y)$, so that

$$|\varphi(\alpha x + y) - \varphi(\alpha' x + y')| \leq 2q((\alpha x + y) - (\alpha' x + y')).$$

Consequently, $L(\varphi) \leq 2$.

By Proposition 8.9.13, φ has an extension f in $\text{Lip}_{od}(X)$ with $L(f) \leq 2$ and $\|f\|_X = \|\varphi\|_Z$, which satisfies all the conditions (i)–(iii) from (8.9.15). □

From this extension result one obtains easily the characterization of best approximation elements.

Proposition 8.9.16 *Let q be a monotone F-norm on a vector space X, Y a subspace of X closed with respect to the topology generated by q, $r \geq 2$, $x \in X \setminus \overline{Y}$ and $y_0 \in Y$. Then y_0 is a best approximation element for x in Y if and only if there exists a function $f \in \text{Lip}_{od}^r(X)$ satisfying the following conditions:*

$$\begin{aligned} &\text{(i) } \|f\|_X = 1, \quad \text{(ii) } f(x) = q(x - y_0), \\ &\text{(iii) } f(y) = 0 \text{ and } f(x + y) = f(x) \text{ for all } y \in Y. \end{aligned} \quad (8.9.16)$$

Proof The necessity part follows from Proposition 8.9.16.

Suppose now that $f \in \text{Lip}^r_{\text{od}}(X)$ satisfies the conditions (i)–(iii) from (8.9.16). Then

$$\frac{q(x - y_0)}{q(x - y)} = \frac{f(x)}{q(x - y)} = \frac{f(x - y)}{q(x - y)} \leq \|f\|_X = 1,$$

that is, $q(x - y_0) \leq q(x - y)$ for all $y \in Y$. $\qquad\square$

Remark 8.9.17 In [624] one proves also that every extreme point of the unit ball of $\text{Lip}^r_{\text{od}}(Y)$ has an extension which is an extreme point of the unit ball of $\text{Lip}^r_{\text{od}}(X)$. Based on this result one obtains characterizations of best approximation elements in terms of the extreme points of the unit ball of $\text{Lip}^r_{\text{od}}(X)$, similar to those given in [637, Ch. I, Theorem 1.12].

Remark 8.9.18 It is shown in [624] that if (X, d) is a metric linear space, then $\text{Lip}_{\text{od}}(X)$ is not complete, provided that the associated F-norm $q(x) = d(x, 0)$, $x \in X$, is monotone.

The proof is based on the equality

$$\text{Lip}_{\text{od}}(X) = \bigcup_{k=1}^{\infty} \text{Lip}^k_{\text{od}}(X),$$

and on the fact that each $\text{Lip}^k_{\text{od}}(X)$ is closed in $\text{Lip}_{\text{od}}(X)$ (see Proposition 8.9.9.4). One shows that the interior of $\text{Lip}^k_{\text{od}}(X)$ is empty for every $k \in \mathbb{N}$, so that, by Baire's category theorem (see Theorem 1.3.19 and Corollary 1.3.18), the space $\text{Lip}_{\text{od}}(X)$ is not complete. See [624] for details.

Remark 8.9.19 The uniqueness of the extension of odd Lipschitz functions is discussed in [498].

Pantelidis [538] proposes another substitute for the dual of a Banach space. For a metric linear space (X, d) with the associated F-norm $q(x) = d(x, 0)$, $x \in X$, one considers the cone

$$E^v := \left\{ f : X \to \mathbb{R} : f \text{ is subadditive, } f(0) = 0 \text{ and } \sup_{x \in X \setminus \{0\}} \frac{|f(x)|}{q(x)} < \infty \right\},$$

called the *dual cone* of X. It is obvious that if X is a normed space with dual X^*, then $X^* \subseteq X^v$ and X^* is the biggest linear subspace of X^v.

The characterization result is the following.

Proposition 8.9.20 *Let (X,d) be a metric linear space with the associated F-norm q, Y a closed subspace of X, $x_0 \in X \setminus Y$ and $y_0 \in Y$. Then y_0 is a best approximation* *element for x_0 in Y if and only if there exists a function $f \in X^\upsilon$ satisfying the following conditions:*

(i) $|f(x) - f(x')| \leq q(x - x')$ *for all* $x, x' \in X$;

(ii) $f(x_0 - y_0) = f(x_0) = q(x_0 - y_0)$;

(iii) $f(x + y) = f(x)$ *for all* $x \in X$ *and* $y \in Y$,

      (or, equivalently, $f|_Y = 0$).

References

1. M.D. Acosta, Denseness of norm attaining mappings. RACSAM **100**(1–2), 9–30 (2006)
2. M.D. Acosta, V. Montesinos, On norm-attaining functionals. Acta Univ. Carolin. Math. Phys. **47**(2), 5–24 (2006)
3. M.D. Acosta, R.M. Aron, D. García, M. Maestre, The Bishop-Phelps-Bollobás theorem for operators. J. Funct. Anal. **254**(11), 2780–2799 (2008)
4. M.D. Acosta, J. Alaminos, D. García, M. Maestre, A variational approach to norm attainment of some operators and polynomials. Acta Math. Sin. (Engl. Ser.) **26**(12), 2259–2268 (2010)
5. M.D. Acosta, J.B. Guerrero, D. García, M. Maestre, The Bishop-Phelps-Bollobás theorem for bilinear forms. Trans. Am. Math. Soc. **365**(11), 5911–5932 (2013)
6. M.D. Acosta, J.B. Guerrero, D. García, S.K. Kim, M. Maestre, Bishop-Phelps-Bollobás property for certain spaces of operators. J. Math. Anal. Appl. **414**(2), 532–545 (2014)
7. M.D. Acosta, J.B. Guerrero, D. García, S.K. Kim, M. Maestre, The Bishop-Phelps-Bollobás property: a finite-dimensional approach. Publ. Res. Inst. Math. Sci. **51**(1), 173–190 (2015)
8. Adimurthi, I.H. Biswas, Role of fundamental solutions for optimal Lipschitz extensions on hyperbolic space. J. Differ. Equ. **218**(1), 1–14 (2005)
9. I. Aharoni, Every separable metric space is Lipschitz equivalent to a subset of c_0^+. Isr. J. Math. **19**, 284–291 (1974)
10. I. Aharoni, J. Lindenstrauss, Uniform equivalence between Banach spaces. Bull. Am. Math. Soc. **84**(2), 281–283 (1978)
11. A.V. Akopyan, A.S. Tarasov, A constructive proof of Kirszbraun's theorem. Mat. Zametki **84**(5), 781–784 (2008) [Russian; English translation in Math. Notes **84**(5–6), 725–728 (2008)]
12. G. Alberti, A. Marchese, On the differentiability of Lipschitz functions with respect to measures in the Euclidean space. Geom. Funct. Anal. **26**(1), 1–66 (2016)
13. G. Alberti, M. Csörnyei, D. Preiss, Structure of null sets in the plane and applications, in *European Congress of Mathematics* (European Mathematical Society, Zürich, 2005), pp. 3–22
14. G. Alberti, M. Csörnyei, D. Preiss, Differentiability of Lipschitz functions, structure of null sets, and other problems, in *Proceedings of the International Congress of Mathematicians*, vol. III (Hindustan Book Agency, New Delhi, 2010), pp. 1379–1394
15. F. Albiac, Nonlinear structure of some classical quasi-Banach spaces and F-spaces. J. Math. Anal. Appl. **340**(2), 1312–1325 (2008)
16. F. Albiac, The role of local convexity in Lipschitz maps. J. Convex Anal. **18**(4), 983–997 (2011)
17. F. Albiac, J.L. Ansorena, Lipschitz maps and primitives for continuous functions in quasi-Banach spaces. Nonlinear Anal. **75**(16), 6108–6119 (2012)

© Springer Nature Switzerland AG 2019
Ş. Cobzaş et al., *Lipschitz Functions*, Lecture Notes in Mathematics 2241,
https://doi.org/10.1007/978-3-030-16489-8

18. F. Albiac, J.L. Ansorena, On a problem posed by M. M. Popov. Studia Math. **211**(3), 247–258 (2012)
19. F. Albiac, J.L. Ansorena, Integration in quasi-Banach spaces and the fundamental theorem of calculus. J. Funct. Anal. **264**(9), 2059–2076 (2013)
20. F. Albiac, J.L. Ansorena, Optimal average approximations for functions mapping in quasi-Banach spaces. J. Funct. Anal. **266**(6), 3894–3905 (2014)
21. F. Albiac, J.L. Ansorena, On Lipschitz maps, martingales, and the Radon-Nikodým property for F-spaces. Mediterr. J. Math. **13**(4), 1963–1980 (2016)
22. F. Albiac, N.J. Kalton, *Topics in Banach Space Theory*. Graduate Texts in Mathematics, vol. 233 (Springer, New York, 2006)
23. F. Albiac, N.J. Kalton, Lipschitz structure of quasi-Banach spaces. Isr. J. Math. **170**, 317–335 (2009)
24. Y. Alber, S. Reich, J.-C. Yao, Iterative methods for solving fixed-point problems with nonself-mappings in Banach spaces. Abstr. Appl. Anal. **2003**(4), 193–216 (2003)
25. J.M. Aldaz, On the derivability of Lipschitz functions. Rend. Circ. Mat. Palermo **50**(1), 213–216 (2001)
26. A.D. Aleksandrov, Almost everywhere existence of the second differential of a convex function and some properties of convex surfaces connected with it. Leningrad. State Univ. Ann. Uchen. Zap. Math. Ser. **6**, 3–35 (1939) [Russian]
27. A.D. Aleksandrov, A theorem on triangles in a metric space and some of its applications. Trudy Mat. Inst. Steklov. **38**, 5–23 (1951) [Russian]
28. S. Alexander, V. Kapovitch, A. Petrunin, Alexandrov meets Kirszbraun, in *Proceedings of the 17th Gökova Geometry-Topology Conference*, Gökova, Turkey, May 31–June 4, 2010 (International Press, Somerville; Gökova: Gökova Geometry-Topology Conferences, 2011), pp. 88–109
29. S. Alexander, V. Kapovitch, A. Petrunin, Alexandrov geometry (2017, Book in progress)) URL: http://anton-petrunin.github.io/book/all.pdf
30. P. Alestalo, D.A. Trotsenko, Plane sets allowing bilipschitz extensions. Math. Scand. **105**(1), 134–146 (2009)
31. P. Alestalo, D.A. Trotsenko, J. Väisälä, The linear extension property of bi-Lipschitz mappings. Sibirsk. Mat. Zh. **44**(6), 1226–1238 (2003)
32. R. Alexander, The circumdisk and its relation to a theorem of Kirszbraun and Valentine. Math. Mag. **57**(3), 165–169 (1984)
33. M.A. Alghamdi, W.A. Kirk, N. Shahzad, Remarks on convex combinations in geodesic spaces. J. Nonlinear Convex Anal. **15**(1), 49–59 (2014)
34. Ch.D. Aliprantis, K.C. Border, *Infinite-Dimensional Analysis - A Hitchhiker's Guide*, 2nd edn. (Springer, Berlin, 1999)
35. Ch.D. Aliprantis, R. Tourky, *Cones and Duality*. Graduate Studies in Mathematics, vol. 84 (American Mathematical Society, Providence, 2007)
36. F.J. Almgren, *Almgren's big regularity paper. Q-valued functions minimizing Dirichlet's integral and the regularity of area-minimizing rectifiable currents up to codimension 2*, ed. by V. Scheffer, J.E. Taylor (World Scientific Publishing, Singapore, 2000)
37. L. Ambrosio, B. Kirchheim, Rectifiable sets in metric and Banach spaces. Math. Ann. **318**(3), 527–555 (2000)
38. L. Ambrosio, P. Tilli, *Topics on Analysis in Metric Spaces* (Oxford University Press, Oxford, 2004)
39. T. Aoki, Locally bounded linear topological spaces. Proc. Imper. Acad. Tokyo **18**, 588–594 (1942)
40. J. Appell, P.P. Zabrejko, *Nonlinear Superposition Operators*. Cambridge Tracts in Mathematics, vol. 95 (Cambridge University Press, Cambridge, 1990)
41. R. Arens, The adjoint of a bilinear operation. Proc. Am. Math. Soc. **2**, 839–848 (1951)
42. R. Arens, Operations induced in function classes. Monatsh. Math. **55**, 1–19 (1951)
43. W. Arendt, Ch. J.K. Batty, M. Hieber, F. Neubrander, *Vector-Valued Laplace Transforms and Cauchy Problems*. 2nd edn. (Birkhäuser, Basel, 2011)

44. R.F. Arens, J. Eells, Jr., On embedding uniform and topological spaces. Pac. J. Math. **6**, 397–403 (1956)

45. A.V. Arkhangel'skij, V.I. Ponomarev, *Fundamentals of General Topology: Problems and Exercises* (D. Reidel Publishing Company, Dordrecht, 1984)

46. D. Ariza-Ruiz, A. Fernández-León, G. López-Acedo, A. Nicolae, Chebyshev sets in geodesic spaces. J. Approx. Theory **207**, 265–282 (2016)

47. R. Aron, V. Lomonosov, After the Bishop-Phelps theorem. Acta Comment. Univ. Tartu. Math. **18**(1), 39–49 (2014)

48. R. Aron, Y.S. Choi, S.K. Kim, H.J. Lee, M. Martín, The Bishop-Phelps-Bollobás version of Lindenstrauss properties A and B. Trans. Am. Math. Soc. **367**(9), 6085–6101 (2015)

49. G. Aronsson, Extension of functions satisfying Lipschitz conditions. Ark. Mat. **6**, 551–561 (1967)

50. G. Aronsson, M.G. Crandall, P. Juutinen, A tour of the theory of absolutely minimizing functions. Bull. Am. Math. Soc. (N.S.) **41**(4), 439–505 (2004)

51. N. Aronszajn, Differentiability of Lipschitzian mappings between Banach spaces. Studia Math. **57**(2), 147–190 (1976)

52. N. Aronszajn, P. Panitchpakdi, Extension of uniformly continuous transformations and hyperconvex metric spaces. Pac. J. Math. **6**, 405–439 (1956) [Correction. Ibid. 7 (1957) 1729]

53. M. Aschenbrenner, A. Fischer, Definable versions of theorems by Kirszbraun and Helly. Proc. Lond. Math. Soc. **102**(3), 468–502 (2011)

54. E. Asplund, Fréchet differentiability of convex functions. Acta Math. **121**, 31–47 (1968)

55. P. Assouad, Espaces de fonctions lipschitziennes. C. R. Acad. Sci. Paris Sér. A-B **281**(2–3), Aii, A107 (1975)

56. P. Assouad, Remarques sur un article de Israel Aharoni sur les prolongements lipschitziens dans c_0 (Isr. J. Math. 19 (1974), 284–291). Isr. J. Math. **31**(1), 97–100 (1978)

57. P. Assouad, Étude d'une dimension métrique liée à la possibilité de plongements dans $\mathbf{R}^n$. C. R. Acad. Sci. Paris Sér. A-B **288**(15), A731–A734 (1979)

58. P. Assouad, Sur la distance de Nagata. C. R. Acad. Sci. Paris Sér. I Math. **294**(1), 31–34 (1982)

59. P. Assouad, Plongements lipschitziens dans $\mathbf{R}^n$. Bull. Soc. Math. Fr. **111**(4), 429–448 (1983)

60. J.-P. Aubin, H. Frankowska, *Set-Valued Analysis* (Birkhäuser Boston Inc., Boston, 1990)

61. D. Azagra, R. Fry, L. Keener, Real analytic approximation of Lipschitz functions on Hilbert space and other Banach spaces. J. Funct. Anal. **262**(1), 124–166 (2012)

62. W.G. Bade, P.C. Curtis, Jr., H.G. Dales, Amenability and weak amenability for Beurling and Lipschitz algebras. Proc. London Math. Soc. **55**(2), 359–377 (1987)

63. K. Ball, Markov chains, Riesz transforms and Lipschitz maps. Geom. Funct. Anal. **2**(2), 137–172 (1992)

64. K. Ball, The Ribe programme. Astérisque **VIII**(352), Exp. No. 1047, 147–159. Séminaire Bourbaki. Vol. 2011/2012. Exposés 1043–1058 (2013)

65. K. Ball, E.A. Carlen, E.H. Lieb, Sharp uniform convexity and smoothness inequalities for trace norms. Invent. Math. **115**(3), 463–482 (1994)

66. W. Ballmann, Lectures on spaces of nonpositive curvature, in *DMV Seminar*, vol. 25 (Birkhäuser, Basel, 1995)

67. Z.M. Balogh, K.S. Fässler, Rectifiability and Lipschitz extensions into the Heisenberg group. Math. Z. **263**(3), 673–683 (2009)

68. Z.M. Balogh, U. Lang, P. Pansu, Lipschitz extensions of maps between Heisenberg groups. Ann. Inst. Fourier (Grenoble) **66**(4), 1653–1665 (2016)

69. G. Basso, Fixed point theorems for metric spaces with a conical geodesic bicombing. Ergod. Theory Dyn. Syst. **38**(5), 1642–1657 (2018)

70. G. Basso, Lipschitz extensions to finitely many points. Anal. Geom. Metr. Spaces **6**(1), 174–191 (2018)

71. B. Beauzamy, *Introduction to Banach Spaces and Their Geometry*, 2nd edn. North Holland Mathematics Studies, vol. 68 [Notas de Matemática, vol. 86] (North-Holland, Amsterdam, 1985)

72. F. Behrouzi, H. Mahyar, Compact endomorphisms of certain analytic Lipschitz algebras. Bull. Belg. Math. Soc. Simon Stevin **12**(2), 301–312 (2005)

73. J.J. Benedetto, W. Czaja, *Integration and Modern Analysis*. Birkhäuser Advanced Texts (Birkhäuser, Basel, 2009)

74. Y. Benyamini, Introduction to the uniform classification of Banach spaces, in *Advanced Courses of Mathematical Analysis* vol. I (World Sci. Publ., Hackensack, 2004), pp. 1–29

75. Y. Benyamini, J. Lindenstrauss, *Geometric Nonlinear Functional Analysis, vol. 1*. American Mathematical Society Colloquium Publications, vol. 48 (American Mathematical Society, Providence, 2000)

76. I.D. Berg, I.G. Nikolaev, Quasilinearization and curvature of Aleksandrov spaces. Geom. Dedicata **133**, 195–218 (2008)

77. I.D. Berg, I.G. Nikolaev, Characterization of Aleksandrov spaces of curvature bounded above by means of the metric Cauchy-Schwarz inequality. Michigan Math. J. **67**(2), 289–332 (2018)

78. M. Bestvina, $\mathbb{R}$-trees in topology, geometry, and group theory, in *Handbook of Geometric Topology* (North-Holland, Amsterdam, 2002), pp. 55–91

79. D.N. Bessis, F.H. Clarke, Partial subdifferentials, derivates and Rademacher's theorem. Trans. Am. Math. Soc. **351**(7), 2899–2926 (1999)

80. C. Bessaga, A. Pełczyński, *Selected Topics in Infinite-Dimensional Topology*. Monografie matematyczne. Tom 58 (PWN, Warszawa, 1975)

81. P. Billingsley, *Convergence of Probability Measures*, 2nd edn., Wiley Series in Probability and Statistics: Probability and Statistics (Wiley, New York, 1999)

82. E. Bishop, D. Bridges, *Constructive Analysis*. Grundlehren der Mathematischen Wissenschaften [Fundamental Principles of Mathematical Sciences], vol. 279 (Springer, Berlin, 1985)

83. E. Bishop, R.R. Phelps, A proof that every Banach space is subreflexive. Bull. Am. Math. Soc. **67**, 97–98 (1961)

84. E. Bishop, R.R. Phelps, The support functionals of a convex set, in *Proceedings of Symposia in Pure Mathematics*, vol. VII (American Mathematical Society, Providence, 1963), pp. 27–35

85. L.M. Blumenthal, *Theory and Applications of Distance Geometry*. 2nd edn (Chelsea Publishing Co., New York, 1970)

86. V.I. Bogachev, B.G. Goldis, Second derivatives of convex functions in the sense of A. D. Aleksandrov in infinite-dimensional measure spaces. Mat. Zametki **79**(4), 488–504 (2006) [Russian; English translation: Math. Notes **79**(3–4), 454–467 (2006)]

87. V.I. Bogachev, A.N. Kalinin, A continuous cost function for which the minima in the Monge and Kantorovich problems are not equal. Dokl. Akad. Nauk **463**(4), 383–386 (2015)

88. V.I. Bogachev, A.V. Kolesnikov, The Monge-Kantorovich problem: achievements, connections, and prospects. Uspekhi Mat. Nauk **67**(5), 3–110 (2012) [Russian; English translation in Russian Math. Surveys **67**(5), 785–890 (2012)]

89. V.I. Bogachev, E. Mayer-Wolf, Some remarks on Rademacher's theorem in infinite dimensions, Potential Anal. **5**(1), 23–30 (1996)

90. V.I. Bogachev, S.A. Shkarin, Differentiable and Lipschitz mappings of Banach spaces. Mat. Zametki **44**(5), 567–583 (1988) [Russian; English translation in Math. Notes **44**(5–6), 790–798 (1988)]

91. V. Bogdan, On Henri Cartan's vectorial mean-value theorem and its applications to Lipschitzian operators and generalized Lebesgue-Bochner-Stieltjes integration theory, arXiv:0910.2277 [math.FA] (2009), p. 35

92. M.C. Boiso, Approximation of Lipschitz functions by Δ-convex functions in Banach spaces. Isr. J. Math. **106**, 269–284 (1998)

93. B. Bollobás, An extension to the theorem of Bishop and Phelps. Bull. London Math. Soc. **2**, 181–182 (1970)

94. D. Bongiorno, Stepanoff's theorem in separable Banach spaces. Comment. Math. Univ. Carolin. **39**(2), 323–335 (1998)

95. D. Bongiorno, Radon-Nikodým property of the range of Lipschitz extensions. Atti Sem. Mat. Fis. Univ. Modena **48**(2), 517–525 (2000)

96. D. Bongiorno, Rademacher theorem in separable Frechet spaces, in *Proceedings of the Fourth International Conference on Functional Analysis and Approximation Theory*, vol. I (Potenza, 2000), no. 68, part I, (2002), pp. 293–301

97. M. Bonk, O. Schramm, Embeddings of Gromov hyperbolic spaces. Geom. Funct. Anal. **10**(2), 266–306 (2000)

98. F.F. Bonsall, J. Duncan, *Numerical Ranges II*. London Mathematical Society Lecture Note Series, vol. 10 (Cambridge University Press, London, 1973)

99. K. Borsuk, Über Isomorphie der Funktionalräume. Bull. Int. Acad. Polon. Sci. Cl. Sci. Math. Nat. Sér. A. Sci. Math. **1933**, 1–10 (1933) [German]

100. M. Borkowski, D. Bugajewski, D. Phulara, On some properties of hyperconvex spaces. Fixed Point Theory Appl., Art. ID 213812, 19 (2010)

101. J.M. Borwein, Continuity and differentiability properties of convex operators. Proc. London Math. Soc. **44**(3), 420–444 (1982)

102. J.M. Borwein, Subgradients of convex operators. Math. Operationsforsch. Statist. Ser. Optim. **15**(2), 179–191 (1984)

103. J.M. Borwein, M. Fabián, On convex functions having points of Gâteaux differentiability which are not points of Fréchet differentiability. Canad. J. Math. **45**(6), 1121–1134 (1993)

104. J.M. Borwein, M. Fabián, On generic second-order Gateaux differentiability. Nonlinear Anal. **20**(12), 1373–1382 (1993)

105. J.M. Borwein, D. Noll, Second order differentiability of convex functions in Banach spaces. Trans. Am. Math. Soc. **342**(1), 43–81 (1994)

106. F. Botelho, J. Jamison, Composition operators on spaces of Lipschitz functions. Acta Sci. Math. (Szeged) **77**(3–4), 621–632 (2011)

107. N. Bourbaki, *Elements of Mathematics. General Topology. Chapters 5–10*. Translated from the French, 2nd edn. (Springer, Berlin, 1998)

108. J. Bourgain, On dentability and the Bishop-Phelps property. Isr. J. Math. **28**(4), 265–271 (1977)

109. J. Bourgain, On Lipschitz embedding of finite metric spaces in Hilbert space. Isr. J. Math. **52**(1–2), 46–52 (1985)

110. J. Bourgain, The metrical interpretation of superreflexivity in Banach spaces. Isr. J. Math. **56**(2), 222–230 (1986)

111. J. Bourgain, Remarks on the extension of Lipschitz maps defined on discrete sets and uniform homeomorphisms, in *Geometrical Aspects of Functional Analysis, Israel Semin. 1985–86.* Lecture Notes in Mathematics, vol. 1267 (Springer, Berlin, 1987), pp. 157–167

112. R.D. Bourgin, *Geometric Aspects of Convex Sets with the Radon-Nikodým Property*. Lecture Notes in Mathematics, vol. 993 (Springer, Berlin, 1983)

113. W.W. Breckner, *Rational s-Convexity. A Generalized Jensen-Convexity* (Presa Universitară Clujeană, Cluj-Napoca, 2011)

114. W.W. Breckner, T. Trif, Equicontinuity and Hölder equicontinuity of families of generalized convex mappings. N. Z. J. Math. **28**(2), 155–170 (1999)

115. W.W. Breckner, A. Göpfert, T. Trif, Characterizations of ultrabarrelledness and barrelledness involving the singularities of families of convex mappings. Manuscripta Math. **91**(1), 17–34 (1996)

116. A. Bressan, A. Cortesi, Lipschitz extensions of convex-valued maps. Atti Accad. Naz. Lincei Rend. Cl. Sci. Fis. Mat. Natur. **80**(7–12), 530–532 (1987)

117. M.R. Bridson, A. Haefliger, *Metric Spaces of Non-Positive Curvature*. Grundlehren der Mathematischen Wissenschaften [Fundamental Principles of Mathematical Sciences], vol. 319 (Springer, Berlin, 1999)

118. N. Brodskiy, J. Dydak, J. Higes, A. Mitra, Assouad-Nagata dimension via Lipschitz extensions. Isr. J. Math. **171**, 405–423 (2009)

119. A. Brøndsted, R.T. Rockafellar, On the subdifferentiability of convex functions. Proc. Am. Math. Soc. **16**, 605–611 (1965)

120. A.M. Bruckner, J.B. Bruckner, B.S. Thomson, *Real Analysis* (Prentice-Hall International, Upper Saddle River, 1997)
121. A. Brudnyi, Yu. Brudnyi, Simultaneous extensions of Lipschitz functions. Uspekhi Mat. Nauk **60**(6), 53–72 (2005)
122. A. Brudnyi, Yu. Brudnyi, Extension of Lipschitz functions defined on metric subspaces of homogeneous type. Rev. Mat. Complut. **19**(2), 347–359 (2006)
123. A. Brudnyi, Yu. Brudnyi, A universal Lipschitz extension property of Gromov hyperbolic spaces. Rev. Mat. Iberoam. **23**(3), 861–896 (2007)
124. A. Brudnyi, Yu. Brudnyi, Linear and nonlinear extensions of Lipschitz functions from subsets of metric spaces. Algebra i Analiz **19**(3), 106–118 (2007)
125. A. Brudnyi, Yu. Brudnyi, Metric spaces with linear extensions preserving Lipschitz condition. Am. J. Math. **129**(1), 217–314 (2007)
126. A. Brudnyi, Yu. Brudnyi, *Methods of Geometric Analysis in Extension and Trace Problems. Volume 1*. Monographs in Mathematics, vol. 102 (Birkhäuser/Springer, Basel, 2012)
127. A. Brudnyi, Yu. Brudnyi, *Methods of Geometric Analysis in Extension and Trace Problems. Volume 2*. Monographs in Mathematics, vol. 103 (Birkhäuser/Springer, Basel, 2012)
128. Yu. Brudnyi, P. Shvartsman, Stability of the Lipschitz extension property under metric transforms. Geom. Funct. Anal. **12**(1), 73–79 (2002)
129. P. Buneman, A note on the metric properties of trees. J. Combin. Theory Ser. B **17**, 48–50 (1974)
130. D. Burago, Y. Burago, S. Ivanov, *A Course in Metric Geometry*. Graduate Studies in Mathematics, vol. 33 (American Mathematical Society, Providence, 2001)
131. Yu. Burago, M. Gromov, G. Perel'man, A. D. Aleksandrov spaces with curvatures bounded below. Uspekhi Mat. Nauk **47**(2), 3–51 (1992) [Russian; English translation in Russian Math. Surveys **47**(2), 1–58 (1992)]
132. H. Busemann, Spaces with non-positive curvature. Acta Math. **80**, 259–310 (1948)
133. H. Busemann, *The Geometry of Geodesics* (Academic Press Inc., New York, 1955)
134. S. Buyalo, V. Schroeder, *Elements of Asymptotic Geometry*. EMS Monographs in Mathematics (European Mathematical Society (EMS), Zürich, 2007)
135. A. Carioli, L. Veselý, Normal cones and continuity of vector-valued convex functions. J. Convex Anal. **20**(2), 495–500 (2013)
136. H. Cartan, *Calcul différentiel* (Hermann, Paris, 1967)
137. H. Cartan, *Differential Calculus* (Hermann, Paris; Houghton Mifflin Co., Boston, Mass., 1971) [Exercises by C. Buttin, F. Rideau and J. L. Verley, Translated from the French]
138. B. Cascales, A.J. Guirao, V. Kadets, A Bishop-Phelps-Bollobás type theorem for uniform algebras. Adv. Math. **240**, 370–382 (2013)
139. V. Caselles, J.-M. Morel, C. Sbert, An axiomatic approach to image interpolation. IEEE Trans. Image Process. **7**(3), 376–386 (1998)
140. V. Caselles, J.-M. Morel, C. Sbert, An axiomatic approach to image interpolation. Trois applications des mathématiques. SMF Journ. Annu., vol. 1998 (Mathematical Society of France, Paris, 1998), pp. 15–38
141. J. Cheeger, Differentiability of Lipschitz functions on metric measure spaces. Geom. Funct. Anal. **9**(3), 428–517 (1999)
142. J. Cheeger, B. Kleiner, On the differentiability of Lipschitz maps from metric measure spaces to Banach spaces, in *Inspired by S.S. Chern*. Nankai Tracts in Mathematics, vol. 11 (World Scientific Publishing, Hackensack, 2006), pp. 129–152
143. J. Cheeger, B. Kleiner, Differentiability of Lipschitz maps from metric measure spaces to Banach spaces with the Radon-Nikodým property. Geom. Funct. Anal. **19**(4), 1017–1028 (2009)
144. M. Chica, V. Kadets, M. Martín, S. Moreno-Pulido, F. Rambla-Barreno, Bishop-Phelps-Bollobás moduli of a Banach space. J. Math. Anal. Appl. **412**(2), 697–719 (2014)
145. M. Chica, V. Kadets, M. Martín, J. Merí, M. Soloviova, Two refinements of the Bishop-Phelps-Bollobás modulus. Banach J. Math. Anal. **9**(4), 296–315 (2015)

146. I. Chiţescu, L. Ioana, R. Miculescu, L. Niţă, Sesquilinear uniform vector integral. Proc. Indian Acad. Sci. Math. Sci. **125**(2), 187–198 (2015)
147. I. Chiţescu, L. Ioana, R. Miculescu, L. Niţă, Monge–Kantorovich norms on spaces of vector measures. Results Math. **70**(3–4), 349–371 (2016)
148. Y.S. Choi, S.K. Kim, H.J. Lee, M. Martín, The Bishop-Phelps-Bollobás theorem for operators on $L_1(\mu)$. J. Funct. Anal. **267**(1), 214–242 (2014)
149. G. Choquet, *Lectures on Analysis. Vol. I: Integration and Topological Vector Spaces. Vol. II: Representation Theory. Vol. III: Infinite Dimensional Measures and Problem Solutions* 3rd printing, ed. by J. Marsden, T. Lance, S. Gelbart (W.A. Benjamin, Inc., Reading, 1976)
150. J.P.R. Christensen, On some measures analogous to Haar measure. Math. Scand. **26**, 103–106 (1970)
151. J.P.R. Christensen, On sets of Haar measure zero in abelian Polish groups. Isr. J. Math. **13**, 255–260 (1972)
152. I. Cîmpean, A remark on the proof of Cobzaş-Mustăţa theorem concerning norm preserving extension of convex Lipschitz functions. Studia Univ. Babeş-Bolyai Math. **57**(3), 325–329 (2012)
153. J.A. Clarkson, Uniformly convex spaces. Trans. Am. Math. Soc. **40**, 396–414 (1936)
154. S. Cobzaş, On the Lipschitz properties of continuous convex functions. Mathematica (Cluj) **21(44)**(2), 123–125 (1979)
155. S. Cobzaş, Lipschitz properties of convex functions, Babeş-Bolyai University, Cluj-Napoca, Research Seminaries, Seminar on Mathematical Analysis, vol. 85, pp. 77–84 (1985)
156. S. Cobzaş, Extreme points in Banach spaces of Lipschitz functions. Mathematica **31(54)**(1), 25–33 (1989)
157. S. Cobzaş, Lipschitz properties for families of convex mappings, in *Inequality Theory and Applications*, vol. I (Nova Science Publishers Inc., Huntington, 2001), pp. 103–112
158. S. Cobzaş, Compactness in spaces of Lipschitz functions. Rev. Anal. Numér. Théor. Approx. **30**(1), 9–14 (2001) [Dedicated to the memory of Acad. Tiberiu Popoviciu]
159. S. Cobzaş, Phelps type duality results in best approximation. Rev. Anal. Numér. Théor. Approx. **31**(1), 29–43 (2002)
160. S. Cobzaş, Adjoints of Lipschitz mappings. Studia Univ. Babeş-Bolyai Math. **48**(1), 49–54 (2003)
161. S. Cobzaş, Lipschitz properties of convex functions. Adv. Oper. Theory **2**(1), 21–49 (2017)
162. S. Cobzaş, *Functional Analysis in Asymmetric Normed Spaces*. Frontiers in Mathematics (Birkhäuser/Springer, Basel, 2013)
163. S. Cobzaş, I. Muntean, Continuous and locally Lipschitz convex functions. Mathematica **18(41)**(1), 41–51 (1976)
164. S. Cobzaş, C. Mustăţa, Norm-preserving extension of convex Lipschitz functions. J. Approx. Theory **24**(3), 236–244 (1978)
165. S. Cobzaş, C. Mustăţa, Extension of Lipschitz functions and best approximation, in *Research on the Theory of Allure, Approximation, Convexity and Optimization*, ed. by E. Popovici (SRIMA, Cluj-Napoca, 1999), pp. 3–21
166. D.L. Cohn, *Measure Theory*, 2nd edn. Birkhäuser Advanced Texts: Basler Lehrbücher (Birkhäuser/Springer, New York, 2013)
167. R.R. Coifman, G. Weiss, *Analyse harmonique non-commutative sur certains espaces homogènes. Étude de certaines intégrales singulières*. Lecture Notes in Mathematics, vol. 242 (Springer, Berlin, 1971) [French]
168. I. Colojoară, On Whitney's imbedding theorem. Rev. Roumaine Math. Pures Appl. **10**, 291–296 (1965) [Correction in the same journal **10**, 1051–1052 (1965)]
169. I. Colojoară, Lipschitz closed embedding of Hilbert-Lipschitz manifolds. Rend. Mat. Appl. **15**(4), 561–568 (1996)
170. D. Cooper, T. Pignataro, On the shape of Cantor sets. J. Differ. Geom. **28**(2), 203–221 (1988)
171. M. Csörnyei, Aronszajn null and Gaussian null sets coincide. Isr. J. Math. **111**, 191–201 (1999)

172. J. Czipszer, L. Gehér, Extension of functions satisfying a Lipschitz condition. Acta Math. Acad. Sci. Hungar. **6**, 213–220 (1955)

173. B. Dacorogna, *Direct Methods in the Calculus of Variations.* 2nd edn. Applied Mathematical Sciences, vol. 78 (Springer, New York, 2008)

174. B. Dacorogna, W. Gangbo, Extension theorems for vector valued maps. J. Math. Pures Appl. **85**(3), 313–344 (2006)

175. L. Danzer, B. Grünbaum, V. Klee, Helly's theorem and its relatives, in *Proceedings of Symposia in Pure Mathematics*, vol. VII (American Mathematical Society, Providence, 1963), pp. 101–180

176. G. David, S. Semmes, *Fractured Fractals and Broken Dreams.* Self-similar Geometry Through Metric and Measure. Oxford Lecture Series in Mathematics and its Applications, vol. 7 (The Clarendon Press, Oxford University Press, New York, 1997)

177. W.J. Davis, T. Figiel, W.B. Johnson, A. Pełczyński, Factoring weakly compact operators. J. Funct. Anal. **17**, 311–327 (1974)

178. M.M. Day, *Normed Linear Spaces.* 3rd edn. Ergebnisse der Mathematik und ihrer Grenzgebiete, Band 21 (Springer, New York, 1973)

179. D.G. de Figueiredo, L.A. Karlovitz, On the extension of contractions on normed spaces, in *Nonlinear Functional Analysis (Proc. Sympos. Pure Math., vol. XVIII, Part 1, Chicago, Ill., 1968)* (American Mathematical Society, Providence, 1970), pp. 95–104

180. D.G. de Figueiredo, L.A. Karlovitz, The extension of contractions and the intersection of balls in Banach spaces. J. Funct. Anal. **11**, 168–178 (1972)

181. K. Deimling, *Nonlinear Functional Analysis* (Springer, Berlin, 1985)

182. K. de Leeuw, Banach spaces of Lipschitz functions. Studia Math. **21**, 55–66 (1961/1962)

183. C. De Lellis, E.N. Spadaro, Q-valued functions revisited. Mem. Am. Math. Soc. **991**, i–v + 79 (2011)

184. G. De Philippis, A. Marchese, F. Rindler, On a conjecture of Cheeger, in *Measure Theory in Non-Smooth Spaces* (De Gruyter Open, Warsaw, 2017), pp. 145–155

185. G.T. Deng, X.G. He, Lipschitz equivalence of fractal sets in $\mathbb{R}$. Sci. China Math. **55**(10), 2095–2107 (2012)

186. J. Deng, Z.-Y. Wen, Y. Xiong, L.-F. Xi, Bilipschitz embedding of self-similar sets. J. Anal. Math. **114**, 63–97 (2011)

187. D. Descombes, Asymptotic rank of spaces with bicombings. Math. Z. **284**(3–4), 947–960 (2016)

188. D. Descombes, U. Lang, Convex geodesic bicombings and hyperbolicity. Geom. Dedicata **177**, 367–384 (2015)

189. D. Descombes, U. Lang, Flats in spaces with convex geodesic bicombings. Anal. Geom. Metr. Spaces **4**, 68–84 (2016)

190. F. Deutsch, S. Li, W. Mabizela, Helly extensions and best approximation, in *Parametric Optimization and Related Topics, III* (Güstrow, 1991). Approx. Optim., vol. 3 (Lang, Frankfurt am Main, 1993), pp. 107–120

191. R. Deville, G. Godefroy, V. Zizler, *Smoothness and Renormings in Banach Spaces.* Pitman Monographs and Surveys in Pure and Applied Mathematics, vol. 64 (Longman Scientific & Technical, Harlow; Copublished in the United States with John Wiley & Sons, Inc., New York, 1993)

192. J. Diestel, *Geometry of Banach Spaces—Selected Topics.* Lecture Notes in Mathematics, vol. 485 (Springer, Berlin, 1975)

193. J. Diestel, *Sequences and Series in Banach Spaces.* Graduate Texts in Mathematics, vol. 92 (Springer, New York, 1984)

194. J. Diestel, J.J. Uhl, Jr., *Vector Measures.* Mathematical Surveys, No. 15 (American Mathematical Society, Providence, 1977) [With a foreword by B. J. Pettis]

195. J. Dieudonné, *Foundations of Modern Analysis.* Enlarged and corrected printing, Pure and Applied Mathematics, Vol. 10-I (Academic Press, New York, 1969)

196. N. Dinculeanu, *Vector Measures*. International Series of Monographs in Pure and Applied Mathematics, vol. 95 (Pergamon Press, Oxford-New York-Toronto; VEB Deutscher Verlag der Wissenschaften, Berlin, 1967)

197. J. Dixmier, Sur un théorème de Banach. Duke Math. J. **15**, 1057–1071 (1948)

198. R.L. Dobrušin, Definition of a system of random variables by means of conditional distributions. Teor. Verojatnost. i Primenen. **15**, 469–497 (1970) [Russian]

199. M. Doré, O. Maleva, A compact null set containing a differentiability point of every Lipschitz function. Math. Ann. **351**(3), 633–663 (2011)

200. M. Doré, O. Maleva, A universal differentiability set in Banach spaces with separable dual. J. Funct. Anal. **261**(6), 1674–1710 (2011)

201. M. Doré, O. Maleva, A compact universal differentiability set with Hausdorff dimension one. Isr. J. Math. **191**(2), 889–900 (2012)

202. C.H. Dowker, On countably paracompact spaces. Can. J. Math. **3**, 219–224 (1951)

203. A.W.M. Dress, Trees, tight extensions of metric spaces, and the cohomological dimension of certain groups: a note on combinatorial properties of metric spaces. Adv. in Math. **53**(3), 321–402 (1984)

204. A. Dress, K.T. Huber, V. Moulton, Metric spaces in pure and applied mathematics. Doc. Math. Extra vol., 121–139 (2001) [Proceedings of the Conference on Quadratic Forms and Related Topics Baton Rouge, LA (2001)]

205. A. Dress, K.T. Huber, V. Moulton, An explicit computation of the injective hull of certain finite metric spaces in terms of their associated Buneman complex. Adv. Math. **168**(1), 1–28 (2002)

206. M. Dubei, E.D. Tymchatyn, A. Zagorodnyuk, Free Banach spaces and extension of Lipschitz maps. Topology **48**(2–4), 203–212 (2009)

207. J. Duda, Metric and w^*-differentiability of pointwise Lipschitz mappings. Z. Anal. Anwend. **26**(3), 341–362 (2007)

208. J. Duda, On Gâteaux differentiability of pointwise Lipschitz mappings. Can. Math. Bull. **51**(2), 205–216 (2008)

209. R.M. Dudley, Convergence of Baire measures. Studia Math. **27**, 251–268, (1966) [Correction: Studia Math. **51**, 275 (1974)]

210. R.M. Dudley, *Probabilities and Metrics. Convergence of Laws on Metric Spaces, with a View to Statistical Testing*, Lecture Notes Series, No. 45 (Matematisk Institut, Aarhus Universitet, Aarhus, 1976)

211. R.M. Dudley, *Real Analysis and Probability*. Cambridge Studies in Advanced Mathematics, vol. 74 (Cambridge University Press, Cambridge, 2002) [Revised reprint of the 1989 original]

212. J. Dugundji, An extension of Tietze's theorem. Pacific J. Math. **1**, 353–367 (1951)

213. J. Dugundji, *Topology*. Reprinting of the 1966 original, Allyn and Bacon Series in Advanced Mathematics (Allyn and Bacon, Inc., Boston, 1978)

214. N. Dunford, J.T. Schwartz, *Linear Operators. I. General Theory*. With the assistance of W.G. Bade, R.G. Bartle. Pure and Applied Mathematics, vol. 7 (Interscience Publishers, New York; Interscience Publishers, London, 1958)

215. A. Dvoretzky, Some results on convex bodies and Banach spaces, in *Proc. Internat. Sympos on Linear Spaces*, Jerusalem, 1960 (Jerusalem Academic Press, Jerusalem; Pergamon, Oxford, 1961), pp. 123–160

216. M. Dymond, O. Maleva, Differentiability inside sets with Minkowski dimension one. Mich. Math. J. **65**(3), 613–636 (2016)

217. M. Edelstein, A.C. Thompson, Contractions, isometries and some properties of inner-product spaces. Nederl. Akad. Wetensch. Proc. Ser. A Indag. Math. **29**, 326–331 (1967)

218. D.A. Edwards, A simple proof in Monge-Kantorovich duality theory. Studia Math. **200**(1), 67–77 (2010)

219. D.A. Edwards, On the Kantorovich-Rubinstein theorem. Expo. Math. **29**(4), 387–398 (2011)

220. V.A. Efremovich, The geometry of proximity, I. Mat. Sb. **31**, 189–200 (1952) [Russian]

221. I. Ekeland, R. Temam, *Convex Analysis and Variational Problems*. Translated from the French, Studies in Mathematics and its Applications, vol. 1 (North-Holland Publishing, Amsterdam-Oxford; Elsevier, New York, 1976)

222. J. Elstrodt, *Maß- und Integrationstheorie*. 7th edn. (Springer, Berlin, 2011) [German]

223. R. Engelking, *General Topology*. 2nd edn. Sigma Series in Pure Mathematics, vol. 6 (Heldermann Verlag, Berlin, 1989) [Translated from the Polish by the author]

224. K. Esmaeili, H. Mahyar, Weighted composition operators between vector-valued Lipschitz function spaces. Banach J. Math. Anal. **7**(1), 59–72 (2013)

225. R. Espínola, On selections of the metric projection and best proximity pairs in hyperconvex spaces. Ann. Univ. Mariae Curie-Skłodowska Sect. A **59**, 9–17 (2005)

226. R. Espínola, A. Fernández-León, Fixed point theory in hyperconvex metric spaces, in *Topics in Fixed Point Theory* (Springer, Cham, 2014), pp. 101–158

227. R. Espínola, M. A. Khamsi, Introduction to hyperconvex spaces, in *Handbook of Metric Fixed Point Theory* (Kluwer Academic Publishers, Dordrecht, 2001), pp. 391–435

228. R. Espínola, A. Nicolae, Continuous selections of Lipschitz extensions in metric spaces. Rev. Mat. Complut. **28**(3), 741–759 (2015)

229. R. Espínola, Ó. Madiedo, A. Nicolae, Borsuk-Dugundji type extension theorems with Busemann convex target spaces. Ann. Acad. Sci. Fenn. Math. 43(1), 225–238 (2018)

230. A. Es-Sahib, H. Heinich, Barycentre canonique pour un espace métrique à courbure négative, in *Séminaire de Probabilités*, XXXIII. Lecture Notes in Math., vol. 1709 (Springer, Berlin, 1999), pp. 355–370

231. L.C. Evans, R.F. Gariepy, *Measure Theory and Fine Properties of Functions*. Revised edn., Textbooks in Mathematics (CRC Press, Boca Raton, 2015)

232. M. Fabian, P. Habala, P. Hájek, V. Montesinos-Santalucía, J. Pelant, V. Zizler, *Functional analysis and infinite-dimensional geometry*. CMS Books in Mathematics/Ouvrages de Mathématiques de la SMC, vol. 8 (Springer, New York, 2001)

233. M. Fabian, P. Habala, P. Hájek, V. Montesinos, V. Zizler, *Banach Space Theory. The Basis for Linear and Nonlinear Analysis*. CMS Books in Mathematics/Ouvrages de Mathématiques de la SMC (Springer, New York, 2011)

234. K.J. Falconer, D.T. Marsh, On the Lipschitz equivalence of Cantor sets. Mathematika **39**(2), 223–233 (1992)

235. S. Fan, Q.L. Guo, L.F. Xi, Lipschitz constant for bi-lipschitz automorphism of self-similar fractal. Progr. Natur. Sci. **16**, 415–420 (2006)

236. H. Federer, *Geometric Measure Theory*. Die Grundlehren der mathematischen Wissenschaften, Band 153 (Springer, New York, 1969)

237. G. Fichtenholz, Note sur les fonctions absolument continues. Bull. Cl. Sci. Acad. R. Belg. **31**(2), 430–443 (1922)

238. G.M. Fichtenholz, On absolutely continuous functions. Mat. Sb. **31**(2), 286–295 (1923) [Russian version of [237]]

239. T.M. Flett, Extensions of Lipschitz functions. J. Lond. Math. Soc. **7**(2), 604–608 (1974)

240. C. Foias, E. Olson, Finite fractal dimension and Hölder-Lipschitz parametrization. Indiana Univ. Math. J. **45**(3), 603–616 (1996)

241. T. Fowler, D. Preiss, A simple proof of Zahorski's description of non-differentiability sets of Lipschitz functions. Real Anal. Exchange **34**(1), 127–138 (2009)

242. M. Fréchet, Sur quelques points du calcul fonctionnel. Rend. Circ. Mat. Palermo **22**, 1–74 (1906) [French]

243. D.H. Fremlin, *Kirszbraun's Theorem*, http://www.essex.ac.uk/maths/people/fremlin/n11706.pdf

244. J. Fried, Open cover of a metric space admits ℓ_∞-partition of unity. Suppl. Rend. Circ. Mat. Palermo **3**(2), 139–140 (1984)

245. Z. Frolík, Existence of ℓ_∞-partitions of unity. Rend. Semin. Mat., Torino **42**(1), 9–14 (1984)

246. G. Gagneux, Approximations lipschitziennes de la fonction signe et schémas semi-discrétisés implicites de problèmes quasi-linéaires dégénérés, in *Publ. Math. Fac. Sci. Besançon*, vol. 8, Besançon (1984), pp. Exp. No. 5 (1984), 16 pp [French]

247. V. Gantmacher, Über schwache totalstetige Operatoren. Rec. Math. Mat. Sbornik, N.S. **7**(49), 301–308 (1940) [German]

248. L.M. García-Raffi, S. Romaguera, E.A. Sánchez Pérez, Extensions of asymmetric norms to linear spaces. Rend. Istit. Mat. Univ. Trieste **33**(1–2), 113–125 (2002)

249. I. Gelfand, Abstrakte Funktionen und lineare Operatoren. Rec. Math. Moscou, n. Ser. **4**, 235–286 (1938) [German]

250. G. Georganopoulos, Sur l'approximation des fonctions continues par des fonctions lipschitziennes. C. R. Acad. Sci. Paris Sér. A-B **264**, A319–A321 (1967)

251. M. Gieraltowska-Kedzierska, F.S. Van Vleck, Fréchet differentiability of regular locally Lipschitzian functions. J. Math. Anal. Appl. **159**(1), 147–157 (1991)

252. M. Gieraltowska-Kedzierska, F.S. Van Vleck, Fréchet vs. Gâteaux differentiability of Lipschitzian functions. Proc. Am. Math. Soc. **114**(4), 905–907 (1992)

253. J.R. Giles, *Convex Analysis with Application in the Differentiation of Convex Functions.* Research Notes in Mathematics, vol. 58 (Pitman, Boston, 1982)

254. J.R. Giles, Fréchet intermediate differentiability of Lipschitz functions on Asplund spaces. Bull. Austral. Math. Soc. **79**(2), 309–317 (2009)

255. J.R. Giles, S. Sciffer, A generic differentiability property of Lipschitz functions on Asplund spaces. J. Nonlinear Convex Anal. **3**(3), 353–363 (2002)

256. J. Goblet, A selection theory for multiple-valued functions in the sense of Almgren. Ann. Acad. Sci. Fenn., Math. **31**(2), 297–314 (2006)

257. J. Goblet, Lipschitz extension of multiple Banach-valued functions in the sense of Almgren. Houst. J. Math. **35**(1), 223–231 (2009)

258. G. Godefroy, On norm attaining Lipschitz maps between Banach spaces. Pure Appl. Funct. Anal. **1**(1), 89–118 (2016)

259. G. Godefroy, De Grothendieck à Naor: une promenade dans l'analyse métrique des espaces de Banach. Gaz. Math. (151), 13–24 (2017) [French; translated into English in Eur. Math. Soc. Newsl. **107**, 9–16 (2018)]

260. G. Godefroy, N.J. Kalton, Lipschitz-free Banach spaces. Studia Math. **159**(1), 121–141 (2003) [Dedicated to Professor Aleksander Pełczyński on the occasion of his 70th birthday]

261. G. Godefroy, N.J. Kalton, G. Lancien, L'espace de Banach c_0 est déterminé par sa métrique. C. R. Acad. Sci. Paris Sér. I Math. **327**(9), 817–822 (1998)

262. G. Godefroy, N.J. Kalton, G. Lancien, Subspaces of $c_0(\mathbf{N})$ and Lipschitz isomorphisms. Geom. Funct. Anal. **10**(4), 798–820 (2000)

263. G. Godefroy, N.J. Kalton, G. Lancien, Szlenk indices and uniform homeomorphisms. Trans. Am. Math. Soc. **353**(10), 3895–3918 (2001)

264. K. Goebel, W.A. Kirk, *Topics in Metric Fixed Point Theory.* Cambridge Studies in Advanced Mathematics, vol. 28 (Cambridge University Press, Cambridge, 1990)

265. M.X. Goemans, Semidefinite programming in combinatorial optimization. Math. Program. **79**(1–3), 143–161 (1997) [Lectures on mathematical programming (ismp97) (Lausanne, 1997)]

266. A. Golbaharan, Weakly compact weighted composition operators on spaces of Lipschitz functions. Positivity **22**(5), 1265–1268 (2018)

267. A. Golbaharan, H. Mahyar, Weighted composition operators on Lipschitz algebras. Houst. J. Math. **42**(3), 905–917 (2016)

268. A. Göpfert, H. Riahi, C. Tammer, C. Zălinescu, *Variational Methods in Partially Ordered Spaces.* CMS Books in Mathematics/Ouvrages de Mathématiques de la SMC, vol. 17 (Springer, New York, 2003)

269. R. Górak, A note on differentiability of Lipschitz maps. Bull. Pol. Acad. Sci. Math. **58**(3), 259–268 (2010)

270. R. Gordon, Riemann integration in Banach spaces. Rocky Mountain J.Math. **21**(3), 923–949 (1991)

271. M. Gromov, Hyperbolic groups, in *Essays in Group Theory.* Mathematical Sciences Research Institute Publications, vol. 8 (Springer, New York, 1987), pp. 75–263

272. B. Grünbaum, On a theorem of Kirszbraun. Bull. Res. Council Israel. Sect. **F 7**, 129–132 (1957/1958)
273. B. Grünbaum, A generalization of theorems of Kirszbraun and Minty. Proc. Am. Math. Soc. **13**, 812–814 (1962)
274. Q.-L. Guo, M. Wu, L.-F. Xi, Lipschitz constant for bi-Lipschitz automorphism on Moran-like sets. J. Math. Anal. Appl. **336**(2), 937–952 (2007)
275. P. Hájek, M. Johanis, Smooth approximations. J. Funct. Anal. **259**(3), 561–582 (2010)
276. P. Hájek, M. Johanis, *Smooth Analysis in Banach Spaces* (de Gruyter, Berlin, 2014)
277. L.G. Hanin, Kantorovich-Rubinstein norm and its application in the theory of Lipschitz spaces. Proc. Am. Math. Soc. **115**(2), 345–352 (1992)
278. L.G. Hanin, On isometric isomorphism between the second dual to the "small" Lipschitz space and the "big" Lipschitz space, in *Nonselfadjoint Operators and Related Topics* (Beer Sheva, 1992). Operator Theory: Advances and Applications, vol. 73 (Birkhäuser, Basel, 1994), pp. 316–324
279. L.G. Hanin, Duality for general Lipschitz classes and applications. Proc. Lond. Math. Soc. **75**(1), 134–156 (1997)
280. L.G. Hanin, An extension of the Kantorovich norm, in *Monge Ampère Equation: Applications to Geometry and Optimization* (Deerfield Beach, 1997). Contemporary Mathematics, vol. 226 (American Mathematical Society, Providence, 1999), pp. 113–130
281. G.H. Hardy, J.E. Littlewood, G. Pólya, *Inequalities*, Cambridge Mathematical Library (Cambridge University Press, Cambridge, 1988) [Reprint of the 1952 edition]
282. M.J. Hirn, E.Y. Le Gruyer, A general theorem of existence of quasi absolutely minimal Lipschitz extensions. Math. Ann. **359**(3–4), 595–628 (2014)
283. J. Heinonen, *Lectures on Analysis on Metric Spaces*. Universitext (Springer, New York, 2001)
284. J. Heinonen, *Lectures on Lipschitz Analysis*. Report. University of Jyväskylä Department of Mathematics and Statistics, vol. 100 (University of Jyväskylä, Jyväskylä, 2005), p. 77
285. J. Heinonen, P. Koskela, N. Shanmugalingam, J.T. Tyson, *Sobolev Spaces on Metric Measure Spaces. An Approach Based on Upper Gradients*. New Mathematical Monographs, vol. 27 (Cambridge University Press, Cambridge, 2015)
286. S. Heinrich, P. Mankiewicz, Applications of ultrapowers to the uniform and Lipschitz classification of Banach spaces. Studia Math. **73**(3), 225–251 (1982)
287. E. Hewitt, K. Stromberg, *Real and Abstract Analysis (A Modern Treatment of the Theory of Functions of a Real Variable)* (Springer, New York-Heidelberg, 1975) [Third printing, Graduate Texts in Mathematics, No. 25]
288. J.-B. Hiriart-Urruty, Extension of Lipschitz functions. J. Math. Anal. Appl. **77**, 539–554 (1980) [English]
289. T.G. Honary, H. Mahyar, Approximation in Lipschitz algebras of infinitely differentiable functions. Bull. Korean Math. Soc. **36**(4), 629–636 (1999)
290. T.G. Honary, H. Mahyar, Approximation in Lipschitz algebras. Quaest. Math. **23**(1), 13–19 (2000)
291. C.D. Horvath, Contractibility and generalized convexity. J. Math. Anal. Appl. **156**(2), 341–357 (1991)
292. S.-T. Hu, On Lipschitz mappings. Port. Math. **7**, 45–49 (1948)
293. M. Huang, Y. Li, On bilipschitz extensions in real Banach spaces. Abstr. Appl. Anal., 9 (2013) [Art. ID 765685]
294. T. Huuskonen, J. Partanen, J. Väisälä, Bi-Lipschitz extensions from smooth manifolds. Rev. Mat. Ibero Am. **11**(3), 579–601 (1995)
295. T. Hytönen, J. van Neerven, M. Veraar, L. Weis, *Analysis in Banach Spaces. Volume I. Martingales and Littlewood-Paley Theory* (Springer, Cham, 2016)
296. J.R. Isbell, Six theorems about injective metric spaces. Comment. Math. Helv. **39**, 65–76 (1964)
297. J.R. Isbell, Three remarks on injective envelopes of Banach spaces. J. Math. Anal. Appl. **27**, 516–518 (1969)

298. E. Ivakhno, V. Kadets, D. Werner, The Daugavet property for spaces of Lipschitz functions. Math. Scand. **101**(2), 261–279 (2007) [Corrigendum: Math. Scand. **104**(2), 319 (2009)]
299. A. Iwanik, Norm attaining operators on Lebesgue spaces. Pac. J. Math. **83**(2), 381–386 (1979)
300. A. Iwanik, On norm-attaining operators acting from $L^1(\mu)$ to $C(S)$. Rend. Circ. Mat. Palermo **2**(Suppl. 2), 147–152 (1982)
301. R.C. James, Super-reflexive Banach spaces. Can. J. Math. **24**, 896–904 (1972)
302. R. Jensen, Uniqueness of Lipschitz extensions: minimizing the sup norm of the gradient. Arch. Ration. Mech. Anal. **123**(1), 51–74 (1993)
303. M. Jiménez-Sevilla, L. Sánchez-González, Smooth extension of functions on a certain class of non-separable Banach spaces. J. Math. Anal. Appl. **378**(1), 173–183 (2011)
304. A. Jiménez-Vargas, Weakly compact composition operators on spaces of Lipschitz functions. Positivity **19**(4), 807–815 (2015)
305. A. Jiménez-Vargas, M.A. Navarro, Hölder seminorm preserving linear bijections and isometries. Proc. Indian Acad. Sci. Math. Sci. **119**(1), 53–62 (2009)
306. A. Jiménez-Vargas, M. Lacruz, M. Villegas-Vallecillos, Essential norm of composition operators on Banach spaces of Hölder functions. Abstr. Appl. Anal. **2011**, 13 (2011)
307. A. Jiménez-Vargas, M. Villegas-Vallecillos, Compact composition operators on noncompact Lipschitz spaces. J. Math. Anal. Appl. **398**(1), 221–229 (2013)
308. A. Jiménez-Vargas, J.M. Sepulcre, M. Villegas-Vallecillos, Lipschitz compact operators. J. Math. Anal. Appl. **415**(2), 889–901 (2014)
309. J.A. Johnson, Banach spaces of Lipschitz functions and vector-valued Lipschitz functions. Trans. Am. Math. Soc. **148**, 147–169 (1970)
310. W.B. Johnson, J. Lindenstrauss, Extensions of Lipschitz mappings into a Hilbert space, in *Conference in Modern Analysis and Probability* (New Haven, Conn., 1982). Contemporary Mathematics. vol. 26 (American Mathematical Society, Providence, 1984), pp. 189–206
311. W.B. Johnson, J. Lindenstrauss (eds.), *Handbook of the Geometry of Banach Spaces*, vol. 1 (North-Holland Publishing Co., Amsterdam, 2001)
312. W.B. Johnson, J. Lindenstrauss (eds.), *Handbook of the Geometry of Banach Spaces*, vol. 2 (North-Holland, Amsterdam, 2003)
313. W.B. Johnson, J. Lindenstrauss, G. Schechtman, Extensions of Lipschitz maps into Banach spaces, in *Israel Seminar on Geometrical Aspects of Functional Analysis (1983/84)* (Tel Aviv University, Tel Aviv, 1984), p. 15
314. W.B. Johnson, J. Lindenstrauss, G. Schechtman, Extensions of Lipschitz maps into Banach spaces. Isr. J. Math. **54**(2), 129–138 (1986)
315. W.B. Johnson, J. Lindenstrauss, G. Schechtman, On Lipschitz embedding of finite metric spaces in low-dimensional normed spaces, in *Geometrical Aspects of Functional Analysis (1985/86)*. Lecture Notes in Mathematics, vol. 1267 (Springer, Berlin, 1987), pp. 177–184
316. W.B. Johnson, J. Lindenstrauss, D. Preiss, G. Schechtman, Almost Fréchet differentiability of Lipschitz mappings between infinite-dimensional Banach spaces. Proc. London Math. Soc. **84**(3), 711–746 (2002)
317. M. Josephy, Composing functions of bounded variation. Proc. Am. Math. Soc. **83**(2), 354–356 (1981)
318. J. Jost, *Nonpositive Curvature: Geometric and Analytic Aspects*. Lectures in Mathematics ETH Zürich (Birkhäuser Verlag, Basel, 1997)
319. M. Jouak, L. Thibault, Equicontinuity of families of convex and concave-convex operators. Can. J. Math. **36**(5), 883–898 (1984)
320. W. Julian, K. Phillips, Constructive bounded sequences and Lipschitz functions. J. Lond. Math. Soc. **31**(3), 385–392 (1985)
321. K.-W. Jun, D.-W. Park, Almost linearity of ϵ-bi-Lipschitz maps between real Banach spaces. Proc. Am. Math. Soc. **124**(1), 217–225 (1996)
322. V.M. Kadets, Lipschitz mappings of metric spaces. Izv. Vyssh. Uchebn. Zaved. Mat. (1), 30–34 (1985)
323. V. Kadets, M. Martín, J. Merí, D. Werner, Lipschitz slices and the Daugavet equation for Lipschitz operators. Proc. Am. Math. Soc. **143**(12), 5281–5292 (2015)

324. V. Kadets, M. Martín, M. Soloviova, Norm-attaining Lipschitz functionals. Banach J. Math. Anal. **10**(3), 621–637 (2016)
325. S. Kakutani, Concrete representation of abstract (M)-spaces. (A characterization of the space of continuous functions.) Ann. Math. (2) **42**, 994–1024 (1941)
326. N.J. Kalton, Curves with zero derivative in F-spaces. Glasg. Math. J. **22**, 19–29 (1981)
327. N.J. Kalton, The existence of primitives for continuous functions in a quasi-Banach space. Atti Semin. Mat. Fis. Univ. Modena **44**(1), 113–117 (1996)
328. N.J. Kalton, Quasi-Banach spaces, in *Handbook of the Geometry of Banach Spaces*, vol. 2 (North-Holland, Amsterdam, 2003), pp. 1099–1130
329. N.J. Kalton, Spaces of Lipschitz and Hölder functions and their applications. Collect. Math. **55**(2), 171–217 (2004)
330. N.J. Kalton, Extending Lipschitz maps into $\mathscr{C}(K)$-spaces. Isr. J. Math. **162**, 275–315 (2007)
331. N.J. Kalton, Extension of linear operators and Lipschitz maps into $\mathscr{C}(K)$-spaces. N. Y. J. Math. **13**, 317–381 (2007)
332. N.J. Kalton, The nonlinear geometry of Banach spaces. Rev. Mat. Complut. **21**(1), 7–60 (2008)
333. N.J. Kalton, G. Lancien, Best constants for Lipschitz embeddings of metric spaces into c_0. Fund. Math. **199**(3), 249–272 (2008)
334. N.J. Kalton, J.H. Shapiro, An F-space with trivial dual and non-trivial compact endomorphisms. Isr. J. Math. **20**, 282–291 (1975)
335. N.J. Kalton, N.T. Peck, J.W. Roberts, *An F-Space Sampler*. London Mathematical Society Lecture Note Series, vol. 89 (Cambridge University Press, Cambridge, 1984)
336. H. Kamowitz, S. Scheinberg, Some properties of endomorphisms of Lipschitz algebras. Studia Math. **96**(3), 255–261 (1990)
337. R. Kannan, C.K. Krueger, *Advanced Analysis on the Real Line*. Universitext (Springer, New York, 1996)
338. L.V. Kantorovich, On the translocation of masses. C. R. (Doklady) Acad. Sci. URSS (N.S.) **37**, 199–201 (1942) [Russian; English translation in J. Math. Sci (NY) **133**(4), 1381–1382 (2006)]
339. L.V. Kantorovich, On a problem of Monge. Uspekhi Mat. Nauk **3**(2), 225–225 (1948) [Russian; English translation in J. Math. Sci (NY) **133**(4), 1383 (2006)]
340. L.V. Kantorovich, G.P. Akilov, *Functional Analysis*. 2nd edn (Pergamon Press, Oxford, 1982) [Translated from the Russian by Howard L. Silcock]
341. L.V. Kantorovich, G.Š. Rubinshtein, On a functional space and certain extremum problems. Dokl. Akad. Nauk SSSR (N.S.) **115**, 1058–1061 (1957)
342. L.V. Kantorovich, G.Š. Rubinshtein, On a space of completely additive functions. Vestn. Leningr. Univ. **13**(7), 52–59 (1958)
343. S. Karamardian, A further generalization of Kirszbraun's theorem, in *Inequalities*, III (Proc. Third Sympos., Univ. California, Los Angeles, Calif., 1969; dedicated to the memory of Theodore S. Motzkin) (Academic Press, New York, 1972), pp. 145–148
344. M.B. Karmanova, Metric Rademacher theorem and the area formula for metric-valued Lipschitz mappings. Vestn. Novosib. Gos. Univ., Ser. Mat. Mekh. Inform. **6**(4), 50–69 (2006) [Russian]
345. M.B. Karmanova, Area and co-area formulas for mappings of the Sobolev classes with values in a metric space. Sibirsk. Mat. Zh. **48**(4), 778–788 (2007) [Russian; English translation in Siberian Math. J. **48**(4), 621–628 (2007)]
346. M. Katětov, On real-valued functions in topological spaces. Fund. Math. **38**, 85–91 (1951) [Correction: Fund. Math. **40**, 203–205 (1953)]
347. S. Keith, Measurable differentiable structures and the Poincaré inequality. Indiana Univ. Math. J. **53**(4), 1127–1150 (2004)
348. S. Keith, A differentiable structure for metric measure spaces. Adv. Math. **183**(2), 271–315 (2004)
349. S. Keith, K. Rajala, A remark on Poincaré inequalities on metric measure spaces. Math. Scand. **95**(2), 299–304 (2004)

350. H.G. Kellerer, Duality theorems and probability metrics, in *Proceedings of the Seventh Conference on Probability Theory* (Braşov, 1982) (VNU Science Press, Utrecht, 1985), pp. 211–220
351. J.L. Kelley, *General Topology* (Springer, New York-Berlin, 1975) [Reprint of the 1955 edition [Van Nostrand, Toronto, Ont.], Graduate Texts in Mathematics, No. 27]
352. M.A. Khamsi, W.A. Kirk, C.M. Yañez, Fixed point and selection theorems in hyperconvex spaces. Proc. Am. Math. Soc. **128**(11), 3275–3283 (2000)
353. S. Khot, A. Naor, Nonembeddability theorems via Fourier analysis. Math. Ann. **334**(4), 821–852 (2006)
354. S.K. Kim, H.J. Lee, Uniform convexity and Bishop-Phelps-Bollobás property. Can. J. Math. **66**(2), 373–386 (2014)
355. S.K. Kim, H.J. Lee, The Bishop-Phelps-Bollobás property for operators from $\mathscr{C}(K)$ to uniformly convex spaces. J. Math. Anal. Appl. **421**(1), 51–58 (2015)
356. B. Kirchheim, Rectifiable metric spaces: local structure and regularity of the Hausdorff measure. Proc. Am. Math. Soc. **121**(1), 113–123 (1994)
357. B. Kirchheim, V. Magnani, A counterexample to metric differentiability. Proc. Edinb. Math. Soc. **46**(1), 221–227 (2003)
358. W.A. Kirk, Hyperconvexity of **R**-trees. Fund. Math. **156**(1), 67–72 (1998)
359. M.D. Kirszbraun, Über die zusammenziehenden und Lipschitzschen Transformationen. Fundam. Math. **22**, 77–108 (1934) [German]
360. E. Kopecká, Bootstrapping Kirszbraun's extension theorem. Fund. Math. **217**(1), 13–19 (2012)
361. E. Kopecká, Extending Lipschitz mappings continuously. J. Appl. Anal. **18**(2), 167–177 (2012)
362. E. Kopecká, S. Reich, Continuous extension operators and convexity. Nonlinear Anal. **74**(18), 6907–6910 (2011)
363. P. Kosmol, Optimierung konvexer Funktionen mit Stabilitätsbetrachtungen. Diss. Math. (Rozprawy Mat.) **140**, (1976)
364. P. Kosmol, *Optimierung und Approximation.* de Gruyter Lehrbuch. [de Gruyter Textbook] (Walter de Gruyter & Co., Berlin, 1991)
365. P. Kosmol, W. Schill, M. Wriedt, Der Satz von Banach-Steinhaus für konvexe Operatoren. Arch. Math. (Basel) **33**(6), 564–569 (1979/80)
366. G. Köthe, *Topological Vector Spaces. I.* Die Grundlehren der mathematischen Wissenschaften in Einzeldarstellungen, vol. 159 (1969) [Translated by D.J.H. Garling]
367. O. Kozhushkina, The Bishop-Phelps-Bollobas Theorem and Operators on Banach Spaces. Thesis (Ph.D.), Kent State University (ProQuest LLC, Ann Arbor, 2014)
368. T. Kuczumow, A. Stachura, Extensions of nonexpansive mappings in the Hilbert ball with the hyperbolic metric. I, II. Comment. Math. Univ. Carolin. **29**(3), 399–402, 403–410 (1988)
369. K. Kuratowski, Quelques problèmes concernant les espaces métriques nonseparables. Fund. Math. **25**, 534–545 (1935) [French]
370. A.G. Kusraev, S.S. Kutateladze, *Subdifferentials: Theory and Applications.* Mathematics and Its Applications, vol. 323 (Kluwer Academic Publishers Group, Dordrecht, 1995) [Translated from the Russian]
371. C. La Russa, Differentiability of Lipschitz maps. Far East J. Math. Sci. (FJMS) **41**(1), 33–44 (2010)
372. T.J. Laakso, Plane with A_∞-weighted metric not bi-Lipschitz embeddable to $\mathbb{R}^N$. Bull. London Math. Soc. **34**(6), 667–676 (2002)
373. G. Lancien, B. Randrianantoanina, On the extension of Hölder maps with values in spaces of continuous functions. Isr. J. Math. **147**, 75–92 (2005)
374. U. Lang, Extendability of large-scale Lipschitz maps. Trans. Am. Math. Soc. **351**(10), 3975–3988 (1999)
375. U. Lang, Injective hulls of certain discrete metric spaces and groups. J. Topol. Anal. **5**(3), 297–331 (2013)

376. U. Lang, B. Pavlović, V. Schroeder, Extensions of Lipschitz maps into Hadamard spaces. Geom. Funct. Anal. **10**(6), 1527–1553 (2000)

377. U. Lang, C. Plaut, Bilipschitz embeddings of metric spaces into space forms. Geom. Dedicata **87**(1–3), 285–307 (2001)

378. U. Lang, T. Schlichenmaier, Nagata dimension, quasisymmetric embeddings, and Lipschitz extensions. Int. Math. Res. Not. (58), 3625–3655 (2005)

379. U. Lang, V. Schroeder, Kirszbraun's theorem and metric spaces of bounded curvature. Geom. Funct. Anal. **7**(3), 535–560 (1997)

380. J.R. Lee, A. Naor, Absolute Lipschitz extendability. C. R. Math. Acad. Sci. Paris **338**(11), 859–862 (2004)

381. J.R. Lee, A. Naor, Extending Lipschitz functions via random metric partitions. Invent. Math. **160**(1), 59–95 (2005)

382. E. Le Gruyer, On absolutely minimizing Lipschitz extensions and PDE $\Delta_\infty(u) = 0$. Nonlinear Differ. Equ. Appl. **14**(1–2), 29–55 (2007)

383. E. Le Gruyer, Minimal Lipschitz extensions to differentiable functions defined on a Hilbert space. Geom. Funct. Anal. **19**(4), 1101–1118 (2009)

384. N. Levine, Remarks on uniform continuity in metric spaces. Am. Math. Mon. **67**, 562–563 (1960)

385. N. Levine, A note on the Lipschitz condition in metric spaces. Proc. Am. Math. Soc. **13**, 314–315 (1962)

386. V.L. Levin, Topics in the duality theory for mass transfer problems, in *Distributions with Given Marginals and Moment Problems* (Prague, 1996) (Kluwer Academic Publications, Dordrecht, 1997), pp. 243–252

387. V.L. Levin, A.A. Miljutin, The problem of the displacement of masses with a discontinuous cost function and the mass formulation of the problem of the duality of convex extremal problems. Uspekhi Mat. Nauk **34**(3), 3–68, 248 (1979)

388. J. Lindenstrauss, On operators which attain their norm. Isr. J. Math. **1**, 139–148 (1963)

389. J. Lindenstrauss, On nonlinear projections in Banach spaces. Mich. Math. J. **11**, 263–287 (1964)

390. J. Lindenstrauss, D. Preiss, Almost Fréchet differentiability of finitely many Lipschitz functions. Mathematika **43**(2), 393–412 (1996)

391. J. Lindenstrauss, D. Preiss, A new proof of Fréchet differentiability of Lipschitz functions. J. Eur. Math. Soc. (JEMS) **2**(3), 199–216 (2000)

392. J. Lindenstrauss, D. Preiss, Fréchet differentiability of Lipschitz functions (a survey), in *Recent Progress in Functional Analysis* (Valencia, 2000). North-Holland Mathematics Studies, vol. 189 (North-Holland, Amsterdam, 2001), pp. 19–42

393. J. Lindenstrauss, D. Preiss, On Fréchet differentiability of Lipschitz maps between Banach spaces. Ann. of Math. **157**(1), 257–288 (2003)

394. J. Lindenstrauss, H.P. Rosenthal, The $\mathscr{L}_p$ spaces. Isr. J. Math. **7**, 325–349 (1969)

395. J. Lindenstrauss, L. Tzafriri, *Classical Banach Spaces*. Vol. 1. *Sequence Spaces*; Vol. 2. *Function Spaces* (Springer, Berlin, 1996) [Reprint of the 1977 and 1979 ed.]

396. J. Lindenstrauss, E. Matoušková, D. Preiss, Lipschitz image of a measure-null set can have a null complement. Isr. J. Math. **118**, 207–219 (2000)

397. J. Lindenstrauss, D. Preiss, J. Tišer, Fréchet differentiability of Lipschitz maps and porous sets in Banach spaces, in *Banach Spaces and Their Applications in Analysis* (Walter de Gruyter, Berlin, 2007), pp. 111–123

398. J. Lindenstrauss, D. Preiss, J. Tišer, Fréchet differentiability of Lipschitz functions via a variational principle. J. Eur. Math. Soc. (JEMS) **12**(2), 385–412 (2010)

399. J. Lindenstrauss, D. Preiss, J. Tišer, *Fréchet Differentiability of Lipschitz Functions and Porous Sets in Banach Spaces*. Annals of Mathematics Studies, vol. 179 (Princeton University Press, Princeton, 2012)

400. N. Linial, Squared l_2 metrics into l_1, in *Discrete Metric Spaces and Their Algorithmic Applications*, Haifa, March 2002, ed. by J. Matoušek (2002)

401. N. Linial, E. London, Y. Rabinovich, The geometry of graphs and some of its algorithmic applications. Combinatorica **15**(2), 215–245 (1995)
402. Ph. Logaritsch, A. Marchese, Kirszbraun's extension theorem fails for Almgren's multiple valued functions. Arch. Math. (Basel) **102**(5), 455–458 (2014)
403. V. Lomonosov, A counterexample to the Bishop-Phelps theorem in complex spaces. Isr. J. Math. **115**, 25–28 (2000)
404. V. Lomonosov, On the Bishop-Phelps theorem in complex spaces. Quaest. Math. **23**(2), 187–191 (2000)
405. V. Lomonosov, The Bishop-Phelps theorem fails for uniform non-selfadjoint dual operator algebras. J. Funct. Anal. **185**(1), 214–219 (2001)
406. M. Llorente, P. Mattila, Lipschitz equivalence of subsets of self-conformal sets. Nonlinearity **23**(4), 875–882 (2010)
407. K. Luosto, Ultrametric spaces bi-Lipschitz embeddable in $\mathbf{R}^n$. Fund. Math. **150**(1), 25–42 (1996)
408. J. Luukkainen, Extension of spaces, maps, and metrics in Lipschitz topology. Ann. Acad. Sci. Fenn. Math. Diss. No. 17 (1978)
409. J. Luukkainen, Assouad dimension: antifractal metrization, porous sets, and homogeneous measures. J. Korean Math. Soc. **35**(1), 23–76 (1998)
410. J. Luukkainen, Lipschitz and quasiconformal approximation of homeomorphism pairs. Topol. Appl. **109**(1), 1–40 (2001)
411. J. Luukkainen, H. Movahedi-Lankarani, Minimal bi-Lipschitz embedding dimension of ultrametric spaces. Fund. Math. **144**(2), 181–193 (1994)
412. J. Luukkainen, E. Saksman, Every complete doubling metric space carries a doubling measure. Proc. Am. Math. Soc. **126**(2), 531–534 (1998)
413. J. Luukkainen, J. Väisälä, Elements of Lipschitz topology. Ann. Acad. Sci. Fenn. Math. **3**(1), 85–122 (1977)
414. M.S. Lyapina, On the Lipschitz constant for a nonisometric bi-Lipschitz transformation of a Cantor set. J. Math. Sci. (N. Y.) **120**(2), 1109–1116 (2004)
415. S. Mabizela, The relationship between Lipschitz extensions, best approximations, and continuous selections. Quaest. Math. **14**(3), 261–268 (1991)
416. P. MacManus, Bi-Lipschitz extensions in the plane. J. Anal. Math. **66**, 85–115 (1995)
417. I.J. Maddox, *Elements of Functional Analysis* (Cambridge University Press, London, 1970)
418. V. Magnani, Lipschitz continuity, Aleksandrov theorem and characterizations for H-convex functions. Math. Ann. **334**(1), 199–233 (2006)
419. V. Magnani, Towards differential calculus in stratified groups. J. Aust. Math. Soc. **95**(1), 76–128 (2013)
420. V. Magnani, T. Rajala, Radon-Nikodym property and area formula for Banach homogeneous group targets. Int. Math. Res. Not. **2014**(23), 6399–6430 (2014)
421. H. Mahyar, Compact endomorphisms of infinitely differentiable Lipschitz algebras. Rocky Mountain J. Math. **39**(1), 193–217 (2009)
422. H. Mahyar, A.H. Sanatpour, Compact and quasicompact homomorphisms between differentiable Lipschitz algebras. Bull. Belg. Math. Soc. Simon Stevin **17**(3), 485–497 (2010)
423. H. Mahyar, A.H. Sanatpour, Quasicompact endomorphisms of Lipschitz algebras of analytic functions. Publ. Math. Debr. **76**(1–2), 135–145 (2010)
424. H. Mahyar, A.H. Sanatpour, Compact composition operators on certain analytic Lipschitz spaces. Bull. Iranian Math. Soc. **38**(1), 85–99 (2012)
425. K. Makarychev, Yu. Makarychev, Union of Euclidean metric spaces is Euclidean. Discrete Anal., Paper No. 14, (2016)
426. O. Maleva, D. Preiss, Cone unrectifiable sets and non-differentiability of Lipschitz functions, arXiv preprint, math.FA, arXiv:1709.04233, 30 (2017)
427. L. Maligranda, V.V. Mykhaylyuk, A. Plichko, On a problem of Eidelheit from The Scottish Book concerning absolutely continuous functions. J. Math. Anal. Appl. **375**(2), 401–411 (2011)

428. P. Mankiewicz, On Lipschitz mappings between Fréchet spaces. Studia Math. **41**, 225–241 (1972)
429. P. Mankiewicz, On the differentiability of Lipschitz mappings in Fréchet spaces. Studia Math. **45**, 15–29 (1973)
430. P. Mankiewicz, On Polish spaces Lipschitz universal for separable metric spaces. Fund. Math. **119**(1), 41–48 (1983)
431. L. Marco, J.A. Murillo, Locally Lipschitz and convex functions. Mathematica (Cluj) **38**(1–2), 121–131 (1996)
432. M.B. Marcus, G. Pisier, Characterizations of almost surely continuous p-stable random Fourier series and strongly stationary processes. Acta Math. **152**(3–4), 245–301 (1984)
433. D.S. Marinescu, A class of functions related to the Lipschitz property, in *Analysis, functional equations, approximation and convexity*. Proceedings of the conference held in honour of Professor Elena Popoviciu on the occasion of her 75th birthday, Cluj-Napoca, Romania, October 15–16, 1999 (Editura Carpatica, Cluj-Napoca, 1999), pp. 160–163
434. G. Marino, When is any continuous function Lipschitzian?. Extracta Math. **13**(1), 107–110 (1998)
435. G. Marino, G. Lewicki, P. Pietramala, Finite chainability, locally Lipschitzian and uniformly continuous functions. Z. Anal. Anwendungen **17**(4), 795–803 (1998)
436. A. Markov, On mean values and exterior densities. Rec. Math. Moscou Ser. **4**, 16–190 (1938)
437. M. Martín, The version for compact operators of Lindenstrauss properties A and B. Rev. R. Acad. Cienc. Exactas Fís. Nat. Ser. A Math. RACSAM **110**(1), 269–284 (2016)
438. P. Mattila, P. Saaranen, Ahlfors-David regular sets and bilipschitz maps. Ann. Acad. Sci. Fenn. Math. **34**(2), 487–502 (2009)
439. J. Matkowski, Functional equations and Nemytskiĭ operators. Funkcial. Ekvac. **25**(2), 127–132 (1982)
440. J. Matkowski, On Nemytskiĭ Lipschitzian operator. Acta Univ. Carolin. Math. Phys. **28**(2), 79–82 (1987), 15th winter school in abstract analysis (Srní, 1987)
441. J. Matkowski, On Nemytskiĭ operator. Math. Japon. **33**(1), 81–86 (1988)
442. J. Matkowski, Uniformly continuous superposition operators in the Banach space of Hölder functions. J. Math. Anal. Appl. **359**(1), 56–61 (2009)
443. J. Matkowski, Uniformly bounded composition operators between general Lipschitz function normed spaces. Topol. Methods Nonlinear Anal. **38**(2), 395–405 (2011)
444. J. Matkowski, M. Wróbel, Uniformly bounded Nemytskij operators generated by set-valued functions between generalized Hölder function spaces. Discuss. Math. Differ. Incl. Control Optim. **31**(2), 183–198 (2011)
445. J. Matkowski, M. Wróbel, Uniformly bounded set-valued Nemytskij operators acting between generalized Hölder function spaces. Cent. Eur. J. Math. **10**(2), 609–618 (2012)
446. J. Matoušek, Extension of Lipschitz mappings on metric trees. Comment. Math. Univ. Carolin. **31**(1), 99–104 (1990)
447. J. Matoušek, E. Matoušková, A highly non-smooth norm on Hilbert space. Isr. J. Math. **112**, 1–27 (1999)
448. P. Mattila, *Geometry of Sets and Measures in Euclidean Spaces. Fractals and Rectifiability*. Cambridge Studies in Advanced Mathematics, vol. 44 (Cambridge University Press, Cambridge, 1995)
449. E. Matoušková, Extensions of continuous and Lipschitz functions. Can. Math. Bull. **43**(2), 208–217 (2000)
450. B. Maurey, Théorèmes de factorisation pour les opérateurs linéaires à valuers dans les espaces L_p. Astérisque **11**, 1–5 (1974)
451. R.D. Mauldin (ed.), *The Scottish Book*, 2nd edn. Mathematics from the Scottish Café with selected problems from the new Scottish Book, Including selected papers presented at the Scottish Book Conference held at North Texas University, Denton, May 1979 (Birkhäuser/Springer, Cham, 2015)
452. S. Mazur, W. Orlicz, Sur les espaces métriques linéaires. I. Studia Math. **10**, 184–208 (1948) [French]

453. S. Mazur, S. Ulam, Über isometrische Abbildungen von normierten Vektorräumen. Ann. Soc. Polon. Math. **10**, 127 (1932) [German]

454. S. Mazur, S. Ulam, Sur les transformations isometriques d'espaces vectoriels, normés. C. R. Acad. Sci., Paris **194**, 946–948 (1932) [French]

455. E.J. McShane, Extension of range of functions. Bull. Am. Math. Soc. **40**, 837–842 (1934) [English]

456. R.E. Megginson, *An Introduction to Banach Space Theory*. Graduate Texts in Mathematics, vol. 183 (Springer, New York, 1998)

457. F. Mémoli, G. Sapiro, P. Thompson, Brain and surface warping via minimizing Lipschitz extensions. IMA Preprint Series, No. 20192 (University of Minnesota, Minneapolis, 2006), pp. 11

458. F. Mémoli, G. Sapiro, P. Thompson, Geometric surface and brain warping via geodesic minimizing Lipschitz extensions, in *1st MICCAI Workshop on Mathematical Foundations of Computational Anatomy: Geometrical, Statistical and Registration Methods for Modeling Biological Shape Variability* (2006), pp. 58–67

459. M. Mendel, A simple proof of Johnson-Lindenstrauss extension, arXiv:1803.03606v2 [math.MG], 2 (2018)

460. M. Mendel, A. Naor, Ramsey partitions and proximity data structures. J. Eur. Math. Soc. (JEMS) **9**(2), 253–275 (2007)

461. M. Mendel, A. Naor, Metric cotype. Ann. Math. **168**(1), 247–298 (2008)

462. M. Mendel, A. Naor, A relation between finitary Lipschitz extension moduli, arXiv:1707.07289 [math.MG], 3 (2017)

463. K. Menger, Untersuchungen über allgemeine Metrik. Math. Ann. **100**, 75–163 (1928) [German]

464. P. Meyer-Nieberg, *Banach Lattices*. Universitext (Springer, Berlin, 1991)

465. E. Michael, Some extension theorems for continuous functions. Pac. J. Math. **3**, 789–806 (1953)

466. E. Michael, Continuous selections. I. Ann. Math. **63**, 361–382 (1956)

467. J.H. Michael, Approximation of functions by means of Lipschitz functions. J. Aust. Math. Soc. **3**, 134–150 (1963)

468. J.H. Michael, Lipschitz approximations to summable functions. Acta Math. **111**, 73–95 (1964)

469. E. Michael, A short proof of the Arens-Eells embedding theorem. Proc. Am. Math. Soc. **15**, 415–416 (1964)

470. E.J. Mickle, On the extension of a transformation. Bull. Am. Math. Soc. **55** (1949), 160–164

471. R. Miculescu, Extensions of some locally Lipschitz maps. Bull. Math. Soc. Sci. Math. Roum. **41**(3), 197–203 (1998)

472. R. Miculescu, Approximating uniformly continuous bounded functions by Lipschitz functions. Rev. Roum. Math. Pures Appl. **44**(2), 253–255 (1999)

473. R. Miculescu, Some applications of Lip-partition of unity. Math. Rep. Bucur. **1**(2), 227–235 (1999)

474. R. Miculescu, Equivalent definitions for Lipschitz compact connected manifolds. Ann. Univ. Bucureşti Mat. **49**(1), 53–62 (2000)

475. R. Miculescu, A LIP immersion of Lipschitz manifolds modelled on some Banach spaces. Bull. Greek Math. Soc. **43**, 99–104 (2000)

476. R. Miculescu, Les fonctions lipschitziennes homotopiques sont Lipschitz homotopiques. Rev. Roum. Math. Pures Appl. **45**(1), 119–122 (2000)

477. R. Miculescu, Approximation of continuous functions by Lipschitz functions. Real Anal. Exchange **26**(1), 449–452 (2000/01)

478. R. Miculescu, Lipschitz approximation of uniformly continuous convex-valued functions. Bull. Greek Math. Soc. **46**, 129–132 (2002)

479. R. Miculescu, Approximations by Lipschitz functions generated by extensions. Real Anal. Exchange **28**(1), 33–40 (2002/03)

480. R. Miculescu, Some observations on generalized Lipschitz functions. Rocky Mountain J. Math. **37**(3), 893–903 (2007)
481. R. Miculescu, A selection of embedding results for Lipschitz manifolds. Ann. Univ. Buchar. Math. Ser. **1**(1), 121–124 (2010)
482. R. Miculescu, C. Mortici, *Funcţii Lipschitz* (Editura Academiei Române, Bucharest, 2004) [Romanian]
483. V.A. Mil'man, Lipschitz extensions of linearly bounded functions. Mat. Sb. **189**(8), 67–92 (1998)
484. G.J. Minty, On the extension of Lipschitz, Lipschitz-Hölder continuous, and monotone functions. Bull. Am. Math. Soc. **76**, 334–339 (1970)
485. M. Moussaoui, Approximations lipschitziennes de multifonctions. Application aux conditions nécessaires d'optimalité. Sém. Anal. Convexe **20**, 34 (1990)
486. H. Movahedi-Lankarani, On the inverse of Mañé's projection. Proc. Am. Math. Soc. **116**(2), 555–560 (1992)
487. H. Movahedi-Lankarani, R. Wells, On bi-Lipschitz embeddings. Port. Math. **62**(3), 247–268 (2005)
488. K. Musiał, Pettis integral, in *Handbook of Measure Theory*, vols. I and II (North-Holland, Amsterdam, 2002), pp. 531–586
489. K. Musiał, Topics in the theory of Pettis integration. Rend. Ist. Mat. Univ. Trieste **23**, 177–262 (1991)
490. C. Mustăţa, On some Chebyshevian subspaces of the normed space of Lipschitz functions. Rev. Anal. Numer. Teor. Aprox. **2**, 81–87 (1973) [Romanian]
491. C. Mustăţa, On the best approximation in metric spaces. Rev. Anal. Numér. Théor. Approx. **4**, 45–50 (1975)
492. C. Mustăţa, Best approximation and unique extension of Lipschitz functions. J. Approx. Theory **19**, 222–230 (1977)
493. C. Mustăţa, Norm preserving extension of starshaped Lipschitz functions. Mathematica (Cluj) **19**, 183–187 (1977)
494. C. Mustăţa, A characterization of semi-Chebyshevian sets in a metric space. Rev. Anal. Numér. Théor. Approx. **7**(2), 169–170 (1978)
495. C. Mustăţa, Extension of bounded Lipschitz functions, in *Seminar of Functional Analysis and Numerical Methods*, Babeş-Bolyai University, Cluj-Napoca, Research Seminaries, vol. 1 (1980), pp. 1–20
496. C. Mustăţa, The extension of starshaped bounded Lipschitz functions. Rev. Anal. Numér. Théor. Approx. **9**(1), 93–99 (1980)
497. C. Mustăţa, On the extension of Hölder functions, in *Seminar of Functional Analysis and Numerical Methods*, Babeş-Bolyai University, Cluj-Napoca, Research Seminaries, vol. 85 (1985), pp. 84–92
498. C. Mustăţa, *On the Unicity of the Extension of Lipschitz Odd Functions*, Seminar on Optimization Theory, Babeş-Bolyai University, Cluj-Napoca, Research Seminaries, vol. 87 (1987), pp. 75–80
499. C. Mustăţa, An application of a theorem of McShane, in *Seminar on Functional Analysis and Numerical Methods*, Babeş-Bolyai University, Cluj-Napoca, Research Seminaries, vol. 89 (1989), pp. 75–84
500. C. Mustăţa, Extension of Hölder functions and some related problems of best approximation, in *Seminar on Mathematical Analysis*, Babeş-Bolyai University, Cluj-Napoca, Research Seminaries, vol. 91 (1991), pp. 71–86
501. C. Mustăţa, Some remarks concerning norm preserving extensions and best approximation. Rev. Anal. Numér. Théor. Approx. **29**(2), 173–180 (2000)
502. C. Mustăţa, Uniqueness of the extension of semi-Lipschitz functions on quasi-metric spaces. Bul. Ştiinţ. Univ. Baia Mare Ser. B Fasc. Mat.-Inform. **16**(2), 207–212 (2000)

503. C. Mustăța, Extensions of semi-Lipschitz functions on quasi-metric spaces. Rev. Anal. Numér. Théor. Approx. **30**(1), 61–67 (2001), Dedicated to the memory of Acad. Tiberiu Popoviciu

504. C. Mustăța, On the extremal semi-Lipschitz functions. Rev. Anal. Numér. Théor. Approx. **31**(1), 103–108 (2002)

505. C. Mustăța, On the extension of semi-Lipschitz functions on asymmetric normed spaces. Rev. Anal. Numér. Théor. Approx. **34**(2), 139–150 (2005)

506. C. Mustăța, Extension and approximation of semi-Lipschitz functions. Ann. Univ. Vest Timiş. Ser. Mat.-Inform. **45**(2), 83–92 (2007)

507. C. Mustăța, Best uniform approximation of semi-Lipschitz functions by extensions. Rev. Anal. Numér. Théor. Approx. **36**(2), 161–171 (2007)

508. C. Mustăța, On a theorem of Baire about lower semicontinuous functions. Rev. Anal. Numér. Théor. Approx. **37**(1), 71–75 (2008)

509. C. Mustăța, Extensions of semi-Hölder real valued functions on a quasi-metric space. Rev. Anal. Numér. Théor. Approx. **38**(2), 164–169 (2009)

510. C. Mustăța, On the approximation of the global extremum of a semi-Lipschitz function. Mediterr. J. Math. **6**(2), 169–180 (2009)

511. C. Mustăța, On the existence and uniqueness of extensions of semi-Hölder real-valued functions. Rev. Anal. Numér. Théor. Approx. **39**(2), 134–140 (2010)

512. C. Mustăța, On the extensions preserving the shape of a semi-Hölder function. Results Math. **63**(1–2), 425–433 (2013)

513. J. Nagata, Note on dimension theory for metric spaces. Fund. Math. **45**, 143–181 (1958)

514. A. Naor, A phase transition phenomenon between the isometric and isomorphic extension problems for Hölder functions between L_p spaces. Mathematika **48**(1–2), 253–271 (2001)

515. A. Naor, An introduction to the Ribe program. Jpn. J. Math. **7**(2), 167–233 (2012)

516. A. Naor, Metric dimension reduction: a snapshot of the Ribe program, arXiv preprint, mathFA, arXiv:1809.02376, 44 (2018) [Proceedings of ICM 2018]

517. A. Naor, Y. Rabani, On Lipschitz extension from finite subsets. Isr. J. Math. **219**(1), 115–161 (2017)

518. A. Naor, Y. Peres, O. Schramm, S. Sheffield, Markov chains in smooth Banach spaces and Gromov-hyperbolic metric spaces. Duke Math. J. **134**(1), 165–197 (2006)

519. I.P. Natanson, *Theory of Functions of a Real Variable* (Frederick Ungar Publishing Co., New York, 1955) [Translated by Leo F. Boron with the collaboration of Edwin Hewitt]

520. A. Navas, An L^1 ergodic theorem with values in a non-positively curved space via a canonical barycenter map. Ergod. Theory Dyn. Syst. **33**(2), 609–623 (2013)

521. A. Nekvinda, L. Zajíček, A simple proof of the Rademacher theorem. Časopis Pěst. Mat. **113**(4), 337–341 (1988)

522. A.B. Németh, On the subdifferentiability of convex operators. J. Lond. Math. Soc. **34**(3), 552–558 (1986)

523. M.M. Neumann, Uniform boundedness and closed graph theorems for convex operators. Math. Nachr. **120**, 113–125 (1985)

524. K.F. Ng, On a theorem of Dixmier. Math. Scand. **29**, 279–280 (1971)

525. V.K. Nguyen, T.N. Nguyen, Lipschitz extensions and Lipschitz retractions in metric spaces. Colloq. Math. **45**(2), 245–250 (1981)

526. V.K. Nguyen, T.N. Nguyen, Extending locally Lipschitz maps with values in an infinite-dimensional nuclear Fréchet space. Bull. Acad. Polon. Sci. Sér. Sci. Math. **29**(11–12), 609–616 (1981)

527. M. Ó Searcóid, *Metric Spaces* (Springer, Berlin, 2006)

528. A.M. Oberman, A convergent difference scheme for the infinity Laplacian: construction of absolutely minimizing Lipschitz extensions. Math. Comp. **74**(251), 1217–1230 (2005)

529. A.M. Oberman, An explicit solution of the Lipschitz extension problem. Proc. Am. Math. Soc. **136**(12), 4329–4338 (2008)

530. S. Ohta, Extending Lipschitz and Hölder maps between metric spaces. Positivity **13**(2), 407–425 (2009)

531. E. Olson, Bouligand dimension and almost Lipschitz embeddings. Pac. J. Math. **202**(2), 459–474 (2002)
532. E.J. Olson, J.C. Robinson, Almost bi-Lipschitz embeddings and almost homogeneous sets. Trans. Am. Math. Soc. **362**(1), 145–168 (2010)
533. M.I. Ostrovskii, *Metric Embeddings. Bilipschitz and Coarse Embeddings into Banach Spaces* (de Gruyter, Berlin, 2013)
534. H. Pajot, Plongements bilipschitziens dans les espaces euclidiens, Q-courbure et flot quasi-conforme, in *Actes du Séminaire de Théorie Spectrale et Géométrie*. Vol. 25 (2008), pp. 149–158 [French]
535. D. Pallaschke, The compact endomorphisms of the metric linear spaces L_φ. Studia Math. **47**, 123–133 (1973)
536. D. Panchenko, *Lecture Notes on Probability Theory*. AMS Open Math Notes: Works in Progress (American Mathematical Society, Providence, 2016)
537. P. Pansu, Métriques de Carnot-Carathéodory et quasiisométries des espaces symétriques de rang un. Ann. of Math. **129**(1), 1–60 (1989)
538. G. Pantelidis, Approximationstheorie für metrische lineare Räume. Math. Ann. **184**, 30–48 (1969) [German]
539. N.S. Papageorgiou, Nonsmooth analysis on partially ordered vector spaces. I. Convex case. Pac. J. Math. **107**(2), 403–458 (1983)
540. N.S. Papageorgiou, Nonsmooth analysis on partially ordered vector spaces. II. Nonconvex case, Clarke's theory. Pac. J. Math. **109**(2), 463–495 (1983)
541. A. Papadopoulos, *Metric Spaces, Convexity and Non-Positive Curvature*. 2nd edn. IRMA Lectures in Mathematics and Theoretical Physics, vol. 6 (European Mathematical Society (EMS), Zürich, 2014)
542. S.H. Park, Quotient mappings, Helly extensions, Hahn-Banach extensions, Tietze extensions, Lipschitz extensions, and best approximation. J. Korean Math. Soc. **29**(2), 239–250 (1992)
543. J. Partanen, J. Väisälä, Extension of bi-Lipschitz maps of compact polyhedra. Math. Scand. **72**(2), 235–264 (1993)
544. K.R. Parthasarathy, *Probability Measures on Metric Spaces* (AMS Chelsea Publishing, Providence, 2005) [Reprint of the 1967 original]
545. N.T. Peck, A note on support points. Bull. Am. Math. Soc. **74**, 1103 (1968)
546. N.T. Peck, Support points in locally convex spaces. Duke Math. J. **38**, 271–278 (1971)
547. J. Pelant, Embeddings into c_0. Topol. Appl. **57**(2–3), 259–269 (1994)
548. J. Pelant, P. Holický, O.F.K. Kalenda, $C(K)$ spaces which cannot be uniformly embedded into $c_0(\Gamma)$. Fund. Math. **192**(3), 245–254 (2006)
549. A. Pełczyński, Projections in certain Banach spaces. Studia Math. **19**(2), 209–228 (1960)
550. P. Pérez Carreras, J. Bonet, *Barrelled Locally Convex Spaces*. North Holland Mathematics Studies, vol. 131 [Notas de Matemática, vol. 113] (North-Holland, Amsterdam, 1987)
551. C. Perez-Garcia, W.H. Schikhof, *Locally Convex Spaces Over Non-Archimedean Valued Fields* (Cambridge University Press, Cambridge, 2010)
552. A.I. Perov, On certain aspects of the general theory of convex functions. Ukrain. Mat. Ž. **18**(3), 129–132 (1966)
553. V.G. Pestov, Free Banach spaces and representations of topological groups. Funct. Anal. Appl. **20**, 70–72 (1986)
554. M. Petrakis, J.J. Uhl, Differentiation in Banach spaces, in *Proceedings of Analysis, Conference, Singapore 1986*. North-Holland Mathematics Studies, vol. 150 (North-Holland, Amsterdam, 1988), pp. 219–241
555. A. Petrunin, *Convex Hull In CAT(0)*, MathOverflow (2014), URL: http://mathoverflow.net/users/1441/anton-petrunin
556. R.R. Phelps, Subreflexive normed linear spaces. Arch. Math. (Basel) **8**, 444–450 (1957)
557. R.R. Phelps, Correction to "Subreflexive normed linear spaces". Arch. Math. **9**, 439–440 (1958)
558. R.R. Phelps, Some subreflexive Banach spaces. Arch. Math. **10**, 162–169 (1959)

559. R.R. Phelps, A representation theorem for bounded convex sets. Proc. Am. Math. Soc. **11**, 976–983 (1960)

560. R.R. Phelps, Uniqueness of Hahn-Banach extensions and unique best approximation. Trans. Am. Math. Soc. **95**, 238–255 (1960)

561. R.R. Phelps, Support cones and their generalizations, in *Proceedings of Symposia in Pure Mathematics*, vol. VII (American Mathematical Society, Providence, 1963), pp. 393–401

562. R.R. Phelps, Weak* support points of convex sets in E^*. Isr. J. Math. **2**, 177–182 (1964)

563. R.R. Phelps, Support cones in Banach spaces and their applications. Adv. Math. **13**, 1–19 (1974)

564. R.R. Phelps, Gaussian null sets and differentiability of Lipschitz map on Banach spaces. Pac. J. Math. **77**(2), 523–531 (1978)

565. R.R. Phelps, *Convex Functions, Monotone Operators and Differentiability*. Lecture Notes in Mathematics, vol. 1364 (Springer, Berlin, 1989)

566. R.R. Phelps, The Bishop-Phelps theorem in complex spaces: an open problem, in *Function spaces* (Edwardsville, IL, 1990). Lecture Notes in Pure and Applied Mathematics, vol. 136 (Dekker, New York, 1992), pp. 337–340

567. R.R. Phelps, *Convex Functions, Monotone Operators and Differentiability*, 2nd edn. Lecture Notes in Mathematics, vol. 1364 (Springer, Berlin, 1993)

568. R.R. Phelps, The Bishop-Phelps theorem in the complex case, Preprint, Lecture at the 27th Winter School, Lhota nad Rohanovem, Czech Republic (1999), p. 6

569. R.R. Phelps, The Bishop-Phelps theorem, in *Ten Mathematical Essays on Approximation in Analysis and Topology*, ed. by J. Ferrera, et al. (Elsevier B. V., Amsterdam, 2005), pp. 235–244

570. Ł. Piasecki, *Classification of Lipschitz Mappings* (CRC Press, Boca Raton, 2014)

571. P.B. Pierce, D. Waterman, On the invariance of classes ΦBV, ΛBV under composition. Proc. Am. Math. Soc. **132**(3), 755–760 (2004)

572. A. Pinamonti, G. Speight, A measure zero UDS in the Heisenberg group, in *Bruno Pini Math. Anal. Semin.*, vol. 2016 (Univ. Bologna, Bologna, 2016), pp. 85–96

573. A. Pinamonti, G. Speight, Structure of porous sets in Carnot groups. Ill. J. Math. **61**(1–2), 127–150 (2017)

574. A. Pinamonti, G. Speight, Porosity, differentiability and Pansu's theorem. J. Geom. Anal. **27**(3), 2055–2080 (2017)

575. A. Pinamonti, G. Speight, A measure zero universal differentiability set in the Heisenberg group. Math. Ann. **368**(1–2), 233–278 (2017)

576. C. Plaut, Metric spaces of curvature $\geq k$, in *Handbook of Geometric Topology* (North-Holland, Amsterdam, 2002), pp. 819–898

577. R.A. Poliquin, V.E. Zizler, Optimization of convex functions on w^*-compact sets. Manuscripta Math. **68**(3), 249–270 (1990)

578. M.M. Popov, On integrability in F-spaces. Studia Math. **110**(3), 205–220 (1994)

579. D. Preiss, Differentiability of Lipschitz functions on Banach spaces. J. Funct. Anal. **91**(2), 312–345 (1990)

580. D. Preiss, G. Speight, Differentiability of Lipschitz functions in Lebesgue null sets. Invent. Math. **199**(2), 517–559 (2015)

581. D. Preiss, J. Tišer, Two unexpected examples concerning differentiability of Lipschitz functions on Banach spaces, in *Geometric Aspects of Functional Analysis* (Israel, 1992–1994). Operator Theory: Advances and Applications, vol. 77 (Birkhäuser, Basel, 1995), pp. 219–238

582. D. Preiss, L. Zajíček, Directional derivatives of Lipschitz functions. Isr. J. Math. **125**, 1–27 (2001)

583. K. Przesławski, D. Yost, Continuity properties of selectors and Michael's theorem. Mich. Math. J. **36**(1), 113–134 (1989)

584. K. Przesławski, D. Yost, Lipschitz retracts, selectors, and extensions. Mich. Math. J. **42**(3), 555–571 (1995)

585. S.T. Rachev, L. Rüschendorf, *Mass Transportation Problems*. Vol. I. *Theory*; Vol. II. *Applications*. Probability and its Applications (New York) (Springer, New York, 1998)

586. S.T. Rachev, R.M. Shortt, Duality theorems for Kantorovich-Rubinstein and Wasserstein functionals. Diss. Math. **299**, (1990)

587. H. Rademacher, Über partielle und totale Differenzierbarkeit von Funktionen mehrerer Variablen und über die Transformation der Doppelintegrale. I. Math. Ann. **79**, 340–359 (1919); II. Math. Ann. **81**, 52–63 (1920) [German]

588. H. Rao, H.-J. Ruan, L.-F. Xi, Lipschitz equivalence of self-similar sets. C. R. Math. Acad. Sci. Paris **342**(3), 191–196 (2006)

589. H. Rao, H.-J. Ruan, Y.-M. Yang, Gap sequence, Lipschitz equivalence and box dimension of fractal sets. Nonlinearity **21**(6), 1339–1347 (2008)

590. H. Rao, H.-J. Ruan, Y. Wang, Lipschitz equivalence of Cantor sets and algebraic properties of contraction ratios. Trans. Am. Math. Soc. **364**(3), 1109–1126 (2012)

591. H. Rao, H.-J. Ruan, Y. Wang, Lipschitz equivalence of self-similar sets: algebraic and geometric properties, in *Fractal Geometry and Dynamical Systems in Pure and Applied Mathematics*. I. Fractals in Pure Mathematics, Contemporary Mathematics, vol. 600 (American Mathematical Society, Providence, 2013), pp. 349–364

592. S. Reich, S. Simons, Fenchel duality, Fitzpatrick functions and the Kirszbraun-Valentine extension theorem. Proc. Am. Math. Soc. **133**(9), 2657–2660 (2005)

593. M. Ribe, On uniformly homeomorphic normed spaces. Ark. Mat. **14**(2), 237–244 (1976)

594. M. Ribe, On uniformly homeomorphic normed spaces. II. Ark. Mat. **16**(1), 1–9 (1978)

595. B. Ricceri, Smooth extensions of Lipschitzian real functions. Proc. Am. Math. Soc. **104**(2), 641–642 (1988)

596. M.A. Rieffel, Lipschitz extension constants equal projection constants, in *Operator Theory, Operator Algebras, and Applications*. Contemporary Mathematics, vol. 414 (American Mathematical Society, Providence, 2006), pp. 147–162

597. F. Riesz, Sur les opérations fonctionnelles linéaires. C. R. Acad. Sci. Paris **149**, 97–977 (1910)

598. A.W. Roberts, D.E. Varberg, *Convex Functions*. Pure and Applied Mathematics, vol. 57 (Academic Press, New York, 1973)

599. A.W. Roberts, D.E. Varberg, Another proof that convex functions are locally Lipschitz. Am. Math. Monthly **81**, 1014–1016 (1974)

600. S.M. Robinson, Z. Robinson, Uniformly continuous partitions of unity of a metric space. Can. Math. Bull. **26**, 115–117 (1983)

601. S. Rolewicz, On a certain class of linear metric spaces. Bull. Acad. Pol. Sci. **5**, 471–473 (1957)

602. S. Rolewicz, A generalization of the Mazur-Ulam theorem. Studia Math. **31**, 501–505 (1968)

603. S. Rolewicz, *Metric Linear Spaces*. 2nd edn. (D. Reidel Publishing Company, Boston, 1985)

604. S. Romaguera, M. Sanchis, Semi-Lipschitz functions and best approximation in quasi-metric spaces. J. Approx. Theory **103**(2), 292–301 (2000)

605. S. Romaguera, M. Sanchis, Properties of the normed cone of semi-Lipschitz functions. Acta Math. Hungar. **108**(1–2), 55–70 (2005)

606. S. Romaguera, J.M. Sánchez-Álvarez, M. Sanchis, On balancedness and *D*-completeness of the space of semi-Lipschitz functions. Acta Math. Hungar. **120**(4), 383–390 (2008)

607. J. Rosenberg, Applications of analysis on Lipschitz manifolds, in *Miniconferences on Harmonic Analysis and Operator Algebras* (Canberra, 1987), Proceedings of the Centre for Mathematical Analysis, Australian National University, vol. 16 (The Australian National University, Canberra, 1988), pp. 269–283

608. A. Rosenfeld, "Continuous" functions on digital pictures. Pattern Recogn. Lett. **4**, 177–184 (1986)

609. A. Roșoiu, D. Frățilă, On the Lipschitz extension constant for a complex-valued Lipschitz function. Stud. Univ. Babeș-Bolyai Math. **53**(1), 101–108 (2008)

610. H.-J. Ruan, Y. Wang, L.-F. Xi, Lipschitz equivalence of self-similar sets with touching structures. Nonlinearity **27**(6), 1299–1321 (2014)

611. W. Rudin, *Principles of Mathematical Analysis*. 3rd edn. International Series in Pure and Applied Mathematics (McGraw-Hill Book Co., New York, 1976)
612. M.E. Rudin, A new proof that metric spaces are paracompact. Proc. Am. Math. Soc. **20**, 603 (1969)
613. W. Rudin, *Real and Complex Analysis*. 3rd edn. (McGraw-Hill Book Co., New York, 1987)
614. W. Rudin, *Functional Analysis*. 2nd edn. International Series in Pure and Applied Mathematics (McGraw-Hill, Inc., New York, 1991)
615. S. Saks, *Theory of the Integral*. 2nd revised edn. English translation by L. C. Young. With two additional notes by Stefan Banach (Dover Publications, Inc., New York, 1964)
616. E. Saksman, Remarks on the nonexistence of doubling measures. Ann. Acad. Sci. Fenn. Math. **24**(1), 155–163 (1999)
617. F. Santambrogio, *Optimal Transport for Applied Mathematicians. Calculus of Variations, PDEs, and Modeling*. Progress in Nonlinear Differential Equations and Their Applications, vol. 87 (Birkhäuser/Springer, Cham, 2015)
618. S.A. Saxon, Some normed barrelled spaces which are not Baire. Math. Ann. **209**, 153–160 (1974)
619. W. Schachermayer, Norm attaining operators on some classical Banach spaces. Pac. J. Math. **105**(2), 427–438 (1983)
620. W. Schachermayer, Norm attaining operators and renormings of Banach spaces. Isr. J. Math. **44**(3), 201–212 (1983)
621. H.H. Schaefer, *Banach Lattices and Positive Operators*. Die Grundlehren der mathematischen Wissenschaften, Band 215 (Springer, New York, 1974)
622. H.H. Schaefer, M.P. Wolff, *Topological Vector Spaces*. 2nd edn. Graduate Texts in Mathematics, vol. 3 (Springer, New York, 1999)
623. E. Schechter, *Handbook of Analysis and Its Foundations* (Academic Press, Inc., San Diego, 1999)
624. K. Schnatz, Nonlinear duality and best approximations in metric linear spaces. J. Approx. Theory **49**, 201–218 (1987)
625. I.J. Schoenberg, On a theorem of Kirzbraun and Valentine. Am. Math. Monthly **60**, 620–622 (1953)
626. S.O. Schönbeck, Extension of nonlinear contractions. Bull. Am. Math. Soc. **72**, 99–101 (1966)
627. S.O. Schönbeck, On the extension of Lipschitz maps. Ark. Mat. **7**, 201–209 (1967)
628. S. Semmes, Bilipschitz embeddings of metric spaces into Euclidean spaces. Publ. Mat. **43**(2), 571–653 (1999)
629. J. Seo, A characterization of bi-Lipschitz embeddable metric spaces in terms of local bi-Lipschitz embeddability. Math. Res. Lett. **18**(6), 1179–1202 (2011)
630. S. Sheffield, Ch.K. Smart, Vector-valued optimal Lipschitz extensions. Comm. Pure Appl. Math. **65**(1), 128–154 (2012)
631. D.R. Sherbert, Banach algebras of Lipschitz functions. Pac. J. Math. **13**, 1387–1399 (1963)
632. D.R. Sherbert, The structure of ideals and point derivations in Banach algebras of Lipschitz functions. Trans. Am. Math. Soc. **111**, 240–272 (1964)
633. S.A. Shkarin, Points of discontinuity of Gateaux-differentiable mappings. Sibirsk. Mat. Zh. **33**(5), 176–185, 224 (1992) [Russian; English translation in Sib. Math. J. **33**(5), 905–913 (1992)]
634. R. Sine, Hyperconvexity and nonexpansive multifunctions. Trans. Am. Math. Soc. **315**(2), 755–767 (1989)
635. R. Sine, Hyperconvexity and approximate fixed points. Nonlinear Anal. **13**(7), 863–869 (1989)
636. I. Singer, *Bases in Banach Spaces. I*. Die Grundlehren der mathematischen Wissenschaften, Band 154 (Springer, New York-Berlin, 1970)
637. I. Singer, *Best Approximation in Normed Linear Spaces by Elements of Linear Subspaces* (Editura Academiei Republicii Socialiste România, Bucharest; Springer, Berlin, 1970)

638. I. Singer, *Bases in Banach Spaces. II* (Editura Academiei Republicii Socialiste România, Bucharest; Springer, Berlin, 1981)

639. A.V. Skorohod, *Integration in Hilbert Space.* Translated from the Russian by K. Wickwire, Ergebnisse der Mathematik und ihrer Grenzgebiete, Band 79. (Springer, New York, 1974)

640. M.A. Sofi, Weaker forms of continuity and vector-valued Riemann integration. Colloq. Math. **129**(1), 1–6 (2012)

641. M.A. Sofi, Some problems in functional analysis inspired by Hahn-Banach type theorems. Ann. Funct. Anal. AFA **5**(2), 1–29 (2014)

642. M.A. Sofi, Nonlinear aspects of certain linear phenomena in Banach spaces, in *Applied Analysis in Biological and Physical Sciences.* ICMBAA, Aligarh, June 4–6, 2015 (Springer, New Delhi, 2016), pp. 407–425

643. M.A. Sofi, Riemann integrability under weaker forms of continuity in infinite dimensional space. arXiv:1612.00642 [math.FA], 18 (2016)

644. J.M. Steele, Certifying smoothness of discrete functions and measuring legitimacy of images. J. Complex. **5**(3), 261–270 (1989)

645. Ch. Stegall, Optimization of functions on certain subsets of Banach spaces. Math. Ann. **236**(2), 171–176 (1978)

646. Ch. Stegall, Optimization and differentiation in Banach spaces. Linear Algebra Appl. **84**, 191–211 (1986)

647. E.M. Stein, *Harmonic Analysis: Real-Variable Methods, Orthogonality, and Oscillatory Integrals.* Princeton Mathematical Series, vol. 43 (Princeton University Press, Princeton, 1993) [With the assistance of Timothy S. Murphy, Monographs in Harmonic Analysis, III]

648. W. Stepanoff, Über totale Differenzierbarkeit. Math. Ann. **90**, 318–320 (1923) [German]

649. V.V. Stepanov, Sur les conditions de l'existence de la différentielle totale. Rec. Math. Moscou (Mat. Sbornik) **32**, 511–527 (1925) [French]

650. N. Teleman, The index of signature operators on Lipschitz manifolds. Inst. Hautes Études Sci. Publ. Math.(58), 39–78 (1983)

651. H. Tong, Book review: a treatise on set topology. Bull. Am. Math. Soc. **54**, 1087–1090 (1948)

652. H. Tong, Some characterizations of normal and perfectly normal spaces. Duke Math. J. **19**, 289–292 (1952)

653. H. Triebel, A new approach to function spaces on quasi-metric spaces. Rev. Mat. Complut. **18**(1), 7–48 (2005)

654. I.G. Tsar'kov, Extension of Hilbert-valued Lipschitz mappings. Vestnik Moskov. Univ. Ser. I Mat. Mekh. (6), 9–16 (1999) [Russian; English translation in Moscow Univ. Math. Bull. **54**(6), 7–14 (1999)]

655. V.A. Tuan, Ch. Tammer, C. Zălinescu, The Lipschitzianity of convex vector and set-valued functions. Top **24**(1), 273–299 (2016)

656. P. Tukia, Lipschitz approximation of homeomorphisms. Ann. Acad. Sci. Fenn. Ser. A I Math. **4**(1), 137–144 (1979)

657. P. Tukia, J. Väisälä, Lipschitz and quasiconformal approximation and extension. Ann. Acad. Sci. Fenn. Ser. A I Math. **6**(2), 303–342 (1981)

658. P. Tukia, J. Väisälä, Bi-Lipschitz extensions of maps having quasiconformal extensions. Math. Ann. **269**(4), 561–572 (1984)

659. Ph. Turpin, Opérateurs linéaires entre espaces d'Orlicz non localement convexes. Studia Math. **46**, 153–165 (1973) [French]

660. J.T. Tyson, Bi-Lipschitz embeddings of hyperspaces of compact sets. Fund. Math. **187**(3), 229–254 (2005)

661. J. Väisälä, Piecewise linear approximation of lipeomorphisms. Ann. Acad. Sci. Fenn. Ser. A I Math. **3**(2), 377–383 (1977)

662. J. Väisälä, Bi-Lipschitz and quasisymmetric extension properties. Ann. Acad. Sci. Fenn. Ser. A I Math. **11**(2), 239–274 (1986)

663. J. Väisälä, Banach spaces and bi-Lipschitz maps. Studia Math. **103**(3), 291–294 (1992)

664. F.A. Valentine, On the extension of a vector function so as to preserve a Lipschitz condition. Bull. Am. Math. Soc. **49**, 100–108 (1943)

665. F.A. Valentine, Contractions in non-Euclidean spaces. Bull. Am. Math. Soc. **50**, 710–713 (1944)
666. F.A. Valentine, A Lipschitz condition preserving extension for a vector function. Am. J. Math. **67**, 83–93 (1945)
667. L.N. Vasershtein, Markov processes over denumerable products of spaces describing large system of automata. Probl. Pereda. Inf. **5**(3), 64–72 (1969) [Russian]
668. A.M. Vershik, The Kantorovich metric: the initial history and little-known applications. J. Math. Sci. N. Y. **133**(4), 1410–1417 (2006)
669. A.M. Vershik, Long history of the Monge-Kantorovich transportation problem. Math. Intell. **35**(4), 1–9 (2013)
670. A.M. Vershik, Monge-Kantorovich problem along years, in *Advances in Economics and Optimization*. Economic Issues, Problems and Perspectives (Nova Sci. Publ., New York, 2014), pp. 7–18
671. A.M. Vershik, S.S. Kutateladze, S.P. Novikov, Leonid Vital'evich Kantorovich (on the 100th anniversary of his birth). Russ. Math. Surv. **67**(3), 589–597 (2012) [Translation from Russian: Uspekhi Mat. Nauk **67**(3), 185–191 (2012)]
672. C. Villani, *Optimal Transport – Old and New*, Grundlehren der Mathematischen Wissenschaften, vol. 338 (Springer, Berlin, 2009)
673. A.L. Vol'berg, S.V. Konyagin, On measures with the doubling condition. Izv. Akad. Nauk SSSR Ser. Mat. **51**(3), 666–675 (1987) [Russian; English translation in Math. USSR-Izv. **30**(3), 629–638 (1988)]
674. Wayne State University, Mathematics Department Coffee Room, Classroom Notes: Every Convex Function is Locally Lipschitz. Am. Math. Mon. **79**(10), 1121–1124 (1972)
675. N. Weaver, *Lipschitz Algebras* (World Scientific Publishing, River Edge, 1999)
676. J.H. Wells, L.R. Williams, *Embeddings and Extensions in Analysis*. Ergebnisse der Mathematik und ihrer Grenzgebiete, Band 84 (Springer, New York, 1975)
677. Z.-Y. Wen, L.-F. Xi, Relations among Whitney sets, self-similar arcs and quasi-arcs. Isr. J. Math. **136**, 251–267 (2003)
678. Z. Wen, Z. Zhu, G. Deng, Lipschitz equivalence of a class of general Sierpinski carpets. J. Math. Anal. Appl. **385**(1), 16–23 (2012)
679. D. Werner, Review to "A. Naor, An introduction to the Ribe program, Jpn. J. Math. 7 (2012), 167–233", ZblMATH, Zbl 1261.46013 (2012)
680. H. Whitney, Analytic extensions of differentiable functions defined in closed sets. Trans. Am. Math. Soc. **36**(1), 63–89 (1934)
681. H. Whitney, On the extension of differentiable functions. Bull. Am. Math. Soc. **50**, 76–81 (1944)
682. S. Willard, *General Topology* (Dover Publications, Inc., Mineola, 2004) [Reprint of the 1970 original Addison-Wesley, Reading]
683. P. Wojtaszczyk, *Banach Spaces for Analysts*. Cambridge Studies in Advanced Mathematics, vol. 25 (Cambridge University Press, Cambridge, 1991)
684. L.-F. Xi, Lipschitz equivalence of self-conformal sets. J. Lond. Math. Soc. **70**(2), 369–382 (2004)
685. L.-F. Xi, Quasi-Lipschitz equivalence of fractals. Isr. J. Math. **160**, 1–21 (2007)
686. L.-F. Xi, Lipschitz equivalence of dust-like self-similar sets. Math. Z. **266**(3), 683–691 (2010)
687. L.-F Xi, H.-J. Ruan, Lipschitz equivalence of generalized {1, 3, 5, }-{1, 4, 5} self-similar sets. Sci. China Ser. A **50**(11), 1537–1551 (2007)
688. L.F. Xi, H.J. Ruan, Lipschitz equivalence of self-similar sets satisfying the strong separation condition. Acta Math. Sinica (Chin. Ser.) **51**(3), 493–500 (2008)
689. L.-F. Xi, H.-J. Ruan, Q.-L. Guo, Sliding of self-similar sets. Sci. China Ser. A **50**(3), 351–360 (2007)
690. L.-F. Xi, Y. Xiong, Ensembles auto-similaires avec motifs initiaux cubiques. C. R. Math. Acad. Sci. Paris **348**(1–2), 15–20 (2010)
691. Y. Xiong, L.-F. Xi, Lipschitz equivalence of graph-directed fractals. Studia Math. **194**(2), 197–205 (2009)

692. Z. Zahorski, Sur l'ensemble des points de non-dérivabilité d'une fonction continue. Bull. Soc. Math. France **74**, 147–178 (1946)

693. L. Zajíček, An elementary proof of the one-dimensional Rademacher theorem. Math. Bohem. **117**(2), 133–136 (1992)

694. C. Zălinescu, A generalization of the Farkas lemma and applications to convex programming. J. Math. Anal. Appl. **66**(3), 651–678 (1978)

695. C. Zălinescu, *Convex Analysis in General Vector Spaces* (World Scientific Publishing, River Edge, 2002)

696. V. Zizler, On some extremal problems in Banach spaces. Math. Scand. **32**, 214–224 (1973)

Index

(X, d, θ) – pointed metric space, 366
$B[x_0, r]$, $B(x_0, r)$ – balls, 20
B_X, B_X', S_X – closed (open) unit ball and
      sphere in a normed space X, 43
$B_p[x, r]$, $B_p(x, r)$ – p-balls in LCS, 39
B_ℓ, B_{sum}, B_{max}, 409
B_{var}, 404
C-full envelope, 4
$L(X, Y)$ – the space of linear operators, 41
$S(x_0, r)$ – sphere, 20
$TM(X, \mathscr{H})$ - totally measurable functions,
      405
$X^\#$ – Lipschitz dual, 393, 519, 548
Y°, $Z_\circ$ – polars, 390
$Y^\perp$ – null space, 542
$Y^\perp$, $Z_\perp$ – null spaces, 390
$Y^\perp$ – the annihilator, 542
$\| \cdot \|_d^\circ$, 437
$\| \cdot \|_{KR}$, 411
$\| \cdot \|_{\text{sum}}$, $\| \cdot \|_{\text{max}}$, 366, 408
$\| \cdot \|_H$ – Hanin norm , 440
$\| \cdot \|_L$, 408
$\complement(A)$ – the complement of A, 69
$\ell^\infty(\Gamma)$, ℓ^∞, 277
$\int \langle f, d\mu \rangle$, 405
$\mathbb{C}$, 1
$\mathbb{F}(X)$, 379
$\mathbb{F}(X)^* = \text{Lip}_0(X)$, 379
$\mathbb{K} = \mathbb{R} \vee \mathbb{C}$, 1
$\mathbb{N}$, 1
$\mathbb{N}_0$, 1
$\mathbb{Q}$, 1
$\mathbb{R}$, 1
$\mathbb{R}$-tree, 263
$\mathbb{R}_+$, 1

$\mathbb{Z}$, 1
$\mathbf{1}$, $\mathbf{1}_X$ – the function identically equal to 1, 55,
      370
$\mathscr{K}(E_1, E_2)$ – the space of compact operators,
      459
$\mathscr{L}(X, Y)$ – the space of continuous linear
      operators, 41
$\mathscr{M}_2$, $\mathscr{M}^k$, $\mathscr{M}_0$, $\mathscr{M}_0^k$, $\mathscr{M}_0^f$ – classes of metric
      spaces, 371
$\mathscr{W}\mathscr{K}(E_1, E_2)$ – the space of weakly compact
      operators, 459
$\text{BLip}(X)$, 367
$\text{BLip}(X, Y)$, 366
$\text{BLip}(X, d^\alpha) = \text{BLip}_\alpha(X)$, 368
$\text{BLip}_\alpha(X)$, 368
$\text{BLip}_{\alpha,0}(X)$, 368
$\text{BLip}_{\alpha,0}(X, Y)$, 368
$\text{BLip}_\alpha(X, Y)$, 368
LCS – locally convex space, 39
Lip-BPB, 528
$\text{Lip}(X)$, 367
$\text{Lip}(X, Y)$, 365
$\text{Lip}_0(X, Y)$, $\text{Lip}_0(X)$, 367
$\text{Lip}_0(X, d^\alpha) = \text{Lip}_{\alpha,0}(X)$, 368
$\text{Lip}_\alpha(X)$, 368
$\text{Lip}_{\alpha,0}(X)$, 368
$\text{Lip}_{\alpha,0}(X, Y)$, 368
$\text{Lip}_\alpha(X, Y)$, 368
$\text{Lip}_{od}(X)$, 549
$\text{Lip}_{x_0}(X)$, 367
$\text{Lip}_{x_0}(X, Y)$, 367
$\text{DA}(X^\#)$, 527
MVT – mean value theorem, 102
$\text{Mol}(X)$ - the space of molecules, 374
RNP – Radon-Nikodým property, 93

© Springer Nature Switzerland AG 2019
Ş. Cobzaş et al., *Lipschitz Functions*, Lecture Notes in Mathematics 2241,
https://doi.org/10.1007/978-3-030-16489-8

LECTURE NOTES IN MATHEMATICS   🐎 Springer

Editors in Chief: J.-M. Morel, B. Teissier;

Editorial Policy

1. Lecture Notes aim to report new developments in all areas of mathematics and their applications – quickly, informally and at a high level. Mathematical texts analysing new developments in modelling and numerical simulation are welcome.

   Manuscripts should be reasonably self-contained and rounded off. Thus they may, and often will, present not only results of the author but also related work by other people. They may be based on specialised lecture courses. Furthermore, the manuscripts should provide sufficient motivation, examples and applications. This clearly distinguishes Lecture Notes from journal articles or technical reports which normally are very concise. Articles intended for a journal but too long to be accepted by most journals, usually do not have this "lecture notes" character. For similar reasons it is unusual for doctoral theses to be accepted for the Lecture Notes series, though habilitation theses may be appropriate.

2. Besides monographs, multi-author manuscripts resulting from SUMMER SCHOOLS or similar INTENSIVE COURSES are welcome, provided their objective was held to present an active mathematical topic to an audience at the beginning or intermediate graduate level (a list of participants should be provided).

   The resulting manuscript should not be just a collection of course notes, but should require advance planning and coordination among the main lecturers. The subject matter should dictate the structure of the book. This structure should be motivated and explained in a scientific introduction, and the notation, references, index and formulation of results should be, if possible, unified by the editors. Each contribution should have an abstract and an introduction referring to the other contributions. In other words, more preparatory work must go into a multi-authored volume than simply assembling a disparate collection of papers, communicated at the event.

3. Manuscripts should be submitted either online at www.editorialmanager.com/lnm to Springer's mathematics editorial in Heidelberg, or electronically to one of the series editors. Authors should be aware that incomplete or insufficiently close-to-final manuscripts almost always result in longer refereeing times and nevertheless unclear referees' recommendations, making further refereeing of a final draft necessary. The strict minimum amount of material that will be considered should include a detailed outline describing the planned contents of each chapter, a bibliography and several sample chapters. Parallel submission of a manuscript to another publisher while under consideration for LNM is not acceptable and can lead to rejection.

4. In general, **monographs** will be sent out to at least 2 external referees for evaluation.

   A final decision to publish can be made only on the basis of the complete manuscript, however a refereeing process leading to a preliminary decision can be based on a pre-final or incomplete manuscript.

   Volume Editors of **multi-author works** are expected to arrange for the refereeing, to the usual scientific standards, of the individual contributions. If the resulting reports can be forwarded to the LNM Editorial Board, this is very helpful. If no reports are forwarded or if other questions remain unclear in respect of homogeneity etc, the series editors may wish to consult external referees for an overall evaluation of the volume.

5. Manuscripts should in general be submitted in English. Final manuscripts should contain at least 100 pages of mathematical text and should always include

   - a table of contents;
   - an informative introduction, with adequate motivation and perhaps some historical remarks: it should be accessible to a reader not intimately familiar with the topic treated;
   - a subject index: as a rule this is genuinely helpful for the reader.
   - For evaluation purposes, manuscripts should be submitted as pdf files.

6. Careful preparation of the manuscripts will help keep production time short besides ensuring satisfactory appearance of the finished book in print and online. After acceptance of the manuscript authors will be asked to prepare the final LaTeX source files (see LaTeX templates online: https://www.springer.com/gb/authors-editors/book-authors-editors/manuscriptpreparation/5636) plus the corresponding pdf- or zipped ps-file. The LaTeX source files are essential for producing the full-text online version of the book, see http://link.springer.com/bookseries/304 for the existing online volumes of LNM). The technical production of a Lecture Notes volume takes approximately 12 weeks. Additional instructions, if necessary, are available on request from lnm@springer.com.

7. Authors receive a total of 30 free copies of their volume and free access to their book on SpringerLink, but no royalties. They are entitled to a discount of 33.3 % on the price of Springer books purchased for their personal use, if ordering directly from Springer.

8. Commitment to publish is made by a *Publishing Agreement*; contributing authors of multiauthor books are requested to sign a *Consent to Publish form*. Springer-Verlag registers the copyright for each volume. Authors are free to reuse material contained in their LNM volumes in later publications: a brief written (or e-mail) request for formal permission is sufficient.

Addresses:
Professor Jean-Michel Morel, CMLA, École Normale Supérieure de Cachan, France
E-mail: moreljeanmichel@gmail.com

Professor Bernard Teissier, Equipe Géométrie et Dynamique,
Institut de Mathématiques de Jussieu – Paris Rive Gauche, Paris, France
E-mail: bernard.teissier@imj-prg.fr

Springer: Ute McCrory, Mathematics, Heidelberg, Germany,
E-mail: lnm@springer.com

Printed in the United States
By Bookmasters